Severe Storms

Someshwar Das · Wei-Kuo Tao
Editors

Severe Storms

Anatomy, Early Warning Systems and Aftermath in Changing Climate Scenarios

Springer

Editors
Someshwar Das
Former Scientist-G/ Advisor
NCMRWF & IMD
Ministry of Earth Sciences
New Delhi, India

Former Professor
Central University of Rajasthan
Ajmer, India

Wei-Kuo Tao
Emeritus Scientist
NASA/ Goddard Space Flight Center
Greenbelt, MD, USA

Chair Professor
National Central University
Taoyuan City, Taiwan

ISBN 978-981-97-7074-8 ISBN 978-981-97-7075-5 (eBook)
https://doi.org/10.1007/978-981-97-7075-5

This Springer imprint is published by the registered company Springer Nature Singapore Pte Ltd.
The registered company address is: 152 Beach Road, #21-01/04 Gateway East, Singapore 189721, Singapore

Foreword

Having 30 years experience working on the development and community support of widely used atmospheric models, MM5 and WRF, I am an ardent believer in shared knowledge, especially in such an important area for humanity as severe storms. I know that Someshwar Das and Wei-Kuo Tao are bringing years of numerical weather modeling accomplishments to gathering chapters of a highly needed Monograph for anyone who wants to sample the forefront of current knowledge in the wide range of topics relevant to observing, understanding, modeling, and forecasting severe storms, and their impacts now and in the future climate.

Those of us who have worked in this area of meteorological research are fully aware of the complexities of every aspect of severe storms; how they develop, their internal dynamics and thermodynamics, their evolution and lifetimes, and their impacts. Furthermore, they differ globally under different environments and climates, and this Monograph gives a sense of that through its strongly international flavor.

Mesoscale convective systems occur in a variety of environments globally, and much research has been done on their climatology and impacts. In mountainous areas, such as the Himalayas, Rockies, Alps, and Andes, orography plays a large role in how these systems develop and interact with slopes and river basins. In flatter areas, organized systems can wreak havoc along long tracks.

Over the past few decades, satellites, radar, aircraft, and ground-based observations have used increasingly sophisticated methods to observe severe storms in terms of their real-time and climatological occurrence and their internal structures. Now casting and warnings rely heavily on a good observing network and real-time communication of observations. Given the impact of severe storms on society, the improvement of national observing networks of radars and mesonets in developing areas of the world is a crucial part of advancements in public safety. Preparedness for disasters such as flooding and severe storms is a large factor in mitigating their impacts. The link between forecasters and emergency managers must be able to reliably communicate forecast risks to the end users. The forecast product is crucial in maintaining the level of trust needed for this to be effective.

Observations of severe storms are also increasingly being used in high-resolution storm-resolving forecast systems as data assimilation techniques have advanced enough to incorporate developed storms into the model's initial conditions via real-time satellite and radar information. Short-range prediction via a rapid-cycling data assimilation system is becoming more common at operational centers and is suited to the short range of deterministic predictability with such stochastic systems. As the range extends beyond a few hours, ensemble forecasts become the main tool to enable forecasters to see what can happen within the growing uncertainty envelope. Ensembles should represent both model and analysis uncertainties to encompass the range of possible outcomes. Other prediction aspects such as hail size or lightning risk that are not directly modeled may make use of offline physical models using the forecast output or machine learning trained on past events and forecasts with the same model.

The development of high-resolution forecast models also benefits from microphysical observations obtained from radar and in situ aircraft. Microphysical properties vary globally, not just due to the thermodynamics of the environment, but also due to the presence of pollutants or dust where local effects need to be considered as these may sometimes strongly modify the storm characteristics. In storm-resolving models, much of the variability among models in simulating severe storms is down to the variability in representing microphysical processes, especially the particle properties and their evolution. Predicting rainfall intensity or hail relies on adequate microphysics schemes. Hail, surface wind, and lightning are dangerous aspects of severe storms and there is now much more understanding of these processes that is helping models to forecast these more reliably.

Climate change is already occurring and has detectable and accelerating trends even over the last 50 years. The global large-scale circulation is changing and will continue to do so for decades into the future with magnitudes of change that just multiply what has happened so far. How this change affects important systems such as the Indian monsoon and extreme rainfall events is a major topic of research. Projections for the future indicate more rainfall and larger extremes but also seasonal changes that remain uncertain. This will of course impact water resources and agriculture as well as urban climates for billions of people and to increasing degrees related to how much and how quickly the climate is permitted to change.

This Monograph has addressed the important topics that I have outlined above by showing current knowledge in the science and some best practices in forecasting related to severe storms. It is a commendable collection of works by experts in all the topics.

Jimy Dudhia
NSF National Center for Atmospheric Research (NCAR)
Boulder, CO, USA

Preface

Severe Storms continue taking their toll on human lives and economy even though there has been considerable advancement in the prediction models. This book is focused on local severe thunderstorms causing intense lightning, strong winds and torrential rainfall, dust storms, hailstorms, tornadoes, and cloudbursts.

The local Severe Storms disrupt the routine life, and have devastating impacts on agriculture, aviation, surface transportation, electricity, power, communication, essential services, and other socio-economic sectors. It is estimated that lightning due to severe storms kills more people than hurricanes, earthquakes, and floods and is the biggest contributor to accidental deaths due to natural causes. A global survey on the mortality and economic losses due to severe storms indicates that over half a million people died and economic losses of US Dollar 521 billion occurred during 1970–2019 (WMO report, 2021). The IPCC (2021) projects an increase in the severity and frequency of severe storms in future due to climate change. As climate shocks become more common, better forecasting and early warning is a critical need for decision-making, such as when to plan and when to flee. According to UN, currently, one-third of the people in the world, especially in the Least Developed Countries and Small Island Developing States, are still not covered by early warning systems.

What causes the local severe storms so furious? How can we predict their intensity, location, and time of occurrence accurately at sufficient lead time? Accurate prediction of severe storms is still a challenge to the Atmospheric Scientists. Such questions are addressed in this book. For understanding a physically evolving system like a thunderstorm, we need to observe the system through various approaches, develop techniques to forecast its evolution in short, medium, and long-time scales, and develop policies for disaster mitigation in the aftermath of the event. These are discussed in three parts containing 26 chapters in the monograph titled *Severe Storms: Anatomy, Early Warning Systems and Aftermath in Changing Climate Scenarios* by leading scientists across the world.

The observational aspects of the severe storms through satellite, radar, aircraft, and ground-based network of stations are discussed in the first part of the book consisting of 8 chapters. The numerical modeling and data assimilation techniques are discussed in the second part consisting of 12 chapters for the development of early warning system, and finally the outlook of the severe storms in a changing climate, their socio-economic impacts, and policies for disaster mitigation are discussed in the third part of the book consisting of 6 chapters.

The first theme consisting of eight chapters focuses on using satellite (TRMM PR, GPM DPR, INSAT-3D/R), ground-based Radar (Doppler/Dual polarization), aircraft, and in situ measurements to improve our understanding of the physical processes (dynamic, thermodynamic and microphysics) associated with severe storms. These observations are for different types of storms and over different geographic locations. Observation of electric field, Maxwell current, and disdrometers also provide the lightning and raindrop size distributions, respectively.

The second theme consists of twelve chapters that focus on numerical model simulations including the impact of the data assimilation on forecasting severe storms. The impact of the model horizontal resolution, hailstorm dynamic, and hail characteristics, atmospheric boundary layer, low- and high-level moisture characteristics, diurnal cycle, monsoon lows, intra-seasonal oscillations, and extreme precipitation events on severe storms are presented. There is one chapter discussing the use of AI in lightning forecasting. All chapters also emphasized the importance of observations (i.e., Mesonet meteorological observations, radar reflectivity and radial velocity, and lidar profiling data, using high spatial and temporal radar-gauge merged precipitation data).

The third theme consists of six chapters that focus on the severe storms in the changing climate and their impact on societal and disaster risk reduction. These papers used observations (gridded rainfall data, satellite), re-analysis, and CMIP6 to address the impact of future climate change on severe storm developments. The solar storm and its characteristics, and cloud-radiation interaction; land-ocean feedback on extreme weather systems are discussed.

We would like to thank all the authors of the book for contributing the chapters that include the state-of-the-art of the subject and the latest findings of their research. We thank Dr. Amitava Bandopadhyay, Director General of the Centre for Science and Technology of the Non-Aligned and Other Developing Countries (NAM S&T Centre) for taking initiative for the publication of the Monograph, and Ms. Jasmeet Kaur Baweja, Programme Officer, NAM S&T Centre for her support throughout the process.

We hope that the Monograph will be a useful resource guide for research scientists, policymakers, and students to understand the "Severe Storms" and minimize the knowledge and technology gaps especially among the scientific community and policymakers for disaster mitigation, thereby minimizing the loss of lives and economy due to Severe Storms.

New Delhi, India — Someshwar Das
Maryland, USA — Wei-Kuo Tao

Introduction

Thunderstorms are among the nature's most spectacular phenomenon. They occur almost everywhere on the earth's surface. Depending on their intensity, they can produce cool showers or torrential rainfall, pleasant downdrafts or damaging winds, hails, lightning strikes, and can cause deaths and loss of properties. It is estimated that at any given time about 2000 thunderstorms are taking place on the globe (http://www.nssl.noaa.gov/education/svrwx101/thunderstorms/). While they occur both during the summer and the winter seasons, their frequency is highest during the pre-monsoon season over the Indian subcontinent. During the pre-monsoon season, severe thunderstorms generally travel from northwest to southeast direction over the east and northeast parts of countries such as India and Bangladesh. Therefore, they are also called the Nor'westers.

The thunderstorms and dust storms severely disrupt the routine human life, transportation, electricity, and other essential services. Livestock are particularly vulnerable to severe storms. Hailstorms can cause heavy damage to crops and vegetation. Secondary hazards like snapping of electric poles due to uprooting of trees, disruption of communication links, etc. are also attributed to thunderstorms. Most of the damages from thunderstorms occur due to strong winds, lightning strikes, hail stones, and flash floods caused by heavy rains. The cloudburst that occurs in the mountains, leads to exceptionally heavy rainfall and sudden flash floods in the streams and rivers, leading to breaching of the banks and overflowing of dams over the plain areas.

While the natural disasters cannot be prevented or controlled, the loss of lives and damage to property can be substantially minimized by issuing accurate forecast/advisories of the impending disasters through field observations, research, and the use of numerical modeling tools. Accurate and advanced advisories can greatly help the disaster mitigation managers to get sufficient lead time to initiate all the necessary preparedness and mitigation actions including evacuation measures which would help in reducing risks due to these disasters. The first step in the development of accurate warning system of severe storms is a proper understanding of their anatomy based on observations from the satellites, radars, aircrafts, and ground-based network of stations. The second step is the development of numerical modeling techniques and data assimilation in the model. The third step is the communication of the forecasts to the public, disaster mitigation agencies, and local administration giving sufficient lead time for disaster preparedness, evacuation, and other mitigation measures.

Considering the scientific importance of the subject and its socio-economic significance, particularly for the countries of the Global South, the Centre for Science and Technology of the Non-Aligned and Other Developing Countries (NAM S&T Centre), New Delhi have brought out this publication titled *Severe Storms—Anatomy, Early Warning Systems and Aftermath in Changing Climate Scenarios*.

The Monograph is divided into three parts comprising chapters contributed by leading experts from various parts of the world. The first two parts of the book present the current understanding of the anatomy and early warning system of severe storms. Further, due to the uncertainties of the effects of climate change on the earth, it is important to deliberate on what the frequency and severity of the severe storms could be in the future. In order to address this, a few chapters have been included in the third part of the book which present the outlook of severe storms in the changing climate scenarios based upon the climate model simulations and preparedness required for mitigation and management of disasters in the aftermath of such storms. The book also touches upon socio-economic implications of severe storms. The book contains 26 chapters contributed by authors from 11 countries and an Epilogue by the Editors.

I express my sincere gratitude to Dr. Jimy Dudhia, NSF National Centre for Atmospheric Research, Boulder, Colorado, USA for writing the "Foreword" of the Monograph.

The leading international scientific and academic publisher—Springer Nature, Singapore has published the Monograph. We are thankful to Dr. Loyola D'Silva, Executive Editor, Springer Nature for his encouragement and kind support; and Mr. Rajasekar Ganesan, Production Editor, Springer Nature, Chennai, India and other staff members of Springer team worldwide especially Ms. Bhagyalakshmee, Production Supervisor for handling all technical and administrative matters related to this publication. I am confident that our association with "Springer" would lead to many more such valuable collaborative endeavors in the future.

I would like to express my gratitude to the Editors, Dr. Someshwar Das, Former Adviser, Ministry of Earth Sciences, Govt. of India, New Delhi & Former Chair Professor of Atmospheric Science, Central University of Rajasthan, India and Dr. Wei-Kuo Tao, Emeritus Scientist, NASA/Goddard Space Flight Center, Greenbelt, Maryland, USA & Chair Professor, National Central University, Taiwan for their guidance, initiatives and efforts, and sharing their time in technical editing of the papers and designing the structure of this book.

In addition, I am grateful to all the experts for accepting our invitation and sparing their valuable time for reviewing the papers published in this book.

I express my gratitude to Mr. Madhusudan Bandyopadhyay, Senior Adviser, NAM S&T Centre for his kind advice and guidance in various stages of planning and execution of this book project. Further, I am thankful to Ms. Jasmeet Kaur Baweja, Senior Programme Officer, NAM S&T Centre for her significant contributions in coordinating this publication project.

I also acknowledge the interest, initiatives and support received from the entire team of the NAM S&T Centre. I am especially thankful to Mr. Pankaj Buttan, Data

Processing Manager; Mr. Rahul Kumra, Assistant Administrative Officer; Mr. Sunil Kumar, Accounts Manager, and Mr. Jayakumaran, Public Relations Manager, NAM S&T Centre for their support in bringing out this publication.

I am sure that this book would be a valuable reference material for the scientists, researchers and other professionals working in the areas of severe storms observations, modeling, and development of prediction and early warning systems.

Amitava Bandopadhyay, Ph.D.
Director General
NAM S&T Centre
New Delhi, India

Contents

Part 2: Modelling, Assimilation, Simulation and Forecasting of Severe Storms

Part 1: Observed Structure of Severe Storms Based on Satellite, Radar, Aircraft & Ground Based Network

Severe Storms Observed by Satellites

Yukari N. Takayabu

Abstract Accumulations of space-borne precipitation radar observations have enabled us to study characteristics of precipitation globally and statistically. Above all, structures of extreme precipitation are studied to renew our conventional assumption that intense precipitation is associated with intense convection. Further studies of wide-spread heavy rainfalls in very moist season in Japan clarified the contrasts between the severe rainfalls resulted from very unstable atmospheric condition and those from very moist atmosphere. The former is related to instability in terms of Convective Available Potential Energy (CAPE), while the latter is related to Moist Available Unstable Layer (MAUL) in very moist atmospheric conditions. Therefore, increase of moisture can increase the frequency of very heavy rainfalls without increasing the atmospheric instability in a conventional sense. In this chapter, we will review a series of studies that led us to restructure the knowledge of severe storms.

Keywords Precipitation characteristics · Extreme precipitation · Space-borne precipitation radar · TRMM PR · GPM DPR · MAUL

1 Introduction

In recent years, frequent wide-area heavy rainfalls causing disastrous floods have been reported in various parts of the world. For the long-term trends in precipitation characteristics over Japan, for example, JMA has made statistics from 51 station data over Japan. Figure 1a shows annual number of days for daily precipitation exceeding 100 mm. We can confirm that in recent 120 years, frequency of heavy precipitation has a significant increasing trend, with a 99% confidence level. On the other hand, numbers of days for daily precipitation over 1.0 mm is significantly decreasing (Fig. 1b), showing light precipitation is decreasing. These trends indicate possible changes in precipitation characteristics.

Y. N. Takayabu (✉)
Atmosphere and Ocean Research Institute, University of Tokyo, Tokyo, Japan
e-mail: yukari@aori.u-tokyo.ac.jp

S. Das and W-K. Tao (eds.), *Severe Storms*,
https://doi.org/10.1007/978-981-97-7075-5_1

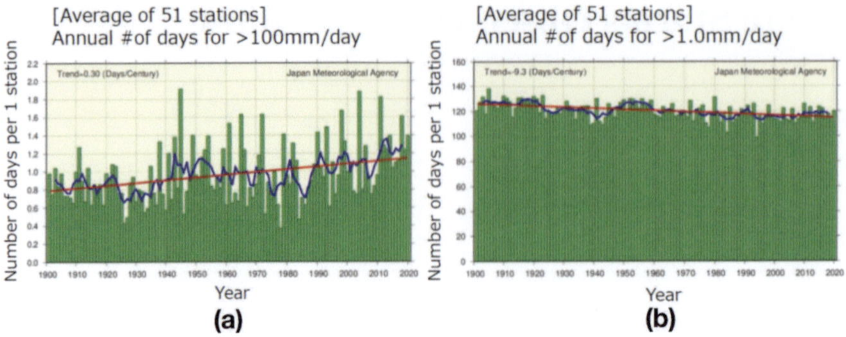

Fig. 1 Statistics showing long-term trends in precipitation characteristics obtained from 51 station data over Japan, made by the Japan Meteorological Agency. **a** Annual number of days for daily precipitation exceeding 100 mm (green bars), and its 5-year running mean (blue curve) and linear trends (red line). In recent 120 years, frequency of heavy precipitation has a significant increasing trend, with a 99% confidence level. **b** Annual number of days for daily precipitation over 1.0 mm. Including light precipitation, the frequency is significantly decreasing. These trends indicate recent changes in precipitation characteristics. Adopted from Climate Change Monitoring Report 2020, JMA

Intergovernmental Panel on Climate Change (IPCC) published the Assessment Report 6 (AR6) in August 2021. About the changes of extreme precipitation over land observed from 1950s, as cited in Fig. 2, it is stated that "The frequency and intensity of heavy precipitation events have increased since the 1950s over most land area for which observational data are sufficient for trend analysis (*high confidence*), and human-induced climate change is *likely* the main driver." However, because of the difficulty of rainfall estimates from small numbers of ground-based measurements, there are many regions, especially over the deep tropics and in southern hemisphere, we cannot obtain enough data to discuss about the trend. Moreover, the mechanism of the increase of heavy rainfalls is not very clear yet.

How do we understand the precipitation change associated with the global warming? Associated with the temperature rise, potential increase of the atmospheric water vapor is estimated as ~7%/K, following the Clausius-Clapeyron (CC) relationship. However, increase of the global mean precipitation is estimated as ~2%/K (Allen & Ingram, 2002). It is attributed to the constraints by the global atmospheric energy budget (Trenberth et al., 2009; Stephens et al., 2012). From the global energy budget viewpoint, total precipitation equals the latent energy moved from the surface to the atmosphere in the form of water vapor and remain there with precipitation. Thus, global total precipitation amount is constrained by the global energy balance, therefore cannot simply increase with the CC rate. On the other hand, it has been shown, that extreme precipitation is increasing with ~7%/K, or more/less, depending on regions, seasons, or on timescales (e.g. Lenderink & Meijgaard, 2008, Lenderink et al., 2011, Utsumi et al., 2011). We still do not sufficiently understand the mechanism in these variations.

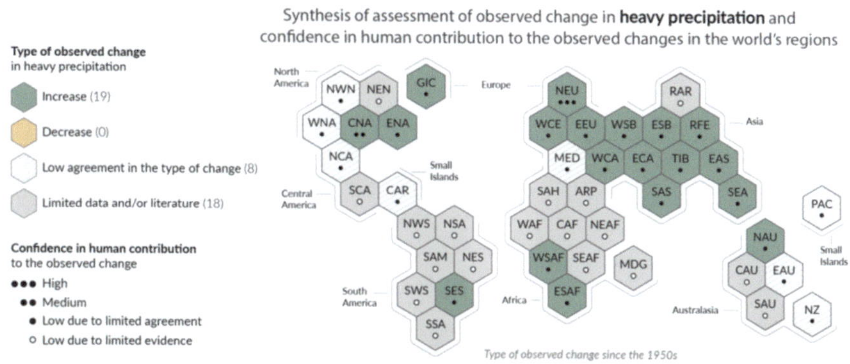

Fig. 2 Synthesis of assessment of observed change in heavy precipitation and confidence in human contribution to the observed changes in the world's regions over land from 1950s. Adopted from Fig. SPM.3 in IPCC, 2021: Summary for Policy Makers. In: *Climate Change 2021: The Physical Science Basis. Contribution of Working Group I to the Sixth Assessment Report of the Intergovernmental Panel on Climate Change*

2 Statistical Studies of Extreme Precipitation Observed from Satellites

To study precipitation globally, satellite observation is a powerful tool. Geostationary satellites have been often used to observe precipitating cloud systems, because of their ability in tracking the systems with high temporal resolutions. However, although visible and infrared frequency (VISIR) observation from the geostationary satellites obtain the cloud information, it cannot directly obtain the precipitation signals.

To capture the precipitation characteristics, ground-based precipitation radar observation has been conventionally utilized since 1970s (e.g. Houze, 1977, 2004; Leary & Houze, 1979). In 1997, the first space-borne precipitation radar (PR) was launched on the Tropical Rainfall Measuring Mission (TRMM) satellite. TRMM PR was a 13.8 GHz Ku-band phased-array radar, which could measure moderate to intense rainfalls in three dimensions with horizontal resolution of ~5 km and vertical resolution of 250 m, with swath width of 215–247 km, from near surface to 20 km altitude. Radar emits the microwave to the target, precipitation particles for precipitation radars, and receives the scattered signal, thus can obtain 3D structure of precipitation. TRMM PR and following Global Precipitation Measurement (GPM) Dual-frequency Precipitation Radar (Kojima et al. 2012; Skofronick-Jackson et al. 2017; Kidd et al. 2020), have continuously collected huge amount of 3D radar observation data of precipitation over almost any part of the globe, covering 35N-35S, and 65N-65S, respectively, now over 25 years together. Figure 3 shows a snapshot of a typhoon captured by the GPM DPR.

Utilizing the level-2 (orbital) products, we identified rainfall events, or continuous volumes of rainfall, captured by TRMM PR or GPM DPR. Figure 4 plots

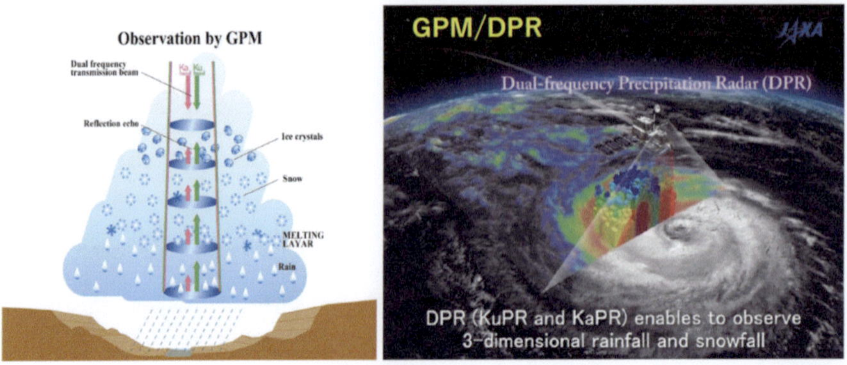

Fig. 3 Images of 3D measurements of Precipitation with GPM Dual frequency Precipitation Radar. Courtesy of JAXA

two examples of rainfall events. By capturing rainfall events in 3D, we can represent "characteristics of rainfalls" with their height, area, stratiform/convective ratio, volumetric rain, and their maximum intensity.

Hamada et al. (2014) accumulated all rainfall events observed by TRMM PR in each 2.5degree × 2.5degree latitude–longitude grid between 35N and 35S, then identified the top 0.1% events (hereafter called the extreme events) in each grid in terms of their maximum rainfall intensity. Figure 5a shows the maximum intensity of the extreme events. It is found that characteristics of extreme rainfall events depend on regions. Figure 5b shows the classification of the extreme rainfall events by their area and maximum intensity into 4 categories. Clear distinction between extreme events over land and over open oceans are found: Over land, extreme rainfalls consist of intense but not very large events, while over open oceans, they are very large but not

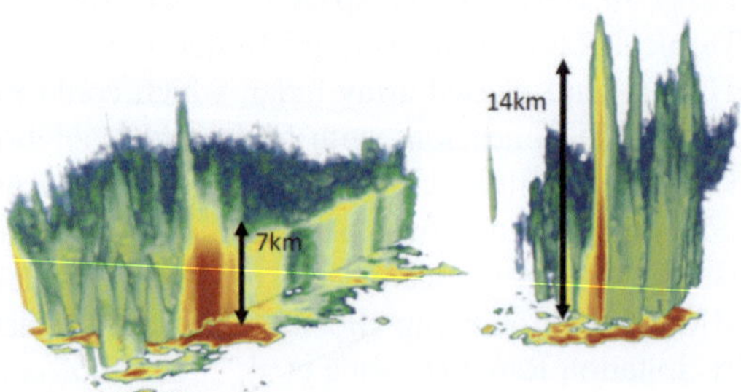

Fig. 4 Examples of rainfall events, or continuous volumes of rainfall, captured by TRMM PR. By capturing rainfall events in 3 dimensions, we can represent "characteristics of rainfalls" with their height, area, maximum intensity, volumetric rainfall, and stratiform/convective ratio. These events correspond to one of extreme rainfall (left) and one of extreme convection (right) over Japan. Courtesy of Prof. Atsushi Hamada

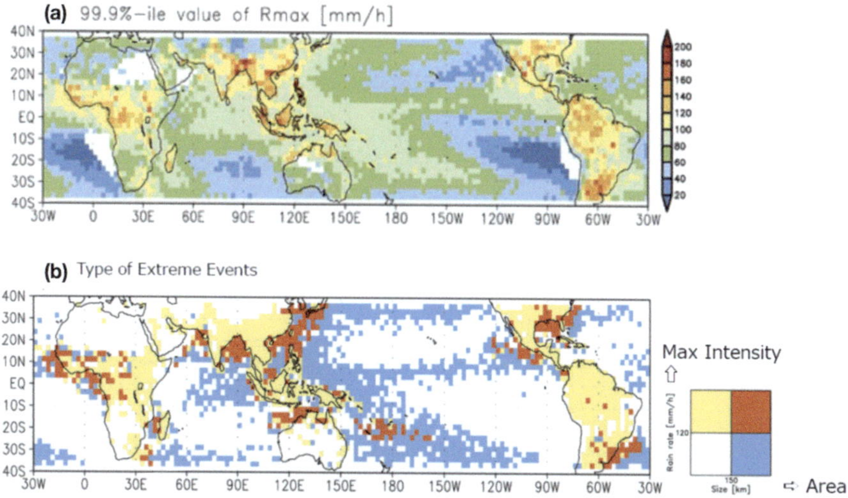

Fig. 5 **a** Maximum rainfall intensity of top 0.1 percentile (extreme) heavy rainfall events in each 2.5degree × 2.5degree longitude-latitude grid. **b** Types of extreme events in each grid categorized in four classes. Yellow shades indicate dominance of very intense (>120 mm hr^{-1}), but not very large (<150 km in length scale) events. Blue shades show very large but not very intense events. Red shades show very intense and very large events. White is others. Adopted from Hamada et al., 2014, J. Climate

very intense. Very large and very intense extreme events are found in the ~1000 km-scale oceanic fringes of continents, or some specific regions over land, such as over central Africa, around the LaPlata River, or over the southern USA.

Hamada et al. (2015) utilized 11 years of TRMM PR observation data and made statistics of top 0.1% extreme rainfall events and extreme convection events at each 2.5degree × 2.5degree latitude–longitude grid. Extreme rainfalls are defined with their maximum near-surface rainfall, and extreme convections are defined with their maximum height of 40dBZ reflectivity. The total number of rainfall events used counts over 85 million events. The plots in Fig. 4 for rainfall events correspond to one of extreme rainfall and one of extreme convection over Japan. Figure 6a and b show reflectivity-height 2D histograms for extreme rainfall events and extreme convection events, respectively, over land for whole TRMM observation region. From the definition, the near surface rainfall intensity is very large for extreme rainfalls, and the rainfall heights are very tall for extreme convection. However, it is very interesting to find that height of extreme rainfall is not quite tall, and intensity of extreme convection is not extreme. So that heaviest rainfall is not often accompanying tallest convection, against our conventional assumptions.

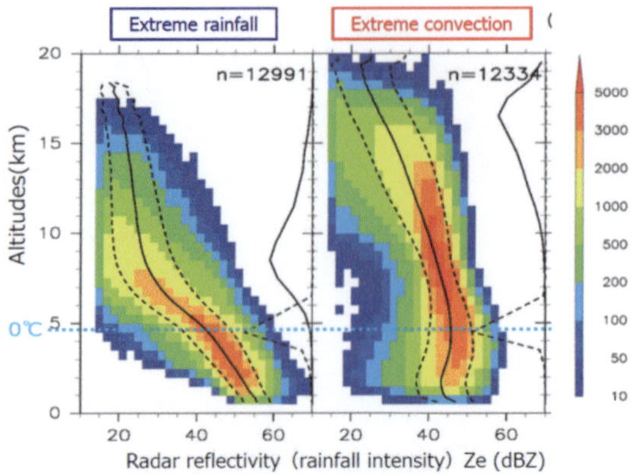

Fig. 6 Two dimensional histograms for top 0.1% extreme rainfall events and extreme convections, respectively, in terms of radar reflectivity and altitudes, at each 2.5 × 2.5 latitude–longitude grid over land from 35N-35S, obtained from 11 years from September 2001 to August 2012 of TRMM PR observations. Extreme rainfalls are defined with their maximum near-surface rainfall, and extreme convections are defined with their max height of 40dBZ reflectivity. The total number of rainfall events used for analysis is over 85 million events. It is shown that heaviest rainfall is not often accompanying tallest convection. From Hamada et al. (2015), Nature Communications

3 Structure and Environments of Heavy Rainfall Events in Summer Rainfall Season in Japan

Three-dimensional rainfall structure of the July 2018 wide-area disastrous rainfall event in Japan, which caused 237 fatalities, was captured by GPM DPR at around 9:30 JST on July 7, 2018, during the heavy rainfall (Fig. 7). It is clearly seen that despite the record-breaking rainfall intensity and amount, rainfall height over 10 km is scarcely observed, which was consistent with the finding by Hamada et al. (2015).

Yokoyama et al. (2020) examined the large-scale environment for this extreme event. Figure 8a shows a Himawari-8 water vapor image. We can see a deep trough which persisted for a few days over the Korean peninsula. Figure 8b shows, potential vorticity at 350 K representing the upper-level trough in black contours, rainfall in color shades, and total precipitable water, or an integrated atmospheric column water vapor amount, in red contours. It is analyzed that this trough dynamically prepared the updraft with very moist atmosphere in its front, then caused the persistent heavy rainfall.

It is reported by Kawano and Suzuki's group (personal communication) that the July 2018 event is scarcely accompanied by lightnings. It is consistent with the GPM DPR observation in Fig. 7 that showed moderate heights of rainfalls for this event. In Tsuji et al. (2020), we compared this July 2018 event with a previous heavy rainfall event in Northern Kyushu in July 2017, when active lightnings were reported. Figure 9 compares 24-h accumulated rainfall for each event. We can see that 2018

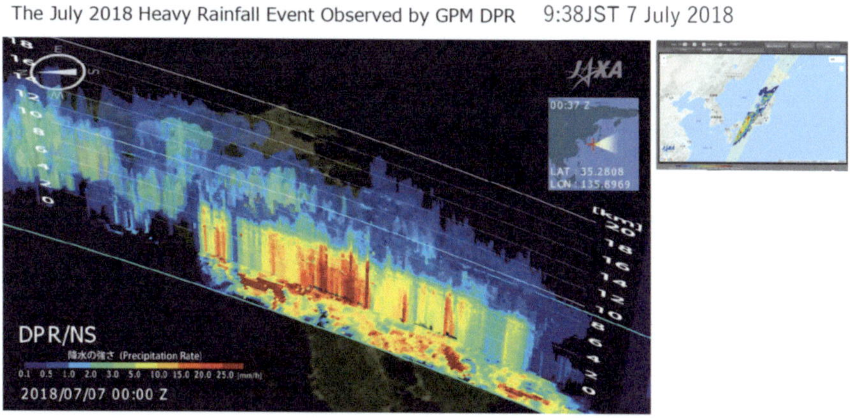

Fig. 7 Three-dimensional precipitation structure captured by GPM DPR at around 9:30 JST on July 7, 2018, during the July 2018 disastrous rainfall event. The right panel shows the GPM DPR scan image of this case. It is clearly seen that despite the record-breaking rainfall amount, rainfall height over 10 km is scarcely observed. Courtesy of JAXA

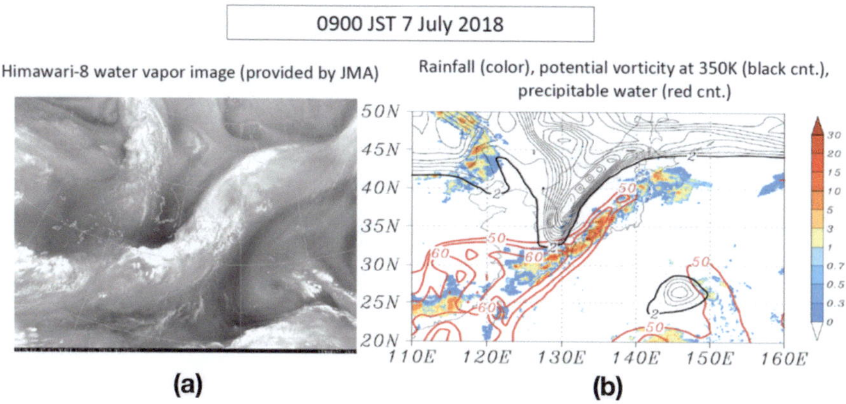

Fig. 8 **a** Himawari-8 water vapor channel image for 00UTC 7 July 2018. Dark area surrounded by while cloud area indicate a deep trough which persisted for a few days over the Korean peninsula. **b** Potential vorticity at 350 K representing the upper-level trough in black contours, hourly rainfall rate obtained from GSMaP in color shades, and total precipitable water, or an integrated atmospheric column water vapor amount, in red contours. Labels on TPW are in mm. From Yokoyama et al. (2020), JMSJ

event had a much wider area coverage, while 2017 event was confined to northern Kyushu.

We then compared the time series of JMA's northern Kyushu ground-based radar data for these cases (Fig. 10). Red curves are for numbers of very tall rain grids with their echo top height (ETH) exceeding 10 km, and blue curves are for moderate-height rain grids, with the ETC < 10 km. We could readily see a distinct contrast between the

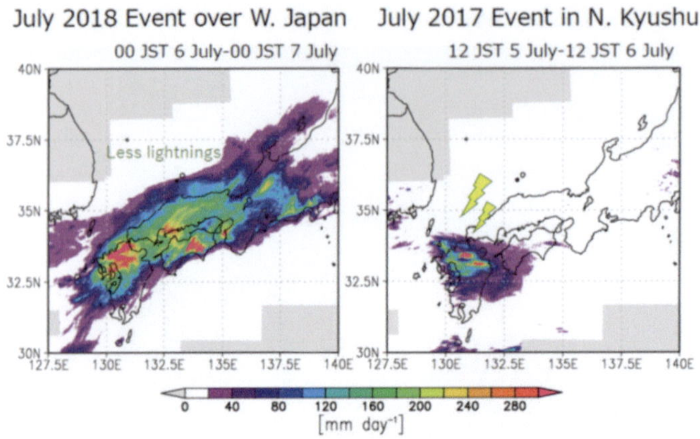

Fig. 9 Twenty four-hour accumulated rainfall from 00JST 6 to 00JST 7 July during the July 2018 heavy rainfall event over western Japan (left), and from 12JST 5 July to 12JST 6 July during the July 2017 heavy rainfall event in northern Kyushu. From Tsuji et al. (2020), JMSJ

two events. For the 2018 case, rainfall is primarily from moderate-height rainfalls, while for the 2017 case, heaviest rainfall is brought by tall rainfall over 10 km. It is also consistent with the fact that more active lightnings are observed in the 2017 case.

Tsuji et al. (2020) further compared the environments of above two cases in anomalies from the climatology. Figure 11a and b show temperature anomalies and moisture anomalies, respectively. Temperature stratification in 2018 case is even more stable than the climatology, while it is more unstable for the 2017 case. On the other hand, atmosphere is more humid than usual in both cases and extremely moister

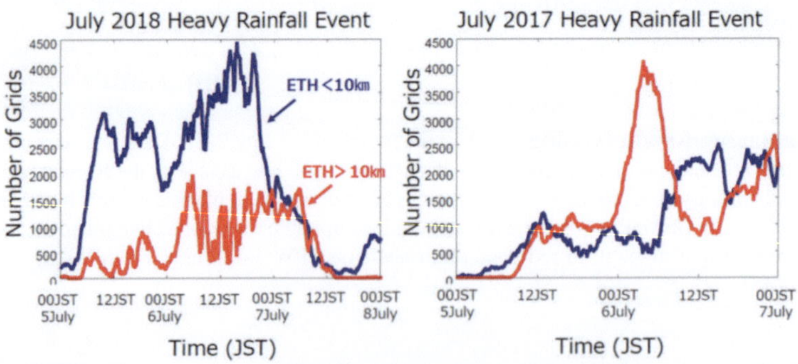

Fig. 10 Time series for very tall rain grid numbers and moderate-height rain grid numbers, for heavy rainfall events in July 2018 (left) and in July 2017 (right). Rainfall height data are obtained from the JMA's Northern Kyushu Radar. Red curves show those for echo top heights larger than 10 km, and blue curves show those for echo top heights smaller than 10 km. From Tsuji et al. (2020), JMSJ

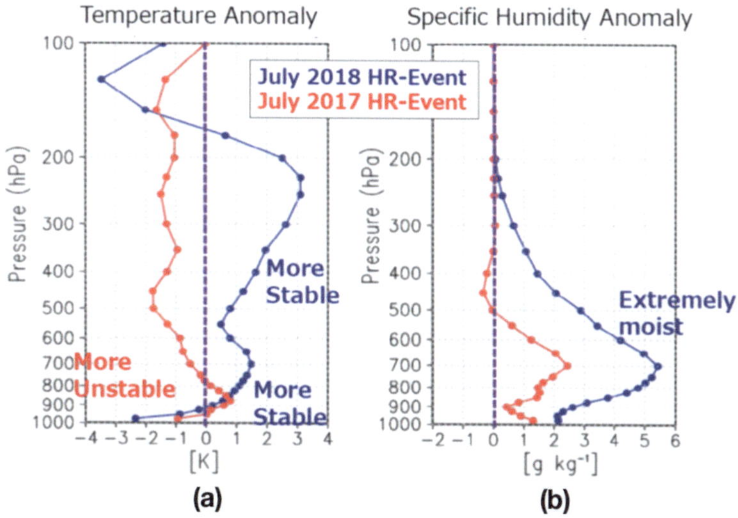

Fig. 11 Vertical profiles of **a** temperature anomaly (K) and **b** specific humidity anomaly (g kg^{-1}) for the 2018 case (blue curves) and 2017 case (red curves). These profiles are averaged over four grids around Fukuoka at 6 UTC 6 July 2018 for the 2018 case and 6 UTC 5 July 2017 for the 2017 case. From Tsuji et al. (2020), JMSJ

in 2018 event. These contrasts well correspond to the statistics for the Extreme Rainfalls and Extreme Convection around Japan observed from space shown in Hamada and Takayabu (2018).

The comparisons were summarized as schematics in Fig. 12. For the 2018 event, a long-lasting rainfall occurred over an extensive area from moderately tall precipitation systems. Compared to the climatology, atmosphere was more stable but extremely moist. The upper tropospheric trough played a role to dynamically induce ascending motion, which provided such environments for a wide-spread heavy rainfall. For the 2017 event, on the other hand, short-term rainfall occurred in a narrow area from extremely tall precipitation systems. Atmosphere was more unstable compared with the climatology. The upper trough played a role to destabilize the atmosphere, which provided environments for a very tall convection with lightnings.

4 A New Synthesized Understandings of Extreme Precipitation

Then, why the rainfall height of the record-breaking rainfall event in 2018 was not very tall? Conventional assumption on heavy rainfall is based on the buoyancy that a near-surface parcel can gain. Convective available potential energy, called CAPE, which is the total work done by buoyancy when the boundary layer parcel is raised to the level of free convection, is often used to measure the instability. It is effective

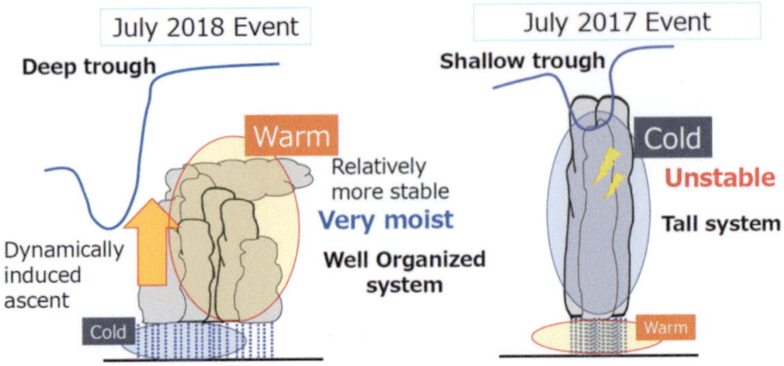

Fig. 12 Schematics of contrasting characteristics of heavy rainfall events in July 2018 and in July 2017. Adopted from Tsuji et al. 2020, JMSJ

to measure the potential for very tall convection with lightening under the condition with warm air below and cold air aloft.

However, another essential instability with very moist atmosphere has not been seriously considered even among many of the experts. It is the moist absolute unstable layer or MAUL (e.g., Bryan & Fritsch, 2000). When the atmosphere is very moist, even if the CAPE is not large, a layer lifting causes a kind of absolute instability to cause convection, and moreover, can be maintained by continuous free tropospheric moisture convergence. Moist absolute instability occurs in the very moist layer lifted upward by, for example, a slant upward velocity in mesoscale systems.

Tsuji et al. (2021) utilized 15 years of JMA's Mesoscale Model (MSM) analysis data to identify 62 heavy rainfall cases, and composited time-series of rainfall and moisture budget averaged over Kyushu-region (Fig. 13). They also diagnosed the MAUL grids over the Kyushu-region and their vertical distribution. Free-tropospheric moisture convergence preceded the rainfall peak and prepared MAUL condition in the lower-to-mid troposphere. The MAUL condition could maintain the heavy precipitation without very tall convection.

An example case for such heavy rainfall and MAUL condition is shown in Fig. 14. The left panel shows a vertical cross-section along the white line in the right figure at the precipitation peak time. Deep and intense inflow layer slantly ascents in the heavy rainfall area. MAUL conditions are satisfied (black circles) around the ascending area. These characteristics are consistent with the schematics of the MCS organized with a layer lifting (Bryan & Fritsch, 2000).

Here, an important finding was that free-tropospheric moisture convergence provides a moist environment before heavy rainfall. Under such very moist environment, an MCS with layer lifting produces intense precipitation without a large instability.

One of the candidates which prepares a very moist condition to realize MAUL, is a phenomenon called Atmospheric River (AR).Atmospheric River is a narrow band of large moisture flux from tropical latitudes to mid-latitudes. It brings deep moist layer in the troposphere (e.g., Ralph et al., 2011, Gimeno et al., 2014). Thus, ARs

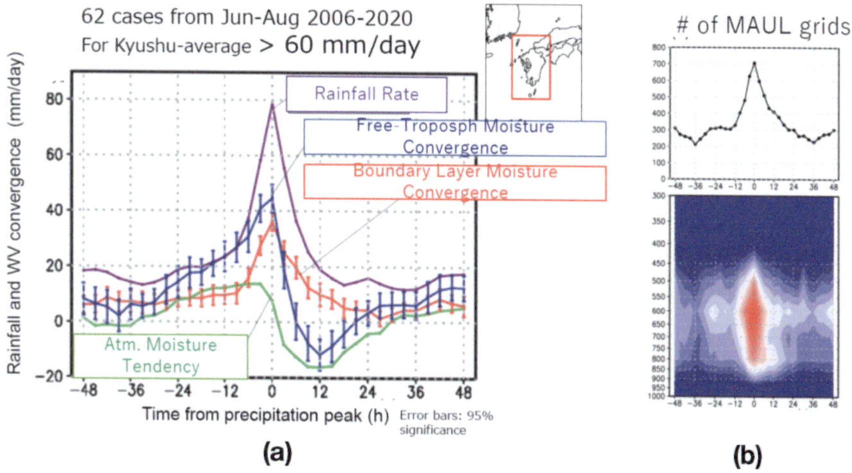

Fig. 13 **a** Composite time series of 62 heavy rainfall cases selected from June to August 2006–2020 over Kyushu region (inset red box) with thresholds of average rainfall >60 mm day^{-1}. Purple curve is for rainfall, Blue curve is for Free-tropospheric moisture convergence, red curve is for BL moisture convergence. The abscissa is hours from the average precipitation peak. **b** Number of MAUL grids diagnosed over the Kyushu-region and its vertical distribution. From Tsuji et al. (2021), GRL

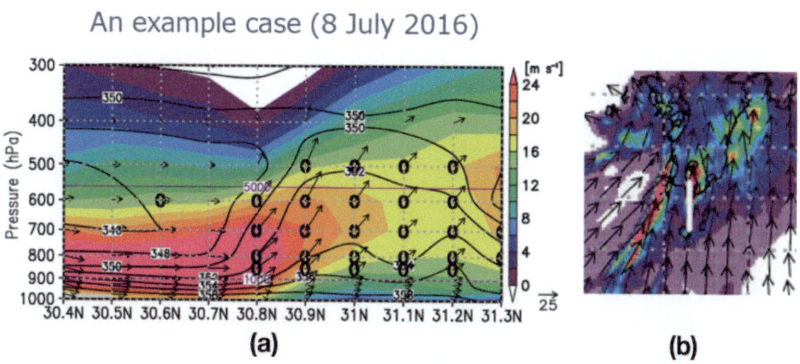

Fig. 14 An example case for heavy rainfall events and MAUL condition on 8 July 2016. Left panel shows a vertical cross-section along the white line in the right figure at the precipitation peak time. Color shows wind velocity along the cross-section, and black contours are equivalent potential temperature. Thick black circles indicate grid points that satisfy a MAUL condition. From Tsuji et al. (2021), GRL

can influence characteristics of rainfall systems, and especially, can affect extreme rainfalls. This phenomenon is famous for causing severe floods in the western coast of the United States, where climate is usually very dry. In moist monsoon regions, like western Japan in the Baiu season, it can prepare very moist atmosphere for MCSs which can maintain MAUL for severe rainfall.

Fig. 15 An image of the water vapor channel of Himawari 8, during the July 2020 severe flood of Kuma River in Kyushu. We can see a wide band of water vapor, or an Atmospheric River, extending from the Indian Ocean to the Baiu Front over Japan and beyond. Adopted from HP of JMA

Figure 15 is an image of the water vapor channel of JMA's geostationary satellite Himawari 8, during the July 2020 severe flood of Kuma River. We can see a wide band of water vapor, or an AR, extending from the Indian Ocean to the Baiu Front over Japan and beyond. Inside this moist band of AR, we could confirm continuous developments of cloud clusters, or MCSs, which covered the Kuma River region and prepared moist condition for heavy rainfalls (Tsuji & Takayabu, 2023).

5 Summary

In recent years, frequency of wide-area heavy rainfalls causing disastrous floods is increasing in Japan, as well as in the world. A significant increasing trend of heavy rainfalls and a decreasing trend of lighter rainfalls in Japan indicate a long-term change of precipitation characteristics. Precipitation transport the energy from the earth's surface to the atmosphere, thus playing a role to close the atmospheric energy budget. We need to utilize climate models to project the future change. However, it is extremely difficult to deal with precipitation characteristics in climate models.

Utilizing accumulated space-borne precipitation radar observations, it was found that extremely heavy rainfalls do not often accompany extremely tall convection. Environments for heavy rainfalls are relatively stable but extremely moist, compared to the climatology. Disastrous wide area heavy rainfall event in July 2018 was captured by the GPM DPR. It was shown from the radar observation, that such extreme rainfall event had moderate heights of precipitation.

By comparing heavy rainfall events in Kyushu in warm season, two types of heavy rainfalls are identified, one associated with instability, another with extremely moist atmosphere. Mesoscale Convective Systems contribute largely to such wide-area heavy rainfalls. The second kind instability 'MAUL' is realized in a very moist environment, thus enabled the occurrence of MCSs to maintain a wide area heavy rainfall. Atmospheric Rivers can prepare such very moist environments for rainfalls.

6 Concluding Message

The most important message in this chapter is that a huge number of data accumulated globally from satellite observations could renew our conventional assumptions and gave us important knowledge about extreme precipitation with statistical evidence. There is an urgent need to quantify the change of precipitation characteristics associated with the global warming for adequate adaptation measure for the sustainable society. To this end, it is essential to understand mechanisms of heavy rainfalls in various environments and their modifications with the climate change. The earth satellite observation provides us with such opportunities to renew our knowledge in physical processes and aid the representation of climate models.

Acknowledgements The author expresses her deep gratitude to late Prof. Tsuyoshi Nitta for showing her its large potential of space-borne precipitation radar observation, at the planning stage of the TRMM satellite. She also acknowledges her collaborators, Prof. Atsushi Hamada in Toyama University, Dr. Chie Yokoyama in the University of Tokyo, and Dr. Hiroki Tsuji in Remote Sensing Technology Center of Japan, for their intensive laboratory discussions in recent years. This research was supported by the Japan Aerospace Exploration Agency (JAXA) Precipitation Measuring Mission (PMM) science, the Environment Research and Technology Development Fund (JPMEERF20192004, JPMEERF20222002) of the Environmental Restoration and Conservation Agency of Japan, and the University of Tokyo through a project "Research hub for the big data analysis of global water cycle and precipitation in changing climate".

References

Allen, M. R., & Ingram, W. J. (2002). Constraints on future changes in climate and the hydrologic cycle. *Nature, 419*, 224–232. https://doi.org/10.1038/nature01092

Bryan, G. H., & Fritsch, J. M. (2000). Moist absolute instability: The sixth static stability state. *Bulletin of the American Meteorological Society, 81*, 1207–1230. https://doi.org/10.1175/1520-0477(2000)081%3c1287:MAITSS%3e2.3.CO;2

Gimeno, L., Nieto, R., Vázquez, M., & Lavers, D. A. (2014). Atmospheric rivers: A mini review. *Frontiers in Earth Science, 2.*

Hamada, A., Murayama, Y., & Takayabu, Y. N. (2014). Regional characteristics of extreme rainfall extracted from TRMM PR measurements abstract. *Journal of Climate, 27*(21), 8151–8169. https://doi.org/10.1175/JCLI-D-14-00107.1

Hamada, A., & Takayabu, Y. N. (2018). Large-scale environmental conditions related to midsummer extreme rainfall events around Japan in the TRMM region. *Journal of Climate, 31*, 6933–6945. https://doi.org/10.1175/JCLI-D-17-0632.1

Hamada, A., Takayabu, Y. N., Liu, C., & Zipser, E. J. (2015). Weak linkage between the heaviest rainfall and tallest storms. *Nature Communications, 6*(6213). https://doi.org/10.1038/ncomms 7213.

Houze, R. A., Jr. (1977). Structure and dynamics of a tropical squall-line system. *Monthly Weather Review, 105*, 1540–1567.

Houze, R. A., Jr. (2004). Mesoscale convective systems. *Review of Geophysics, 42*, RG4003. https://doi.org/10.1029/2004RG000150.

Kidd, C., Takayabu, Y. N., Skofronick-Jackson, G., Huffman, G. J., Braun, S., Kubota, T., & Turk, J. (2020). The Global Precipitation Measurement (GPM) mission. In V. Levizzani, C.

Kidd, D. Kirschbaum, C. Kummerow, K. Nakamura, & F. Turk (Eds.), *Satellite Precipitation Measurement. Advances in Global Change Research* (Vol. 67). Springer. https://doi.org/10.1007/978-3-030-24568-9_1.

Kojima, M., Miura, T., Furukawa, K., Hyakusoku, K., Ishikiri, T., Kai, H., Iguchi, T., Hanado, H., & Nakagawa, K. (2012). Dual-frequency precipitation radar (DPR) development on the global precipitation measurement (GPM) core observatory. In: *Proceedings of the SPIE 8528, Earth observing missions and sensors: Development, Implementation, and characterization II*, 85281A, 9 November 2012. https://doi.org/10.1117/12.976823

Leary, C. A., & Houze, R. A., Jr. (1979). The structure and evolution of convection in a tropical cloud cluster. *Journal of Atmospheric Science, 36*.

Lenderink, G., & van Meijgaard, E. (2008). Increase in hourly precipitation extremes beyond expectations from temperature changes. *Nature Geoscience, 1*(8), 511–514. https://doi.org/10.1038/ngeo262

Lenderink, G., Mok, H. Y., Lee, T. C., & vanOldenborgh, G. J. (2011). Scaling and trends of hourly precipitation extremes in two different climate zones—HongKong and The Netherlands. *Hydrology and Earth System Sciences, 15*, 3033–3041. https://doi.org/10.5194/hess-15-3033-2011

Ralph, F. M., Neiman, P. J., Kiladis, G. N., Weickmann, K., & Reynolds, D. W. (2011). A multiscale observational case study of a Pacific atmospheric river exhibiting tropical extratropical connections and a mesoscale frontal wave. *Monthly Weather Review, 139*, 1169–1189.

Skofronick-Jackson, G., Petersen, W. A., Berg, W., Kidd, C., Stocker, E. F., Kirschbaum, D. B., Kakar, R., Braun, S. A., Huffman, G. J., Iguchi, T., Kirstetter, P. E., Kummerow, C., Meneghini, R., Oki, R., Olson, W. S., Takayabu, Y. N., Furukawa, K., & Wilheit, T. (2017). The Global Precipitation Measurement (GPM) mission for science and society. *Bulletin of the American Meteorological Society, 98*, 1679–1695. https://doi.org/10.1175/BAMS-D-15-00306.1

Stephens et al. (2012). An update on earth's energy balance in light of the latest global observations. *Nature Geoscience*, 691–696.

Trenberth, K. E., Fasullo, J. T., & Kiehl, J. (2009). Earth's global energy budget. *BAMS, 2009*, 311–323.

Tsuji, H., & Takayabu, Y. N. (2023). A hierarchical structure of the heavy rainfall event over Kyushu in July 2020. *Journal of the Meteorological Society of Japan, 101*, 271–287. https://doi.org/10.2151/jmsj.2023-017

Tsuji, H., Takayabu, Y. N., Shibuya, R., Kamahori, H., & Yokoyama, C. (2021). The role of free-tropospheric moisture convergence for summertime heavy rainfall in western Japan. *Geophysical Research Letters, 48*, e2021GL095030. https://doi.org/10.1029/2021GL095030.

Tsuji, H., Yokoyama, C., & Takayabu, Y. N. (2020). Contrasting features of the July 2018 heavy rainfall event and the 2017 Northern Kyushu rainfall event in Japan. *Journal of the Meteorological Society of Japan, 98*. Special Edition on Extreme Rainfall Events in 2017 and 2018. https://doi.org/10.2151/jmsj.2020-045.

Utsumi, N., Seto, S., Kanae, S., Maeda, E. E., & Oki, T. (2011). Does higher surface temperature intensify extreme precipitation? *Geophysical Research Letters, 38*, L16708.

Yokoyama, C., Tsuji, H., & Takayabu, Y. N. (2020). The effects of an upper-tropospheric trough on the Heavy Rainfall Event in July 2018 over Japan, *98*, 235–255. Special Edition on Extreme Rainfall Events in 2017 and 2018. https://doi.org/10.2151/jmsj.2020-013.

Satellite Observations of Severe Storms: Indian Perspectives

Ashim Kumar Mitra, Pratiksha Dubey, Kavita Navria, and Chandan Mishra

Abstract The Imager and Sounder for enhanced meteorological observations are the main meteorological payloads of the multipurpose geosynchronous satellite INSAT-3D/R. INSAT-3D/R carries 6 channel imager and 19 channel sounder payloads. INSAT-3D/R offers fine resolution vertical profiles over India and the adjacent area, along with other polar satellite soundings. For use in numerical weather product models and day-to-day weather forecasting, these derived products' accuracy is crucial. For the NWP model assimilation cycle for a tropical nation like India, where high-impact convective outbreaks are frequent, good quality high-density observations on both a spatial and temporal scale are needed. Given the significance of satellite data, precise estimation of satellite observations may be the only practical alternative approach for locations with a lack of data. We now have an exclusive perspective on meteorological occurrences in the Indian subcontinent because of the existence of multi-spectral imager channels in INSAT3D/3DR and now INSAT3DS satellite with a staggered pattern of temporal frequency at every 15 min. In this study, Real-time Analysis of Products & Information Dissemination (RAPID), a web-based quick visualization and analysis tool for INSAT satellite data on a real- time basis has been introduced for identification of pre-monsoon severe weather events. The combination of channels using red–green–blue (RGB) composites of INSAT3D/3DR satellites and its physical significant value contents are presented. The solar reflectance and brightness temperatures are the major components of the RGB composite. The threshold technique has been developed separately for both the RGB products of the year (of March to June) of data for the detection of the thunderstorms prior to the events. A validation analysis was conducted using Forecast Demonstration Project (FDP) of Storm Bulletins for pre-monsoon weather systems prepared by India Meteorological Department (IMD) and RADAR observations, showing that the RGBs have a reasonable agreement with ground-based observations and RADAR data. This threshold technique yields a very good probability of thunderstorm detection more than 95 and 92% with acceptable false alarm conditions less than 5 and 10%

A. K. Mitra (✉) · P. Dubey · K. Navria · C. Mishra
Satellite Meteorology Division, India Meteorological Department, New Delhi 110003, India
e-mail: ashimmitra@gmail.com

S. Das and W-K. Tao (eds.), *Severe Storms*,
https://doi.org/10.1007/978-981-97-7075-5_2

for Day Microphysics (DMP) and Night Microphysics (NMP) respectively. Furthermore, limitations of these RGB products are also highlighted, and future scope of refinement of these products in perspective of upcoming Indian geostationary satellite missions is proposed. The thresholds techniques are found useful for day-to-day weather forecasting and being used operationally.

.

1 Introduction

Initial stage of instability is a very important factor in nowcasting and short-term forecasting. This is basically due to the ascent of large quantities of warm air, lifting forces and surface heating during summer time (Mitra et al., 2012). Generally, in weather forecasting, it is very useful to estimate when and where the initialization of instability is going to develop. For the purpose of forecasting severe weather, it is useful to predict the initial stage when atmospheric instability produces significant convection (Chrysoulakis et al., 2003Boyden (1963), George (1960), Rackliff (1962), Showalter (1953)).

The Thunderstorm activities over the Indian region starts from early March onward with the advancement of season in India. The thunderstorm frequency is higher during the pre-monsoon season (March–May) due to highly unstable atmosphere of high temperatures prevailing at surface levels. During pre-monsoon months (March–May), there are as many as 35–45 days of thunderstorm occurrences over northeast India (Science report, 2005) whereas over the northwest India the thunderstorm activity is the highest in the month of March to April, because of associated western disturbances. During the winter and pre-monsoon seasons in India Hail is observed (Das et al., 2014). Nizzamuddin (1993) has reported about 29 hail days per year (analysis of IMD reports from 1982 to 1989) of moderate to severe intensity. Eliot found that, out of 597 hailstorms in India, 153 yielded hailstones of 3-cm diameter or greater. The current operational NWP models are incapable of simulating the thunderstorm progress due to the dependency of thunderstorm on mesoscale and synoptic scale processes that act with boundary layer that is influenced by the prevailing synoptic regime.

In the current scenario, the effect of severe weather events particularly thunderstorms over the Indian subcontinent are high and it has not been studied exclusively by any means of thresholds criteria using satellite data in both spatial and temporal resolution. With the rapidly evolving technologies in the field of remote sensing, it is desirable to evolve such a algorithms, which can provide better problem-solving scheme. Therefore, satellite based thunderstorm monitoring have becomes popular for its enhanced spatial, spectral and higher temporal resolution for near real-time thunderstorm detection and monitoring. The INSAT-3D indigenous satellite has launched

Table 1 General specification of INSAT-3D imager

Spectral band	Wave length (μm)	Ground resolution (km)
VIS	0.55–0.75	1
SWIR	1.55–1.70	1
MIR	3.80–4.00	4
WVP	6.50–7.10	8
TIR 1	10.3–11.3	4
TIR 2	11.5–12.5	4

on 26th July 2013, and is provided a new opportunity to the Indian meteorologists and users. It has 6-imager channels to provide imaging capability of the earth disc from geostationary altitude in one visible (0.52–0.77 μm) and five infrared channels; 1.55–1.70 μm (SWIR), 3.80–4.00 μm (MIR), 6.50–7.10 μm (water vapour), 10.3–11.3 μm (TIR-1) and 11.5–12.5 μm (TIR-2) bands. The ground resolution at the sub-satellite point is nominally 1 km × 1 km for visible and SWIR bands, 4 km × 4 km for one MIR and both TIR bands and 8 km × 8 km for WV band. The Specification of INSAT-3D Imager is given in Table 1. The detailed information of INSAT-3D can be found on http://satellite.imd.gov.in/dynamic/INSAT3D_Catalog.pdf.

In the current paper, the author initiative is concerning to weather forecast, particularly for the to detect initial of stages of instability/convection developments such as thunderstorms using new scheme of RGB‘s with INSAT-3D satellite in IMD.

An on-line INSAT-3D visualization tool, RAPID have been developed by Space Application Center (SAC), India Space Research Organization (ISRO), Ahmadabad in consultation with IMD and made operational at India Meteorological Department (IMD), New Delhi. The web address is: http://rapid.imd.gov.in and shown in Fig. 1. The RAPID provides features of interest to the scientific community with no specific operating system/software/library/compiler required on the desktop. It can run on any machine and access through any browsers. It has all the features from overlaying map boundaries (World Coastline, State, District Boundaries, and Gridlines) to probing the digital INSAT-3D data on the fly in real- time. On spot computation of time-series, vertical profile, transect, area measurement, distance measurement over the pixel can be done.

In the current paper, an approach representing much of the physical information content of the INSAT-3D geostationary satellite using red–green–blue (RGB) composites image for identification of thunderstorms with a threshold is attempted. The solar reflectance component of the shortwave thermal Infrared (1.6 μm) (SWIR), visible (0.5 μm) and thermal IR channels (10.8 μm) representing cloud microstructure is known as DMP RGB and the BT differences between 10.8, 12.0 and 3.9 μm is known as NMP RGB. The thresholds technique has been developed separately for both the RGB products of the year (2015–16 and 2016–17 of March to June) of data for the identification of the thunderstorms. Overall 250 cases of thunderstorms

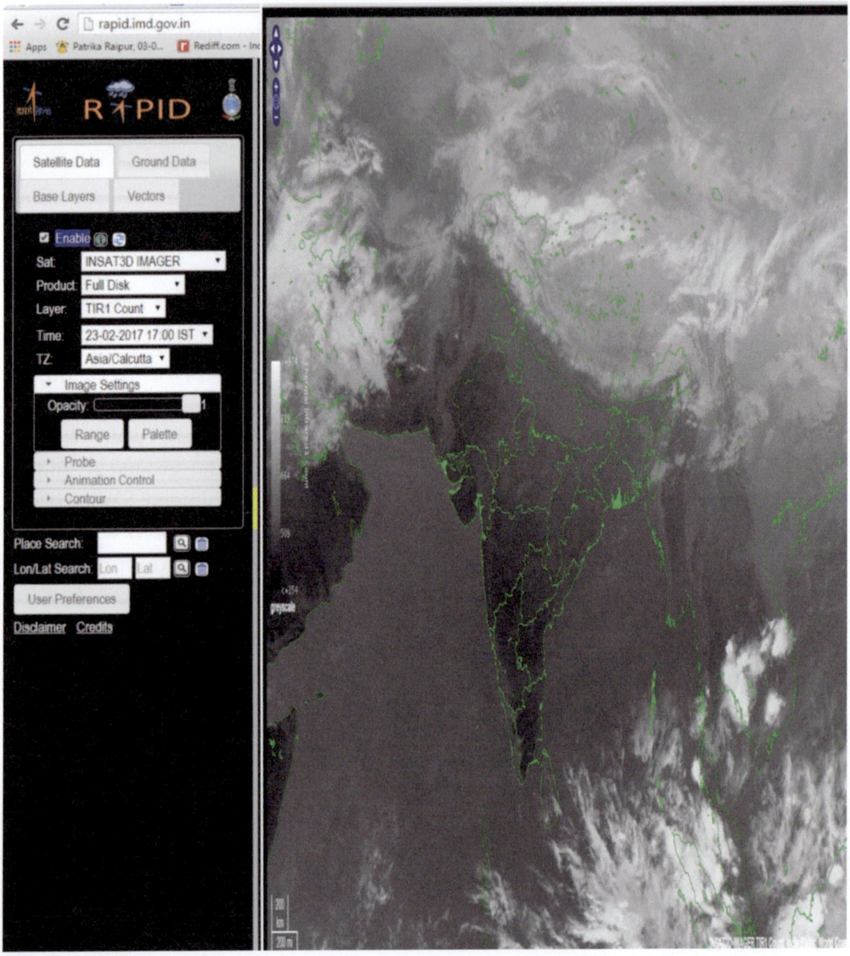

Fig. 1 RAPID on IMD website

including hail and squalls with RGB thresholds have been carried out for March to June 2018 against FDP Storm Bulletins of IMD (http://nwp.imd.gov.in/fdp_now/).

2 Methodology

2.1 Formation of RGB Microphysics Using INSAT-3D (Day Time and Night Time)

Since four decades, INSAT weather satellites have provided a vital tool for observing all types of weather phenomenon such as monsoon, fog, thunderstorms, tropical convections etc. India established its own geostationary satellite program, known as INSAT, in early 1980's which has enabled operational forecasters to continuously monitor the development of hazardous weather. Satellite imagery has shown that tropical convection is multi-scalar in nature ranging from isolated cumulus to mesoscale convective systems (MCSs) or mesoscale convective complexes (MCCs) (Tyagi et al., 2012). However, the extract, distil nature of the clouds and its characteristics, and its interpretation for forecasters to understand the weather and utilization, is still poses a challenge. The Red, Green, Blue or RGB processing offers a unique solution from different signature of spectral channels into meaningful inference for severe mesoscale phenomena than any individual IR and VIS grayscale images or traditional single-channel color enhancement techniques. In this article, the Lensky and Rosenfeld (2008) RGB technique (operational in EUMESAT) have been fine-tuned according to INSAT-3D multispectral channels specification and implemented to INSAT Meteorological Data Processing System (IMDPS), IMD, New Delhi. The solar reflectance and brightness temperatures are the major components of the RGB composite. The technique uses a meaningful RGB combination for a qualitative as well as quantitative analysis of the cloud microphysics of INSAT-3D Imager channels.

2.2 Day Time Microphysics Color Scheme

Table 2 shows an RGB composition of the DMP color scheme along with the spectral signature of several cloud types: 0.5, 1.6 and 10.8 μm in the red–green–blue beams and can be used only in day time. Figure 2 shows the INSAT-3D RGB composition of DMP imagery of 04-04-2015, 0400 UTC. In DMP RGB, snow and ice clouds appear red because they strongly absorb in 1.6 μm; small particle of ice cloud appears orange (western disturbance) in the imagery whereas the ice cloud with large particle (Mature Cumulonimbus (CB) cell) identified with the greater red color. Snow on the ground appears as full red because its grains are usually larger than cloud ice particles. Lensky and Rosenfeld (2008) and Rosenfeld and Lensky (1998) has very well demonstrated that the ice particles that form by mixed phase process in a supercooled water cloud grow quickly to much larger sizes than crystals forming by vapor deposition in ice-only clouds. This helps to separate convective precipitating clouds from non-precipitating or layer ice clouds. Due to the larger reflectance in the 1.6 and 0.5 μm, the desert appear bright cyan color. The 10.8 μm brightness temperature modulates

Table 2 Shows an RGB composition of the DMP color scheme

Beam	Channel	Range	Gamma
Red	VIS (0.55–0.75 um)	0... +100%	1.0
Green	SWIR (1.67 um)	0... +60%	1.0
Blue	IR (10.8 um)	+203... +323 °K	1.0

the blue. Ocean appears dark blue because of high thermal emission. The major applications of this color scheme are analysis of different cloud types, initial stages of convection, maturing stages of thunderstorm, identification of snow area, and the detection of fires. An ideal case of DMP colors scheme along with its physical values on 27-05-2016, 0400 UTC is shown by RAPID in Fig. 3. From the figure, it can be easily demarcated the low/medium/CB clouds and snow. The physical value of visible albedo for low/medium/CB clouds and snow are almost same i.e., approximately 50% but SWIR albedo is different; as in the case of snow is very low compared to low/medium/CB clouds. This is because of strong absorption in 1.6μzm. For mature CB cell, the large particle of ice cloud, lower the temperature and appears less than 220 K compared higher temperatures of low/medium/snow (above 250 K).

2.3 *Night Time Microphysics Color Scheme*

Table 3 shows an RGB composition of the NMP color scheme along with the spectral signature of several; cloud types: brightness temperature differences between TIR2-TIR1, TIR-1-MIR, and TIR1 in the red green blue beams respectively. and can be used only at nighttime only. NMP is used to monitor the evolution of night-time clouds/low stratus. It can also be used in the detection of fires, low-level moisture concentration and cloud classifications. Figure 4 shows the INSAT-3D RGB composition of NMP imagery of 23-04-2014, 2130 UTC. In NMP RGB, the BT difference between the 12.0 and 10.8 μm channels (12.0–10.8) is a measure for the clouds' opaqueness (Inoue, 1987) and it is displayed in red whereas the BT difference between 10.8 and 3.9 μm channels (10.8–3.9) modulates the green beam. The BT distinction in these two bands recognizes stratus clouds comprising of water droplets versus cirrus clouds and also cloud-free areas. Lensky and Rosenfeld (2003a) showed how sensitive is the BT difference between 10.8 and 3.9 μm channels to particle size, and used this information to delineate precipitation (Lensky and Rosenfeld, 2003b). The large difference between these two channels will indicate clouds with small drops. An ideal case of NMP colors scheme along with its physical values on 14-06-2016, 1730 UTC is shown by RAPID in Fig. 5. From the figure, it can be easily demarcated the low/medium and CB clouds during nighttime. It can be noted that the physical value of different classes, color schemes and its interpretation of low/medium clouds are different in day and nighttime microphysics. The tops of cold Cb radiation emitted from 10.8 μm is very low, resulting in low temperature (<220 K)

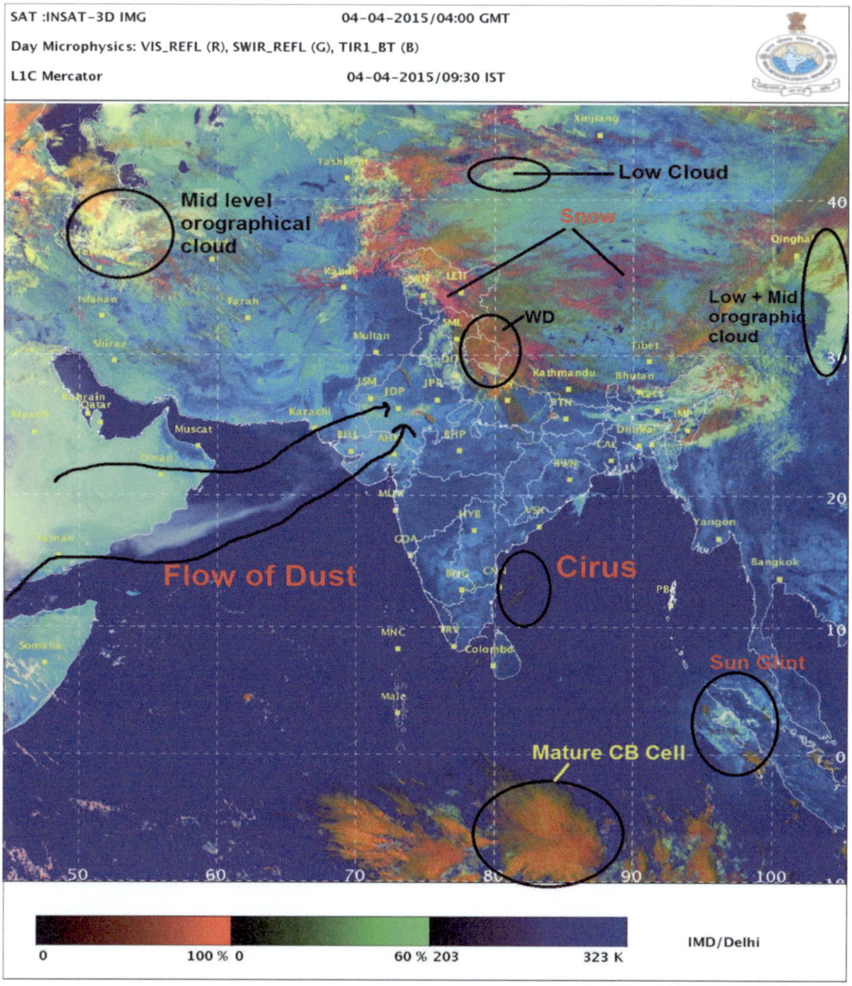

Fig. 2 INSAT-3D RGB composition of the DMP color scheme: 1.6, 0.5, and 10.8 μm reflectance/temperatures in the red green and blue beams respectively from 4 April 2015, 04:00 UTC

and showing sprinkled color at the mature Cb tops. Therefore mature Cb clouds at night seen in sprinkled orange-red color as compared to low and medium clouds.

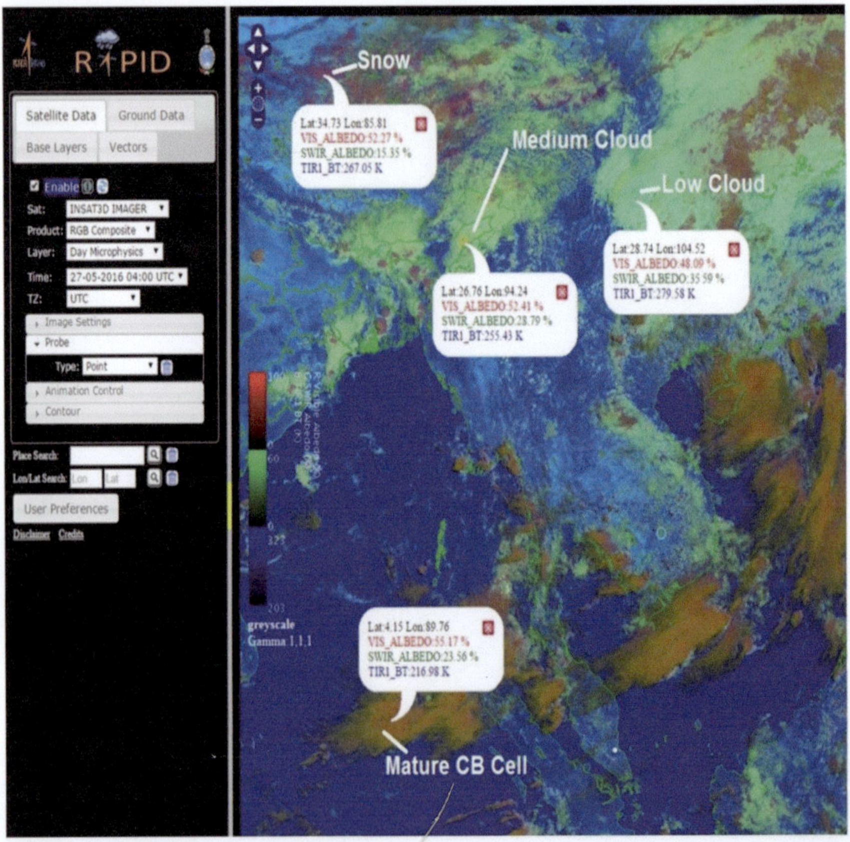

Fig. 3 INSAT-3D RGB composition of the DMP color scheme along with its physical values on 27-05-2016, 0400 UTC

Table 3 Shows an RGB composition of the NMP color scheme

Beam	Channel	Range	Gamma
Red	IR 12.0 um—IR 10.0 um (TIR2—TIR1)	−4… +2 K	1.0
Green	IR 10.8 um—IR 3.9 um (TIR1—MIR)	−4… +6 K	1.0
Blue	IR 10.8 um (TIR1)	+243… +293 K	1.0

3 Threshold Technique for Identifying Thunderstorm in DMP and NMP Prior to 3-h

The general methodology of thunderstorm detection using satellite involves identification of some threshold radiance or brightness temperature at different wavelengths. The channel TIR2 is referred to as the Split-Window IR channel and is centered at

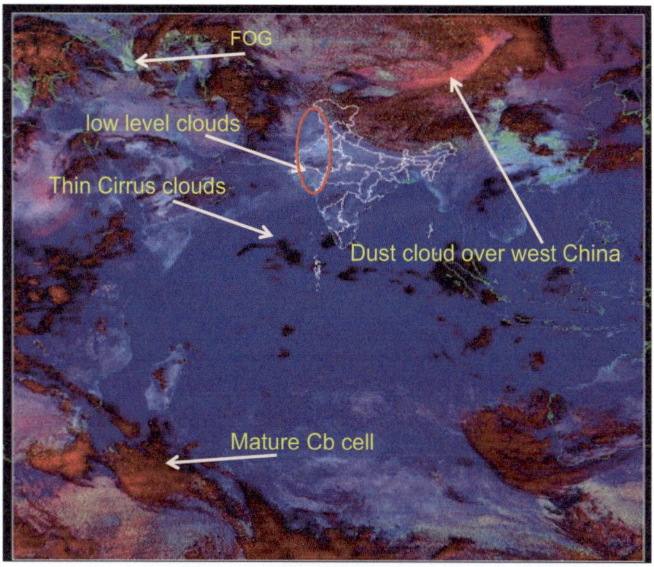

Fig. 4 INSAT-3D RGB composition of the NMP color scheme: 10.8, 12.0 and 3.9 μm reflectance/temperatures in the red green and blue beams respectively from 23-04-2014, 2130 UTC

approximately 12 microns whereas the channel TIR1 is the more standard IR channel and is centered at approximately 11 microns. At these two wavelengths, optically thin clouds emit different amounts of energy. Optically opaque clouds emit the same amount of energy at these wavelengths. By subtracting these two channels, regions of thin clouds (high, thin cirrus) can be distinguished properly. In contrast to TIR, INSAT-3D MIR channel is sensitive to reflected and emitted thermal energy at about 3.9 microns. This reflected energy relies on the relative points between the sun, clouds, the position of the satellite and reflecting objects for example, ocean, land, deserts, snow ground, ice particles and water droplets in clouds. The MIR channel can possibly segregate clouds composed of water versus those made out of ice particles. In the process, clouds composed of water reflect considerably more MIR energy than clouds composed of ice or snow-covered ground. The measure of reflection is dependent upon the span of the water drops and ice particles. During night, reflection is not adding to the estimations and any contrasts amongst water and ice/snow are purely based on emission.

The INSAT-3D RGB microphysics scheme is extremely valuable to the extent cloud analysis, mesoscale convections are concerned. In this scheme, water clouds that do not precipitate seem white, due to the fact that cloud drops are little, while large drops that are typical to precipitating clouds appear pink, in view of the low reflectance at 3.9 μm channel (Lensky & Rosenfeld, 2003a; Rosenfeld & Woodley, 2000). Supercooled water clouds appear more pink shading because of the lower temperature. Cold and thick clouds with tops made out of of large ice particles, e.g. CB tops, appear brown shading. In order to determine the thresholds criteria for

Fig. 5 INSAT-3D RGB composition of the NMP color scheme along with its physical values on 14-06-2016, 1730 UTC

identification of thunderstorms in RGB of DMP and NMP images, the composition of channels along with its physical values have been studied and analyzed from two years of (2015–16 and 2016–17 of March to June). The thunderstorms case including hails and squalls using RGB thresholds have been validated against FDP Storm Bulletins of IMD. The DMP and NMP thresholds along with its resolution and time of use for thunderstorm identification prior to the event (1- to 3-h) have been listed in the Table 4.

4 INSAT3D/3DR Sounder Derived Total Precipitable Water (TPW) and a Case Tudy

The atmospheric sounder payload has been introduced for the first time in Indian satellites, namely INSAT3D and INSAT3DR, to observe vertical profiles of temperature and humidity in the atmosphere over the Indian sub-continent. Continuous upper level temperature and moisture profiles with a temporal resolution of 30 min

Table 4 DMP and NMP thresholds along with its pixel resolution and time of use for thunderstorms

Product	Pixel resolution	Time	Source	Use (conditions)
Night time RGB (3.9, 10.8 μm)	1–4 km	13–02 UTC (of next day) half hourly	RAPID rapid.imd.gov.in	1. TIR2-TIR1 BT DIFF > must be negative and −5 to −2 or up to 1.5 K 2. TIR1-MIR BT must be −27 K to −1, if positive = Ice/snow clouds 3. TIR1 BT < 254 K ===> **Thunderstorm with rain** (i) TIR2-TIR1 BT DIFF > becoming positive and −2 to 2.5 K (ii) TIR1-MIR BT must be −28 K to 15 (iii) TIR1 BT < 208 K ===> **Thunderstorm with hail**
Day Time RGB (TIR, SWIR, Vis)	1–4 km	02:30 to 12:30 UTC half hourly	RAPID rapid.imd.gov.in	(i) TIR1 BT rate of change is (in consecutive hours) −5 to −15 K (ii) VIS 26–55% (increasing trend) (iii) SWIR < 28% (decreasing trend) and (iv) VIS > SWIR ==> **Thunderstorm with Rain** (i) TIR1 BT rate of change (in consecutive hours) is −5 to −38 K (ii) VIS 15–59% (increasing trend) (iii) SWIR < 24% (decreasing trend) and (iv) VIS > SWIR ==> **Thunderstorm with hail**

and a spatial resolution of 10×10 km can be obtained using the staggering mode. The INSAT3D/3DR sounder comprises eighteen channels in the shortwave infrared (SWIR), middle infrared (MIR), and longwave infrared (LIR) spectrums in addition to one visible spectral channel. Additional information about the INSAT3D sounder is available in Mitra et al. (2015).

In order to examine the possible impact of TPW, hourly GNSS TPW and INSAT-3D sounder generated TPW were analyzed for thunderstorm occurrences that happened in Shimla, Trivandrum, and Dibrugarh in June 2017 and displayed in Fig. 8. The IMD GNSS instrument was selected due to its ability to cover the entirety of India, its redundancy, availability, and access to multiple satellites. The example studies show that during thunderstorm outbreaks, the TPW derived from INSAT-3D closely matches the TPW observations made by GNSS. It can be observed that the value of TPW increases quickly—in less than one to three hours—prior to rainstorms, thunderstorms, and hailstorms. Together with other meteorological parameters, instability indices utilizing INSAT-3D TPW and mesoscale activities are readily identifiable and useful for weather forecasting.

Conversely, the subsequent case studies validate the application of TPW derived from INSAT-3D prior to an event and can be regarded as one of the precursors for convective processes. These days, one of the most important input parameters for numerical weather prediction used for nowcasting during severe weather events is TPW.

5 Lightning Detection Instruments in India

The forecasting of cloud to ground (CG) lightning flashes is of great importance. Because of lighting is a high-impact weather phenomenon that causes deaths and injuries in India every year, especially in the warm seasons from May to September. It also causes significant property damage and economical loses. The ideal tools for detecting thunderstorms are weather radars,

lightning detectors and satellite imagery. For this purpose, first time Lightning Location Networks (LLN) an auxiliary data set overlaid with weather radars reflectivity (dBZ) and INSAT-3D infrared satellite imagery retrieved from the India Meteorological Department (IMD) Network. The whole of India is covered by weather radars updated every 10 min and LLN observations updated every 15 min. Furthermore, the time difference between INSAT-3D satellite and LLN data will be around 1 h, due to the satellite scan strategy. The initial motivation is to explore the possibility of defining indicators for the occurrence of lightning activity, because inspection of individual case studies has shown a strong relation between the regions of cold brightness temperature and significant lightning activity. The overall compared results showed that the greater threshold value for reflectivity > 30 dBZ echo detected is the best indicator for predicting the beginning of lightning activity. In addition, we also observed in the operational analysis that the most of the cases with flash initiations have INSAT-3D satellite the coldest brightness temperature (−30 to −

50 °C) and this work is specifically applicable for data sparse regions, where radar systems are not available. An overall comparison of weather radar reflectivity and INSAT-3D satellite with respect to LLN ground-based data is an extremely valuable tool and will be highly beneficial for the nowcasting of severe weather events.

Lightning Location Network (LLN) data used in IMD are collected by the Indian Institute of Tropical Meteorology (IITM),Pune, India. Lightning information is stored in ASCII files, containing data for each lightning flash, such as the date-time of occurrence, its location in latitude and longitude, the polarity and strength of the electric current, and an indicator telling whether it is intracloud (IC) or CG. In addition, LLN can provide real-time detection, high detection efficiency and broad coverage, and has great potential in providing severe weather prediction and warning updated every 15 min interval. The LLN comprises more than 100 sensors over various parts of the country. The LLN detection efficiency is 90% and the associated location accuracy is 500 m varying slightly by region to region.

6 Results and Discussion

The occurrence of thunderstorms reported by IMD observations of FDP STORM bulletins is analyzed with RGB products i.e., DMP and NMP, based on the threshold criteria of given in the Table 4. Over all 16 cases of thunderstorm have been taken in their most intense phases during pre-monsoon season and shown in Table 5. On average, the total time duration of an intense thunderstorm was 2–3 h. It is important to mention that, in this study, prior to the thunderstorm events was considered as the time when convection appearance was not seen in the satellite imageries. Once the convection initiates, the RGB features within the initial hours is crucial to identify the location and increasing warning lead times. Traditionally, the satellite instruments only observe the uppermost cloud tops, revealing no information about the internal structure of the storm (Setvak et al., 2008, 2010). Taking full advantage of the high temporal sampling of INSAT-3D satellite and its spectral channels capability, particularly the four channels TIR, MIR, SWIR and Visible, this may be taken as a measure of convective activity.

Figure 6a and b illustrates the all the 16 cases of thunderstorms from initiation to dissipation stage of different cities all over the country. The evolution of the initiation and maturing stage of a thunderstorm is very clear when analyzed using the proposed RGB approach. In the day time, DMP, the Visible and SWIR pixel values were showing the opposite trend i.e., higher the value of visible reflection value (Visible > SWIR, by 70–80%), than SWIR and rate of change in the TIR1 pixel in the consecutive hours are representing the stronger convective activity. During maximum convection activity, the value of each parameter of RGB proved to be good indicators of differences in intensity from one case to another. In the night time, NMP, Fig. 6b, the BT difference between TIR1 and MIR, plays the major role to identify the area of location of convective activities along with TIR1 temperature. Differencing the MIR data from the TIR data produces negative values for clouds composed of

Table 5 Thunderstorm events during Pre-monsoon 2018

	Date	Occurrences of thunderstorms/ hails from FDP STORM bulletin (in UTC)	Place & central position (°N/ °E)	Weather reported from observations	Organization of the cells (Isolated single cells/ multiple cells/ convective regions/ squall lines) with height of 20 dBZ echo top and maximum reflectivity from IMD RADAR observation	INSAT-3D RGB composition *prior to 3-h*
DMP	27.05.2019	0520–0524	Lucknow	Thunderstorm and rain	Single cells with average height of 7.5 KM. Echo tops: 6.5 KM with maximum reflectivity: **48 dBZ**	Visible = 41% SWIR = 21% TIR(K) = 252
	20.04.2018	0105–0230	Purnia	Thunderstorm and rain	Multiple Cell. Echo Top: 13.0 KM with maximum reflectivity: **50 dBZ**	Visible = 46% SWIR = 27% TIR(K) = 212
	22.04.2018	1105–1150	Meerut	Thunderstorm and rain	Isolated/ multiple weak to moderate echoes were seen from DWR Lucknow, at 1050 UTC. With maximum reflectivity: **50 dBZ**	Visible = 52% SWIR = 25% TIR(K) = 253

(continued)

Table 5 (continued)

Date	Occurrences of thunderstorms/ hails from FDP STORM bulletin (in UTC)	Place & central position (°N/ °E)	Weather reported from observations	Organization of the cells (Isolated single cells/ multiple cells/ convective regions/ squall lines) with height of 20 dBZ echo top and maximum reflectivity from IMD RADAR observation	INSAT-3D RGB composition *prior to 3-h*
23.04.2018	0702–0730	Agartala	Thunderstorm and rain	Multiple cells were seen from DWR at 1000 UTC with maximum height 14 km with maximum reflectivity: **40 dBZ**	Visible = 58% SWIR = 27% TIR(K) = 250
03.05.2018	1015–1100	Bangalore	Thunderstorm and rain	Isolated/ multiple echoes were seen in DWR Chennai composite at 1010 h IST indicated significant isolated convective activity over Karnataka. Maximum height 10–15 km with maximum reflectivity: **>55dBZ**	Visible = 46% SWIR = 28% TIR(K) = 258

(continued)

Table 5 (continued)

	Date	Occurrences of thunderstorms/ hails from FDP STORM bulletin (in UTC)	Place & central position (°N/ °E)	Weather reported from observations	Organization of the cells (Isolated single cells/ multiple cells/ convective regions/ squall lines) with height of 20 dBZ echo top and maximum reflectivity from IMD RADAR observation	INSAT-3D RGB composition *prior to 3-h*
	16.05.2018	0630–0652	Sundernagar	Thunderstorm and rain	Significant convection were indicated DWR composite at 1720 h IST over south Himachal Pradesh and adjoining Uttrakhand	Visible = 43% SWIR = 28% TIR(K) = 256
	20.04.2018	1135–1155	Jagdalpur	Thunderstorm and hailstorm with hail diameter 0.3–0.5 cm	Isolated convection were seen DWR composite at 1610 h IST over east Telangana adjoining Chhattisgarh	Visible = 46% SWIR = 28% TIR(K) = 258

(continued)

Table 5 (continued)

	Date	Occurrences of thunderstorms/hails from FDP STORM bulletin (in UTC)	Place & central position (°N/°E)	Weather reported from observations	Organization of the cells (Isolated single cells/multiple cells/convective regions/squall lines) with height of 20 dBZ echo top and maximum reflectivity from IMD RADAR observation	INSAT-3D RGB composition *prior to 3-h*
	20.05.2018	1030–1045	Phalodi	Thunderstorm and Hailstorm with hail diameter 0.2–0.4 cm	Isolated/multiple echoes were seen in DWR Delhi, Patiala & Jaipur. Maximum height 10–12 km with maximum reflectivity: **50 dBZ**	Visible = 54% SWIR = 24% TIR(K) = 236
NMP	28.05.2018	1140–0010	Bareilly	Thunderstorm and rain	Strong multiple convection were indicated DWR composite at 1640 h IST over Uttarakhand	TIR2-TIR1(K) = −2.06 TIR2-MIR(K) = 3.6 TIR1(K) = 249
	29.05.2018	2310–0100	Jaipur	Thunderstorm and rain	Scattered convection were indicated DWR composite at 1720 h IST over NE Rajasthan	TIR2-TIR1(K) = −0.16 TIR2-MIR(K) = 2.1 TIR1(K) = 244

(continued)

Table 5 (continued)

	Date	Occurrences of thunderstorms/ hails from FDP STORM bulletin (in UTC)	Place & central position (°N/ °E)	Weather reported from observations	Organization of the cells (Isolated single cells/ multiple cells/ convective regions/ squall lines) with height of 20 dBZ echo top and maximum reflectivity from IMD RADAR observation	INSAT-3D RGB composition *prior to 3-h*
	29.05.2018	1310–1445	Kota	Thunderstorm and rain	Scattered convection were indicated DWR composite at 1720 h IST over Rajasthan	TIR2-TIR1(K) = −2.4 TIR2-MIR(K) = −11 TIR1(K) = 249
	09.05.2018	1237–1330	Akola	Thunderstorm and rain	Scattered significant convection were indicated DWR composite at1640hrs IST over Southeast Madhya Pradesh, Vidarbha, Extreme east Maharashtra. With maximum reflectivity: **50 dBZ**	TIR2-TIR1(K) = −0.5 TIR2-MIR(K) = −6 TIR1(K) = 252

(continued)

Table 5 (continued)

Date	Occurrences of thunderstorms/hails from FDP STORM bulletin (in UTC)	Place & central position (°N/°E)	Weather reported from observations	Organization of the cells (Isolated single cells/multiple cells/convective regions/squall lines) with height of 20 dBZ echo top and maximum reflectivity from IMD RADAR observation	INSAT-3D RGB composition *prior to 3-h*
07.05.2018	2300–0130	Kanyakumari	Thunderstorm and rain	Multiples significant convection were indicated DWR composite at 1630 IST over Tamilnadu including interior Kerala. With maximum reflectivity: **48dBZ**	TIR2-TIR1(K) = −2.9 TIR2-MIR(K) = −7.7 TIR1(K) = 243
28.04.2018	0240–0320	Guwahati	Thunderstorm and rain	Scattered significant convective activities observed DWR composite at 1610 UT Cover NE Tripura & Guwahati. With maximum reflectivity: **45dBZ**	TIR2-TIR1(K) = −0.9 TIR2-MIR(K) = −10 TIR1(K) = 255

(continued)

Table 5 (continued)

	Date	Occurrences of thunderstorms/ hails from FDP STORM bulletin (in UTC)	Place & central position (°N/ °E)	Weather reported from observations	Organization of the cells (Isolated single cells/ multiple cells/ convective regions/ squall lines) with height of 20 dBZ echo top and maximum reflectivity from IMD RADAR observation	INSAT-3D RGB composition *prior to 3-h*
	22.04.2018	2228–2231	Coimbatore	Thunderstorm and hailstorm with hail	Significant convection were indicated DWR composite at 1620 h IST over north Tamilnadu isolated/ multiple strong echoes were seen in echo tops: >15KM with maximum reflectivity: **>55dBZ**	TIR2-TIR1(K) = −0.34 TIR2-MIR(K) = 29 TIR1(K) = 233
	28.04.2018	2315–2325	Bikaner	Thunderstorm and hailstorm with hail	Scattered significant convective activities observed DWR composite at 1610 UTC over NE Rajasthan. With maximum reflectivity: **48dBZ**	TIR2-TIR1(K) = 0.9 TIR2-MIR(K) = −16 TIR1(K) = 247

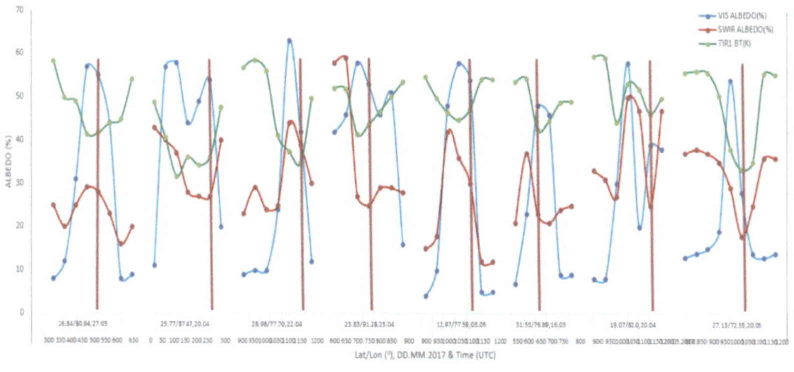

(a) DMP

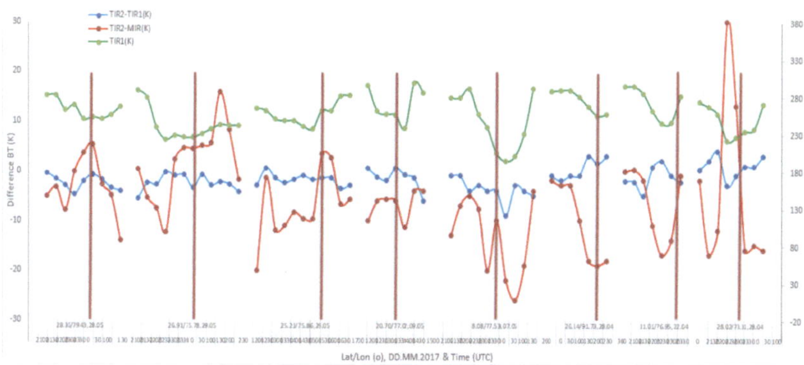

(b) NMP

Fig. 6 **a** and **b** illustrates the all the 16 cases of thunderstorms from initiation to dissipation stage

water and near-zero or positive values for ice/snow clouds. The combined analysis of TIR2-TIR1, TIR1-MIR and TIR1 data, showed notable differences in extreme and less severe cases. This routine analysis of RGB parameters was very useful as a diagnostic and prognostic tool for the following few hours of the thunderstorms.

6.1 Validation with Actual Observations and Radar Data

Figure 7 displays a comparison between radar and NMP RGB satellite imageries of 23rd May 2018. It can be seen from the Kolkata DWR Composite at 11:21 UTC indicating strong convection over East & North East Jharkhand adjoining West Bengal. This was squall line echo (Max dBZ 55–60 and height > 15 km). From RAPID, the RGB imagery was also showing the significant convective clouds over the same area. The largest radar reflectivity is seen in areas near NMP pixels. The NMP parameters

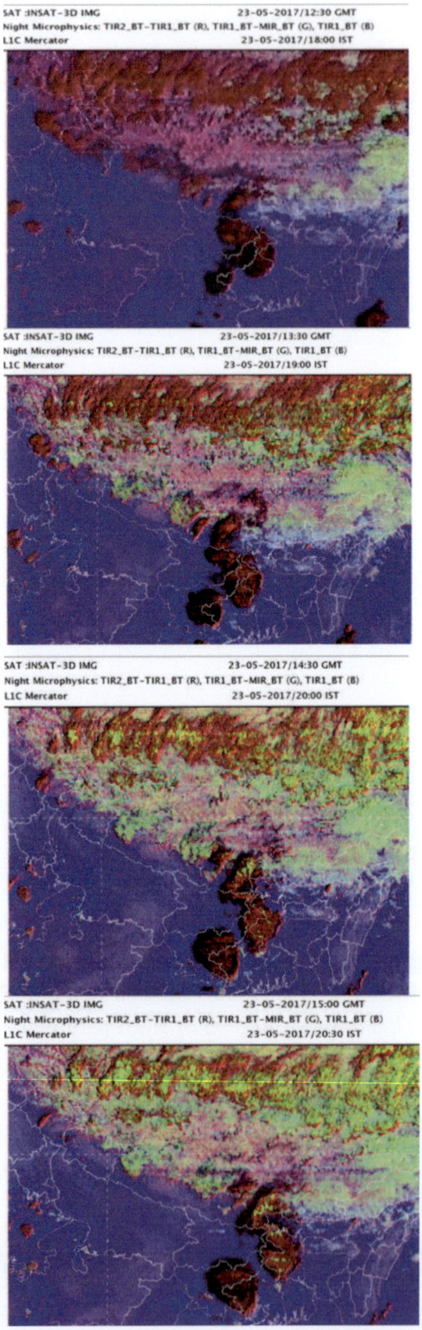

Fig. 7 Night time RGB and radar reflectivity (dBZ) for 23rd May 2018

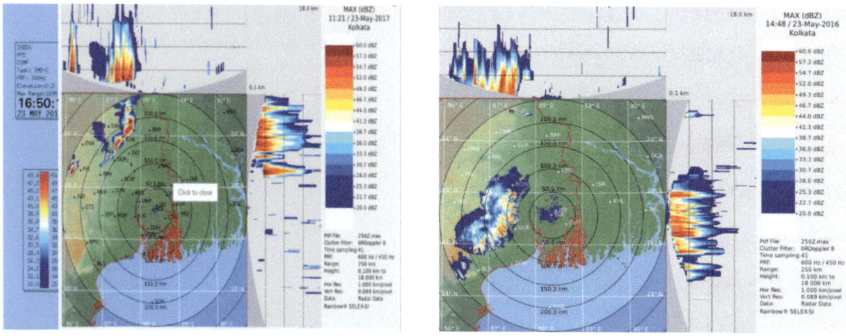

Fig. 7 (continued)

at 12:30 UTC was, $TIR2\text{-}TIR1(K) = -1.26; TIR2\text{-}MIR(K) = -12.3; TIR1(K) = 249$, meeting the thresholds criteria of a severe thunderstorm. The associated convective clouds in the NMP images were suggested the occurrence of a thunderstorm over Kolkata and accordingly warning were issued by the IMD‘s FDP STORM bulletin of 23rd May 2018 (*FDP STORM Bulletin No.79 (23-05-2018)*), and IMD Kolkata Office has issued the Nowcast Bulletin. This event was very well forecasted. The thunderstorm was reported at between 14:30 and 15:00 UTC. Multiple cells are shown by radar which transformed into big cells with a maximum reflectivity of 58.5 dBz with height > 18 km. This exercise suggests that thunderstorms events can be more precisely estimated by satellite using RGB technique of every half an hour interval, with the stronger cores of reflectivity highlighted by radar parameter. Both the instruments were in good agreement in terms of delineating areas of deep convection (Fig. 7).

A quantitative validation of RGB threshold technique has been carried for both, DMP and NMP product and shown in Table 6. The radar data was used to evaluate convection evolution. It can be seen that most of the occurrences of convective activity period were estimated very well by the both the RGB schemes. It can be seen from the Table 6, the percent of detection using DMP threshold is 95%, whereas NMP threshold is 92%. The false detections are 7 and 9%, for DMP and NMP respectively. This may be due to the many factors such as high viewing angle scan during early morning and late evening. Over the hilly region such as Sub- Himalayan West Bengal, Western Ghats, Nilgiris in southern parts of the country where surface heating has rapidly intensified. Hence, localized factor plays an important role along with synoptic forcing for any convection formation. The navigation of the satellite image may also add some false alarms.

However, INSAT-3D RGB scheme provides important precursors for the study of convective formations specifically for severe weather such as thunderstorm and their growth prior to the event. The synoptic information would not provide the internal structure for the study of different life stages and intensity in the development of mesoscale convection in the pre-monsoon season over the Indian region. Even though the lead time of detection of severe weather is very low, the RGB scheme has provided

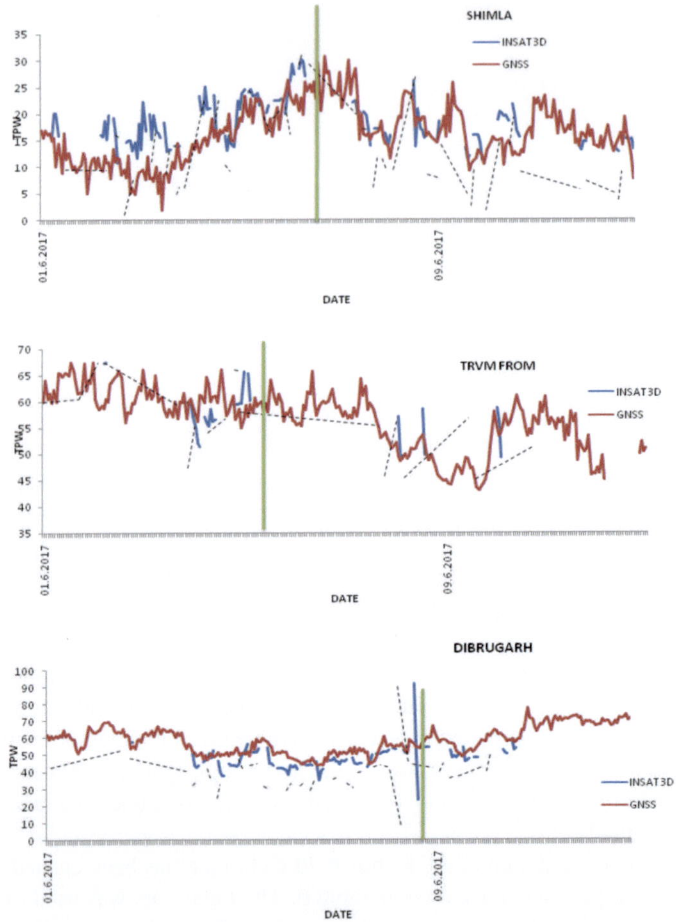

Fig. 8 INSAT-3D & GNSS TPW and hourly realized thunderstorms events at Shimla, Trivandrum and Dibrugarh in1st to 10th June 2017). Green line is the time of occurrence of thunderstorms. Dotted lines are cloudy scene of INSAT3D

Table 6 Statistics of DMP and NMP observed using threshold criteria

	Day micro physics	Night micro physics
Total no. of observations	116	134
Number of miss	3%	5%
No. of hits	97%	95%
Percent of detection (POD)	94%	93%
Percent of false detection (due to navigation)	7%	9%
Critical success index	90%	87%

vital information with regard to development of mesoscale convection over the region and demonstrated very well for nowcasting of mesoscale convection from half-hourly observations from INSAT-3D RGB.

7 Limitations of These RGB Products

The availability of multispectral Imager channels in INSAT-3D satellite from its predecessor, provide us a new way to look at weather events such as different types of clouds and its microphysics. The most effective way to utilize these products is in the use of RGB combinations such as DMP and NMP, in the RAPID tool. The different RGB combinations have the relative advantage for observing different phenomena. At present at IMDPS, the DMP and NMP scheme is applied only on Imager.

At present at IMDPS, the DMP and NMP scheme is applied only on Imager. In DMP, water and ice cloud which is more sensitive to the different cloud microstructure, some time lacks the valuable information of cloud top temperature and reduced the delectability of thin clouds in the DMP. In the orographic region, the development of CB cell also sometimes give rise to false alarm because of rising air motion originating at high level (can be called orographic wave clouds) from the surface and misleading in the NMP. The Lensky and Rosenfeld (2003a) technique is basically applied on rich spectral information of Spinning Enhanced Visible and Infrared Imager (SEVIRI) on board METEOSAT Second Generation (MSG, 11 bands spectral information)—the European geostationary satellite operated by EUMETSAT, which has multispectral capability for cloud microphysical characterization and aerosol-cloud interaction detection. In the current INSAT-3D channels, we have used only 6-band spectral information imager. In the RAPID tool, the large reflectance of the CB and lower level clouds is sometimes not detectable because of unavailability of 0.8 and 1.6 μm in the DMP color scheme. The difference between the 1.6 and 0.6 μm of solar reflectance play a major role for discriminating the ice clouds and surface. The large negative difference between the 1.6 and 0.6 μm indicates ice clouds and a much larger positive difference between the 1.6 and 0.6 μm is typical for the surface. Another missing channel in the INSAT-3D satellite is an 8.7 μm channel, which is most sensitive to clouds' particle effective radius in the night. This particular channel can also be used in the sand and dust discrimination. The RGB scheme has also some has limitations during dawn to dusk period.

To improve the forecasting accuracy for the detection of severe weather and deep convection phenomena, a more flexible and powerful observing strategy is also needed. A rapid scan strategy is required during pre-monsoon season over the Indian region. Purdom (1996a, b) and Levizzani and Setvák (1996) has discussed the benefits of very rapid scan imagery (5 min interval) from GOES-8 for the study of explosive convection and demonstrated the great impact of rapid scan down to 30 s repetition time for the analysis of the genesis and evolution of squall lines and their fine structure. Test on INSAT-3DR satellite at every 5-min interval has been conducted over the Indian region. Next season, this will be implemented in the IMDPS.

8 Conclusions and Future Scopes

In this study, we have presented the day and nighttime RGB scheme using RAPID visualization tool for INSAT-3D satellite data on a real-time basis for identification of weather events. The solar reflectance and brightness temperatures are the major components of the RGB composite from INSAT-3D spectral channels. The thresholds technique have been developed separately for both the RGB products of year (2016–17 and 2017–18 of March to June) of data for the detection of the thunderstorms prior to the events. Both the RGBs i.e., DMP and NMP have reasonable agreement with ground-based observations and RADAR data. This threshold technique yields a very good probability of thunderstorm detection more than 94 and 93% with acceptable false alarm conditions less than 7 and 9% for DMP and NMP respectively The threshold's techniques are found to be very useful for day-to-day weather forecasting and being used by forecasters and user community operationally.

In order to provide us with meteorological occurrences throughout the Indian area every 15 minutes, work is also underway to apply the DMP and NMP color scheme over the 19 channels of the recently deployed INSAT-3DS satellite and the INSAT-3DR Sounder. For future Indian geostationary satellite GISAT-2 missions, these products will also undergo additional refinement. An Indian geo-imaging satellite called GISAT-2 can deliver photos rapidly during severe weather emergencies. GISAT-2 will have a resolution of 50m to 1.5km. Future 4th generations of INSAT satellites are planned to be equipped with more advanced payloads.

Acknowledgements *Authors are very much grateful to SAC, Ahmadabad team for their technical, software expertise and implementation of 'RAPID' tool at IMDPS, New Delhi. We specially thank Shri Ghansyam Sanger and Nitesh Kausik, scientists of SAC for consulting with IMD and their valuable suggestions while developing 'RAPID'. The first author thanks to I. M. Lensky and D. Rosenfeld for CAPSAT tool information.*

References

Boyden, C. J. (1963). A simple instability index for use as a synoptic parameter. *Meteorological Magazine, 92*, 198–210.

Chrysoulakis, N., Spiliotopoulos, M., Domenikiotis, C., & Dalezios, N. (2003). Towards monitoring of regional atmospheric instability through MODIS/AQUA images. In *Proceedings of the International Symposium* (pp. 7–9). Volos.

Das, S., Mohanty, U. C., Tyagi, A., Sikka, D. R., Joseph, P. V., Rathore, L. S., Habib, A., Baidya, S., Sonam, K., & Sarkar, A. (2014). The SAARC STORM: A coordinated field experiment on severe thunderstorm observations and regional modeling over the South Asian Region. *Bulletin of the American Meteorological Society, 95*, 603–617.

George, J. J. (1960). *Weather forecasting for Aeronautics* (pp. 409–415). Academic.

Inoue, T. (1987). An instantaneous delineation of convective rainfall areas using split window data of NOAA-7 AVHRR. *Journal of Meteorological Society of Japan, 65*, 469–481.

Lensky, I. M., & Rosenfeld, D. (2003a). Satellite-based insights into precipitation formation processes in continental and maritime convective clouds at nighttime. *Journal of Applied Meteorology, 42*, 1227–1233.

Lensky, I. M., & Rosenfeld, D. (2003b). A night rain delineation algorithm for infrared satellite data based on microphysical considerations. *Journal of Applied Meteorolgy, 42*, 1218–1226.

Lensky, I. M., & Rosenfeld, D. (2008). Clouds-Aerosols-Precipitation Satellite Analysis Tool (CAPSAT). *Atmospheric Chemistry and Physics, 8*, 6739–6753.

Levizzani, V., & Setvák, M. (1996). Multispectral, high-resolution satellite observations of plumes on top of convective storms.

Mitra, A. K., Sharma, A. K., Bajpai, I., & Kundu, P. K. (2012). An atmospheric instability derived with MODIS profile using real-time direct broadcast data over the Indian region. *Natural Hazards, 63*(2), 1007–1023.

Nizzamuddin, S. (1993). Hail occurrences in India. *Weather*, https://doi.org/10.1002/j.1477-8696.1993.tb05847.x.

Purdom, J. F. W. (1996a). One minute interval imaging of atmospheric phenomena using NOAA's new generation of geostationary satellites. In *Preparation for Eighth Conference on Satellite Meteorology and Oceanography, AMS* (pp. 164–167).

Purdom, J. F. W. (1996b). Advanced atmospheric studies using GOES-8/9 multichannel imagery. In *Proceedings of 1996 Meteorological Satellite Data Users' Conference, EUMETSAT EUM P* (Vol. 19, pp. 77–86).

Rackliff, P. G. (1962). Application of an instability index to regional forecasting. *Meteorite Magazine, 91*, 113–120.

Rosenfeld, D., & Lensky, I. M. (1998). Satellite-based insights into precipitation formation processes in continental and maritime convective clouds. *Bulletin of the American Meteorological Society, 79*, 2457–2476.

Rosenfeld, D., & Woodley, W. L. (2000). Deep convective clouds with sustained supercooled liquid water down to -37.5 °C. *Nature, 405*, 440–442. https://doi.org/10.1038/35013030

Science Plan. (2005). Severe Thunderstorms Observations and Regional Modeling (STORM) programme. Department of Science and Technology, Government. of India.

Showalter, A. K. (1953). A stability index for thunderstorm forecasting. *Bulletin of the American Meteorological Society, 34*, 250–252.

Setvak, M., Lindsey, D. T., Rabin, R. M., Wang, P. K., & Demeterova, A. (2008). Indication of water vapor transport into the lower stratosphere above midlatitude convective storms: Meteosat second generation satellite observations and radiative transfer model simulations. *Atmospheric Research, 89*, 170–180.

Setvak, M., Lindsey, D. T., Novak, P., Wang, P. K., Radová, M., Kerkmann, J., Grasso, L., Su, S.-H., Rabin, R. M., Šťástka, J., & Charvát, Z. (2010). Satellite-observed cold-ring-shaped features atop deep convective clouds. *Atmospheric Research, 97*, 80–96. https://doi.org/10.1016/j.atmosres.2010.03.009

Tyagi, A., Sikka, D. R., Goyal, S., & Bhowmik, M. (2012). A satellite based study of pre-monsoon thunderstorms (Nor'westers) over eastern India and their organization into mesoscale convective complexes. *MAUSAM, 63*, 1, 29–54.

Mitra, A. K., Bhan, S. C., Sharma, A. K., Kaushik, N., Parihar, S., Mahandru, R., Kundu, P.K. (2015) INSAT-3D vertical profile retrievals at IMDPS. New Delhi: A preliminary evaluation in Mausam, 66, 4, 687–694.

Structures of Severe Storms Observed by Dual Polarization Doppler Weather Radar

V. Chandrasekar and Patrick C. Kennedy

Abstract Many severe storm phenomena have horizontal length scales in the mesoscale range (on the order of 10–100 km). In the warm season, these relatively small-scale severe weather elements include damaging, thunderstorm-driven winds, rainfall intensities that hamper transportation and eventually overwhelm water drainage systems, and concentrated hailfalls. Cold season severe weather conditions are typically associated with high snowfall rates and significant accumulations of snow and ice. Weather radar provides an effective means of observing mesoscale severe weather systems in all seasons. Advances in weather radar technology have expanded the scope of information available for the real time recognition of severe weather. Radial velocity measurements obtained with Doppler-capable hardware provide considerable insight on three-dimensional storm motion fields. Dual polarization observations have allowed significant improvements in the identification of hydrometeor types and estimations of precipitation rates. In this chapter, we present examples of weather radar data collected during a variety of severe weather conditions.

Keywords Severe thunderstorm · Tornado · Flash flood · Winter storm · Hurricane · Doppler/dual polarization weather radar

1 Introduction

The nature of severe storms varies by climatological area. For example, hail and tornadoes in continental areas, hurricanes in tropics blizzards in polar regions, etc. In this chapter we have categorized severe storms based on the primary surface weather phenomena that they produce. To support real time identification of severe weather

V. Chandrasekar (✉) · P. C. Kennedy
Colorado State University, Fort Collins, CO 80523, USA
e-mail: chandra@colostate.edu

V. Chandrasekar
Finnish Meteorological Institute, Helsinki, Finland

S. Das and W-K. Tao (eds.), *Severe Storms*,
https://doi.org/10.1007/978-981-97-7075-5_3

patterns, we primarily present images of basic polar coordinate radar data fields of the type that are typically available from current technology Dual Polarization and Doppler weather radars. A number of these example cases are a subset of examples that have been posted at the CSU-CHILL radar facility: http://www.chill.colostate.edu/w/Articles.

2 Summary of Dual Polarization Observations

A Dual-polarized weather radar measures the amplitude and phase of echoes for one or more polarizations. The amplitude is proportional to the size, composition, and number of scatterers. The phase is related to the distance to the scatterers, their velocity, and scattering properties. There are three common Doppler radar observations which are estimated, namely: equivalent reflectivity factor (Ze), mean Doppler velocity (Vr), and Doppler spectrum width (σv). With dual polarization using an alternate transmission and simultaneous receive mode (Chandrasekar et al., 2023), three additional variables are estimated: differential reflectivity (Zdr), differential phase shift (Φdp) and copolar correlation (ρhv). In some radars, linear depolarization ratio (LDR) and cross polar correlation are also available.

Reflectivity is commonly used to designate the intensity of the precipitation in weather radar data displays, it is the basic measurement observed by the radar. For coherent radars, the mean Doppler velocity and the Doppler spectrum width can also be measured. These additional measurements complement the microphysics information provided by the radar reflectivity with the capability of sensing the atmospheric dynamics. Similar to Doppler radar measurements, the polarimetric radar observations rely on measurements of the amplitude and phase of the signal from two orthogonal (horizontal and vertical) polarizations. (The terms observations and measurements are used interchangeably here). Both the propagating and the scattered waves can be described by their complex amplitudes. Several radar parameters can be derived based on the magnitude of the backscattered wave at horizontal and vertical polarizations, their phase, or both. The most straightforward additional measurement that a polarimetric radar can provide is the differential reflectivity Zdr, i.e., the ratio of the equivalent reflectivities at horizontal and vertical polarizations (Zh and Zv respectively). Note that Zdr, just like reflectivity, is often expressed in logarithmic scale (dB).

Zdr reveals information about the shape, composition, and density of the hydrometeors. For any given particle shape, Zdr varies depending on the composition of the particle (i.e., water, ice, or mixed phase), because of the different relative permittivity of the material (Chandrasekar et al., 2023). For example, oblate raindrops will have a larger Zdr than solid ice particles with the same shape. Low density aggregates such as snowflakes will have even smaller Zdr values (typically close to 0 dB), due to the small effective relative permittivity of low bulk density, diluted ice-air mixtures. The differential phase shift Φdp (i.e., the phase difference between the received signals at horizontal and vertical polarization) can be highly informative about the composition and concentration of the particles along the propagation path. The copolar correlation coefficient between the horizontal and vertical polarization signals (ρhv) provides a measure of the similarity of the horizontal and vertical scattering cross section of the objects. A higher value of ρhv (close to 1) is typically observed in rain as a result of the consistency of the raindrops shape and fall behavior. Lower correlation values indicate a greater variability in the population of scatterers. For example, reduced ρhv is normally observed in the melting layer (coexistence of liquid, solid, and mixed phase particles with varying shape), or from non-meteorological echoes such as tornado debris, insects, birds. Finally, for radars that transmit one polarization at a time and receive both the co-polar and cross-polar signals, an additional parameter can be estimated. For example, if the radar transmits a horizontal polarization, the linear depolarization ratio LDR is given by the ratio of the power received on the vertical polarization (cross-polar signal) to the power received on the horizontal polarization (co-polar signal). Like ρhv, LDR can be a good indicator of regions where a mixture of precipitation types occurs (typically when the radar resolution volume is populated with randomly oriented particles). In other instances, LDR can be used to detect preferential orientation of the hydrometeors, such as the more vertical alignment of ice crystals that develops in the presence of strong electric fields in thunderstorms (Chandrasekar et al., 2023).

3 Wind

3.1 Thunderstorm Outflow

In the mature and dissipating thunderstorm life cycle stages, downdrafts, typically characterized by relatively cold air temperatures, reach the surface and generate outflowing winds (Byers & Braham, 1948). Convergence along the leading edge of these outflows (the gust front) often collects radar-detectable scatterers into fine line echoes (Wakimoto, 1982). The strength and size scale of these thunderstorm outflows varies significantly from case to case from microbursts (4 km or less outflow diameter) to regional scale outflows (derecho echo systems affecting 100 km scale areas (Fujita, 1981). Monitoring low elevation angle radial reflectivity and velocity data patterns is critical to the identification of high wind threats in convective weather situations.

Figure 1a shows a 2.3° elevation angle PPI scan taken by the CSU-CHILL radar during an episode of damaging surface winds in northeastern Colorado. The fine line echo associated with the gust front had passed the radar and was located within the 10 km range ring to the northeast of the radar. The thunderstorm line generating this outflow was beyond the 30 km range ring southwest of the radar. The reflectivity levels in this convective line were decreasing. At this time, the overall reflectivity pattern was not particularly threatening in terms of echo intensity or organization level.

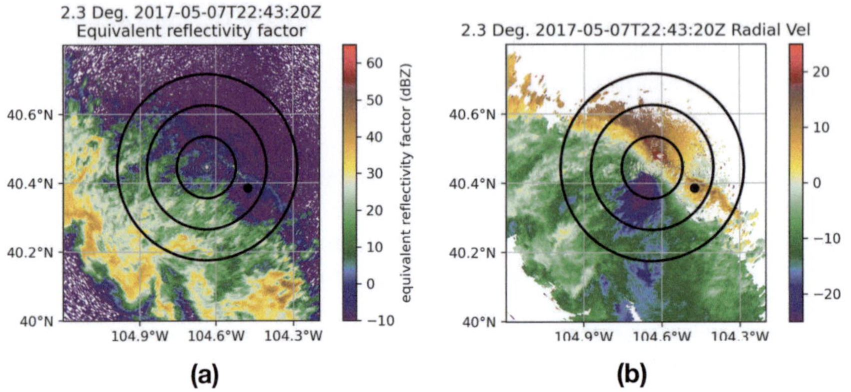

Fig. 1 **a** CSU-CHILL radar S-band reflectivity. Range rings are 10, 20, 30 km. **b** Radial velocity values (ms^{-1}) corresponding to Fig. 1a. The black dot marks a spotter report of 75 mph (34 ms^{-1})

Figure 1b shows the radial velocity data corresponding to Fig. 1a. Inbound (negative sign) radial velocity values reach magnitudes of ~25 ms^{-1} just southwest of the radar. These strong (dark blue) radial velocities only extend out to approximately 20 km range. This is due to the strongest winds being confined to low height levels. The existence of high winds was confirmed by a volunteer weather observer located at the black dot approximately 15 km southeast of the radar. This observer estimated the wind speed to 75 mph (34 ms^{-1}) with significant blowing dust and the downing of a tree. These conditions were noted within two minutes of the radar data time.

3.2 Tornado

Tornadoes have long been recognized as the source of the Earth's highest observed surface wind speeds (Wurman et al., 2014). Due to the extreme damage capabilities of these winds, the identification and tracking of tornadoes has been a goal from the early efforts to apply World War II radar technology to weather observations. The first recorded success in the recognition of a tornadic echo on a radar display took place on 9 April 1953 near Champaign Illinois, using a 3 cm wavelength US Navy surplus radar that was continuously scanning at a low elevation angle (Fig. 2). The track of the hook-shaped appendage emanating from the southwestern portion of an intense, isolated parent thunderstorm echo was found to be well correlated with the extreme wind damage path on the ground and public tornado sightings (Huff et. al., 1954; Fujita, 1958). Over subsequent years, observations of the hook echo signature have been found to be a useful radar reflectivity pattern for the real time identification of tornadoes (Forbes, 1981). However, it should be noted that tornadoes can also develop from echo systems that do not contain well-defined hook echoes. For example, tornadogenesis can occur when developing thunderstorm updrafts interact

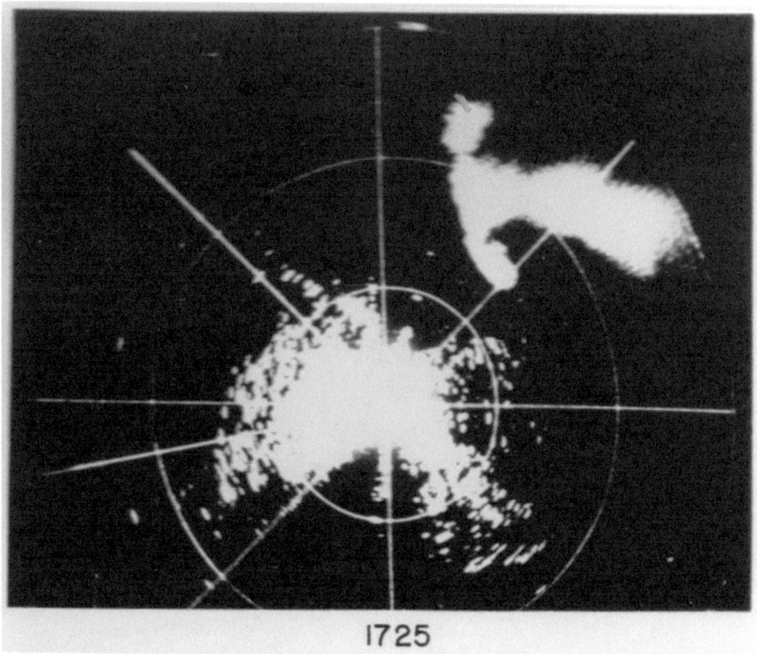

Fig. 2 Hook echo located 15 nautical miles northeast of the Champaign Illinois airport on 9 April 1953 at 1725 local time. Data collected by photographing an analog cathode ray tube display of received power from a low elevation PPI scan. Range rings at 10 nautical mile intervals, irregular bright echo near the radar is ground clutter

with surface convergence lines (Wakimoto, 1989), in association with larger scale "bookend circulations" in bow echo/squall line systems (Atkins & St. Laurent, 2009), and in hurricane rainbands (Gentry, 1983; Nowotarski et al., 2021).

Continuing severe thunderstorm research revealed that hook echo was associated with the existence of an intense, mesoscale cyclonic circulation. Hydrometeors swept along by this mesocyclone formed the overall low-level hook pattern. The three-dimensional airflows in these rotating "supercell" thunderstorms were found to persist over time scales 2–3 times longer than the ~45 to 60-min lifetimes of non-severe airmass thunderstorms (Browning, 1964).

As Doppler radar measurements became available in the 1970s, radial velocity patterns composed of azimuthally adjacent regions of inbound and outbound (opposite sign) velocities were found to be very useful for identifying areas of rotation. In tornadic supercell thunderstorms, two distinct circulation scales were found: The larger mesocyclone with a diameter on the order of 10 km, and the much smaller (few km)/few range gates scale tornado vortex signature (TVS) associated the tornado (Brown et. al., 1978).

Figure 3a shows the reflectivity pattern observed in a tornadic supercell thunderstorm viewed by the 11 cm wavelength CSU-CHILL radar. The red box marks the

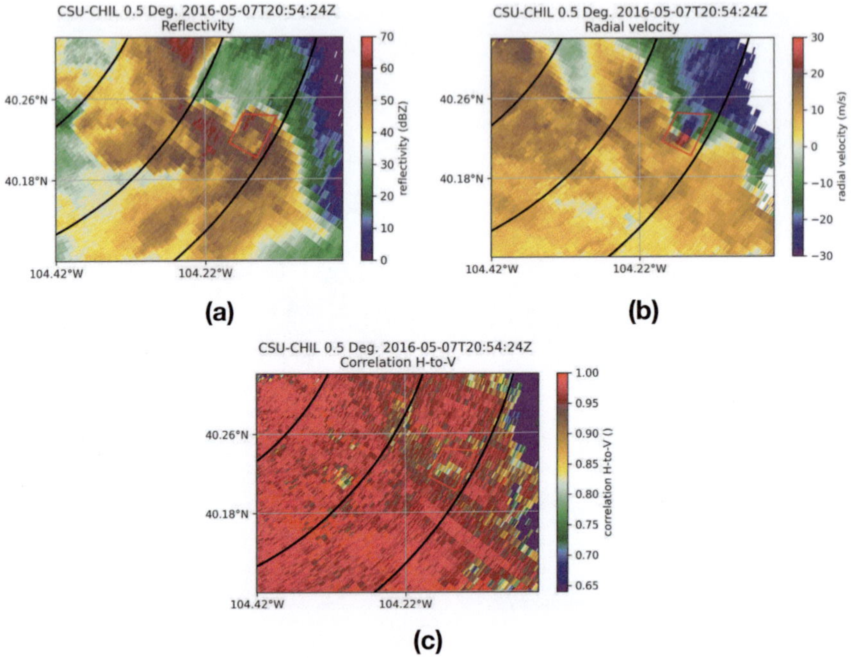

Fig. 3 **a** 7 May 2016 tornado case S-band reflectivity. Range rings 30, 40, 50 km. **b** 7 May 2016 radial velocity (ms^{-1}). Outbound motions have positive sign, inbound motion is negative. **c** ρ_{hv} data in the 7 May 2016 case

tornado signature area. The southern portion of the parent supercell's >60 dBZ reflectivity core is located along the northern edge of the plot. This hook echo also contains a reduced reflectivity "eye" within the boxed area. These eye features apparently are due centrifugal forces acting on larger diameter radar targets in intense, small radius circulations (Wakimoto & Martner, 1992).

Figure 3b shows the corresponding radial velocity field. A tornado vortex signature (TVS) exists inside the boxed area. This is identified by the strong (~$\pm$25 ms^{-1}) inbound/outbound range gate velocity maxima that occur on adjacent radials. The surrounding dark blue/inbound velocities located to the northeast of the TVS and dark brown/outbound velocities to the southwest are indicative of the larger scale mesocyclone.

The radial velocity information, especially as demonstrated by the TVS, is clearly useful in locating the maximum tornado threat location. (Storm spotter reports confirmed a multi-vortex tornado on the ground associated with the TVS at this time). As with all radar measurements, the ability to detect near-surface tornado signatures decreases with increased range as the above ground height of the radar sample volumes increases. The probability of detecting low level hook and radial velocity couplets is appreciably lower at ranges beyond ~70 km.

Dual polarization data provides additional, confirmatory tornado identification information. The correlation coefficient between co-polar horizontally and vertically polarized radar returns (ρ_{hv}) is significantly reduced when the pulse volume contains irregularly shaped and randomly oriented debris raised from the surface by the tornado (Ryzhkov et. al., 2003). For reference, hydrometeor echoes generally have ρ_{hv} levels above 0.95 while correlation from debris returns is often below ~0.85. As a result, the presence of these low correlation levels in association with the hook echo and TVS signatures can be used as the basis for the issuance of tornado warnings even in the absence of human spotter reports. Figure 3c shows an example of the tornado debris signature (TDS) within the boxed region.

The presence of Mie scattering and mixtures of different types of particles are important factors in reducing correlation levels (Bringi & Chandrasekar, 2001). Since Mie scattering develops sooner at shorter wavelengths (i.e., 5 and 3 cm), radars operating at wavelengths shorter than 10 cm may observe lower ρ_{hv} in debris echoes, improving the visibility of the TDS pattern.

As noted earlier, tornadoes can also develop in association with non-supercell thunderstorms (i.e., thunderstorms without a mid-level mesocyclone or a low-level hook echo, etc.) Fig. 4 shows the combined 0.5° elevation angle radial velocity and reflectivity patterns during the development of a non-supercell tornado that formed ~22 km northeast of the Denver Colorado WSR-88D radar (KFTG). At the initial time (2319 UTC, Fig. 4a), the general afternoon boundary layer echo pattern contained a well-defined cyclonic shear line to the northeast of the radar. This shear line is an example of the Denver convergence/vorticity zone (DCVZ) flow field that develops when southeasterly low level synoptic flow interacts with the local terrain (Brady & Szoke, 1989). Small scale rotation features (X = 13, Y = 15 in Fig. 4a) often develop along this shear line. Ten minutes later (2329 UTC, Fig. 4b), the expanding reflectivity contours associated with a thunderstorm that was rapidly developing above the 0.5° scan surface appeared near X = 18, Y = 15 km. This storm's low-level convergence acted to enhance local circulation along the shear line. This resulted in the appearance of a well-defined cyclonic rotation signature at X = 15, Y = 18 km at 2339 UTC. Several spotter reports confirmed the existence of a tornado associated with this rotation. When thunderstorms develop close to preexisting shear lines in the boundary layer flow, forecasters should be alert to the possibility of the development of non-supercell tornadoes.

3.3 Microburst

Downdrafts in convective systems generate rapidly evolving surface outflows on a variety of horizontal scales. Small diameter outflows with a separation of 4 km or less between the maxima in the diverging near surface winds have been designated microbursts (Fujita, 1981). When these small-scale downdraft/diverging outflow air motion fields occur within a few km of airport runways, landing and departing aircraft may encounter serious difficulties in maintaining safe flight paths during low altitude

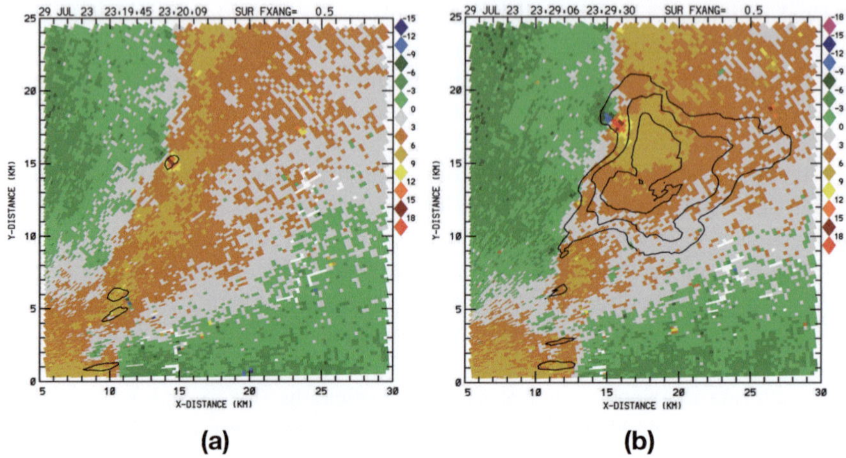

Fig. 4 **a** Color fill is radial velocity (ms^{-1}, away motion positive) observed in a 0.5° PPI scan made by the NWS WSR-88D radar near Denver, Colorado at 2319 UTC on 29 July 2023. The radar is the origin for the axes distance labels. Solid black contour is 25 dBZ. **b** As in 4a except time is 2329 UTC. Reflectivity contours are 25, 35, 45 dBZ

microburst encounters (Fujita & Caracena, 1977). In addition to aviation hazards, the localized, rapidly strengthening surface winds associated with microbursts often pick up movable surface objects and raise dust clouds. The sudden onset of these conditions can have dangerous impacts on outdoor gatherings, highway driving conditions, etc.

Low elevation Doppler radar observations can be used to identify the divergent radial velocity pattern produced when downdrafts spread out as they approach the surface. Figure 5 shows a composite plot of a microburst centered near X = 8, Y = −12.5 km (~15 km range) in a 1.5° PPI scan done by the CSU-CHILL radar. The two range rings pass through the local maxima in outbound and inbound radial velocities, the 4 km spacing between these centers indicates that the outflow is at the upper end of the microburst size scale this time. The location of the thunderstorm generating this outflow is shown by the gray shaded areas where reflectivities of 53 dBZ or greater exist. It should be noted that these fairly high reflectivity levels are not necessary for the development of microbursts, precipitation shafts that completely evaporate during their descent towards the surface ("dry microbursts") can also generate significant outflowing surface wind speeds (Wilson et. al., 1984).

An example of a younger microburst with a smaller diameter is shown in Fig. 6. The two range rings are 25 and 27 km. The bulk of the high reflectivity associated with this microburst cell is within a diameter of 2–3 km (Fig. 6a). The corresponding radial velocity maxima in the divergence pattern are only separated by the 2 km range increment (Fig. 6b). Assuming a groundspeed of ~65 ms^{-1}, a typical value for a jet transport aircraft flying near the ground during landing or takeoff, the radial velocity divergence feature would be completely transited in ~30s. The resultant

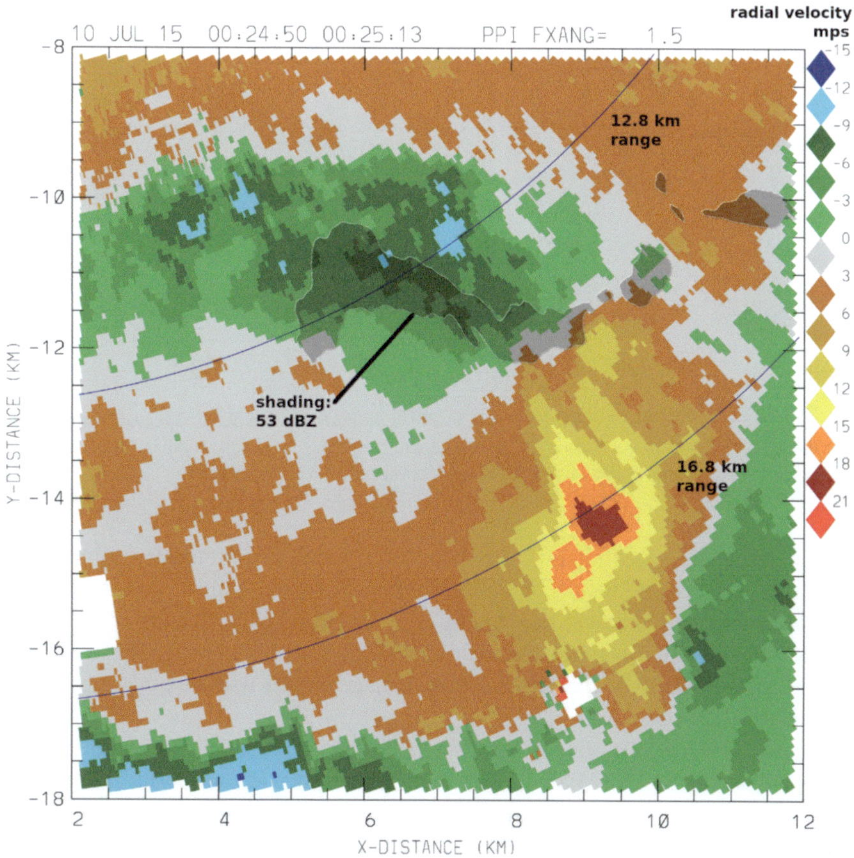

Fig. 5 CSU-CHILL X-band data from a 1.5° elevation PPI scan at 0025 UTC on 10 July 2015

sudden loss of visibility and onset of significant airspeed fluctuations compromise the maintenance of safe landing and takeoff altitude profiles.

3.4 Squall Line/Derecho

When individual thunderstorms become organized into quasi-circular complexes (mesoscale convective systems, MCS (Maddox, 1980)), or into extensive linear systems (squall lines and derechos Johns & Hirt, 1987), the surface area threatened by strong outflow winds increases significantly from the region affected by single thunderstorms. The upward mass flux in the leading portions of these thunderstorm complexes typically drives an extensive area of stratiform precipitation in the trailing portions of the echo system. The melting and evaporation/sublimation experienced

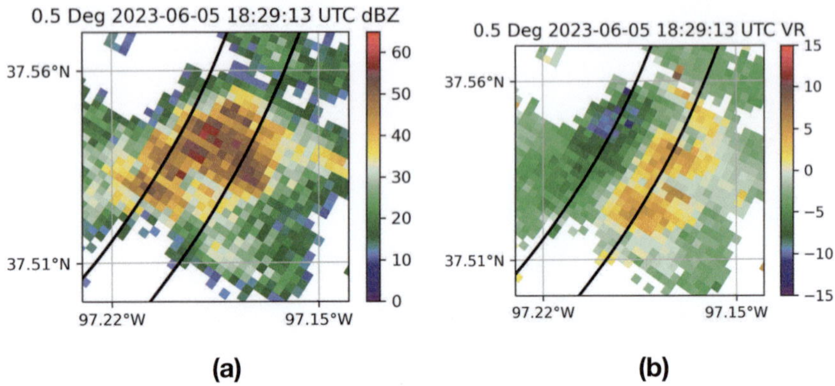

Fig. 6 **a** Reflectivity (dBZ) observed in a 0.5° PPI scan by the NWS WSR-88D radar at Wichita KS (KICT) at 1829 UTC on 5 June 2023. Range rings are 25 and 27 km. **b** As in 6a except color fill is radial velocity (ms^{-1}, positive values are motion away from the radar.)

by this descending precipitation generates mesoscale environmental cooling. This stratiform area cooling results in the development of a rear inflow jet that descends and reaches the surface near the leading edge of the complex (Davis et al., 2004). The development of this mesoscale rear inflow jet promotes and sustains potentially damaging outflow winds that can result in damage swaths with length scales on the order of 100's of km.

Figure 7a shows the reflectivity data collected at 0.5° elevation angle by the National Weather Service WSR-88D radar at Minneapolis Minnesota during an event that produced widespread reports of wind damage collected during the previous 90-min period (black dots). The primary reflectivity pattern feature is the essentially continuous line ~50 dBZ echo mass that curves across most of the 240 km diameter of the outer range ring. This line was advancing rapidly towards the northeast. This convective line has been in existence long enough to generate the broad region of ~25–45 dBZ echo in the trailing stratiform precipitation region.

The corresponding radial velocity pattern is shown in Fig. 7b. The strongest velocities (~40 ms^{-1}) are found along the leading edge of the outflow, generally in the direction of the overall convective system motion. Linear thunderstorm systems that assume this outward-bowed configuration are particularly likely to generate high surface wind speeds (Davis et al., 2004). MCS and squall line systems generally have lifetimes that span several hours. Thus, monitoring radar data from "upstream" sites is useful for recognizing the threat that an approaching mesoscale convective system may present to a local forecast area.

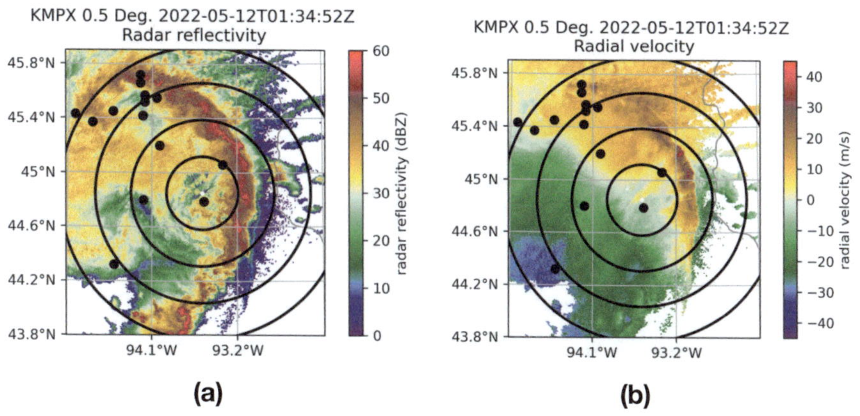

Fig. 7 **a** Minneapolis MN WSR-88D reflectivity data in a 0.5° PPI scan at 0135 UTC on 12 May 2022. Range rings: 30, 60, 90, 120 km. **b** Radial velocities (ms^{-1}) corresponding to Fig. 7a

4 Hail

4.1 Damaging Hail

Historically, when only echo intensity data was available, the presence of hail was inferred based on the observation of reflectivity levels above a threshold level, (often ~53 dBZ), measurement of great thunderstorm echo top height (Donaldson, 1961), and quantification of the vertical extent of high reflectivities (Greene & Clark, 1972). Dual polarization radar's capability to sense hydrometeor shape and orientation characteristics allows significant refinement in hail detection. The aerodynamic forces acting on hailstones generally cause the stones to gyrate through an appreciable range of orientation angles (Knight & Knight, 1970). In contrast, the net effects of aerodynamic pressures and surface tension forces cause raindrops, especially larger diameter drops, to assume a horizontally flattened cross sectional shape. The difference in the preferred orientations of hailstones and raindrops is detectable by radars using the dual linear (horizontal and vertical) polarization basis. These radars measure the co-polar received signal levels for both the horizontal and vertical polarization returns. The typical horizontal orientation of oblate raindrops will generate a horizontally polarized return signal that is several dB stronger than the vertically polarized returns from the same pulse volume. Since differential reflectivity (Zdr) is based on the ratio of the H and V returns:

$$\text{Zdr} = 10\log_{10}(\text{Zh}/\text{Zv})$$

Zdr will be a positive quantity from raindrops, while the fairly equal H and V returns from tumbling hailstones will result in a Zdr near 0 dB (Wakimoto & Bringi, 1988, Fig. 9).

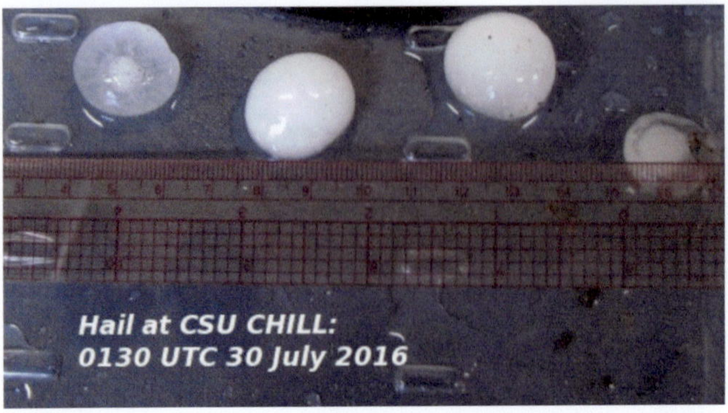

Fig. 8 Hailstones collected after the storm passage on 30 July 2016. The top ruler scale is in mm, lower (inverted) scale is in inches

A hailstorm that passed over the CSU-CHILL radar site during the early UTC hours of 30 July 2016 left the hailstones shown in Fig. 8. Figure 9 shows example 0.9° elevation angle PPI data collected as the storm subsequently moved off to the south of the radar. The reflectivity contained a well-defined >60 dBZ core (red irregular box area along the 30 km range ring in Fig. 9a). The Zdr field from this same sweep (Fig. 9b) shows the near 0 dB values associated with randomly oriented hailstones in the red outlined area. In contrast, positive Zdr prevails in the surrounding echo areas where oblate raindrops are the primary hydrometeor type.

The measurement of Linear Depolarization Ratio (LDR) requires the use of an alternating transmit polarization sequence and a receiver system that allows both the co-polar and cross-polar received signal levels to be measured. Many dual polarization radars simultaneously transmit H and V polarization, which prevents measurement of the cross-polar component. When LDR is available, it provides another dual polarization indication of the presence of hail (Herzegh & Jameson, 1992). LDR is based on the ratio of the cross-pol and co-pol return signal levels, where the cross-polar signal is typically defined by the vertically polarized received signal level following the transmission of a horizontally polarized pulse, and the co-polar signal is the usual H received level following an H pulse transmission. Hailstone shapes are typically not perfectly spherical, especially at larger diameters (Knight, 1986). When the hailstone gyrations bring the largest ice particle dimension to an appreciable angle (~30–40°) away from the polarization plane of the incident microwave radiation, the canted hailstone will cause a measurable portion of the backscattered radiation will be directed into the cross-polar plane. This depolarization is magnified when hailstone has a high liquid water content, especially on its outer surface. These factors typically increase the LDR from hail to levels greater than ~−25 dB, while the LDR from the more stably oriented raindrops is usually below −25 dB (Bringi & Chandrasekar, 2001). An example of this hail LDR enhancement is shown in Fig. 9c where the highest depolarization levels are closely associated with low Zdr hail area marked in red. The additional regions with LDR values slightly above

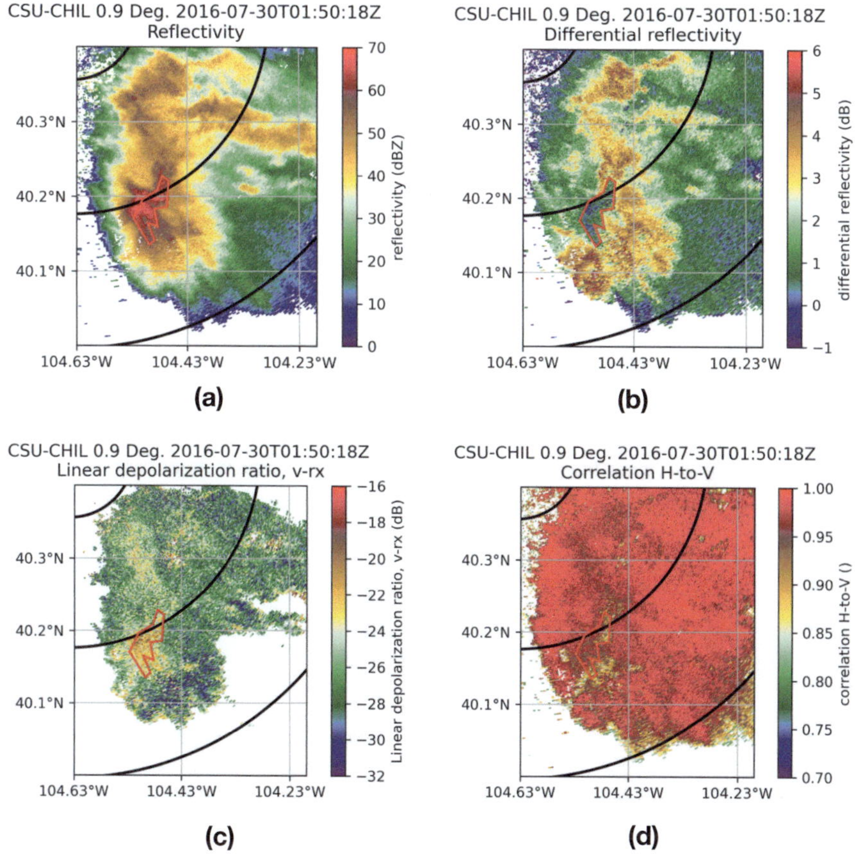

Fig. 9 **a** CSU-CHILL reflectivity data at 0.9° elevation angle on 30 July 2016. Range rings: 10, 30, 50 km. **b** Differential reflectivity data corresponding to Fig. 9a. **c** CSU-CHILL S-Band LDR corresponding to Fig. 9a. **d** CSU-CHILL co-polar correlation (ρ_{hv}) corresponding to Fig. 9a

the −25 dB reference level (white and light-yellow colors) probably are due to low concentrations of hailstones mixed with heavy rain.

The magnitude of the correlation coefficient between the co-pol H and V received signals (ρ_{hv}) is measurably lowered as the diversity of hydrometeor shapes, orientations and thermodynamic phases varies in the radar pulse volume (Balakrishnan & Zrnic, 1990). This behavior makes regions of reduced ρ_{hv} useful in identifying melting zones where a mixture of liquid drops and larger, still partially frozen particles coexist (Zrnic et al., 1993). In a similar manner, areas where hailstones and raindrops are approximately equally mixed in terms of backscattered power generation result in local ρ_{hv} reductions. Figure 9d shows this pattern in the 30 July 2016 CSU-CHILL S-band data. The lowest correlations occur near or outside of the boxed hail region that was based on the low Zdr and enhanced LDR features. It is likely that the minimum ρ_{hv} zones mark the areas where approximately equal rain hail mixtures exist; correlation increases in the core of the red-marked area where the precipitation composition shifts towards a pure hail composition.

Dual polarization indications of hail become more difficult to recognize when radar wavelength is less than 10 cm. In the heavy, oblate drop shape rain that often occurs in hail-producing thunderstorms, Zdr values are generally lowered due to differential attenuation. With sufficiently long beam paths, differential attenuation can bias Zdr down to the ~0 dB levels classically associated with tumbling hailstones. In contrast, resonance effects produce a several dB positive Zdr bias at C-band (5 cm wavelength) when 4.5–5 mm diameter raindrops are present (Kumjian & Ryzhkov, 2008). As a result, C-band radars can display distinctly positive Zdr from some mixed hail and rain areas. Finally, extensive heavy rain areas can cause complete signal extinction at shorter wavelengths (Allen et al., 1981), reducing the regions where severe weather can be observed. Operational users of radars operating at less than 10 cm wavelength should take advantage of attenuated-corrected products and be aware of the fundamental limitations of shorter wavelength radars.

5 Heavy Rain: Flooding

5.1 Training Echo Flooding

Surface rain accumulation is fundamentally the product the instantaneous rainfall rate and the precipitation duration. The resultant spatial distribution of the rainfall is controlled by the combination of the precipitation system's structural evolution and echo area's motion with time (Doswell et al., 1996). Finally, the capability for this integrated, evolving rainfall pattern to produce flooding conditions is critically sensitive to the local drainage and soil characteristics. For example, in the steep drainages with the rocky, thin soil conditions that are typical of many mountainous continental locations, the rain deposited by a single thunderstorm system can cause significant localized flash flooding (Maddox et al., 1978).

Historically, radar estimates of rainfall rate were based only on reflectivity. Natural variations in raindrop size distributions cause considerable uncertainties in these power-based rain rate estimates (Wilson & Brandes, 1979). The accuracy of radar estimation of the rainfall rate has been improved with the availability of dual polarization data. For example, Zdr can be used to characterize the median drop diameter in pure rain areas (i.e., areas without frozen or partially melted hydrometeors Bringi & Chandrasekar, 2001). The range accumulation of differential propagation phase shift (Φdp) can be related to the concentration of oblate particles along the beam path (Jameson, 1985). Rain rate can be directly estimated from the local slope of Φdp with respect to range (Kdp). (The insensitivity of Φdp to randomly oriented hailstones materially reduces the effect of hail on this rain rate estimate). Current technology rain mapping techniques make use of dual polarization data for removal of non-precipitation echo, hydrometeor classification, and selection of the locally optimal rain rate calculation method (Wolf et al., 2019).

When multiple thunderstorms affect the same stream drainage in a short time period, an enhanced threat of flooding often results. Often these individual storms do not present threatening characteristics on any single radar scan. However, their combined effects over a several hour period can be quite serious. The flash flooding that affected portions of Ft. Collins Colorado on the evening of 28 July 1997 provides an example of sequence of "training" storms producing flooding when they repeatedly affected the same stream drainage. A detailed analysis of this event (Petersen et al. 1999, Fig. 15) identified three increasingly intense reflectivity maxima that affected the Spring Creek drainage during the evening of the flood. Figure 10a–f show the low-level reflectivity and specific propagation differential phase (Kdp) data collected by the CSU-CHILL radar during these three precipitation pulses. In all of these figure panels, an irregular red line depicts the path of Spring creek through the flood area. The first two rain fall maxima (0046 and 0147 UTC) primarily affected the headwaters of the creek. The third and most intense pulse at 0346 UTC (Fig. 10e and f) with reflectivities above 50 dBZ and Kdp levels approaching 5° km^{-1}, maximized further downstream and more directly over the creek bed in which water levels had already risen from the earlier rainfall. Five fatalities occurred when the rapid development of deep, fast-moving flood swept through a mobile home park adjacent to the creek just downstream of the black dot in the Fig. 10 panels.

In practice, awareness of the current synoptic weather pattern and local forecast provides critical anticipation for the possible issuance of later flood warnings. The presence of an air mass with a high freezing level, an abnormally high precipitable water content and light vertically averaged wind speeds can support the development of slow-moving areas of heavy convective rainfall. Storm intensity and motion/structural evolution characteristics should be closely monitored in these environmental conditions.

It is apparent that unlike some severe weather phenomenon, such as tornadoes and microbursts, it is difficult for a single PPI image of basic radar data moments to clearly identify a flood threat. Processed data products that integrate rain rate estimations over time to provide an evolving map of rain accumulation are generally more useful. Using independent, real time ground truth observations (telemetered rain gauges, storm spotters, etc.) is also crucial in evaluating the accuracy of radar rainfall estimation products in each particular situation.

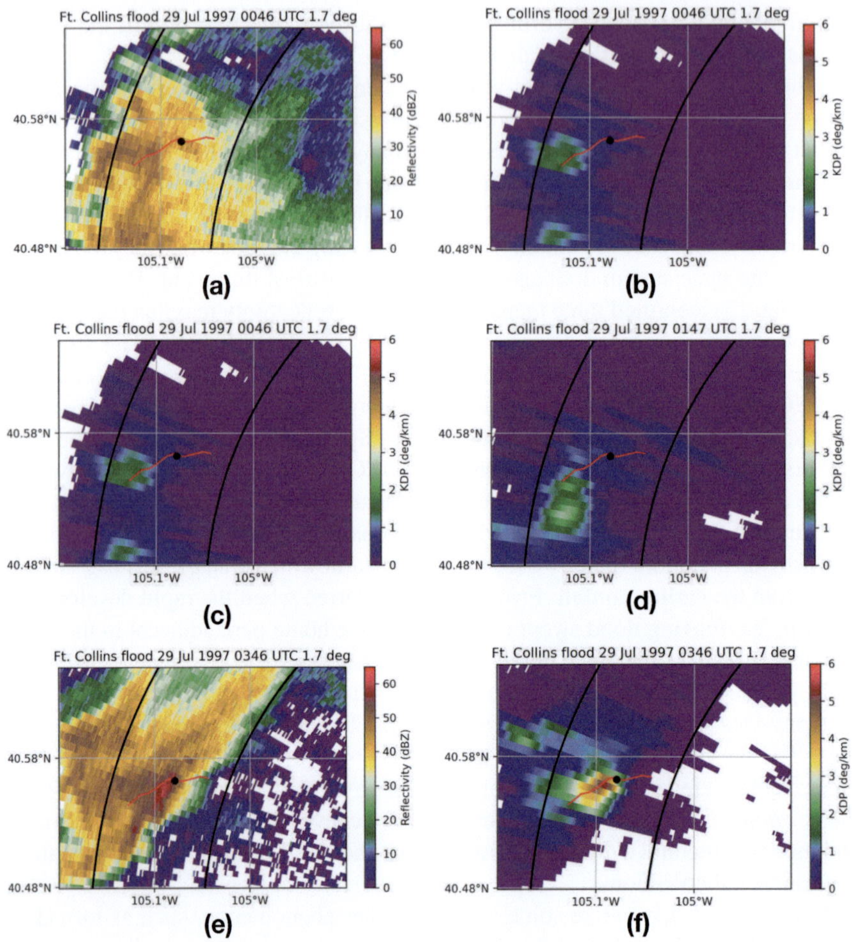

Fig. 10 **a** CSU-CHILL reflectivity data in a 1.7° PPI scan during the first precipitation pulse during the Fort Collins flash flood event. Red line: Spring Creek from its headwaters (on the left) to the general end of the primary flood zone (on the right). Black dot marks the site where fatalities occurred. Range rings = 35 and 45 km. **b** Specific differential propagation phase corresponding to Fig. 10a. **c** CSU-CHILL reflectivity data during the second precipitation pulse. (See Fig. 10a for additional figure details). **d** Specific differential propagation phase during the second precipitation pulse. **e** Reflectivity data during the third, most intense precipitation pulse of the Fort Collins flash flood event (See Fig. 10a for additional figure details). **f** Specific propagation phase corresponding to Fig. 10e

6 Snow

6.1 *Large Scale East Coast Storm*

During the cold season, convective activity decreases and synoptic scale, migratory low-pressure systems become more significant in the generation of mid-latitude precipitation. This is less true in mountainous areas where orographically-driven rising air motions and precipitation generation occurs relatively independently from the dynamically induced processes in synoptic scale flow features. Under suitable atmospheric temperature and moisture profiles, this cold season precipitation can arrive at the surface in frozen or freezing states, with significant impacts on most transportation modes. In the eastern region of the United States, these winter systems present forecasting challenges and contain significant mesoscale precipitation structures (Maglaras et al., 1995). Weather radar's capability monitor these mesoscale features has been recognized for many years (Sanders & Bosart, 1985).

Figure 11 provides an example of a deep, intense low-pressure system that generated widespread heavy snow and high winds over the greater Boston, MA area on 29 January 2022. At the time nearest to the WSR-88D data shown in the figure (1623 UTC), the surface observation taken at the Boston Logan Airport (KBOS, black dot near X = 15, Y = 40 km with respect to the radar) reported heavy snow and freezing fog with the prevailing visibility reduced to 0.25 statute miles (~400 m) and northerly winds gusting to 19 ms^{-1}. The surface air temperature was −8 °C. In the 3.1° elevation data presented in Fig. 11a, the beam is just exiting the top of the surface-based maximum reflectivity layer at a height of ~2.9 km AGL near KBOS. The corresponding ρ_{hv} PPI data (not shown) does not contain any localized minima, implying that melting is not active, and the precipitation is entirely frozen. The increase of reflectivity towards the surface indicates that the snow particles are aggregating into larger diameter crystal collections during their descent to the ground.

The corresponding radial velocity pattern is shown in Fig. 11b. Near the ground, maximum velocities exceed 30 ms^{-1}, the downward mixing of these strong north-northeasterly winds contributes to the high winds and associated blowing snow reported at the surface. With increasing range/beam height, the wind direction rotates to southeasterly direction at echo top (~5 km MSL). This veering wind profile reflects the warm air advection and implied synoptic scale rising motion that develops ahead of mid-tropospheric troughs (Carlson, 1980). This overall radial velocity pattern suggests that precipitation system remains well organized and that high impact surface conditions are likely to continue for at least the next ~12 h.

A dual polarization confirmation of the active snow growth processes in this winter storm is shown by the corresponding Kdp field shown in Fig. 11c. As noted earlier, positive Kdp develops as the concentration of horizontally oriented hydrometeors increases. The preferred growth of dendritic ice crystals that maximizes at ~−12.5 °C under saturated conditions sustains a population of quasi-horizontally oriented ice particles in winter storms. When the concentration of these particles becomes large enough, detectible (few tenths of degree) S-band Kdp magnitudes result (Kennedy

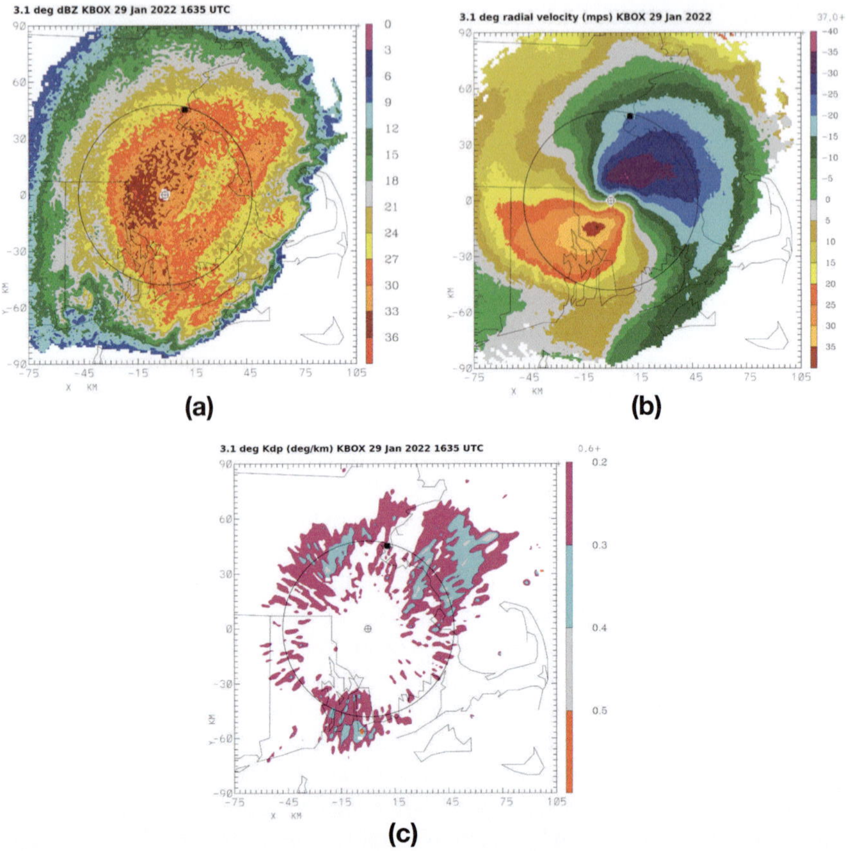

Fig. 11 **a** Reflectivity at 3.1° elevation from the Boston MA, WSR-88D radar at 1635 UTC on 29 January 2022. Single range ring at 48 km indicates where this PPI sweep intercepts the ~−12.5° environmental temperature height (2.8 km MSL). Black dot along the northern limit of the ring marks the Boston airport (KBOS). **b** Radial velocities corresponding to Fig. 11a. **c** One-way specific differential propagation phase in the PPI sweep shown in panels b and c

and Rutledge, 2011). After generation, these individual dendritic crystals tend to aggregate as they descend. The transition to the irregular, more randomly oriented aggregate shapes cause the positive Kdp signature to fade at lower heights. Figure 11c shows the resultant tendency for Kdp to maximize in a roughly ring-shaped pattern centered in the vicinity of the −12 to −15 °C environmental temperature level. The observation of this elevated Kdp layer indicates that active ice crystal growth is in progress and that heavy snow is likely to continue at the surface. As per the horizontal wind profile indicated in Fig. 11b, the ice crystals that are rapidly growing in the positive Kdp areas may be expected to follow trajectories that gradually curve from an east-northeastward direction in the generation zone to a rapid motion towards the south-southwest in the strong near-surface layer winds. Thus, the heavy snow

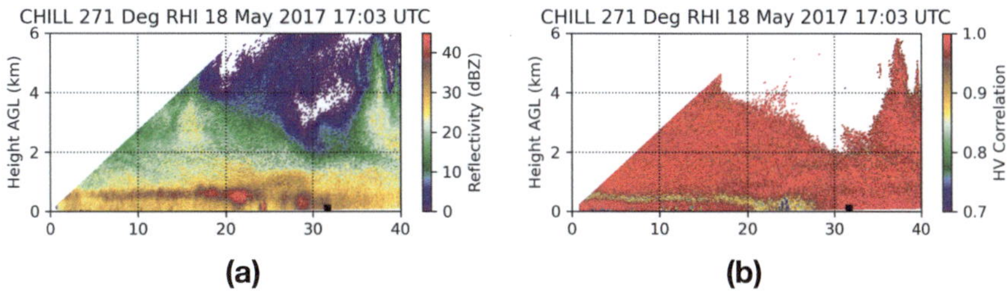

Fig. 12 **a** CSU-CHILL S-band RHI scan through a melting layer. Black square marks the range of the Fort Collins—Loveland airport (KFNL) automated surface weather observing station. At 1656 UTC KFNL was reporting light snow and fog with an air temperature of +1° C. At the same time, surface observations near the radar site at Greeley (KGXY) reported light rain and a temperature of +5 °C. **b** Corresponding correlation coefficient data. The mixture of frozen and liquid hydrometeors in the melting layer produces a local ρ_{hv} reduction. On this azimuth, the melting zone reaches the surface between the locations where snow (KFNL, black square) and rain (the radar site) were being reported

observed at KBOS is probably associated with the large positive Kdp area aloft that is located generally east and north of the airport.

In the preceding 29 January 2022 Boston area example, surface precipitation was entirely snow. However, areas of mixed phase or liquid precipitation occur in many cold season events. Local reductions in ρhv are generally useful in the identification of melting areas where mixed hydrometeor phases coexist. When the melting is confined to the near surface height layer, which is a fairly common situation, the radar beam will overshoot the reduced ρhv signature at longer ranges. An example of this situation is shown in the Fig. 12 RHI plots. Under these conditions, surface weather observations can help refine radar-depicted precipitation patterns. Surface temperature can be an especially critical factor when a reduced ρhv layer aloft implies that liquid precipitation is reaching the surface. If the surface temperatures are below 0 °C, the melted precipitation below the ρhv minimum will freeze upon impact, allowing a layer of ice to develop on many ground surfaces. Radar antenna height relative to surrounding terrain is also a critical factor in identifying near-surface melting. When the radar data is collected from an elevated antenna site, the melting layer cannot be observed if it is located below the lowest PPI scan surface. The utilization of dual polarization radar data to improve the understanding of cold season precipitation remains an active area of current research (McMurdie et al., 2022).

7 Hurricane

Tropical storm systems, particularly hurricanes, are responsible for a significant portion of damaging winds and heavy rainfall events in lower latitudes (Rezapour & Baldock, 2014). Satellite data provides the most comprehensive means of observing the genesis and path of these systems. As these tropical storms approach

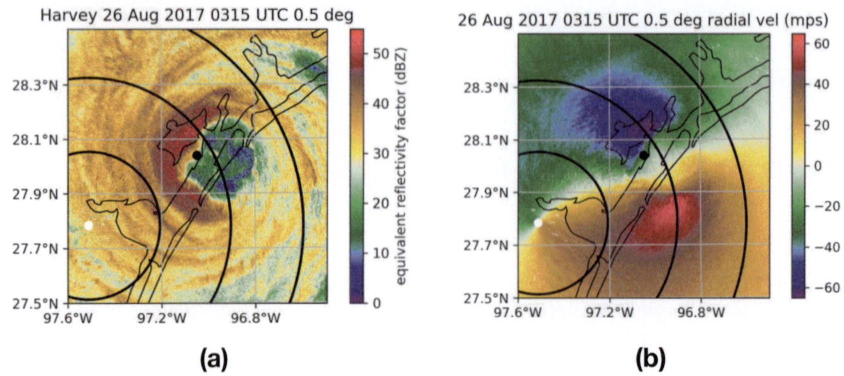

Fig. 13 **a** Hurricane Harvey Corpus Christi WSR-88D reflectivity. Range rings: 30, 60, 90 km. **b** De-aliased radial velocities (ms^{-1}) corresponding to Fig. 13a. **c** Specific differential propagation phase from the Corpus Christi Texas WSR-88D data corresponding to Fig. 13a

coastal regions, many operational land-based weather radars are able to provide dual polarization data during the critical landfall stage.

Since wind speeds and rain rates maximize in the immediate eyewall area, radar observations of this feature are important. Figure 13a shows the eyewall observed by the Corpus Christi WSR-88D radar as major hurricane Harvey reached the south Texas coast at 0315 UTC on 26 August 2017. The circular reflectivity structure is helpful in refining the storm center's location and in mapping the details of the eyewall and spiral band structures. Fernandez-Caban et al. (2019), with the aid of short-range data collected by two truck-mounted mobile radars, documented mesovortex structures that appeared to be associated with localized extreme (~80 ms^{-1}) wind speeds in the inner eyewall of hurricane Harvey at the time shown in Fig. 13a.

Figure 13b shows the associated de-aliased, ground relative radial velocity field at 0.5° elevation. The intense (~$\pm$60 ms^{-1}) mesoscale wind field associated with the hurricane core is evident. Although not present in these data, small, TVS-scale rotation signatures are crucial in identifying tornadoes that tend to develop in the spiral bands in the northeast quadrant of the hurricane (Nowotarski et al., 2021).

A snapshot of the Kdp field in Hurricane Harvey is shown in Fig. 13c. Maximum values of ~4° km^{-1} are found in small regions of the eyewall. Due to the generally smaller drop diameters in tropical rain, and the lack of melting ice particles at near surface levels, Kdp magnitudes in tropical storms are characteristically smaller versus the peak values observed in midlatitude, continental convective rain (Thompson et. al., 2018). Once the associated adjustments have been made to the rain rate coefficients used in the radar-based rain rate estimators, both conventional R(Z) and dual polarization, the resultant time-evolving rain accumulation maps are critical to monitoring the flooding threat posed by land falling hurricanes (Wolf et al., 2019).

8 Flooding Tropical Rain

While hurricanes are relatively rare events in terms of frequency of occurrence at a given point, convective rain showers regularly produce a significant contribution to annual rainfall in many tropical regions. When synoptic weather conditions promote the frequent passage of tropical rain showers over the same area, flooding may develop. Such a situation developed on 8 March 2021 when the southern extension of a mid-latitude trough system affected the Hawaiian Islands. As this trough and frontal system approached the islands from the northwest, multiple episodes of rain showers, some of heavy intensity moved inland from the Pacific Ocean. One area of widespread, destructive flooding during this event was reported in the greater Hilo area. Figure 14a shows the reflectivity pattern at a representative point in this multiday rainfall period. Most of the showers have maximum reflectivities in the 40 dBZ range, with isolated cores exceeding 50 dBZ. The corresponding differential reflectivity field is shown in Fig. 14b. While the Zdr values become more positive in the higher reflectivity areas, the differential reflectivity levels generally remain below ~2.5 dB. This relationship is further illustrated in the Zdr versus Z scatter plot shown in Fig. 15. On a global basis, the warm rain processes that are typically active in tropical rain produce drop size distributions with smaller median diameters, and consequently lower Zdr, than mid-latitude rain showers (Bringi et al., 2003; Fig. 6; Thompson et al., 2018). The numerical coefficients used in "standard" mid-latitude dual polarization rain rate estimators often need to be adjusted for optimal performance in the tropical rainfall regime.

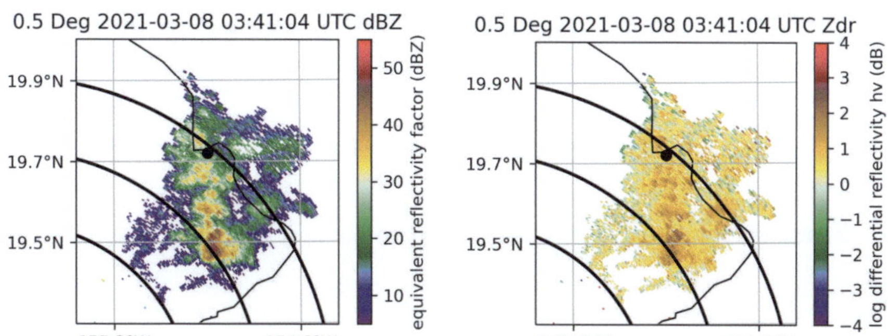

Fig. 14 **a** Reflectivity data from the South Shore WSR-88D radar on the island of Hawaii at 0341 UTC on 8 March 2021. Range rings are 50, 70, and 90 km; elevation angle is 0.5°. Black dot is Hilo airport (HPTO). Surface observations at the airport included peak rain rates of 44 mm h^{-1} during the 03 UTC hour. **b** As in 14a except data field is differential reflectivity.

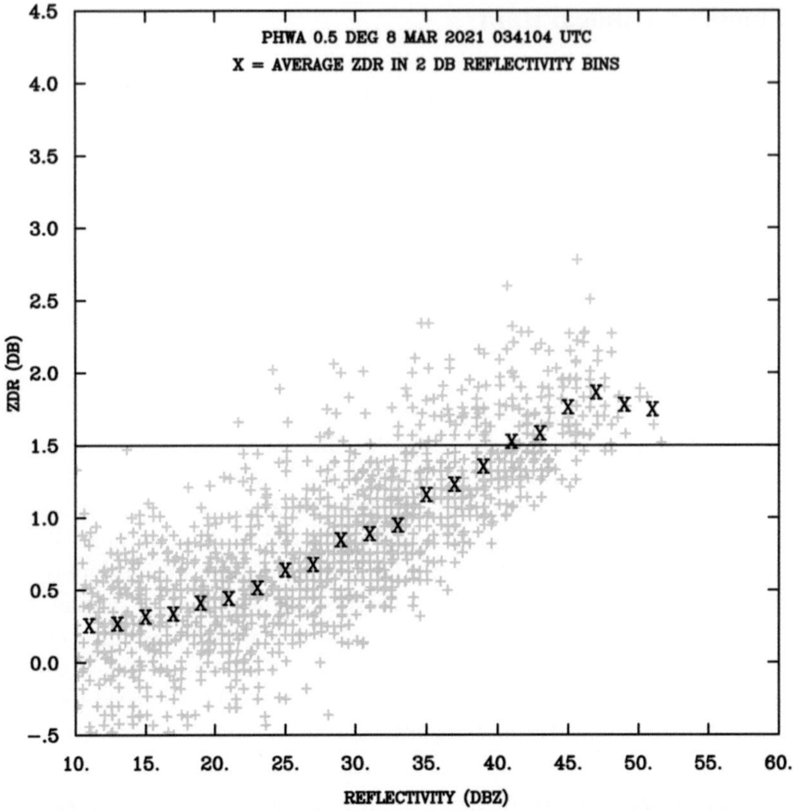

Fig. 15 Scatter plot of the Zdr versus Reflectivity range gate values from the PPI scans shown in Fig. 14. Black X's mark the average Zdr in each 2 dB wide reflectivity bin. The shallow positive slope of these bin averages shows the modest increase in Zdr with increasing reflectivity that is typical of tropical rain. For reference, a horizontal line is shown at the +1.5 dB Zdr level

9 Summary

Experience developed over the last seven decades has firmly established weather radar's capabilities for the recognition of many forms of hazardous conditions. These abilities have been enhanced as radar measurement technology expanded to include radial velocity and dual polarization fields. The real time distribution of data from both traditional national-scale radar surveillance networks, as well as more specialized local networks that observe specific urban areas, now support a wide variety of data fusion techniques. The resultant data sets will support continuing advancements in the assimilation of radar data into numerical weather prediction models. It is also likely that current research into machine learning techniques can be applied to weather radar data sets to further improve the anticipation and recognition of severe weather. Weather radar observations will continue to provide crucial support for improving the interactions between the atmosphere's behavior and societal decision support systems such as watches and warnings.

Acknowledgements The research reported here is from the long-term efforts of the authors at Colorado State University. We acknowledge all the students, staff, and collaborators who have contributed to this material. We acknowledge support from multiple funding agencies over the years including NSF, NASA, NOAA, DOE, DOD, and Colorado State University. Special recognition is given to the CSU-CHILL radar facility staff for their dedicated support in the collection of high-quality data.

References

Allen, R. H., Burgess, D. W., & Donaldson, R. J. (1981). Attenuation problems associated with a 5 cm radar. *Bulletin of the American Society, 62*, 807–810.

Atkins, N. T., & St. Laurent, M. (2009). Bow echo mesovorticies. Part I: Processes that influence their damaging potential. *Monthly Weather Review, 137*, 1497–1513.

Balakrishnan, N., & Zrnic, D. S. (1990). Use of polarization to characterize precipitation and discriminate large hail. *Journal of Atmospheric Science, 47*, 1525–1540.

Brady, R. H., & Szoke, E. J. (1989). A case study of nonmesocyclone tornado development in Northeast Colorado: Similarities to waterspout formation. *Monthly Weather Review, 117*, 843–856.

Bringi, V. N., & Chandrasekar, V. (2001). *Polarimetric Doppler Weather Radar: Principles and Applications* (636pp.). Cambridge University Press.

Bringi, V. N., Chandrasekar, V., Hubbert, J., Gorgucci, E., Randeu, W. L., & Schoenhuber, M. (2003). Raindrop size distribution in different climatic regimes from disdrometer and dual-polarized radar analysis. *Journal of Atmospheric Science, 60*, 354–365.

Brown, R. A., Lemon, L. R., & Burgess, D. W. (1978). Tornado detection by pulsed Doppler radar. *Monthly Weather Review, 106*, 29–38.

Browning, K. A. (1964). Airflow and precipitation trajectories within severe local storms which travel to the right of the winds. *Journal of Atmospheric Science 21*, 634–639.

Byers, H. R., & Braham, R. R. (1948). Thunderstorm structure and circulation. *Journal of Meteorology, 5*, 71–86.

Carlson, T. N. (1980). Airflow through midlatidude cyclones and the comma cloud pattern. *Monthly Weather Review, 108*, 1498–1509.

Chandrasekar, V., Beauchamp, R. M., & Bechini, R. (2023). *Introduction to Dual Polarization Weather Radar: Fundamentals, Applications, and Networks* (496pp.). Cambridge University Press.

Davis, C., Atkins, N., Bartels, D., Bosart, L., Coniglio, M., Bryan, G., Cotton, W., Dowell, D., Jewett, B., Johns, R., Jorgensen, D., Knievel, J., Knupp, K., Lee, W.-C., McFarquhar, G., Moore, J., Przybylinski, R., Rauber, R., Smull, B., Trapp, R., Trier, S., Wakimoto, R., Weisman, M., and Ziegler, C. (2004). The bow echo and MCV experiment. *Bulletin of the American Society, 85*, 1075–1093.

Donaldson, R. J. (1961). Radar reflectivity profiles in thunderstorms. *Journal of Meteorology, 18*, 292–305.

Doswell, C. A., Brooks, H. E., & Maddox, R. A. (1996). Flash flood forecasting: An ingredients-based methodology. *Weather and Forecasting, 11*, 560–581.

Fernandez-Caban, A. A., Alford, M. J., Bell, M. I., Biggerstaff, G. D., Carrie, B., Hirth, K., Kosiba, B. M., Phillips, J. L., Schroeder, S. M., Waugh, E., Williford, J. W., & Masters, F. J. (2019).

Observing hurricane Harvey's eyewall at landfall. *Bulletin of the American Meteorological Society, 100*, 759–775.
Forbes, G. S. (1981). On the reliability of hook echoes as tornado indicators. *Monthly Weather Review, 109*, 1457–1466.
Fujita, T. T. (1958). Mesoanalysis of the Illinois tornadoes of 9 April 1953. *Journal of Meteorology, 15*, 288–296.
Fujita, T. T. (1981). Tornadoes and downbursts in the context of generalized planetary scales. *Journal of Atmospheric Science, 38*, 1511–1534.
Fujita, T. T., & Caracena, F. (1977). An analysis of three weather-related aircraft accidents. *Bulletin of the American Meteorological Society, 58*, 1164–1181.
Gentry, C. R. (1983). Genesis of tornadoes associated with hurricanes. *Monthly Weather Review, 111*, 1793–1805.
Greene, D. R., & Clark, R. A. (1972). Vertically integrated liquid water—A new analysis tool. *Monthly Weather Review 100*, 548–552.
Herzegh, P. H., & Jameson, A. R. (1992). Observing precipitation through dual-polarization radar measurements. *Bulletin of the American Society, 73*, 1365–1374.
Huff, F. A., Hiser, H. W., & Bigler, S. G. (1954). Study of an Illinois tornado using radar, synoptic weather and field survey data. Report of Investigation of State of Illinois. No 22.
Jameson, A. R. (1985). Microphysical interpretation of multiparameter radar measurements in rain. Part III: Interpretation and measurement of propagation differential phase shift between orthogonal linear polarizations. *Journal of Atmospheric Science, 42*, 607–614.
Johns, R. H., & Hirt, W. D. (1987). Derechos: Widespread convectively induced windstorms. *Weather and Forecasting, 2*, 32–49.
Knight, N. C. (1986). Hailstone shape factor and its relation to radar interpretation of hail. *Journal of Climate and Applied Meteorology, 25*, 1956–1958.
Kennedy, P. C., & Rutledge, S. A. (2011). S-Band dual-polarization radar observations of winter storms. *Journal of Applied Meteorology and Climotology, 50*, 844–858.
Knight, C., & Knight, N. (1970). The falling behavior of hailstones. *Journal of Atmospheric Science, 27*, 672–681.
Kumjian, M. R., & Ryzhkov, A. V. (2008). Polarimetric signatures in supercell thunderstorms. *Journal of Applied Meteorology and Climotology, 47*, 1940–1961.
Maddox, R. A. (1980). Mesoscale convective complexes. *Bulletin of the American Meteorological Society, 61*, 1374–1400.
Maddox, R. A., Hoxit, L. R., Chappell, C. F., & Caracena, F. (1978). Comparison of meteorological aspects of the Big Thompson and Rapid City flash floods. *Monthly Weather Review, 106*, 375–389.
Maglaras, G. J., Waldstreicher, J. S., Kocin, P. J., Gigi, A. F., & Marine, R. A. (1995). Winter weather forecasting throughout the Eastern United States. Part I: An overview. *Weather and Forecasting., 10*, 5–16.
McMurdie, Heymsfield, G. M., Yorks, J. E., Braun, S. A., Skfronick-Jackson, G., Rauber, R. M., Yuter, S., Colle, B., McFarquhar, G. M., Poellot, M., Novak, D. R., Lang, T. J., Kroodsma, R., McLinden, M., Oue, M., Kollias, P., Kumjian, M. R., Greybush, S. J., Heymsfield, A. J., Finlon, J. A., McDonald, V. L., & Nicholls, S. (2022). Chasing snowstorms. *Bulletin of the American Meteorological Society, 103*, 1243–1269.
Nowotarski, C. J., Spotts, J., Edwards, R., Overpeck, S., & Woodall, G. R. (2021). Tornadoes in hurricane Harvey. *Weather and Forecasting, 36*, 1598–1609.
Petersen, W.A., Carey, L. D., Rutledge, S. A., Knievel, J. C., Doesken, N. J., Johnson, R. H., McKee, T. B., Vonder Harr, T. & Weaver J. F. (1999). Mesoscale and radar observations of the Fort Collins flash flood of 28 July 1997. *Bulletin of the American Society, 80*, 191–215.
Rezapour, M., & Baldock, T. E. (2014). *Classification of hurricane hazards: The importance of rainfall. Weather and Forecasting, 29*, 1319–1331. https://doi.org/10.1175/WAF-D-14-00014.1
Ryzhkov, A. V., Schuur, T. J., & Burgess, D. W. (2003). Polarimetric Tornado detection. *Journal of Applied Meteorology, 44*, 557–570.

Sanders, F., & Bosart, L. F. (1985). Mesoscale structure in the megalopolitan snowstorm, 11–12 February 1983. Part II: Doppler radar study of the New England Snowband. *Journal of Atmospheric Science, 42*, 1398–1407.

Thompson, E. J., Rutledge, S. A., Dolan, B., Thurai, M., & Chandrasekar, V. (2018). Dual-polarization radar rainfall estimation over tropical oceans. *Journal of Applied Meteorology and Climatology, 57*, 755–775.

Wakimoto, R. M. (1982). The life cycle of thunderstorm gust fronts as viewed with Doppler radar and rawinsonde data. *Monthly Weather Review, 110*, 1060–1082.

Wakimoto, R. M. (1989). Non-supercell Tornadoes. *Monthly Weather Review, 117*, 1113–1140.

Wakimoto, R. M., & Bringi, V. N. (1988). Dual-polarization observations of microbursts associated with intense convection: The 20 July storm during the MIST project. *Monthly Weather Review, 116*, 1521–1539.

Wakimoto, R. M., & Martner, B. E. (1992). Observations of a Colorado tornado Part II: Combined photogrammetric and Doppler radar analysis. *Monthly Weather Review, 120*, 522–543.

Wilson, J. W., & Brandes, E. A. (1979). Radar measurement of rainfall—A summary. *Bulletin of the American Society, 60*, 1048–1058.

Wilson, J. W., Roberts, R. D., Kessinger, C., & McCarthy, J. (1984). Microburst wind structure and evaluation of Doppler radar for airport wind shear detection. *Journal of Applied Meteorology and Climatology, 23*, 898–915.

Wolf, D. B., Petersen, W. A., Tokay, A., Marks, D. A., & Pippitt, J. L. (2019). Assessing dual-polarization radar estimates of extreme rainfall during hurricane Harvey. *Journal of Atmospheric and Oceanic Technology, 36*, 2501–2520.

Wurman, J., Kosiba, K., Robinson, P., & Marshall, T. (2014). The role of multiple-vortex tornado structure in causing storm researcher fatalities. *Bulletin of the American Meteorological Society, 95*, 31–45.

Zrnic, D. S., Balakrishnan, N., Ziegler, C. L., Bringi, V. N., Aydin, K., & Matejka, T. (1993). Polarimetric signatures in the stratiform region of a mesoscale convective system. *Journal of Applied Meteorology, 32*, 678–693.

Detection of Severe Storms by Doppler Weather Radar

Devendra Pradhan

Abstract Doppler Weather Radar (DWR) has been utilized during last 30 years for weather related phenomena and especially for Thunderstorms and Severe Convective activities associated with thunderstorms. Nowcasting (Short Term Weather Forecast) is possible with the "High Resolution" data received from DWR. Strong winds, Microburst, Tornado and Hailstorms are the severe weather activities which can be monitored and predicted at least few hours in advance with continuous monitoring by DWR. As Severe Weather Activities like Mesoscale Convective systems (MCS) greatly influence the aviation sector (public and Defense), utilization of DWR data is of immense use in the aviation sector. DWR measured Radar reflectivity (Z)—a measure of intensity of precipitation, Radial Velocity (V)—the velocity with which the sample volume moves (towards or away from radar) and the Spectrum Width (W)—a measure of turbulence within the sample volume, are quite useful in tracking and monitoring the severity and direction of movement of the severe storms in the range of 250 km. With the introduction of Dual Polarization DWRs, a great improvement has been noticed in the prediction of Hailstorms and Heavy rainfall associated with the severe storms during a short duration of time. This facility may be utilized in the short-term alert of the "Cloudburst" kind of the systems in the Hilly regions where Cloud Burst is very commonly observed during the occurrences of Western Disturbance (WDs) in the Northen sector of India during winter and early summer season. DWR has become an indispensable tool for weather observations world wide and being used in Hydrology (Heavy rainfall, Flood forecast, Water lever in the sub-catchment areas), research, avalanche prediction and Defense purposes. So far, in India 36 DWRs have been installed and 21 more Radars are in the process of installation, thus covering the whole country under an umbrella of 57 Radars in near future.

Keywords Doppler weather radar (DWR) · Meso-scale convective systems (MCS) · Dual-pol DWR · Thunderstorm

D. Pradhan (✉)
India Meteorological Department (Retired), New Delhi, India
e-mail: pradhandev1960@gmail.com

S. Das and W-K. Tao (eds.), *Severe Storms*,
https://doi.org/10.1007/978-981-97-7075-5_4

1 Introduction

After the implementation of Doppler Principle in EM waves and technological inclusion in Weather Radars, these radars have got another direction of observation with radial velocity and Spectral Width base products in addition to radar Reflectivity. Such radars now termed as Doppler weather Radars (DWR) are being used for several purposes viz. monitoring, tracking and predicting the severe weather phenomena (Thunderstorm, Hailstorm, Tornado, Cyclones, Heavy rainfall, cloudburst, gusting winds etc.) for disaster management as well as for missile launching. This is well known fact that for accurate testing of missiles and launch of satellites, development of thunderstorm associated with strong wind is an important factor and therefore, DWR plays an important role for missile and satellite launch missions.

Several studies have been performed by the weather scientists for the prediction of thunderstorms & hailstorms in the different parts of India and worldwide. Initial studies were carried by the weather scientists using data from analog radars. This data were captured using achieved using Polar diagrams and Cameras by taking pictures of the PPI & RHI scopes. Subsequently, after the technological developments towards digital technology and computers, weather researchers got a new approach Desai (1950), Mull et al (1955) and Rai Sircar (1957) in their studies using analog weather radar observations for thunderstorms in Gangetic west Bengal region, mentioned that thunderstorms approach most frequently from west to northwest directions. These thunderstorms usually develop during 1200 UTC and 1500 UTC and travel with a speed of 50–60 km/h. Very strong echoes were observed by analog radar from these tall and intense convective clouds and had high potential of heavy rainfall. The vertical extent of such thunderstorm clouds found to be exceeding 18 km in many cases but on few occasions crossed 20 km. These deadly thunderstorms which are usually originated in the North-West sector of Kolkata are commonly known as Nor'westers based on their direction of approach.

Das et al. (1957) analyzed the radar observations of two violent Nor'westers occurred during April 1955 using data of X-Band radar installed at the Meteorological Office of Dumdum Airport. They found that these thunderstorms are not associated with any well-marked frontal activity. Koteshwaram and Srinivasan (1958) conducted a significant study for the occurrence of severe thunderstorms in the pre-monsoon season and correlated with the synoptic factors favourable for their formation. Koteswaram and De (1959) studied the hourly precipitation echo maps during the occurrence of south-easterly, southerly or south-westerly thunderstorms using analog radar located at Kolkata. The radar provided very useful inputs for structural studies of radar echoes. De (1959) analyzed the radar echoes of the severe weather storms in the eastern part of India using the weather radar during pre-monsoon season. De and Sen (1961) studied the pre-monsoon thunderstorms, bright band phenomena over Calcutta and neighborhood using the analog radar located at Dumdum airport. Chaudhury (1961) after studying the pre-monsoon thunderstorms in Assam, Tripura and Manipur concluded that thunderstorm activity was relatively profound both in morning and afternoon in the area with convergence at 1.5 km level. Sen and Sengupta

(1961) investigated thunderstorm activity occurred on May 21, 1959 over Kolkata which was associated with tornadic characteristics. They found that the tornado formed in the lee of the squall-head is due to vigorous lifting of warm moist air by influx of cold air spreading from the squall. De and Rakshit (1961) performed Radar studies on the formation of cumulus clouds near Calcutta during monsoon season using Kolkata radar. De (1962) studied the "Bright band" Phenomena over Kolkata and surroundings by Analog radar installed at dumdum airport Kolkata. De (1963) studied the movement of pre-monsoon squall lines over Gangetic West Bengal observed by analog radar located at Dumdum airport Kolkata. In these studies, analysis of the formation of squall lines, their speed of movement and their dimensions were performed. Krishna Rao (1966), compiled many cases of severe thunderstorms occurred in various parts of India and concluded that thunderstorms have different characteristics in the different parts but most severe activities occur in the eastern and north eastern regions of India. Narayanan and Krishnamurthy (1966) conducted a radar study of pre-monsoon thunderstorms over Bombay Airport. Mukherjee et al. (1972) studied the "Norwesters" around Calcutta (now Kolkata) and found that the tops of Cb cloud associated with the thunderstorm reach to a very high altitude sometimes crossing Tropopause. Chatterjee and Prakash (1990) performed radar study of thunderstorms around Delhi during summer monsoon season. Since inception of operational Doppler weather radar in 1988, DWR has proved its utilization worldwide in the study and analysis of severe weather phenomena like cyclones, thunderstorm, squalls, hailstorm and tornadoes. Due to immense applications and product generating capabilities, DWR has become an indispensable instrument in the field of Nowcasting. Doviak et al. (1979) in their book, compiled all basic and applied information regarding Doppler Weather Radars, basic theory and applications. This book is taken as the refence book for all Radar Meteorologists. Suresh and Bhatnagar (2004) using data from DWR Chennai conducted case studies on the occurrences of two hailstorms during May 2002 around Chennai and suburbs and concluded that DWR generated products like Max_Z and VIL are good predictors of the hailstorm identification. Mukhopadhyay et al. (2005) studied two severe Nor'westers in April 2003 over Kolkata, using Doppler radar observations and satellite imagery. Suresh et al. (2005a) also conducted studies on rain rate estimation using DWR Chennai data. Suresh et al. (2005b) performed a case study on the clear air echoes from the atmospheric boundary Layer over Chennai. Pradhan and Sinha (2005) conducted a case study for two thunderstorm genesis over Kolkata during the month of June and discussed about their severity in this region. Sinha and Pradhan (2006) performed an analysis of a "Super Cell" thunderstorm over Kolkata and surroundings using DWR data and established that the rare occurrences of such thunderstorms which lasted for more than 8 h and had generated two tornadoes on its way is only detectable using DWR. Many more approaches have been put forward for the prediction and studies of the thunderstorms in this region. Chaudhary (2006) adopted the technique of "Ampliative reasoning" to view the prevalence of severe thunderstorms in Kolkata and surroundings. Using frequency domain analysis for the Identification of the level of downdraft formation during severe thunderstorms Chaudhuri (2008a) conducted a study and attempted to establish the fact that the method is quite useful in the prediction of such thunderstorms in the pre-monsoon period. Chaudhuri (2008b) also analyzed the preferred type of cloud in the genesis of severe thunderstorms using soft computing approach. Chaudhuri and Middey (2009) made an intensive analysis of applicability of bipartite graph model for thunderstorms forecast over Kolkata. Chaudhuri (2010) in another study, made use of convective energies in forecasting severe thunderstorms with one hidden layer neural net and variable learning rate back propagation algorithm. Chaudhuri and Middey (2011) conducted study on the nowcasting of severe thunderstorms with graph spectral distance and entropy estimation technique. Arora and Srivatava (2010) utilized Aerostat DWR owned by Indian Air Force in nowcasting of convective phenomena in the north Indian region and concluded that DWR provides a good prediction for the convective phenomena. Pradhan et al. (2012) conducted an intensive study on the severe thunderstorms formed in the Gangetic West Bengal region during pre-monsoon period of 2005–2007 using DWR Kolkata images. A technique for Nowcasting of severe thunderstorms has been suggested by the authors. As no range of convective indices for thunderstorm prediction was available in this region, threshold values of five convective parameters such as CAPE, CINE, BRN, LI & VGP have been suggested by the authors.

Many attempts have also been made by the Indian scientists for the prediction of thunderstorms & hailstorms in this region using mesoscale models. In some of these studies, DWR derived radial velocity and radar reflectivity data has been used and significant improvements were observed in such studies. Abhilash et al. (2007a, b) utilized DWR derived wind for simulating the intensity and propagation of rain bands associated with meso-scale convective complexes using MM5—3 DVAR system. This was the first attempt in India in which radial velocity from DWR Kolkata and Chennai was used in the MM5 model. The study showed a remarkable improvement in the prediction of severe thunderstorms using radial velocity data from DWR compared to the results without DWR data. In another study by Abhilash et al. (2007a, b) DWR observations were assimilated in MM5 model for the prediction of rainfall associated with mesoscale convective systems. Chatterjee et al. () conducted a study of three cases of pre monsoon thunderstorms using MM5 model and compared the results with the DWR Kolkata imageries. It was shown in the study that mesoscale model is capable of capturing the maximum intensity associated with the severe thunderstorms formed in the pre-monsoon period. In another study by Chatterjee et al. () for simulation of weather data in MM5 model for the prediction of Hailstorm, it was stated that after modifying the cloud microphysics schemes suitable to the environmental conditions of India, it is possible to have improved prediction of hailstorms. The results were compared with the actual occurrence of hailstorm observed by DWR Kolkata. Shrivastava et al. (2010) simulated the high pact convective events over Indian region using ARPS model with assimilation of radar reflectivity and radial velocity data from DWR. The results were very encouraging and confirmed the capabilities of regional scale models for the prediction of thunderstorms.

2 Radars

RADAR an acronym of ***"Radio Detection and Ranging"*** is a system for the detection and location of reflecting objects which are capable of reflecting Electromagnetic Waves. It has been employed to detect targets on or above the ground, in the air, in the space, under water etc. Radar works on the principle of "**Echo Sounding**" principle. When a radio wave emitted by a Radio transmitter meets any object, a part of the energy contained in the radiation is reflected, a part is transmitted through, a part is scattered in all directions and a part is absorbed by the material of the object. The basic principles of Radar Meteorology are described by several authors such as Battan (1973) and Atlas (1984) in their studies and text books. The basic theory and working of various kinds of radars used for different applications is compiled by Skolnik (1991) in details. A suitably placed radar receiver receives the reflected or scattered radiation. From the time of travel of the radiation and the direction from which it is received, the position of the object is determined. The measurement of the received energy and monitoring of the various characteristics of the received radiation enable determination of several characteristics of the target.

2.1 Weather Radars

Radars used for the detection of clouds, precipitation, ice, snow, winds and other hydrometeors are termed as "**Weather Radar**". The targets are basically an ensemble of "water drops" may be in the form of water, ice or snow present in the clouds, which back scatter the electromagnetic waves after incidence.

Weather radars are typically "**Pulsed Radars**" which transmit energy in the form of pulses of suitable width and interval (Fig. 1). Another reason of using pulsed radar for weather detection is that only one antenna may be used for both transmission and reception purpose. During the high time of the pulse, the transmission of the energy and between two pulses the reception of the back scattered energy takes place.

After the development of digital technology and the computers, data processing, archival and analysis became easier and fast. Offline processing of the data and research have been intensively increased. In case of Digital radars, received signals after back scattered from the targets are converted in to digital signals using A/D converters and then analyzed through computers. A new era has emerged in the field of weather forecasting by the advent of digital technology. Fast Fourier Transformation (FFT) and Pulse Pair Processing (PPP) techniques adopted in the digital radars have given a boost to the image processing from weather radars.

Digital radars may again be classified as "**Non-Doppler Radars**" and "**Doppler Radars**". The Non-Doppler radars are those which provide only quantitative estimates of the precipitation contents in digital form but no information is available for velocity of the clouds or any other parameter related with the velocity and its derived products.

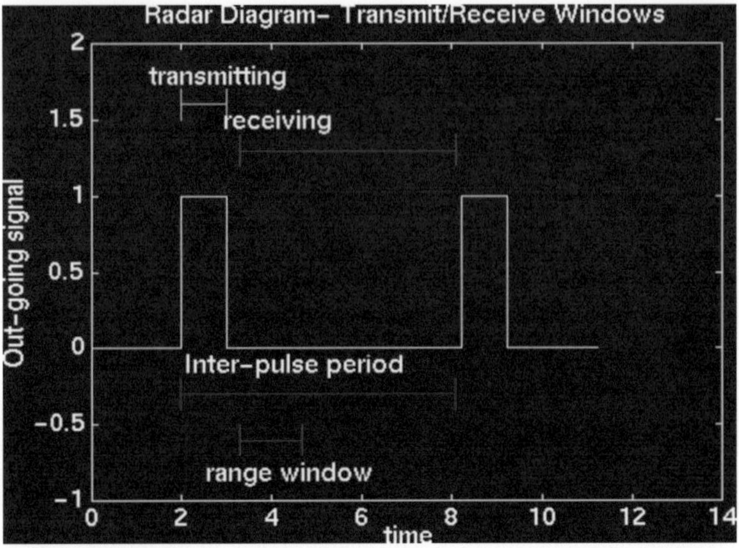

Fig. 1 Principle of pulsed radar

Weather Radars working on the principle of "***Doppler Effect***" applied to the Electromagnetic Waves are known as **Doppler Weather Radars**. These radars provide the Reflectivity (precipitation contents) as well as the radial component of the velocity of the target. Many derived products are obtained from Reflectivity and Radial Velocity data using standard algorithms developed by meteorologists and scientists. These products are very useful in the research and development purpose and help in short term weather forecasting (Nowcasting) for aviation and non-aviation purposes in the event of Thunderstorm and Hailstorm occurrences. Suresh (2006) utilized the DWR data for nowcasting the genesis and movement of convective storms over Chennai region along with the NWP (Numerical weather Prediction) models and upper air observations. Pradhan (2006) studied the structure of a super Cell thunderstorm by DWR installed at Kolkata (India). The analysis of Super Cell storm was done using DWR data and thermodynamic conditions for its evolution, structure & movement. Most of the ground based weather radars operate in S, C & X bands for cloud and precipitation purposes whereas Ku, K, Ka and mm frequency bands are used for "Cloud Dynamics Research" studies and in space borne radars (such as radars used in satellites being small in diameter and size). Satellite TRMM (Tropical Rainfall Measuring Mission) was deployed with precipitation radar operating in Ka band (35 GHz). Table 1 shows the wavelength spectrum for various meteorological applications.

Table 1 Wavelengths used for meteorological applications

Band	Frequency	Wavelength	Applications
VHF	30–300 MHz	10–1 m	Wind profilers
UHF	300–1000 MHz	1–0.3 m	Wind profilers
L	1–2 GHz	30–15 cm	Military applications
S	2–4 GHz	15–8 cm	Clouds study, heavy rainfall and cyclone prediction
C	4–8 GHz	8–4 cm	Clouds study and aviation
X	8–12 GHz	4–2.5 cm	Clouds prediction
Ku, K & K	12–40 GHz	2.5–1.7 cm	Cloud microphysics

2.2 *Principle of Doppler Weather Radar*

"Doppler Effect" in electromagnetic waves states that whenever there is a relative motion between the source of the electromagnetic waves and the target, then the reflected signal from the target gets a change in the frequency or phase as compared to the transmitted signal. This change in the frequency is known as "Doppler Shift" and is directly proportional to the relative velocity between target and the source of electromagnetic waves.

2.2.1 Maximum Unambiguous Velocity

The maximum velocity a Doppler radar can detect correctly or unambiguously is given by the velocity which produces a phase shift of π radians. This is also called the Nyquist frequency or Nyquist velocity.

Mathematically we can express this as

$$V_{max} = f_{max}\lambda/2 \quad (1)$$

where the maximum frequency

$$f_{max} = PRF/2 \quad (2)$$

and PRF is the Pulse Recurrence Frequency of the radar (Number of pulses transmitted per second).

Thus, the maximum unambiguous velocity detectable by a Doppler radar is

$$V_{max} = PRF\,\lambda/4 \quad (3)$$

It says that to detect high velocities, we must use long wavelengths, large PRF or both.

2.2.2 Maximum Unambiguous Range

The maximum range up to which a Doppler radar can detect a target correctly or unambiguously is known as Maximum Unambiguous Range. As the electromagnetic radiation travels at the speed of light, the time taken for a signal to go out and return back from a target is

$$T = 2r/c \tag{4}$$

where c is the speed of light, r is the range and T is the time. The "2" accounts for the distance out and back from the target.

The time T between pulses is thus

$$T = 1/PRF \tag{5}$$

Now with given T, we can determine the maximum range a radar signal can travel and return before the next pulses is sent out. This is simply

$$R_{max} = c\,T/2 \tag{6}$$

$$R_{max} = c/(2 \cdot PRF) \tag{7}$$

2.2.3 Doppler Dilemma

The combination of maximum unambiguous velocity and maximum unambiguous range forms two constraints which must be considered in choosing the PRF for use with a Doppler Radar. Unfortunately, PRF appears in the both the Eqs. (3) and (7), but in the numerator of one and in the denominator of the other. By solving both equations for PRF and equating them, we find that

$$V_{max}R_{max} = c \cdot \lambda/8 \tag{8}$$

As c (Speed of EM waves) and λ (wavelength of EM Waves) are constant for a specified radar, the product of V_{max} R_{max} is a constant. This can be written as

$$V_{max}R_{max} = \text{Constant} \tag{9}$$

This expression shows that V_{max} is inversely proportional to R_{max} and termed as "Doppler Dilemma" because it is difficult to choose a single PRF to optimize both V_{max} & R_{max} simultaneously.

For measuring high velocities, PRF should be high but for the same PRF, the value of R_{max} decereases. Similarly, for getting long range of observation, PRF should be low but that gives a reduced value of Velocity range measurement. Therefore, for getting a good range of Velocity and Obserbvational Range, an optimum value of PRF may be selected as a compromise.

3 Base Products Derived from DWR

3.1 Radar Reflectivity Factor (Z)

This product is a measure of the precipitation in the sample volume of a cloud system. The quantitative estimate of the precipitation contents, rainfall rate and other features of the system may be understood from this product. Based on the precipitation contents in the cloud, the strength of the back scattered signal is presented in the form of Radar Reflectivity factor (Z) expressed in the units of mm^6/m^3. Usually, the range of Radar Reflectivity Factor varies from a very small value of the order of 10^{-2} to a very large value in the order 10^8 and therefore it is convenient to express Z in the logarithmic scale denoted as dBZ given by

$$\text{dBZ} = 10\log_{10}\left\{\text{Z}/\left(\text{mm}^6/\text{m}^3\right)\right\} \tag{10}$$

$$\text{Z} = \sum \text{D}^6 \text{per unit volume}$$

where 'D' is the diameter of the rain drops. Reflectivity in dBZ is presented in the images in pseudo colors generated by the software.

3.2 Mean Radial Velocity (V)

This product is an indirect measurement of mean wind velocity of the system towards or away from radar using Doppler Principle discussed earlier. As the radar is able to measure only the radial component of the wind velocity (along the radar beam), the velocity is referred as "Radial Velocity".

The radial velocity towards the radar is considered to be **Negative** and represents the actual wind velocity if the system is approaching the radar along the radar beam. On the other hand, the radial velocity is taken as **Positive** if the radial velocity is measured away from the radar in the outwards direction. The radial velocity will be **zero** if the wind is in the perpendicular direction of the radar beam. Figure 2 shows PPI_V image of a thunderstorm indicating the radial velocity associated with the system.

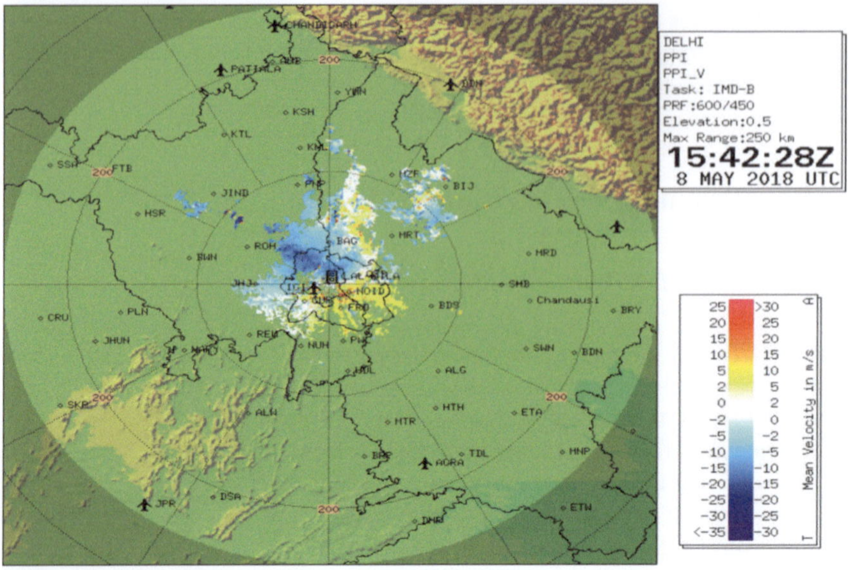

Fig. 2 PPI image for radial velocity

3.3 Spectrum Width (W)

According to Rinehart (1999) a Doppler radar produces a single velocity for the entire sample volume known as Mean velocity or the radial velocity by processing all the returned signals from the sample volume. However, velocities of the individual hydrometeors vary from mean velocity and show the variability within the sample volume. The spectrum width (σ) is the square root of the Variance is defined as the average departure of individual velocities from the mean velocity of a sample and expressed as

$$\sigma^2 = \sum (\mathrm{V_i} - \mathrm{V_{av}})^2/(\mathrm{N} - 1)$$

where V_i is the velocity of an individual scatterer, V_{av} and N is the number of targets in the sample. In short, σ indicates how variable the velocities are in the sample.

This product shows the dispersion of the velocities of the individual particles with reference to the mean velocity of the system in the form of a spectrum of velocities and is a measure of turbulence in the system. The unit of measurement is m/s and is shown by the pseudo colors for the presentation in the Doppler Weather Radar images.

4 Derived Products from Base Products

Various products are derived from the base products (Z, V & W) based on standard algorithms suggested by Weather Researchers and Scientists. Some of the derived products are:

4.1 Reflectivity (Z) Products

(a) PPI (Plan Position Indicator)
(b) Maximum Reflectivity
(c) CAPPI (Constant Altitude Plan Position Indicator)
(d) PCAPPI (Pseudo CAPPI)
(e) SRI (Surface Rainfall Intensity)
(f) PAC (Precipitation Accumulation)
(g) N-Hr Precipitation
(h) EBAS (Base of the Cloud)
(i) ETOP (Top of the Cloud).

4.2 Radial Velocity (V) Products

As the radial component of the velocity is measured by DWR, products are derived from this velocity:

(a) Plan Position Indicator (PPI)
(b) Maximum Reflectivity (MAX)
(c) Constant Altitude Plan Position Indicator (CAPPI)
(d) Pseudo CAPPI (PCAPPI)
(e) Velocity Azimuth Display (VAD)
(f) Uniform Wind Technique (UWT)
(g) Volume Velocity Processing (VVP)
(h) Wind Shear products: Azimuthal Shear (AZS), Elevation Shear (ELS), Radial Shear (RDS), Horizontal Shear (HZS), Vertical Shear (VCS), 3 D Shear (3DS).

4.3 Phenomena Detection and Warning Products

(a) Hail Detection (HHW)
(b) Gust Detection (GUST)
(c) Tornado Vortex Detection (TVS)
(d) Cloud Motion Vector (CMV)
(e) Meso Cyclone System Detection (MCS).

Table 2 Matrix of dual pole DWR products

Type of hydrometeor	Reflectivity (dBZ)	Correlation coefficient	Differential reflectivity (dB)	Specific differential phase (deg/km)
Big drops	30–50	0.92–1.01	2.5–7.0	0.0–2.0
Hail and hail rain mixed	45–60	0.75–1.0	−0.3–4.5	−2.0–10.0
Graupel	30–50	0.92–1.01	−0.3–2.2	−2–2.0
Dry snow	10–35	0.97–1.01	−0.3–1.5	−1.0–0.5
Wet snow	25–50	0.88–0.985	−0.5–3.0	−1.0–0.5
Ice crystals	10–25	0.95–1.01	−1.0–5.0	−1.0–0.5

Advancement in DWR technology

During last few years many upgradations and new technologies of DWRs have been emerged which provide better nowcasting and improvement in the quality of data. One such upgraded version is Dual Polarization Radar (Dual Pol or Polarimetric radar) which deploys the transmission of two pulses with different polarizations (Horizontal and Vertical). These pulses may be transmitted and received simultaneously (STAR) or may be sent and received alternately. The products derived from Dual Pol radars are described briefly as follows:

1. **Correlation Coefficient (CC)**: It is a measure of the correlation between the returns from Horizontally & Vertically Polarized waves after scattering from a Hydrometeor. This product provides distinction between raindrops and hails.
2. **Differential Reflectivity (ZDR)**: Differential Reflectivity measures the absolute difference between the horizontal and vertical reflectivity pulse, as opposed to CC which looked at their correlation value.
3. **Specific Differential Phase (KDP)**: This is a measure of difference in two way attenuation for the horizontal and vertical pulses in a pulse volume.

These products, in addition to the single polarization products are used in the better nowcasting of rainfall estimates. (Table 2) Shows the matrix of possible combinations of various parameters derived from Dual Pol DWRs.

5 Data and Methodology

India Meteorological Department has installed DWRs in the various bands such as X, C and S-Band at different locations in India. S Band DWRs (range of observation 400 km) are mainly located at coastal stations for monitoring and tracking of Tropical Cyclones in Indian Ocean (Bay of Bengal and Arabian Sea). Along the East coast of India, Kolkata, Visakhapatnam, Chennai, Machilipatnam, Karaikal and Paradip whereas along the West Coast, Kochi, Panaji, Mumbai, Bhuj. Few S Band DWRs are also located in the plains to monitor thunderstorm activities. C Band DWRs are located at New Delhi and Jaipur for monitoring severe weather phenomena in the range of 250 km. X band DWRs are located at Srinagar, Jammu, Mukteshwar, Kufri and few more places in J&K, Himachal Pradesh and Uttarkahand to cover the NW sector of the Himalayan region. Another 11 no of C-band and 8 no of X band DWRs are also planned by IMD to install at different locations to intensify the network and to cover the gap areas in the plains. Initially installed DWRs were singly polarized (Horizontal Polarization) whereas recently installed radars and proposed radars in near future are Dual Polarized after the installation of these radars, the network will have more than 55 DWRs in the various bands. In addition to IMD, other agencies such as ISRO, Indian Navy, Indian Air Force and Avalanche Research Institute (Defense) have also procured some DWRs for weather monitoring for specific purposes and applications.

6 Capabilities of Doppler Weather Radars in Detecting the Severe Weather Phenomena

Doppler weather radars have been utilized for detection, monitoring and predicting the severity of various severe weather phenomena in terms of heavy rainfall, strong winds and other parameters related to. The identification of tornadoes, super cell storms, hail storms, gust fronts, downbursts and fronts is presently accomplished by standard computer algorithms as well as human intervention by meteorologists by monitoring and interpreting displayed information in the images. Apart from computer algorithms, forecasters play a critical role in the interpretation of the radar information presented on two and three dimensional images. Doppler radar products such as divergence, vertical and horizontal velocities, micro bursts, storm movement vectors and precipitation amounts are more successfully obtained by computer but still benefit substantially from human input. Effective Doppler interpretation in real-time requires a human–machine mix, with the computer expected to have a greater share of the load as automatic interpretation techniques involved. Some of these are described below.

6.1 Thunderstorm Initiation

Wilson and Schreiber (1986) have shown that thunderstorms typically initiate along boundary layer convergence lines that are visible on Doppler radars. Wilson and Mueller (1993) have demonstrated how the monitoring of these boundary layer-convergence lines can be used to successfully prepare very short period forecasts of thunderstorm initiation. Thunderstorm develops mainly due to intense convection. It is the towering cumulus or the cumulonimbus clouds of the convective origin and high vertical extent that are capable of producing heavy rainfall, thunder, lightning, hail and squall line. Usually thunderstorm is generated from cumulonimbus (Cb) cloud.

The months of April and May account for most of the thunderstorm development in several parts of India. A majority of these thunderstorms are accompanied by squalls and the frequency of squalls increases from northwest to southeast. The squalls are mostly from northwesterly direction. The highest wind speed recorded in these squalls reached up to 140–150 km/h. A definite time sequence had been observed for occurrence and movement of these Norwesters by various authors in India. They generally develop over Bihar plateau, southeast Madhya Pradesh or West Orissa and travel east or southeastwards towards Gangetic West Bengal or head Bay of Bengal. They do not advance more than 130–160 km into the sea. Desai (1950) mentioned that thunderstorms approach most frequently from west to northwest and their occurrence takes place during 1200 UTC and 1500 UTC with a movement speed of 50–60 km/h. The intensity of echoes received by analog radars were quite strong and showed qualitatively the potential of heavy rainfall. The vertical extent of such thunderstorm clouds exceeded 18 km in many cases but on few occasions crossed 20 km.

6.2 Thunderstorm Detection and Issue of Nowcasting

The short-term weather forecasting is known as '**Nowcasting**'. According to weather scientists, prediction of weather phenomena from 1 to 3 h is considered as 'Nowcasting'. Sometimes, the period of Nowcasting may be extended up to 6 h but the accuracy of Nowcasting decreases gradually with increase in the duration of the prediction.

After the development of Doppler Weather Radars, the predictability of such short term and severe phenomena like thunderstorms, hailstorm, wind squall, tornado, dust storm, heavy rainfall has been improved in addition to the precision of forecast. Microburst detection can also be performed using the velocity products of DWR. Researchers use Doppler radars to obtain detailed understanding of storm and clear-air boundary layer wind fields. Using techniques originally proposed by Lhermitte (1968) and refined by Armijo (1969) detailed three-dimensional fields of air motion are obtained by combining the Doppler radial wind components from two or more radars. Using multiple Doppler techniques to obtain the full wind field, the approach is impractical for operational applications because of the large number of radars required. Fortunately, experiences with single Doppler radar in research and quasi-operational experiments have demonstrated numerous applications for warning and short-period forecasting. Pradhan et al. (2012) developed a technique for nowcasting the thunderstorm and a range of convective indices for thunderstorm prediction in Gangetic West Bengal (India) using Doppler Weather Radar and upper air data in the surroundings of Kolkata (WB).

6.3 Classification of Storms

Based on the DWR observations of many thunderstorms over and surrounding Kolkata during 2003–2015, author has proposed a classification criteria for different types of thunderstorms in the following categories.

6.3.1 Classification Based on Structure, Life Span and Speed of Movement: Single Cell Storm

A single cloud cell moving independently without combining with any other cell has been considered as a single cell storm. Usually, the life cycle of such storms is between 30 min to 1 h as observed by DWR observations in India, however, it was found that many single cell thunderstorms lived for more than 1 h on some occasions. Figure 3 is the image of a single Cell thunderstorm observed by DWR Kolkata.

Multiple cell storms:

This storm system consists of a series of evolving cells. Cells typically form on or near the storm periphery at 10–15 min intervals and each cell eventually becomes the dominant cell of the storm complex, building to higher levels as it approaches and

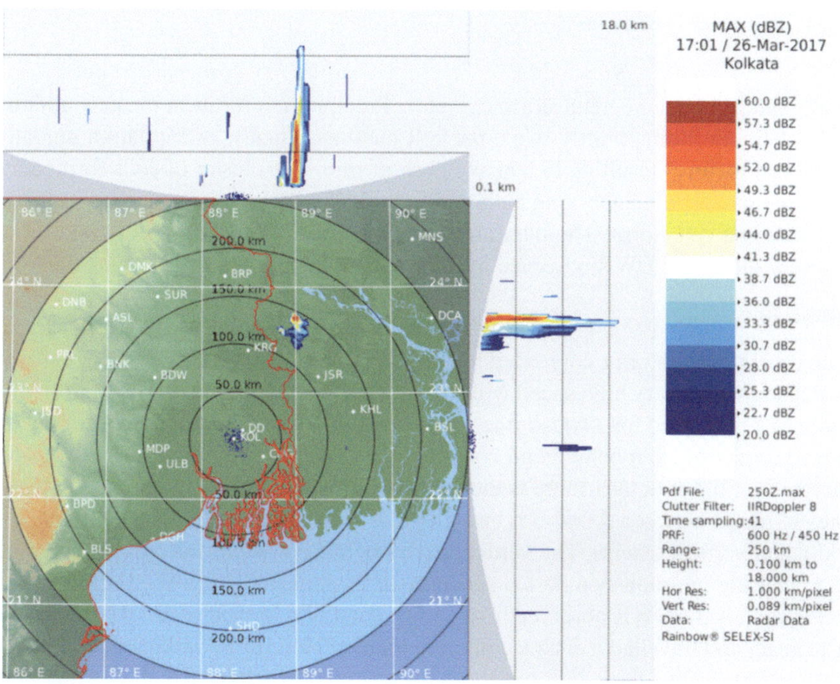

Fig. 3 Single cell thunderstorm

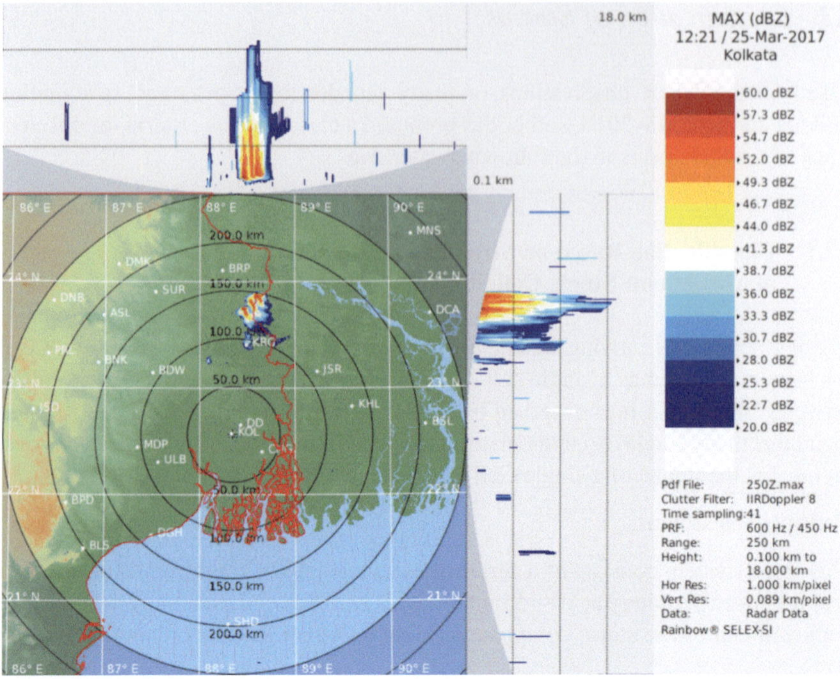

Fig. 4 Multiple cell thunderstorm

finally merges with the main storm complex. Precipitation forms in the new updraft and is held aloft temporarily while the cell matures. Finally, precipitation unloads as the cell matures resulting in a heavy gush of rain. The cluster of cells formed by the combination of neighboring cells and the whole system moves together is taken as a multiple cell storm. The duration of such storms has been found to be 1–3 h. Figure 4 shows the DWR observed multiple Cell thunderstorm.

Squall line

A chain of thunderstorms connected together and moving like a single entity, usually associated with a very high speed wind. As per the literature available, the minimum width to length ratio for a squall line should be 1:10. Squall lines may be straight or curved (concave from behind and convex in the direction of movement). More the curvature of the structure, more is the severity of the system. The life span of squall lines is found between 3 and 6 h based on their structure and dimensions (length, width and vertical extent). The vertical extent of these systems may approach up to 16 km during pre-monsoon period specially in Eastern and North Western India.

Figure 5 is the DWR observed image of a squall line formed in the NW direction of Kolkata and travelled a distance of more than 300 km from Jharkhand to Kolkata and further travelled through Bangladesh for several hundreds of Km. It was reported by Airport Authority of India, Kolkata that 10 flights were diverted to nearby airports

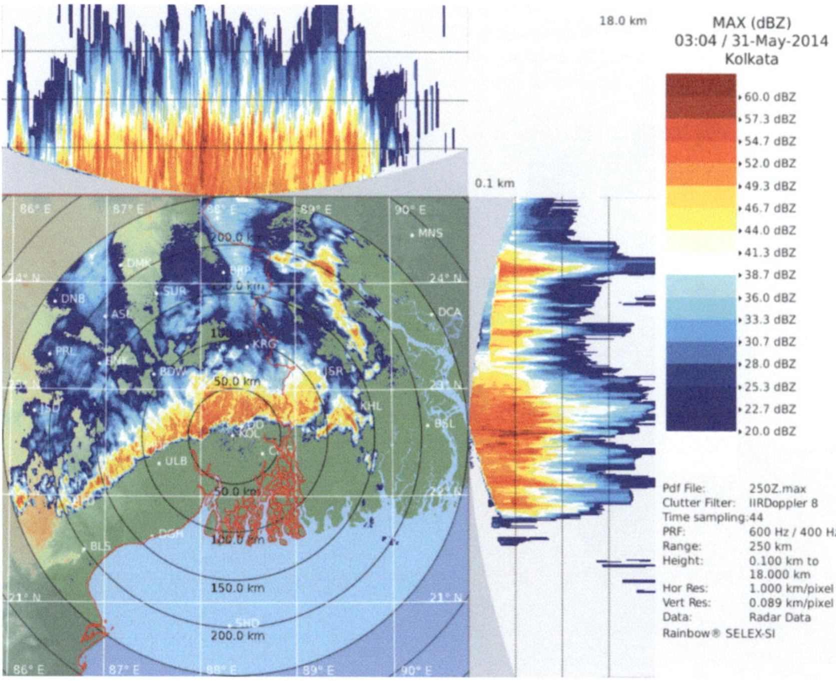

Fig. 5 Squall line formation over Kolkata

and about 15 flights were delayed because of severe thunderstorm approaching to airport in the morning hours. The squall line (Fig. 5) arrived over Kolkata airport at 0830 Hrs. IST and due to dense clouds associated with Cb clouds in the squall line, complete darkness was seen over entire Kolkata for more than 45 min. Strong winds of the order of 70 km/h were recorded by IMD Observatory located at Kolkata airport.

The maximum reflectivity associated with the squall line is exceeding 60 dBZ. The image also shows the vertical extent of the associated clouds. This may be seen that height of some of the clouds is exceeding 16 km and possessing a very high reflectivity > 55 dBz showing a high potential of heavy rainfall and strong wind speed due to downbursts.

Super cell thunderstorm

Supercell storms are larger, intense and persistent and normally produce more severe weather than other type of thunderstorms. A highly organized internal circulation that reaches a nearly steady state enables the super cell to propagate continuously. Usually associated with Hailstorms and Tornadoes. The life of such storms is usually 6–8 Hrs from development to dissipation and sometimes more. Such thunderstorms move with average speed of 70–80 km/h. Figure 6 depicts a Super Cell thunderstorm seen by DWR Kolkata on 12th march 2003. The cell was developed in the NW sector

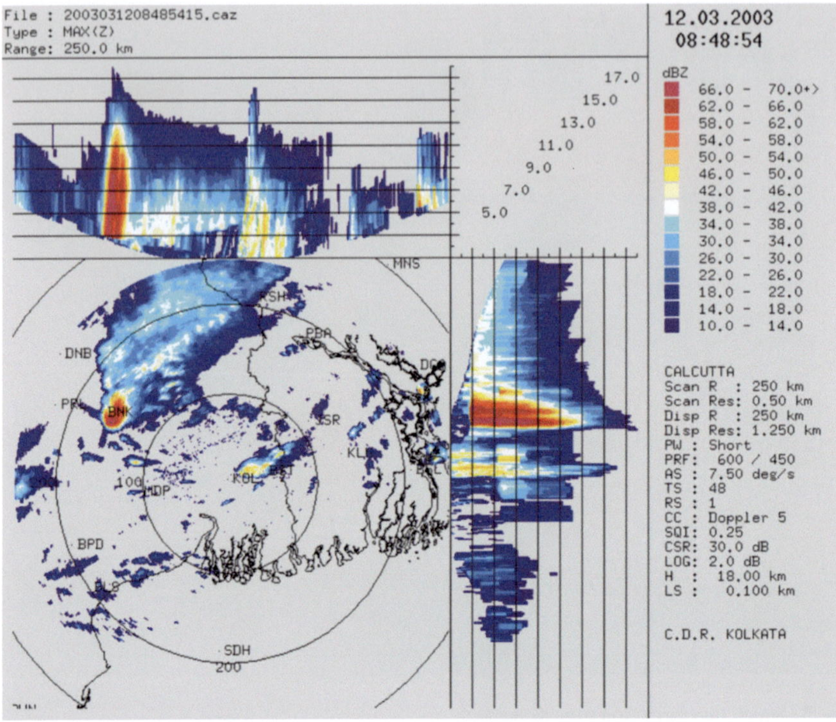

Fig. 6 Super cell thunderstorm

of Kolkata at a distance of about 200 km (Purulia) as a single cell and travelled more than 400 km in the SEly direction. It was reported in the newspapers that two Tornadoes were created by this thunderstorm on its way (Near Bankura & Kolkata) which damaged a lot of properties. Very strong winds were also noticed by the people in its way and few people were also reported died due to tornadic activities of the cell.

6.3.2 Classification Based on DWR Estimated Reflectivity and Vertical Extent

Author has classified the storms on the basis of DWR observations taken in India. A large volume of the data generated by the DWRs in India has been analyzed in this study. Following are the major categories of clouds and thunderstorms:

Severe thunderstorm

Thunderstorms in which the height of 50 dBz reflectiv is more than 16 km are defined as "**Severe**". Usually, multiple cell clouds generate severe thunderstorms, however, the possibility of single cell thunderstorms to become severe may not be ruled out.

It has been observed from the rainfall data that such thunderstorms provide heavy rainfall at the rate of 30 mm/h or more. Line squalls are also put in the category of severe thunderstorms.

Moderate thunderstorm

Thunderstorms in which the height of 50 dBz reflectivity lies within 12 and 16 km are defined as "**Moderate**". The study of the rainfall data revealed that such thunderstorms produce rainfall in the range of 5–30 mm/h.

Weak thunderstorm

Thunderstorms having the maximum reflectivity level below 40 dBz have been considered as "**weak**" thunderstorms. The vertical extent of weak thunderstorms may reach up to 10 km where the reflectivity is not more than 20 dBz. Such thunderstorms may be accompanied with thunder and lightning but provide a rainfall in the range of 1–5 mm/h.

Very weak thunderstorm

Thunderstorms in which the maximum reflectivity lies below 30 dBz have been considered as "**very weak**" thunderstorms. The vertical extent of such thunderstorms does not exceed 6 km and they dissipate very quickly but produce thunder and lightning for a short interval. A drizzle or very light rain has been observed with such thunderstorms.

7 Current Severe Thunderstorm Detection

Most of the weather radars are used to detect severe thunderstorms. The simplest signatures are maximum reflectivity determined from the PPI display and echo top height measured from the Max display. In addition, a whole spectrum of other signatures, some derived empirically but never verified, is employed by various segments of the severe storm community. Maximum low-level reflectivity and echo top height are useful to remove weak echoes from consideration, but not as reliable signatures since they produce too many false alarms. Detection of strong echo (50 dBZ) at elevated heights (>8 km) indicates a possible severe storm, especially a large hail producer. Threshold values vary with latitude and with season. Such a storm, with strong echo at high levels, is also more likely to produce strong winds or a tornado, but skill for that application is considerably less. Figure 7 shows the severe thunderstorm formation in the NW direction at 1434 UTC on 18th April 2011 showing an intermediate stage of a squall line formation (line echo wave patterns and bow or comma-shaped echoes).

The squall line is an important echo shape which is a useful detector of damaging winds. The length of the squall line was more than 200 km and it created winds of the magnitude 90 km and more at many places while travelling from NW direction towards Kolkata. The system arrived Kolkata at 1544 UTC but at about 1500 UTC,

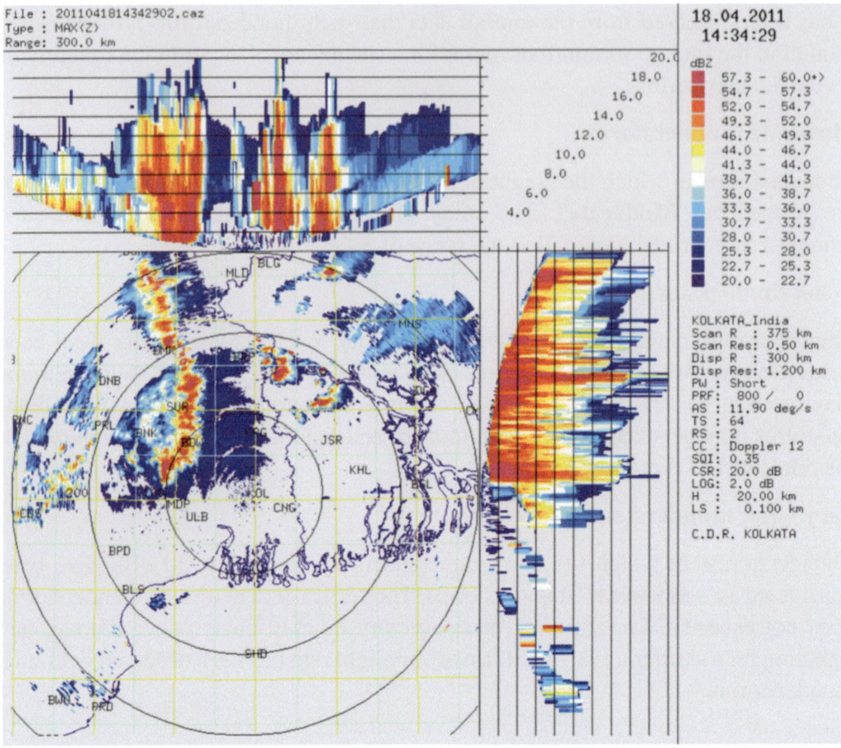

Fig. 7 Squall line formation

winds of the magnitude 90 km were recorded by Met Office located at the Dumdum airport Kolkata.

Some tornadoes may occur in conjunction with the gust fronts or rotating comma-heads of these storms. Three-dimensional echo structure is a good signature for all types of severe weather. A storm that possesses a strong core aloft, a weak echo region (WER) and an echo top over the WER will probably produce damaging wind and hail. If the weak echo region becomes bounded (BWER) or a hook like echo appears in low levels, the storm may produce a tornado, particularly as the WER or BWER collapses.

7.1 *Near Future Severe Thunderstorm Detection*

With Doppler capability, several new wind-field signatures become available for severe storm detection. Mesocyclones can be detected and their potential for producing tornadoes can be monitored. In addition, under the appropriate conditions, signatures for the smaller Tornadic Vortex (TVS) can be seen and tracked.

Together these two signatures (mesocyclone and TVS) provide good skill in identifying tornadoes, particularly the large and severe kind. Weak, ephemeral tornadoes with small circulations may escape detection by Doppler radars. Mesocyclones are also useful in finding the rotating updrafts and strong winds of supercell storms and provide aid in monitoring long track events.

When the direction of low-level damaging wind aligns with radar radials, the true wind magnitudes will be sensed, otherwise, only the radial component of the wind will be measured. Assuming that the angle between the wind and the radial is small, the damaging wind will be detected by its large magnitude. If the storm producing the winds is too far away from the radar, the strongest flows may be below the radar horizon and not detected. The gust front associated with damaging winds may show up as an enhanced band in one or more of the Doppler displays (reflectivity, mean velocity, and spectrum width). Thus, single-Doppler radars are useful in identifying windstorms, but not all damaging winds within range of the radars can be detected.

7.2 *Rainfall Forecast*

Radar measures the reflectivity (Z) in terms of dBZ associated with a cloud which may be converted (theoretically) in the rainfall rate (R) following an empirical relation between Z & R and known as Z-R relation given by

$$Z = A\,R^{b} \tag{11}$$

where A and b are constants which have different values for different kinds of the rains, different seasons and different places. Initially, Marshal & Palmer (1948) gave the values of A as 200 and b as 1.6 and therefore the relation

$$Z = 200\,R^{1.6} \tag{12}$$

is also known as Marshal Palmer relation.

Therefore, to issue the rainfall forecast, it is necessary to know the values of A and b over an area in addition to the Radar Reflectivity (Z). The reflectivity derived product SRI (Surface Rainfall Intensity) provides good estimation of the rainfall rate from radar. The rainfall forecast is possible but requires proper validation of the DWR products with the actual rainfall recorded by a network of several Self Recording Rain Gauges (SRRG) available within the shorter range of radar (usually 100–120 km). Pradhan and Talukdar (2016) performed a study over Kolkata for the validation of SRI product and determination of Z-R constants A and b.

The unit of SRI is mm/h as it provides the rainfall rate converted from Reflectivity measured by radar (Fig. 8).

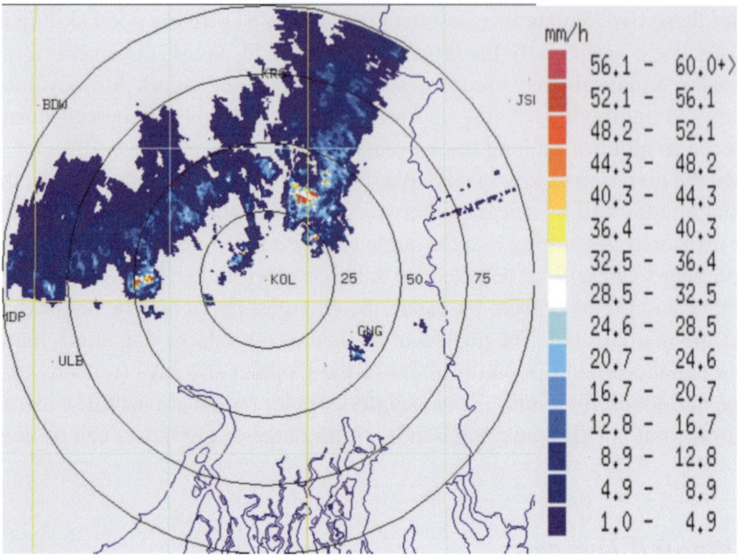

Fig. 8 SRI product of 07th May 2010 over Kolkata

7.3 Wind Squall Warning

The wind squalls are the high speed winds generated by the severe thunderstorms during its dissipation phase due to strong downdrafts. The speed of the squall may be measured using the PPI (velocity) data. However, the exact prediction of the wind squall is quite difficult but a good estimation may be done by the forecasters based on the structure of the cloud system forming a thunderstorm. This has been observed from DWR images that the large vertical extent (more than 12 km) of the severe thunderstorm and sufficiently high reflectivity (>55 dBz) gives rise to high speed wind squalls of the order of 50–60 kmph or more. Now a days, audio-visual warning is also generated automatically by the DWR through software controls and suitably defining the threshold for the Wind Squall.

Figure 9 shows the Time Series of the Vertical Profile of the horizontal winds derived from the Radial Velocity measured by Doppler Weather Radar. The product provides winds at different levels in the range of 50 km around DWR. Although the range of this product is only 50 km and the height is 10 km, the winds remain almost same over 100 km (being planar) and therefore, the wind speed and direction estimates cab be done at the various level above ground. Backing and veering of the winds can also be estimated from this product. During the cyclone observations, “No wind zone” in the eye of the cyclone is very well seen by this product. The winds at middle level (between 3 and 5 km) provide the expected direction of movement of thunderstorm.

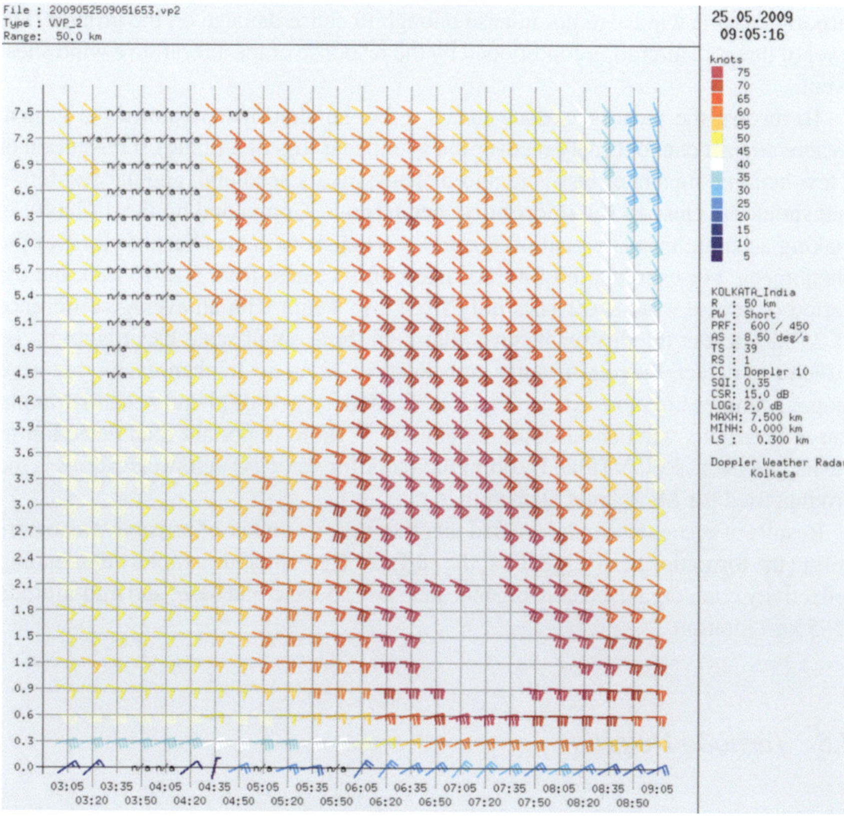

Fig. 9 Vertical profile of the horizontal winds

7.4 *Microburst Forecast*

Initially let us understand the meaning of "**Downburst**". It is defined as a "*Strong downdraft that induces an outburst of damaging wind on or near the ground*". Downbursts are mostly associated with intense, severe storms but they also have been observed in the high plains regions where sub-cloud moisture contents are higher. When the horizontal dimensions of a thunderstorm's density current are small but velocities are intense, the downdraft and its associated outflow are called "**Microburst**". More specifically, a microburst has been defined as "**a diverging outflow that exhibits anywhere a wind speed difference of at least 10 ms^{-1} across a distance of no more than 4 km**". Experiments indicate that intense small-scale divergent flow is more common than previously thought and, as with vortices, it is the smaller-scale downdrafts that produce the most hazardous shear. Strong divergence of horizontal flow can cause low-flying aircraft to markedly depart from their intended flight altitude, which can be disastrous on approach to and departure from airports, even if winds are not intense enough to cause damage on the ground. The level of threat to aircraft is conditioned by the response of the aircraft to a wind shear event.

To resolve the most hazardous shears a few hundred meters above the ground (where aircraft cannot timely recover the significant loss of altitude), a resolution of a few hundred meters or less is required. Thus, radars need high angular resolution and should be **close to the microburst occurrence**. Compounding the difficulty in making accurate measurement of hazardous wind shear is the short life time of the phenomena. For example, lifetimes of microbursts range from 5 to 15 min, but the period of severe wind shear lasts only from 2 to 4 min with an average difference of 25 m/s across the divergent flow. Although these small-scale hazards might be difficult to detect, they are usually embedded in larger-scale phenomena that have longer lifetimes are more easily detected. Thus, detection of phenomena that might cause hazardous wind shear could be equally or more important than detection of the hazard itself. PPI_V (Plan Position Indicator for velocity) at low elevations is the product used for Microburst Detection.

Results of recent research revealed that there are a number of possible precursors before the formation of divergence at the surface. These include a rapidly descending reflectivity core, organized convergence near and above cloud base, and mid-altitude (2–5 km) rotation.

7.5 *Tornado Detection*

Of all the atmospheric storms, tornadoes are the most severe and probably the least understood. Because of their unpredictability and high energy density, tornadoes represent an awesome challenge to the scientific community. Anemometers falling directly on the path of the storm seldom survive to give instrumental evidence and micro-barographs respond too sluggishly to the strong pressure gradient and rapidly travelling vortex and therefore, almost all the data on tornadoes have been gathered from visual and photographic records and damage surveys. A tornado is a violent, rotating column of air which is in contact with both the surface of the earth and a cumulonimbus cloud or, in rare cases, the base of a cumulus cloud. The word "tornado" originates from a Spanish word "Tronada" which means a Thunderstorm. Tornado is the fiercest meteorological phenomena usually associated with thunderstorms. According to Battan (1973), tornado is the most violent meteorological phenomena usually associated with thunderstorm. Because of the lack of suitable observational techniques, it has been very difficult to learn a great deal about tornadoes. A Tornado may be defined as a concentrated vortex or revolving storm with a vertical or near vertical axis which is endowed with great power of destruction because of its strong updraft and high rotational velocity and enormous pressure defect inside. It is visible in the sky as a narrow funnel usually below the base of a thunderstorm sometimes reaching the surface but often remaining suspended in the air. It wields its power of destruction only when the base of the funnel happens to be near the earth's surface. Tornadoes passing over the sea, large lake or wide river are often found to raise water like a pillar—these are known as "**Water Spouts**" However, these are less intense than tornadoes, far more frequent and form readily from even cumulus congestus cloud lines. Tornadoes, funnel clouds and waterspouts are atmospheric vortices.

In India, a tornado is a rare experience in the lives of many. About 72% of the reported tornadoes in India have occurred in northeast India and Bangladesh. About 76% of the tornadoes in India occur during March to May, the most favored month being April. According to the studies performed by many authors, Gupta and Ghosh (1980), Bhattacharya and Banerjee (1980), it was established that more number of tornadoes have occurred in the afternoon and evening. Asnani (1985) and Goldar et al. (2001) have given reviews on tornadoes of India and given exhausted lists of references on the subject. Gupta and Ghosh (1982) have described the reported cases of tornadoes in the Indian subcontinent. Considerable efforts have been spent in seeking characteristic radar echo forms that might positively identify a tornado or water-spout. Most of the tornadoes are generated from extremely violent thunderstorms of great vertical development but on rare occasions they are observed with stratified clouds and without associated precipitation or radar echo. Many of the massive thunderstorms associated with tornadoes give rise to radar echoes that have a hooked shaped protuberance in the right rear quadrant of the storm (when facing in the direction toward which the storm is moving). The hook is believed to be the echo from a huge mass of water droplet that is swept out of the mother cloud by the vortex and not an echo from the tornado funnel itself. Unfortunately, the size of the hook is so small that it cannot be distinguished at long ranges owing to beam-width distortion and similar protuberances occur with many echoes from thunderstorms which have no tornadoes associated with them. Based on the several studied conducted on the generation of Tornadoes, following echo features seen in the DWR images may be considered as precursors of a tornado associated with a severe thunderstorm:

(a) Unusually great vertical extent (echo top exceeds the reported tropopause);
(b) An echo which moves in a direction to the right (looking downwind) of the direction of other nearby echoes;
(c) When two echoes merge into one.
(d) Hook shaped echoes seen in the PPI (reflectivity and velocity) images.

All of the Doppler signatures contribute directly to tornado detection, however, not all may be seen in every Tornado. It is easy to justify the contributions of mesocyclone and TVS signatures because they are measures of the rotation. Gust front signatures contribute positively because of the knowledge that wind shear lines are breeding grounds of tornadoes through convergence and stretching of existing vertical vorticity. High-level divergence is a measure of updraft strength that plays a critical part in tornado formation through either stretching or tilting of appropriate horizontal or vertical vorticity. Even downburst signatures can contribute indirectly to tornado formation when outflow shear and convergence are involved in producing local vorticity extremes.

Doppler velocity data has significantly improved the radar meteorologist's ability to identify severe storms. In particular, mesoscale vorticity signatures in the Doppler radial velocity field can be used to identify super cell storms. By monitoring the magnitude of this mesocyclone, tornado warnings can be issued. This mesocyclone signature is a couplet of closed Doppler velocity contours of opposite sign spaced in azimuth by about 4–15 km. In some conditions a tornado may produce a very large beam to beam shear called Tornado Vortex signature (TVS).

Small, short duration tornadoes often occur in the absence of a super cell storm, these non-supercell tornadoes can only be detected within 40 km of radar. These tornadoes discussed by Wakimoto and Wilson (1989) & Brady and Szoke (1989) can occur when pre-existing boundary layer mesocyclones (diameter < 4 km) along convergence lines rapidly intensify when they become co-located with an intense updraft of a rapidly developing convective storm. Forecasts of non-supercell tornadoes are possible by monitoring boundary layer convergence lines for mesocyclones and rapidly developing cells. A warning can be issued when the mesocyclone begins to strengthen and grow vertically when collocated with a cell rapidly developing overhead.

A case study was performed by the author (unpublished) using DWR generated for a tornado occurred at around 1130 UTC (1700 IST) over Kendrapara (Orissa) on 31st March 2009 (Fig. 10) which created a lot of destruction along a straight path of more than 5 km in length and about 1 km in width (Fig. 11). Although the place was about 240 km South West from Kolkata it was tracked by the DWR Kolkata. Of course, being at a large distance the lower part of the cloud could not be observed by DWR but the reflectivity images capturing the upper part of the cloud are observed. Author attempted to see the pre-cursor related to the origin of the Tornado on the basis of the DWR images. Figure 12a shows the vertical cross section of the Tornado cloud depicting the downward moving funnel created by the strong winds ejected out from the base of the cloud at (1102 UTC). Figure 12b shows further descent of the Tornado Funnel towards the base of the cloud creating an orifice between the strong reflectivity clouds (Reflectivity more than 60 dBZ) at (1122 UTC). The vertical cross section of the thunderstorm cloud (Fig. 12c) shows a "Funnel" shaped structure at a height of 10 km, which is clearly visible coming out from the centre of the Cb cloud which might be the probable initiation of the "Tornado" descending downwards towards the base of the cloud and to emerge out from the base of the cloud. Another vertical cross section at 1122 UTC (Fig. 12d) shows further descent of the same "Funnel" at the height of about 6 km above the ground. This is to be noted that DWR observations were taken every 20 min interval that time (during 2009), so actual observation was somewhere about 20 min prior to the time of generation of images by DWR. The observation shows that funnel descended 4 km in 20 min (downdraft speed is about 12 km/h i.e. 3.3 m/s). The actual orifice of ejection may be quite lower than the funnel shown in these images. The lower part of the cloud is not observed being the large distance of the thunderstorm from DWR Kolkata. The Disaster Management Odisha reported the time of occurrence of Tornado at 1640 Hrs (IST) which is well in agreement with the time of observations (around 1110 UTC) by DWR Kolkata.

Fig. 10 Tornado occurrence on 31st March 2009 in Orissa (India) (Courtesy: Rajkanika Tornado Disaster Management, Government of Odisha)

Fig. 11 Destruction caused by the Tornado (Courtesy: Rajkanika Tornado Disaster Management, Government of Odisha)

Author tried to pick up the **Pre-Cursors of tornado** despite the larger distance of the system from radar. The alert period was hardly 20–30 min before the actual occurrence of the tornado.

7.6 Hail Warning

Hail of damaging size is most common in continental areas in temperate latitudes, and because of its exceptional hazard to aircraft as well as the damage done on the ground, it would be valuable to have positive indication of its presence from the radar PPI (Z & V) pictures of the typical clouds. Hail is, of course, associated with violent convection and the radar echoes which have been positively identified as

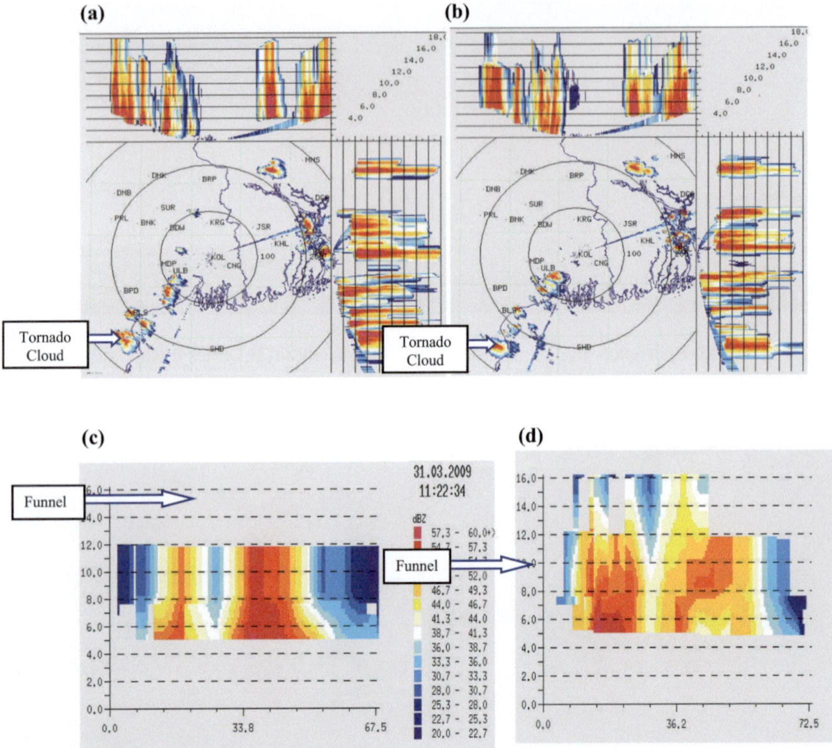

Fig. 12 **a** DWR image at 1102 UTC. **b** DWR image at 1122 UTC. **c** Vertical cross-section at 1102 UTC. **d** Vertical cross-section at 1122 UTC

associated with hail have shown the sharp-edged high intensity echoes characteristic of precipitation arising from convection. In many cases, hail has been associated with protuberance or fingers from the edges of the bright convection-type echoes with perhaps a V-shaped notch between the fingers. The Hail warning product is used for the prediction of hails within a thunderstorm. Based on the user selectable criterion this warning is issued. For example:

1. Reflectivity interval (45–55 dBZ)
2. Layer height (2–10 km).

When the above criteria is satisfied, warning will be issued by the radar. However, this may not be always true because above conditions are to be tested with several cases of thunderstorms in the different seasons. Figure 13 depicts a warning for probable hailstorm on 30.04.2003 at 1036 UTC to the south of Kolkata (India).

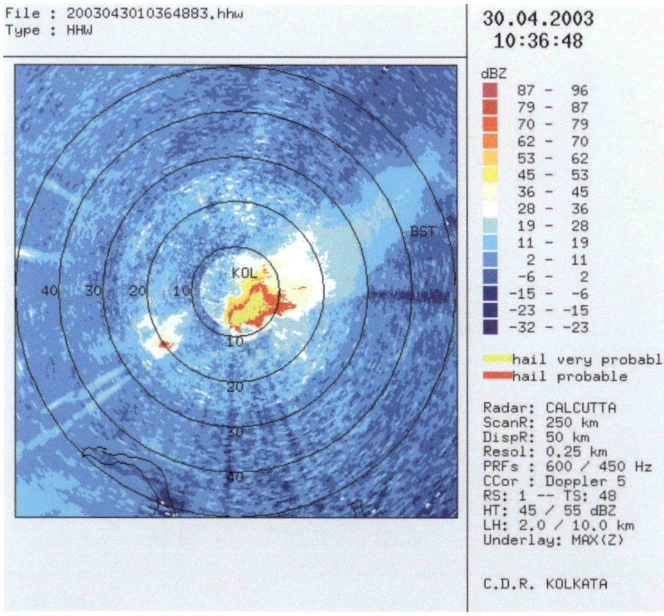

Fig. 13 Hail warning

In reallity, a hailstorm was reported on that day during 1030–1100 UTC in the surroundings of Kolkata (India) of hail size about 3 cm in diameter. Therefore, importance of this product may not be ignored by the forecasters. However, the product validation is to be done by taking many such events recorded by the DWR and must be compared with the ground truths. Figure 14 shows the typical signatures as Hail Spikes observed on 10.03.2006. Hail Storm created the hails upto 2 cm in diameter around and in Kolkata that day.

7.7 DWR Images of Thunderstorms

Few images from DWR have been introduced in the text below showing some typical severe weather events observed by DWR Kolkata (Fig. 15).

(i) **Development of the Line squall observed on 03.06.2004 (Fig. 16)**
(ii) **Super cell Thunderstorm observed on 12.03. 2003 (Fig. 17).**

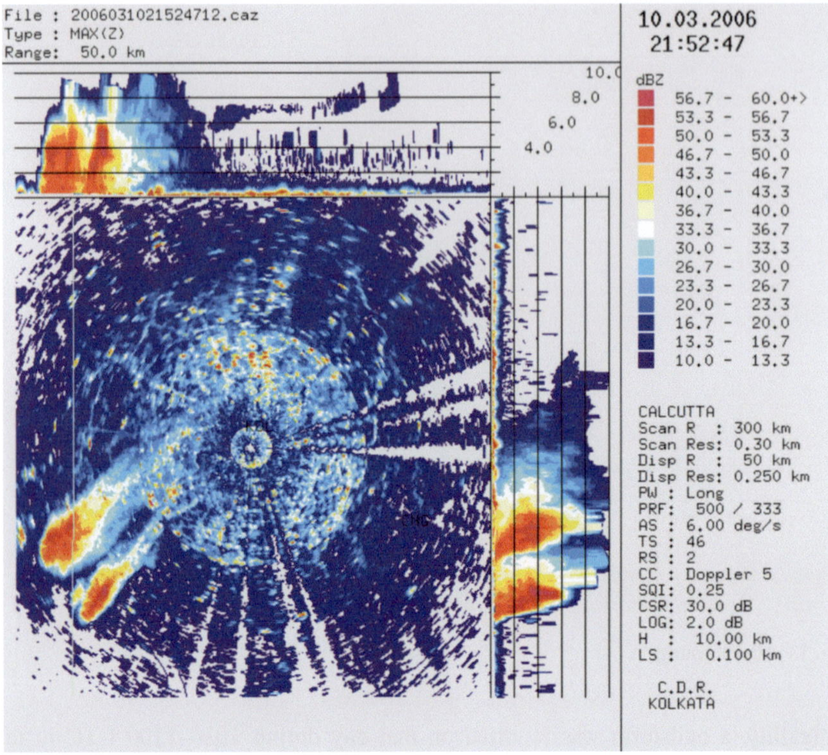

Fig. 14 Hail signatures (hail spikes)

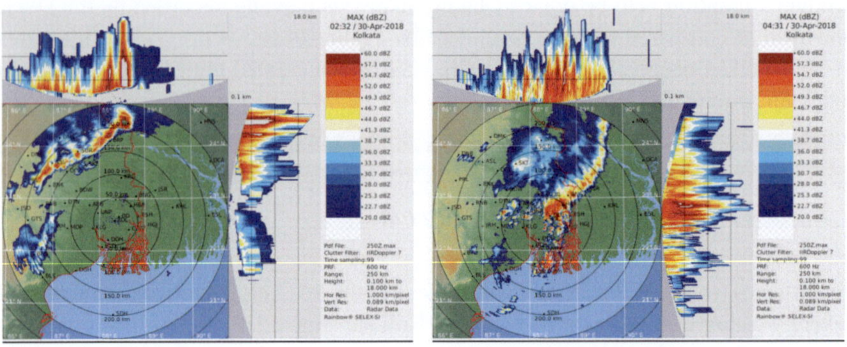

(a) (b)

Fig. 15 a Squall line movement. b Squall line movement

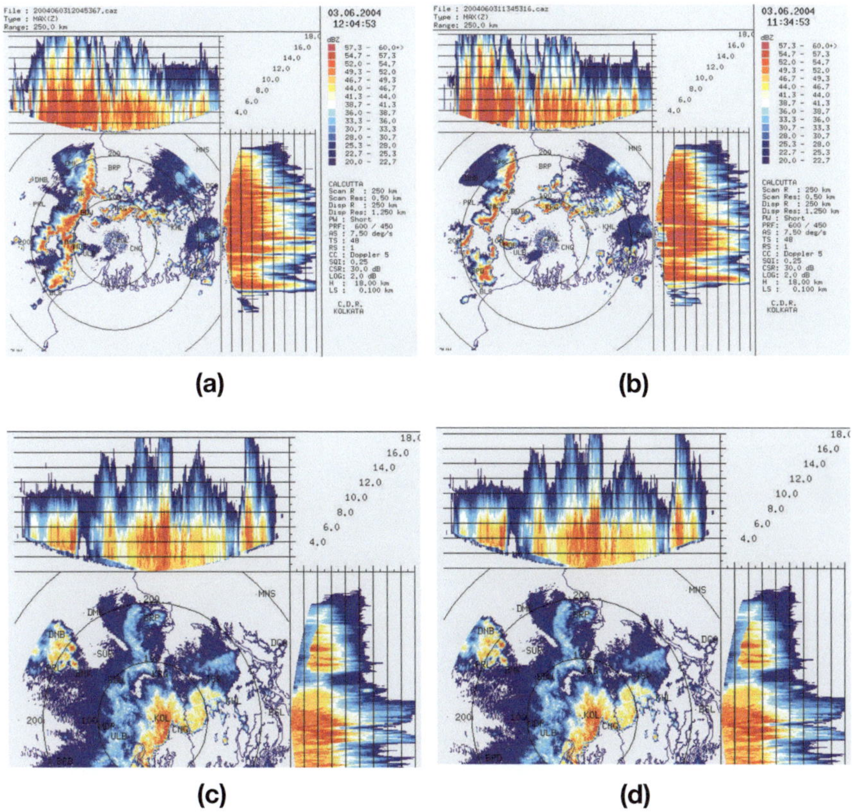

Fig. 16 **a** 1134 UTC. **b** 1204 UTC. **c** 1304 UTC. **d** 1404 UTC

8 Conclusions

Doppler weather Radar has become an indispensable tool for monitoring the severe weather phenomena with a very high accuracy. Although, DWR suffers from many limitations such as short range of observation, difference in the estimated rainfall and actual rainfall, DWR plays a vital role in the Nowcasting of Wind Squall, thunderstorm, Hail storm, detection of MCS and associated tornado. A large network of DWRs world wide has improved the nowcasting very significantly. In the recent developments during last 10 years, Polarimetric Radars (Dual Polarization Radars) i.e. (Horizontal and Vertical Polarizations of electromagnetic waves) are being utilized for better rainfall classification and prediction of the Hailstorms categorically distinguished from Thunderstorms without hails.

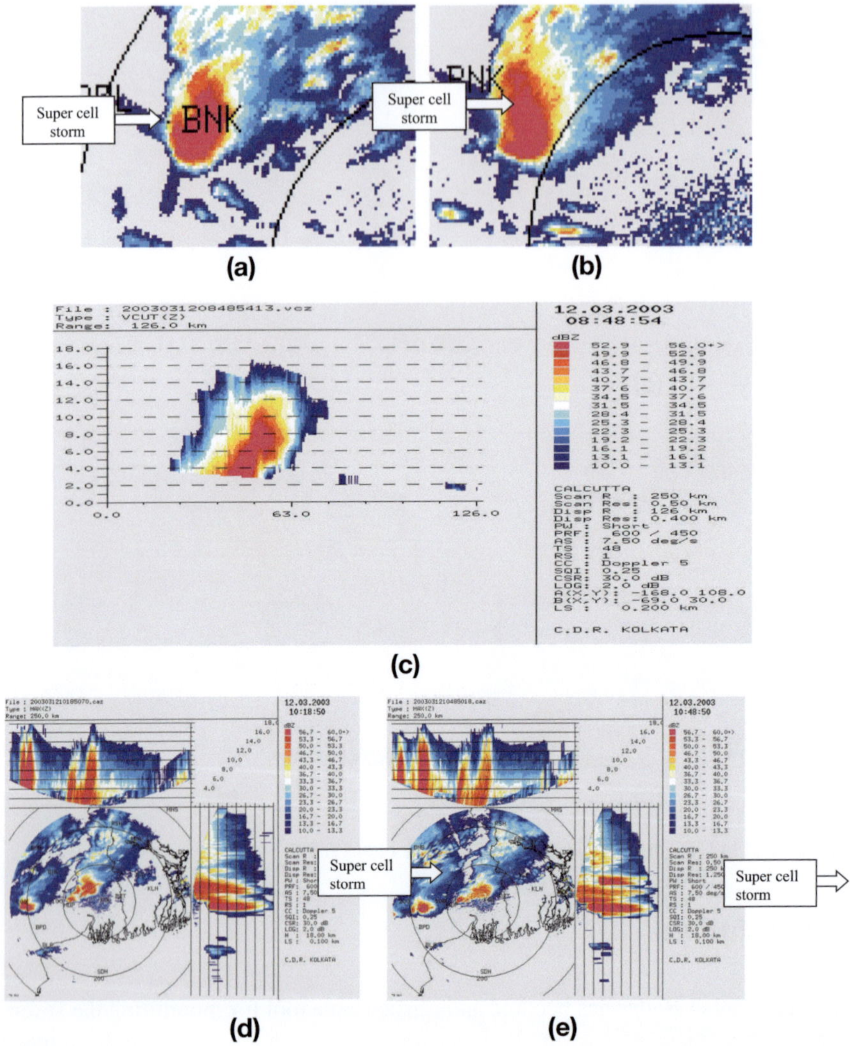

Fig. 17 **a** 0848 UTC at 260 km distance. **b** 0918 UTC 220 km distance. **c** Vertical Cross section of the super cell storm on 12.03.2003. **d** 1018 UTC at 200 km distance. **e** 1048 UTC at 150 km distance

References

Abhilash, S., Das, S., Kalsi, S. R., Dasgupta, M., Mohankumar, K., George John, P., Banerjee, S. K., Thampi, S. B., Pradhan, D. (2007a). Impact of Doppler radar wind in simulating the intensity and propagation of rainbands associated with meso-scale convective complexes using MM5—3 DVAR system. In *PAGEOPH* (p. 164).

Abhilash, S., Das, S., Kalsi, S. R., Das gupta, M., Mohankumar, K., George John, P., Banerjee, S. K., Thampi, S. B., & Pradhan, D. (2007b). Assimilation of Doppler weather radar observations in

a mesoscale model for the prediction of rainfall associated with mesoscale convective systems. *Journal of Earth System Sciences, 116*, 4.
Arora, P. K., & Srivastava, T. P. (2010). Utilization of aerostat doppler weather radar in nowcasting of convective phenomena. *Mausam, 61*(1), 95–104.
Armijo, L. (1969). A theory for the determination of wind and precipitation velocities with Doppler radars. *Journal of Atmospheric Sciences, 26*, 570–573.
Asnani, G. C. (1985). Tornado—A review. *Vayumandal*, 97–133.
Atlas, D. (1984). Highlights of the Symposium on the multiple parameter radar measurements of the precipitation: Personal reflections. *Radio Science, 19*(1), 238–242.
Battan, L. J. (1973). *Radar Observations of the Atmosphere* (Chaps. 12, 17, 20a, 21a, 23a, 27a, 29a, 29b, 30a, 323pp.). University of Shikago Press.
Bhattacharya, P. K., & Banerjee, S. K. (1980). Pre-monsoon Tornadoes over West Bengal during April 1977. *Vayumandal, 10*, 14–18.
Brady & Szoke. (1989). Observations of a Colorado Tornado. Part I: Mesoscale environment and Tornado genesis. *Monthly Weather Review, 120*, 497–520.
Chatterjee, P., Pradhan, D., & De, U. K. (2008a). Simulation of local severe storms by Mesoscale model MM5. *Indian Journal of Radio and Space Physics, 37*, 419–433.
Chatterjee, P., Pradhan, D., & De, U. K. (2008b). Simulation of hailstorm event using mesoscale model MM5 with modified cloud microphysics scheme. *Annales Geophysicae, 26*, 3545–3555.
Chatterjee, R. N., & Prakash, P. (1990). Radar study of thunderstorms around Delhi during summer monsoon season. *Mausam, 41*(2), 161–165.
Chaudhary, S. (2006). Ampliative reasoning to view the prevalence of severe thunderstorms. *Mausam, 57*, 523–526.
Chaudhuri, S. (2008a). Identification of the level of downdraft formation during severe thunderstorms: a frequency domain analysis. *Meteorology and Atmospheric Physics.*
Chaudhuri, S. (2008b). Preferred type of cloud in the genesis of severe thunderstorms—A soft computing approach. *Atmospheric Research, 88*, 149–156.
Chaudhuri, S. (2010). Convective energies in forecasting severe thunderstorms with one hidden layer neural net and variable learning rate back propagation algorithm. *Asia-Pacific Journal of Atmospheric Sciences, 46*(2), 173–183.
Chaudhuri, S., & Middey, A. (2009). Applicability of bipartite graph model for thunderstorms forecast over Kolkata. In *Advances in Meteorology* (pp. 1–12).
Chaudhuri, S., & Middey, A. (2011). Nowcasting thunderstorms with graph spectral distance and entropy estimation. *Meteorological Applications., 18*(2), 238–249.
Chaudhury, A. K. (1961). Pre-monsoon thunderstorms in Assam, Tripura and Manipur. *Indian Journal of Meteorology & Geophysics, 12*(1), 33–40.
Das, P. M., De, A. C., & Gangopadhyaya, M. (1957). Movements of two Nor'westers of West Bengal. A radar study. *Indian Journal of Meteorology & Geophysics, 8*(3), 399–406.
De, A. C. (1959). An unusually high Nor'wester radar cloud. *Indian Journal of Meteorology & Geophysics, 10*(3), 359–362.
De, A. C. (1962). Bright band phenomena over Calcutta and neighbourhood. *Indian Journal of Meteorology and Geophysics, 13*, pp. 143–146.
De, A. C. (1963). Movement of pre-monsoon squall lines over Gangetic West Bengal as observed by radar at Dumdum airport. *Indian Journal of Meteorology and Geophysics, 14*(1), 37–45.
De, A. C., & Rakshit, D. K. (1961). Radar observations on the formation of cumulus clouds near Calcutta during monsoon season. *Indian Journal of Meteorology and Geophysics, 12*(1), 1–16.
De, A. C., & Sen, S. N. (1961). A radar study of pre-monsoon thunderstorms (Norwesters) over Gangetic West Bengal. *Indian Journal of Meteorology and Geophysics, 12*(1), 51–55.
Desai, B. N. (1950). Mechanism of Nor'westers of Bengal. *Indian Journal of Meteorology & Geophysics, 1*(1), 74–76.
Doviak, R. J., Zrnik, D. S., & Sirmans, D. S. (1979). Doppler weather radar. *Proceedings of the IEEE, 67*, 1522–1553.

Goldar, R. N., Banerjee, S. K., & Debnath, G. C. (2001). Tornado in India and neighbourhood. Pre-published Scientific Report No. 2/2001, India Meteorological Department, RMC Calcutta (West Bengal).

Gupta, H. N., & Ghosh, S. K. (1980). North Delhi Tornado of 17th March 1978. *Mausam, 31*(1), 93–100.

Gupta, H. N., & Ghosh, S. K. (1982). Reported cases of Tornadoes in Indian subcontinent. *Vayumandal, 12*, 57–60.

Koteshwaram, P., & Srinivasan, V. (1958). Thunderstorms over Gangetic West Bengal in the pre-monsoon season and the synoptic factors favourable for their formation. *Indian Journal of Meteorology and Geophysics, 9*(3), 301–312.

Koteswaram, P., & De, A. C. (1959). Study of pre-monsoon thunderstorms over Gangetic West Bengal by radar. *Indian Journal of Meteorology & Geophysics, 10*(2), 275–282.

Krishna Rao, P. R. (1966). Thunderstorm studies in India—A review. *Indian Journal of Meteorology and Geophysics, 17*(1), 3–13.

Lhermitte, R. M. (1968). Turbulent air motion as observed by Doppler radar. In *Proceedings of 13th Conference on Radar Meteorology* (pp 498–503). Montreal AMS Boston.

Marshall, J. S., & Palmer, W. M. (1948). The distribution of raindrops with size. *Journal of Meteorology, 5*, 165–166.

Mukherjee, A. K., Rakshit, D. K., & Chaudhary, A. K. (1972). On the very high tops of Cb cloud around Calcutta. *Indian Journal of Meteorology and Geophysics, 23*(2), 241–245.

Mukhopadhyay, P., Singh, H. A. K., & Singh, S. S. (2005). Two severe Nor'westers in April 2003 over Kolkata, India, using Doppler radar observations and satellite imagery. *Weather, 60*(12), 343–353.

Mull, S., Gangopadhyaya, M., & George, C. A. (1955). A review of some layer analyses in relation to prediction of Nor'westers. *Indian Journal of Meteorology & Geophysics, 6*(1), 5–30.

Narayanan, V., & Krishnamurthy, G. (1966). A radar study of pre-monsoon thunderstorms over Bombay Airport. *Indian Journal of Meteorology and Geophysics, 17*(2), 237–244.

Pradhan, D. (2006). Supercell storm at Kolkata, India and Neighborhood—Analysis of thermodynamic conditions, evolution, structure & movement. *Indian Journal of Radio & Space Physics, 35*, 270–279.

Pradhan, D., & Sinha, V. (2005). Thunderstorm Genesis over Kolkata—A case study. *Vatavaran, AFAC Journal of Meteorology, 29*, 20–27.

Pradhan, D., & Talukdar, N. (2016). Evalution of ZR constants for Kolkata (India) and surroundings using Doppler weather radar and rain gauge data for pre-monsoon period. *AFAC Journal of Meteorology, 40*(2), 31–41.

Pradhan, D., Singh, U. V., & De, U. K. (2012)*:* Development of Nowcasting technique and evaluation of convective indices for thunderstorm prediction in Gangetic West Bengal (India) using Doppler Weather radar and upper air data. *Mausam, 63*(2), 299–318.

Rai Sircar, N. C. (1957). On the forecasting of nor'westers in Gangetic West Bengal. *Indian Journal of Meteorology & Geophysics, 8*(1), 21–32.

Rinehart, R. E. (1999). *Radar for Meteorologists* (3rd edn.). Rinehart Publications, P.O. Box 6124, grand Forks, ND 58026-6124, USA.

Sen, S. N., & Sengupta, J. R. (1961). A pre-monsoon thunderstorm with tornadic characteristics at Calcutta. *Indian Journal of Meteorology & Geophysics, 12*(1), 41–48.

Sinha, V., & Pradhan, D. (2006). Analysis of super cell thunderstorm over Kolkata. *Indian Journal of Radio & Space Physics, 35*, 270–279.

Skolnik, M. (1991). *Radar Handbook* (2nd edn.). McGraw hill Publishing Company.

Srivastava, K., Roy Bhowmik, S. K., & Sen Roy, S. (2010). Simulation of high impact convective events over Indian region by ARPS model with assimilation of Doppler weather radar radial velocity and reflectivity. *Atmósfera, 23*(1), 53–73.

Suresh, R., & Bhatnagar, A. K. (2004). Unusual hailstorm during May 2002 in Chennai and suburbs—A study using data from a single Doppler weather radar. *Mausam, 55*(4), 655–670.

Suresh, R., Ravichandran, P. K., Gupta, J. P., Thampi, S. B., Sundaram, S. K., & Rao, P. R. (2005a). On optimum rain rate estimation from a pulsed Doppler weather radar at Chennai. *Mausam, 56*(2), 433–446.

Suresh, R., Arindam, V., Rao, P. R., & Bhatnagar, A. K. (2005b). Clear air echoes from the atmospheric boundary layer over Chennai—A case study using s-band Doppler weather radar. *Mausam, 56*(2), 447–466.

Suresh, R. (2006). Nowcasting the genesis and movement of convective storms. *VATAVARAN, AFAC Journal of Meteorology, 30*(2), 73–81.

Union Conference on Radar Meteorology Report No. SFCSI–COMM (281pp.) (Chaps. 12, 29a).

Wakimoto & Wilson. (1989). Non-supercell tornadoes. *Monthly Weather Review, 117*, 1113–1140.

Wilson, J. W., & Schreiber, W. E. (1986). Initiation of convective storms at radar observed boundary layer convergence lines. *Monthly Weather Review, 114*, 2516–2536.

Wilson, J. W., & Mueller, C. K. (1993). Nowcast of thunderstorm initiation and evolution. *Weather Forecasting, 8*, 113–131.

Weather Radar Networks and Their Use in Monitoring Severe Storms in the Northeast Indian Subcontinent: Status and Expectations

Hideyuki Kamimera and Yusuke Yamane

Abstract This chapter discusses weather radar networks and their use in monitoring severe storms in the northeast Indian subcontinent, mainly in Bangladesh and seven states in northeast India. This chapter focuses on two types of severe storms in the region, tornadoes and the torrential rains and flash floods. As for tornadoes that cause damage and losses in Bangladesh, the feasibility of building an operational system on their predictions in the country is discussed. It is possible to build the system by combining a technique of predicting temporal variations in latent instability of the atmosphere using a numerical model and a technique of detecting convective clouds and/or mesocyclones using weather radar observations. As for torrential rains and flash floods over the basin of the Meghna River, flowing through three countries, Bangladesh, India, and Myanmar, it is possible to detect, monitor, and predict them by combining weather radars around the basin and rain gauges in a coverage of the radars.

Keywords Severe storms · Indian subcontinent · Weather radars · Disaster risk reduction

1 Introduction

This chapter discusses weather radar networks and their use in monitoring severe storms in the northeast Indian subcontinent, mainly in Bangladesh and seven states in northeast India. Basic investigations on weather radar networks in the region have been made in the Borderless Radar Information Networking over South and Southeast Asia (BRAIN) project (Kamimera, 2020) after its launch in 2018, as a

H. Kamimera (✉)
National Research Institute for Earth Science and Disaster Resilience, Tsukuba, Japan
e-mail: kamimera@bosai.go.jp

Y. Yamane
Tokoha University, Shizuoka, Japan

S. Das and W-K. Tao (eds.), *Severe Storms*,
https://doi.org/10.1007/978-981-97-7075-5_5

part of activities in the Asian Precipitation Experiment (AsiaPEX) project (Terao et al., 2023). According to the investigation in the BRAIN project, an international network of weather radars in the northeast Indian subcontinent is recognized as shown in Fig. 1. These radars can be utilized in monitoring severe storms in the region. A brief introduction to weather radar and its applications for rainfall estimation and weather prediction is provided in Sect. 2. Although there are many types of severe storms defined in earlier studies (e.g., Das, 2022), this chapter focuses on two types of severe storms in the region, tornadoes and the torrential rains and flash floods. As for tornadoes it is discussed in Sect. 3; as for the torrential rains and flash floods in Sect. 4.

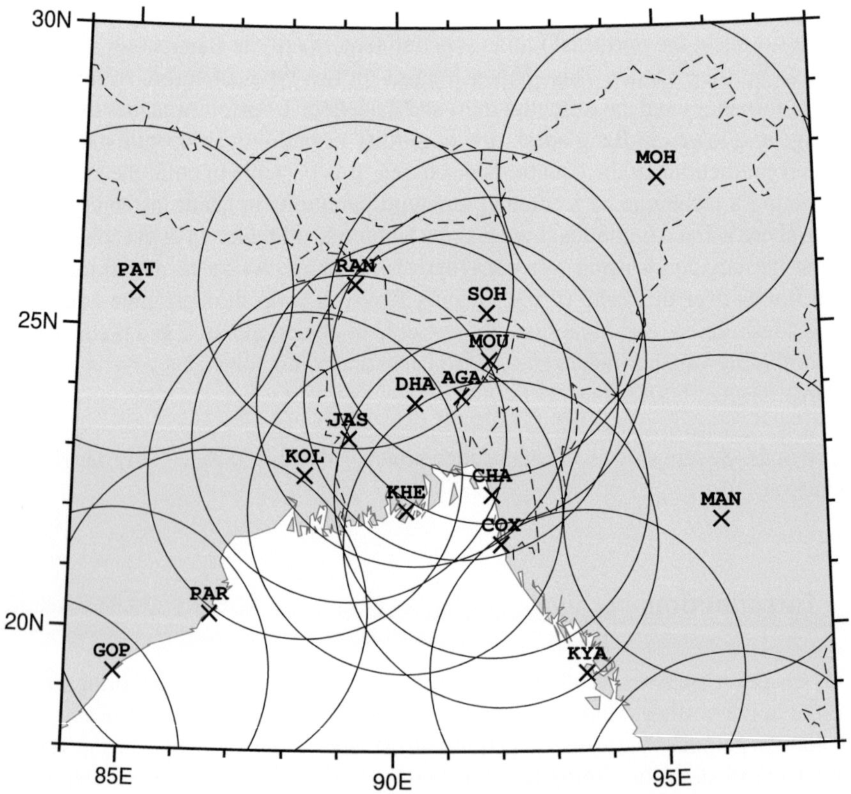

Fig. 1 Network of weather radars in the northeast Indian subcontinent. Five sites of the Bangladesh Meteorological Department (BMD): Rangpur (RAN), Khepupara (KHE), Dhaka (DHA), Moulvibazar (MOU), and Cox's Bazar (COX). Two sites of the Bangladesh Air Force (BAF): Jashore (JAS) and Chattogram (CHA). Seven sites of the India Meteorological Department (IMD): Gopalpur (GOP), Patna (PAT), Paradeep (PAR), Kolkata (KOL), Agartala (AGA), Sohra (SOH), and Mohanbari (MOH). Two sites of the Department of Meteorology and Hydrology (DMH) in Myanmar: Kyaukpyu (KYA) and Mandalay (MAN) (one more site at Yangon, but outside the box)

2 Weather Radar and Its Applications

Radar is an abbreviation for "radio detection and ranging", which is an electronic system that generates electromagnetic waves in the transmitter, radiates them into space via antenna, receives the scattered signal returning from the target, and measures the position, movement of the target, etc. (Fukao and Hamazu, 2013).

The diameters of precipitation particles are between several μm and several mm. Thus, electromagnetic waves of wavelengths of several mm (millimeter wave) to several cm (micro wave) are adopted in the radar measurements of precipitation particles. For larger precipitation particles such as raindrops, electromagnetic waves of about 2.5–15 cm wavelengths are adopted. Electromagnetic waves of 2.5–3.75 cm wavelengths (8–12 GHz) are categorized as X-band of radar frequency bands; 3.75–7.5 cm wavelengths (4–8 GHz) are as C-band; and 7.5–15 cm wavelengths (2–4 GHz) are as S-band. Shorter wavelengths are suitable for short range and high space resolution measurements; longer wavelengths are for long range and low space resolution measurements.

Weather radars directly measure the received signal power, the mean Doppler shift (or frequency), and the Doppler frequency spectrum width. The echo intensity and radar reflectivity factor are deduced from the received signal power and the relevant radar system parameters in consideration of power attenuation of the radar wave on the way to and back from scatterers. The mean Doppler shift is caused by the radial velocity of scatterers, and the Doppler frequency spectrum width is caused by the velocity fluctuation of scatterers. The Doppler velocity and the Doppler velocity spectrum width are derived, respectively, from the mean Doppler shift and the Doppler frequency spectrum width in reference to the transmitted wave frequency.

Radar reflectivity factor Z ($\mathrm{mm}^6\ \mathrm{m}^{-3}$) is the sixth moment of drop size distribution (DSD), and expressed as

$$Z = m_6 = \int_0^{\infty} D^6 N(D)\mathrm{d}D, \tag{1}$$

where D (mm) is the diameter of a raindrop, and $N(D)$ ($\mathrm{m}^{-3}\ \mathrm{mm}^{-1}$) is the number of raindrops per unit volume and unit size interval. A variety of functions have been proposed for the DSD, the following (modified) gamma distribution (Ulbrich, 1983) is known as the most popular one.

$$N(D) = N_0 D^{\mu} \exp(-\Lambda D), \tag{2}$$

where N_0 ($\mathrm{m}^{-3}\ \mathrm{mm}^{-1-\mu}$) is an intercept parameter; μ is a shape parameter (dimensionless); and Λ (mm^{-1}) is a slope parameter. Substituting (2) into (1), Z can be expressed as

$$Z = N_0 \Lambda^{-7-\mu}\Gamma(7+\mu), \tag{3}$$

where $\Gamma(x)$ is the complete gamma function.

When $\mu = 0$ in (2), $N(D)$ becomes an exponential distribution as follows;

$$N(D) = N_0 \exp(-\Lambda D). \tag{4}$$

This is the same with the DSD proposed by Marshall and Palmer (1948). They reported the following parameters N_0 and Λ;

$$N_0 = 8 \times 10^3 \ [\mathrm{m}^{-3}\mathrm{mm}^{-1}], \tag{5}$$

$$\Lambda = 4.1R^{-0.21} \ [\mathrm{mm}^{-1}], \tag{6}$$

where R (mm h^{-1}) is rainfall rate. It is known that both N_0 and Λ are dependent on R as shown in earlier studies (e.g., Ulbrich, 1983). Thus, N_0 and Λ are, in general, expressed as

$$N_0 = a_{N_0} R^{b_{N_0}}, \tag{7}$$

$$\Lambda = a_\Lambda R^{b_\Lambda}. \tag{8}$$

Substituting (7) and (8) into (3), a power-law Z–R relationship is derived as follows;

$$Z = aR^b, \tag{9}$$

$$a = a_{N_0} {a_\Lambda}^{-7-\mu} \Gamma(7 + \mu), \tag{10}$$

$$b = b_{N_0} - (7 + \mu) b_\Lambda. \tag{11}$$

As shown in (9)–(11), rainfall rate R can be estimated from radar reflectivity factor Z when all the DSD parameters, a_{N_0}, b_{N_0}, a_Λ, b_Λ and μ, are known. For example, when $\mu = 0$, using (5) and (6) the following Z–R relationship is obtained;

$$Z = 296R^{1.47}. \tag{12}$$

Radar observations of such radar reflectivity and radial velocity are quite useful not only for a better understanding of severe storms, such as supercell storms and tornadoes, but also for the improvement of nowcasting and/or short-range numerical weather prediction with radar data assimilation.

Assimilating weather radar observations of reflectivity and radial velocity into numerical weather prediction models, the performance of short-range predictions can be improved. For example, following an earlier work of Caumont et al. (2010), Japan Meteorological Agency (JMA) estimates relative humidity from reflectivity and assimilates it into a model in operational meso-scale and local analyses in Japan (e.g., Ikuta, 2021; JMA, 2023). According to Seko et al. (2004), Ishikawa and Koizumi (2006), and JMA (2023), JMA is doing the assimilation of radial velocity data also for improving the model performance. Such radar data assimilation, that is, a fusion of radar data and a model, or further, inter-complementary integration of them with satellite data is effective for improving the performance of short-range predictions.

3 Tornadoes

3.1 Climatological Aspects of Tornadoes in the Indian Subcontinent

3.1.1 What is Tornado?

Tornado is one of the most destructive meteorological phenomena in the world associated with developed cumulonimbus. Tornado has significant strong vortex in the vertical under the developed cumulonimbus, and often causes heavy damages to human and property. It is well known that tornadoes occur in the United States of America the most frequently across the world. In the United States of America, the averaged frequency of tornadoes is around 1,200 per year, and the averaged death by tornadoes is about 60 per year (National Oceanic and Atmospheric Administration, 2023).

It is well known that there are environmental conditions favorable for the development of cumulonimbus associated with tornadoes. All cumulonimbus is not necessary associated with tornadoes. Cumulonimbus generated in the environmental suitable for the development tends to produce tornadoes. We will describe in detail about the environmental situations of the development of cumulonimbus leading to the occurrence of tornadoes in Sect. 3.2.

3.1.2 Distribution of Tornadoes Across the World

Fujita (1973) showed the distribution of tornadoes across the world, and tornadoes occur most frequently in the United States of America. He also showed that tornado relatively frequently occur in the European, Japan, Australia, Argentina, and the northeastern part of the Indian subcontinent. In contrast, the occurrence of tornadoes is rare in the tropical area. Tornadoes frequently occur in the middle latitude area. This indicates that the instability of the atmosphere can be easily increased in the middle latitude area, where the warm air in the tropical region conflicts the cold air in the pole area.

3.1.3 Climatological Aspects of Tornadoes Across the Indian Subcontinent

Tornadoes occur the most frequently across the northeastern part of the Indian subcontinent in the South Asia as mentioned before. In this subsection, we will show climatological aspects of tornadoes in the Indian subcontinent based on Peterson and Mehta (1981, 1995).

Peterson and Mehta (1981, 1995) show the spatial distribution of tornado and monthly frequency of tornado in the Indian subcontinent, respectively. Peterson and Mehta (1995) analyzed 73 tornado events during 1838–1993 based on the investigation using literature reports such as newspapers. They show that tornadoes significantly concentrate in the northeastern part of the Indian subcontinent centered at Dhaka. Table 1 shows the monthly frequency of tornado across the Indian subcontinent shown in Peterson and Mehta (1995). This figure shows that tornado occur frequently during March to May peaking in April. In this area, the season from March to May is called "Pre-monsoon season". The season from June to September is rainy season with the highest rainfall in the year and called "Monsoon season". After monsoon season, the season from October to November is called "Post monsoon season". Tornado is occurred the most frequent in the pre-monsoon season. The occurrence of tornadoes is significantly decreasing in the monsoon season. After the monsoon season, tornado occurrence is slightly increasing in the post monsoon season.

In the monsoon season, convection is active with the highest rainfall through the year. However, the most of convections is not associated with severe local storms such as tornadoes. This indicates that the instability of the atmosphere is significantly increasing favoring for the development of cumulonimbus especially in the pre-monsoon season than the monsoon season. This point will be described in detail in Sect. 3.2. The occurrence of tornado is slightly increasing in the post monsoon season, and this indicates the instability of the atmosphere is slightly enhanced in the post-monsoon season compared to the monsoon season, but this mechanism has not been clarified.

Table 1 Monthly frequency of tornadoes across the Indian subcontinent shown in Peterson and Mehta (1995)

Jan	Feb	Mar	Apr	May	Jun	Jul	Aug	Sep	Oct	Nov	Dec
2	4	8	30	10	2	1	3	5	6	2	0

3.2 *Convective Parameters Associated with Severe Local Storms in Bangladesh*

As mentioned before, tornadoes concentrate in and around Bangladesh relatively higher in the Indian subcontinent. In Bangladesh, not only tornadoes but also other types of severe local storms such as hail and lightning frequently occur during the pre-monsoon season (Yamane et al. (2010a)). Why do severe local storms including tornadoes frequently occur in and around Bangladesh during the pre-monsoon season? To understand this question, the characteristics of convective parameters such as CAPE (Convective Available Potential Energy) in Bangladesh are mentioned in this subsection. A convective parameter is an index for identifying and forecasting the environmental situations favorable for the outbreak of deep convections with severe local storms. In this subsection, the atmospheric conditions in Bangladesh during the pre-monsoon is significantly favorable for the outbreak of severe local storms based on analysis using convective parameters. And, the efficiency of convective parameters for the forecasting of severe local storms in Bangladesh is also described.

3.2.1 Evaluation of Environmental Situations of Severe Local Storms Using Convective Parameters Across the Indian Subcontinent

Potential instability and vertical wind shear are important factors for the outbreak of deep convective clouds associated with severe local storms. Brooks et al. (2003) showed that tornadoes likely occur in the environment with large magnitude of potential instability and vertical wind shear.

Yamane and Hayashi (2006) showed that the frequency of large value of CAPE and vertical wind shear is significantly higher in and around Bangladesh during the pre-monsoon season, which leads to frequent occurrences of severe local storms over this region during this season. In contrast, CAPE and vertical wind shear are significantly lower all over the Indian subcontinent during the monsoon season. In the monsoon season, although convective activity with heavy rainfall is active in this region, deep convections associated with severe local storms do not tend to be occurred because the environment is not suitable for the occurrence of developed deep convections leading to the generation of severe local storms.

3.2.2 Characteristics of Convective Parameters Associated with Severe Local Storms in Bangladesh

Yamane et al. (2010b) compared convective parameters of the day with the damage reports of severe local storms (hereafter, referred to as SLSD) with those of the day without the damage reports of severe local storms (hereafter, referred to NSLSD). Table 2 shows that the mean and median values of indices regarding to potential instability (KI, TT, CAPE) in the SLSD is higher than those in the NSLSD (For SSI and LI, the mean and median values in the SLSD are lower than those in the NSLSD). Therefore, this study indicates that convective parameters regarding to potential instability of the atmosphere is relatively effective for discriminating between the SLSD and the NSLSD in Bangladesh during the pre-monsoon season.

The PW in the SLSD is higher than that in the NSLSD. Yamane et al. (2010b) showed that southerly wet inflow from the Bay of Bengal is significantly enhanced in the lower layer peaking at 500 m Above Ground Level in the SLSD compared with the NSLSD. Thus, the amount of water vapor in the atmosphere in SLSD is larger than in NSLSD. This study indicates that the PW is an effective parameter for discriminating between SLSD and NSLSD. Table 2 shows that the mean and median values of vertical wind shear related parameters in SLSD are higher than in NSLSD. However, the differences between SLSD and NSLSD of wind shear related parameters are not clear compared with the parameters related to potential instability.

Table 2 Statistics on convective parameters on severe local storm days (SLSD) and non-severe local storm days (NSLSD) during the pre-monsoon season in Bangladesh (Yamane et al. (2010b))

	SLSD (59 days)			NSLSD (135 days)		
	Mean	Median	SD	Mean	Median	SD
KI	27.6	29.0	6.3	19.6	24.6	16.1
TT	68.0	66.9	12.5	65.9	64.0	14.6
SSI	0.8	0.7	3.5	3.9	2.8	5.1
LI	−0.2	0.1	3.2	3.6	2.6	5.4
PW	38.2	38.5	5.4	32.7	32.6	10.2
CAPE	1363	1170	1283	799	254	1107
CIN	322	300	239	209	224	202
SHEAR	16.7	16.5	6.6	16.2	16.3	7.0
MS0-1 km ($\times 10^{-3}$)	20.0	19.2	7.7	18.8	18.0	6.0
MS0-2 km ($\times 10^{-3}$)	16.7	16.2	5.4	15.6	15.0	4.0
MS0-3 km ($\times 10^{-3}$)	15.0	14.6	4.0	14.5	14.1	3.6
MS0-4 km ($\times 10^{-3}$)	14.3	13.9	3.1	14.1	13.7	3.3
SREH	148	115	131	105	94	104
VGP0-1 km	0.59	0.58	0.43	0.38	0.27	0.38
VGP0-2 km	0.51	0.46	0.37	0.31	0.23	0.31
VGP0-3 km	0.46	0.42	0.33	0.29	0.23	0.28
VGP0-4 km	0.44	0.42	0.30	0.28	0.22	0.27
EHI	1.32	0.43	1.87	0.57	0.03	1.04
BRN	16.1	7.4	21.3	36.6	2.2	298.9

KI: K index **TT:** Total Total **SSI:** Showalter's Stability Index **LI:** Lifted Index **PW:** Precipitable Water **CAPE:** Convective Available Potential Energy **CIN:** Convective Inhibition **SHEAR:** The magnitude of the vector difference between the winds at the surface and 500 hPa. **MS:** Mean Shear **SREH:** Storm Relative Environmental Helicity **VGP:** Vorticity Generation Parameter **EHI:** Energy Helicity Index **BRN:** Bulk Richardson Number

It is well known that SREH is an effective parameter for the forecasting of supercell associated with mesocyclone. The SREH in SLSD is higher than in NSLSD. VGP, EHI and BRN are well known as parameters of combinations of potential instability and vertical wind shear. Table 2 shows that the mean and median values of the VGP and EHI in SLSD are larger that in NSLSD. The mean values of BRN in SLSD is smaller than in NSLSD (BRN value is smaller for the environment favorable for the outbreak of supercell). In the above analysis, it was confirmed that LI, PW, CAPE, SSI, VGP, KI, EHI have statistical significant differences between SLSD and NSLSD with the 99% confidence level.

Table 3 shows the HSS (Hidke's skill score) of convective parameters associated with the outbreaks of severe local storms during the pre-monsoon soon season in Bangladesh. The HSS is a skill score for evaluating the ability for prediction of likelihood of rare phenomena such as tornadoes. Table 3 shows that the HSS of the LI is the highest among all parameters. The threshold with the highest HSS for the LI is −2.6 K. Thus, this study indicates that the LI is the most effective parameter for the prediction of severe local storms in Bangladesh during the pre-monsoon season. In addition, it is likely that we can predict the occurrence of severe local storm when we obtain the value of LI less that −2.6 K. Table 3 shows that the HSS values of parameters related to potential instability is relatively higher than those of parameters related to vertical wind shear. This indicates that potential instability is more important factor for the generation of severe local storms during the pre-monsoon season in Bangladesh. In addition, the large value of the HSS of the PW indicates that the amount of water vapor in the atmosphere is important for generation of convective clouds associated with severe local storms. In particular, the highest value of the HSS of the LI indicates that the water vapor in the lowest layer is significantly crucial for the development of deep convection associated with severe local storms because the LI reflects the amount of water vapor in the lower layer.

Table 3 The Heidke Skill Score (HSS) and thresholds yielding the best HSS for each convective parameter (Yamane et al., 2010b)

	HSS	Threshold
LI	0.29	−2.6
PW	0.29	31.9
CAPE	0.24	1121
SSI	0.24	0.7
VGP0-2 km	0.23	0.67
KI	0.21	29.6
VGP0-1 km	0.20	0.77
VGP0-3 km	0.17	0.70
VGP0-4 km	0.17	0.13
EHI	0.12	0.01

3.3 *Early Warning System and Forecasting of Severe Local Storms in Japan Using Convective Parameters and Radar Observational Data*

This subsection states the forecasting and early warning system for gusty wind such as tornadoes using convective parameters and doppler radar observation in Japan based on JMA website (JMA, 2024).

Numerical weather prediction system of JMA simulates convective parameters such as CAPE and EHI. Radar echo data are used to detect developing convective clouds, which is likely to be associated with severe local storms. The region with higher values of convective parameters such as CAPE and EHI and developing convective clouds can be forecasted as the place with relatively higher possibility of severe local storm occurrences.

Doppler radar observation is also utilized to detect the vortex of mesocyclone associated with supercell. The pair of approaching doppler velocity into a radar site and going away doppler velocity from a radar site indicates the existence of the vortex of mesocyclone. Tornado tends to be occurred in the region of mesocyclone detected with the doppler radar observations. When the vortex of mesocyclone is detected with higher values of convective parameters predicted with numerical weather model and the indication of high convective area such as intensive radar echo over a certain region, the region has the high possibility of the occurrence of gusty wind such as tornado and downburst.

The region with the high value of convective parameters has the large potential for developing of cumulonimbus clouds such as supercell. Developed convective clouds are likely to occur in the region with higher radar echo intensity observed with weather radar and the high value of convective parameters predicted with numerical simulations. In addition, if the vortex of mesocyclone is detected with doppler radar in such region, supercell with gusty wind such as tornado tends to be occurred there. JMA developed the danger index for gust to diagnosis the situation with gusty wind such as tornado. Higher value of the danger index for gust is found in the region with higher radar echo intensity and larger value of convective parameters predicted with numerical simulation. JMA predicts the possibility of occurrence of gusty wind by utilizing the combination of the danger index for gust and detection of mesocyclone vortex detected with the doppler radar.

JMA categorizes the possibility of the occurrence of tornadoes into two levels and produces the distribution of the levels of possibility of the occurrence of tornadoes up to 1 h ahead updating every 10 min. JMA issues a tornado warning to the area with high level of the possibility of the occurrence of tornadoes. A tornado warning is reported in JMA website, TV, radio and so on. Japanese people can prepare for mitigating damage of tornadoes based on a tornado warning in advance.

However, Japanese people, who utilizes tornado warning to mitigate the damages of tornadoes, is not so many unfortunately. Perception for the damage of tornado and tornado warning is low. Although tornado caused devastating damage to life and property, Japanese people thinks that they may be not damaged by tornadoes because tornado is significantly rare. Moreover, false alarm (an event forecast that is associated with no event observed) is frequent because it is difficult that we precisely predict the occurrence of tornadoes even by current technology. This problem is likely to cause low perception of Japanese people for tornado warning.

Although tornado causes devastating damages to human and property, if we can predict the coming of tornadoes, we can reduce damages of tornado. Forecasting and early warning system of tornado often tends to result in false alarm. However, the cost of preparing for a tornado that may not come is much smaller than the cost of damage. We should prepare for tornado warning to minimize the damage admiring the false alarm.

3.4 Proposal Ideas of Forecasting and Early Warning System of Severe Local Storms Utilized Convective Parameters and Continuously Observational Radar Data in Bangladesh

This subsection proposes the instruction of early warning system based on the combination of detection of mesocyclone utilized doppler radar and gusty wind index calculated from convective parameters predicted with numerical simulation into Bangladesh. Bangladesh does not conduct this kind of forecasting and early warning system and the instruction of JMA method mentioned above is significantly meaningful for Bangladesh to migrate damages of tornado and severe local storms.

In Bangladesh, there are vulnerable houses commonly in rural areas, and these houses can be easily collapsed by high wind such as tornadoes. Debris of collapsed houses are blown off at huge high speed and often cause terrible and fatal damage to a human body. In rural area, there are many houses constructed with tin plates. Debris of tin plates blown off by high wind are so dangerous and often cut human bodies.

To construct robust house and avoid using dangerous construction materials which can be easily blown off by high winds, is important for mitigating the damages of gusty wind such as tornadoes, and we suggest that the government should support financially for the construction of houses and buildings with high resistance against high wind.

In addition, it is also important to construct the early warning system for quickly taking actions for mitigating the damages of high wind. The early warning system for gusty wind such as tornadoes in Japan as mentioned above will be helpful for introduction of an early warning system in Bangladesh, because recently, numerical weather prediction has been already utilized, and the distributions of temperature, wind and precipitation are predicted using a numerical model in Bangladesh Meteorological Department (BMD). Therefore, it is not difficult for BMD to predict convective parameters using the numerical model. Yamane et al. (2010b) shows that convective parameters are effective tools for forecasting of severe local storms, and some kinds of parameters can discriminate SLSD and NSLSD as mentioned before.

Figure 2 shows that the radar echo image observed by BMD radar in Dhaka on 22 March 2013. On this day, a tornado occurred in Brahmanbaria district, which is located in the eastern part of Bangladesh. The occurrence point of the tornado is indicated by the circle with "T" in Fig. 2. It can be seen that the tornado is occurred near hook echo pattern with high rainfall rate. Akter and Ishikawa (2014) shows that the value of EHI is relatively higher over the regional area included the occurrence point of the hook echo mentioned above. Figure 3 shows the distribution of EHI at 06UTC (12 BST, Bangladesh Standard Time) on 22 March 2013 calculated with JRA-55 (Japanese 55-year Reanalysis) produced by JMA. Figures 2 and 3 show that the storm cloud associated with the tornado with hook echo pattern is occurred over the area with the larger value of EHI (EHI > 1.0). The result of this case analysis suggests that the prediction method based on the combination of convective parameters such as EHI and radar observation data is likely to be effective in predicting severe weather such as tornadoes in Bangladesh. As mentioned above, BMD has already utilized numerical model in the weather forecasting. Therefore, they easily can predict the distribution of convective parameters using numerical modeling.

The problem to be improved for constructing the early warning system in BMD is to make a continuous radar observation. Some doppler radar systems of BMD are not active now. And BMD does not make a continuous radar observation (the radar is working for 1 h and then stopping for 2 h). Reliable detection of mesocyclones, which abruptly occurs, needs continuous radar observations. Various factors such as the shortage of electric power and human resources in BMD may cause that they cannot make a continuous radar observation. To solve the problem which prevents making a continuous radar observation is significantly crucial for constructing the

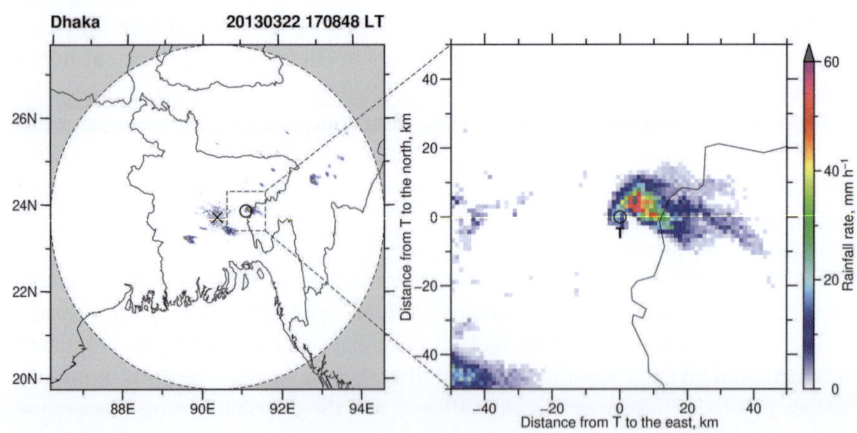

Fig. 2 Radar echo image at 1708 BST (Bangladesh Standard Time) on 22 March 2013 observed by BMD radar in Dhaka. Circle with "T" indicates the location of tornado occurrence

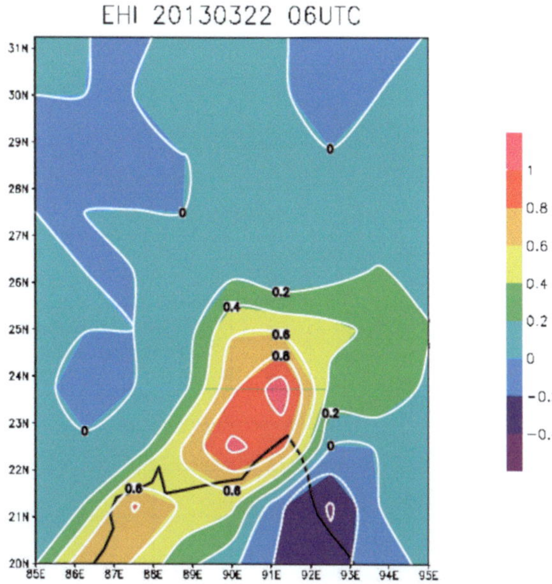

Fig. 3 Distribution of EHI (Energy Helicity Index) at 06 UTC (12 BST) on 22 Marth 2013 calculated with JRA-55 (Japanese 55-year Reanalysis)

early warning system for drastically mitigating the damages of high wind such as tornadoes in Bangladesh.

4 Torrential Rains and Flash Floods

The same as other severe storms, it is expected that existing weather radars will be utilized for detecting, monitoring, and predicting torrential rains and succeeding flash floods, which appear on a small and short scale in space and time. This section discusses the detection, monitoring, and prediction of the torrential rains and flash floods in the Meghna River basin in the northeast Indian subcontinent.

The Meghna River is an international, transboundary river flowing through three countries, Bangladesh, India, and Myanmar (Fig. 4). Bhairab Bazar is one of the most important points for hydrological observations in the lower basin of the river. The catchment area of Bhairab Bazar is 64,501 km^2, according to the HydroSHEDS data set (version 1, Lehner et al., 2008). India (43,379 km^2) and Myanmar (784 km^2) cover about 68% of the catchment area, whereas Bangladesh covers the remaining 32% (20,338 km^2). Over the catchment area, spatio-temporal distributions of convective cloud activities and succeeding torrential rains are not well known.

It is not easy to obtain accurate and detailed rainfall data from rain gauge observations only or satellite observations only for such large areas in general. Therefore, integrating rain gauge and satellite data, or adjusting satellite-based rainfall estimates

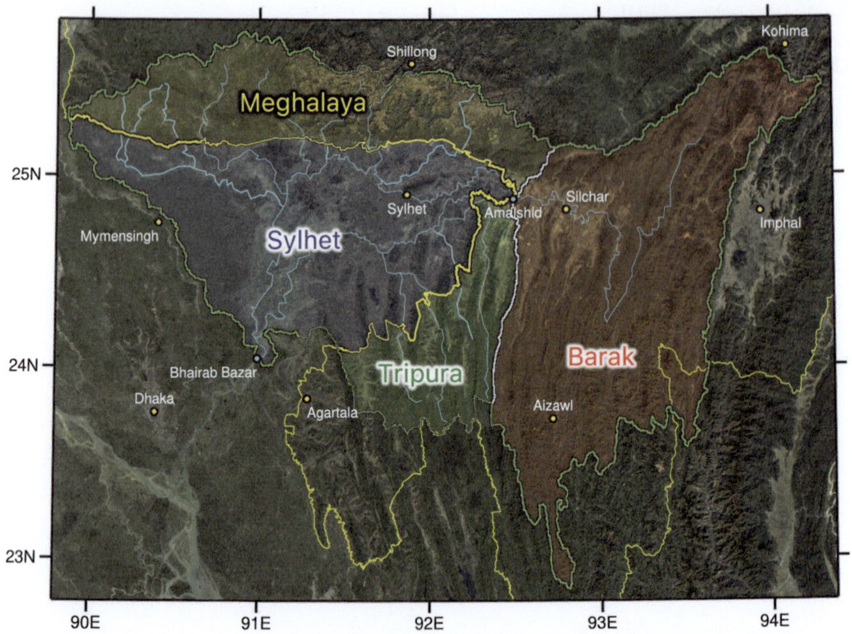

Fig. 4 Map of the Meghna River basin. Green line shows the boundary of the basin (the catchment of Bhairab Bazar); light blue line shows rivers; and yellow line shows national borders. Circle filled by light blue shows the location of two hydrological stations at Amalshid and Bhairab Bazar. The basin is divided into four sub-regions: Barak (the catchment of Amalshid, red), Meghalaya (yellow), Tripura (green), and Sylhet (blue)

using rain gauge observations in other words, various satellite-based rainfall products have been produced. However, it is not easy to know rainfall over the basin even using such rainfall products. There is a preliminary study that reveals this difficulty.

Kamimera et al. (2022) divided the basin into four sub-regions, Barak, Meghalaya, Tripura, and Sylhet, as shown in Fig. 4 and compared five products of satellite-based rainfall estimates, TRMM/PR, CHIRPS, GSMaP-G, TMPA-G, and MSWEP, over each sub-region. As a result of the comparison, they showed that there are large variances in rainfall estimates over all the four sub-regions, that is, there are large uncertainties over all the four sub-regions, as shown in Fig. 5. And also, they showed that among the four sub-regions, the variance in rainfall estimates is particularly high over the Meghalaya region. This is an important fact that cannot be overlooked for monitoring and predicting the flash floods caused by torrential rains over the Meghalaya region.

As shown above, it is difficult to know rainfall over the Meghna River basin using only rain gauges and/or satellite-borne sensors. Therefore, there are great expectations for quantitative precipitation estimation (QPE) and even quantitative precipitation forecasting (QPF) based on weather radar observations. Radar-based QPE has been applied in many countries including Japan, Vietnam, and Indonesia, even

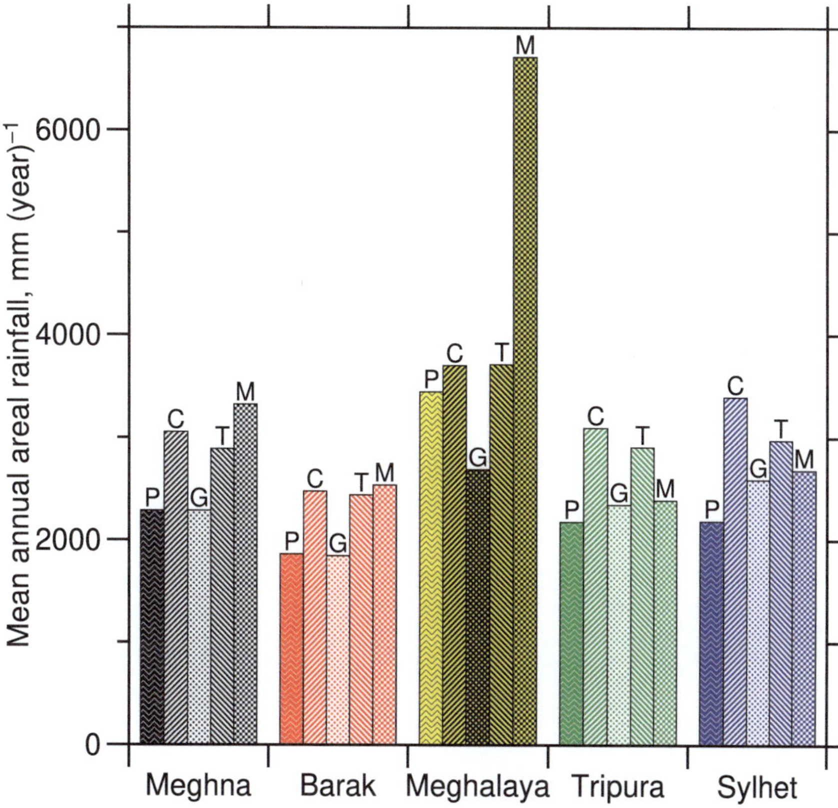

Fig. 5 Mean annual areal rainfall over the entire basin of the Meghna River (the catchment of Bhairab Bazar, black) and four sub-regions of the basin, Barak (red), Meghalaya (yellow), Tripura (green), and Sylhet (blue), for five products of satellite-based rainfall estimates, TRMM/PR (P), CHIRPS (C), GSMaP-G (G), TMPA-G (T), and MSWEP (M), basically for the eight-year period from 2009 to 2016, or for the 16-year period from 1998 to 2013 only for the TRMM/PR product

if it focuses only on Asian countries. And then it has played an important role not only in fundamental studies such as a study to elucidate rainfall mechanisms (e.g., Kamimera et al., 2012), but also in applied studies such as flood predictions (e.g., Kamimera et al., 2015) and water hazard mapping (e.g., Kamimera, 2022).

Also in the Meghna River basin, radar-based QPE can be possible by combining various observations from two countries, upstream India and downstream Bangladesh. Figure 6 shows the locations of weather radars around the basin and rain gauges in/around a coverage of the radars. At Agartala and Mohanbari in northeast India, two radars have been operated by the India Meteorological Department (IMD) and they cover the entire basin of the Meghna River. According to the investigation in the BRAIN project (Kamimera, 2020), these two radars have continuous observations basically at 10-min intervals for the coverage with a radius (range) of roughly

300 km (Fig. 7). Thus, combining with the data of rain gauges in the coverage, radar-based QPE and its validation may be feasible.

There are two types of rain gauge observations, experimental and operational observations, in the region. An experimental observation has been made by the South Asia Hydro-Meteoro-Climatological Observation Network (SOHMON) research team for about 18 years from 2006 to the present. Operational observations have been made by the IMD for every 15 min in real time in Indian part. For Bangladeshi

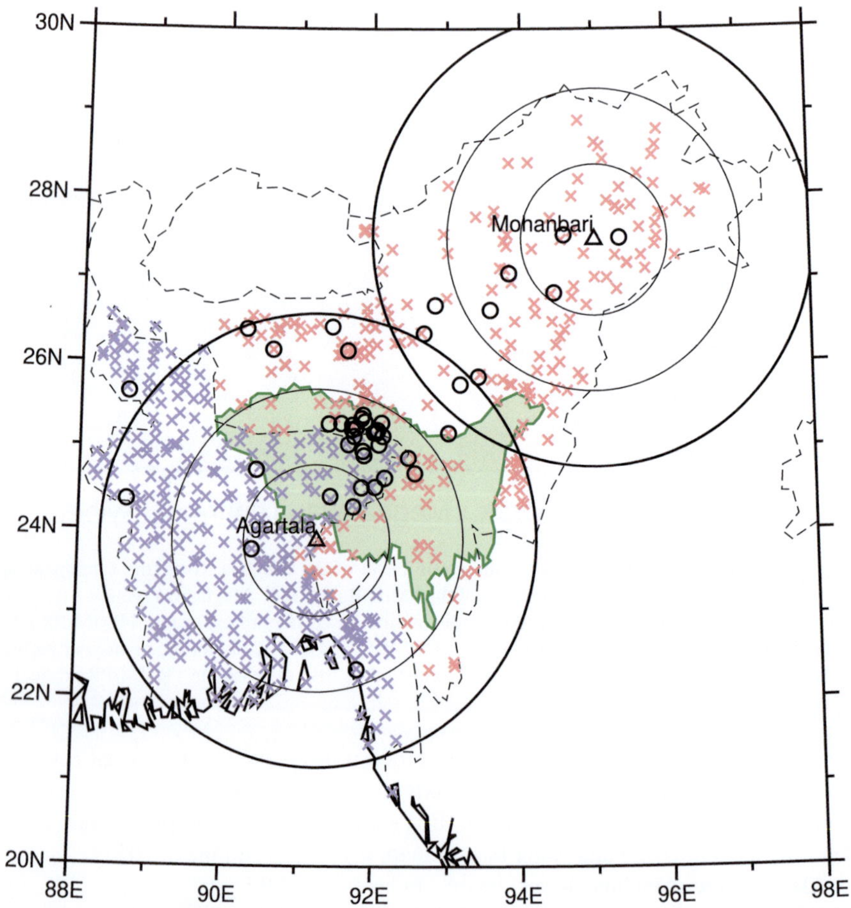

Fig. 6 Map of the Meghna River basin (the catchment of Bhairab Bazar, green-shaded area), two weather radars (triangles) at Agartala and Mohanbari in northeast India, and rain gauges (small open circles and crosses) in/around a coverage of the two radars. The radars are operated by the India Meteorological Department (IMD). Range rings of each radar are for 100, 200, and 300 km. Small open circles are rain gauges of the South Asia Hydro-Meteoro-Climatological Observation Network (SOHMON). Red and blue crosses are rain gauges operated by the IMD and the Bangladesh Water Development Board (BWDB) respectively

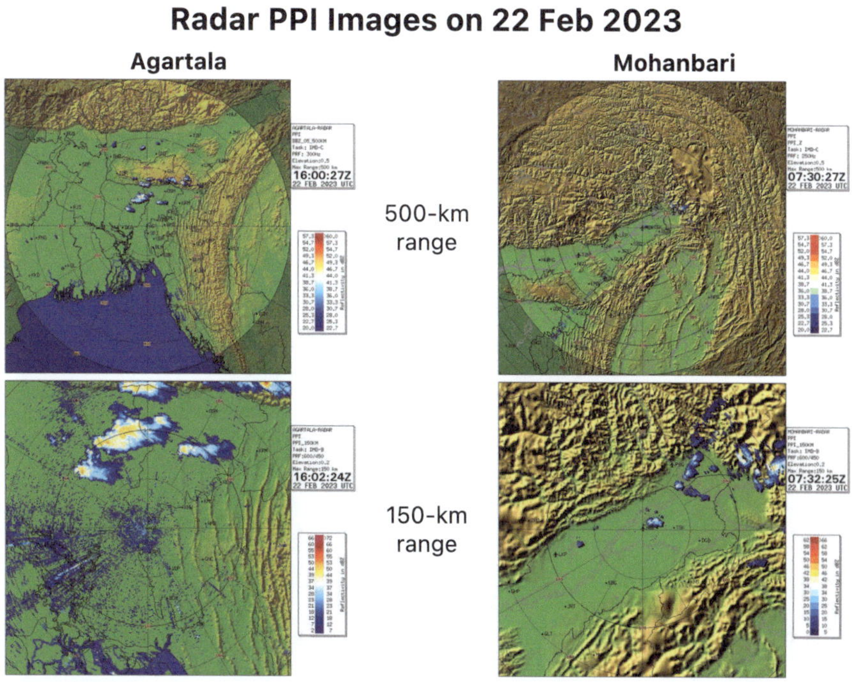

Fig. 7 Plan position indicator (PPI) images on 22 February 2023 for two weather radars at Agartala and Mohanbari in northeast India. The radars are operated by the India Meteorological Department (IMD)

part, daily data by the Bangladesh Water Development Board (BWDB) are available, but it is not in real time.

Using the above data, a relationship between reflectivity measured by a weather radar (Z) and rainfall on the ground measured by rain gauges (R) is empirically determined in advance as a Z–R relationship. Since the time intervals of the Z and R values are different, the Z–R relationship is determined after aggregating the Z and R values every 30 or 60 min. Using the determined Z–R relationship, Z values can be converted to R values. This simple method has been applied in Vietnam (Kamimera et al., 2011), Indonesia (Kamimera et al., 2012), and Cambodia (Kamimera et al., 2023), and high performance of rainfall estimates has been confirmed in each country.

Figure 8 shows a geographical pattern of the probability of appearance of radar reflectivity echoes (POA) by the IMD Agartala radar for the 6.5-month period from February 18 to August 31, 2023 in the pre-monsoon and monsoon seasons of the region. In the radar coverage, there are almost no shadow areas created by obstacles blocking radar beams, such as topography, trees, or buildings. A large area of high frequency is dominant over around the State of Meghalaya, India. It seems that rainfall phenomena during the pre-monsoon season are correctly observed by the radar. And thus, the radar can be utilized in QPEs or QPFs.

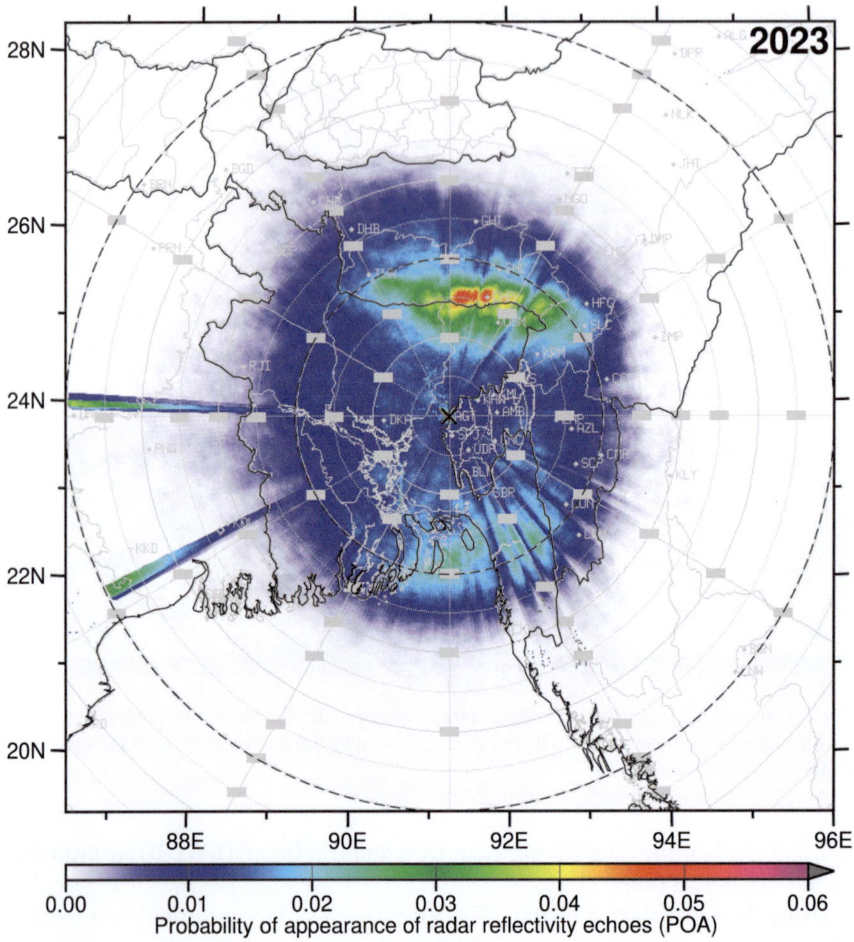

Fig. 8 Geographical pattern of the probability of appearance of radar reflectivity echoes (POA) by the India Meteorological Department (IMD) Agartala radar for the 6.5-month period from February 18 to August 31, 2023. Black solid line shows shorelines or national borders. Cross at the center shows the position of the radar. Black dashed line shows a 200-km or 500-km range ring of the radar

5 Summary

As discussed for tornadoes in Sect. 3 and the torrential rains and flash floods in Sect. 4, utilizing existing weather radars can be a viable, effective solution to detect, monitor, and predict such types of severe storms in the northeast Indian subcontinent.

As for tornadoes that cause damage and losses in Bangladesh, the feasibility of building an operational system on their predictions in the country has been discussed. It is possible to build by combining a technique of predicting temporal variations in latent instability of the atmosphere using a numerical model and a technique of detecting convective clouds and/or mesocyclones using weather radar observations. Because weather prediction using a numerical model has already been implemented, it is not difficult to compute indices of latent instability from the results of the prediction. Weather radar observations, which are discontinuous in time at present, should be obtained continuously in the future operation.

As for torrential rains and flash floods over the basin of the Meghna River, flowing through three countries, Bangladesh, India, and Myanmar, it is possible to detect, monitor, and predict them by combining weather radars around the basin and rain gauges in a coverage of the radars. In this regard, it is important to strengthen not only international cooperation but also domestic, inter-organizational cooperation through open dialogues among stakeholders.

References

Akter, F., & Ishikawa, H. (2014). Synoptic features and environmental conditions of the tornado outbreak on March 22, 2013 at Brahmanbaria in the east-central region of Bangladesh. *Natural Hazards, 74*, 1309–1326. https://doi.org/10.1007/s11069-014-1252-y

Brooks, H. E., Lee, J. W., & Craven, J. P. (2003). The spatial distribution of severe thunderstorm and tornado environments from global reanalysis data. *Atmospheric Research, 67–68*, 73–94.

Caumont, O., Ducrocq, V., Wattrelot, E., Jaubert, G., & Pradier-Vabre, S. (2010). 1D+3DVar assimilation of radar reflectivity data: A proof of concept. *Tellus A, 62*(2), 173–187. https://doi.org/10.1111/j.1600-0870.2009.00430.x

Das, S. (2022). Challenges in predicting extreme weather events over the South Asian region. In A. S. Unnikrishnan, F. Tangang, R. J. Durrheim (Eds.), *Extreme Natural Events* (pp. 51–106). Springer. https://doi.org/10.1007/978-981-19-2511-5_3

Fujita, T. (1973). Tornadoes-The mystery of a vortex, Kyouritsu Shuppan (228pp.) (in Japanese).

Fukao, S., & Hamazu, K. (2013). *Radar for Meteorological and Atmospheric Observations.* Springer. https://doi.org/10.1007/978-4-431-54334-3

Ikuta, Y. (2021). Radar data assimilation in numerical weather prediction models. In S. Michaelides (Ed.), *Precipitation Science: Measurement, Remote Sensing, Microphysics, and Modeling* (pp. 743–756). Elsevier. https://doi.org/10.1016/B978-0-12-822973-6.00012-3

Ishikawa, Y., & Koizumi, K. (2006). Doppler radar wind data assimilation with the JMA Meso 4D-Var. In *Research Activities in Atmospheric and Oceanic Modelling.* CAS/JSC Working Group on Numerical Experimentation Report No. 36 (WMO/TD No. 1347) (pp. 1-11–1-12).

Japan Meteorological Agency (2023). *Outline of the Operational Numerical Weather Prediction at the Japan Meteorological Agency.*

Japan Meteorological Agency (2024). About severe gusty wind such as tornado (in Japanese). Retrieved November 20, 2024, from https://www.jma.go.jp/jma/kishou/know/toppuu/tornado0-0.html.

Kamimera, H. (2020). BRAIN: Borderless Radar Information Networking over South and Southeast Asia. In *Book of Abstracts, ERAD 2020: European Conference on Radar in Meteorology and Hydrology* (p. 55).

Kamimera, H. (2022). BRAIN and its hydrological application. In *Book of Abstracts, ERAD 2022: 11th European Conference on Radar in Meteorology and Hydrology* (p. 146).

Kamimera, H., Khairul, I. M., Rasmy, M., & Murata, F. (2022, December 10, 11, 15). Intercomparison and hydrological validation of satellite-based rainfall estimates over the Meghna River basin. In *International Conference on Climate Change 2022 (ICCC 2022)*, Dhaka, Bangladesh.

Kamimera, H., Mori, S., Yamanaka, M. D., & Syamsudin, F. (2012). Modulation of diurnal rainfall cycle by the Madden–Julian oscillation based on one-year continuous observations with a meteorological radar in west Sumatera. *SOLA, 8*, 111–114. https://doi.org/10.2151/sola.2012-028

Kamimera, H., Ngo-Duc, T., Le, V. X., Matsumoto, J., & Iwami, Y. (2015). Rain-gauge, radar and satellite rainfall intercomparison in central Vietnam. In *Program and Proceedings of the Fourth International Workshop on Climatic Changes and Evaluation of Their Effects on Agriculture in Asian Monsoon Region* (pp. 43–46).

Kamimera, H., Ngo-Duc, T., Masuda, K., Le, V. X., Nguyen, T. T. T., & Matsumoto, J. (2011). Building a radar-raingauge combined gridded rainfall data set over central Vietnam for the heavy rainfall event in November 2007. *Scientific and Technical Hydro-Meteorological Journal, 601*, 43–47.

Kamimera, H., Tsujimoto, K., Hak, M., Ryna, O., Monichoth, S. I., & Youtray, K. (2023, February 18–22). Radar-based rainfall estimation over Cambodia. In *9th International Conference on Flood Management (ICFM9)*, Tsukuba, Japan.

Lehner, B., Verdin, K., & Jarvis, A. (2008). New global hydrography derived from spaceborne elevation data. *Eos, Transactions, American Geophysical Union, 89*(10), 93–94. https://doi.org/10.1029/2008EO100001

Marshall, J. S., & Palmer, W. M. (1948). The distribution of raindrops with size. *Journal of Meteorology, 5*(4), 165–166. https://doi.org/10.1175/1520-0469(1948)005<0165:TDORWS>2.0.CO;2

National Oceanic and Atmospheric Administration (2023). The online tornado FAQ. Retrieved March 10, 2023, from https://www.spc.noaa.gov/faq/tornado/.

Peterson, R. E., & Mehta, K. C. (1981). Climatology of tornadoes of India and Bangladesh. *Archives for Meteorology, Geophysics, and Bioclimatology Series B, 29*, 345–356.

Peterson, R. E., & Mehta, K. C. (1995). Tornadoes of the Indian subcontinent. In *Proceedings of the 9th International Wind Engineering Conference* (Vol. VI, pp. 2077–2081).

Seko, H., Kawabata, T., Tsuyuki, T., Nakamura, H., Koizumi, K., & Iwabuchi, T. (2004). Impacts of GPS-derived water vapor and radial wind measured by Doppler radar on numerical prediction of precipitation. *Journal of the Meteorological Society of Japan, 82*(1B), 473–489. https://doi.org/10.2151/jmsj.2004.473

Terao, T., Kanae, S., Fujinami, H., Das, S., Dimri, A. P., Dutta, S., Fujita, K., Fukushima, A., Ha, K.-J., Hirose, M., Hong, J., Kamimera, H., Kayastha, R. B., Kiguchi, M., Kikuchi, K., Kim, H. M., Kitoh, A., Kubota, H., Ma, W., … Matsumoto, J. (2023). AsiaPEX: Challenges and prospects in Asian precipitation research. *Bulletin of the American Meteorological Society, 104*(4), E884–E908. https://doi.org/10.1175/BAMS-D-20-0220.1

Ulbrich, C. W. (1983). Natural variations in the analytical form of the raindrop size distribution. *Journal of Climate and Applied Meteorology, 22*(10), 1764–1775. https://doi.org/10.1175/1520-0450(1983)022<1764:NVITAF>2.0.CO;2

Yamane, Y., & Hayashi, T. (2006). Evaluation of environmental conditions for the formation of severe local storms across the Indian subcontinent, *Geophysical Research Letters, 33*, LI7806, https://doi.org/10.1029/2006GL026823.

Yamane, Y., Hayashi, T., Dewan, A. M., & Akter, F. (2010a). Severe local convective storms in Bangladesh: Part I. Climatology. *Atmospheric Research, 95*, 400–406.

Yamane, Y., Hayashi, T., Dewan, A. M., & Akter, F. (2010b). Severe local convective storms in Bangladesh: Part II. Environmental conditions. *Atmospheric Research, 95*, 407–418.

Cloud Microphysical Parameters Observed in Convective Clouds

Thara Prabhakaran, Sachin Patade, Gayatri Kulkarni, Soumya Samanta, Neelam Malap, and P. Murugavel

Abstract Convective clouds are ubiquitous over the tropics and midlatitudes, and they are part of aggregated convection. The convection is initially driven by the boundary layer eddies and the fair-weather cumulus, which gets favored by the vertical ascent. The deep convective clouds in the Indian monsoon environment, as inferred from the in situ measurements, ground-based polarimetric radar and numerical simulations are a major topic of discussion in this chapter. The environmental conditions, the growth/decay of cumulus clouds, and processes involved in forming warm and cold rain, cloud organization are discussed. The in situ measurements are focussed on the continental convective clouds and their sensitivity to the large-scale moisture content and sub-cloud aerosols that are critically investigated with high-resolution models as well. Polarimetric radars can give a key information on the microphysical processes and organization of clouds and rainfall types. It is shown from existing literature that key cloud dynamics and microphysics observations give pathways for improving the emerging global cloud-resolving models.

1 Introduction

Convective clouds are formed due to the vertical ascent of air parcels which are generally associated with environmental instability. Convection is driven by rapid surface heating or when the surface heats up more than surrounding regions. Other mechanisms include the ascent along orographic escarpments and frontal convection. Convective clouds play an integral part in the diurnal cycle of precipitation over land. They are vital in determining the earth's radiation budget, hydrological cycle, and global circulation. Convective clouds are more prominent in mid to low latitude and

T. Prabhakaran (✉) · G. Kulkarni · S. Samanta · N. Malap · P. Murugavel
Pune, India
e-mail: Thara@tropmet.res.in

S. Patade
Lund, Sweden

S. Das and W-K. Tao (eds.), *Severe Storms*,
https://doi.org/10.1007/978-981-97-7075-5_6

are deepest in the tropics. The warm cloud depth is much higher in tropical convection when compared to midlatitude.

Several controlling factors influence the convective clouds such as surface-atmosphere exchanges, boundary layer convection, microphysical and dynamical feedbacks, organization of convection and mesoscale circulations, moisture advection, etc. Solar heating during the day generates boundary layer convection, which mixes the surface properties with that of the cloud base. The convective clouds initiate as small cumuli with a high degree of entrainment and constantly form and disappear as the boundary layer grows during the daytime over land. Few of those clouds last longer and become candidates for congestus or cumulonimbus clouds in the afternoon hours. These individual clouds organize and aggregate to form convective clusters. These cloud systems may contribute to the most extreme precipitation (Feng et al., 2021) due to their areal coverage and longevity.

There are major gaps in our understanding that in several instances, the diurnal cycle of precipitation over land is not represented in the models, especially with respect to intensity, timing, and location (Tang et al., 2022; Ban et al., 2021; Jucker et al., 2020; Covey et al., 2018). A multitude of aspects contributes to the gap areas related to the life cycle of convection due to the land surface, boundary layer processes, precipitation formation, factors attributing to deep convection, and further growth from deep convective cells to Mesoscale Convective Systems (MCS). The physical processes associated with these clouds are the convective updrafts and downdrafts and the impacts due to cold pools in the lower atmosphere. Observations of different kinds, such as water vapor, atmospheric dynamics, surface fluxes, and precipitation, are needed to understand the linkages and processes within the boundary layer and cloud layer. There is a strong positive feedback between the land surface moisture and the occurrence of moist convection (Taylor et al., 2011; Findell Kirsten & Elthair, 2003). Also, the large tropospheric water column favors deeper convection (Neelin et al., 2009; Zhuang et al., 2017). The transformation of cumulus clouds to deep, convective clouds happens through several interactive and nonlinear processes. Identifying favorable environmental conditions and feedback processes for the cloud organization and formation of cloud clusters is trivial. Clouds have radiative as well as physical transformative effects due to precipitation. As precipitation starts, it cools the boundary layer, and often a cold pool develops with associated organization features. The subsequent events may entirely change the boundary layer thermodynamics.

The boundary layer is characterized by thermals/cells and further transformation to roll-type circulations that, favor the growth of shallow cumulus and transform to deep clouds and precipitation. Generally, the area fraction of downdrafts in the boundary layer is higher than that of updrafts in the boundary layer. The updrafts are very narrow and sharp and form canters of convection. Studies have also shown (Feng et al., 2015) that the intersection of different cold pools has a dramatic influence on the formation of secondary convection and precipitation. The model intercomparison projects are a way to increase basic understand to parameterization development.

Convection and its associated self-aggregation is addressed with model intercomparison projects (Wing et al., 2018) to get a measure of climate sensitivity. The organization of convective clouds can trigger MCS, which have a more dramatic impact on the precipitation over a large area (Rowe and Houze, 2015). The MCS contributes a significant part of rainfall e.g. 30–70% of warm season rainfall over the east of the Rocky Mountains (Fritsch et al., 1986; Nesbitt et al., 2008) and 50–60% of tropical rainfall (Yuan and Houze, 2010). The 100 years of research on the subject gives an in-depth understanding of the features of MCS (Houze, 2018a, b), and most of such observations were reported from the mid-latitudes. These studies show that the intensification of convective clouds depends on updraft/downdraft characteristics, environmental interaction, and microphysical processes.

Besides the challenges associated with our understanding of the dynamic conditions of convective clouds, many cloud microphysical processes in convective clouds are poorly understood. Recent advancements in cloud microphysical observations have improved our understanding of various warm microphysical processes in convective clouds, including droplet nucleation, condensational growth, and evaporation. Comparatively, our understanding of microphysical processes that involve the formation of ice is limited and incomplete in many aspects.

In this chapter, we deal with three aspects that are least understood regarding convective clouds and their microphysical processes in monsoon environments. The three topics are (a) Entrainment effects on the deep convection, (b) microphysical processes leading to the precipitation, and (c) the impact of aerosol on deep convection. It is also understood that combining observations and numerical simulations give insights into these grey areas. The major thrust is given to the microphysical observations over the Indian subcontinent from in situ measurements in convective clouds.

2 Field Campaigns

Typically field campaigns are conducted with a focussed theme or a specific science problem and gather data from various types of investigations in different facets of that problem. The hot tower hypothesis (Riehl & Malkus, 1958) argues that the Hadley circulation is driven by several deep convective cores, which are undiluted. The undiluted core was contrary to inferences from several field campaigns (e.g., LeMone & Zipser, 1980; Igau et al., 1999). It is demonstrated that the parcel required additional buoyancy, probably from latent heating above freezing level, to reach the upper troposphere. The cloud top entrainment and evaporation of cloud particles may contribute to negative buoyancy, detrimental to cloud growth. Since then, several field campaigns have been conducted to address the intricate details of deep convection with the help of detailed observations with C-band radar (CACTI; Varble et al., 2021; CAIPEEX; Samanta et al., 2021) and with airborne in situ measurements (Gayatri et al., 2017; Prabha et al., 2011). A comprehensive list of various field campaigns was conducted to study the deep convective clouds, and their microphysics is detailed in

Table 1. It may be noted that the continental cloud observations are less compared to their marine counterpart.

Though there were other field experiments such as SWAMI, ENCOMPASS, etc. conducted over India, they were not focussed on convective clouds. Observations of the microphysics of convective clouds are conducted with the help of in situ probes with the help of airborne platforms. The different types of instruments can be part of an airborne campaign, as described in Table 2. The cloud droplet probe, precipitation imaging probe, cloud imaging probe, and the 2D stereo probes can give the complete size distribution of cloud droplets from 2 microns to 6 mm raindrops. The more advanced probes are used to characterize ice particles in the cloud. The polarization probes can characterize the shape of particles. The ice particle habits and their existence at different altitudes give information on the key processes for their formation, evolution, transformation, or decay. Remote sensing instruments such as radars give insight into the evolutionary features of convective clouds. Satellite observations have shown further insights globally, such as Zhao et al. (2023), who showed that enhancement in the aerosol loading increases ice particle size and ice water content over the tropical convergence zones and decreases that in the subtropical subsidence regions. The aerosol effect influences the cloud cover, while there is a weak impact on the cloud top height. It may be noted that cloud top height is more determined by the dynamical and thermodynamical processes. In the present chapter, we illustrate in situ observations and inferences derived from radars.

3 Environmental Conditions

Environmental conditions suitable for triggering deep convective clouds, their vertical growth and organization into mesoscale convective clusters is a topic of significant relevance in the forecast and nowcast of extreme precipitation. Zhang and Klein (2010) gave compelling evidence that the environment plays an important role in the transition to deep convection. The thermodynamics and the lifting mechanisms in the boundary layer due to the impact of cold pools (Tompkins, 2001) extend the availability of convective available potential energy (CAPE) during later hours of the day. They are not captured in large-scale numerical models and have a significant impact on the diurnal cycle of precipitation. The conditions of high mid-tropospheric humidity favor the formation of deep convective clouds (Chepfer et al., 2019; Zhang & Klein, 2013), as may be noted from observations in Fig. 1, where shallow convective cloud (when clouds are not taller than 4 km) and deep convective cloud regimes are identified with a microwave radiometer profiler observations of liquid water content and the vertical profile of relative humidity from radiosonde observations conducted in the pre-convective environment. The shallow convective regime has several dry layers, and the overall relative humidity is below 80%, unlike in the deep convective cases. Investigating the diurnal cycle of CAPE and the convective temperature (CT), both triggering convection parameters in the shallow/deep cloud. The hysteresis curves (Fig. 2) for the evolution of shallow and

Table 1 Different field experiments on convective clouds

Field campaign	Aim	Reference link
Deep Convective Clouds and Chemistry (DC3) Platform NASA DC-8 NCAR GV DLR Falcon	Thunderstorms, including their dynamical, physical, and electrical structures and their impact on the chemical composition of the troposphere	https://doi.org/10.1175/BAMS-D-13-00,290.1
Mid-latitude Continental Convective Clouds Experiment (MC3E)	To provide the complete characterization of convective cloud systems, precipitation	https://doi.org/10.1175/BAMS-D-14-00,228.1
GoAmazon2014/5 field campaign	To improve understanding of how land-atmosphere processes affect tropical hydrology and climate	https://doi.org/10.1073/pnas.1719842115
ACRIDICON–CHUVA Campaign: Amazonia Using the New German Research Aircraft HALO	Studying Tropical Deep Convective Clouds and Precipitation over Amazonia Using the New German Research Aircraft HALO	https://doi.org/10.1175/BAMS-D-14-00,255.1
2017 StratoClim campaign	Deep-convective influence on the upper troposphere–lower stratosphere composition in the Asian monsoon anticyclone region:	https://doi.org/10.5194/acp-20-12,193-2020
European High-Altitude Ice Crystals (HAIC) and North American HIWC projects NASA)/FAA HIWC-RADAR I campaign	In situ microphysical properties of deep convective clouds	https://doi.org/10.4271/01-14-02-0007
Ice in clouds experiment—tropical	Measurements of ice nucleation to predict the number of ice particles forming by nucleation mechanisms	https://data.eol.ucar.edu/project/ICE-T
HIBISCUS campaign	To investigate the impact of deep convection on the Tropical Tropopause Layer (TTL) and the lower stratosphere	https://acp.copernicus.org/articles/11/2309/2011/acp-11-2309-2011.pdf
DYNAMO field campaign	improve understanding of the Madden–Julian Oscillation as well as to improve prediction skill of climate and weather models	https://doi.org/10.1002/2014MS000335
Barbados Cloud Observatory (BCO) German High Altitude and Long Range Research Aircraft (HALO)	Understand cloudiness, circulation, and climate at windward edge of Barbedos	https://journals.ametsoc.org/supplemental//journals/bams/97/5/bams-d-14-00247.1.xml/10.1175_BAMS-D-14-00247.2.pdf

(continued)

Table 1 (continued)

Field campaign	Aim	Reference link
Experiment of Sea Breeze Convection, Aerosols, Precipitation and Environment (ESCAPE)	Investigate sea breeze related convection; characterization of life cycle of convective systems using two aircraft (NCAR-C130 and SPEC Learjet), surface-based fixed and mobile measurements	https://doi.org/10.1175/BAMS-D-21-0104.1
Tracking Aerosol Convection Interactions Experiment (TRACER) Air Quality (TAQ) campaign	Investigate the life cycle of the convective clouds with a comprehensive suite of remote sensing and in situ instrumentation, uses ARM mobile facility instruments	https://doi.org/10.1175/BAMS-D-21-0104.1
CACTI	Deep Convection Initiation, Growth, and Environments in the Complex Terrain of Central Argentina	https://doi.org/10.1175/MWR-D-21-0237.1
CAIPEEX	Monsoon clouds over the Indian subcontinent	Prabha et al. (2011), Kulkarni et al. (2012), Prabhakaran et al., (2023)

deep clouds have dramatic differences. In the shallow cloud case, the development of CAPE is gradual and attains a lower maximum value than the deep convection. The removal of CAPE is a gradual process in the shallow cumulus compared to the deep case. Although CAPE is sustained in the later hours of the day, the convective temperatures are rather higher in shallow convection compared to deep convection. The deep convection indeed needs only a lower CT compared to the shallow convection. The lower CT and high CAPE in the deep convection are likely drivers for further convective developments in the later part of the day (probably due to the incidence of CP, stabilizing the lower atmosphere) through the influence of cold pools moistening and cooling the lower atmosphere, as well as triggering convection along its boundaries.

4 Entrainment and Mixing

The entrainment can be defined as the mixing of droplet free environmental air into the cloud. Entrainment results in dilution of the cloud parcel (evaporation of cloud droplets and depletion of cloud water content) and reduction in its buoyancy. The conditions of dry/wet regimes have high/low boundary layers, cloud bases, lifting condensation levels, and convective available potential energy (CAPE). A theoretical relationship between Bowen's ratio (ratio of sensible to latent heat flux) and lifting condensation level (LCL) height was noted from the observations and simulations

Table 2 Various parameters and related instruments used in the CAIPEEX field experiment

Parameters	Expression/comment	Instrument
Cloud drop effective radius	$r_{eff} = \int f(r)r^3dr / \int f(r)r^2dr$	Cloud droplet probe
Liquid water content (LWC) (derived from DSD f(r), where r is the radius of the droplets)	$LWC = \int f(r)r^3dr$	
Number concentration of droplets	$N = \int f(r)dr$	
Liquid water content (direct measurements)		Hot wire probe
Ice water content (derived from ice Particle Size Distribution (PSD))	Various mass dimensional relationships are used to estimate IWC	Cloud imaging probe, 2DS probe, precipitation imaging probe
Ice number concentration	50–2000 μm, Particles smaller than 200 um are omitted to overcome effects from shattering	
Raindrop mean volume diameter	100–6200 μm	Precipitation imaging probe
Cloud condensation nuclei		Cloud condensation nuclei counter
Ice nucleating particle	INP are usually measured using continuous flow diffusion chamber or filter method	Commercial ice nuclei counter, aerosol collected on filters activated in cloud chambers
Typical gamma function used to represent particle size distributions	$N(D) = N_0 D^{\mu} \exp(-\lambda D)$ where N_0 is the intercept, λ is the slope and μ is the dispersion of size distribution, for exponential size distribution $\mu = 0$	Particle size distribution observed using imaging probes

(Thomas et al., 2018), indicating a strong relationship between the surface fluxes in deciding the cloud bases. The Thermodynamics of the environment has a key role. In a dry environment, droplet-free air entering the cloud may evaporate cloud droplets and reduce the width of droplet size distribution. The moistened region surrounds the cloud core in a convective cloud as a protective cover. Glenn and Kruger (2014) showed subsiding zones of drier air near the edges through evaporation in the downdraft of deep convective clouds from a large eddy simulation (LES) model. These subsiding zones lead to a shell of cold air around the convective cores and shells characterize negative buoyancy as a result of mixing and entrainment and evaporation of cloud parcels.

In situ, measurements give great details of the cloud properties. CAIPEEX; (Kulkarni et al., 2012) observations give first-time detailed and in situ insights into pre-monsoon and monsoon clouds that grow over continental monsoon regions. In

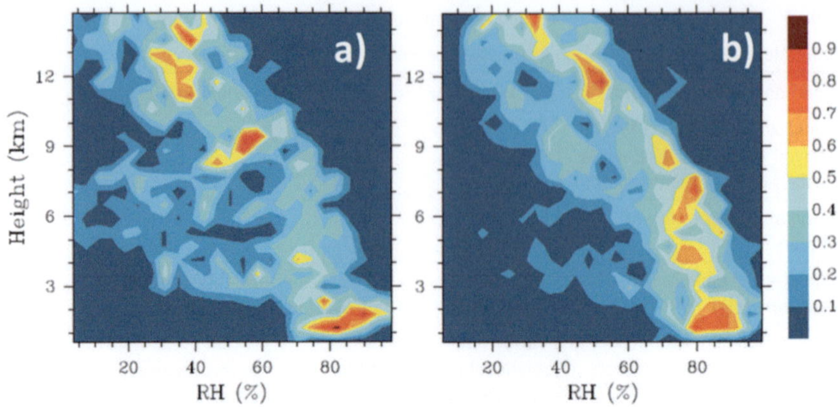

Fig. 1 Vertical variation of relative humidity during **a** shallow cloud and **b** deep convective cloud observations. The probability of occurrence is presented from CAIPEEX observations

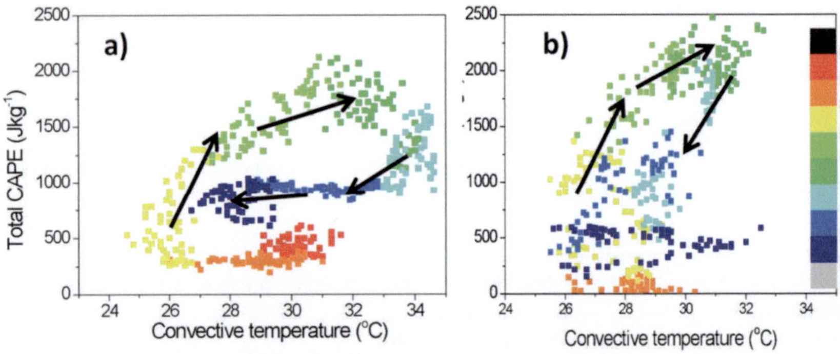

Fig. 2 CAPE and convective temperature during the diurnal cycle (color indicates the time of day in local time) of **a** shallow cloud and **b** deep convective cloud over the Indian peninsula from CAIPEEX radiosonde observations. The hysteresis curve is indicated by arrows

a pre-monsoon environment and monsoon break conditions with a dry environment, the droplet size distribution (DSD) was narrow. As the wet monsoon conditions set in, the DSD broadened (Prabha et al., 2012). The DSD in the pre-monsoon/monsoon clouds has high/low droplet number concentrations in the dry/wet environments depending on the high/low aerosol number concentrations. This difference drives a narrower droplet spectrum skewed to small droplet sizes (<20 um). This is due to the activation of several aerosols, evaporation of large cloud droplets, and the suppressed collisions/coalescence leading to many small droplets. The pre-monsoon DSDs are impacted significantly by the entrainment of dry air, as indicated in Prabha et al. (2012). The droplet spectrum is subjected to homogeneous and inhomogeneous mixing in the cloud core and cloud edges, depending on the entrainment time scales

(Bera et al., 2016b). The entrainment in the deep cumulus was observed at different altitudes, and the cloud core was always well-defined with near adiabatic zones.

Nair et al. (2012) found monsoon clouds to be significantly diluted due to entrainment mixing. The study has developed parameterization for the effective radius from CAIPEEX observations in convective clouds. Bera et al. (2019) found stronger cloud-base updrafts in the pre-monsoon clouds, which could introduce a high supersaturation and activate more aerosol particles. Adiabatic fraction measures how diluted the cloud parcel is with respect to the adiabatically growing cloud, that is not mixing with the environment. A high Adiabatic fraction and low entrainment rate in the polluted clouds are especially noted (Bera et al., 2016a). Meanwhile, the relative dispersion (ratio of the droplet spectral width to the mean radius of droplets), used to represent droplet spectral broadening, highly depends on the airmass characteristics, which advect different aerosols at the cloud base.

A sample cloud pass observation from airborne observation shown in Fig. 3 illustrates DSD behaviour at the cloud core and cloud edges, the variation of Cloud droplet number concentration (CDNC), liquid water content (LWC), and droplet effective radius (reff) over the pass. The LWC is maximum near the cloud core and reduces near the cloud edges. A few large drops are seen in the core where maximum LWC is found. The reff and CDNC do not change significantly across the cloud core, while there is a drastic reduction in CDNC at the cloud edges, detailing the evaporation of cloud droplets. There are also flanks of the cloud with enhanced CDNC (with low LWC) at regions of new triggers for cloud formation. These are likely the new clouds generated at the intersection of the subsiding zone and the moist environment.

An example of in situ cloud measurements of liquid water content (a) and effective radius (b) in a monsoon cloud with a cloud base at 800 m is shown in Fig. 4. The color indicates the temperature corresponding to each cloud pass at different altitudes. The scatter at each altitude indicates the spatial variation of properties of clouds with very diluted and undiluted cloud parcels. The effective radius increases with height and becomes maximum at 7.5 km. The level at which the effective radius crosses 12

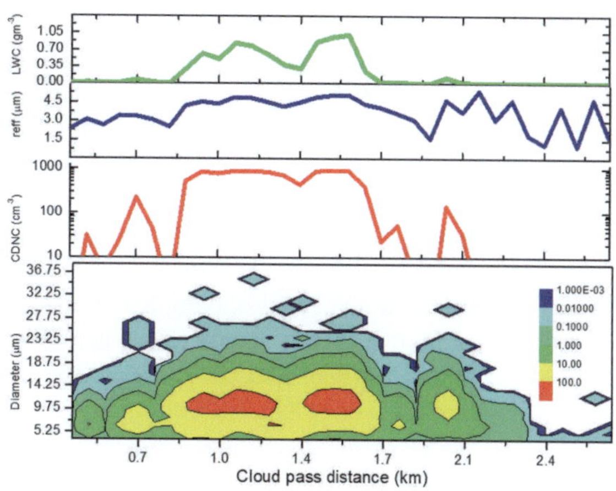

Fig. 3 A sample cloud pass over 2.5 km level observations at 2.2 km above mean sea level conducted over the Indian west coastal areas in a convective cloud during CAIPEEX. The figure shows from top to bottom; liquid water content (LWC, gm^{-3}), droplet effective radius (reff), Cloud droplet number concentration (CDNC), and cloud DSD across the pass

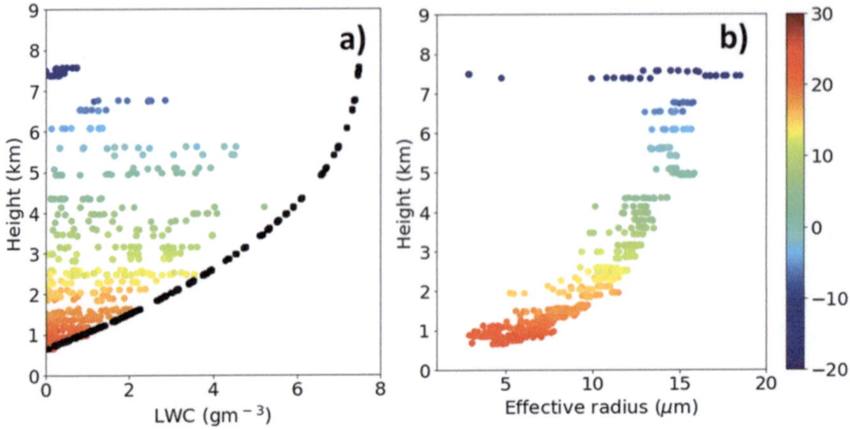

Fig. 4 Vertical variation of cloud liquid water content **a** and effective radius of cloud droplets observed during CAIPEEX observations **b**. Temperature (in °C) observed at each level is indicated with a color map. The black symbol in panel **a** indicates the adiabatic liquid water content. Near adiabatic parcels in the cloud falls close to the adiabatic curve

microns (5 km) is considered to be the level of collision coalescence to form large drops as centers of raindrops. This is the level at which the maximum liquid water content is found in the cloud. This is also level at which melting of ice hydrometeors may happen. The collisions will remove liquid water through raindrops or make avenues for freezing drops below 0 °C. The LWC decreases monotonically with height thereafter and becomes minimum at 8 km in this example. In the instance that cloud base is higher elevated, the cloud droplets will be too small at the freezing level and large enough drops may not cross the freezing level for ice multiplication processes.

The inset figure is indicating different zones within the cumulus clouds identified with the help of clustering methods from LES snapshot (Fig. 5). The colors indicate regions from the cloud core to the environment. These zones are identified with clustering on the LWC, droplet number concentration, and vertical velocity from the LES fields. Blue and black lines are the linear and nonlinear fits between the entrainment rate and the adiabatic fraction. Small dots are LES results and large dots are CAIPEEX observations (derived from Malap et al., 2021). When the clouds are shallow (non precipitating cloud is considered), in the form of cumulus, they experience more dilution due to higher buoyancy and entrainment rates, both due to the strong vertical development and due to the subsiding motions. The dilution (LWC/LWC_{max}) and entrainment rate in cumulus clouds using an LES and is compared (Fig. 5) with aircraft observations (for core when LWC is greater than 0.05 gm^{-3} and edge when LWC is less than 0.05 gm^{-3}). The different cloud segments from LES are separated based on the turbulent fields of liquid water content, number concentration, and vertical velocity (Malap & Prabha, 2023). The inset figure shows an example of the classification of dilution regimes for a continental cumulus cloud. The core of

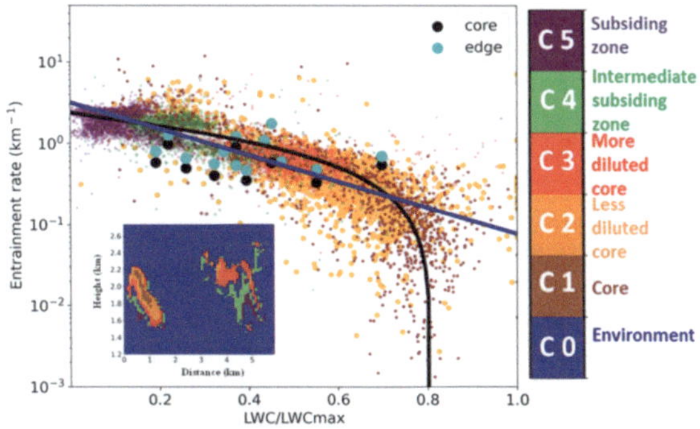

Fig. 5 Relationship between dilution (LWC/LWCmax) and entrainment rate in cumulus clouds using LES and compared with observations (for core and edge, derived from Malap & Prabha, 2023)

the cloud has the highest LWC, which is reduced at the diluted segments. The core also has the highest adiabatic fraction and lowest entrainment rate. The core is also protected from the surrounding diluted regions due to its highest adiabatic fraction. These are important for the further development of clouds to form deep cumuli.

5 Microphysics of Convective Clouds

The precipitation formation in convective clouds culminates several complex microphysical processes. Understanding these complex processes is essential as these processes can alter atmospheric heating profiles through latent heat exchange. The cloud microphysical properties result in precipitation formation and can modify many cloud macrophysical properties, such as cloud cover, and thickness, affecting the radiation budget. The interplay between cloud microphysical properties and cloud dynamics is not well understood in convective as well as in all major common cloud types. We illustrate the warm cloud and cold cloud properties with a focus on the relevant processes.

(a) **Warm microphysical properties of convective clouds**

Major microphysical processes below freezing level are nucleation of cloud drops, growth by condensation, evaporation, raindrop formation through collision coalescence, raindrop formation, drops breakup, evaporation, and ice particle melting.

(i) **Cloud condensation Nuclei (CCN) activation, droplet nucleation**

The accumulation mode particles serve as a major source of cloud condensation nuclei. The convective clouds are associated with strong updrafts (2–15 m/s). Such strong updrafts result in higher supersaturation which can activate much smaller aerosol particles (~100 nm Aitken mode). The number concentration of CCN near the cloud base determines the total number of cloud drops (at higher altitudes, aerosol also may play a role as ice nuclei and only a few observations are available over the Indian region). CCN concentration is measured using a commercially available

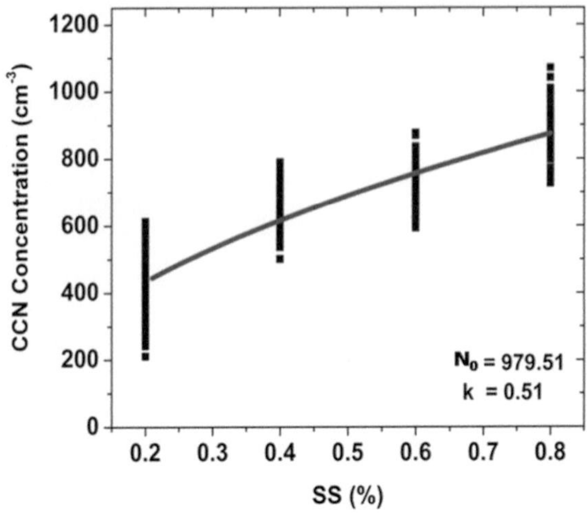

Fig. 6 Measured CCN concentration by a CCN counter during CAIPEEX. The CCN concentration is plotted as a function of supersaturation (derived from Gayatri et al., 2017). Each square corresponds to a CCN measurement below the cloud base at various supersaturations. The solid line indicates the exponential fit to the observations. Bars indicate the variation due to spatial variability near the cloud base

CCN counter. During CAIPEEX, CCN concentration was measured using a Droplet Measurement Technology (DMT, Inc.) CCN counter mounted on research aircraft. CCN measurements were conducted below a few meters of cloud base at different water supersaturations. Figure 6 shows the CCN activity spectrum near the cloud base of a convective cloud observed during CAIPEEX. The observed number concentration of CCN in this case (N0) was 979 cm − 3 at 1% supersaturation. An empirical formulation $N = Noe^{-k}$ was fitted (Fig. 6) to the observed CCN spectrum. It gives the intercept (No) and slope (k) values that can be used in bulk microphysics schemes to represent CCN activation.

Prabha et al. (2011) investigated the properties of DSD in convective clouds using in situ measurements during the Cloud Aerosol Interaction and Precipitation Enhancement Experiment (CAIPEEX). The results showed that the DSD in these clouds was bimodal, with the first mode occurring at a smaller size range and the second mode at a larger size range. The first mode was dominated by the droplets activated near the cloud base. In-cloud nucleation takes place in a highly supersaturated environment even above the cloud base, which gives rise to the second mode of the DSD. The observational evidence suggests that the in-cloud nucleation at the elevated layers is the fundamental mechanism for bi-modal DSDs, especially in monsoon clouds.

We may notice (Fig. 7) that the cloud base in a super continental case is located at around 3.4 km, while over the continental region at 2.7 km and over the marine regime at 500 m. The size of symbols indicates the effective radius of droplets. Several small droplets exist at all different levels in the super continental case, while in the continental case only up to the middle of the cloud, while in the marine case, small droplets are only near the cloud base.

The vertical variation of different microphysical parameters of DSDs, above the cloud base such as the mean radius, and spectral width (Liu & Daum, 2000) of the

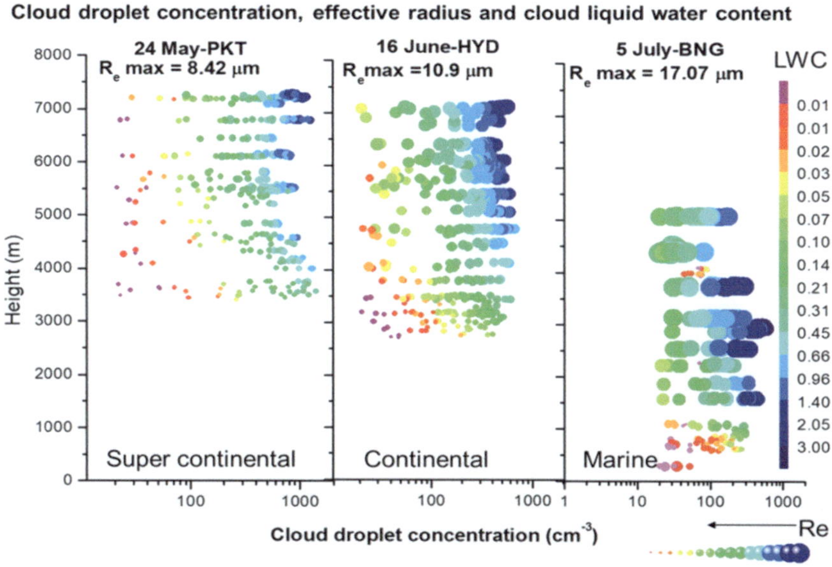

Fig. 7 Examples of cloud microphysical parameters under three different environments: Over the super continental (pre-monsoon time over Pathankot), continental (transition time over Hyderabad), and over a marine (Monsoon time over Bengaluru) airmass dominated over three different maximum effective radii of 8.42 um, 10.9 um, d 17.07 um. Minimum size scale is 3 microns and maximum is 15 micron

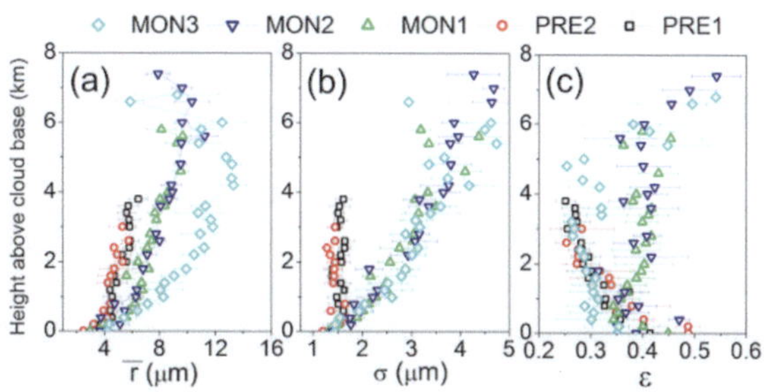

Fig. 8 Mean radius **a**, the spectral width **b** and relative dispersion **c** of the droplet size distribution for clean monsoon (MON3), high CCN (MON2), and for moderate CCN (MON1) and polluted pre-monsoon (PRE1 and PRE2) for convective clouds over the Indian monsoon region. Error bars indicate the spatial variation of each parameter at a specific altitude (adapted from Prabha et al., 2012)

droplet in three different CCN conditions of monsoon (Prabha et al., 2012) are shown in Fig. 8.

The droplet mean radius is <10 um in the cases with moderate and high CCN. Meanwhile, only the cleanest condition showed droplet growth above 10 um. The spectral width of DSD is not significantly different in all three monsoon conditions.

The contrasting behavior of mean radius and spectral width in the pre-monsoon condition is especially noted with nearly constant values with altitude. These clouds are characterised by a protected adiabatic core with a cooler and moist environment around it with a depleted number of droplets and liquid water content. The droplet spectra remained narrow with altitude. In an extreme scenario of PRE1, where droplet radius is reduced with height above the cloud base, indicating the evaporating cloud tops. The relative dispersion (ratio of spectral width and mean radius) is 0.3 in the pre-monsoon while it exceeds 0.4 in the polluted monsoon clouds, where droplet growth due to collision coalescence is suppressed. Typically, a constant value of 0.3 is used in the numerical models, and it can vary with height significantly mainly due to the broadening of droplet spectra towards the smaller sizes in the polluted environments. The droplet activation above cloud bases as well as evaporation of droplets above cloud bases are noted. There were considerable regional differences in the droplet growth rates with height (Morwal et al., 2012). Clouds observed above the elevated aerosol layers showed High droplet concentrations (200–1400 cm^{-3}) and small effective radii were observed in the clouds observed above elevated aerosol layers (Padmakumari et al., 2013).

(ii) **Growth of droplets, collision coalescence, effective radius profile, etc.**

The growth of cloud droplets through condensation and collision coalescence is crucial in precipitation formation in convective clouds. First rain formation is determined largely by the basic condensation and collisions mechanisms in ascending adiabatic volumes (Khain et al., 2013). During the active monsoon season, convective clouds have a low altitude cloud base (around 1.5 km) when compared with pre-monsoon conditions. The freezing level (4.5–5 km) is located around 4.5–5 km. When compared with pre-monsoon conditions, The distance between the cloud base and freezing level (warm layer depth) is higher during the active monsoon season. It provides a longer time for cloud drops to grow in updrafts through condensation and collision coalescence. This is evident from the higher effective radius in monsoon clouds (see Fig. 9). The pre-monsoon clouds have a relatively higher cloud base which results in a lower warm layer depth. In PRE clouds, cloud droplets have limited time to grow into raindrops. In addition to lower warm layer depth, higher polluted conditions in PRE clouds lead to smaller cloud droplet sizes.

(iii) **Supercooled liquid water**

The amount of liquid present in clouds above freezing level is called supercooled liquid water. Supercooled liquid can affect ice microphysical processes through various mechanisms, including cloud/rain drop freezing, and ice particle growth by riming. The convective clouds with broad drop size distribution near the warm cloud base can support rapid precipitation through collision coalescence. They may also support the production of supercooled liquid water in the clouds. Laboratory experiments and theory suggest a 10μm cloud droplet freezes homogeneously near −39 °C. Airborne observations in deep convective clouds by Rosenfeld and Woodley (2000) showed that most condensed water remains liquid at temperatures as cold as −

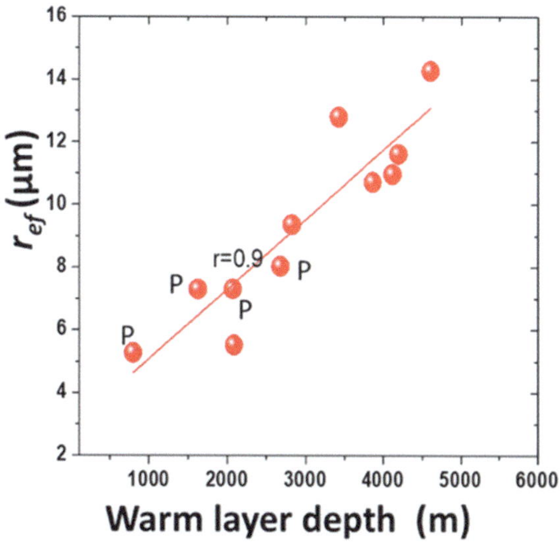

Fig. 9 The relationship between the effective radius at the freezing level and warm layer depth, from several observations during CAIPEEX from Patade et al., 2019. Pre-Monsoon cases are denoted by "P"

37.5 °C. This can happen due to a lack of potential ice nuclei in the upper troposphere, which leads to the inhibition of heterogeneous ice nucleation. The observations of ice phase at −14 °C were also noted during CAIPEEX observations near foothills and the likelihood of dust aerosol acting as potential ice nuclei were proposed (Padmakumari et al., 2013). The in situ measurements also showed that cloud tops were glaciated at −17 °C (Prabha et al., 2012) in this region.

An example of vertical profiles of LWC and IWC in different clouds growing in the coastal area and within 50 km inland from the coast, with equally polluted conditions, is shown in the Fig. 10. The figure shows that a considerable amount of supercooled liquid was present in coastal clouds, which grow over a moist polluted condition. The difference in the cloud base height introduces a different warm layer depth and, thus, a different vertical microstructure; more vertical growth and ice in the coastal cloud compared to the inland cloud. Details of cloud microphysics and aerosol interaction in these clouds can be found in Varghese et al. (2021).

Depending on the updraft speeds and the depth of the warm layer, the supercooled liquid can be seen in higher altitudes even across the homogeneous freezing level. The relationship between the supercooled liquid and rapid glaciation depends on factors such as drop temperature, composition, electrical charge, turbulence, etc. Other secondary ice production mechanisms, such as fragmentation during ice-ice collision may play an important role. It has been noted that with a low and warm cloud base, the drop size distribution is broad, and there is a likelihood of the formation of a deep warm layer depth and the formation of drizzle and raindrops in the convective cores with updrafts (Khain et al., 2013).

The Fig. 11 compares LWC from various observations and the convective cloud during CAIPEEX. It is particularly noted that monsoon clouds have a large liquid water content next to tropical deep convection and hail storms. However, the raindrop

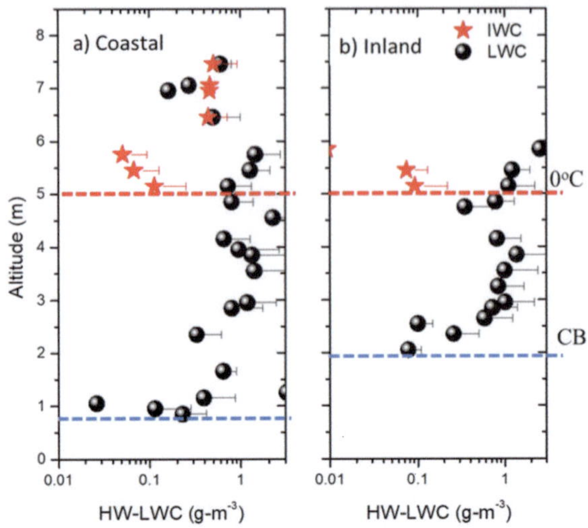

Fig. 10 Liquid water content (from hot wire anemometer) and ice water content estimated from the CIP-PIP probes measured inside convective clouds during CAIPEEX **a** from a coastal region and **b** from 50 km inland. Note the difference in the cloud base (CB) heights (blue dashed line)

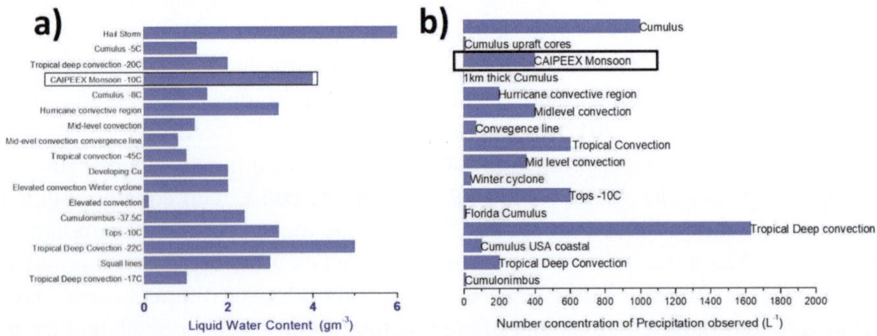

Fig. 11 Liquid water content and number concentration of raindrops in the different clouds

number concentration is much less than several other examples worldwide, likely from the collision-coalescence of drops.

(b) **Ice and mixed-phase microphysics process**

Globally, ice particles play a crucial role in modulating precipitation and radiation transfer. This will eventually affect the hydrological cycle and climate. Based on satellite data, Field and Heymsfield (2015) found that globally 50% of rain events are linked to snow production in clouds. A major part of precipitation in convective clouds is associated with ice and mixed-phase cloud microphysics processes. In a stratiform region of convective clouds melting of ice particles contributes significantly to the surface precipitation.

In mixed-phase clouds, ice particles and supercooled liquid droplets coexist. Mixed-phase clouds are associated with many phase transition processes, including condensation and evaporation of cloud drops, deposition and sublimation of ice particles, ice particle growth by riming, raindrop freezing, and melting of ice particles below freezing level. In addition, there are other microphysical processes active in mixed-phase clouds, such as primary ice nucleation, secondary ice formation, collisions between different ice particle habits etc. The understanding of ice and mixed-phase processes in convective clouds is poor as compared to other cloud types due to challenges in collecting observations in a strong updraft environment.

(i) **Primary ice nucleation, Ice nuclei measurements**

Primary ice nucleation involves the formation of ice particles through heterogeneous nucleation and homogeneous freezing. The formation of ice particles between 0 °C and about −38 °C occurs through heterogeneous ice nucleation and requires the presence of aerosol particles, called ice-nucleating particles (INPs). Whereas, homogeneous freezing does not require any INPs. Homogeneous nucleation generally occurs at temperatures colder than −38 °C through the spontaneous freezing of water drops and hygroscopic aerosol particles. Heterogeneous ice nucleation can occur in four ways involving various insoluble or partially insoluble INP: (a) Deposition mode, in which there is the direct deposition of water vapor on the surface of INP to form ice crystals, (b) Immersion freezing mode, where INP immersed INP initiate ice formation (c) Condensation freezing in which INP first act as CCN to form drop and immediately induces freezing d) the contact mode in which ice initiates in droplets when their surfaces come in contact with INP. Additional details about ice nucleation modes can be found in Kanji et al. (2017). The relative importance of each ice nucleation mode is still unclear. Particles that may serve as INPs include mineral dust, soil dust and primary biological aerosol particles. In addition, aerosols of anthropogenic origin such as soot, metallic aerosols and solid ammonium sulphate can also act as INP (Kanji et al., 2017). When compared with CCN activation, large uncertainties exist in our current knowledge of ice nucleation rates due to limited airborne observations and laboratory experiments.

Like CCN influences the liquid clouds, an increase in INP can cause the cloud glaciation effect. The increase in INP leads to the formation of more ice crystals through primary nucleation. These ice crystals can further grow by diffusional growth or riming. This rapid conversion of liquid into ice may lead to a reduction in cloud optical thickness. It is also hypothesized that this can produce more precipitation.

In most numerical models, the ice nucleation parameterization schemes rely on the empirical relationship between INP concentration and temperature and/or supersaturation over ice (e.g. Fletcher, 1962; Meyers et al., 1992). A disadvantage of these parameterization schemes is that they do not include explicit dependencies on the physical and chemical characteristics of aerosol particles. Demott et al. (2010) showed that the INP concentrations could be related to temperature and the number concentrations of aerosol particles larger than 0.5 μm in diameter. This new parameterization scheme strongly altered cloud liquid and ice water distributions in a global climate model compared to widely used parameterizations based only on temperature or supersaturation. Thus, the empirical parametrization for heterogeneous ice nucleation that includes the diversity of physical and chemical characteristics of INPs can improve the microphysical processes in mixed-phase clouds (e.g. Phillips et al., 2008, 2013). However, the INPs data included in most of the cloud models are highly uncertain. Consequently, the atmospheric models need to be constrained and confirmed by observational data of INPs from different geographical locations to have a better estimation of their effect on clouds.

The INPs in the atmosphere can be measured using a continuous flow diffusion chamber mounted on the research aircraft. A more sophisticated IN counter (Spectrometer Ice Nuclei by Droplet Measurement Technologies) is available commercially and can give continuous INP measurements at various ice supersaturations. However, mounting and operating these chambers on research aircraft is challenging, and hence in many campaigns, continuous INP measurements are conducted at the surface. In a few field campaigns (e.g. CAIPEEX Patade et al., 2014), aerosols were collected on filter papers and were analyzed in cloud chambers to estimate their concentrations in the atmosphere. The observed number concentration of INP as a function of ice supersaturation or temperature was then utilized to propose a new parameterization of ice nucleation for monsoon clouds. The maximum concentration of depositional ice nuclei measured during CAIPEEX was about 5 per litre (Patade et al., 2014).

(ii) **Secondary ice production and ice enhancement**

The observed concentrations of ice crystals in clouds often exceed the concentration of INPs by many orders of magnitude. This discrepancy is associated with secondary ice production (SIP), which produces new ice crystals in the presence of pre-existing ice without involving INPs. Based on laboratory experiments and aircraft observations, several possible mechanisms were proposed to explain the SIP of ice crystals and are well documented in the literature (e.g., Korolev & Leisner, 2020). The most important SIP mechanisms are

1. Fragmentation during Ice-ice collision
2. Fragmentation during riming
3. Droplet fragmentation during freezing
4. Fragmentation during sublimation

The collision of Ice and snow particles with cloud droplets results in the freezing of cloud droplets on their surface. This process is called riming and results in the formation of graupel or hail. Patade et al. (2016) has found the existence of both small and large cloud droplets, needles, columns, and graupel particles within the temperature zone of -3 to -8 °C. These observations strongly suggest that the mixed phase monsoon clouds conditions were favorable for secondary ice production by riming splintering (also called the Hallet-Mossop process ;Hallet & Mossop, 1974). The images from cloud imaging probe collected during CAIPEEX were often encountered with broken ice particles that might have resulted from ice-ice collisions (See Fig. 12).

The secondary ice particles produced by these SIP mechanisms can initiate the freezing of additional drops by contact nucleation, or they can also grow into larger crystals by vapor diffusion. The SIP can affect the precipitation formation, cloud lifetime, cloud electrification, and radiative properties of clouds through the modulation of ice particle concentration. Despite their importance, the physical basis of many SIP mechanisms is not fully understood.

Model simulations have revealed that, in deep continental convective clouds, SIP through ice-ice collision is dominant among all SIP mechanisms (Patade et al., 2022; Waman et al., 2022). When implemented in the global climate model, SIP altered the global annual average shortwave, longwave, and net cloud forcing by 2.1, −1.0, and 1.1 W m^{-2}, respectively (Zhao & Liu, 2021).

(iii) **Growth of ice particles**

Once activated, ice particles can grow by various mechanisms including riming, deposition, aggregation as shown in Fig. 12. After initiation, ice crystals can further grow into macroscopic pristine ice crystals through diffusional growth as long as ice supersaturation is maintained. The collision of an ice crystal with another ice crystal results in the formation of aggregated snowflakes. This process of ice particle growth is called aggregation and is more efficient at higher temperatures due to higher sticking efficiency.

Aerosol particles can affect ice particle growth by riming through changes in drop size distribution. The cloud lifetime effect in mixed-phase clouds may also exist in the same way as in the **cloud lifetime effect** of liquid clouds. This is manifested as narrow droplet size distributions, fewer ice particles of precipitation size, increasing cloud lifetime, and decreasing precipitation (Lance et al., 2011). Smaller droplet sizes led to less efficient riming and reduced snow production (Borys et al., 2003).

The schematic Fig. 13 shows the various types of hydrometeors in the convective cloud observed during CAIPEEX. The presence of numerous pristine ice crystals

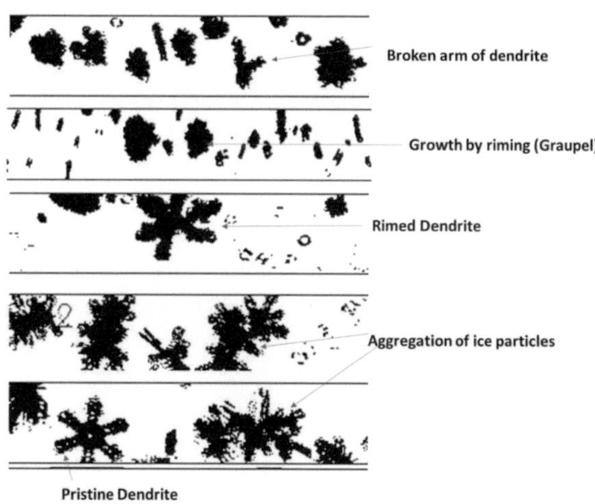

Fig. 12 Ice particle images showing their possible growth mechanisms. The ice particle images were collected in deep convective clouds using a cloud imaging probe mounted on research aircraft during the CAIPEEX experiment

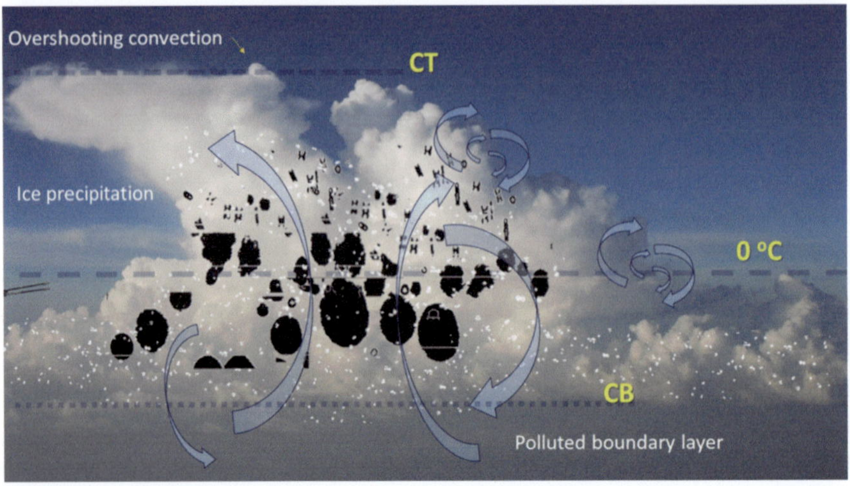

Fig. 13 Pictorial representation of different hydrometeors found in deep cumulus clouds. The blue arrows indicate cloud scale motions and turbulent motions near the cloud edges. While the core updrafts sustain cloud growth, the subsiding motions and the entertainment would deplete the cloud (CB: Cloud Base, CT: Cloud Top; Freezing level is indicated by 0 °C temperature line)

like plates and columns can be noticed in the updraft. Ice particle growth by riming results in graupel formation and is observed frequently in monsoon clouds. Larger raindrops can be observed below freezing level.

The CAIPEEX measurement showed evidence of changes in ice particle growth by riming associated drop size distribution as shown in Fig. 14 where probability distribution function of area ratio and diameter of each ice particle is plotted for polluted pre-monsoon (P-PRE), polluted monsoon (P-MON) and clean monsoon (C-MON) clouds. The particles with a higher area ratio (>0.7) have more or less spherical morphology. Particles with an area ratio lower than 0.5 show complex structures and are non-spherical. The MON cloud showed more ice particles with a higher Area ratio which is a measure of ice particle morphology. The PRE clouds were growing in dry and polluted environments when compared with MON clouds which were growing in moist and relatively less polluted environments. Ice particle growth by riming was more efficient in MON clouds than in PRE clouds.

In mixed-phase cloud ice and water coexist which is a thermodynamically unstable state. At water subsaturation, ice can grow at the expense of liquid as the saturation vapor pressure over ice is less than that over water. This process is known as the Wegener–Bergeron–Findeisen (WBF) process (Bergeron, 1935; Findeisen, 1938; Wegener, 1911).

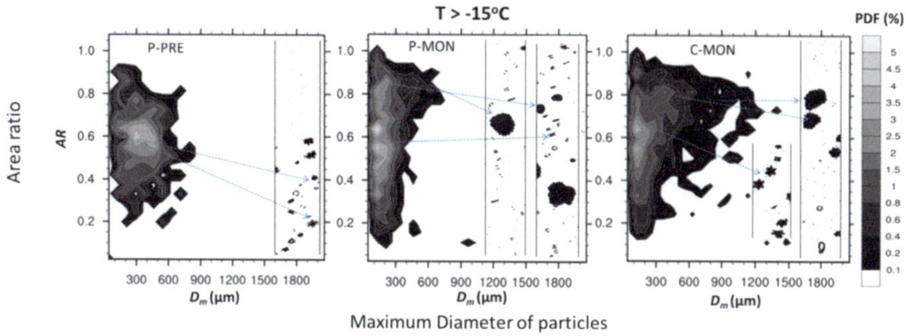

Fig. 14 Probability distribution functions of area ratio and diameter of ice particles as observed in polluted pre-monsoon (P-PRE), polluted monsoon (P-MON), and clean monsoon (C-MON) conditions (adapted from Patade et al., 2019)

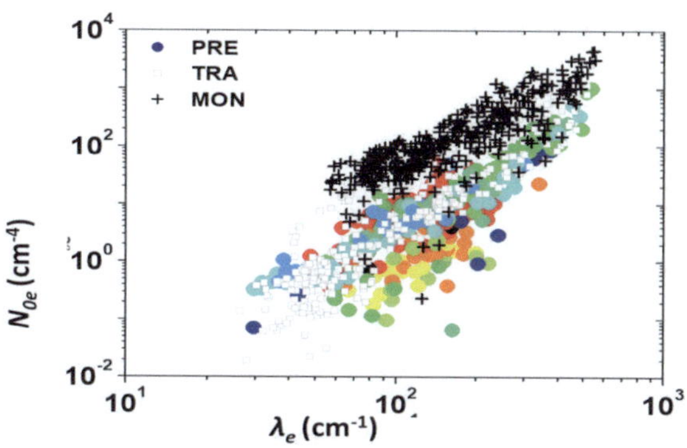

Fig. 15 The relationship between intercept (N_0) and slope (λ) of the exponential size distribution for IN PRE, TRA, and MON clouds (Adopted from Patade et al., 2015)

6 Properties of Particle Size Distributions

Accurate representation of particle size distribution is vital for any bulk microphysical scheme. The bulk microphysics schemes usually use exponential or gamma distributions to represent particle size distributions. A comprehensive analysis of particle size distributions from in situ airborne observations during CAIPEEX is presented by Patade et al. (2015). They fitted exponential and gamma distributions to the observed particle size distributions. All cloud samples considered for analysis (Fig. 15) were taken during the developing stage of continental convective clouds during premonsoon (PRE), transition (TRA), and monsoon (MON) periods at temperatures from 25 to −22 °C.

There is a significant difference in the environmental conditions in which PRE, TRA, and MON clouds grew. The pre-monsoon clouds develop in the dry boundary layer and polluted environment, whereas Monsoon clouds develop in the humid boundary layer and less polluted environment. The variable environmental conditions resulted in changes in particle size distribution in these clouds. Figure 15 shows the relationship between intercept (N_0) and slope (λ) derived from fitted exponential size distributions to observed Particle Size Distribution (PSD). These relationships are important to constrain the snow size distributions in large-scale models that use single-moment bulk microphysical schemes. Figure 15 shows a considerable difference in PSD properties of PRE and MON clouds. For a fixed value of λ, the values of No are always higher in monsoon clouds. This indicates the higher particle number (rain and ice) concentration in monsoon clouds.

A significant difference in PSD properties in PRE and MON clouds is associated with large differences in the ice initiation and growth processes between these cloud regimes. These results showed that in addition to dynamics, aerosol and moisture are important for modulating ice microphysical processes in PRE and MON clouds due to the modification of cloud drop size distribution. In general, the monsoon clouds showed considerably higher total particle number concentrations (N_i) and higher ice water content (IWC) than the pre-monsoon clouds. This suggests significant differences in the ice production and growth processes between these clouds.

7 Aerosol-Cloud Interaction in Convective Clouds

It has long been recognized that aerosols can serve as cloud condensation nuclei (CCN) and alter the microphysical characteristics of clouds. According to a National Academy of Science report, "The greatest uncertainty about the aerosol climate forcing—is probably the indirect effect of aerosols on clouds" (NRC, 2005). The fourth assessment report of the Intergovernmental Panel on Climate Change (IPCC, 2007) notes that the interactions between aerosols and clouds are still poorly understood and that even a slight alteration in cloud characteristics brought on by aerosol forcing can have a sizable impact when added up over extensive time and space scales. In the recent IPCC (2022) report, the understanding of aerosol direct effect has been reported with high confidence and is attributed to the understanding from several field campaigns and satellite observations.

8 CCN Effect on Cloud Microphysical Characteristics

In addition to aerosol contributing to direct effect by absorbing and reflecting radiation, they (act as cloud condensation nuclei) determine the initial cloud droplet number concentration, albedo, precipitation formation, and lifetime of clouds, known as indirect effects. A high concentration of aerosols leads to numerous smaller cloud particles that reflect more solar radiation to space, leading to climate cooling. This effect is known as the 1st indirect effect (e.g. Ramaswamy et al., 2001) or cloud albedo effect (e.g. Lohmann & Feichter, 2005), or Twomey effect (e.g. Twomey, 1977). This increase in CCN also increases the cloud drop number concentration but reduces the mean cloud droplet size. This suppresses the collision-coalescence process and thus delays the precipitation formation mechanism, thereby prolonging cloud lifetime and further increasing the reflection of solar radiation, referred to as the 2nd indirect effect (e.g. Ramaswamy et al., 2001) or cloud lifetime effect (e.g. Lohmann & Feichter, 2005) or Albrecht effect (e.g. Albrecht, 1989). The response of deep convective clouds to increase in aerosol concentration is poorly understood as the presence of mixed-phase (co-existence of ice and supercooled liquid) makes it complicated to understand the interaction between different processes. Rosenfeld and Woodley (2000) suggested that a reduction in cloud drop size in polluted deep clouds suppresses precipitation and delays the onset of freezing. They showed the presence of supercooled drops at temperatures as low as −37.5 °C. This study was supported by a cloud-resolving modelling study (Khain et al., 2001) which showed that if the size of cloud droplets were small, then supercooled drops could be simulated at −37.5 °C. When these droplets freeze homogeneously below this temperature, associated latent heat could enhance the updrafts and increase the cloud tops. This effect is known as a thermodynamic effect. Another aerosol indirect effect connecting to deep convective clouds is the "riming indirect effect" (Lohmann and Feitcher, 2005). The reduction in mean cloud droplet diameter due to increased CCN concentration reduces the efficiency of droplet accretion onto an existing ice crystal, this decreases riming and precipitation rate in polluted mixed-phase clouds.

(a) **Aerosol invigoration effect**

The clouds developing in a polluted environment tend to grow taller and cause stronger thunderstorms compared to the clouds in clean conditions. Various processes have been proposed to explain precipitation enhancement in deep convective clouds. The positive sign of precipitation changes due to an increase in aerosol loading (while there is uncertainty regarding the type of aerosols) is attributed to the convective invigoration mechanism, that is, an increase in convective updrafts and an increase in cloud size. Several studies suggest that increased aerosol concentration will lead to the invigoration of deep convection (Rosenfeld et al., 2008; review paper by Altaratz et al., 2014; Gayatri et al., 2017). Rosenfeld et al. (2008) suggested that the increased aerosol number concentration suppresses the onset of the collision—coalescence process and delays the formation of precipitation to greater heights (Albrecht, 1989; Rosenfeld, 1999, 2000). Thus more cloud water reaches the freezing level and releases more latent heat due to enhanced freezing. This additional latent heat increases updrafts and hence buoyancy (Seifert & Beheng, 2006; Rosenfeld et al., 2008) along with the formation of large ice hydrometeors due to the availability of supercooled water. As a result of the increase in updrafts, there is an enhancement of mass flux which brings more water to the mixed phase part of the cloud, further releasing more latent heat. This creates positive feedback and produces deeper clouds (Khain et al., 2005; Teller and Levin, 2006; van den Heever et al., 2006; Li et al., 2011). Using CAIPEEX aircraft observations and numerical simulation, Gayatri et al. (2017) showed that aerosol invigoration effects could enhance precipitation (by 20% in a single convective cell) in deep convective clouds, and the increase in precipitation is more pronounced in humid environments (Gayatri et al., 2021).

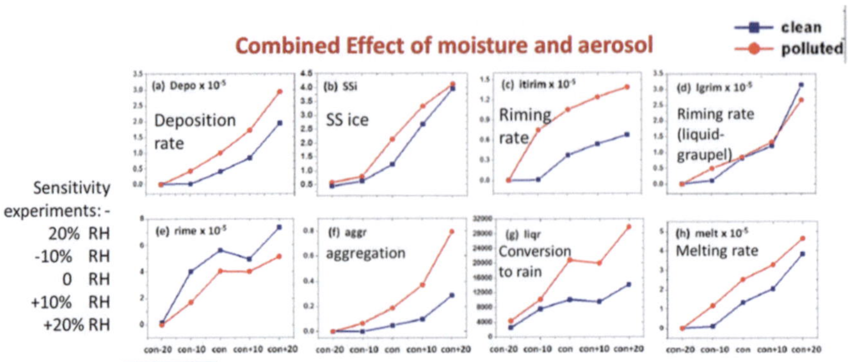

Fig. 16 Population mean (Wang, 2005) of different microphysical processes rates; **a** deposition rate (g/kg/s), **b** super saturation over ice (in %), **c** Conversion ice/snow to ice/snow by riming (g/kg/s), **d** Conversion of liquid to graupel by riming (g/kg/s), **e** total riming rate (g/kg/s), **f** aggregation rate (/kg/s), **g** liquid conversion to rain rate (/kg/s), and **h** melting rate (g/kg/s) for several relative humidity (RH) sensitivity experiments (Adapted from Gayatri et al., 2021)

(b) **Combined effect of aerosol and moisture on cloud properties**

In addition to aerosol forcing, the question of invigoration or suppression of clouds/precipitation may strongly depend on the environmental conditions such as atmospheric instability and environmental humidity (Rosenfeld, 1999; Khain et al., 2008; Khain and Lynn, 2009) and wind shear (Lee et al., 2008, 2012; Fan et al., 2009; Storer and van den Heever, 2013). Many previous studies highlighted the importance of relative humidity (RH) as a key factor affecting cloud microphysics and dynamics (Lynn et al., 2007; Khain et al., 2008; Lee, 2011; Cui et al., 2011; Kalina et al., 2014; Li et al., 2017).

Different microphysical processes due to change in aerosol concentration and moisture are shown in Fig. 16. The cloud development is dramatically affected by the presence of high moisture. In both dry and moist environments, the warm phase has the same response to the increase in CCN concentration. Numerous smaller drops narrow the drop sizes distribution spectra and suppress the collision coalescence process. However, the increased condensation releases large amounts of latent heat and increases updrafts which leads to an increase in supercooled liquid water. In moist environments, clouds have low cloud bases compared to dry environments and large amounts of cloud water are transported above freezing level. These increases drop freezing, vapour deposition and other mixed-phase processes like riming and aggregation. Due to this, the ice particle spectra (ice, snow and graupel) broaden in moist environments and cause more precipitation via melting of the cold phase. In a humid environment, the increase in precipitation with a rise in the CCN is more substantial. In a dry environment, all processes are suppressed hence less precipitation at the surface.

(c) **Effect on rainfall**

Konwar et al. (2012) showed the linear relationship between the cloud droplet number and the level of warm rain initiation. The raindrop forms at the undiluted part of the cloud (Khain et al., 2013) as inferred from the aircraft observations and supporting turbulence simulations. On the large scale, Maheshkumar et al. (2013) using a comprehensive set of reanalysis and in situ CAIPEEX observations over the Western Ghats indicated that the moist adiabatic/multi-level inversion stratifications during the high/low rainfall spells. Shallow and deep convective clouds coexist in moist adiabatic stratification regimes. The low updraft rates provide the sufficient time required for warm rain processes to produce rainfall from shallow clouds. However, these different airmass characteristics bring different aerosol types into the study area. Under different large-scale conditions, contrasting precipitation microphysical characteristics were found (Sreenath et al., 2021). The high aerosol conditions showed narrow precipitation drop spectra, while low pollution showed wide precipitation spectra.

9 Deep Cloud Vertical Structure from Radar Observations

Meteorological Radars are very useful for continuous observations of cloud populations, as well as their vertical structure. The single polarization weather radar only provides conventional radar parameters, i.e., reflectivity (Z_H), radial velocity (V_R), spectral width, etc. which are only useful to study the bulk properties of clouds. Z_H is the fraction of transmitted power returned to the radar receiver after hitting the hydrometeors within clouds. It provides useful insights into the size and concentration of the hydrometeors (Doviak et al., 1979). V_R provides a radial wind field and/or fall velocity of the hydrometeors (Browning and Wexler, 1968; Keppas et al., 2018).

In recent times, the development of dual-polarization weather radar started a new era in the field of cloud microphysics study. The key microphysical signatures identified from polarimetric radar are listed in Table 3. In addition to the conventional radar parameters, the dual polarization weather radars provide different polarimetric parameters, i.e., differential reflectivity (Z_{DR}), differential phase shift (Φ_{DP}), specific differential phase (K_{DP}), co-polar correlation coefficient (ρ_{HV}), etc. Z_{DR} is the logarithmic ratio of the reflectivity factor between the horizontal (H) and vertical (V) polarization and it provides information about the shape, orientation, density, and composition/phase of the hydrometeors (Kumjian, 2013). Φ_{DP} is the additional phase difference acquired between H and V channels when these two signals propagate through non-spherical/oblate raindrops and are useful for quantitative precipitation estimation (Kumjian, 2013). K_{DP} indicates the differential phase shift per unit distance along the radial direction and is a useful indicator for locating the region of heavy precipitation (Kumjian, 2013). ρ_{HV} provides information about the contribution of each target in a radar sampling volume to the overall H and V-polarization

Table 3 Important microphysical signatures inferred from polarimetric radar

Microphysical signatures	Authors
Z_{DR} column: narrow, vertically extended region of positive Z_{DR} values extend above 0 °C isotherm in a deep convective storm which forms due to the transport of supercooled liquid droplets by the strong updraft	Hall et al. (1984), Illingworth et al. (1987), Conway and Zrnić (1993), MacGorman et al. (2008), Kumjian et al. (2014a)
Specific differential phase (K_{DP}): potential identifier of heavy precipitation	Zrnić and Ryzhkov (1996), Ryzhkov and Zrnić (1996), Kumjian (2013)
An increase in Z_{DR} and K_{DP} associated with a reduced ρ_{HV} near the cloud top indicates the presence of anisotropic pristine ice crystals	Kumjian et al. (2014b), Ryzhkov et al. (2016)
Local K_{DP} maxima (~0.15° –0.4° km^{-1}) near −15 °C temperature regime is associated with the higher concentration of dendritic particles (diameter range ~ 0.8–1.2 mm) due to the enhancement of depositional growth and/or onset of aggregation. The passage of these local K_{DP} maxima was also associated with increased surface precipitation	Kennedy and Rutledge (2011), Bechini et al. (2013), Schneebeli et al. (2013), Moisseev et al. (2015), Kumjian and Lombardo (2017)
Sharp vertical gradient of Z_H, enhancement in Z_{DR} (up to 1.5–2 dB) and reduction of ρ_{HV} above the freezing level (at −10 to −15 °C temperature regime) indicate the dendritic growth and onset of aggregation	Kennedy and Rutledge (2011), Ryzhkov et al. (2016)
An increase of Z_H and a decrease of Z_{DR} below the (dendritic growth zone) DGZ indicates riming and/or aggregation. Z_{DR} value of dry aggregated snow is slightly positive (~0.25 dB) compared to rimed snow (close to zero)	Vogel and Fabry (2018), Ryzhkov et al. (2016)
Melting Layer/Bright Band: melting layer/ bright band is associated with enhanced Z_H, Z_{DR}, and low ρ_{HV}	Ryzhkov et al. (2016)
Saggy melting layer/bright band: heavily rimed particles fall faster due to their higher density and descend to a greater distance before complete melting, compared to non-rimed snow, causing a downward extension (local sagging) of the melting layer/bright band leading to an increase in its depth	Kumjian et al. (2016), Ryzhkov et al. (2016)
Enhanced Z_H and Z_{DR} below the melting layer indicate the presence of larger and denser particles below the melting layer (falling from above), melting and subsequently becoming big drops in due course	Kumjian et al. (2016)

(continued)

Table 3 (continued)

Microphysical signatures	Authors
Size Sorting: difference in sedimentation rate owing to the hydrometeor mass (i.e., heavier particles fall faster) produces an area of a sparsely concentrated large drop, causing an enhancement in Z_{DR} and a decrease in Z_H over this region	Kumjian and Ryzhkov (2012)
Refreezing: freezing of the supercooled liquid droplets into ice pellets causes an enhancement of Z_{DR} and K_{DP} and a decrease of Z_H and ρ_{HV} within the low-level subfreezing air	Kumjian et al. (2013), Ryzhkov et al. (2016)

signals and is a very useful parameter to distinguish between meteorological ($\rho HV > 0.8$) and non-meteorological ($\rho HV < 0.8$) targets (Kumjian, 2013).

Some studies have further extended the potential of dual-polarization radar by developing hydrometeor identification (HID) algorithm using several polarimetric radar parameters (e.g., Vivekanandan et al., 1999; Liu & Chandrasekar, 2000; Dolan & Rutledge, 2009; Dolan et al., 2013; etc.). These algorithms are able to classify the hydrometeors into different pre-defined categories. The evolution of various hydrometeors during the lifecycle of a cloud cluster during the Indian Summer Monsoon is illustrated by Samanta et al. (2021), using dual-polarization radar observations from the CAIPEEX Phase-IV campaign. An example of the vertical structure of a convective cloud and associated hydrometeor distribution is shown in Fig. 17. The deep convective cloud is able to reach 14 km and an anvil developing with ice crystals. The cloud has a core region with graupel and the melting layer is visible with the wet snow, which melts below this level to form large raindrops. A few graupel and hail are also seen below the melting layer. There are a few hail reaching the surface. The vertical ice indicated in the cold region of the cloud are possible centres of electrical effects. Aggregates are most abundantly seen and can be due to several processes from accretion, riming, etc. The vertical structure of deep convective clouds noted was similar to the in situ observations. However, there are several processes to be unravelled and the lifecycle of a storm needs to be also investigated. It may be specifically noted that these clouds have significantly less cloud width (<10 km) and are able to give precipitation of 10–40 mm over a short period.

A combination of strong updrafts, supercooled liquid water content, large ice crystal and graupel number concentration contribute to lightning (Saunders et al., 2006). Thus the charging mechanisms in clouds are supported by the strong updrafts, which lift water content to greater depths to form these precursors for lightning flashes. The indicators of lightning in the radar observations are from the vertical ice signature. A combination of radar and lightning location sensors can give detailed information.

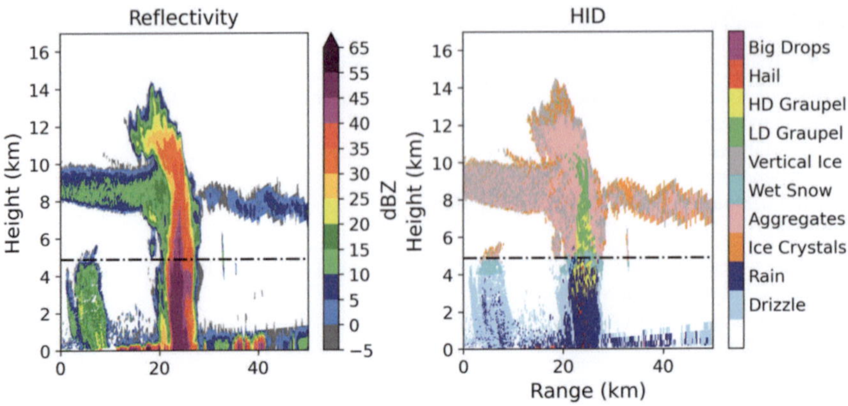

Fig. 17 An example of deep convective cloud observation using polarimetric radar during CAIPEEX IV. **a** reflectivity and **b** hydrometer identification showing big drops, hail, high-density graupel, low-density graupel, vertical ice, wet snow, aggregates, ice crystals, rain, and drizzle. The horizontal line indicates the 0 °C level

10 Organization of Convection

Convective clouds often organize under favorable dynamical and/or synoptic conditions and give rise to a contiguous area of precipitation of several hundreds of kilometres. Depending on the morphology and intensity of these cloud clusters, they are often termed Mesoscale Convective Systems (MCS) (Houze, 2004), Mesoscale Convective Complex (MCC) (Maddox, 1980), etc. These clusters of convective clouds are often responsible for very heavy rainfall over a shorter time span over any location, causing flood-like situations. Recent studies (IPCC, 2021) showed that the frequency of intense precipitation events increases in the present scenario of global climate change, which poses a threat to the life and livelihood of millions of people around the world.

The coexistence of warm and mixed-phase cloud processes is seen in organized convective systems and they possess a complex dynamical-microphysical coupling, which primarily affects the vertical distribution of momentum, heat, and water vapour more rigorously compared to shallow and/or deep convective clouds. To date, it is not very clear how individual deep convective cells of a horizontal dimension of ~10 km organize themselves in a group of clouds with a horizontal dimension of ~100 km (Houze, 2018).

The organization of clouds depends on several factors. Orography plays an important role in initiation (Orville, 1965) and the organization of convection. During 'North American Monsoon Experiment' (NAME) (Higgins et al., 2006) it was noticed that topography has a significant influence on the diurnal variability of rainfall (Gochis et al., 2007; Lang et al., 2007) as well as the organization of cloud systems (Rowe et al., 2012). During the NAME experiment, it was also frequently observed that convections were initiated over high terrain, started organizing when it moved off the higher terrain and became well organized over lower terrain (Rowe et al., 2012). A simulation study of tropical MCSs in the absence of ice processes indicated less organized convection as a result of associated dynamical modulations which eliminate the surface gust front (Liu et al., 1997). Analysing several cases of NAME experiment, Rowe et al. (2012) also suggested that there is a possible dependency of cloud organization on the microphysical processes as in most of the cases; prominent organization occurred in the presence of precipitation-sized ice, rear inflow, and outflow boundaries.

The cold pool seems to be another controlling factor in the organization of convection (Rowe et al., 2012). The strength of the cold pool is mainly controlled by the microphysical processes through convective downdrafts by means of evaporation of precipitation in the sub-saturated environment and evaporation as well as melting of ice (Knupp & Cotton, 1985; Srivastava, 1987). Robe and Emanuel (2001) showed that wind shear at a lower level also has a pronounced effect on the orientation of an organized cloud band; the band of convection tends to organize itself into an arc, broadly perpendicular to the shear.

The organization of convection is studied quantitatively using parameters, e.g., Simple Convective Aggregation Index (SCAI) (Tobin et al., 2012), Organization Index (I_{org}) (Tompkins & Semie, 2017), Convection Organization Potential (COP) (White et al., 2018), etc. Recent studies focused on understanding the organization of convection using satellite, radar, and numerical simulations. Samanta et al. (2023) reported that the organization of convection in numerical simulations highly depends on the choice of the microphysics schemes. This may be noted in Fig. 18 that the SCAI is significantly different between the models and observations. Although some schemes give somewhat comparable features with observations during the initiation of storms, the remainder of its lifecycle is simulated very differently from the observation. The basic problem is that models are simulating significantly large areas of convection compared to observations. Though there are improvements in the bin microphysics scheme (Khain et al., 2015), there is still a lot of room for improvement. The cloud processes are still under-resolved in the models especially the deep convective clouds and their fundamental understanding needs to be improved (Morrison et al., 2020).

11 The Importance of Deep Convective Clouds in Warming Climate

Having known that nearly 50% of the precipitation around the globe is coming from the mesoscale convective clusters (Feng et al., 2021), the organization of convection to be investigated in details in a warming climate. The impact of global warming on precipitation is very direct; increased heating leads to more evaporation and surface drying and more drought conditions. For every 1 °C change in temperature, the water holding capacity of air increases by 7%, leading to more water vapor in the

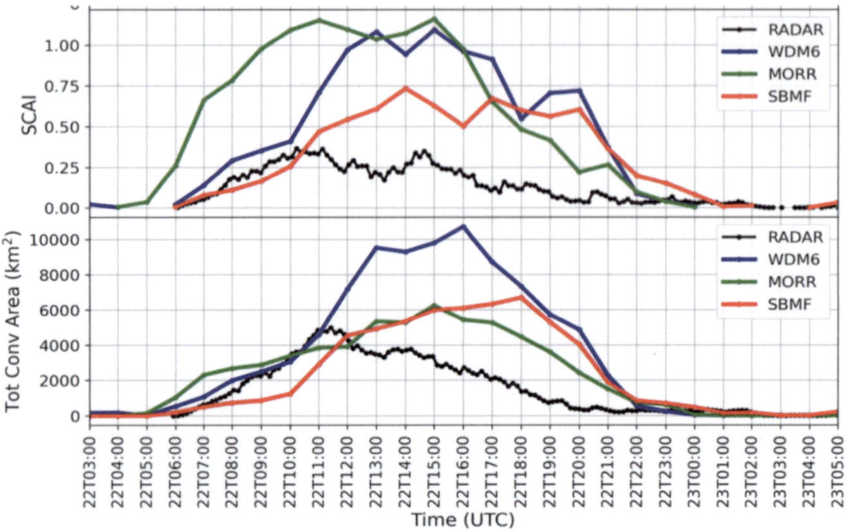

Fig. 18 The simple convective aggregation index and total convective area from radar observations and from the three microphysics schemes

atmosphere, which is a trigger for intense storms. Romps (2011) indicated that several properties of clouds including the water vapor gradient, cloud pressure depth and cloud velocity could change in the warming environment.

Convective self-aggregation is a spontaneous organization of scattered clouds to isolated convective clusters (Holloway et al., 2017). How organized convection and climate change interact? The changing convective organization is one of the mechanisms, which could help understand the variations in extreme precipitation, through dynamical interactions (Pendergrass 2020). The problem of self-aggregation of convection in the warming environment is a challenging problem, identified in recent times (Wing et al., 2017) and is an active area of research. Bao and Sherwood (2018) showed that the daily extreme precipitation is higher in the convectively aggregated state due to persistence. This adds further complexity to dealing with the topic.

Singh and O'Gorman (2014) indicated the important role of ice and mixed-phase processes in the precipitation extremes and especially the fall velocity of hydrometers explains the sensitivity of different parameterization schemes. The study unequivocally indicates that microphysics plays a key role in determining the response of convective extreme precipitation to global warming. Seiki et al. (2022) detail global cloud-resolving models and the improvements they can make by constraining and optimizing cloud optical properties using satellite simulators. The datasets developed for GCM validations such as CALIPSO-GOCCP (Chepfer et al., 2010). However, the hydrometeor classification from CALIOP is limited to nonprecipitating particles near the cloud tops due to the attenuation issues. The recent developments with Global Precipitation Measurement dual-frequency precipitation radar (GPM-DPR) satellite observations, hydrometeor classes are identified as snow, rain, and mixed-phase clouds using Ku/Ka-band radar echo and the dual-frequency ratio (Liao et al., 2020). The GPM-DPR also gives a global picture of the mesoscale convective clusters and indicates that nearly 50% of precipitation in the globe is from MCS (Feng et al., 2021).

The combination of Cloud Profiling radar (CPR) and DPR can give details of cloud hydrometeors (Castella et al., 2017). The ground validation of satellite data is a key aspect, where confidence in the cloud microphysics datasets could be developed. This requires major synergic ground deployments of different types of radars such as C-band polarimetric radars, X-band phased array radars for fast scanning, Ka and Wband radar, and wind profilers together with satellite data. An example of such deployments is ULTra-sIte for Measuring Atmosphere of Tokyo metropolitan Environment (ULTIMATE) site together with satellite simulators and cloud-resolving GCM (Satoh et al., 2022). The use of such data in Global cloud-resolving models will be the key.

Feng et al. (2023) tracked deep convective clouds and the MCS in the global cloud-permitting models. They showed that lifetime, speed of MCS, rainfall volume, and cloud shield area are well captured in the models. The cloud top heights and convective rainfall intensity over land are hugely overestimated in the model, while stratiform rainfall amount and area are underestimated. The model evaluations and intercomparison experiments would give key insights into the process that is relevant, key shortcomings of the model, and the controlling factors of the life cycle of organized systems in a framework that can unravel the dynamics-microphysics interactions.

Acknowledgements CAIPEEX is funded by the Ministry of Earth Sciences, Government of India and conducted by IITM, Pune. Several datasets used in this chapter was from CAIPEEX and contributions from several colleagues are acknowledged.

References

Albrecht, B. A. (1989). Aerosols, cloud microphysics and fractional cloudiness. *Science, 245*(4923), 1227–1230. https://doi.org/10.1126/science.245.4923.1227

Altaratz, O., Koren, I., Remer, L. A., & Hirsch, E. (2014). Review: Cloud invigoration by aerosols-coupling between microphysics and dynamics. *Atmospheric Research, 140–141*, 38–60. https://doi.org/10.1016/j.atmosres.2014.01.009

Ban, N., et al. (2021). The first multi-model ensemble of regional climate simulations at kilometer-scale resolution, part I: Evaluation of precipitation. *Climate Dynamics, 57*, 275–302. https://link.springer.com/article/10.1007/s00382-021-05708-w.

Bao, J., & Sherwood, S. C. (2018). The role of convective self-aggregation in extreme instantaneous vs. daily precipitation. *Journal of Advances in Modeling Earth Systems, 11*(1), 19–33. https://doi.org/10.1029/2018MS001503

Bechini, R., Baldini, L., & Chandrasekar, V. (2013). Polarimetric radar observations in the ice region of precipitating clouds at C-band and X-band radar frequencies. *Journal of Applied Meteorology and Climatology, 52*(5), 1147–1169. https://doi.org/10.1175/JAMC-D-12-055.1

Bera, S., Pandithurai, G., & Prabha, T. V. (2016a). Entrainment and droplet spectral characteristics in convective clouds during transition to monsoon. *Atmospheric Science Letters, 17*(4), 286–293. https://doi.org/10.1002/asl.657

Bera, S., Prabha, T. V., & Grabowski, W. W. (2016b). Observations of monsoon convective cloud microphysics over India and role of entrainment-mixing. *Journal of Geophysical Research Atmospheres, 121*, 9767–9788. https://doi.org/10.1002/2016JD025133

Bera, S., Prabha, T. V., Malap, N., Patade, S., Konwar, M., Murugavel, P., & Axisa, D. (2019). Thermodynamics and microphysics relation during CAIPEEX-I. *Pure and Applied Geophysics, 176*(1), 371–388. https://doi.org/10.1007/s00024-018-1942-6

Bergeron, T. (1935). On the physics of clouds and precipitation. Proces Verbaux de l'Association de Météorologie. *International Union of Geodesy and Geophysics*, 156–178.

Borys, R. D., Lowenthal, D. H., Cohn, S. A., & Brown, W. O. J. (2003). Mountaintop and radar measurements of anthropogenic aerosol effects on snow growth and snowfall rate. *Geophysical Research Letters, 30*(10). https://doi.org/10.1029/2002GL016855.

Browning, K. A., & Wexler, R. (1968). The determination of kinematic properties of a wind field using doppler radar. *Journal of Applied Meteorology and Climatology, 7*(1), 105–113. https://doi.org/10.1175/1520-0450(1968)007%3c0105:TDOKPO%3e2.0.CO;2

Castella, D., Panegrossi, G., Sanò, P., Marra, A. C., Dietrich, S., Johnson, B. T., & Kulie, M. S. (2017). Evaluation of the GPM-DPR snowfall detection capability: Comparison with CloudSat-CPR. *Atmospheric Research, 197*, 64–75. https://doi.org/10.1016/j.atmosres.2017.06.018

Chepfer, H., Bony, S., Winker, D., Cesana, G., Dufresne, J. L., Minnis, P., Stubenrauch, C. J., & Zeng, S. (2010). The GCM-oriented CALIPSO cloud product (CALIPSO-GOCCP). *Journal of Geophysical Research-Atmospheres, 115*(D4), https://doi.org/10.1029/2009JD012251.

Chepfer, H., Brogniez, H., & Noel, V. (2019). Diurnal variations of cloud and relative humidity profiles across the tropics. *Science and Reports, 9*, 16045. https://doi.org/10.1038/s41598-019-52437-6, www.nature.com/articles/s41598-019-52437-6.

Conway, J. W., & Zrnić, D. S. (1993). A study of embryo production and hail growth using dual-doppler and multiparameter radars. *Monthly Weather Review, 121*(9), 2511–2528. https://doi.org/10.1175/1520-0493(1993)121%3c2511:ASOEPA%3e2.0.CO;2

Covey, C., Doutriaux, C., Gleckler, P. J., Taylor, K. E., Trenberth, K. E., & Zhang, Y. (2018). High-frequency intermittency in observed and model-simulated precipitation. *Geophysical Research Letters, 45*(22), 12514–12522. https://doi.org/10.1029/2018GL078926

Cui, Z., Carslaw, K. S., & Blyth, A. M. (2011). The coupled effect of mid-tropospheric moisture and abundance on deep convective cloud dynamics and microphysics. *Atmosphere, 2*(4), 222–241. https://doi.org/10.3390/atmos2030222

DeMott, P. J., Prenni, A. J., Liu, X., Kreidnweis, S. M., Petters, M. D., Twohy, C. H., Richardson, M. S., Eidhammer, T., & Rogers, D. C. (2010). Predicting global ice nuclei distributions and their impact on climate. *Proceedings of the National Academy Sciences, 107*, 11217–11222.

Dolan, B., & Rutledge, S. A. (2009). A theory-based hydrometeor identification algorithm for X-band polarimetric radars. *Journal of Atmospheric and Oceanic Technology, 26*(10), 2071–2088. https://doi.org/10.1175/2009JTECHA1208.1

Dolan, B., Rutledge, S. A., Lim, S., Chandrasekar, V., & Thurai, M. (2013). A robust C-band hydrometeor identification algorithm and application to a long-term polarimetric radar dataset. *Journal of Applied Meteorology and Climatology, 52*(9), 2162–2186. https://doi.org/10.1175/JAMC-D-12-0275.1

Doviak, R. J., Zrnic, D. S., & Sirmans, D. S. (1979). Doppler weather radar. *Proceedings of the IEEE, 67*(11), 1522–1553. https://doi.org/10.1109/PROC.1979.11511

Fan, J., Yuan T., Comstock J. M., Ghan S., et al. (2009). Dominant role by vertical wind shear in regulating aerosol effects on deep convective clouds. *Journal of Geophysical Research-Atmospheres, 114*(D22). https://doi.org/10.1029/2009JD012352.

Feng, Z., Hagos, S., Rowe, A. K., Burleyson, C. D., Martini, M. N., & De Szoeke, S. P. (2015). Mechanisms of convective cloud organization by cold pools over tropical warm ocean during the AMIE/DYNAMO field campaign. *Journal of Advances in Modeling Earth Systems, 7*(2), 357–381. https://doi.org/10.1002/2014MS000384

Feng, Z., Leung, L. R., Hardin, J., Terai, C. R., Song, F., & Caldwell, P. (2023). Mesoscale convective systems in DYAMOND global convection-permitting simulations. *Geophysical Research Letters, 50*, e2022GL102603. https://doi.org/10.1029/2022GL102603.

Feng, Z., Leung, L. R., Liu, N., Wang, J., Houze, R. A., Li, J., et al. (2021). A global high-resolution mesoscale convective system database using satellite-derived cloud tops, surface precipitation, and tracking. *Journal of Geophysical Research: Atmospheres, 126*(8). https://doi.org/10.1029/2020JD034202.

Field, P. R., & Heymsfield, A. J. (2015). Importance of snow to global precipitation. *Geophysical Research Letters, 42*(21), 9512–9520. https://doi.org/10.1002/2015GL065497

Findeisen, W. (1938). Kolloid-meteorologische Vorgängebei Neiderschlags-bildung. *Meteorologische Zeitschrift, 55*, 121–133.

Findell, K. L., & Eltahir, E. A. B. (2003). Atmospheric controls on soil moisture-boundary layer interactions. Part II: Feedbacks within the continental United States. *Journal of Hydrometeorology, 4*(3), 570–583. https://doi.org/10.1175/1525-7541(2003)004%3c0570:ACOSML%3e2.0.CO;2

Fletcher, N. H. (1962). *The Physics of Rain Clouds*. Cambridge Univ. Press.

Fritsch, J. M., Kane, R. J., & Chelius, C. R. (1986). The contribution of mesoscale convective weather systems to the warm-season precipitation in the United States. *Journal of Climate Applied Meteorology, 25*, 1333–1345. https://doi.org/10.1175/1520-0450(1986)025%3c1333:TCOMCW%3e2.0.CO;2

Gayatri, K., Patade, S., Fan, J., & Prabhakaran, T. (2021). Pathways of precipitation formation in different thermodynamic and aerosol environments over the Indian Peninsula. *Atmospheric Research, 266*. https://doi.org/10.1016/j.atmosres.2021.105934

Gayatri, K., Patade, S., & Prabha, T. V. (2017). Aerosol cloud interaction in deep convective clouds over the Indian peninsula using spectral (bin) microphysics. *Journal of Atmospheric Science, 74*(10), 3145–3166. https://doi.org/10.1175/JAS-D-17-0034.1

Glenn, I. B., & Krueger, S. K. (2014). Downdrafts in the near cloud environment of deep convective updrafts. *Journal of Advances in Modeling Earth Systems, 6*, 1–8. https://doi.org/10.1002/2013MS000261

Gochis, D. J., Watts, C. J., Garatuza-Payan, J., & Cesar-Rodriguez, J. (2007). Spatial and temporal patterns of precipitation intensity as observed by the NAME event rain gauge network from 2002 to 2004. *Journal of Climate, 20*(9), 1734–1750. https://doi.org/10.1175/JCLI4092.1

Hall, M. P. M., Goddard, J. W. F., & Cherry, S. M. (1984). Identification of hydrometeors and other targets by dual-polarization radar. *Radio Science, 19*(01), 132–140. https://doi.org/10.1029/RS019i001p00132

Hallett, J., & Mossop, S. C. (1974). Production of secondary ice crystals during the riming process. *Nature, 249*, 26–28. https://doi.org/10.1038/249026a0

Higgins, W., et al. (2006). The NAME 2004 field campaign and modeling strategy. *Bulletin of the American Meteorological Society, 87*(1), 79–94. https://doi.org/10.1175/BAMS-87-1-79

Holloway, C. E., Wing, A. A., Bony, S., et al. (2017). Observing convective aggregation. *Surveys in Geophysics, 38*, 1199–1236. https://doi.org/10.1007/s10712-017-9419-1

Houze, R. A., Jr. (2004). Mesoscale convective systems. *Reviews of Geophysics, 42*(4), 1–43. https://doi.org/10.1029/2004RG000150

Houze, R. A., Jr. (2018a). 100 years of research on mesoscale convective systems. *Meteorological Monographs, 59*(1), 17.1-17.54. https://doi.org/10.1175/AMSMONOGRAPHS-D-18-0001.1

Houze, R. A., Jr. (2018b). 100 years of research on mesoscale convective systems, University of Washington, Seattle, and Pacific Northwest National Laboratory, Richland, Washington. *American Meteorological Society, 59*(1), 171–1754. https://doi.org/10.1175/AMSMONOGRAPHS-D-18-0001.1

Igau, R. C., LeMone, M. A., & Wei, D. (1999). Updraft and downdraft cores in TOGA COARE: Why so many buoyant downdraft cores? *Journal of Atmospheric Science, 56*, 2232–2245.

Illingworth, A. J., Goddard, J. W. F., & Cherry, S. M. (1987). Polarization radar studies of precipitation development in convective storms. *Quarterly Journal of the Royal Meteorological Society, 113*(476), 469–489. https://doi.org/10.1002/qj.49711347604

IPCC. (2007). *Climate Change (2007): The Physical Science Basis; Contribution of Working Group I to the Fourth Assessment Report of the Intergovernmental Panel on Climate Change—Summary for Policymakers*. https://wedocs.unep.org/20.500.11822/30763.

IPCC, 2022: Climate Change 2022: Impacts, Adaptation, and Vulnerability. Contribution of Working Group II to the Sixth Assessment Report of the Intergovernmental Panel on Climate Change [H.-O. Pörtner, D.C. Roberts, M. Tignor, E.S. Poloczanska, K. Mintenbeck, A. Alegría, M. Craig, S. Langsdorf, S. Löschke, V. Möller, A. Okem, B. Rama (eds.)]. Cambridge University Press. Cambridge University Press, Cambridge, UK and New York, NY, USA, 3056 pp., https://doi.org/10.1017/9781009325844.

Jucker, M., Lane, T. P., Vincent, C. L., Webster, S., Wales, S. A., Louf, V. (2020). Locally forced convection in sub kilometre-scale simulations with the Unified Model and WRF. *Quarterly Journal of the Royal Meteorological Society, 146*(732), 2981–3621. https://rmets.onlinelibrary.wiley.com/doi/epdf/10.1002/qj.3855.

Kalina, E. A., Friedrich, K., Morrison, H., & Bryan, G. H. (2014). Aerosol effects on idealized supercell thunderstorms in different environments. *Journal of the Atmospheric Sciences, 71*(12), 4558–4580. https://doi.org/10.1175/JAS-D-14-0037.1

Kanji, Z. A., et al. (2017). *Overview of Ice Nucleating Particles in Ice Formation and Evolution in Clouds And Precipitation: Measurement and Modeling Challenges* (Vol. 58, pp. 1.1–1.33). American Meteorological Society.

Kennedy, P. C., & Rutledge, S. A. (2011). S-band dual-polarization radar observations of winter storms. *Journal of Applied Meteorology and Climatology, 50*(4), 844–858. https://doi.org/10.1175/2010JAMC2558.1

Keppas, S. C., Crosier, J., Choularton, T. W., & Bower, K. N. (2018). Microphysical properties and radar polarimetric features within a warm front. *Monthly Weather Review, 146*(7), 2003–2022. https://doi.org/10.1175/mwr-d-18-0056.1

Khain, A. & Lynn, B. (2009). Simulation of a supercell storm in clean and dirty atmosphere using weather research and forecasting model with spectral SBM microphysics. *Journal of Geophysical Research—Atmospheres, 114*(D19). https://doi.org/10.1029/2009JD011827.

Khain, A., Prabha, T. V., Benmoshe, N., Pandithurai, G., & Ovchinnikov, M. (2013). The mechanism of first raindrops formation in deep convective clouds. *Journal of Geophysical Research—Atmospheres, 118*, 9123–9140. https://doi.org/10.1002/jgrd.50641

Khain, A., Roseenfeld, D., & Pokrovsky, A. (2005). Aerosol impact on the dynamics and microphysics of deep convective clouds. *Quarterly Journal of the Royal Meteorological Society, 131*, 2639–2663. https://doi.org/10.1256/qj.04.62

Khain, A. P., Benmoshe, N., & Pokrovsky, A. (2008). Factors determining the impact of aerosols on surface precipitation from clouds: Attempt of classification. *Journal of the Atmospheric Sciences, 65*(6), 1721–1748. https://doi.org/10.1175/2007JAS2515.1

Khain, A. P., Pinsky, M. B., Shapiro, M., & Pokrovsky, A. (2001). Graupel-drop collision efficiencies. *Journal of Atmospheric Science, 58*, 2571–2595. https://doi.org/10.1175/1520-0469(2001)058%3c2571:CROSGA%3e2.0.CO;2

Khain, A. P., et al. (2015). Representation of microphysical processes in cloud resolving models: Spectral (bin) microphysics versus bulk parameterization. *Reviews of Geophysics, 53*, 247–322. https://doi.org/10.1002/2014RG000468

Knupp, K. R., & Cotton, W. R. (1985). Convective cloud downdraft structure: An interpretive survey. *Reviews of Geophysics, 23*(2), 183–215. https://doi.org/10.1029/RG023i002p00183

Konwar, M., Maheshkumar, R. S., Kulkarni, J. R., Freud, E., Goswami, B. N., & Rosenfeld, D. (2012). Aerosol control on depth of warm rain in convective clouds. *Journal of Geophysical Research—Atmospheres, 117*(D13), https://doi.org/10.1029/2012JD017585.

Korolev, A., & Leisner, T. (2020). Review of experimental studies of secondary ice production. *Atmospheric Chemistry and Physics, 20*, 11767–11797. https://doi.org/10.5194/acp-20-11767-2020

Kulkarni, J. R., et al., (2012). The Cloud Aerosol Interactions and Precipitation Enhancement Experiment (CAIPEEX): Overview and preliminary results. *Current Science, 102*(3), 413–425. https://www.jstor.org/stable/24083888.

Kumjian, M. (2013). Principles and applications of dual-polarization weather radar. Part I: Description of the polarimetric radar variables. *Journal of Operational Meteorology, 1*(19), 226–242. https://doi.org/10.15191/nwajom.2013.0119.

Kumjian, M. R., & Lombardo, K. A. (2017). Insights into the evolving microphysical and kinematic structure of Northeastern U.S. winter storms from dual-polarization Doppler radar. *Monthly Weather Review, 145*(3), 1033–1061. https://doi.org/10.1175/MWR-D-15-0451.1

Kumjian, M. R., & Ryzhkov, A. V. (2012). The impact of size sorting on the polarimetric radar variables. *Journal of the Atmospheric Sciences, 69*(6), 2042–2060. https://doi.org/10.1175/JAS-D-11-0125.1

Kumjian, R., Khain, A. P., Benmoshe, N., Ilotoviz, E., Ryzhkov, A. V., & Phillips, V. T. J. (2014a). The anatomy and physics of ZDR columns: Investigating a polarimetric radar signature with a spectral bin microphysical model. *Journal of Applied Meteorology and Climatology, 53*(7), 1820–1843. https://doi.org/10.1175/JAMC-D-13-0354.1

Kumjian, M. R., Mishra, S., Giangrande, S. E., Toto, T., Ryzhkov, A. V., & Bansemer, A. (2016). Polarimetric radar and aircraft observations of saggy bright bands during MC3E. *Journal of Geophysical Research—Atmospheres, 121*(7), 3584–3607. https://doi.org/10.1002/2015JD024446

Kumjian, M. R., Rutledge, S. A., Rasmussen, R. M., Kennedy, P. C., & Dixon, M. (2014b). High-resolution polarimetric radar observations of snow-generating cells. *Journal of Applied Meteorology and Climatology, 53*(6), 1636–1658. https://doi.org/10.1175/JAMC-D-13-0312.1

Kumjian, M. R., Ryzhkov, A. V., Reeves, H. D., & Schuur, T. J. (2013). A dual-polarization radar signature of hydrometeor refreezing in winter storms. *Journal of Applied Meteorology and Climatology, 52*(11), 2549–2566. https://doi.org/10.1175/JAMC-D-12-0311.1

Lance, S., Shupe, M. D., Feingold, G., Brock, C. A., Cozic, J., Holloway, J. S., Moore, R. H., Nenes, A., Schwarz, J. P., Spackman, J. R., Froyd, K. D., Murphy, D. M., Brioude, J., Cooper, O. R., Stohl, A., & Burkhart, J. F. (2011). Cloud condensation nuclei as a modulator of ice processes in Arctic mixed-phase clouds. *Atmospheric Chemistry and Physics, 11*, 8003–8015. https://doi.org/10.5194/acp-11-8003-2011

Lang, T. J., Ahijevych, D. A., Nesbitt, S. W., Carbone, R. E., Rutledge, S. A., & Cifelli, R. (2007). Radar-observed characteristics of precipitating systems during NAME 2004. *Journal of Climate, 20*(9), 1713–1733. https://doi.org/10.1175/JCLI4082.1

Lee, S. S. (2011). Dependence of aerosol-precipitation interactions on humidity in a multiple-cloud system. *Atmospheric Chemistry and Physics, 11*(5), 2179–2196. https://doi.org/10.5194/acp-11-2179-2011

Lee, S. S., Donner, L. J., Phillips, V. T. J., & Ming, Y. (2008). Examination of aerosol effects on precipitation in deep convective clouds during the 1997 ARM summer experiment. *Quarterly Journal of the Royal Meteorological Society, 134*(634), 1201–1220. https://doi.org/10.1002/qj.287

Lee, S. S., Feingold, G., & Chuang, P. Y. (2012). Effect of aerosol on cloud-environment interactions in trade cumulus. *Journal of the Atmospheric Sciences, 69*(12), 3607–3632. https://doi.org/10.1175/JAS-D-12-026.1

LeMone, M. A., & Zipser, E. J. (1980). Cumulonimbus vertical velocity events in GATE. Part I: Diameter, intensity and mass flux. *Journal of Atmospheric Science, 37*, 2444–2457.

Li, Z., Niu, F., Fan, J., Liu, Y., Rosenfeld, D., & Ding, Y. (2011). Long-term impacts of aerosols on the vertical development of clouds and precipitation. *Nature Geoscience, 4*(12), 888–894. https://doi.org/10.1038/Ngeo1313

Li, Z., Rosenfeld, D., & Fan, J. (2017). Aerosol and their impact on radiation, clouds, precipitation, and severe weather events. *Oxford Research Encyclopedia of Environmental Science*. https://doi.org/10.1093/acrefore/9780199389414.013.126.

Liao, L., Meneghini, R., Tokay, A., Kim, H. (2020). Assessment of Ku- and Ka-band dual-frequency radar for snow retrieval. *Journal of the Meteorological Society of Japan*. https://doi.org/10.2151/jmsj.2020-057.

Liu, C., Moncrieff, M. W., & Zipser, E. J. (1997). Dynamical influence of microphysics in tropical squall lines: A numerical study. *Monthly Weather Review, 125*(9), 2193–2210. https://doi.org/10.1175/1520-0493(1997)125%3c2193:DIOMIT%3e2.0.CO;2

Liu, H., & Chandrasekar, V. (2000). Classification of hydrometeors based on polarimetric radar measurements: Development of fuzzy logic and neuro-fuzzy systems, and in situ verification. *Journal of Atmospheric and Oceanic Technology, 17*(2), 140–164. https://doi.org/10.1175/1520-0426(2000)017%3c0140:COHBOP%3e2.0.CO;2

Liu, Y. & Daum, P. H. (2000). Spectral dispersion of cloud droplet size distributions and the parameterization of cloud droplet effective radius. *Geophysical Research Letters, 27*(13), 1903–1906. https://doi.org/10.1029/1999GL011011.

Lohmann, U., & Feichter, J. (2005). Global indirect aerosol effects: A review. *Atmospheric Chemistry and Physics, 5*(3), 715–737. https://doi.org/10.5194/acp-5-715-2005

Lynn, B., Khain, A., Rosenfeld, D., Woodley, W. L. (2007). Effects of aerosols on precipitation from orographic clouds. *Journal of Geophysical Research—Atmospheres, 112*(D10). https://doi.org/10.1029/2006JD007537.

MacGorman, D. R., et al. (2008). TELEX the thunderstorm electrification and lightning experiment. *Bulletin of the American Meteorological Society, 89*(7), 997–1014. https://doi.org/10.1175/2007BAMS2352.1

Maddox, R. A. (1980). Meoscale convective complexes. *Bulletin of the American Meteorological Society, 61*(11), 1374–1387. https://twister.caps.ou.edu/MM2005/Maddox_1980.pdf.

Maheskumar, R. S., Narkhedkar, S. G., Morwal, S. B., Padmakumari, B., Kothawale, D. R., Joshi, R. R., Deshpande, C. G., Bhalwankar, R. V., & Kulkarni, J. R. (2013). Mechanism of high rainfall over the Indian west coast region during the monsoon season. *Climate Dynamics*, 1–17. https://doi.org/10.1007/s00382-013-1972-9

Malap, N., & Prabha, T. V. (2023). Entrainment rates in the cloud zones of continental shallow cumulus. *Atmospheric Research, 286*, 106679. https://doi.org/10.1016/j.atmosres.2023.106679

Malap, N., Prabha, T. V., & Karipot, A. (2021). Impact of middle atmospheric humidity on boundary layer turbulence and clouds. *Journal of Atmospheric Solar-Terrestrial Physics, 215*, 105553. https://doi.org/10.1016/j.jastp.2021.105553

Meyers, M. P., DeMott, P. J., & Cotton, W. R. (1992). New primary ice-nucleation parameterizations in an explicit cloud model. *Journal of Applied Meteorology, 31*, 708–721. https://doi.org/10.1175/1520-0450(1992)031%3c0708:NPINPI%3e2.0.CO;2

Moisseev, D. N., Lautaportti, S., Tyynela, J., & Lim, S. (2015). Dual-polarization radar signatures in snowstorms: Role of snowflake aggregation. *Journal of Geophysical Research-Atmospheres, 120*(24), 12644–12665. https://doi.org/10.1002/2015JD023884

Morrison, H., van Lier-Walqui, M., Fridlind, A. M., Grabowski, W. W., Harrington, J. Y., Hoose, C., et al. (2020). Confronting the challenge of modeling cloud and precipitation microphysics. *Journal of Advances in Modeling Earth Systems, 12*, e2019MS001689. https://doi.org/10.1029/2019MS001689

Morwal, S. B., Maheskumar, R. S., Padma Kumari, B., Kulkarni, J. R., & Goswami, B. N. (2012). Cloud microphysical properties over Indian monsoon regions during CAIPEEX-2009. *Journal of Atmospheric Solar-Terrestrial Physics, 81–82*, 76–85.

Nair, S., Sanjay, J., Pandithurai, G., Maheskumar, R. S., & Kulkarni, J. R. (2012). On the parameterization of cloud droplet effective radius using CAIPEEX aircraft observations for warm clouds in India. *Atmospheric Research, 108*, 104–114.

National Research Council (NRC). (2005). *Radiative Forcing of Climate Change: Expanding the Concept and Addressing Uncertainties*. The National Academies Press.

Neelin, J. D., Peters, O., & Hales, K. (2009). The transition to strong convection. *Journal of Atmospheric Science, 66*, 2367–2384.

Nesbitt, S. W., Gochis, D. J., & Lang, T. J. (2008). The diurnal cycle of clouds and precipitation along the sierra madre occidental observed during NAME-2004: Implications for warm season precipitation estimation in complex terrain. *Journal of Hydrometeorology, 9*(4), 728–743. https://doi.org/10.1175/2008JHM939.1

Orville, H. D. (1965). A numerical study of the initiation of cumulus clouds over mountainous terrain. *Journal of the Atmospheric Sciences, 22*(6), 684–699. https://doi.org/10.1175/1520-0469(1965)022<0684:ANSOTI>2.0.CO;2.

Padmakumari, B., Maheskumar, R. S., Harikishan, G., Morwal, S. B., Prabha, T. V., & Kulkarni, J. R. (2013). In situ measurements of aerosol vertical and spatial distributions over continental India during the major drought year 2009. *Atmospheric Environment, 80*(2013), 107–121.

Pendergrass, A. G. (2020). Changing degree of convective organization as a mechanism for dynamic changes in extreme precipitation. *Current Climate Change Reports, 6*, 47–54. https://doi.org/10.1007/s40641-020-00157-9

Patade, S., Gayatri, K., Patade, S., Deshmukh, A., Dangat, P., Axisa, D., Fan, J. W., Pradeepkumar, P., & Prabha, T. V. (2019). Role of liquid phase in the development of ice phase in monsoon clouds: Aircraft observations and numerical simulations. *Atmospheric Research, 229*, 157–174. https://doi.org/10.1016/j.atmosres.2019.06.022

Patade, S., Nagare, B., Wagh, S., Maheskumar, R. S., Prabha, T. V., & Kumar, P. P. (2014). Deposition ice nuclei observations over the Indian region during CAIPEEX. *Atmospheric Research, 149*, 300–314.

Patade, S., Prabha, T. V., Axisa, D., Gayatri, K., & Heymsfield, A. (2015). Particle size distribution properties in mixed-phase monsoon clouds from in situ measurements during CAIPEEX. *Journal of Geophysical Research-Atmospheres, 120*(19), 10418–10440. https://doi.org/10.1002/2015JD023375

Patade, S., Shete, S., Malap, N., Gayatri, K., & Prabha, T. V. (2016). Observational and simulated cloud microphysical features of rain formation in the mixed phase clouds observed during CAIPEEX. *Atmospheric Research*. https://doi.org/10.1016/j.atmosres.2015.09.018

Patade, S., Waman, D., Deshmukh, A., Gupta, A. K., Jadav, A., Phillips, V. T. J., Bansemer, A., Carlin, J., & Ryzhkov, A. (2022). The influence of multiple groups of biological ice nucleating particles on microphysical properties of mixed-phase clouds observed during MC3E. *Atmospheric Chemistry and Physics, 22*(18), 12055–12075. https://doi.org/10.5194/acp-22-12055-2022

Phillips, V. T. J., DeMott, P. J., & Andronache, C. (2008). An empirical parameterization of heterogeneous ice nucleation for multiple chemical species of aerosol. *Journal of Atmospheric Science, 65*, 2757–2783.

Phillips, V. T. J., Demott, P. J., Andronache, C., Pratt, K. A., Prather, K. A., Subramanian, R., & Twohy, C. (2013). Improvements to an empirical parameterization of heterogeneous ice nucleation and its comparison with observations. *Journal of Atmospheric Science, 70*, 378–409. https://doi.org/10.1175/JAS-D-12-080.1

Prabha, V. T., Patade, S., Pandithurai, G., Khain, A., Axisa, D., Pradeep Kumar, P., Maheshkumar, R. S., Kulkarni, J. R., & Goswami, B. N. (2012). Spectral width of premonsoon and monsoon clouds over Indo-Gangetic valley during CAIPEEX. *Journal of Geophysical Research-Atmospheres, 117*(D20), https://doi.org/10.1029/2011JD016837.

Prabha, T. V., Khain, A., Maheshkumar, R. S., Pandithurai, G., Kulkarni, J. R., Konwar, M., & Goswami, B. N. (2011). Microphysics of premonsoon and monsoon clouds as seen from in situ measurements during the Cloud Aerosol Interaction and Precipitation Enhancement Experiment (CAIPEEX). *Journal of the Atmospheric Sciences, 68*(9), 1882–1901. https://doi.org/10.1175/2011JAS3707.1

Prabhakaran, T., Murugavel, P., Konwar, M., Malap, N., Gayatri, K., Dixit, S., ... & Nanjundiah, R. (2023). CAIPEEX - Indian cloud seeding scientific experiment, *Bulletin of the American Meteorological Society, 104*, E2095–E2120, https://doi.org/10.1175/BAMS-D-21-0291.1

Ramaswamy, V., Boucher, O., Haigh, J., et al. (2001). Radiative forcing of climate change. In: J. Houghton et al. (Eds.), *Climate Change, 2001: Contribution of Working Group I to the Third Assessment Report of IPCC* (pp. 349–416). Cambridge University Press. https://www.ipcc.ch/site/assets/uploads/2018/03/TAR-06.pdf.

Riehl, H., & Malkus, J. S. (1958). On the heat balance in the equatorial trough zone. *Geophysica, 6*, 503–538.

Robe, F. R., & Emanuel, K. A. (2001). The effect of vertical wind shear on radiative–convective equilibrium states. *Journal of the Atmospheric Sciences, 58*(11), 1427–1445. https://doi.org/10.1175/1520-0469(2001)058%3c1427:TEOVWS%3e2.0.CO;2

Romps, D. M. (2011). Response of tropical precipitation to global warming. *Journal of Atmospheric Science, 68*, 123–138. https://doi.org/10.1175/2010JAS3542.1

Rosenfeld, D. (1999). TRMM observed first direct evidence of smoke from forest fires inhibiting rainfall. *Geophysical Research Letters, 26*(20), 3105–3108. https://doi.org/10.1029/1999GL006066

Rosenfeld, D. (2000). Suppression of rain and snow by urban and industrial air pollution. *Science, 287*(5459), 1793–1796. https://doi.org/10.1126/science.287.5459.1793

Rosenfeld, D., & Woodley, W. L. (2000). Deep convective clouds with sustained supercooled liquid water down to _37.5_C. *Nature, 405*(6785), 440–441. https://doi.org/10.1038/35013030

Rosenfeld, D., Lohmann, U., Raga, G. B., O'Dowd, C. D., Kulmala, M., Fuzzi, S., Reissell, A., & Andreae, M. O. (2008). Flood or drought: How do aerosols affect precipitation? *Science, 321*(5894), 1309–1313. https://doi.org/10.1126/science.1160606

Rowe, A. K., & Houze, R. A., Jr. (2015). Cloud organization and growth during the transition from suppressed to active MJO conditions. *Journal of Geophysical Research Atmospheres, 120*, 10 324-10 350. https://doi.org/10.1002/2014JD022948

Rowe, A., Rutledge, S. A., & Lang, T. J. (2012). Investigation of microphysical processes occurring in organized convection during NAME. *Monthly Weather Review, 140*(7), 2168–2187. https://doi.org/10.1175/MWR-D-11-00124.1

Ryzhkov, A., Zhang, P., Reeves, H., Kumjian, M., Tschallener, T., Trömel, S., & Simmer, C. (2016). Quasi-vertical profiles-A new way to look at polarimetric radar data. *Journal of Atmospheric and Oceanic Technology, 33*(3), 551–562. https://doi.org/10.1175/JTECH-D-15-0020.1

Ryzhkov, A., & Zrnić, D. (1996). Assessment of rainfall measurement that uses specific differential phase. *Journal of Applied Meteorology and Climatology, 35*(11), 2080–2090. https://doi.org/10.1175/1520-0450(1996)035%3c2080:AORMTU%3e2.0.CO;2

Samanta, S., Murugavel, P., Gurnule, D., Rao, Y. J., Vivekanandan, J., & Prabha, T. V. (2021). The life cycle of a stationary cloud cluster during the Indian summer monsoon: A microphysical investigation using polarimetric C-band radar. *Monthly Weather Reviews, 149*(11), 3761–3780. https://doi.org/10.1175/MWR-D-20-0274.1

Samanta, S., Prabha, T. V., Murugavel, P., & Suneetha, P. (2023). Morphological and microphysical characteristics associated with the lifecycle of a stationary cloud cluster during the Indian Summer Monsoon: A comparative study with numerical simulations and radar observation. *Atmospheric Research, 281*(106464). https://doi.org/10.1016/j.atmosres.2022.106464.

Satoh, M., Matsugishi, S., Roh, W., Ikuta, Y., Kuba, N., Seiki, T., Hashino, T., & Okamoto, H. (2022). Evaluation of cloud and precipitation processes in regional and global models with ULTIMATE (ULTra-sIte for Measuring Atmosphere of Tokyo metropolitan Environment): A case study using the dual-polarization Doppler weather radars. *Progress in Earth and Planetary Science, 9*(1), 51. https://doi.org/10.1186/s40645-022-00511-5

Saunders, C. P. R., Bax-norman, H., Emersic, C., Avila, E. E., & Castellano, N. E. (2006). Laboratory studies of the effect of cloud conditions on graupel/crystal charge transfer in thunderstorm electrification. *Quarterly Journal of Royal Meteorological Society, 132*, 2653–2673. https://doi.org/10.1256/qj.05.218

Schneebeli, M., Dawes, N., Lehning, M., & Berne, A. (2013). High-resolution vertical profiles of X-band polarimetric radar observables during snowfall in the Swiss Alps. *Journal of Applied Meteorology and Climatology, 52*(2), 378–394. https://doi.org/10.1175/JAMC-D-12-015.1

Seifert, A., & Beheng, K. D. (2006). A two-moment cloud microphysics parameterization for mixed-phase clouds. Part 2: Maritime vs. continental deep convective storms. *Meteorology and Atmospheric Physics, 92*, 67–82. https://doi.org/10.1007/s00703-005-0113-3

Seiki, T., Roh, W., & Satoh, M. (2022). Cloud microphysics in global cloud resolving models. *Atmosphere-Ocean, 60*(3–4), 477–505. https://doi.org/10.1080/07055900.2022.2075310

Singh, M. S., & O'Gorman, P. A. (2014). Influence of microphysics on the scaling of precipitation extremes with temperature. *Geophysical Research Letters, 41*(16), 6037–6044. https://doi.org/10.1002/2014GL061222

Sreenath, A. V., Malap, N., Abhilash, S., Kulkarni, G. U., & Prabhakaran, T. (2021). Precipitation processes over Indian region under different environmental conditions from in situ measurements. *Atmospheric Research, 262*, 105775. https://doi.org/10.1016/j.atmosres.2021.105775

Srivastava, R. C. (1987). A model of intense downdrafts driven by the melting and evaporation of precipitation. *Journal of the Atmospheric Sciences, 44*(13), 1752–1774. https://doi.org/10.1175/1520-0469(1987)044%3c1752:AMOIDD%3e2.0.CO;2

Storer, R. L., & van den Heever, S. C. (2013). Microphysical processes evident in aerosol forcing of tropical deep convective clouds. *Journal of the Atmospheric Sciences, 70*(2), 430–446. https://doi.org/10.1175/JAS-D-12-076.1

Tang, S., et al. (2022). Long-term single-column model intercomparison of diurnal cycle of precipitation over midlatitude and tropical land. *Quarterly Journal of the Royal Meteorological Society, 148*(743), 641–669. https://doi.org/10.1002/qj.4222

Taylor, C., Gounou, A., Guichard, F., Harris, P. P., Ellis, R. J., Couvreux, F., & De Kauwe, M. (2011). Frequency of Sahelian storm initiation enhanced over mesoscale soil-moisture patterns. *Nature Geoscience, 4*, 430–433. https://www.nature.com/articles/ngeo1173.

Teller, A. & Levin Z. (2006). The effects of aerosols on precipitation and dimensions of subtropical clouds: a sensitivity study using a numerical cloud model. *Atmospheric Chemistry and Physics, 6*(67). https://doi.org/10.5194/ACP-6-67-2006.

Thomas, L., Malap, N., Grabowski, W. W., Dani, K., & Prabha, T. V. (2018). Convective environment in pre-monsoon and monsoon conditions over the Indian subcontinent: The impact of surface forcing. *Atmospheric Chemistry and Physics, 18*(10), 7473–7488. https://doi.org/10.5194/acp-18-7473-2018.

Tobin, I., Bony, S., & Roca, R. (2012). Observational evidence for relationships between the degree of aggregation of deep convection, water vapor, surface fluxes, and radiation. *Journal of Climate, 25*(20), 6885–6904. https://doi.org/10.1175/jcli-d-11-00258.1

Tompkins, A. M. (2001). Organization of tropical convection in low vertical wind shears: The role of water vapor. *Journal of the Atmospheric Sciences, 58*(6), 529–545. https://doi.org/10.1175/1520-0469(2001)058%3c0529:OOTCIL%3e2.0.CO;2

Tompkins, A. M., & Semie, A. G. (2017). Organization of tropical convection in low vertical wind shears: Role of updraft entrainment. *Journal of Advances in Modeling Earth Systems, 9*(2), 1046–1068. https://doi.org/10.1002/2016ms000802

Twomey, S. (1977). The influence of pollution on the shortwave albedo of clouds. *Atmospheric Environment, 8*(12), 1251–1256. https://doi.org/10.1016/0004-6981(74)90004-3

Van den Heever, Carrio S. C., G., Cotton W. R., DeMott P. J., Prenni A. J. (2006). Impacts of nucleating aerosol on Florida convection. Part I: Mesoscale simulations. *Journal of the Atmospheric Sciences, 63*, 1752–1775. https://vandenheever.atmos.colostate.edu/vdhpage/papers/vandenHeever-etal-JAS-2006.pdf.

Varble, A. C., et al. (2021). Utilizing a storm-generating hotspot to study convective cloud transitions: The CACTI experiment. *Bulletin of the American Meteorological Society, 102*, E1597–E1620. https://doi.org/10.1175/BAMS-D-20-0030.1

Varghese, M., Prabha, T. V., Patade, S., Gayatri, K., Safai, P. D., & Axisa, D. (2021). Characteristics of CCN activation and cloud microphysics over the east coast of India during the Northeast Monsoon onset. *Atmospheric Research, 257*, 105589. https://doi.org/10.1016/j.atmosres.2021.105589

Vivekanandan, J., Zrnic, D. S., Ellis, S. M., Oye, R., Ryzhkov, A. V., & Straka, J. (1999). Cloud microphysics retrieval using S-Band dual-polarization radar measurements. *Bulletin of the American Meteorological Society, 80*(3), 381–388. https://doi.org/10.1175/1520-0477(1999)080%3c0381:CMRUSB%3e2.0.CO;2

Vogel, J. M., & Fabry, F. (2018). Contrasting polarimetric observations of stratiform riming and nonriming events. *Journal of Applied Meteorology and Climatology, 57*(2), 457–476. https://doi.org/10.1175/JAMC-D-16-0370.1

Waman, D., Patade, S., Jadav, A., Deshmukh, A., Gupta, A. K., Phillips, V. T. J., Bansemer, A., & DeMott, P. J. (2022). Dependencies of four mechanisms of secondary ice production on cloud-top temperature in a continental convective storm. *Journal of the Atmospheric Sciences, 79*(12), 3375–3404. https://doi.org/10.1175/JAS-D-21-0278.1

Wang, C., (2005). A model study of the response of tropical deep convection to the increase of CCN concentration. 1. Dynamics and microphysics. *Journal of Geophysical Research-Atmospheres, 110*(D21). https://doi.org/10.1029/2004JD005720.

Wegener, A. (1911). *Thermodynamik der Atmosphäre* (331pp.). Leipzig.

White, B. A., Buchanan, A. M., Birch, C. E., Stier, P., & Pearson, K. J. (2018). Quantifying the effects of horizontal grid length and parameterized convection on the degree of convective organization using a metric of the potential for convective interaction. *Journal of the Atmospheric Sciences, 75*(2), 425–450. https://doi.org/10.1175/jas-d-16-0307.1

Wing, A. A., Emanuel, K., Holloway, C. E., & Muller, C. (2017). Convective self-aggregation in numerical simulations: A review. *Surveys in Geophysics, 38*(6), 1173–1197.

Wing, A. A., Reed, K. A., Satoh, M., Stevens, B., Bony, S., & Ohno, T. (2018). Radiative-convective equilibrium model intercomparison project. *Geoscientific Model Development, 11*, 793–813. https://doi.org/10.5194/gmd-11-793-2018

Yuan, J., & Houze, R. A., Jr. (2010). Global variability of mesoscale convective system anvil structure from A-train satellite data. *Journal of Climate, 23*, 5864–5888. https://doi.org/10.1175/2010JCLI3671.1

Zhang, Y., & Klein, S. A. (2010). Mechanisms affecting the transition from shallow to deep convection over land: Inferences from observations of diurnal cycle collected at the ARM southern great plains site. *Journal of Atmospheric Science, 67*, 2943–2959. https://doi.org/10.1175/2010JAS3366.1

Zhang, Y., & Klein, S. A. (2013). Factors controlling the vertical extent of fair-weather shallow cumulus clouds over land: Investigation of diurnal-cycle observations collected at the ARM southern great plains site. *Journal of Atmospheric Science, 70*, 1297–1315. https://doi.org/10.1175/JAS-D-12-0131.1

Zhao, C., Yang, Y., Chi, Y., Sun, Y., Zhao, X., Letu, H., & Xia, Y. (2023). Recent progress in cloud physics and associated radiative effects in China from 2016 to 2022. *Atmospheric Research, 293*, 106899. https://doi.org/10.1016/j.atmosres.2023.106899

Zhao, X., & Liu, X. (2021). Global importance of secondary ice production. *Geophysical Research Letters, 48*(11), https://doi.org/10.1029/2021GL092581.

Zhuang, Y., Fu, R., Marengo, J. A., & Wang, H. (2017). Seasonal variation of shallow-to-deep convection transition and its link to the environmental conditions over the Central Amazon. *Journal of Geophysical Research: Atmospheres, 122*(5), 2649–2666. https://doi.org/10.1002/2016JD025993

Zrnić, D. S., & Ryzhkov, A. (1996). Advantages of rain measurements using specific differential phase. *Journal of Atmospheric and Oceanic Technology, 13*(2), 454–464. https://doi.org/10.1175/1520-0426(1996)013%3c0454:AORMUS%3e2.0.CO;2

Electrical Characteristics of Thunderstorms Over India

V. Gopalakrishnan, S. D. Pawar, and Konatham Prasanna

Abstract Electrical, microphysical, and dynamic characteristics in a cloud occur almost simultaneously inside thunderclouds and interact with each other and understanding this interaction is essential to recognize several phenomena such as cloud bursts, rain gush, and tornadoes and better prediction of such events. Many fields and laboratory experiments have been conducted in the USA and Europe to study the electrical characteristics associated with thunderstorm. However, very few studies have been conducted in India to understand the electrical properties of thunderclouds and their interaction with dynamics and micro-physics of cloud. Studies conducted over Pune and other locations in India have shown that some thunderstorms forming over India have inverted polarity charge structure. This inverted polarity charge structure can be attributed to increased aerosol concentration or high liquid water content in mixing layer. Laboratory and field experiments also have shown that electric forces in thundercloud can significantly affect rain formation processes in thunderclouds. Here, we review some important studies conducted to understand the electrical characteristics of thunderstorms over India.

1 Introduction

A vast majority of clouds develop and dissipate in the Earth's atmosphere without producing precipitation or lightning. However, they can still be weakly electrified. As the updraft increases to a few tens of centimeters per second, the electrification intensifies in clouds. Clouds get strongly electrified when the updraft increases to several meters per second and they rapidly grow in both vertical and horizontal directions and develop precipitation. Strong convective cumulonimbus clouds are the source of most of the lightning in the Earth's atmosphere. Satellite observations show that occurrence of thunderclouds and the lightning discharges are far more common over land than over oceans (Orville & Spencer, 1979). Estimates made on

V. Gopalakrishnan · S. D. Pawar (✉) · K. Prasanna
Indian Institute of Tropical Meteorology, Pune, India
e-mail: pawar@tropmet.res.in

S. Das and W-K. Tao (eds.), *Severe Storms*,
https://doi.org/10.1007/978-981-97-7075-5_7

the basis of climatological observations, satellite data and inferences from electrical observations indicate that there are between 1000 and 2000 storms active at the any given time, covering 1–10% of the earth's surface (Ccopa et al., 2021).

Inspite of a large number of theoretical, laboratory and field studies conducted to investigate the electrical nature of thunderstorms, the mechanism causing thunderstorm electrification and charge distribution yet remain an age-old unresolved problem. The main reason for such a status is because thunderclouds are large, complex and short-lived phenomena, and to understand their electric structure, they should be studied in detail for their whole life time (Krehbiel, 1986). Many meteorological parameters such as dew-point depression, air temperature, vertical stability of atmosphere, liquid water and ice content and aerosol concentration affect the electrification processes and charge structure inside thunderclouds (Williams, 1989; Williams et al., 1989, 1991, 1992, 2002). Increased ionization rate in the lower atmosphere also has been found to be related with lightning frequency of thunderstorm (Israelsson et al., 1986). However, still there is a lack of understanding on the role of electrical forces in modifying the dynamics and microphysics of the thundercloud.

Wilson (1920) first proposed a positive dipole structure i.e., positive charge in the upper part and negative charge in the lower part of a thundercloud. Wilson's (1920) conclusion was based on his observations of electric field at ground surface below and at different distances from thundercloud. On the contrary, Simpson (1927) proposed a negative dipole structure, from his measurements of the charge on rain beneath the central portion of thundercloud, which he found positive most of the time. Subsequent measurements of charge and electric field inside thunderclouds by Simpson and Scrase (1964) and Robinson and Simpson (1941) confirmed that the upper positive and the main negative were the dominant charge regions. In addition, however, most of the thunderstorms showed the existence of a positive charge pocket below the main negative charge region. Williams (1989) has critically reviewed many measurements of the electric field and charge on precipitation particles made beneath or inside thunderclouds and concluded that a tri-pole structure is the correct representation of thunderstorm charge structure. The charges distributed inside a thundercloud are the main cause of lightning. However, the charges outside thundercloud such as corona charge in the sub-cloud layer below thundercloud and screening layer above cloud top certainly play an important role in cloud electrification; their role in cloud electrification process is not fully understood.

Recent observations of Wiens et al. (2005) and Rust et al. (2005) during Severe Thunderstorm Electrification and Precipitation Study (STEPS) suggest that some of the severe thunderstorms can have inverted polarity charge structure. In one super cell thunderstorm in Northwestern Kansas during STEPS, Wiens et al. (2005) observed that the charge structure for the most time of the storm was dominated by midlevel region of positive charge. In three storms that occurred during STEPS, Rust et al. (2005) found the positive and negative charge regions at altitudes where respectively negative and positive charge would be found in normal polarity storms. Thus, they concluded that these storms had anomalous and inverted polarity electrical structure.

Evolutions of electrical, microphysical, and dynamic characteristics in a cloud occur almost simultaneously in a cloud and influence each other. For example, the shape, growth, breakup, evaporation characteristics of water drops are strongly influenced if the drops are charged or they are falling through electric fields (e.g. Taylor, 1964, Ausman and Brook, 1967, Richards and Dawson, 1971 et al., 1985; Kamra and Ahire, 1989; Kamra et al., 1991; Bhalwankar Rohini et al., 2004). Laboratory experiments of Kamra et al. (1993) show that water drops can produce corona in much lower values of electric field if the electric field through which they are falling are horizontal rather than vertical. Theoretical calculations have confirmed such elongation of the drop along the direction of electric field. Kamra (1982, 1985) has shown that electrification in storms can modify the distribution of drops and thus create accumulation zones of liquid water content inside storms. Recent observations of Bhalwankar and Kamra (2007) show that water drops falling in horizontal electric field can breakup more readily than those falling in vertical electric field. This will cause wider size distribution and thus can lead to larger growth rates of water drops in those regions of cloud where the direction of electric field is vertical rather than horizontal.

Theoretical calculations of Kamra (1982) and Bala and Kamra (1991) show that storm electrification can modify the time and intensity of the occurrence of downdraft. Vonnegut (1960) has suggested that intense activity of lightning discharges frequently reported to occur inside the tornado funnel can be driven by strong winds. This was confirmed by a simulation experiment conducted by Vonnegut (1960).

Storm electrification has been reported to strongly enhance the collision and coalescence characteristics of cloud drops. Laboratory experiments of Telford et al. (1955), Woods and Mason (1965), Beard and Pruppacher (1968, 1971) and theoretical models of Beard and Grover (1974), Grover and Pruppacher (1985) support such enhancements. Observations of rain gush phenomenon below thunderclouds are one of the strongest manifestations of such an enhancement in the collision and coalescence efficiencies of water drops (Moore et al., 1961). Presence of a Doppler radar or a dense network of rain gauge stations and a network for the lightning location are ideally suited to study the phenomenon of rain-gush. A better understanding of the rain gush phenomenon can certainly help in evaluating the efficiency of electrical forces in weather modification efforts.

Lightning discharges are known to be the source for the production of NOx in the lightning channels. Global production of NOx by lightning discharges has been estimated to be 0.8–8 tg/yr (Borucbi and Chameides, 1984) and is considered to cause substantial destruction of ozone. Production of ozone associated with point discharge current that occurs at sharp points raised above the ground surface, below thunderstorms can add substantially to the ozone budget in the lower atmosphere.

Here, we briefly present our observations of electrical characteristics of thunderstorms observed mainly over Pune and laboratory experiments conducted at Indian Institute of Tropical Meteorology (IITM, Pune) to understand the role of electrical forces on rain characteristics.

2 Climatology of Thunderstorms Over India

Because of its position close to the equator, most parts of India experience thunderstorm activity some or the other time of a year. A study by Tyagi (2007) on climatology of thunderstorms over Indian region using 30 years IMD data (1950–1980) shows that North-eastern part of India, some part of Kerala, some part of Jammu-Kashmir and Bangladesh experience thunderstorm for more than 80 days per year. More recently Murugavel et al. (2012) studied the climatology of lightning activity over Indian region during different seasons using TRMM lightning data for 18 years. According to this study, during pre-monsoon lightning activity is highest over North-eastern part of India and Bangladesh. Further, the Foot hills of Himalayas, east coast of India and Kerala coast also show very high lightning activity during this season. During monsoon, some parts of Jammu-Kashmir and adjoining parts of Pakistan show very high lightning activity. The foot hills of Himalayas and east coast of India also show high lightning activity during this season. The north-eastern part of India is known to experience very severe thunderstorms during the pre-monsoon season, locally known as 'Nor'wester'. The airflow over Gangetic plains and the adjoining areas has a shallow layer of moist southerlies/southwesterlies from the Bay of Bengal near the ground and dry westerlies of continental origin above. Availability of high Convective Available Potential Energy (CAPE) is one of the favorable conditions for outbreak of severe Nor'wester. Using monthly data of thunderstorms from a large number of stations Manohar et al. (1999) have studied the latitudinal variation of thunderstorms during the pre-monsoon (March–May) and monsoon season (June-September) over India. They examined the seasonal variation of monthly thunderstorm days in every 5° interval latitude-belts from 8°N to 30°N and observed that the monthly thunderstorm days show systematic changes in their signals of semi-annual oscillation. These changes are a function of latitude and season and appear to be consistent with the seasonal migration of the Inter tropical Convergence Zone and solar heating of the Indian landmass.

Figure 1 shows the seasonal as well as annual climatology of lightning distribution over India for the period 1996–2013. TRMM satellite data was used for to derive the climatology. As seen from the Figure maximum lightning is seen over north-western parts of India (>50 flashes/km^2/year) and the other over north-eastern parts of India (>40 flashes/km^2/year) and also along east coast of India. Some parts of Kerala also experience high lightning during pre-monsoon period. The third hotspot region is over southern Peninsular India with a maximum concentration over Kerala during the pre-monsoon season. Kamra and Ramesh Kumar (2021) report that the characteristics of lightning associated with the pre-monsoon seasons over northeast India differ from the monsoon season lightning conditions. The high lightning activity over northwest India during the monsoon period is attributed to the high thunderstorm activity associated with intense convergence in the boundary layer, cyclonic vorticity above the boundary layer, and deep organized convection over this region. As stated above, during pre-monsoon period very high number of lightnings are observed due to excessive surface heating and associated convection.

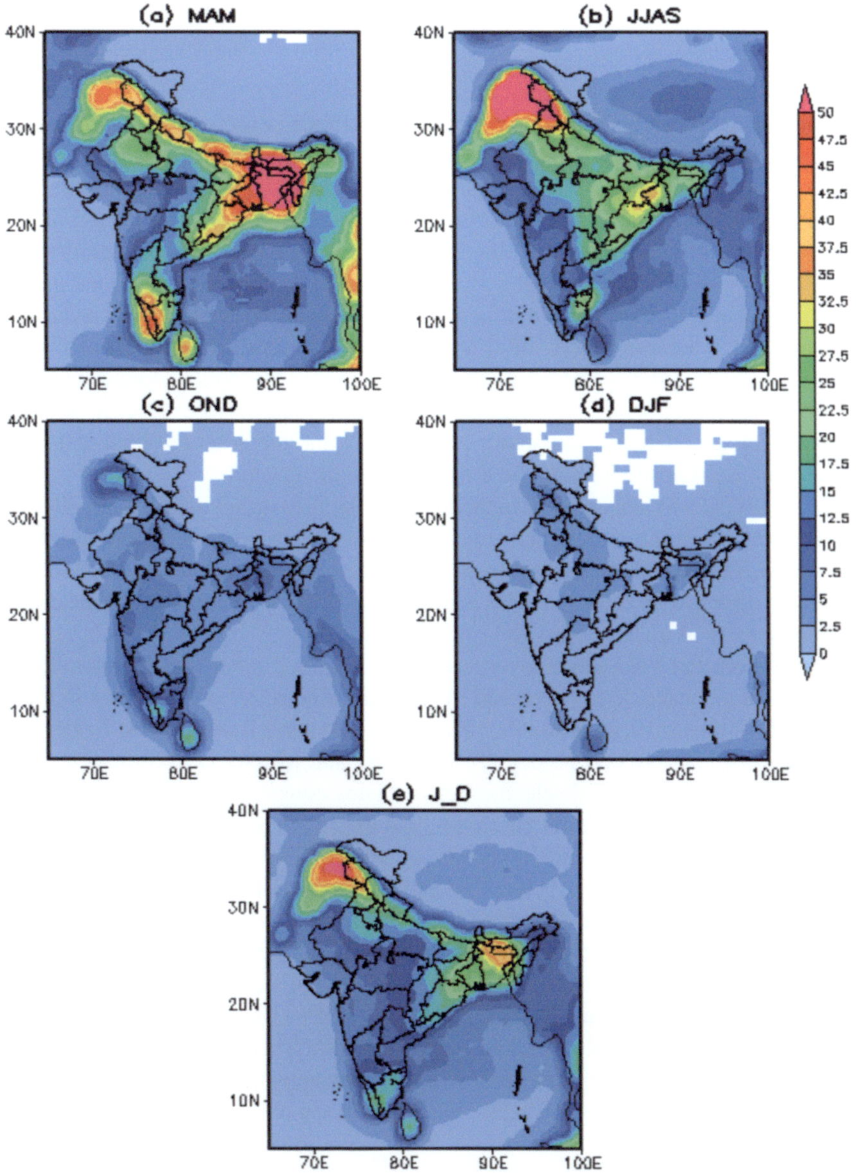

Fig. 1 Climatological mean of lightning flash rate density (flashes/Km^2/year) over India during the period 1996–2013 for i(a) MAM, ii(a) JJAS, iii(a) OND, iv(a) DJF, and v(a) J-D (annual)

3 Charge Structure of Thunderstorms Over India

Electric field and Maxwell current measurements:

Indigenously designed and fabricated a.c field mill is used for measurement of vertical component of the atmospheric electric field at the Earth's surface. The field-mill consists of two stators sectors that are periodically shielded and exposed to the atmospheric electric field with rotor sectors. The rotor sectors are erected on the shaft of an a. c. synchronous motor. The rotor is grounded using a mercury cup at the other end of motor. The stators are connected to the inverting inputs of two operational amplifiers. The magnitude of the charge induced on the stators is directly proportional to the intensity of the atmospheric electric field. Field-mill is calibrated by directly placing it in a uniform electric field produced by applying known voltages across two parallel plates separated by a distance. It can measure electric fields of upto $\pm$ 12 kV m^{-1} with a sensitivity of 15 V m^{-1} and has response time of 10 ms. Observations are made with the field-mill flush to the ground.

Maxwell current density is measured directly by measuring the current collected by an isolated 1 m^2 aluminum plate mounted flush with the ground on four porcelain insulators. The plate is connected to an electrometer with a resistor of 1000 MΩ in parallel with a 100 pF capacitor. This instrument is designed to measure the Maxwell current density of up to $\pm$ 5 nA/m^2.

Surface electric field and recovery curves of the electric field between two lightning discharges have often been used to study the electrification processes in thunderclouds (Wilson, 1920; Wormell, 1953; Livingston and Krider, 1978; Krider and Musser, 1982; Devere and Krider, 1991). Krider and Musser (1982) observed that the rate of change of surface electric field remains constant between two lightning discharges even though the electric field, both at ground and aloft, undergoes large changes in amplitude and sometimes even in polarity. These observations lead them to conclude that the charge generating processes operating in thunderclouds are field-independent. Pawar and Kamra (2004, 2007, 2009) also have used the surface electric field and the Maxwell current density measured below thunderclouds at Pune (India) to study the electrical characteristics of thunderstorms occurring in that region. Studies by Pawar and Kamra (2004, 2007, 2009) have clearly shown that the surface measurements of electric field and Maxwell current density can be effectually used to study the charge structure and charging mechanisms inside thundercloud. A schematic diagram of charge structure that is generally observed in normal thunderstorms is shown in Fig. 2. As shown in Fig. 2, in a normal thundercloud, positive charge is observed in the upper portion of cloud and negative charge below it. Small pockets of positive charge called as lower positive charge centers (LPCC) also have been observed in many earlier studies.

In India, earliest measurements of electric field were made by Banerjee (1930) at Colaba observatory on the west coast of India by using water dropper method. His observations, made during 18 thunderstorms, showed that most of these thunderclouds were having negative charge in the front part, positive charge in the center

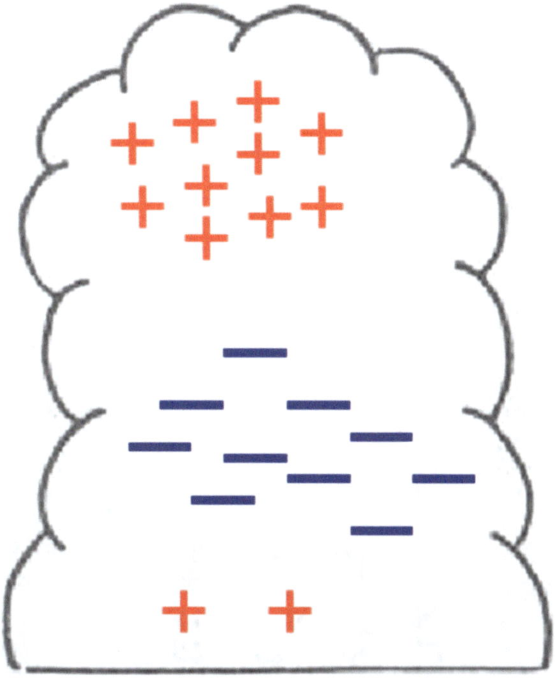

Fig. 2 Schematic diagram of charge structure inside normal thunderstorm

part and negative again in the rear part. His observations of electric field during pre-monsoon seasons showed extensive regions of positive charge in the lower portions of thunderstorms. (Banerji, 1932, 32). Based on the observations of electric field made beneath thunderclouds, Pawar and Kamra (2002) reported that the thunderstorms forming over Pune produce positive electric field changes after lightning discharges. This indicates the removal of positive charge from overhead; and recovery curves of electric field between two lightning discharges show buildup of positive charge inside the cloud. From these observations Pawar and Kamra (2002, 2009) suggested that some thunderstorms forming over Pune have strong and widespread lower positive charge centers. Subsequent measurements by Pawar et al. (2017) over Pune, Pawar et al. (2014) and Gopalakrishnan et al. (2011) over Kharagpur, Pawar et al. (2010) over Guwahati has clearly shown that strong and widespread positive charge exists in some thunderstorms forming over some parts of India..

Recovery curves of electric field between two lightning discharges and electric field changes induced by lightning discharges are used to identify the charge structure of the thunderstorm. Electric field changes observed during a normal thunderstorm with positive dipole (positive charge in the upper portion of cloud and negative charge below it) and a thunderstorm with strong positive charge region in the lower portion are illustrated in the Fig. 3. For a normal thunderstorm, all electric field changes are negative (removing negative charge from overhead) and recovery curves show building up of negative charge overhead; whereas, electric field changes and recovery

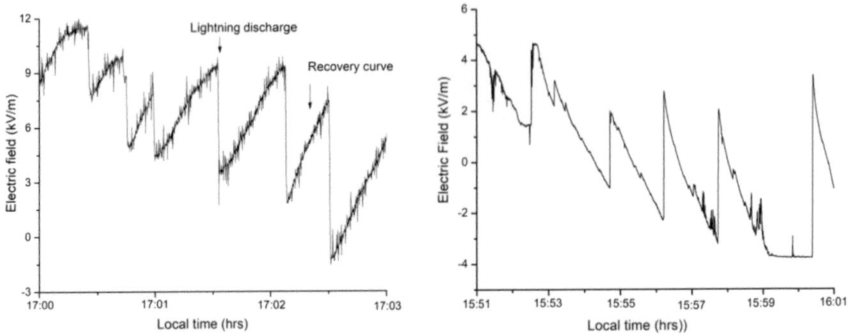

Fig. 3 Electric field records of two thunderstorms observed at Pune plotted on extended time scale **a** for inverted polarity charge structure and **b** for normal polarity charge structure

curves are opposite, indicating the presence of a strong positive charge region in the lower portion of the cloud.

From the measurements of surface electric field and Maxwell current density below thunderstorms over Pune, Pawar and Kamra (2009) proposed three type of charge structures over this region. Figure 4 shows the charge structure proposed by Pawar and Karma (2009). Some thunderstorms get formed with normal polarity or positive dipole type charge structure (positive charge in the upper portion and negative charge below it) and some thunderstorms get formed with tripole structure, having negative charge in the middle portion and positive charge above and below it. However, thunderstorms with negative dipole or strong positive charge in the lower portion of cloud and negative charge above it are also observed. These thunderstorms with negative dipole are termed as inverted polarity charge structures.

Thunderstorms with inverted polarity charge structures have also been reported by Gopalakrishnan et al. (2011) and Pawar et al. (2014) in the eastern part of India (Khargapur). North-eastern states are very much prone to experience severe thunderstorms, locally known as *'kalbaisakki'*, during pre-monsoon season. Measurements made at Guwahati reveal that these severe thunderstorms exhibit short active phases. During this period, lightning frequency jumps to even 60–80 flashes per second (Fig. 5). These observations are in sharp contrast to thunderstorms over Pune that produce only 10–15 flashes per 5 min. Observations show that thunderstorms over eastern parts of India generally exhibit high lightning flash rates.

A study by Gangane et al. (2023) has shown that during summer, over North and North western parts of India, thunderstorms associated with dust storms usually show a high percentage of positive CG discharges. Another study by Ghosh et al. (2023), using ground-based observation of Lightning Location Network (LLN) over the Indian region suggests that during the local break period in summer monsoon seasons, CG lightning of positive polarity increases dramatically relative to the pre-monsoon days. They have suggested that in the thunderstorms formed in the wet season, the non-inductive charge separation happens in the presence of larger mean cloud liquid water in the mixed phase of storms. This facilitates the positive charging

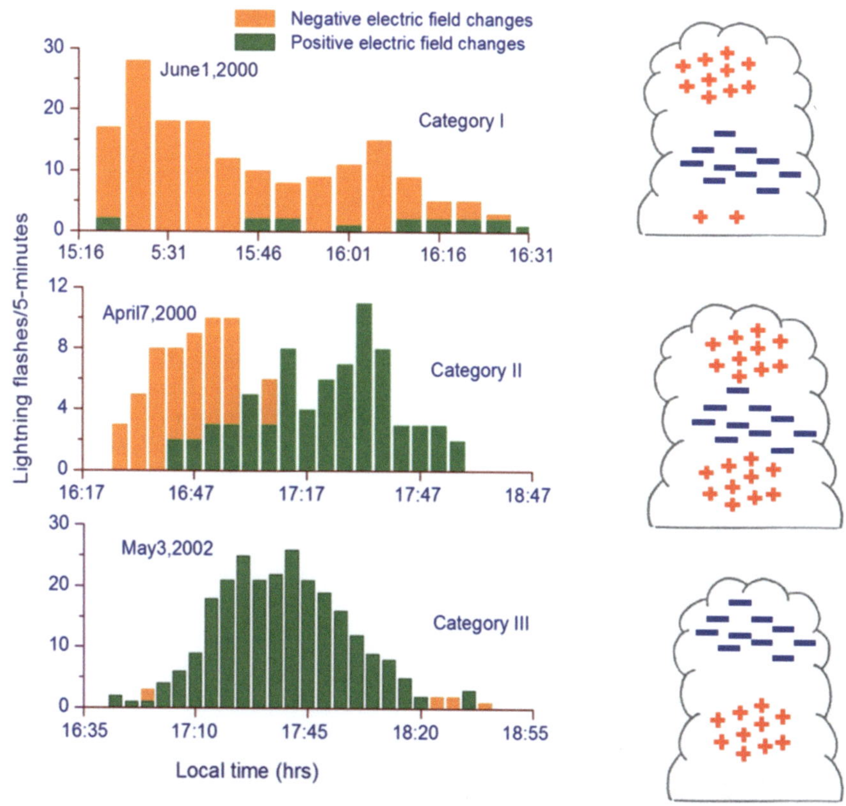

Fig. 4 Number of lightning-induced field changes of positive and negative polarities occurring in a period of every 5-min intervals along with possible charge structure in threethunderstorms storms (Adapted from Pawar & Kamra, 2009)

of graupel particles, thereby producing a large amount of positive CG lightning. This observation is found to be consistent with the laboratory results of Takahasi (1978).

Williams et al. (2005) hypothesized that the inverted polarity charge structure in thunderstorms is the result of superlative liquid water content in the mixed phase region. Environmental conditions such as large instability, atmosphere drier than normal, higher than normal cloud base height, and increased aerosol concentrations can transport higher liquid water content in the mixed phase region and therefore are considered as favorable environmental conditions for the formation of inverted polarity thunderstorms. However, many observations clearly show that all the thunderstorms fulfilling all the above conditions do not form with inverted polarity charge structure. Some recent observations show that some small thunderstorms, without strong updraft, can have inverted polarity charge structures (Qie et al., 2005; Pawar and Kamra, 2009). In organized convection such as supercells and mesoscale convective systems (MCS), many studies have reported multiple charge regions inside and outside of the updraft regions (Stolzenburg et al., 1998; Lang et al., 2000;

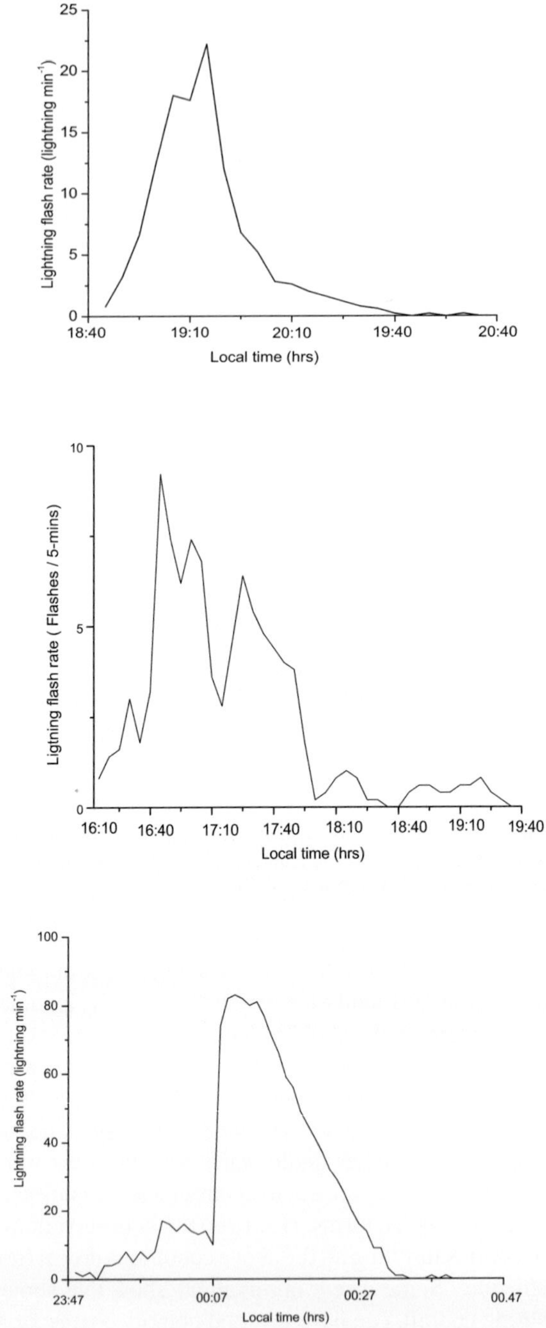

Fig. 5 Electric field records of three thunderstorms observed at **a** Kharagpur on May 11, 2009 **b** Pune on May 22, 2008 and **c** Guwahati on April, 20, 2007

Bruning et al., 2010). Pawar et al. (2011, 2017) attributed the inverted polarity charge structures observed over Indian region to increased aerosol concentration.

4 Laboratory Studies to Investigate the Effect of Electric Forces on Rain Formation Processes in Thunderclouds

A water drop falling at its terminal velocity experiences many forces such as force due to surface tension, hydrostatic and aerodynamic. The balance between them determines the shape and stability of the drop. Water drops, that are larger than a critical size, are hydro-dynamically unstable and break into smaller droplets. Kamra et al. (1991) developed a vertical wind tunnel to study the effect of electrical forces on rain drop characteristics (Fig. 6). Raindrop size distribution (RSD) is mainly determined by the breakup processes and limits the maximum size of the raindrops. Laboratory simulation experiments performed by Kamra et al., (1991) using a vertical wind tunnel show water drops in free-fall may grow up to around 6 mm diameter before they spontaneously break up depending on the level of turbulence in different wind tunnels. Sometimes the collision of two drops produces a large drop due to coalescence. Kamra et al. (1993), Bhalwankar and Kamra (2007) by wind tunnel experiments have shown that a strong electrical force acting on the drops located in high-ambient electric field regions of a thundercloud is a possible factor that effects the drop breakup. Laboratory experiments by Richards and Dawson (1971), Levine (1971), Rasmussen et al., (1985), Chuang and Beard (1990), Coquillat et al. (2003) also have shown that electric field plays important role in determining the DSD in thunderclouds. Strong electric fields that exist in thunderclouds may even cause deformed water drops to produce corona discharge. Many studies consider direction of electric field inside the thunderstorm to be vertical (Fitzerald & Byers, 1962; Marshall & Rust, 1991); however, inclined electric field is also reported in storms with large stratified regions (Krehbiel et al., 2000). Many in-cloud measurements show existence of horizontal electric field also. Maximum measured electric field of 430 kVm^{-1} in thunder clouds was observed to be in horizontal direction (Winn et al., 1974).

Kamra et al. (1993) and Bhalwankar and Kamra (2007) conducted extensive studies in wind-tunnel to study the effect of electric field on breakup of water drops under various conditions.

Bhalwankar and Karma (2007) observed that the electric field in thunderclouds, vertical or horizontal, impacts the distortion of the drop that results into the breakup and initiation of corona from raindrops. The freely suspended drops of various sizes in a vertical wind tunnel and distortion characteristics are studied when electric field 3 $kVcm^{-1}$ is applied in both vertical as well as horizontal direction. The changes in drops are continuously photographed using a movie camera. The changes observed in size and shapes of the drops when electric field is applied are shown in Fig. 7. As shown in this figure, as the size of the suspended drop increases, distortion also

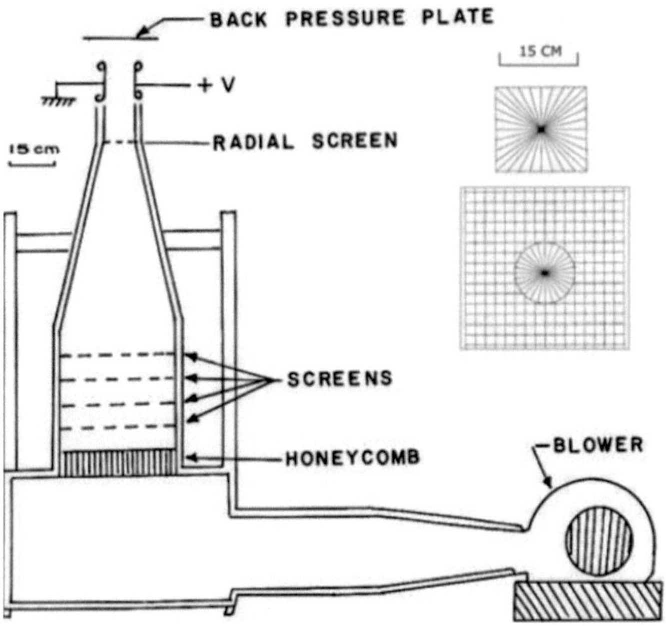

Fig. 6 A schematic drawing of a vertical wind tunnel and the cross-wire screen/electrode used to create a velocity well (Adapted from Bhalwankar & Kamra, 2007)

increases; the drops elongate in the direction of applied electric field and distortion is more when electric field is applied in the horizontal direction. It is observed that the drops break up more readily if the applied electric field is in horizontal direction rather than vertical. Therefore, this study suggests that the size distribution of drops is expected to be wider, and therefore the drop growth is likely to be faster in those regions of the cloud where vertical rather than horizontal electric fields prevail.

Thunderstorms which develop over big cities in a polluted environments may contain drops that are polluted with chemical impurities. As electrical discharges occurring on the raindrop's surface generally initiate the propagation of lightning discharges in thunderclouds, an experiment is performed by Bhalwankar and Kamra to estimate the effect of polluted water drops on the shape, distortion, and oscillation of the drops. Distilled as well as polluted water drops subjected to horizontal electric fields were suspended in a small vertical wind tunnel and shape characteristics were monitored from a movie camera. Water drops polluted with sulphate and/or nitrate salts exhibited lesser oscillation when suspended with zero electric field. However, oblateness is more for polluted drops subjected to horizontal electric field and it is observed that degree of distortion increases with the increase in electric field. Higher distortion of polluted drops could lead to trigger and propagate a discharge in lower electric fields, significantly contributing to the enhancement of lightning. Results suggest a considerable difference in distortion, breakup and corona emission characteristics of drops in polluted clouds and unpolluted clouds. They also observed that the

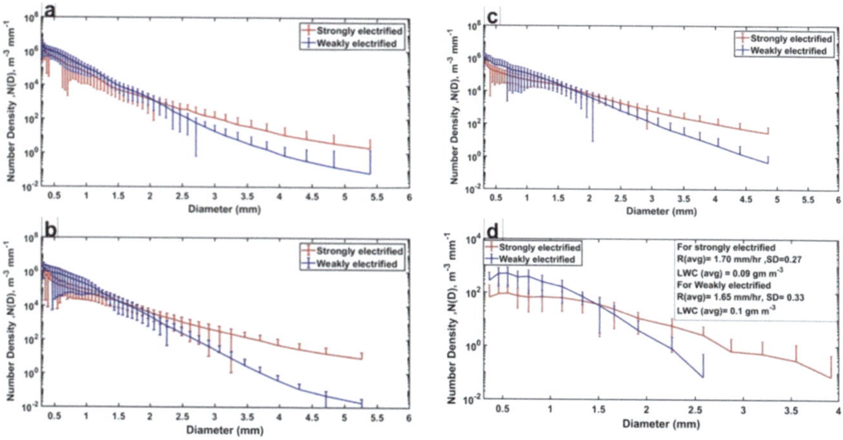

Fig. 7 Composite raindrop size distribution (six events in each composite) at selected altitudes for strongly electrified and weakly electrified clouds as observed by MRR at **a** 2,400 m, **b** 1,200 m, **c** 600 m, and **d** at surface observed by JW disdrometer (From Mudiar et al., 2018)

threshold required for sustained discharge propagation is reduced in polluted clouds. These factors are likely to modify the lightning activity in polluted environments on a regional or even on a global scale. Hence, these results are particularly important in the current context of increasing pollution over many cities and occurrence of severe thunderstorms with higher lightning frequency.

5 Observational Evidences of Effect of Electric Forces on Cloud Microphysics

Even though many laboratory experiments and theoretical studies have confirmed the role of electric field in modifying the raindrop size distribution (RSD),there are very few verified attempts to quantify the same by observation in tropical clouds. None of the earlier studies presented any evidence for electrical forces acting inside the thundercloud modifying microphysical characteristics. Studies by Mudiar et al. (2018) provide such observational evidence for modification in microphysical properties of thunderclouds by electrical field. They have examined therain-drop size distribution (RSD) of strongly electrified and weakly electrified stratiform rain events at different levels using micro rain radar and at surface by disdrometer. Presence or absence of lightning is used for classification of electrified and non-electrified events. Observations indicate RSD at all heights are considerably different for strongly electrified events and weakly electrified events. This observed difference in RSD is due to electric field and surface charge of the raindrop that effect collision-coalescence process and breakup characteristics of raindrops. (Fig. 7).

Piepgrass et al. (1982) have established that a good correlation exists between lightning frequency and rainfall. Precipitation lagged the lightning by 4–9 min. Their observations indicate that the association between lightning frequency and rainfall with some minutes time lag may be caused due to lightning-induced growth of raindrops below the melting layer rather than the enhancement in precipitation in the mixed phase region of cloud. A study by Mudiar et al. (2021), presents the evidence for influence of strong electric field and rainfall. They have shown how overhead lightning discharges could modify observed rain intensity due to enhanced collision-coalescence growth of raindrops. As seen from Fig. 8, increase in rainfall rate is observed 2–4 min after strong lightning. Mudiar et al. (2021) have clearly established the correlation between surface rain intensity and lightning rate that signify the role of electrical forces in modification of intensity of surface precipitation.

6 Conclusions

Observations of electric field associated with thunderstorms over Pune and other places reveal many interesting facets of the electrical nature of thunderstorms occurring over a tropical region like India. Some characteristics are very unique and different from observations made over other parts of the world. Observations of Maxwell current made over many years at Pune can be grouped into three categories. (I) Maxwell current remains positive throughout the life of the storm; (II) remains positive in the initial and final stages and changes to negative during middle stage of the thunderstorm and (III) remains negative throughout the lifetime of the thunderstorm. Observations on recovery curves during lightning, also exhibit some novel features. During dissipation stage, lightning discharges show an almost constant value of the pre-discharge surface electric field. A new phenomenon of the 'inverted EOSO' in a storm with inverted polarity electric dipole has been observed.

Many observations also suggest presence of tripolar charge structure of thunderclouds and in initiating or triggering an intracloud or cloud-to-ground lightning discharge, lower positive charge center (LPCC) plays a dominant role. Such tripolar charge structure or inverted thunderstorms are also observed over Kharagpur which produce many positive CG lightning.

Our simulation studies conducted using vertical wind tunnels show that electrical forces inside thunderstorms play a dominant role in modifying the microphysical characteristics of thunderclouds and in their initiation and strengthening. Results also show the changes in raindrop size distribution due to combined effect of increase in aerosol loading/pollution and electric field inside the clouds formed due to such aerosol loading. These observations clearly demonstrate how electric field in thunderclouds could change rainfall and its characteristics.

It is to be noted that above conclusions are drawn from measurements made at a few places only over India. These observations are sparse. Measurements of electric field at more places, especially over those that witness severe thunderstorms could surely help in our understanding of the effect of electric fields on rainfall

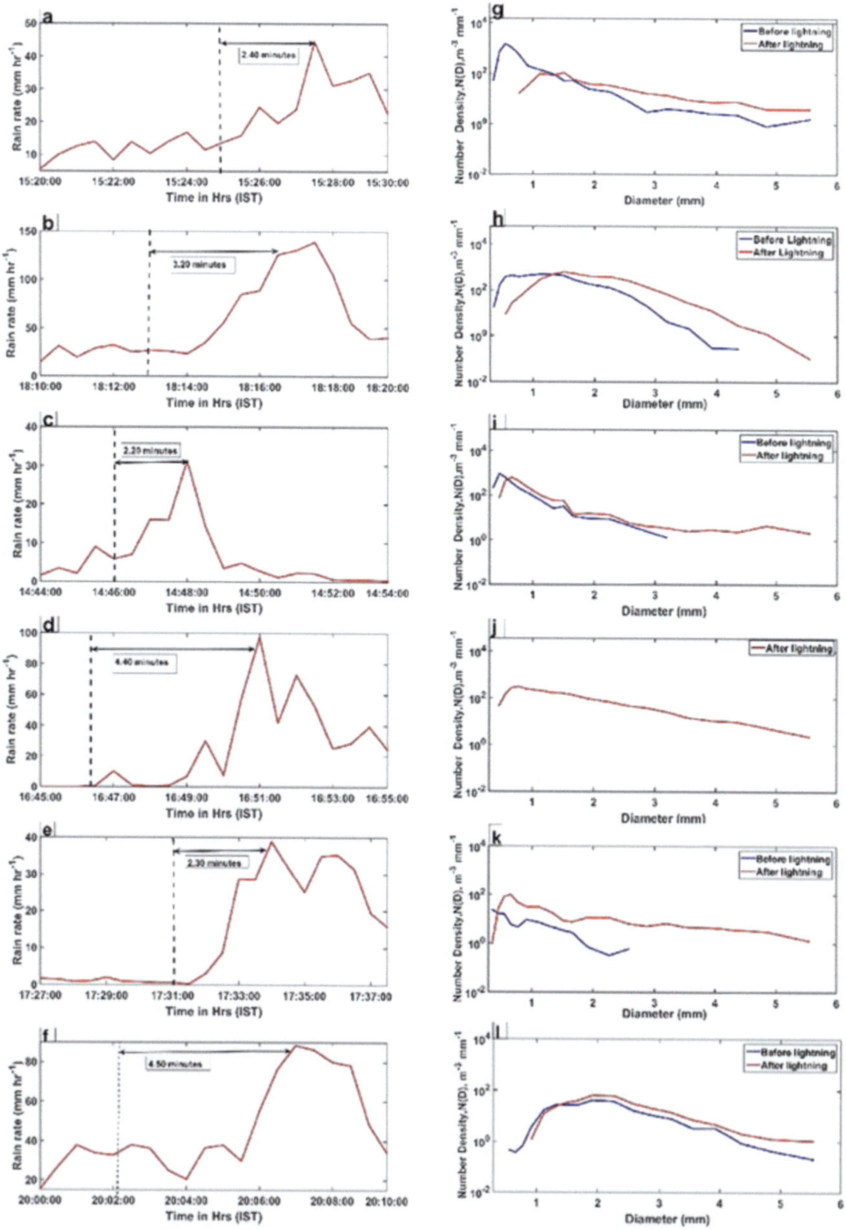

Fig. 8 Rain Gushes associated with some of the cloud-to-ground (CG) lightning flashes observed and their corresponding raindrop size distribution (RDSDs). Vertical dot lines denote the timing of lightning (From Mudiar et al., 2021)

characteristics. This understanding can possibly improve the simulation of tropical rainfall in numerical models.

References

Ausman, E. L., & Brook, M. (1967). Distortion and disintegration of water drops in strong electric fields. *Journal of Geophysical Research, 72*, 6131–6135.

Bala, S. S., & Kamra, A. K. (1991). Change in precipitation drag due to thunderstorm electrification. *Atmospheric Research, 26*, 377.

Banerjee, S. K. (1930). The electric field of overhead thunderclouds. *Quarterly Journal of the Royal Meteorological Society, 56*, 305.

Banerji, S. K. (1932). The electric field of overhead thunderstorms. *Philosophical Transactions of the Royal Society of London, Series A, 231*, 1–27.

Bhalwankar, R. V., Kamra, A. K. (2007). Wind tunnel investigation of the deformation of water drops in the vertical and horizontal electric fields. *Journal of Geophysical Research, 112*(D10215), 1–9.

Bhalwankar Rohini, V., Kamra, A. K., & Sathe, A. B. (2004). The evaporation of the charged and uncharged water drops suspended in a wind tunnel. *Proceedings of the Indian Academy of Science (Earth & Planetary Science), 13*, 129.

Beard, K. V., & Grover, S. N. (1974). Numerical collision efficiencies for small raindrops colliding with micron size particles. *Journal of Atmospheric Science, 31*, 543.

Beard, K. V., & Pruppacher, H. R. (1968). An experimental test of theoretically calculated collision efficiencies of cloud drops. *Journal of Geophysics, 73*, 6407.

Beard, K. V., & Pruppacher, H. R. (1971). A wind tunnel investigation of collection Kernels for small water drops in air. *Quarterly Journal of the Royal Meteorological Society, 97*, 243.

Borucbi, & Chameides. (1984). Lightning estimates of the rates of energy dissipation and nitrogen fixation. *Review of Geophysics, 22*, 364.

Bruning, E. C., Rust, W. D., MacGorman, D. R., Biggerstaff, M. I., Schuur, T. J. (2010). Formation of charge structures in a supercell. *Monthly Weather Review, 138*, 3740–3761.

Ccopa, J. A., Tacza, J., Raulin, J. P., & Morales, C. A. (2021). Estimation of thunderstorms occurrence from lightning cluster recorded by WWLLN and its comparison with the 'universal' Carnegie curve. *Journal of Atmospheric and Solar-Terrestrial Physics, 221*, 105682. https://doi.org/10.1016/j.jastp.2021.105682

Chuang, C., & Beard, K. V. (1990). A numerical model for the equilibrium shape of electrified raindrops. *Journal of the Atmospheric Sciences, 47*, 1374–1389.

Coquillat, S., Combal, B., & Chauzy, S. (2003). Corona emission from raindrops in strong electric fields as a possible discharge initiation: Comparison between horizontal and vertical field configurations. *Journal of Geophysical Research: Atmospheres, 108*(D7), 4205, ACL 1-16. https://doi.org/10.1029/2002JD002714

Deaver, L. E., & Krider, E. P. (1991). Electric fields and current densities under small Florida thunderstorms. *Journal of Geophysical Research, 96*, 22273–22281.

Fitzerald, D. R., & Byers, H. R. (1962). *Aircraft electrostatic measurement instrumentation and observations of cloud electrification.* Report contract AF 19(604)2189, University of Chicago.

Gangane, A., Pawar, S. D., Gopalakrishnan, V., Saikrishnan, K. C. (2023). Effect of dust particles on lightning flash rate and polarity of dust storms over India. *Natural Hazards, 115*, 2505–2529. https://doi.org/10.1007/s11069-022-05651-x.

Ghosh, R., Pawar, S. D., Hazra, A., Wilkinson, J., Mudiar, D., & Domkawale, M. A., et al. (2023). Seasonal and regional distribution of lightning fraction over Indian subcontinent. *Earth and Space Science, 10*, e2022EA002728. https://doi.org/10.1029/2022EA002728.

Gopalakrishnan, V., Pawar, S. D., Murugavel, P., & Johare, K. P. (2011). Electrical characteristics of thunderstorms in the Eastern part of India, *Journal of Atmospheric and Solar Terrestrial Physics, 73*, 1876–1882. https://doi.org/10.1016/j.jastp.2011.04.022.

Grover, S. N., & Pruppacher, H. R. (1985). The effect of vertical turbulent fluctuations in the atmosphere on the collection of aerosol particles by cloud drops. *Journal of Atmospheric Science, 42*, 2305.

Israelsson, S., Schutte, T., Pisler, E., & lundquist, S. (1987). Increased occurrence of lightning flashes in Sweden during 1986. *Journal of Geophysical Research, 92*, 10,996–10998.

Kamra, A. K. (1982). Possible effect of the cloud electrification on the downdraft in thunderstorms. *Geophysical Research Letters, 9*, 235.

Kamra, A. K. (1985). Modification of some microphysical properties in the highly electrified regions of thundercloud. *Archives for Meteorology, Geophysics, and Bioclimatology, Series A, 34*, 37.

Kamra, A. K., & Ahire, D. V. (1989). Wind tunnel studies of the shape of charged and uncharged water drops in the absence or presence of an electric field. *Atmospheric Research, 23*, 117.

Kamra, A. K., Bhalwankar, R. V., & Sathe, A. B. (1991). Spontaneous breakup of charged and uncharged water drops freely suspended in a wind tunnel. *Journal of Geophysical Research, 96*, 17159.

Kamra, A. K., Bhalwankar, R. V., & Sathe, A. B. (1993). The onset of disintegration and corona in water drops falling at terminal velocity in horizontal electric field. *Journal of Geophysical Research, 98*, 12901.

Kamra, A. K., Ramesh Kumar, P. (2021). Regional variability in lightning activity over South Asia, International Journal of Climatology, 41, 625–646. https://doi.org/10.1002/joc.6641

Krehbiel, P. R. (1986). The electrical structure of thunderstorms. In *The Earth's Electrical Environment* (pp. 90–113). National Academy Press.

Krehbiel, P. R., Thomas, R. J., Rison, W., Hamlin, T., Harlin, J., & Davis, M. (2000). GPS-based mapping system reveals lightning inside storms. *Eos Transactions AGU, 81*(3), 21–25. https://doi.org/10.1029/00EO000014

Krider, E. P., & Musser, J. A. (1982). Maxwell currents under thunderstorms. *Journal of Geophysical Research, 87*, 11171–11176.

Lang, T. J., Rutledge, S. A., Dye, J. E., Venticinque, M., Laroche, P., & Defer, E. (2000). Anomalously low negative cloud-to-ground lightning flash rates in intense convective storms observed during STERAO-A. *Monthly Weather Review, 128*, 1601–1673.

Levine, N. E. (1971). Disruption of charged water drops in an external electric field. *Journal of Geophysical Research, 76*, 5079–5100. https://doi.org/10.1029/JC076i021p05097

Livingston, J. M., & Krider, E. P. (1978). Electric fields produced by Florida thunderstorms. *Journal of Geophysical Research, 83*, 385–401.

Manohar, G. K., Kandalgaonkar, S. S., & Tinmaker, M. I. R. (1999). Thunderstorm activity over India and the Indian southwest monsoon. *Journal of Geophysical Research, 104*, 4169–4188.

Marshall, T. C., & Rust, W. D. (1991). Electric field soundings through thunderstorms. *Journal of Geophysical Research, 96*, 22297–22306.

Moore, C. B., Vonnegut, B., Stein, B. A., & Survilas, H. J. (1961). Observations of electrification and lightning in warm clouds. *Journal of Geophysical Research, 65*, 1907.

Mudiar, D., Pawar, S. D., Hazra, A., Konwar, M., Gopalakrishnan, V., Srivastava, M. K., Goswami, B. N. (2018). Quantification of observed electrical effect on the raindrop size distribution in tropical clouds. *Journal of Geophysical Research, 123*, 1–18. https://doi.org/10.1029/2017JD028205.

Mudiar, D., Pawar, S. D., Hazra, A., Gopalakrishnana, V., Lal, D. M., Kaustav, C., Domkawale, M. A., Srivastava, M. K., Goswami, B. N., William, E. (2021). Lightning and precipitation: The possible electrical modification of observed raindrop size distributions. *Atmospheric Research, 259*, 105663, 1–17. https://doi.org/10.1016/j.atmosres.2021.105663.

Murugavel, P., Pawar, S. D., Gopalakrishnan, V. (2012). Trends of convective available potential energy over the Indian region and its effect on rainfall. *International Journal of Climatology, 32*, 1362–1372. https://doi.org/10.1002/joc.2359.

Orville, R. E., & Spencer, D. W. (1979). Global lightning flash frequency. *Monthly Weather Review, 107*, 934–943.

Pawar, S. D., & Kamra, A. K. (2002). Recovery curves of the surface electric field after lightning discharges occurring between the positive charge pocket and negative charge centre in a thundercloud. *Geophysical Research Letters, 29*, 2108–2111.

Pawar, S. D., & Kamra, A. K. (2004). Evolution of lightning and the possible initiation/triggering of lightning discharges by the lower positive charge center in an isolated thundercloud in the tropics. *Journal of Geophysical Research, 109*, D02205. https://doi.org/10.1029/2003JD003735

Pawar, S. D., & Kamra, A. K. (2007). End-of-storm-oscillation in tropical air-mass thunderstorms. *Journal of Geophysical Research, 112*(D03204). https://doi.org/10.1029/2005JD006997

Pawar, S. D., & Kamra, A. K. (2009). Maxwell current density characteristics below isolated thunderstorms in tropics. *Journal of Geophysical Research, 114*(D04208), 1–12.

Pawar, S. D., Murugavel, P., & Gopalakrishnan, V. (2010). Anomalous electric field changes and high flash rate beneath a thunderstorm in northeast India. *Journal of Earth System Science, 119*, 617–625.

Pawar, S. D., Murugavel, P., & Gopalkrishnan, V. (2011). New particle formation by ion-induced nucleation during dissipation stage of thunderstorm. *Journal of Earth System Science, 120*, 843–850.

Pawar, S. D., Gopalkrishnan, V., Murugavel, P., Sinkevich, A., & Lal, D. M. (2014). Effects of environmental conditions on inducing charge structures of thunderstorms over Eastern India. *Earth, Planets and Space, 66*, 1–23. https://doi.org/10.1186/1880-5981-66-54

Pawar, S. D., Gopalakrishnan, V., Murugavel, P., Veremey, N. E., & Sinkevich, A. A. (2017). Possible role of aerosols in the charge structure of isolated thunderstorms. Atmospheric Research, 183, 331–340. https://doi.org/10.1016/j.atmosres.2016.09.016

Piepgrass, M. V., Krider, E. P., Moore, C. B. (1982). Lightning and surface rainfall during Florida thunderstorms. *Journal of Geophysical Research, 87*, 11,193–11,201.

Qie, X., Zhang, T., Chen, C., Zhang, G., Zhang, T., & Wei, W. (2005). The lower positive charge center and its effect on lightning discharges on the Tibetan Plateau. *Geophysical Research Letters, 32*, L05814. https://doi.org/10.1029/2004GL022162

Rasmussen, R., Walcek, C., Pruppacher, H. R., Mitra, S. K., Lew, J., Levizzani, V., Wang, P. K., & Barth, U. (1985). A wind tunnel investigation of the effect of an external, vertical electric field on the shape of electrically uncharged rain drops. *Journal of Atmospheric Science, 42*, 1647–1652.

Richards, C. N., & Dawson, G. A. (1971). The hydrodynamic instability of water drops falling at terminal velocity in vertical electric fields. *Journal of Geophysical Research, 76*, 3445–3455.

Rust, W. D., MacGorman, D. R., Bruning, E. C., Weiss, S. A., Krehbiel, P. R., Thomas, R. J. Rison, W., Hamline, T., & Harlin, J. (2005). Inverted polarity electrical structures in thunderstorms in the Severe Thunderstorm Electrification and Precipitation Study (STEPS). *Atmospheric Research, 76*, 247–271.

Simpson, G. C. (1927). The mechanism of thunderstorm. *Proceedings of the Royal Society of London, Series A, 114*, 376–401.

Simpson, G. C., & Robinson, G. D. (1941). The distribution of electricity in thunderclouds, 2. *Proceedings of the Royal Society of London, Series A, 177*, 281–329.

Simpson, G. C., & Scrase, F. J. (1937). The distribution of electricity in thunderclouds. *Proceedings of the Royal Society of London, Series A, 161*, 309–352.

Stolzenburg, M., Rust, W. D., & Marshall, T. C. (1998). Electrical structure in thunderstorm convective regions. 2. Isolated storms. *Journal of Geophysical Research, 103*, 14079–14096.

Taylor, G. (1964). The disintegration of water drops in an electric field. *Proceedings of the Royal Society of London, a280*, 383–397.

Takahasi, T. (1978). Riming electrification as a charge generation mechanism in thunderstorms. *Journal of Atmospheric Science, 35*, 1536–1548.

Telford, J. W., Thorndike, N. S., & Bowen, E. G. (1955). The coalescence between small water drops. *Quarterly Journal of the Royal Meteorology Society, 81*, 241.

Tyagi, A. (2007). Thunderstorm climatology over Indian region. *Mausam, 58*, 189–212.

Vonnegut, B. (1960). Electrical theory of tornadoes. *Journal of Geophysical Research, 65*, 203–212.

Wiens, K. C., Rutledge, S. A., & Tessendorf, S. A. (2005). The 29 June 2000 suppercell observed during STEPS. Part II: Lightning and charge structure. *Journal of the Atmospheric Sciences, 62*, 4151–4177.

Williams, E. R. (1989). The tripole structure of thunderstorms. *Journal of Geophysical Research, 94*, 13151–13167.

Williams, E. R., Weber, M. E., & Oville, R. E. (1989). The relationship between lightning type and convective state of thunderclouds. *Journal of Geophysical Research, 94*, 13,213–13,220.

Williams, E. R., Zhang, R., & Rydock, J. (1991). Mixed-phase microphysics and cloud electrification. *Journal of Atmospheric Science, 48*, 2195.

Williams, E. R., Rutledge, S., Geotis, S., Renno, N., Ramussen, E., & Rickenback, T. (1992). Radar and electrical study of tropical hot towers. *Journal of Atmospheric Science, 49*, 1386–1395.

Williams, E., Rosenfeld, D., Madden, N., Gerlach, J., Gears, N., Atkinson, L., Dunnemann, N., Frostrom, G., Antonio, M., Biazon, B., Camargo, R., Franca, H., Gomes, A., Lima, M., Machado, R., Manhaes, S., Nachtigall, L., Piva, H., Quintiliano, W., Machado, L., Artaxo, P., Roberts, G., Renno, N., Blakeslee, R., Bailey, J., Boccippio, D., Betts, A., Wolff, D., Roy, B., Halverson, J. et al. (2002). Contrasting convective regimes over the Amazon: implications for cloud electrification. *Journal of Geophysical Research, 107*(D20):LBA 50-1–LBA 50-19. https://doi.org/10.1029/2001JD000380

Williams, E. R., Mustak, V., Rosenfeld, D., Goodman, S., & Boccippio, D. (2005). Thermodynamic conditions favorable to superlative thunderstorm updraft, mixed phase microphysics and lightning flash rate. *Atmospheric Research, 76*, 288–306.

Wilson C. R. T. (1920). Investigations on lightning discharges and on the electric field of thunderstorms. *Philosophical Transactions A, 221*, 73–115.

Winn, W. P., Schwede, G. W., & Moore, C. B. (1974). Measurements of electric fields in thunderclouds. *Journal of Geophysical Research, 79*, 1761–1767. https://doi.org/10.1029/JC079i012p01761

Woods, J. D., & Mason, B. J. (1965). The effect of electric charges upon collisions between equal-size water drops in air. *Quarterly Journal Royal Meteorological Society, 91*, 353–855.

Wormell, T. W. (1953). Atmospheric electricity: Some recent trends and problems. *Quarterly Journal of the Royal Meteorological Society, 79*, 3–50.

Raindrop Size Distributions of Indo-Pacific Tropical Cyclones and Their Hydro-Meteorological Implications

Pay-Liam Lin, Balaji Kumar Seela, and Jayalakshmi Janapati

Abstract Tropical cyclones (TCs) cause tremendous loss of properties and life through their gusty winds and slashing rain. Better precipitation estimations and modeling simulations of TCs can be achieved through an enhanced understanding of the microphysical properties, particularly the raindrop size distributions of the TCs. This chapter documents raindrop size distributions of Indo-Pacific tropical cyclones measured at three locations using two types of disdrometers. The differences in the raindrop size distributions (RSDs) between the Indian and Pacific Ocean tropical cyclones and their possible rationale are detailed. Also, the dissimilarities of the north Indian Ocean (NIO) TCs measured at the coastal and inland stations and their possible microphysical attributes are documented. Furthermore, the RSDs of Western Pacific (WP) TCs measured at an Oceanic station in the WP are also detailed. The raindrop size distribution and rainfall kinetic energy relations that are handy in the rain retrieval algorithm of ground-based/remote sensing radar and rainfall erosivity studies are also reported in this chapter.

1 Introduction

The Indo-Pacific region is chiefly prone to severe weather events like thunderstorms, tropical cyclones, and seasonal flash floods. Among different forms of severe weather events, tropical cyclones (TCs) with heavy to extremely heavy rainfall and gusty

P.-L. Lin (✉)
Institute of Atmospheric Physics, Department of Atmospheric Sciences, National Central University, Zhongli District, Taoyuan City, Taiwan
e-mail: tliam@pblap.atm.ncu.edu.tw

B. K. Seela · J. Janapati
Department of Atmospheric Sciences, National Central University, 300 Zongda Road, Zhongli District, Taoyuan City, Taiwan
e-mail: seelabalaji@pblap.atm.ncu.edu.tw

J. Janapati
e-mail: lakshmijaya2009@pblap.atm.ncu.edu.tw

S. Das and W-K. Tao (eds.), *Severe Storms*,
https://doi.org/10.1007/978-981-97-7075-5_8

winds can cause severe damage to life and the economy. Extreme rain from TCs also results in floods, rainfall erosion, soil erosion, debris flow, and landslides (Chen et al., 2019a; Chen et al., 2013b, 2015; Dadson et al., 2003). In order to better understand and mitigate the TCs, there is an urgent need for detailed investigations on the raindrop size distributions (RSDs) to apprehend the precipitation estimation algorithms and cloud modeling simulations of these TCs (Chang et al., 2009; Chen et al., 2013a; Mcfarquhar et al., 2015; Tokay et al., 2008; Wu et al., 2021; Xia et al., 2020; Zhang et al., 2019). The RSD is a significant precipitation parameter that can offer cloud and rain microphysics information (Rosenfeld and Ulbrich, 2003). The knowledge of RSDs can advance the cloud modeling parameterizations and rainfall estimation algorithms of remote sensing instruments (Janapati et al., 2021; Janapati et al., 2019; Liao and Meneghini, 2019; Liao et al., 2020; Mcfarquhar et al., 2015).

Apart from the meteorological implications, the RSD information also plays a vital role in hydrological studies, especially in estimating the rainfall kinetic energy and rainfall erosivity. Because of the expensive experimental setups, hydrological researchers rely on indirect measurements of rainfall kinetic energy. Such approaches require the empirical relations between rainfall kinetic energy and rainfall intensity. The raindrop size information obtained from the ground-based disdrometer can serve the purpose of establishing the relationships between the rainfall kinetic energy and the rainfall intensity, thereby allowing to evaluate rainfall erosivity factor (or R-factor) (Fornis et al., 2005; Kinnell, 1981; Salles et al., 2002; van Dijk et al., 2002). The relationship between the rainfall kinetic energy and intensity was documented in different forms like linear, exponential, logarithmic, and power (Kinnell, 1981; Rosewell, 1986; Sempere-Torres et al., 1992; Uijlenhoet and Stricker, 1999; Wischmeier and Smith, 1958). These kinetic energy relations vary with geographical location, local climate, and precipitation microphysics (Kinnell, 1981; Rosewell, 1986; van Dijk et al., 2002).

The above-mentioned implications of the RSDs in the field of meteorology and hydrology necessitate investigating the microphysical characteristics of TCs (Chakravarty, et al., 2020; Chen et al., 2019b; Hopper et al., 2020; Janapati et al., 2021; Janapati et al., 2020b; Janapati et al., 2017; Kumar & Reddy, 2013; Kumari et al., 2014; Lee et al., 2019; Seela et al., 2021; Seela et al., 2022; Suh et al., 2016; Tokay et al., 2008). With this brief introduction, Sect. 2 provides the data and methods. Disparities in the RSDs of Indian and Pacific Ocean tropical cyclones are provided in Sect. 3. The RSDs of NIO TCs measured at the coastal and inland stations in South India are detailed in Sect. 4. The RSDs of WP TCs over the oceanic station (Palau Islands) are provided in Sect. 5. Finally, the findings drawn from the Indo-Pacific RSDs analysis are summarized in the section (Sect. 6).

2 Data and Methods

The raindrop information of the Indo-Pacific TCs was measured with the aid of two disdrometers: the Parsivel and Joss-Waldvogel disdrometer (JWD). The geographic map of the Indo-Pacific region with the locations of two kinds of disdrometers, where the tropical cyclones measurements were carried out, is illustrated in Fig. 1. The red circle denotes the location of JWD, and the blue circles indicate the location of Parsivel disdrometers.

2.1 *JWD (Joss–Waldvogel Disdrometer)*

The JWD can measure the size information of raindrops (ranging from 0.3 to 5.3 mm) once they hit the conical surface (Styrofoam cone) area of 50 cm^2 (Joss and Waldvogel, 1969). The JWD stores the raindrop size information in 20-size binds. With this 20-size bin information, the RSD and other rain integral parameters can be estimated. Based on the JWD's raindrop count information, the raindrop size distribution (*N(D),* m^{-3} mm^{-1}), *R* (rainfall intensity in mm h^{-1}), Z (radar reflectivity in dBZ), N_t (total number concentration in $m^{-3)}$, and *LWC* (liquid water content in gm^{-3}) and hydrological parameters like kinetic energy content (KE_{mm}, J m^{-2} mm^{-1}), kinetic energy expenditure (KE_{time}, J m^{-2} h^{-1}) can be estimated (Fornis et al., 2005; Janapati et al., 2021, 2023; Salles et al., 2002; Tokay et al., 2008; Ulbrich, 1983; van Dijk et al., 2002).

$$N(D_i)\left(m^{-3}mm^{-1}\right) = \sum_{i=1}^{20} \frac{n_i}{A \times \Delta t \times V(D_i) \times \Delta D_i} \tag{1}$$

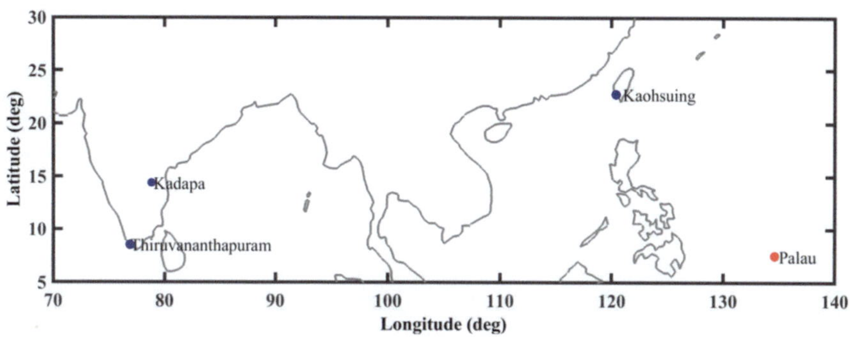

Fig. 1 Indo-Pacific geographic map with the locations of JWD (red filled circle) and Parsivel disdrometer (blue filled circles)

where, n_i, A (=0.005 m^2), Δt (=60 s), D_i (mm), ΔD_i (mm)), $V(D_i)$ (m s^{-1}) are the number of raindrops counted in size bin i, sampling area, sampling time, drop diameter for the size bin i, diameter interval for the drop size bin i, and raindrops terminal velocity in the ith channel, respectively. The terminal velocity of raindrops at each channel is estimated using the below equation (Gunn and Kinzer, 1949).

$$V(D_i) = 9.65 - 10.3 \times \exp(-0.6 \times D_i) \tag{2}$$

$$Z(\text{dBZ}) = 10 \times log_{10}\left(\sum_{i=1}^{20} N(D_i) D_i^6 \Delta D_i\right) \tag{3}$$

$$LWC\left(\text{g m}^{-3}\right) = \frac{\pi}{6} 10^{-3} \rho_w \sum_{i=1}^{20} N(D_i) D_i^3 \Delta D_i \tag{4}$$

$$N_t\left(\text{m}^{-3}\right) = \sum_{i=1}^{20} N(D_i) \Delta D_i \tag{5}$$

The nth-order moment (M_n) of raindrop size distribution can be expressed as

$$M_n = \sum_{D_{min}}^{D_{max}} D_i^n N(D_i) \Delta D_i \tag{6}$$

The mass-weighted mean diameter, D_m, can be expressed as

$$D_m(mm) = \frac{M_4}{M_3} \tag{7}$$

The one-min RSD samples are fitted to gamma function as given below (Ulbrich, 1983)

$$N(D) = N_0 D^{\mu} \exp(-\Lambda D) \tag{8}$$

where D(mm) is the drop diameter, $N(D)$ is the number of drops per unit volume per unit size interval, $N0$ is the number concentration parameter, μ is the shape parameter, and Λ is the slope parameter,

$$\Lambda = \frac{(\mu + 4) M_3}{M_4} \tag{9}$$

$$\mu = \frac{(11 \times G - 8) + \sqrt{G \times (G + 8)}}{2 \times (1 - G)} \tag{10}$$

$$G = \frac{M_4^3}{M_6 M_3^2} \tag{11}$$

The normalized intercept parameter, N_w, can be expressed as (Bringi et al., 2003)

$$N_w\left(m^{-3}mm^{-1}\right) = \frac{4^4}{\pi \rho_w}\left(\frac{10^3 LWC}{D_m^4}\right) \tag{12}$$

The σ_m (mass spectrum standard deviation, mm) can be estimated using (Thurai et al., 2014; Williams et al., 2014)

$$\sigma_m(mm) = \left[\frac{\sum_{i=1}^{20}(D_i - D_m)^2 N(D_i) D_i^3 \Delta \mathrm{D_i}}{\sum_{i=1}^{20} N(D_i) D_i^3 dD}\right]^{1/2} \tag{13}$$

$$KE_{time} = \left(\frac{\pi}{12}\right)\left(\frac{1}{10^6}\right)\left(\frac{3600}{\Delta t}\right)\left(\frac{1}{A}\right)\sum_{\mathrm{i=1}}^{20} n_i D_i^3 V(D_i)^2 \tag{14}$$

$$KE_{mm} = \frac{KE_{time}}{I} \tag{15}$$

2.2 *Parsivel Disdrometer*

Parsivel disdrometer is a laser (wavelength 650 nm and length × width × thickness = 180 × 30 × 1 mm) disdrometer, which can estimate the size (liquid and solid drops of diameters, 0.2 – 5 mm and 0.2 – 25 mm, respectively) and velocity (0.2 – 20 $\mathrm{ms^{-1}}$) of the precipitating particles (Löffler-Mang and Joss, 2000; Yuter et al., 2006). The particle count information is stored in 32 × 32 raindrop size and velocity matrix. Based on the Parsivel raindrop count information, the meteorological (*N(D)* in$\mathrm{m^{-3}\ mm^{-1}}$, *R in* $\mathrm{mm\ h^{-1}}$, Z in dBZ, N_t in $\mathrm{m^{-3}}$, and *LWC* in $\mathrm{gm^{-3}}$) and hydrological (KE_{mm} and KE_{time}) parameters are estimated using the below-mentioned equations.

$$N(D_i) = \sum_{i=1}^{32}\sum_{j=1}^{32}\frac{n_{ij}}{A_{eff}\,\Delta t\, V(D_i)\Delta D_i} \tag{16}$$

where '*i*' and '*j*' denotes the drop and velocity size bin number, n_{ij} corresponds to total particle/raindrops in size bin '*i*' and velocity bin '*j*'. Δt (sampling time = 60 s), A_{eff} ($\mathrm{m^2}$) [Effective area = $10^{-6} \times L\left(B - \frac{D_i}{2}\right)m^2$], L = 180 mm, B = 30 mm.

$$Z(dBZ) = 10log_{10}\sum_{i=1}^{32}\sum_{j=1}^{32} D_i^6 \frac{n_{ij}}{A_{eff}\,\Delta t\, V(D_i)} \tag{17}$$

$$R\left(mmh^{-1}\right) = 6\pi \times 10^{-4} \sum_{i=1}^{32} \sum_{j=1}^{32} D_i^3 \frac{n_{ij}}{A_{eff}\, \Delta t} \tag{18}$$

$$N_t\left(m^{-3}\right) = \sum_{i=1}^{32} \sum_{j=1}^{32} \frac{n_{ij}}{A_{eff}\, \Delta t V\left(D_i\right)} \tag{19}$$

$$LWC\left(gm^{-3}\right) = \frac{\pi}{6} \times 10^{-3} \rho_w \sum_{i=1}^{32} \sum_{j=1}^{32} D_i^3 \frac{n_{ij}}{A_{eff}\, \Delta t V\left(D_i\right)} \tag{20}$$

$$\boldsymbol{KE_{time}}\left(\boldsymbol{Jm^{-2}h^{-1}}\right) = \left(\frac{\pi}{12}\right)\left(\frac{1}{10^6}\right)\left(\frac{3600}{\boldsymbol{t}}\right)\left(\frac{1}{\boldsymbol{A_{eff}}}\right) \sum_{i=1}^{32} \boldsymbol{n_i D_i^3 [V(D_i)]^2} \tag{21}$$

$$\boldsymbol{KE_{mm}}\left(\boldsymbol{Jm^{-2}mm^{-1}}\right) = \left(\frac{\boldsymbol{KE_{time}}}{\boldsymbol{R}}\right) \tag{22}$$

Apart from the ground-based disdrometers data, the convective available potential energy (CAPE, J kg^{-1}), water vapor, vertical air temperature profiles, and relative humidity values for the ERA-interim and cloud top temperature (CTT) from the MODIS satellite are also used.

3 Distinctions in the RSDs of Indian and Pacific Ocean TCs

Similarities or dissimilarities between the Indian and Pacific Ocean TCs RSDs are understood using Parsivel disdrometer measurements (Fig. 2). The mean RSDs of Indian and Pacific Ocean TCs are illustrated in the right panel of Fig. 2. The Pacific Ocean TCs have a larger mean rain rate (R, mm h^{-1}), D_m (mm), and smaller N_w (mm^{-1} m^{-3}) than the Indian Ocean. The raindrop concentration of mid-size (1–3 mm) and large drops (>3 mm) is higher for TCs of the Pacific Ocean than the Indian Ocean. A higher concentration of small drops and lower concentration of mid- and large-size drops in TCs of the Indian Ocean result in lower D_m than TCs of the Pacific Ocean.

To further look into the dissimilarities between Indian and Pacific Ocean TCs, the RSDs of both Oceanic TCs are segregated into stratiform and convective types using the classification criteria of Bringi et al. (2003). The mean raindrop size distribution, D_m, and $\log_{10}N_w$ for the Indian and Pacific Ocean TCs for different precipitation types are illustrated in Fig. 3. Segregation of RSD spectra into different precipitation types showed 66.67% (33.33%) of RSDs as stratiform (convective type) type for Pacific Ocean TCs, and 82.35% (17.65%) of RSDs as stratiform (convective) type for Indian Ocean TCs. Even after separating the Indian and Pacific Ocean TCs into different precipitation types, the Pacific Ocean TCs demonstrated a higher concentration of mid- and large-size drops. Irrespective of the Oceanic region, the stratiform

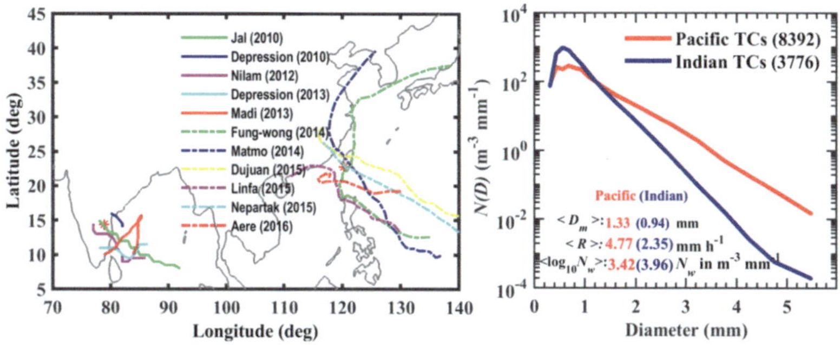

Fig. 2 Indian and Pacific Ocean TCs tracks (left panel) and mean raindrop size distribution (right panel). From Janapati et al. (2020b)

precipitation showed nearly exponential distribution. The convective regimes exhibited broad distributions, which can be attributed to the collisional breakup of the large drops in convective rain (Hu and Srivastava, 1995a). The distribution of mean D_m and $\log_{10}N_w$ values for different precipitations of Indian and Pacific Ocean TCs shows that the convective regimes have higher mean D_m and $\log_{10}N_w$ values than stratiform regimes (Fig.3 right panel). On the other hand, TCs from the Pacific Ocean show larger D_m values than that of the Indian Ocean in both precipitations. Smaller D_m and larger $\log_{10}N_w$ in Indian Ocean TCs suggest that the Indian Ocean TCs are related to the melting of small graupel/rimed ice particles, and the Pacific Ocean TCs with large snowflakes melting.

The μ-Λ relations afford in indulgent the RSD features and in estimating the polarimetric radar rain parameters based on the constrained-gamma distribution (Brandes et al., 2004b; Cao et al., 2008b; Zhang et al., 2001a, 2003a, 2006a). These

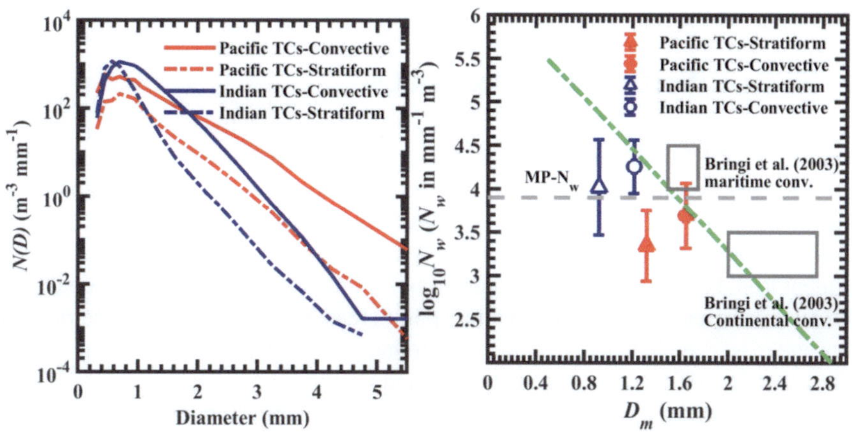

Fig. 3 Stratiform and convective mean raindrop size distribution (left panel) and mean D_m and $\log_{10}N_w$ values. From Janapati et al. (2020b)

relations were found to vary with region and rain type (Seela et al., 2018; Tang et al., 2014; Zhang et al., 2003a), necessitating to investigate the region's representative μ-Λ relations. Distributions plots of μ and Λ for the Indian and Pacific Ocean TCs are depicted in Fig. 4a and b. The relationship between μ and Λ is established for both oceanic TCs using the RSD samples of rain rate > 5 mm h^{-1}, $\mu > 20$, and $\Lambda > 20$ mm^{-1} (Cao et al., 2008b; Zhang et al., 2003a). Estimated μ-Λ relations clearly demonstrated distinctions between the Indian and Pacific Ocean TCs. Furthermore, the D_m and standard deviation of mass spectrum (σ_m) relations ($\sigma_m - D_m$ relations), which are handy in improving precipitation estimation algorithms of ground-based and remote sensing radars, are also established for the Indian and Pacific Ocean TCs (Fig. 4c and d). While establishing the $\sigma_m - D_m$ relations, RSD samples with raindrop count > 50, Z > 10 dBZ, rainfall rate > 5 mm h^{-1}, and Dm > 0.5 mm are considered (Williams et al., 2014). The established $\sigma_m - D_m$ relations for the Indian and Pacific Ocean TCs are, respectively, found to be $\sigma_{m=}$ 0.29 ${D_m}^{1.71}$ and $\sigma_{m=}$ 0.36 ${D_m}^{1.35}$.

Conceivable mechanisms responsible for the differences in the Indian and Pacific Ocean TCs are tried to understand with the convective available potential energy (CAPE, J kg^{-1}), water vapor (Kg m^{-2}), and cloud top temperature (CTT) values for the analyzed TCs over the disdrometer sites. The CAPE, water vapor (CTT) values are found to be relatively higher (lower) for the Pacific Ocean TCs (Figs. 5 and 6) than the Indian Ocean TCs. Dissimilarities in these parameters (CAPE, water vapor, and CTT) signify the strong convection associated with vigorous updrafts and downdrafts in the Pacific Ocean TCs, which can lead to the growth of precipitating particles by aggregation and riming process in the cold region and by collision-coalescence process in the warm regions. Furthermore, relatively higher (lower) air temperature (relative humidity) in the Pacific Ocean TCs indicates (figure not shown) the possibility for the evaporation of small drops, leading to more large size drops in TCs from the Pacific Ocean.

4 RSDs of North Indian Ocean TCs

The RSDs of NIO TCs are measured using Parsivel disdrometers installed at the coast (Thiruvananthapuram) and inland (Kadapa) stations located in south India. The tracks of four TCs measured at the coastal and four TCs at the inland station are illustrated in Fig. 7a, and their mean raindrop concentrations are shown in Fig. 7b. It is evident from Fig. 7b that the mid- and large-size drops (>1 mm) are predominant for the NIO TCs measured at Thiruvananthapuram than Kadapa. Moreover, the rainfall rate and mass-weighted mean diameter (normalized intercept parameters) are higher (lower) at the coastal station than at the inland station.

The rain retrieval algorithms of GPM DPR use the relationships between the differential frequency ratio (DFR in dBZ: differences in the radar reflectivities measured at Ku- and Ka-band frequency in dBZ) and the mass-weighted mean diameter (DFR-D_m relations) (Liao and Meneghini, 2005). However, these DFR-D_m relations are

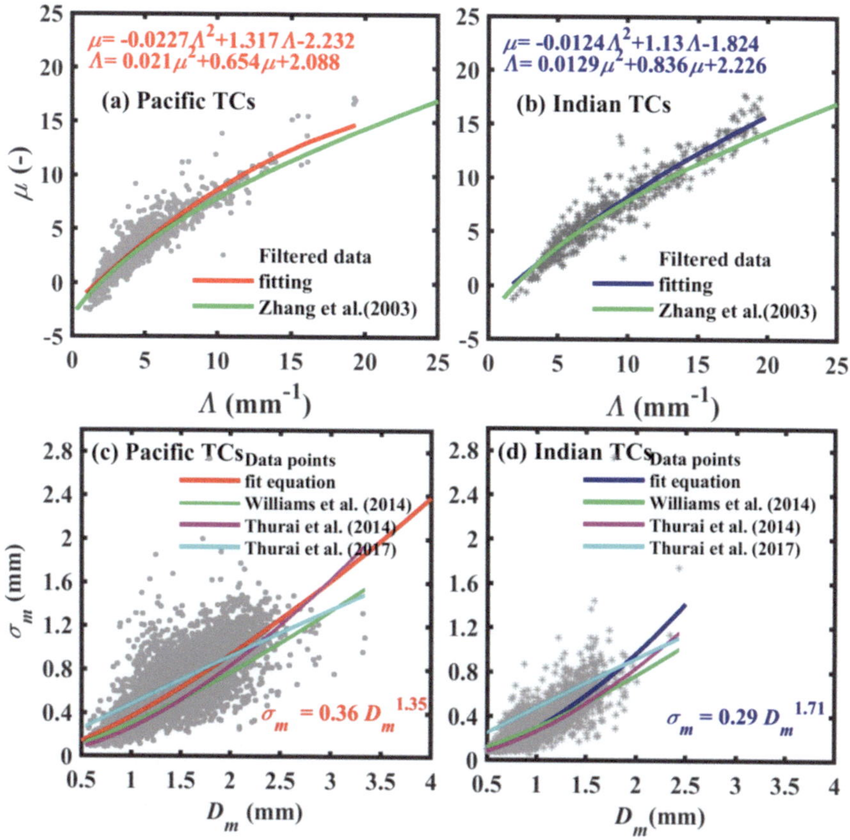

Fig. 4 $\mu - \Lambda$ and $\sigma_m - D_m$ relations for Indian and Pacific Ocean tropical cyclones. From Janapati et al. (2020b)

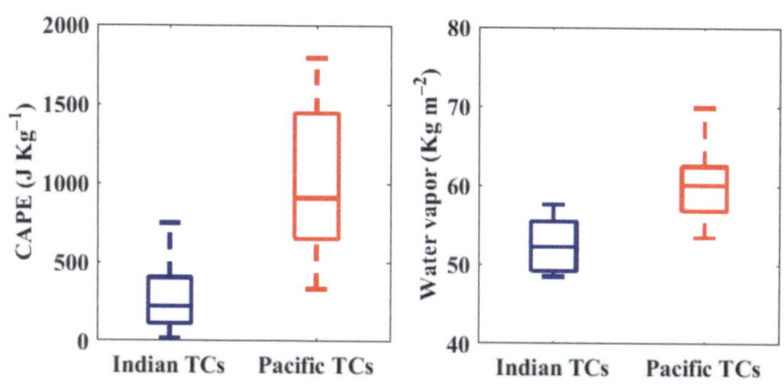

Fig. 5 Box plots of convective available potential energy (CAPE, J Kg^{-1}) and water vapor for the Indian and Pacific Ocean TCs. From Janapati et al. (2020b)

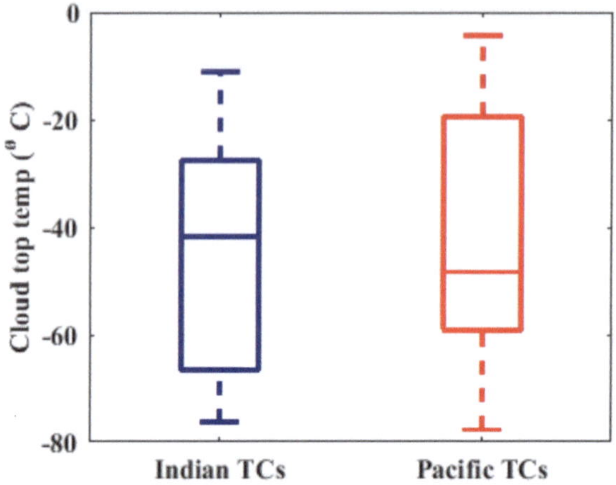

Fig. 6 Box plots of cloud top temperature (CTT) for the Indian and Pacific Ocean TCs. From Janapati et al. (2020b)

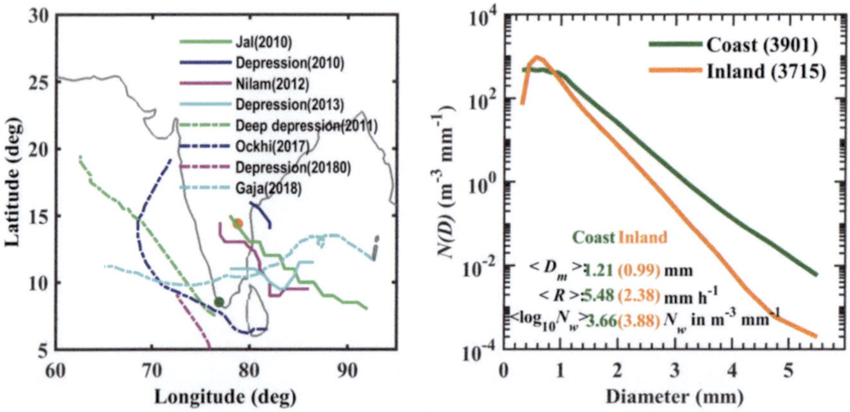

Fig. 7 Tracks of NIO TCs measured at the coastal (green filled circle) and inland stations (orange filled circle) (left panel) and their mean raindrop size distribution (right panel). From Seela et al. (2021)

constrained by the dual solution problem, especially for the lower D_m values (Liao and Meneghini, 2005; Liao et al., 2014). As an alternative approach for the rain retrieval algorithms of the GPM DPR, previous researchers suggested the relationships among D_m, Z_{ku}, and Z_{ka}. Hence, here we establish the $D_m - Z_{Ku}/Z_{Ka}$, and $\log_{10}N_w - D_m$ relations. The radar reflectivity at Ku- and ka-band frequencies for the NIO TCs are estimated using the T-matrix simulations (with 20 °C temperature assumption) applied to the RSDs of NIO TCs (Brandes et al., 2002; Mishchenko et al., 1996). The scatter plots of $D_m - Z_{Ku}/Z_{Ka}$, and $\log_{10}N_w - D_m$ and their polynomial fit equation are depicted in Fig. 8. The established empirical relations demonstrated substantial variations between the NIO TCs measured in the coastal and inland stations, which signifies the adaptation of region-specific empirical relations.

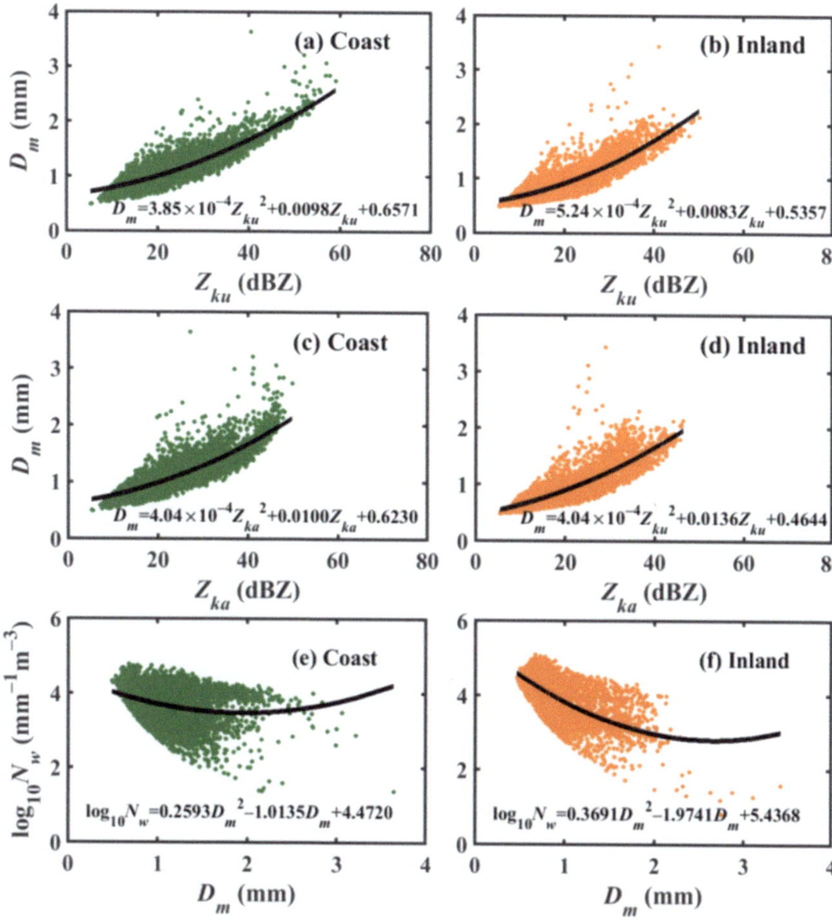

Fig. 8 D_m–Z_{ku}/Z_{ka}, and $\log_{10}N_w$–D_m relations for the NIO TCs measured at the coastal and inland stations. *From* Seela et al. (2021)

The kinetic energy (KE) of the raindrops/rainfall plays a significant role in understanding the soil erosion caused by the rain, also known as rainfall erosivity (Wischmeier and Smith, 1958). As the direct measurement of rainfall KE requires expensive instrumentation, the previous researchers adopted the indirect estimation of KE through the RSD information (Fornis et al., 2005; Jayawardena and Rezaur, 2000; Kinnell, 1981). The empirical relations between the rainfall KE and rainfall rate (KE-*R* relations) can help to estimate the KE at the locations (where there were no RSD measurements) with rain gauge measurements (Janapati et al., 2019; Verstraeten et al., 2006). Moreover, while estimating the rainfall KE for the harsh locations with no ground-based rainfall measurements, the relationship between the rainfall KE and mass-weighted mean diameter (KE-D_m relations) can offer the opportunity to estimate the KE using the D_m values estimated from GPM DPR measurements. As these rainfall KE relations were found to differ from one region to another and from one weather system to another (Janapati et al., 2019, 2020a; Nanko et al., 2016; van Dijk et al., 2005), here, we tried to establish these relations for the NIO TCs measured at the coastal and inland station. The scatter plots of rainfall KE (KE_{time} and KE_{mm}) with rainfall rate and mass-weighted mean diameter and their polynomial fit equations are illustrated in Fig. 9. These relations demonstrated the disparities between the NIO TCs measured at the coastal and inland stations.

To reveal the possible rationale for the differences in the RSDs for NIO TCs measured at the coastal and inland stations, the CAPE, water vapor, air temperature, and relative humidity profiles are demonstrated in Figs. 10 & 11. Higher CAPE and water vapor values for the NIO TCs measured at the coastal station depict intense convection with strong updrafts and downdrafts (collision-coalescence and

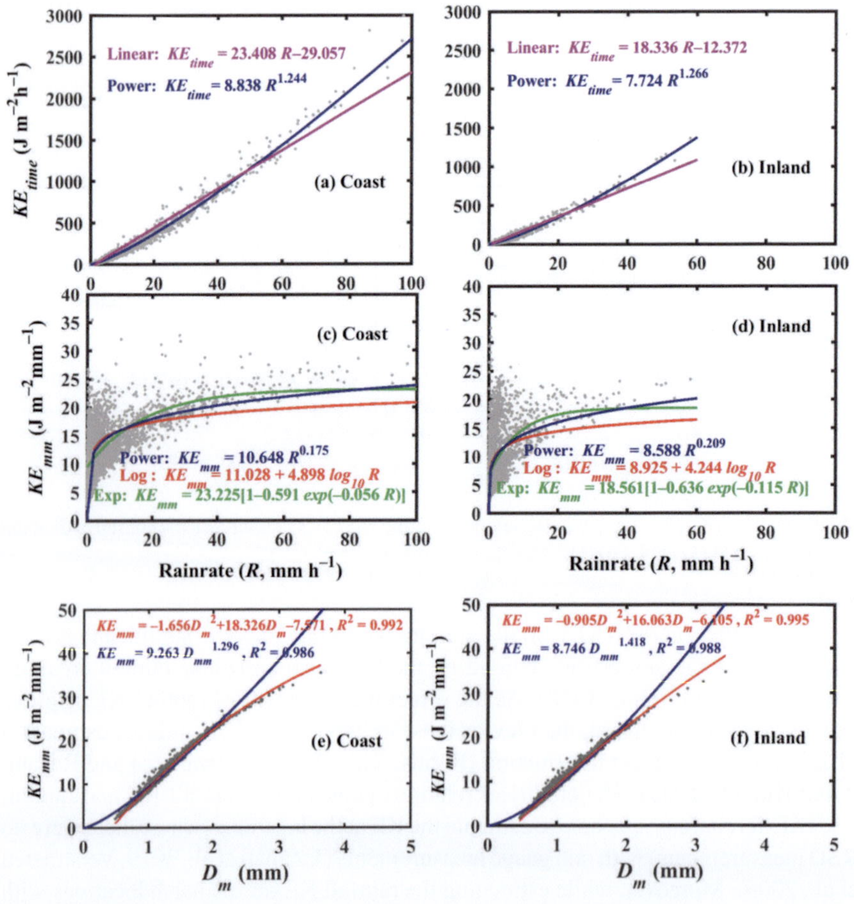

Fig. 9 KE_{time}/KE_{mm}–R, and KE_{mm}–D_m relations for the NIO TCs measured at the coastal and inland station. *From* Seela et al. (2021)

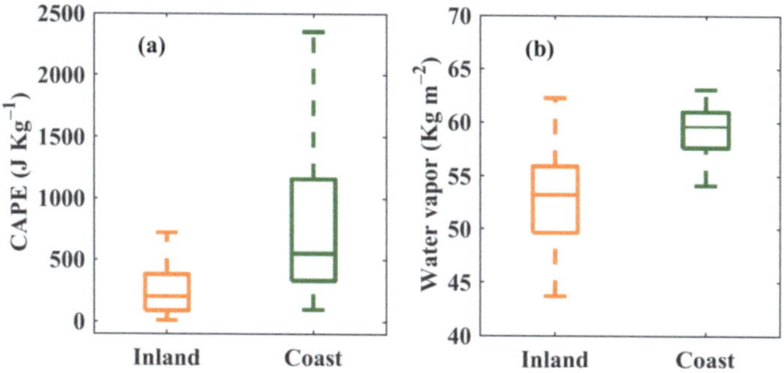

Fig. 10 Box plots of CAPE and water vapor for the NIO TCs measured at the coastal and inland stations. *From* Seela et al. (2021)

drop sorting processes), leading to the growth of the precipitating particles through the enhanced vapor deposition and aggregation in the cold (warm) regions of the TCs cloud systems. A loft suspension of the small drops by the intense updrafts provides adequate time for the growth of the mid-size drops via collision and coalescence processes, resulting in large-size drops in the NIO TCs measured at the coastal station than the inland station (Atlas et al., 2000; Hu and Srivastava, 1995b). The possibility for the small-drops evaporation, which was generated via collision-breakup mechanisms, at the coastal station can be confirmed with relatively higher temperatures and lower relative humidity values (Fig. 11). Henceforth, the above mechanism could result in a large drop in the NIO TCs measured at the coastal station.

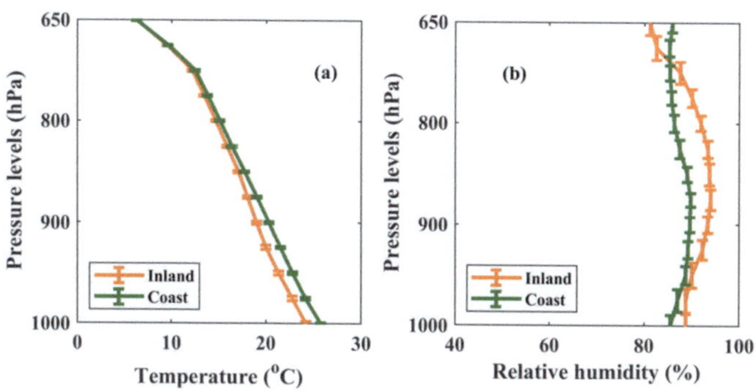

Fig. 11 Air temperature and relative humidity profiles for the NIO TCs measured at the coastal and inland stations. *From* Seela et al. (2021)

5 Raindrop Size Distributions of Pacific Ocean Tropical Cyclones

The RSDs of the WP TCs measured at the Palau islands (Fig. 12, red circle in Fig. 1) offer the opportunity to understand the microphysical attributes of the WP TCs over the oceanic area. The track information of 14 WP TCs measured with the JWD installed at the Palau islands, and their stratiform/convective precipitation mean raindrop size distribution, and variations of D_m with $\log_{10}N_w$ are illustrated in Fig. 13. The RSDs of the WP TCs measured at the Palau islands are segregated into stratiform and convective type by adopting the modified methodology of (Bringi et al., 2003). Especially mean and standard deviation of 10 consecutive samples are >5 mm h^{-1} and 1.5 mm h^{-1}, respectively, such samples are considered convective, otherwise stratiform. This classification results in 80% of RSD samples into the stratiform type and 20% RSD samples into convective type, contributing to 33 and 67% rainfall, respectively. The stratiform & convective mean RSDs of WP TCs showed clear distinctions with an exponential distribution in stratiform precipitation and broader distribution in convective precipitation. The higher D_m and $\log_{10}N_w$ values in convective precipitation can be related to the broader distributions of the RSD spectrum that arises due to the increased collision-breakup processes in convective precipitation (Hu and Srivastava, 1995b).

The inability to directly measure the rainfall amounts by the operational radars urges the adaptation of relationships between the radar reflectivity (Z) & rainfall rate (*R*) (known as the *Z–R relations*). These relations can be achieved with the aid of RSD information. Furthermore, the coefficient and exponent values of the *Z–R relations* ($Z = AR^b$) can also be used to reveal the microphysical attributes of a given precipitation type. The larger coefficient (A) infers the large drops. The coefficient value greater than one denotes the domination of the collision-coalescence (size-controlled) process, while equal to unity represents the domination of collision, coalescence, and breakup (number-controlled) processes in a given precipitation (Atlas et al., 1999; Atlas and Williams, 2003; Steiner et al., 2004). The Z–R relations exhibited diversity with the geographic region, precipitation type, and weather conditions (Janapati et al., 2020b; Rosenfeld and Ulbrich, 2003; Seela et al., 2017; Ulbrich and Lee, 2002). As a result, the Z–R relations for the total, stratiform and convective precipitation of WP TCs are also evaluated (Fig. 11). Furthermore, relationships between the shape and slope parameters (μ–Λ relations) can help to find out the variabilities of the RSDs. These relations have further implications in quantitative precipitation estimation algorithms of ground-based radars and bulk microphysics schemes (Brandes et al., 2003, 2004a; Cao et al., 2008a; Milbrandt and Yau, 2006; Ulbrich, 1983; Zhang et al., 2006b; Zhang et al., 2001b; Zhang et al., 2003b). For instance, these relations can be used in the constrained gamma distribution to reduce the rainfall estimation errors of the polarimetric radars (Zhang et al., 2003b). Hence, the μ–Λ relations for stratiform, convective, and total precipitation of WP TCs are also evaluated (Fig. 15). Established RSD relations (Z–*R and* μ–Λ relations) indicate apparent disparities between the stratiform and convective precipitation (Figs. 14 and 15).

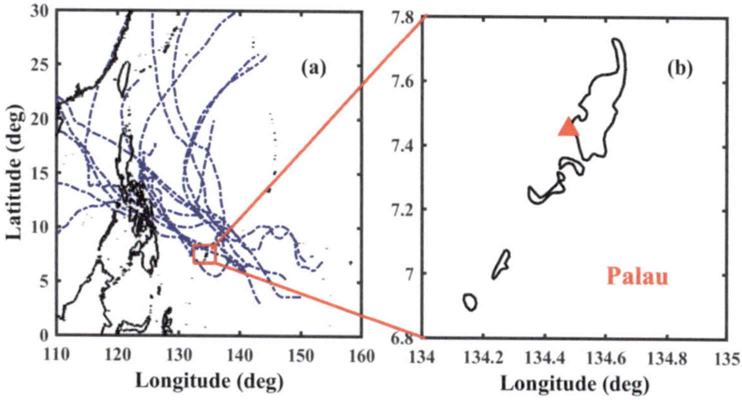

Fig. 12 Geographic Map of Palau islands and tracks of WP TCs measured with JWD (red filled triangle). *From* Seela et al. (2022)

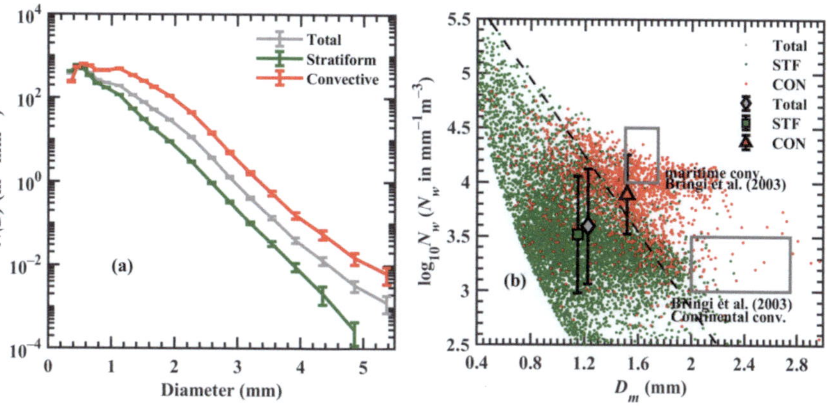

Fig. 13 Mean raindrop size distribution and distribution between Dm and log10Nw for WP TCs' total, stratiform, and convective precipitation. *From* Seela et al. (2022)

As mentioned in Sect. 4, rainfall KE and intensity relations can help to estimate the rainfall KE at the locations with the rain gauge measurements. The distribution of rainfall KE (KEtime and KEmm) with rainfall rates for total, stratiform, and convective precipitations is illustrated in Fig. 16. An increase in KE_{time} with rainfall rate can be seen for stratiform and convective precipitation. However, a sharp and gradual rise of KE_{mm} can be seen for rainfall intensities less than 20 mm h^{-1} and greater than 20 mm h^{-1}, respectively. The KE-I relations estimated in different forms are also depicted in Fig. 16. By taking advantage of GPM DPR D_m measurements, one can calculate the rainfall KE using KE-D_m relations. Hence, the WP TCs RSDs are also used to establish KE-D_m relations. The distribution of rainfall KE with Dm for total, stratiform, and convective precipitations of WP TCs are depicted in

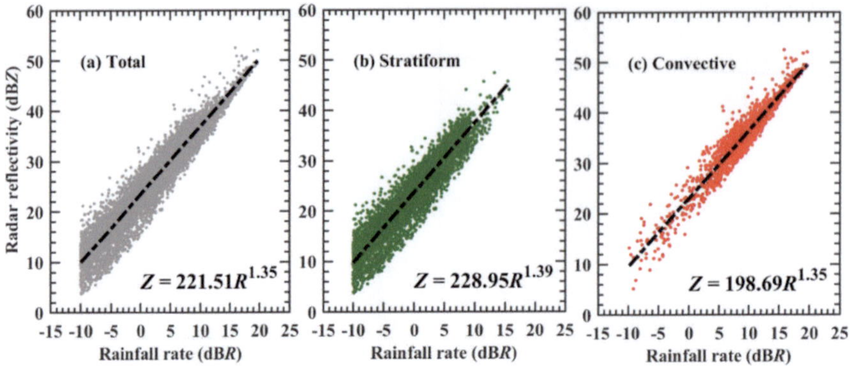

Fig. 14 Radar reflectivity and rainfall rate (*Z–R*) relations for WP TCs' total, stratiform, and convective precipitation. *From* Seela et al. (2022)

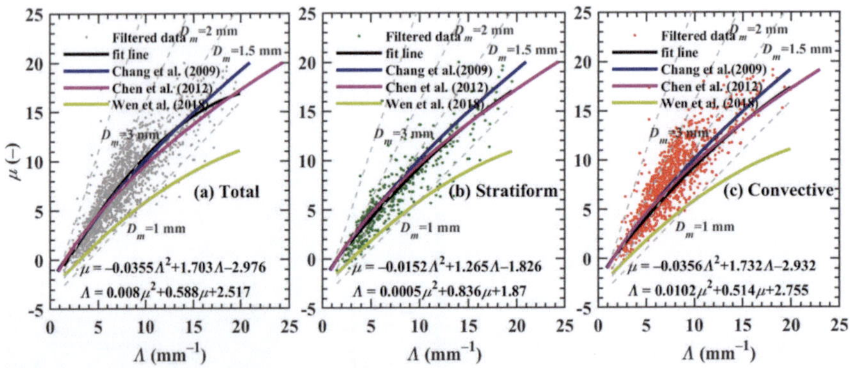

Fig. 15 Shape–slope relations for WP TCs' total, stratiform, and convective precipitations. *From* Seela et al. (2022)

Fig. 17. The estimated second-degree polynomial expression of KE-D_m relations are also shown in Fig. 17. These relations showed a distinction between stratiform and convective precipitation. Along with the estimated fit lines, the KE-D_m relation of WP TCs measured over Taiwan islands is also shown with the blue dotted line in Fig. 17 (Janapati et al., 2021). A negligible difference in the KE-D_m relations of WP TCs measured at the Palau and Taiwan islands demonstrated the nearly identical microphysical attributes of WP TCs over these two islands.

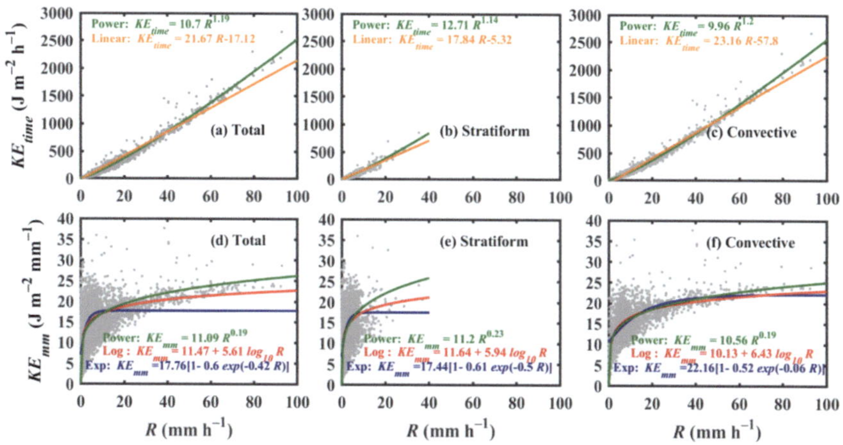

Fig. 16 KE_{time}–R and KE_{mm}–R relations for WP TCs' total, stratiform, and convective precipitations. *From* Seela et al. (2022)

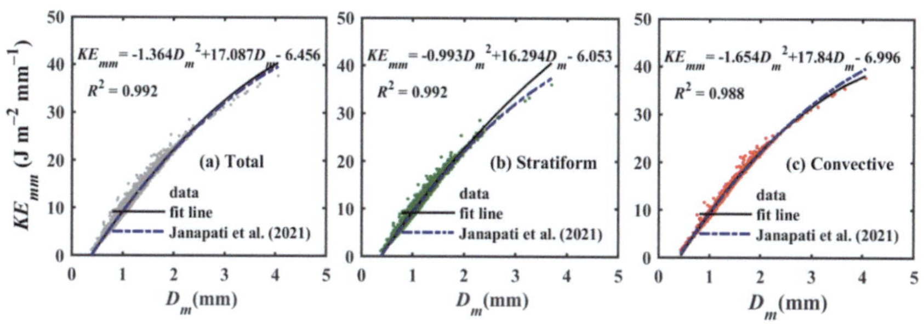

Fig. 17 Rainfall kinetic energy (KE_{mm}) and mass-weighted mean diameter (D_m) relations for WP TCs' total, stratiform and convective rainfall. *From* Seela et al. (2022)

6 Summary

This chapter tried to provide the raindrop size distributions (measured with ground-based disdrometers) of tropical cyclones (TCs) from two oceanic regions (the north Indian Ocean: NIO and the west Pacific Ocean: WP) and their variation with geography and precipitation types. Intercomparison of Indian and Pacific Ocean tropical cyclones demonstrated that the mass-weighted mean diameters are higher in WP TCs due to more mid- and large-size drops. A significant number of bigger drops in the WP TCs can be attributed to the association of intense precipitating clouds with surplus moisture in the WP TCs. Apart from the disparities in the RSDs between one oceanic region to other, regional variations were also observed for a given oceanic TCs. The RSDs of NIO TCs measured at the coastal and inland stations revealed a relatively large number of mid- and large-size drops at the coastal station than at the inland station. As the advancement of TCs in the inland region can cutoff the moisture supply from the ocean, strong convection in the TCs can reduce once they enter the inland area. On the other hand, intense cloud systems of TCs measured at the coastal station can result in bigger drops than at the inland station. The TCs measured at an oceanic station (Palau islands) in the WP also demonstrated distinctions in the RSDs between stratiform and convective precipitation. With the aid of RSDs information of TCs measured with the ground-based disdrometer, meteorological empirical (D_m-Zka/Zku, μ-Λ, and Z-R) relations are established. These established relations are handy for improving ground-based and remote-sensing radar precipitation estimation. Apart from the RSD relations, the RSDs of Indo-Pacific TCs are used to establish the hydrological relations (KE_{time}–R, KE_{mm}–R, and KE_{mm}–D_m), and these relations could help in estimating the improved rainfall KE and erosivity, particularly for the Indo-Pacific TCs.

References

Atlas, D., Ulbrich, C. W., Marks, F. D., Jr., Amitai, E., & Williams, C. R. (1999). Systematic variation of drop size and radar-rainfall relations. *Journal of Geophysical Research: Atmospheres, 104*, 6155–6169.

Atlas, D., et al. (2000). Partitioning tropical oceanic convective and stratiform rains by draft strength. *Journal of Geophysical Research: Atmospheres, 105*, 2259–2267.

Atlas, D., & Williams, C. R. (2003). The anatomy of a continental tropical convective storm. *Journal of the Atmospheric Sciences, 60*, 3–15.

Brandes, E. A., Zhang, G., & Vivekanandan, J. (2002). Experiments in rainfall estimation with a polarimetric radar in a subtropical environment. *Journal of Applied Meteorology, 41*, 674–685.

Brandes, E. A., Zhang, G., & Vivekanandan, J. (2003). An evaluation of a drop distribution-based polarimetric radar rainfall estimator. *Journal of Applied Meteorology, 42*, 652–660.

Brandes, E. A., Zhang, G., & Vivekanandan, J. (2004a). Comparison of polarimetric radar drop size distribution retrieval algorithms. *Journal of Atmospheric and Oceanic Technology, 21*, 584–598.

Brandes, E. A., Zhang, G., & Vivekanandan, J. (2004b). Comparison of polarimetric radar drop size distribution retrieval algorithms. *Journal of the Seismological Society of Japan, 21*, 584–598.

Bringi, V. N., et al. (2003). Raindrop size distribution in different climatic regimes from disdrometer and dual-polarized radar analysis. *Journal of the Atmospheric Sciences, 60*, 354–365.

Cao, Q., et al. (2008a). Analysis of video disdrometer and polarimetric radar data to characterize rain microphysics in Oklahoma. *Journal of Applied Meteorology and Climatology, 47*, 2238–2255.

Cao, Q., et al. (2008b). Analysis of video disdrometer and polarimetric radar data to characterize rain microphysics in Oklahoma. *Journal of Applied Meteorology and Climatology, 47*, 2238–2255.

Chakravarty, K., et al. (2021). Characteristics of precipitation microphysics during Tropical Cyclone Nisarga (2020) as observed over the orographic region of Western Ghats in the Indian subcontinent. *Atmospheric Research*, 105861.

Chang, W.-Y., Wang, T.-C.C., & Lin, P.-L. (2009). Characteristics of the raindrop size distribution and drop shape relation in typhoon systems in the Western Pacific from the 2D video disdrometer and NCU C-band polarimetric radar. *Journal of Atmospheric and Oceanic Technology, 26*, 1973–1993.

Chen, B., Yang, J., & Pu, J. (2013a). Statistical characteristics of raindrop size distribution in the meiyu season observed in Eastern China. *Journal of the Meteorological Society of Japan Series II, 91*, 215–227.

Chen, C.-W., et al. (2019). Assessing landslide characteristics in a changing climate in Northern Taiwan. *CATENA, 175*, 263–277.

Chen, Y.-C., Chang, K.-T., Chiu, Y.-J., Lau, S.-M., & Lee, H.-Y. (2013b). Quantifying rainfall controls on catchment-scale landslide erosion in Taiwan. *Earth Surface Processes and Landforms, 38*, 372–382.

Chen, Y.-C., Chang, K.-T., Lee, H.-Y., & Chiang, S.-H. (2015). Average landslide erosion rate at the watershed scale in Southern Taiwan estimated from magnitude and frequency of rainfall. *Geomorphology, 228*, 756–764.

Chen, Y., Duan, J., An, J., & Liu, H. (2019b). Raindrop size distribution characteristics for tropical cyclones and meiyu-baiu fronts impacting Tokyo, Japan. *Atmosphere-Basel, 10*, 391.

Dadson, S. J., et al. (2003). Links between erosion, runoff variability and seismicity in the Taiwan orogen. *Nature, 426*, 648.

Fornis, R. L., Vermeulen, H. R., & Nieuwenhuis, J. D. (2005). Kinetic energy-rainfall intensity relationship for Central Cebu, Philippines for soil erosion studies. *Journal of Hydrology, 300*, 20–32.

Gunn, R., & Kinzer, G. D. (1949). The terminal velocity of fall for water droplets in stagnant air. *Journal of Meteorology, 6*, 243–248.

Hopper, L. J., Schumacher, C., Humes, K., & Funk, A. (2020). Drop-size distribution variations associated with different storm types in Southeast Texas. *Atmosphere-Basel, 11*, 8.

Hu, Z., & Srivastava, R. C. (1995a). Evolution of raindrop size distribution by coalescence, breakup, and evaporation: Theory and observations. *Journal of Atmospheric Science, 52*, 1761–1783.

Hu, Z., & Srivastava, R. C. (1995b). Evolution of raindrop size distribution by coalescence, breakup, and evaporation: Theory and observations. *Journal of the Atmospheric Sciences, 52*, 1761–1783.

Janapati, J., Seela, B. K., Lin, P.-L., Lee, M.-T., Joseph, E. (2020a). Microphysical features of typhoon and non-typhoon rainfall observed in Taiwan, an island in the northwest Pacific. In *Hydrology and Earth System Sciences Discussions* (pp. 1–44).

Janapati, J., Seela, B. K., Lin, P.-L., Lee, M.-T., & Joseph, E. (2021). Microphysical features of typhoon and non-typhoon rainfall observed in Taiwan, an island in the northwestern Pacific. *Hydrology and Earth System Sciences, 25*, 4025–4040.

Janapati, J., Seela, B. K., Lin, P.-L., Wang, P. K., Kumar, U. (2019). An assessment of tropical cyclones rainfall erosivity for Taiwan. *Scientific Reports, 9*, 15862.

Janapati, J., et al. (2017). A study on raindrop size distribution variability in before and after landfall precipitations of tropical cyclones observed over southern India. *Journal of Atmospheric and Solar-Terrestrial Physics, 159*, 23–40.

Janapati, J., et al. (2020b). Raindrop size distribution characteristics of Indian and Pacific Ocean tropical cyclones observed at India and Taiwan sites. *Journal of the Meteorological Society of Japan Series II, 98*, 299–317.

Janapati, J., et al. (2023). An assessment of rainfall kinetic energy functional relationships with GPM DPR. *Journal of Hydrology, 617*, 128754.

Jayawardena, A. W., & Rezaur, R. B. (2000). Measuring drop size distribution and kinetic energy of rainfall using a force transducer. *Hydrological Processes, 14*, 37–49.

Joss, J., & Waldvogel, A. (1969). Raindrop size distribution and sampling size errors. *Journal of the Atmospheric Sciences, 26*, 566–569.

Kinnell, P. I. A. (1981). Rainfall intensity-kinetic energy relationships for soil loss prediction 1. *Soil Science Society of America Journal, 45*, 153–155.

Kumar, S. B. & Reddy, K. K. (2013). Rain drop size distribution characteristics of cyclonic and north east monsoon thunderstorm precipitating clouds observed over Kadapa (14.47°N, 78.82°E), tropical semi-arid region of India. *Mausam, 64,* 35–48.

Kumari, N., Kumar, S. B., Jayalakshmi, J., & Reddy, K. K. (2014). Raindrop size distribution variations in JAL and NILAM cyclones induced precipitation observed over Kadapa (14.47 o N, 78.82 o E), a tropical semi-arid region of India. *Indian Journal of Radio and Space Physics, 43*, 57–66.

Lee, M.-T., Lin, P.-L., Chang, W.-Y., Seela, B. K., & Janapati, J. (2019). Microphysical characteristics and types of precipitation for different seasons over North Taiwan, *Journal of the Meteorological Society of Japan. Series II*.
Liao, L., & Meneghini, R. (2005). A study of air/space-borne dual-wavelength radar for estimation of rain profiles. *Advances in Atmospheric Sciences, 22*, 841–851.
Liao, L., & Meneghini, R. (2019). Physical evaluation of GPM DPR single- and dual-wavelength algorithms. *Journal of Atmospheric and Oceanic Technology, 36*, 883–902.
Liao, L., Meneghini, R., Iguchi, T., & Tokay, A. (2020). Characteristics of DSD bulk parameters: Implication for radar rain retrieval. *Atmosphere-Basel, 11*, 670.
Liao, L., Meneghini, R., & Tokay, A. (2014). Uncertainties of GPM DPR rain estimates caused by DSD parameterizations. *Journal of Applied Meteorology and Climatology, 53*, 2524–2537.
Löffler-Mang, M., & Joss, J. (2000). An optical disdrometer for measuring size and velocity of hydrometeors. *Journal of Atmospheric and Oceanic Technology, 17*, 130–139.
Mcfarquhar, G. M., Hsieh, T.-L., Freer, M., Mascio, J., & Jewett, B. F. (2015). The characterization of ice hydrometeor gamma size distributions as volumes in N0–λ–μ phase space: Implications for microphysical process modeling. *Journal of the Atmospheric Sciences, 72*, 892–909.
Milbrandt, J. A., & Yau, M. K. (2006). A multimoment bulk microphysics parameterization. Part IV: Sensitivity experiments. *Journal of the Atmospheric Sciences, 63*, 3137–3159.
Mishchenko, M. I., Travis, L. D., & Mackowski, D. W. (1996). T-matrix computations of light scattering by nonspherical particles: A review. *Journal of Quantitative Spectroscopy and Radiative Transfer, 55*, 535–575.
Nanko, K., Moskalski, S. M., & Torres, R. (2016). Rainfall erosivity–intensity relationships for normal rainfall events and a tropical cyclone on the US southeast coast. *Journal of Hydrology, 534*, 440–450.
Rosenfeld, D., & Ulbrich, C. W. (2003). Cloud microphysical properties, processes, and rainfall estimation opportunities. *Meteorological Monographs, 52*, 237–258.
Rosewell, C. J. (1986). Rainfall kinetic energy in Eastern Australia. *Journal of Applied Meteorology and Climatology, 25*, 1695–1701.
Salles, C., Poesen, J., & Sempere-Torres, D. (2002). Kinetic energy of rain and its functional relationship with intensity. *Journal of Hydrology, 257*, 256–270.
Seela, B. K., Janapati, J., Lin, P.-L., Wang, P. K., & Lee, M.-T. (2018). Raindrop size distribution characteristics of summer and winter season rainfall over North Taiwan. *Journal of Geophysical Research: Atmospheres, 123*, 11,602–611,624.
Seela, B. K., et al. (2017). A comparison study of summer season raindrop size distribution between Palau and Taiwan, two islands in Western Pacific. *Journal of Geophysical Research: Atmospheres, 122*, 11,787–711,805.
Seela, B. K., et al. (2021). Raindrop size distributions of North Indian Ocean tropical cyclones observed at the coastal and inland stations in South India. *Remote Sensing, 13*, 3178.
Seela, B. K., et al. (2022). Raindrop size distribution characteristics of the Western Pacific tropical cyclones measured in the Palau Islands. *Remote Sensing, 14*, 470.
Sempere-Torres, D., Salles, C., Creutin, J. D., & Delrieu, G. (1992). Quantification of soil detachment by raindrop impact: performance of classical formulae of kinetic energy in mediterranean storms. In *Erosion and Sediment Transport Monitoring Programmes in River Basins (Proceedings of the Oslo Symposium, August 1992)* (pp. 115–124). IAHS Publications.
Steiner, M., Smith, J. A., & Uijlenhoet, R. (2004). A microphysical interpretation of radar reflectivity-rain rate relationships. *Journal of the Atmospheric Sciences, 61*, 1114–1131.
Suh, S. H., You, C. H., & Lee, D. I. (2016). Climatological characteristics of raindrop size distributions in Busan, Republic of Korea. *Hydrology and Earth System Sciences, 20*, 193–207.
Tang, Q., Xiao, H., Guo, C., & Feng, L. (2014). Characteristics of the raindrop size distributions and their retrieved polarimetric radar parameters in Northern and Southern China. *Atmospheric Research, 135–136*, 59–75.

Thurai, M., Williams, C. R., & Bringi, V. N. (2014). Examining the correlations between drop size distribution parameters using data from two side-by-side 2D-video disdrometers. *Atmospheric Research, 144*, 95–110.
Tokay, A., Bashor, P. G., Habib, E., & Kasparis, T. (2008). Raindrop size distribution measurements in tropical cyclones. *Monthly Weather Review, 136*, 1669–1685.
Uijlenhoet, R., & Stricker, J. N. M. (1999). A consistent rainfall parameterization based on the exponential raindrop size distribution. *Journal of Hydrology, 218*, 101–127.
Ulbrich, C. W. (1983). Natural variations in the analytical form of the raindrop size distribution. *Journal of Climate and Applied Meteorology, 22*, 1764–1775.
Ulbrich, C. W., & Lee, L. G. (2002). Rainfall characteristics associated with the remnants of tropical storm Helene in upstate South Carolina. *Weather and Forecasting, 17*, 1257–1267.
van Dijk, A. I. J. M., Bruijnzeel, L. A., & Rosewell, C. J. (2002). Rainfall intensity-kinetic energy relationships: A critical literature appraisal. *Journal of Hydrology, 261*, 1–23.
van Dijk, A. I. J. M., Meesters, A. G. C. A., Schellekens, J., & Bruijnzeel, L. A. (2005). A two-parameter exponential rainfall depth-intensity distribution applied to runoff and erosion modelling. *Journal of Hydrology, 300*, 155–171.
Verstraeten, G., Poesen, J., Demarée, G., & Salles, C. (2006). Long-term (105 years) variability in rain erosivity as derived from 10-min rainfall depth data for Ukkel (Brussels, Belgium): Implications for assessing soil erosion rates. *Journal of Geophysical Research, 111.*
Williams, C. R., et al. (2014). Describing the shape of raindrop size distributions using uncorrelated raindrop mass spectrum parameters. *Journal of Applied Meteorology and Climatology, 53*, 1282–1296.
Wischmeier, W. H., & Smith, D. D. (1958). Rainfall energy and its relationship to soil loss. *Transactions, American Geophysical Union, 39*, 285–291.
Wu, D., et al. (2021). Evaluation of microphysics schemes in tropical cyclones using polarimetric radar observations: Convective precipitation in an outer rainband. *Monthly Weather Review, 149*, 1055–1068.
Xia, Q., et al. (2020). Quantification of precipitation using polarimetric radar measurements during several typhoon events in Southern China. *Remote Sensing, 12*, 2058.
Yuter, S. E., Kingsmill, D. E., Nance, L. B., & Löffler-Mang, M. (2006). Observations of precipitation size and fall speed characteristics within coexisting rain and wet snow. *Journal of Applied Meteorology and Climatology, 45*, 1450–1464.
Zhang, G., Sun, J., & Brandes, E. A. (2006a). Improving parameterization of rain microphysics with disdrometer and radar observations. *Journal of Atmospheric Science, 63*, 1273–1290.
Zhang, G., Sun, J., & Brandes, E. A. (2006b). Improving parameterization of rain microphysics with disdrometer and radar observations. *Journal of Atmospheric Science, 63*, 1273–1290.
Zhang, G., Vivekanandan, J., & Brandes, E. (2001a). A method for estimating rain rate and drop size distribution from polarimetric radar measurements. *IEEE Transactions on Geoscience and Remote Sensing, 39*, 830–841.
Zhang, G., Vivekanandan, J., & Brandes, E. (2001b). A method for estimating rain rate and drop size distribution from polarimetric radar measurements. *IEEE Transactions on Geoscience and Remote Sensing, 39*, 830–841.
Zhang, G., Vivekanandan, J., Brandes, E. A., Meneghini, R., & Kozu, T. (2003a). The shape-slope relation in observed gamma raindrop size distributions: Statistical error or useful information? *Journal of the Seismological Society of Japan, 20*, 1106–1119.
Zhang, G., Vivekanandan, J., Brandes, E. A., Meneghini, R., & Kozu, T. (2003b). The shape-slope relation in observed gamma raindrop size distributions: Statistical error or useful information? *Journal of Atmospheric and Oceanic Technology, 20*, 1106–1119.
Zhang, Y., et al. (2019). Analysis of dual-polarimetric radar variables and quantitative precipitation estimators for landfall typhoons and squall lines based on disdrometer data in Southern China. *Atmosphere-Basel, 10*, 30.

Part 2: Modelling, Assimilation, Simulation and Forecasting of Severe Storms

Analysis and Numerical Modeling of Linear-Shaped Mesoscale Convective Systems

Tetsuya Takemi and Nanami Naka

Abstract A linear-shaped mesoscale convective system (MCS) is one of the major forms of organized deep convection. A moving linear-shaped MCS is known as a squall line. This chapter focuses on how the development and intensity of linear-shaped MCSs are affected by environmental stability and humidity conditions. Numerical experiments with idealized settings indicated that the intensity of convective updrafts within simulated linear-shaped MCSs becomes weaker under tropical/oceanic conditions than under midlatitude/continental conditions. The dependence of the convective intensity on the free tropospheric humidity appears differently in environments with different temperature lapse rate. The analysis on the climatological characteristics of quasi-stationary MCSs in Japan, as a case of MCSs that develop in humid climate regions, indicated that middle-level relative humidity controls the development of such MCSs and also that linear-shaped MCSs preferably occur with a certain amount of vertical shear. High-resolution modeling of a linear-shaped MCS demonstrated that the horizontal resolution of about 100 m is able to capture the successive development of convective cells on the upstream side of a linear-shaped MCS and the stationarity of the MCS as a whole through representing more precisely a complex and complicated nature of topography and explicitly resolving convective motion.

Keywords Mesoscale convective system · Squall line · Linear-shaped mesoscale convective system · Convective organization · Environmental control · Mesoscale meteorological modeling

T. Takemi (✉) · N. Naka
Disaster Prevention Research Institute, Kyoto University, Gokasho, Uji, Kyoto 611-0011, Japan
e-mail: takemi@storm.dpri.kyoto-u.ac.jp

S. Das and W-K. Tao (eds.), *Severe Storms*,
https://doi.org/10.1007/978-981-97-7075-5_9

1 Introduction

Deep cumulus clouds organize as mesoscale systems when appropriate environmental conditions develop. Mesoscale convective systems (MCSs) can occur in various climate regions of the world and sometimes cause severe weather such as strong winds, heavy rainfalls, and hailfalls. Understanding and predicting the processes related to MCSs and resulting severe weather are challenging issues in mesoscale meteorology and weather forecasting. Furthermore, environmental changes under future global warming are anticipated to worsen the threat of severe weather resulting from MCSs. Assessing the impacts of climate change is also critically important to design adaptation strategy and measure against severe weather in a future warmed climate.

MCSs develop in various forms. The morphology of MCSs can be classified as a linear and a circular type; a linear-type MCS is called as a squall line, while a circular-type MCS as a cloud cluster or mesoscale convective complex (Maddox, 1980). Among these MCSs, squall lines have been extensively investigated in the literature. With the development of precipitation radar networks, the convective-scale processes and organization modes of squall lines have been able to be examined in detail. A pioneering work of Bluestein and Jain (1985) demonstrated, through examining the radar images of squall lines in the central plains of the United States, that the distinguishing mode of squall line organization was classified into 4 types: broken line; back building; broken areal; and embedded areal. The environmental properties that distinguish these squall-line modes were found to be vertical shear in the lower troposphere and convective instability.

From theoretical and numerical studies, vertical shear in the troposphere has been shown to determine the organization mode of MCSs (Weisman & Klemp, 1982; Rotunno et al., 1988; Fovell & Ogura, 1989; Weisman, 1992; Robe & Emanuel, 2001; Weisman & Rotunno, 2004). Observationally, LeMone et al. (1998) indicated that the intensity of vertical shears at different height levels (i.e., low-level vs. middle-level) characterizes the direction and shape of squall lines in tropical oceanic regions. Parker and Johnson (2000) demonstrated that the magnitude and direction of the tropospheric vertical shear affects the shape and evolution of midlatitude squall lines with stratiform precipitation. The organization mode of squall lines was further investigated for those occur in east China and found to form generally in a broken-line mode with a trailing-stratiform pattern (Meng et al., 2013). Unuma and Takemi (2016a) examined the morphology of meso-β-scale, quasi-stationary convective clusters (Unuma & Takemi, 2016b) and found that the vertical shear in the lower troposphere is a distinguishing factor which determines the shape of quasi-stationary convective clusters as a linear or a circular shape. In this way, the role of vertical shear has been recognized to significantly affect the organization mode and morphology of MCSs, especially squall lines.

Compared with the role of shear, the effects of convective instability have not well been explored. In the moist atmosphere, the amount of water vapor influences the stability conditions and hence should be considered in delineating the stability control of MCSs. A study that investigated the sensitivity to convective instability was conducted by Weisman and Klemp (1982) who numerically showed that the mode of convective storms can be categorized as a multi-cellular or super cellular type depending on the stability and shear conditions. In their study, a low-level moisture content was systematically changed in order to set various stability conditions under a prescribed vertical shear. It should be noted that in Weisman and Klemp (1982) the continental, midlatitude region was assumed in setting the vertical profiles of temperature, humidity, and winds; therefore, strong cold pools are generated and play a major role in developing and maintaining MCSs under moderate-to-strong shear conditions. For tropical MCSs, Barnes and Sieckman (1984) pointed out that fast-moving squall lines which develop in drier middle-level environments are stronger than slow-moving counterparts, but, on the other hand, LeMone et al. (1998) suggested that a drier condition in the free troposphere below the 500-hPa level shorten the lifetimes of slow-moving lines. Thus, there are arguments on how tropical squall lines respond to the changes in the moisture. Actually, a study with the use of two-dimensional numerical experiments indicated that the squall-line properties vary significantly with changes in free-tropospheric humidity and shear profiles (Lucas et al., 2000). The impacts of environmental stability conditions on the structure and organization of MCSs should be systematically investigated.

With this recognition, a series of studies by Takemi (2006, 2007a, 2007b, 2010, 2014) conducted numerical experiments of linear-shaped MCSs, i.e., squall lines, under idealized environmental conditions with the vertical profiles of wind, temperature, and moisture being systematically varied and investigated how squall lines develop in response to these different conditions of shear and stability. In this chapter, reviewing those studies, we will firstly describe and summarize how the environmental stability and shear conditions affect and determine the development and organization of squall lines. Secondly, we will present the characteristics of MCSs that develop during the warm season over Japan, as an example of those occurring in humid climate regions. Quasi-stationary MCSs are specifically focused on. Then, we demonstrate the representation of a stationary, linear-shaped MCS in high-resolution numerical modeling from a case study on an extreme-rain-producing system over Japan. Finally, we will give concluding remarks and summarize this chapter.

2 Idealized Modeling of Squall Lines in Various Stability Conditions

2.1 Problem Setting

The strength of convective updraft depends on the tropospheric potential instability, according to the conservation law of mechanical energy in which potential energy can be converted to kinetic energy. The potential instability for the convective motion in the atmosphere can be diagnosed in terms of convective available potential energy

(CAPE). Zipser and LeMone (1980) and Jorgensen and LeMone (1989) showed that the strength of convective updraft in thunderstorms is stronger in the midlatitude case than in the tropical case. In Jorgensen and LeMone (1989), the updraft strength was compared for cases in the tropics, sub-tropics, and midlatitude regions and was qualitatively estimated via the value of CAPE. From the conservation law of mechanical energy, CAPE for an air parcel at the surface will be converted to the kinetic energy whose velocity corresponds to the magnitude of updraft at the level of neutral buoyancy. Although in the real atmosphere the mixing and diabatic processes will dilute the instability for an air parcel at the surface, the maximum updraft speed (w_{max}), which would be achieved for the surface air parcel if lifted to the level of neutral buoyancy, can be diagnosed in terms of CAPEas follows:

$$w_{\text{max}} = \sqrt{2\text{CAPE}}. \tag{1}$$

Note that, from a quantitative perspective, Eq. (1) neglects the pressure gradient force and turbulent mixing and thus should be regarded as a general guide to compare the updraft strength under different CAPE conditions. By examining the vertical change of excess temperature for an air parcel from the ambient environment, Lucas et al. (1994a) noted that the area of the positive excess temperature is "skinny" over the ocean but is "fat" over the continental region of the United States. Figure 1 indicates how the environmental temperature affects the value of CAPE through the changes in the magnitude of the excess temperature. The skinny profile indicated by the green line in Fig. 1 suggests that the smaller buoyancy for a lifted air parcel over the ocean is easily diluted with the environment though mixing (i.e., entrainment) when compared with a larger buoyancy over the continent in which the fat profile indicated by the blue line in Fig. 1 is seen. In other words, updraft air parcels in midlatitude, continental convective clouds will be less diluted with their environments and can be more accelerated by positive buoyancy than the air parcels in tropical, oceanic convection.

Therefore, the magnitude of buoyancy (or positive temperature anomaly from the background state) has to be carefully examined in order to assess the updraft intensity. The excess temperature from the background depends on the temperature lapse rate of the background state. In drier climate regions such as midlatitude, continental regions, the temperature lapse rate becomes closer to dry-adiabat. On the other hand, in more humid climate regions such as tropical, oceanic regions, the lapse rate becomes closer to moist-adiabat. Such changes in the temperature lapse rate in various climate regions will lead to the difference in the intensity of updrafts in convective clouds. Xu and Randall (2001) showed that the largest difference between the tropical and midlatitude convection is seen for the strongest vertical velocity. McCaul et al. (2005) demonstrated that stronger updrafts in cooler environments are due to larger positive buoyancy from fusion processes that start at lower altitudes than in warmer environments. By keeping CAPE unchanged in various tropospheric temperatures in numerical experiments, James et al. (2006) indicated that convective clouds in warmer, moister environments, which are effective in reducing the evaporation rate, generate weaker cold pools, which determines the three-dimensional

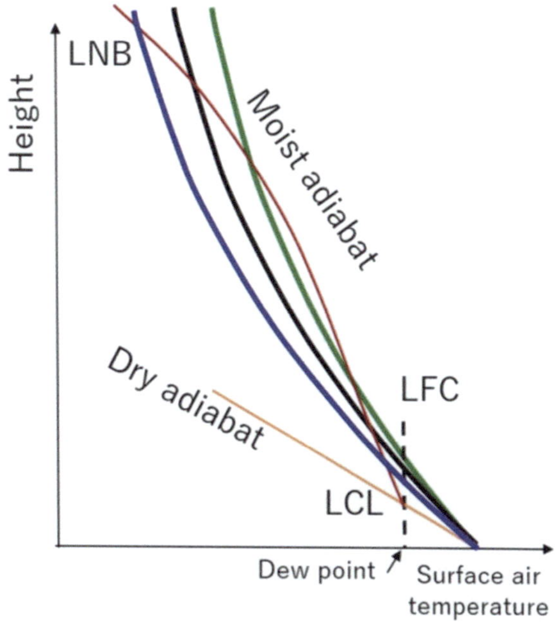

Fig. 1 The conceptual diagram showing how the surface air parcel obtains the positive temperature excess from the environment through the adiabatic lifting. The surface air parcel is lifted firstly along the dry adiabatic line (the orange line) up to the lifting condensation level (LCL) and secondly along the moist adiabatic line (the red line) through the level of free convection (LFC) up to the level of neutral buoyancy (LNB). The environmental temperature profiles are denoted as the green, black, and blue lines

structure of convective lines. By conducting numerical experiments, Houston and Niyogi (2007) further examined the influence of the temperature lapse rate above the level of free convection (LFC) on the development of cumulus convection and showed that stronger convection is produced under conditions with a larger temperature lapse rate.

In this way, the difference in the environmental temperature lapse rate is an important viewpoint to diagnose the strength of convective updraft and thus the organization and intensity of MCSs. Since the buoyancy depends not only on the properties of a low-level air parcel but also on the environmental temperature lapse rate, the vertical distribution of buoyancy is critically important for accelerating the air parcel upward (Lucas et al., 1994b; Takemi & Satomura, 2000; Takemi, 2010; Takemi, 2014). Therefore, the motivation of the idealized modeling of squall lines here is to investigate the sensitivity of the simulated squall lines to environmental stability conditions. In the followings, we will describe the design of numerical experiments and summarize the experimental results. We will then give implications for squall lines in the tropical, oceanic and the midlatitude, continental regions.

2.2 Experimental Design

In the studies of Takemi (2006, 2007a, 2007b, 2010, 2014), an idealized experimental approach was employed in order to focus on the dynamics of squall lines, as was done in Rotunno et al. (1988), Weisman et al. (1988), Fovell and Ogura (1989), Weisman (1992), Weisman and Rotunno (2004), etc. Here we specifically explain the experimental design employed in Takemi (2014).

The numerical model used here was the Weather Research and Forecasting (WRF) model, the Advanced Research WRF (ARW) version 3.1.1 (Skamarock et al., 2008). To conduct idealized experiments, the model was configured in a Cartesian computational domain of 300 km in the line-normal direction (east–west, referred to as x axis) by 60 km in the line-parallel direction (north–south, y axis) by 23 km (vertical, z axis), having minimum physics processes under horizontally homogeneous base states without the Coriolis force. The model physics here included cloud microphysics and turbulence mixing but ignored surface fluxes and atmospheric radiation. The cloud microphysics scheme employed here was the Goddard microphysics scheme (Tao et al., 1989; Tao and Simpson, 1993) which includes prognostic equations of the mixing ratios of water vapor, cloud water, rainwater, cloud ice, snow, and graupel, and the turbulence mixing was parameterized with the Deardorff scheme (1980) that solves a prognostic equation of turbulent kinetic energy to determine the eddy viscosity coefficient with the modification by Takemi and Rotunno (2003). The horizontal grid spacing was set to be 500m, which was considered to be within a convection-resolving range. The vertical grid spacing was stretched with height, varied in the range of 45–900 m. The lateral boundary conditions were periodic at the north and south boundaries and open at the east and west boundaries. The lower boundary was free slip, as we ignored the surface fluxes for the present idealized numerical experiments. In the upper layer, a Rayleigh-type damping layer was imposed within the 5-km layer from the rigid-lid upper boundary to prevent unwanted reflections at the upper boundary.

The base-state conditions were set to represent a tropical, oceanic environment and a midlatitude, continental environment. The temperature profile of a tropical, oceanic environment was prescribed with sounding data obtained for squall line over the tropical western Pacific during the field observation of the Tropical Ocean Global Atmosphere-Coupled Ocean Atmosphere Response Experiment (TOGA-COARE) in February 1993 (Trier et al., 1996). This environment will be hereinafter referred to as TROPICS. On the other hand, the midlatitude, continental profile was given with an analytical function representative of the environment for midlatitude convective storms (Weisman & Klemp, 1982). For the moisture profile, the data shown in Trier et al. (1996) were used to determine the TROPICS environment. For the moisture profile of the midlatitude environment, the analytical sounding of Weisman and Klemp (1982) was used and was divided into two types: a moist type (the original Weisman-Klemp humidity profile, referred to as MIDLATM) and a dry type (reduced humidity from the Weisman-Klemp profile, referred to as MIDLATD). The idea to define the dry type profile MIDLATD was to keep the vertical distribution of CAPE unchanged between the TROPICS and the MIDLATD environment. The vertical profiles of the base-state temperature and relative humidity in the cases of TROPICS, MIDLATD, MIDLATM are exhibited in Fig. 2.

The vertical profile of base-state winds was prescribed as having a uni-directional shear profile in a certain layer and a constant wind profile above the shear layer.

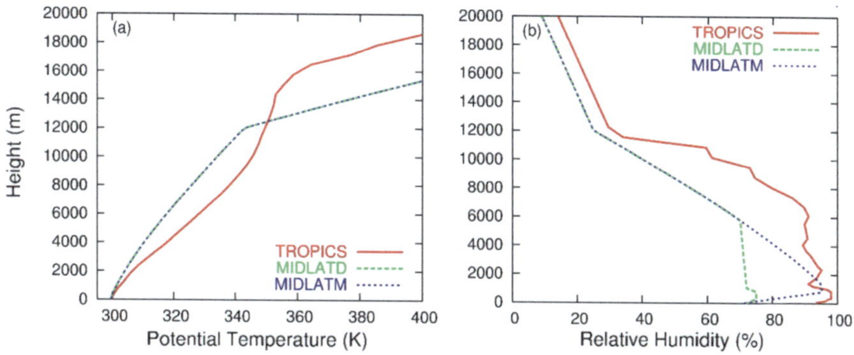

Fig. 2 The base-state vertical profiles of **a** temperature and **b** relative humidity in the cases of TROPICS (red line), MIDLATD (green dashed line), and MIDLATM (blue dotted line). ©2014 Elsevier. Used with permission

The prescribed shear layer was set either in the lower layer between 0 and 2.5 km (referred to as low-level shear) or in the middle layer between 2.5 and 5 km (referred to as middle-level shear) or in the deep layer between 0 and 5 km (referred to as deep shear). The magnitude of shear was chosen as the wind speed difference of 5 m s^{-1} and 15 m s^{-1} within the depth of 2.5 km (i.e., the low-level and middle-level shears) and the difference of 10 m s^{-1} within the depth of 5 km (i.e., the deep shear).

To initiate convection, we set either warm or cold linear-shaped potential temperature anomaly, aligned in the y direction placed at the central point of the x axis, having a peak value of +3 K for the warm thermal and -3 K for the cold pool with an elliptical shape in the $x - z$ cross section. A series of numerical experiments initialized by the warm line thermal and the cold pool was referred to as W-type and C-type, respectively. The conceptual diagram of the present numerical experiments is shown in Fig. 3.

Fig. 3 A conceptual diagram of the idealized experiments of squall lines

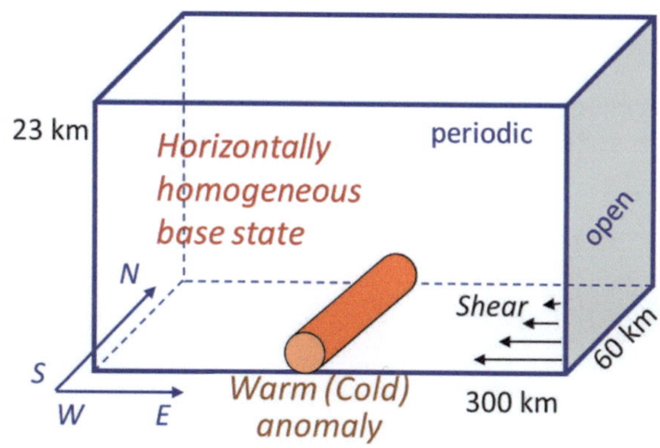

2.3 Results

In this subsection, we demonstrate the results from the above-mentioned numerical experiments, which were fully described in Takemi (2014).

The features of the updrafts within the simulated squall lines under different stability conditions in the lower layer, strong shear case are compared in Fig. 4. For the temporal averaged updraft speed, the both area-means and area-maxima are significantly stronger in the MIDLAT environment than in the TROPICS environment, except the case MIDLATD-W in which the convection did not develop with the warm thermal initialization in this drier case. The area size of stronger updraft ($w \geq 1$ m s^{-1}) is much larger in MIDLAT than in TROPICS (except the MIDLATD-W case). The low-level strong shear seems to distinguish more clearly the differences between the MIDLAT and TROPICS cases than the low-level weak shear cases (see Fig. 9 in Takemi, 2014). This is because the interaction between the low-level shear and a cold pool effectively strengthens the upward motion in cases which generate well-defined cold pools, according to the theory of Rotunno et al. (1988). The smaller fraction of strong updraft in the tropical environment seen in Fig. 4b is consistent with an observational evidence (LeMone & Zipser 1980) and modeling studies (Tao et al., 1987, 2022).

In the elevated shear case shown in Fig. 5, only the case MIDLATM-W case generates stronger and extensive updrafts within the simulated squall line. If a shear layer is elevated apart from the surface, the interaction between the low-level shear and the cold pool is less effective in order to sustain updrafts and hence convective systems, and therefore, the difference in the environmental stability will control more significantly the features of convective updrafts. Since the vertical profiles of CAPE between the TROPICS and MIDLATD environments are similar (see Fig. 2), the updrafts features appear to be similar between these environments.

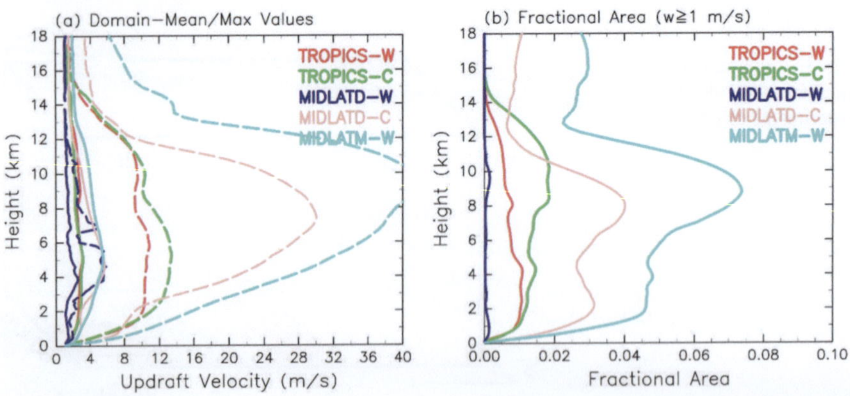

Fig. 4 The temporal averaged vertical profiles of **a** the mean (solid lines) and the maximum (dashed lines) updraft speed over the updraft grids and **b** the fractional area of updraft grids over the computational area for the case of the low-level, strong shear (15 m s^{-1} in the 0–2.5 km layer)

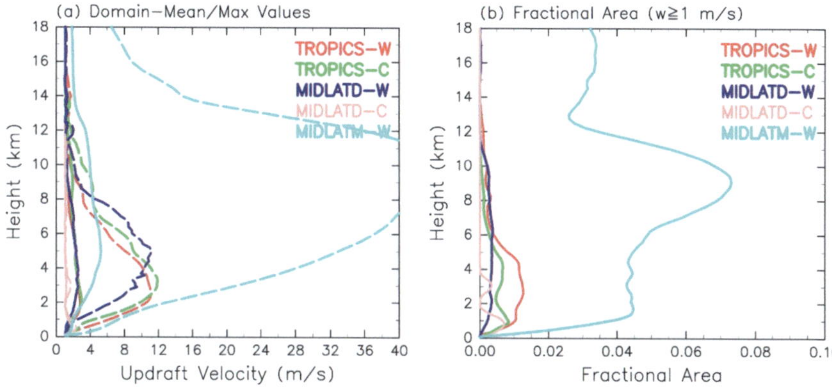

Fig. 5 The same as Fig. 4, except for the case of the elevated, strong shear (15 m s^{-1} in the 2.5–5 km layer)

The updraft features for all the sensitivity experiments are summarized in Fig. 6 for the means and the standard deviations obtained from the time series of the maximum updraft speed within the computational domain. In this figure, the simulated outputs over the whole computational domain from the simulated time period from 2 to 6 h were used. As seen in Figs. 4 and 5, the MIDLATM-W case produced the strongest updrafts among all the stability conditions for all the shear profiles. The maximum updraft is generally stronger in MIDLATD than in TROPICS under the low-level shear conditions; this feature seem to be consistent with the observations of updrafts whose peaks are much higher in the midlatitude than in the tropics (Zipser & LeMone, 1980; Jorgensen & LeMone, 1989; Igau et al., 1999). On the other hands, under the elevated shear conditions, the maximum updrafts in TROPICS are comparable to or slightly greater than those in MIDLATD. As described previously for Fig. 5, convection in TROPICS is as active as or more active than that in MIDLATD, because convection in the midlatitude environments becomes much weaker in MIDLATD than in MIDLATM in the elevated shear.

Figure 7 shows the mean precipitation intensity averaged in space over the computational area and in time during the period from 2 to 6 h. The standard deviation calculated from the time series of area-mean precipitation intensity is also included in this figure. Note that the spatiotemporal mean precipitation intensity is proportional to the total accumulation of precipitation over the computational area during the specified time frame. Consistent with the features shown in Figs. 4 and 5, MIDLATM-W produces the largest amount of precipitation for all the shear cases. On the other extreme, there is virtually no precipitation in the MIDLATD-W condition for every shear profile; furthermore, the lower troposphere is quite dry, and hence precipitation will evaporate before reaching the ground surface. Although the updraft appears to be strong in terms of its peak value, precipitation is not produced correspondingly. It was previously identified that the statistics of maximum updraft and mean precipitation do not necessarily agree with each other (Weisman & Rotunno, 2004).

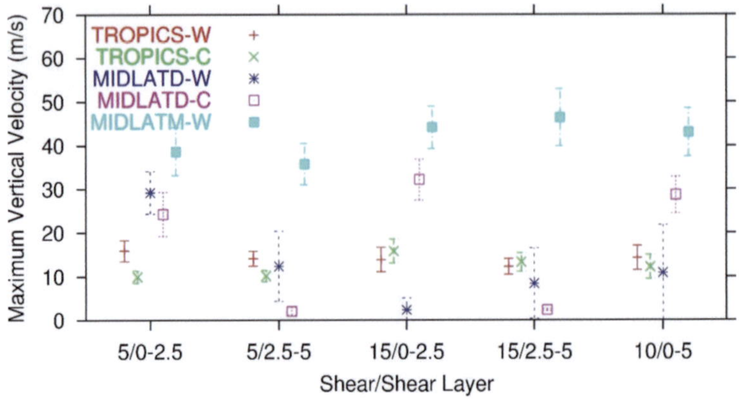

Fig. 6 The means (symbols) and standard deviations (bars) of the maximum updraft speed in the computational domain during 2 and 6 h for all the numerical experiments. Reproduced from Takemi (2014). ©2014 Elsevier. Used with permission

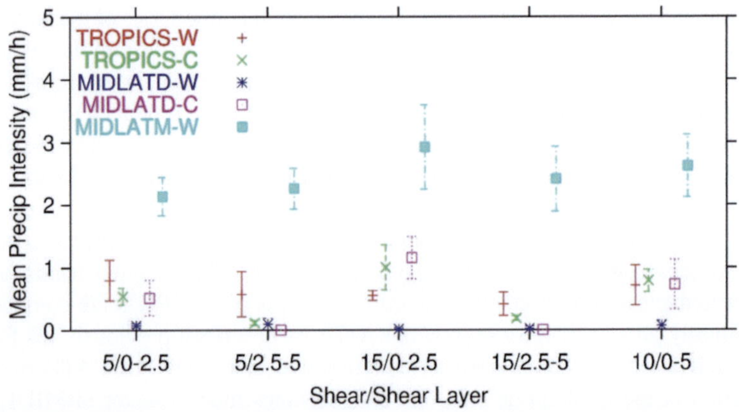

Fig. 7 The mean (symbols) and standard deviation (bars) from the time series of spatially averaged precipitation intensity over the computational area during the time period of 2 and 6 h. Reproduced from Takemi (2014). ©2014 Elsevier. Used with permission

In the MIDLATD-C case, precipitation is produced if the shear layer is based at the surface. On the other hand, in the TROPICS cases, the sensitivity to the initialization method as seen in the MIDLATD cases is not so significant. Depending on the height and intensity of the shear layer, the mean precipitation in the warm and the cold initialization varies, but the variation in TROPICS is less sensitive to the shear profile, compared with the results in the MIDLATD cases.

Takemi (2007a) proposed that the temperature lapse rate in the convectively unstable layer (which basically corresponds to the lower-to-the middle troposphere) well explains the relationship between the squall-line intensity and the stability condition. This lapse rate is written as:

$$\Gamma = -\frac{T(z=z(\theta_{e\,\min}))-T(z=z(\theta_{e\,\max}))}{z(\theta_{e\,\min})-z(\theta_{e\,\max})}, \tag{2}$$

where T is temperature, and $z(\theta_{emin})$ and $z(\theta_{emax})$ are the heights having the minimum and maximum equivalent potential temperatures, respectively. Note that both θ_{emin} and θ_{emax} appear within the convectively unstable layer which is below $z(\theta_{emin})$. With Eq. (2), the lapse rate in the convectively unstable layer is quantified. Takemi (2007b) found that the rainfall amount generated by MCSs scaled well with the lapse rate defined by Eq. (2). Furthermore, the vertical distribution of CAPE for each air parcel that originated at a different height was shown to strongly depend on the temperature lapse rate (Takemi, 2010). In the environment with a smaller lapse rate, the depth having a substantial amount of CAPE becomes shallower even though the CAPE value for the surface-based air parcel is not changed. If the depth having some CAPE values becomes deeper, air parcels not only at the lowest level but also at elevated levels in the lower troposphere will contribute to the upward motion. Updraft enhancement is definitely beneficial for the development of MCSs. The importance of the vertical distribution of CAPE in developing MCSs was emphasized by Takemi and Satomura (2000). They investigated the mechanism for the maintenance of squall lines in a dry environment and found that air parcels at elevated levels in the boundary layer are potentially unstable (i.e., having non-zero CAPE values) with short distances to their levels of free convection (LFCs) and hence are raised easily above LFCs to become convective cells. From a number of sensitivity experiments, Takemi (2010) found that the difference in the buoyancy profile under different temperature lapse rates can be interpreted as the difference in the vertical distribution of CAPE for surface-based and elevated air parcels. The vertical distribution of CAPE was also recognized to play a key role in the development of nocturnal convective systems (Parker, 2008).

3 Linear-Shaped Convective Systems in a Humid Condition

3.1 Observed Features of Linear-Shaped Convective Systems in Japan

The morphology and environmental properties of MCSs have been widely investigated as described in Sect. 1. However, most studies have paid attention to cases that occurred over continental, drier/semi-arid climate regions and over tropical oceanic regions. In contrast, less studies have conducted analyses on cases occurred in Asian monsoon regions. The work of Meng et al. (2013), briefly mentioned in Sect. 1, should be noted as a study examining cases occurred in East Asia. They used 2-year radar data to identify three patterns of the structural evolution of squall lines in east China and found that the squall lines in east China tend to form in a moister environment (and hence a larger amount of precipitable water vapor) with weaker vertical shear than the United States counter-parts like those shown in Bluestein and Jain (1985). They also showed that the largest instability and the most humid environment are seen in the squall lines with leading stratiform regions in east China, which contrast to the cases in the United States where trailing-stratiform-type squall lines display the most unstable and humid condition. The dominance of the leading-stratiform type in China may be due to the influences of synoptic-scale flow patterns such as upper-level trough and cyclonic circulation. In the following, the climatology of linear-shaped convective systems over Japan will be described, as an example of those occurring in humid climate regions.

Climatological analysis on convective systems over Japan was first conducted by Unuma and Takemi (2016b) who examined the composites of operational weather radars throughout the Japanese main islands. Despite anticipated limitation of the spatial coverage over the Japanese land area, long-term dataset for 8 years enabled to detect robust features of mesoscale precipitating systems during warm seasons (i.e., from May to October). Stationary and slow-moving systems were specifically chosen, which were termed as quasi-stationary convective clusters (QSCCs). An automated detection algorithm which was originally developed by Shimizu and Uyeda (2012) to identify an individual convective cell was used to extract QSCCs. To diagnose the environmental conditions for the occurrence of mesoscale precipitating systems, upper-air sounding data observed nearby stations were used. It was found that the typical horizontal scale of QSCCs is about 20 km, indicating that they are meso-β-scale phenomena. The comparison of the environmental conditions between the QSCC and no-rain cases indicated that the relative humidity at low- and middle-levels controls the development of QSCCs; specifically, the QSCCs favorably occur under conditions of higher humidity in the middle troposphere. The study suggests that the middle-tropospheric moistening before the convection development plays an important role in generating QSCCs.

Unuma and Takemi (2016a) further investigated the morphology of QSCCs and found that most QSCCs take a linear shape throughout the areas where QSCCs frequently occur (Fig. 8). The comparison of the environmental conditions for linear-shaped and circular-shaped QSCCs shows that CAPE and shear-related parameters are statistically significant in distinguishing the difference in the environmental conditions. Bulk Richardson number, which is a non-dimensional parameter combining CAPE and vertical shear, also separates significantly the environmental difference; the bulk Richardson number for the linear-shaped QSCCs is 32.8, significantly lower than that for the circular QSCCs, and is comparable to the value for the back-building squall lines in the United States (Bluestein & Jain 1985). Considering that the majority of QSCCs takes a linear shape, a mechanism for generating linear-shaped QSCCs is suggested to be a back-building process.

The orientation angles of linear-type QSCCs over Japan are demonstrated in Fig. 9. Linear-type QSCCs are oriented mostly in the southwest-northeast direction, followed by those oriented in the west–east direction. These dominant directions are determined primarily by monsoonal winds during the summer season in East Asia. In general, propagating squall lines as observed in the continental region of the

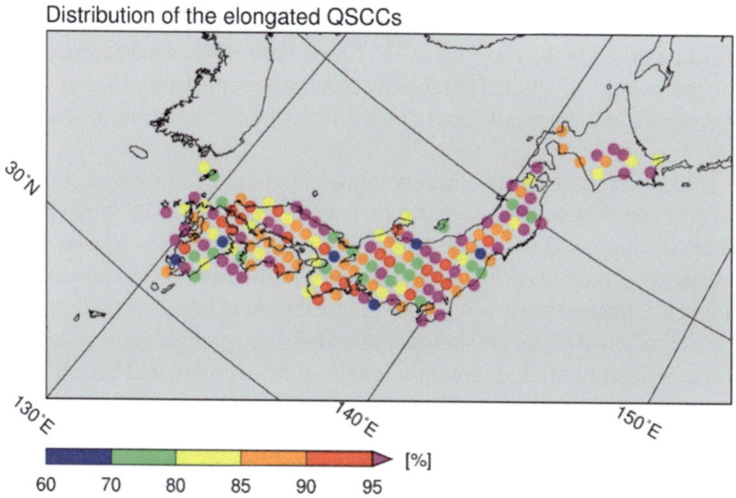

Fig. 8 The frequency distribution of the percentage of the number of linear-type quasi-stationary convective clusters to the total number of the quasi-stationary convective clusters over the Japanese islands. Reproduced from Unuma and Takemi (2016b). ©2016 The Meteorological Society of Japan. Used with permission

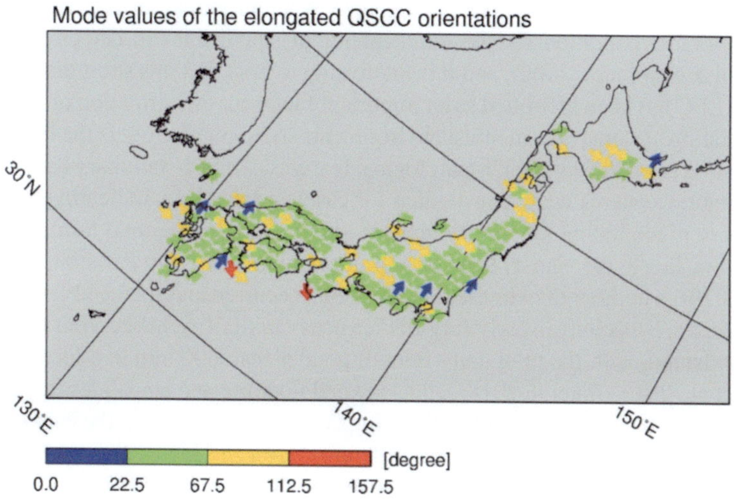

Fig. 9 The distribution of the mode values of the orientation angle for the linear-type quasi-stationary convective clusters over the Japanese islands. Reproduced from Unuma and Takemi (2016a). ©2016 The Meteorological Society of Japan. Used with permission

United States are oriented not so much in the dominant wind direction but rather in the shear direction. On the other hand, the linear-type QSCCs over Japan develop in environments where directional shear is not so large as in the United States. Therefore, dominant winds like the monsoonal flows affected the orientation of linear QSCCs in Japan.

The research progress and current understanding of stationary precipitating systems over Japan have been recently reviewed in Kato (2020). Special focus was paid to quasi-stationary band-shaped precipitation systems that produced heavy rainfalls (i.e., the total rainfall of 200 mm or greater), termed as "senjo-kousuitai". Senjo-kousuitai is a Japanese word and its direct translation is linear-shaped precipitation area. The definition of the quasi-stationary band-shaped precipitation systems in Kato (2020) is a banded rainfall area with a length of 50–300 km and a width of 20–50 km, generated by successive development of convective cells within similar areas. Major processes for the formation of such linear precipitation area are a broken-line type and a back-building type similar to those identified for squall lines in the United States (Bluestein & Jain, 1985). Kato (2020) listed 6 conditions favorable for the generation of "senjo-kousuitai": a larger amount of moisture flux at the 500-m height; short distance of LFC for an air parcel originated at the 500-m height; higher relative humidity at the 700- and 500-hPa levels; higher storm-relative helicity; a larger upward velocity at the 700-hPa level; and higher LNB for a parcel at the 500-m height. A short displacement distance of an air parcel to LFC was found to be a favorable condition for the maintenance of squall lines in dry environments (Takemi & Satomura, 2000), and it is interesting to note that this short displacement toward LFC has been identified as an important factor for the formation of mesoscale precipitating systems also in moist environments. Also pointed out is the importance of moisture at middle levels. Climatological features of quasi-stationary band-shaped precipitation systems were investigated for cases in East Asia, indicating that those stationary precipitating systems contribute significantly to the total rainfall amount in western Japan, the Nansei Islands, and East China Sea (Goto & Satoh, 2022).

In 2010s, a number of extreme rainfall events occurred and spawned severe disasters in Japan. Especially in early July 2017, a heavy rainfall event occurred in northern Kyushu Island, with the total daily rainfall greater than 600 mm in a localized area (Kato et al., 2018; Takemi, 2018), and spawned flooding and landslides in a concentrated region, having significant societal impacts. One year later in the event in early July 2018, heavy rainfalls developed wide areas over the Japanese islands, particularly in the western and central parts of Japan, and flooding, inundation, and landsides caused a large number of human losses and devastating damages to social infrastructures, agriculture, economic activities, etc. In response to these extreme events, a number of studies have been conducted. A part of the outcomes from such studies was briefly summarized in Takemi (2021).

The July 2017 heavy rainfall case is considered to be caused by a typical case of a linear-shaped MCS (Kato et al., 2018; Takemi, 2018). Naka and Takemi (2023) showed that the presence of moist absolutely unstable layer (MAUL) (Bryan & Fritsch, 2000) provides a favorable condition for the rapid development of convection, which was considered to contribute to the intensification of rainfalls. MAUL was shown to be maintained by continuous moisture flux in the lower troposphere. Because MAUL is characterized as having a saturated state, a process that would create a saturated state should be carefully investigated. Figure 10 shows the areas of heavy rainfall and thick MAUL and the west–east and vertical cross section along the 32.9° N latitude. In Fig. 10a, it is seen that a deep MAUL appeared at this time and a heavy rainfall of 60 mm or greater developed in the following 3 h. Figure 10b indicates that relative humidity at the same time with that shown in Fig. 10a is seen to gradually increase from the west to the east in the lower troposphere before the development of the heavy rainfall of 60 mm or greater. Besides, the vertical gradient of equivalent potential temperature is generally negative in the lower to middle troposphere. Therefore, if a saturated layer appears within the layer having negative vertical gradient of equivalent potential temperature, the saturated layer becomes MAUL. The flow pattern shown in Fig. 10b seems to be very similar to the conceptual of slab convective overturning in which MAUL is present (Bryan & Fritsch, 2000). Actually, low-level saturated inflow toward MCSs was noted in Kingsmill and House (1999). Such view is considered to lead to concepts of layer lifting (Mechem et al., 2002) and slabular convection (James et al., 2005). The presence of MAUL is considered to enhance convective motion and hence contribute to the organization and evolution of MCSs. Moistening processes of the lower to middle troposphere is important to understand mechanisms for the formation and maintenance of MAUL and the impact of MAUL in intensifying convection through, for example, layer lifting or slab overturning.

In the followings, the numerical modeling of the July 2017 event will be described.

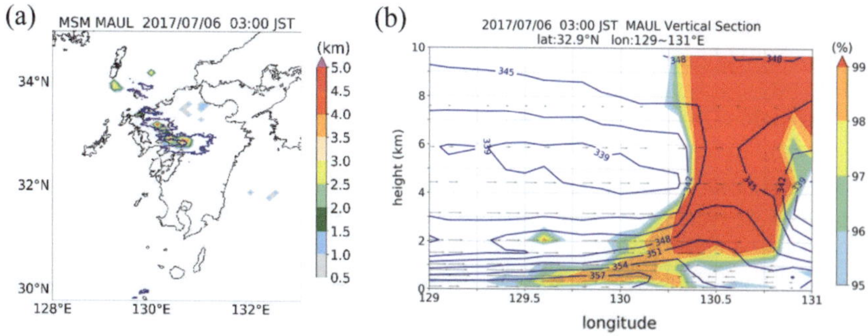

Fig. 10 **a** The horizontal distribution of the thickness of the MAUL (in km, color shading) and the accumulated rainfall of 60 mm for the 3 h afterward (blue contour) in the area around Kyushu at 1800 UTC 5 July 2017. Reproduced from Naka and Takemi (2023). ©The Author(s). Published by the Meteorological Society of Japan under a Creative Commons Attribution 4.0 International (CC BY 4.0) license. **b** The vertical cross section of relative humidity (color shading), equivalent potential temperature, and winds vectors projected on this section along the 32.9° N latitude

3.2 High-Resolution Modeling of Linear-Shaped Convective Systems: A Case Study

In this subsection, a heavy rainfall case that occurred in northern Kyushu, Japan on 5–7 July 2017, referred to as the July 2017 Northern Kyushu Heavy Rainfall event, was chosen. This case produced extreme rainfalls concentrated in a limited region in northern Kyushu and spawned damaging floods and landslides. This case has recently been investigated by Kato et al. (2018), Takemi (2018), Kawano and Kawamura (2020), Tsuji et al. (2020), Ito et al. (2021), Nayak and Takemi (2021), Unuma and Takemi (2021), and Sato et al. (2022). Here we focus on the high-resolution modeling of a linear-shaped stationary MCS by precisely representing a complex and complicated topography in numerical simulations.

Takemi (2018) conducted convection-resolving simulations of a linear-shaped stationary MCS that generated the July 2017 Northern Kyushu Heavy Rainfall in order to examine whether a numerical simulation is able to represent the stationarity of the extreme-rain-producing convective system and to quantitatively reproduce the extreme amount of rainfall. The numerical model used was the WRF/ARW model, version 3.6.1 (Skamarock et al., 2008). The physics parameterizations used in the numerical simulations were a double-moment cloud microphysics scheme for cloud and precipitation processes (Morrison et al., 2009), the Yonsei University scheme for boundary-layer mixing (Hong et al., 2006), a Monin–Obukhov similarity-based scheme for surface fluxes (Jiménez et al., 2012), and the RRTMG scheme for short-wave and long-wave radiation (Iacono et al., 2008). No cumulus parameterization was used. The two-way nesting technique was employed in order to cover a large domain including the Japanese islands and to resolve convective motion and complex topography at a sufficient resolution. We set four computational domains: the outer-most domain (Domain 1) covers most of the Japanese islands and the surrounding regions within the area of 2025 km by 2025 km at the 4.5-km horizontal grid spacing under the Lambert conformal projection on the plane crossing the latitudes of 30° N and 60° N, the second domain (Domain 2) Kyushu Island in the area of 450 km by 450 km at the 1.5-km grid, the third domain (Domain 3) the northern half of Kyushu Island in the 220 km by 200 km area at the 500-m grid, and the innermost domain (Domain 4) focuses northern Kyushu in the area of $(100\ \text{km})^2$at the 167-m grid (Fig. 11). In the domains having the horizontal grid spacing of 500 m (Domain 3) and 167 m (Domain 4), the model topography was generated with the use of a high-resolution digital elevation model (DEM) dataset having a 50-m resolution available from the Geospatial Information Authority of Japan (GSI) to reproduce precisely a complex and complicated feature of the topography in the region. Figure 12 exhibits the difference in the elevation of the model topography generated from the 50-m resolution DEM data of GSI and generated from the global 30-s data, GTOPO30, of the United States Geological Survey (USGS). It is clearly indicated that the higher and lower elevations in the domain are represented higher and lower, respectively, in the case of GSI DEM data than in the case of the USGS data, suggesting that detailed features of the complex topography in the region are represented more precisely in the GSI DEM data than in the USGS data. In other words, the topography generated with the USGS data are smoothly represented.

Figure 13 shows the simulated radar reflectivity at the 2-km height in Domain 4 from 0600 to 0900 UTC on 5 July 2017, which well reproduced the observed radar reflectivity (Takemi, 2018). Around the center of the domain, a linear-shaped MCS oriented in the east–west direction, indicated by areas of strong radar reflectivity, is reproduced. The linear MCS seems to be stagnated in a similar area. From the simulated fields, it was found that individual convective cells continuously developed in the western part of the linear rainfall area and moved eastward. This process is regarded as a back-building type in the definition of Bluestein and Jain (1985). The successive development of convective cells kept the precipitating system stationary and generated strong rainfalls in the same area, resulting in a linear-shaped area of extreme rainfall.

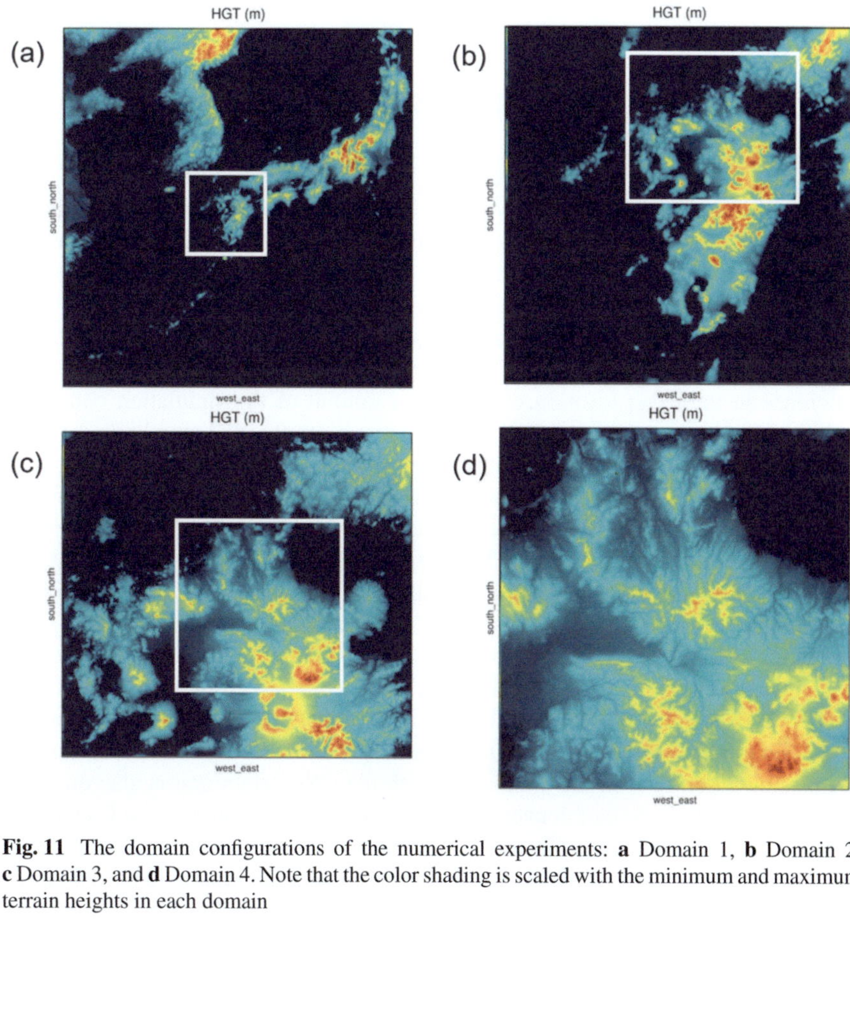

Fig. 11 The domain configurations of the numerical experiments: **a** Domain 1, **b** Domain 2, **c** Domain 3, and **d** Domain 4. Note that the color shading is scaled with the minimum and maximum terrain heights in each domain

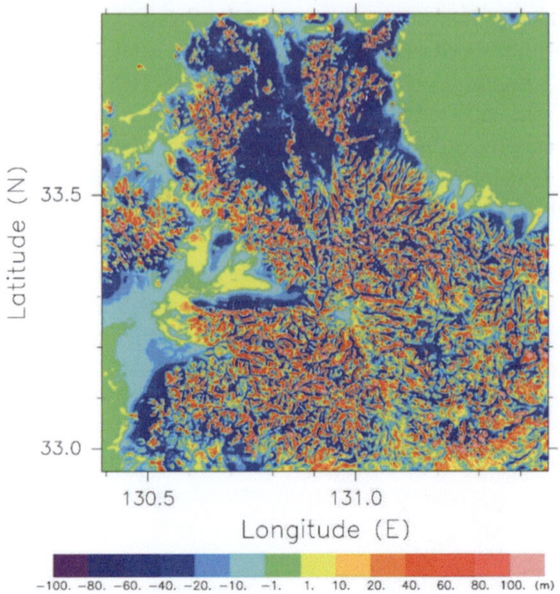

Fig. 12 The difference of thesurface elevation height (m) generated from the GSI from that generated from the G30 DEM data in Domain 4. Reproduced from Takemi (2018). ©The Author(s). Published by the Meteorological Society of Japan under a Creative Commons Attribution 4.0 International (CC BY 4.0) license

As a result of the successive development of convective cells and the formation of a linear-shaped stationary precipitating system, the accumulated rainfall exceeds 600 mm (Fig. 14a). Among the raingauge stations in northern Kyushu, the heaviest rainfall was measured at the Asakura station: the maximum accumulated rainfalls for 24, 48, and 72 h were 545.5 mm (by 1140 JST 6 July), 600.5 mm (by 1040 JST 7 July), and 616.0 mm (by 0600 JST 7 July), respectively. Although there is a slight difference in the exact location of the heavy rainfall of the simulation compared with the observation, the area with over 500 mm rainfall in the simulation extends longitudinally, and the pointwise maximum rainfall in the simulation amounts to 633 mm. In this way, the performance of the 167-m-resolution simulation is highly evaluated in representing the observed features of the back-building, linear-shaped MCS and the resulting linear-shaped heavy rainfall area.

Figure 14 further compares the representation of the total rainfall amount in the simulations with the GSI 50-m-mesh DEM data and with the GTOPO30 data. The simulation with the GTOPO30 terrain (Fig. 14b) indicates the total rainfall amount of at most 400 mm and a smoothed rainfall area, compared to the simulation with the GSI terrain (Fig. 14a). Takemi (2018) indicated that this reduced rainfall amount in the simulation with the GTOPO30 terrain is due to weaker low-level forcing in this simulation than in the simulation with the GSI terrain and is also due to a more propagating feature of convective clouds in the simulation with the GTOPO30 terrain. Thus, it is clearly indicated that the decreased frequency of stronger rains and the less stationarity of the convective system make the accumulated rainfall in the limited region smaller. As Yoshizaki et al. (2000) showed that even small and low mountains are able to generate an organized rainband under a moist convectively

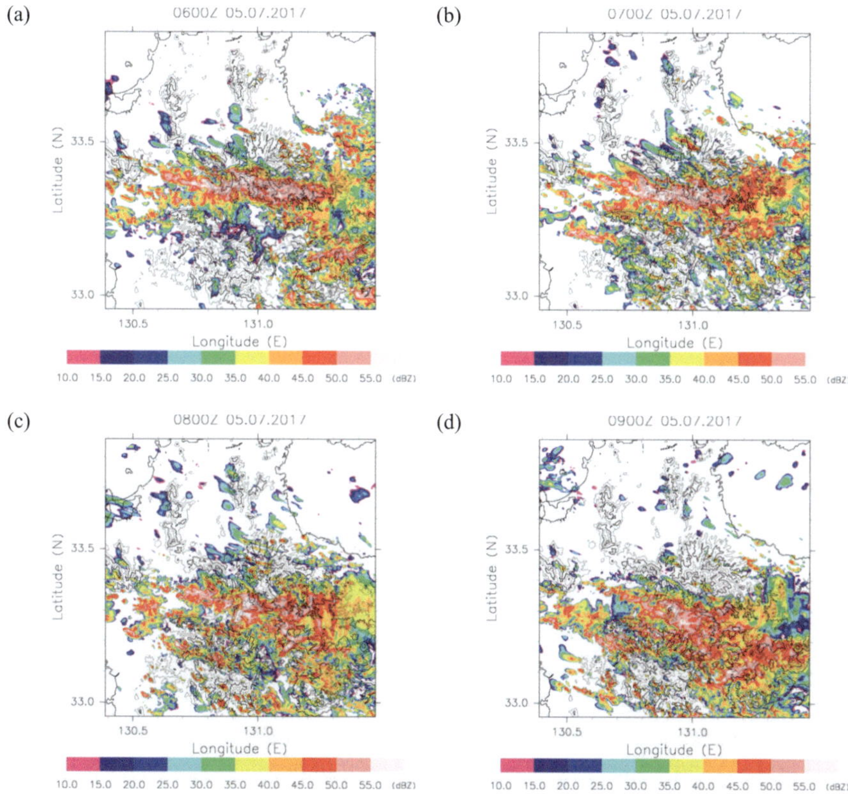

(c) 0800 UTC, and (d) 0900 UTC on 5 July 2017.

Fig. 13 The simulated radar reflectivity field at the 2-km height in Domain 4 at **a** 0600 UTC, **b** 0700 UTC, **c** 0800 UTC, and **d** 0900 UTC on 5 July 2017

unstable condition, the present analysis shown in Fig. 14 emphasizes a possible role of fine-scale terrain features in generating and maintaining stationary MCSs and hence leading to a concentrated area of rainfall.

The impact of the fine-scale terrain representation on the simulation of rainfall amount is further examined by focusing on how the two-way-nested inner domains quantitatively affect the rainfall. For this purpose, we conducted three additional numerical experiments in which inner computational domains were excluded: (1) Domain 4 was excluded; (2) Domain 3 and 4 were excluded; and (3) Domain 2, 3, and 4 were excluded. In these sensitivity experiments, the model topography in Domain 3 was generated with the use of the GSI's 50-m-mesh DEM data. The rainfall distributions in the area corresponding to the Domain 4 size when Domain 4, Domain 3–4, and Domain 2–4 are excluded are shown in Fig. 15a–b, respectively. With the 500-m resolution (Fig. 15a), the rainfall is seen to be concentrated around the center of the domain, similar to the feature shown in Fig. 14a but generally reduced. In

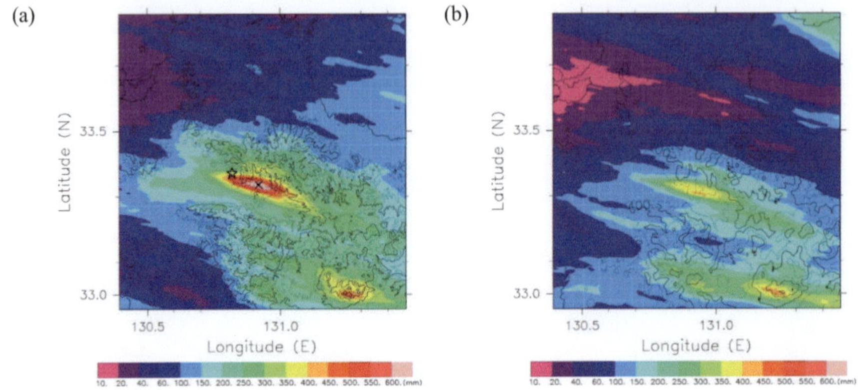

Fig. 14 The accumulated rainfall from 0300 JST 5 July to 0900 JST 7 July 2017 in Domain 4 simulated in the experiment with the topography generated from **a** the GSI's 50-m-mesh DEM data and **b** the GTOPO30 data. In **a**, the contour lines indicate the surface elevation at the 200-m interval, and the star and the cross marks denote the locations of the Asakura grid point and the point of the maximum accumulated rainfall, respectively. Reproduced from Takemi (2018). ©The Author(s). Published by the Meteorological Society of Japan under a Creative Commons Attribution 4.0 International (CC BY 4.0) license

comparison to the rainfall shown in Fig. 14b, a larger amount of rainfall is reproduced in Fig. 15a, which suggests that a higher-resolution terrain dataset has an advantage in quantitatively representing the rainfall amount. With the resolution reduced to 1.5 km (Fig. 15b) and 4.5 km (Fig. 15c), the total rainfall becomes significantly smaller and the linear shaped feature of the rainfall area is lost. The features shown in Fig. 15 strongly support the importance of the terrain representation.

4 Summary and Concluding Remarks

This chapter firstly presented the results of the idealized numerical experiments of squall lines and demonstrated how the tropospheric stability and humidity affect the development and intensity of squall lines. As discussed in Sect. 2.1, the idea in focusing on the stability condition, i.e. temperature lapse rate, is that the magnitude of the positive buoyancy depends on the environmental temperature lapse rate. The buoyancy is generally smaller in the tropical, oceanic environment than in the midlatitude, continental environments. Smaller buoyancy leads to weaker acceleration of updrafts, resulting in weaker convection and hence weaker intensity of MCSs. The numerical experiments clearly indicated that the intensity of updrafts is much weaker in tropical/oceanic squall lines than in midlatitude/continental squall lines. The tropospheric humidity also characterizes how the lower to middle troposphere is unstable from a viewpoint of the vertical distribution of CAPE whose value is

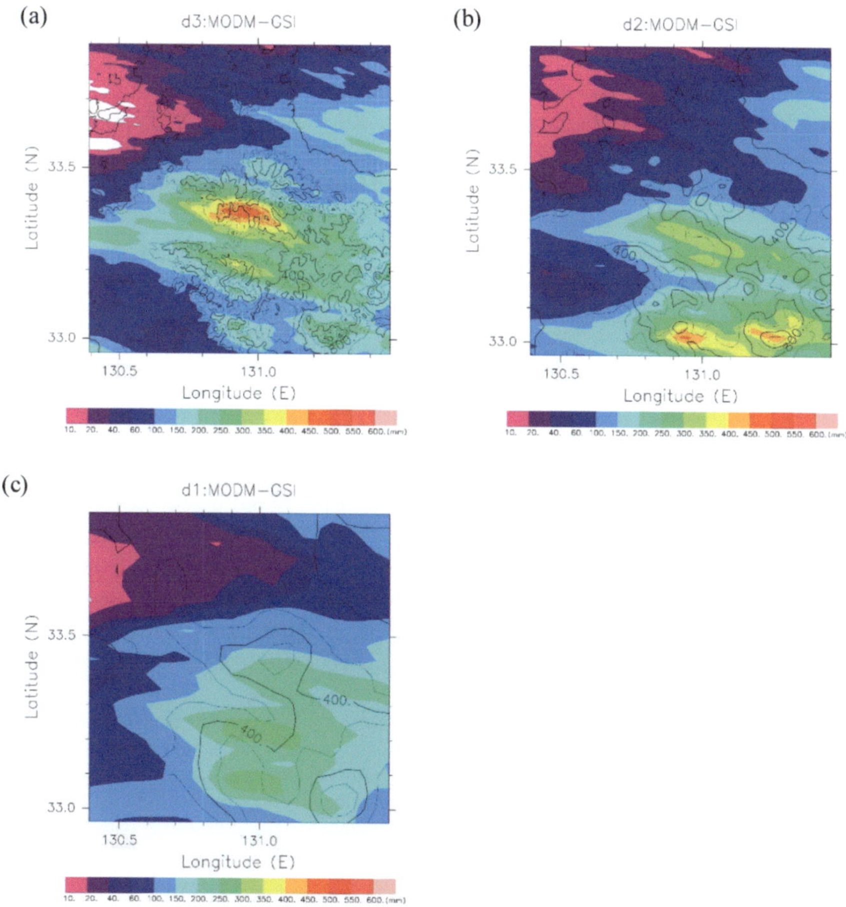

Fig. 15 The accumulated rainfall from 0300 JST 5 July to 0900 JST 7 July 2017 simulated **a** with Domain 4 excluded, **b** with Domain 3 and 4 excluded, and **c** with Domain 2, 3, and 4 excluded

assessed for air parcels originated not only at the surface level but also at elevated levels in the lower troposphere.

We then describe the geographical characteristics of the occurrence of quasi-stationary MCSs over the Japanese islands in the warm season, as an example of the MCSs under humid conditions by using the operational radar network in Japan. From the upper-air sounding data, it was found that the humidity in the middle troposphere well distinguish the environmental properties whether quasi-stationary MCSs develop. Higher humidity at middle levels (i.e., the layer of 700–500 hPa) is favorable for the generation of the MCSs. With a certain amount of vertical shear in the low-to-middle levels, linear-shaped MCSs preferably occur at everywhere over the Japanese islands. A back-building process is regarded as a key mechanism for the maintenance of the linear-shaped MCSs.

Furthermore, high-resolution modeling of a linear-shaped MCS is described. Special attention is paid to the precise reproduction of fine-scale, complex topography in high-resolution simulations. A heavy rainfall case that occurred in the northern part of the Kyushu Island in Japan in early July 2017 was chosen, because this heavy rainfall was shown to be caused by a well-defined, linear-shaped stationary MCS that generated a linear-shaped rainfall area (Kato, 2020). The high-resolution simulation at the horizontal grid spacing of 167 m with the model topography produced by a 50-m-mesh DEM dataset well represents the successive development of convective cells on the upstream side of a linear-shaped MCS and the stationarity of the linear-shaped MCS. Key issues in this high-resolution modeling are to represent more precisely the complex and complicated nature of topography and to explicitly resolve convective motion. In other words, not only resolving convective-scale motions but also representing complex topography is critically important in high-resolution modeling of stationary-type, linearly organized MCSs.

A notable feature for the generation of linear-shaped MCSs in humid environments is a very humid state not only in the lower troposphere but also in the middle-to-upper troposphere. Therefore, the processes leading to such humid conditions should be investigated in more detail. As indicated in Fig. 10b, moistening of the middle and upper troposphere will be critically important. A multi-scale analysis of linear-shaped MCSs and their environmental properties is required to reveal mechanisms for the moistening of the free troposphere and the generation of linear-shaped MCSs.

Acknowledgements We would like to thank Drs. Someshwar Das and Wei-Kuo Tao who are the editors of this book for encouraging me to write this chapter and being patient and supportive for my writing. The comments from the reviewers are greatly acknowledged in improving the initial version of the manuscript. We would like to acknowledge the funding supports from JSPS Kakenhi 20H00289 and 21H01591 and also by the MEXT-Program for the advanced studies of climate change projection (SENTAN) Grant Number JPMXD0722678534.

References

Bluestein, H. B., & Jain, M. H. (1985). Formation of mesoscale lines of precipitation: Severe squall lines in Oklahoma during the spring. *Journal of Atmospheric Science, 42*, 1711–1732.

Bryan, G. H., & Fritsch, J. M. (2000). Moist absolute instability: The sixth static stability state. *Bulletin of the American Meteorological Society, 81*, 1207–1230.

Deardorff, J.W. (1980). Stratocumulus-capped mixed layers derived from a three-dimensional model. *Boundary-Layer Meteorology, 18*, 495–527.

Fovell, R. G., & Ogura, Y. (1989). Effect of vertical wind shear on numerically simulated mul-ticell storm structure. *Journal of Atmospheric Science, 46*, 3144–3176.

Goto, Y., & Satoh, M. (2022). Statistical analysis of "Senjo-Kousuitai" in East Asia and characteristics of associated large-scale circulations in the Baiu season. *SOLA, 18A*, 15–20.

Hong, S.-Y., Noh, Y., & Dudhia, J. (2006). A new vertical diffusion package with an explicit treatment of entrainment processes. *Monthly Weather Review, 134*, 2318–2341.

Houston, A. L., & Niyogi, D. (2007). The sensitivity of convective initiation to the lapse rate of the active cloud-bearing layer. *Monthly Weather Review, 135*, 3013–3032.

Iacono, M. J., Delamere, J. S., Mlawer, E. J., Shephard, M. W., Clough, S. A., & Collins, W. D. (2008). Radiative forcing by long-lived greenhouse gases: Calculations with the AER radiative transfer models. *Journal of Geophysical Research, 113*, D13103. https://doi.org/10.1029/2008JD009944

Igau, R. C., LeMone, M. A., & Wei, D. (1999). Updraft and downdraft cores in TOGA COARE: Why so many buoyant downdraft cores? *Journal of Atmospheric Science, 56*, 2232–2245.

Ito, J., Tsuguchi, H., Hayashi, S., & Niino, H. (2021). Idealized high-resolution simulations of a back-building convective system that causes torrential rain. *Journal of Atmospheric Science, 78*, 117–132.

James, R. P., Fritsch, J. M., & Markowski, P. M. (2005). Environmental distinctions between cellular and slabular convective lines. *Monthly Weather Review, 133*, 2669–2691.

James, R. P., Markowski, P. M., & Fritsch, J. M. (2006). Bow echo sensitivity to ambient moisture and cold pool strength. *Monthly Weather Review, 134*, 950–964.

Jiménez, P. A., Dudhia, J., González-Rouco, J. F., Navarro, J., Montávez, J. P., & García-Bustamante, E. (2012). A revised scheme for the WRF surface layer formulation. *Monthly Weather Review, 140*, 898–918.

Jorgensen, D. P., & LeMone, M. A. (1989). Vertical velocity characteristics of oceanic convection. *Journal of Atmospheric Science, 46*, 621–640.

Kato, T. (2020). Quasi-stationary band-shaped precipitation systems, named "senjo-kousuitai", causing localized heavy rainfall in Japan. *Journal of the Meteorological Socety of Japan, 98*, 485–509.

Kato, R., Shimose, K., & Shimizu, S. (2018). Predictability of precipitation caused by linear precipitation systems during the July 2017 Northern Kyushu Heavy Rainfall Event using a cloud-resolving numerical weather prediction model. *JOurnal of Disaster ReSearch, 13*, 846–859.

Kawano, T., & Kawamura, R. (2020). Genesis and maintenance processes of a quasi-stationary convective band that produced record-breaking precipitation in northern Kyushu, Japan on 5 July 2017. *Journal of the Meteorological Socety of Japan, 98*, 673–690.

Kingsmill, D. E., & House, R. A., Jr. (1999). Thermodynamic characteristics of air flowing into and out of precipitating convection over the west Pacific warm pool. *Quarterly Journal of Royal Meteorological Society, 125*, 1209–1229.

LeMone, M. A., & Zipser, E. J. (1980). Cumulonimbus vertical velocity events in GATE. Part I: Diameter, intensity and mass flux. *Journal of Atmospheric Science, 37*, 2444–2457.

LeMone, M. A., Zipser, E. J., & Trier, S. B. (1998). The role of environmental shear and thermodynamic conditions in determining the structure and evolution of mesoscale convective systems during TOGA COARE. *Journal of Atmospheric Science, 55*, 3493–3518.

Lucas, C., Zipser, E. J., & LeMone, M. A. (1994a). Vertical velocity in oceanic convection off tropical Australia. *Journal of Atmospheric Science, 51*, 3183–3193.

Lucas, C., Zipser, E. J., & LeMone, M. A. (1994b). Convective available potentialenergy in the environment of oceanic and continental clouds: Correction and comments. *Journal of Atmospheric Science, 51*, 3829–3830.

Lucas, C., Zipser, E., & Ferrier, B. S. (2000). Sensitivity of tropical west Pacific oceanic squall lines to tropospheric wind and moisture profiles. *Journal of Atmospheric Science, 57*, 2351–2373.

Maddox, R. A. (1980). Mesoscale convective complexes. *Bulletin of the American Meteorological Society, 61*, 1374–1387.

McCaul, E. W., Jr., Cohen, C., & Kirkpatrick, C. (2005). The sensitivity of simulated storm structure, intensity, and precipitation efficiency to environmental temperature. *Monthly Weather Review, 133*, 3015–3037.

Mechem, D. B., House, R. A., Jr., & Chen, S. S. (2002). Layer inflow into precipitating convection over the western tropical Pacific. *Quarterly Journal of Royal MeteorOlogical Society, 128*, 1997–2030.

Meng, Z., Yan, D., & Zhang, Y. (2013). General features of squall lines in east China. *Monthly Weather Review, 141*, 1629–1647.

Morrison, H., Thompson, G., & Tatarskii, V. (2009). Impact of cloud microphysics on the development of training stratiform precipitation in a simulated squall line: Comparison of one- and two-moment schemes. *Monthly Weather Review, 137*, 991–1007.

Naka, N., & Takemi, T. (2023). Characteristics of the environmental conditions for the occurrence of recent extreme rainfall events in northern Kyushu, Japan. *SOLA, 19A*, 9–16.

Nayak, S., & Takemi, T. (2021). Atmospheric driving mechanisms of extreme precipitation events in July of 2017 and 2018 in western Japan. *Dynamic of Atmospheres and Oceans, 93*, 101186.

Parker, M. D. (2008). Response of simulated squall lines to low-level cooling. *Journal of Atmospheric Science, 65*, 1323–1341.

Parker, M. D., & Johnson, R. H. (2000). Organizational modes of midlatitude mesoscale convective systems. *Monthly Weather Review, 128*, 3413–3436.

Robe, F. R., & Emanuel, K. A. (2001). The effects of vertical wind shear on radiative-convective equilibrium states. *Journal of Atmospheric Science, 58*, 1427–1445.

Rotunno, R., Klemp, J. B., & Weisman, M. L. (1988). A theory for strong, long-lived squall lines. *Journal of Atmospheric Science, 45*, 463–485.

Sato, Y., Hayashi, S., & Hashimoto, A. (2022). Difference in the lightning frequency between the July 2018 heavy rainfall event over central Japan and the 2017 northern Kyushu heavy rainfall event in Japan. *Atmos. Sci. Lett., 23*, e1067.

Shimizu, S., & Uyeda, H. (2012). Algorithm for the identification and tracking of convective cells based on constant and adaptive threshold methods using a new cell-merging and -splitting scheme. *Journal of the Meteorological Socety of Japan, 90*, 869–899.

Skamarock, W. C., Klemp, J. B., Dudhia, J., Gill, D. O., Barker, O. M., Duda, M. G., Huang, X. Y., Wang, W., Powers, J. G. (2008). *A description of the Advanced Research WRF version 3* (Vol. 475, 113 pp.) (NCAR Technical Note).

Takemi, T. (2006). Impacts of moisture profile on the evolution and organization of midlatitude squall lines under various shear conditions. *Atmospheric Research, 82*, 37–54.

Takemi, T. (2007a). A sensitivity of squall line intensity to environmental static stability under various shear and moisture conditions. *Atmospheric Research, 84*, 374–389.

Takemi, T. (2007b). Environmental stability control of the intensity of squall lines under low-level shear conditions. *Journal of Geophysical Research, 112*, D24110.

Takemi, T. (2010). Dependence of the precipitation intensity in mesoscale convective systems to temperature lapse rate. *Atmospheric Research, 96*, 273–285.

Takemi, T. (2014). Convection and precipitation under various stability and shear conditions: Squall lines in tropical versus midlatitude environment. *Atmospheric Research, 142*, 111–123.

Takemi, T. (2018). Importance of terrain representation in simulating a stationary convective system for the July 2017 Northern Kyushu Heavy Rainfall case. *SOLA, 14*, 153–158.

Takemi, T. (2021). Editorial for the special edition on extreme rainfall events in 2017 and 2018. *Journal of the Meteorological Socety of Japan, 99*, 1145–1147.

Takemi, T., & Rotunno, R. (2003). The effects of subgrid model mixing and numerical filtering in simulations of mesoscale cloud systems. *Monthly Weather Review, 131*, 2085–2101.

Takemi, T., & Satomura, T. (2000). Numerical experiments on the mechanisms for the development and maintenance of long-lived squall lines in dry environments. *Journal of Atmospheric Science, 57*, 1718–1740.

Tao, W.-K., & Simpson, J. (1993). The Goddard cumulus ensemble model. Part I: Model description. *Terrestial, Atmospheric and OceanIc Sciences, 4*, 35–72.

Tao, W.-K., Simpson, J., & Soong, S.-T. (1987). Statistical properties of a cloud ensemble: A numerical study. *Journal of Atmospheric Science, 44*, 3175–3187.

Tao, W.-K., Simpson, J., & McCumber, M. (1989). An ice-water saturation adjustment. *Monthly Weather Review, 117*, 231–235.

Tao, W.-K., Iguchi, T., Lang, S., Li, X., Mohr, K., Matsui, T., van den Heever, S. C., & Braun, S. (2022). Relating vertical velocity and cloud and precipitation properties: A numerical modeling study. *Journal of Advances in Modeling Earth Systems*, e2021MS002677.

Trier, S. B., Skamarock, W. C., LeMone, M. A., Parsons, D. B., & Jorgensen, D. P. (1996). Structure and evolution of the 22 February 1993 TOGA COARE squall line: Numerical simulations. *Journal of Atmospheric Science, 53*, 2861–2886.

Tsuji, H., Yokoyama, C., & Takayabu, Y. N. (2020). Contrasting features of the July 2018 heavy rainfall event and the 2017 northern Kyushu Rainfall Event in Japan. *Journal of the Meteorological Socety of Japan, 98*, 859–876.

Unuma, T., & Takemi, T. (2016a). A role of environmental shear on the organization mode of quasi-stationary convective clusters during the warm season in Japan. *Scientific Online LettErs on the Atmosphere, 12*, 111–115.

Unuma, T., & Takemi, T. (2016b). Characteristics and environmental conditions of quasi-stationary convective clusters during the warm season in Japan. *Quarterly Journal of Royal Meteorological Society, 142*, 1232–1249.

Unuma, T., & Takemi, T. (2021). Rainfall characteristics and their environmental conditions during the heavy rainfall events over Japan in July of 2017 and 2018. *Journal of the Meteorological Socety of Japan, 99*, 165–180.

Weisman, M. L. (1992). The role of convectively generated rear-inflow jets in the evolution of long-lived mesoconvective systems. *Journal of Atmospheric Science, 49*, 1826–1847.

Weisman, M. L., & Klemp, J. B. (1982). The dependence of numerically simulated convective storms on vertical wind shear and buoyancy. *Monthly Weather Review, 110*, 504–520.

Weisman, M. L., & Rotunno, R. (2004). "A theory for strong, long-lived squall lines" revisited. *Journal of Atmospheric Science, 61*, 361–382.

Weisman, M. L., Klemp, J. B., & Rotunno, R. (1988). Structure and evolution of numerically simulated squall lines. *Journal of Atmospheric Science, 45*, 1990–2013.

Xu, K. M., & Randall, D. A. (2001). Updraft and downdraft statistics of simulated tropical and midlatitude cumulus convection. *Journal of Atmospheric Science, 58*, 1630–1649.

Yoshizaki, M., Kato, T., Tanaka, Y., Takayama, H., Shoji, Y., & Seko, H. (2000). Analytical and numerical study of the 26 June 1998 orographic rainband observed in western Kyushu, Japan. *Journal of the Meteorological Socety of Japan, 78*, 835–856.

Zipser, E. J., & LeMone, M. A. (1980). Cumulonimbus vertical velocity events in GATE. Part II: Synthesis and model core structure. *Journal of Atmospheric Science, 37*, 24582–25469.

Modelling Hail Microphysical Processes in Severe Storms

Adrian M. Loftus and Stephen E. Lang

Abstract The occurrence of hail in many regions across the globe poses a significant threat to property and agriculture, yet most Numerical Weather Prediction (NWP) models use simplified microphysical schemes that do not explicitly simulate the complex processes involved with hail production. Observations and theoretical research on hail and hailstorms over the past 60+ years have greatly advanced our knowledge of the myriad of factors responsible for hail formation and growth within deep convection, ultimately leading to improvements in hail detection through remote sensing and representation of hail processes in increasingly sophisticated microphysical schemes. This review paper provides an overview of hail and hailstorm characteristics, hail production mechanisms, and examples of various microphysical schemes currently employed by cloud-resolving and mesoscale models for use in hail research. Continued advancements in hail modeling are expected to help improve remote sensing retrievals of hail as well as enhance understanding of potential hail trends in future climate scenarios.

1 Motivation and Purpose

Unlike tornadoes and tropical cyclones, highly variable events that may or may not be problematic in any given year, damage from hailstorms is significant every year in many mid-latitude regions across the globe. In the US, for example, hail damage to structures, automobiles, aircraft, livestock, and crops results in insurance claims averaging over $10 billion per year according to the Insurance Information Institute (2020), with similar losses likely in many other regions as well (Allen et al., 2019). Although large hail (>19 mm) is largely responsible for structural insurance claims and occurs throughout the U.S. (Brown et al., 2015), agricultural losses are driven by

A. M. Loftus (✉) · S. E. Lang
NASA Goddard Space Flight Center, Greenbelt, MD, USA
e-mail: Adrian.m.loftus@nasa.gov

S. E. Lang
Science Systems and Applications, Inc., Lanham, MD, USA

S. Das and W-K. Tao (eds.), *Severe Storms*,
https://doi.org/10.1007/978-981-97-7075-5_10

a wider size distribution and density of hail, depending on the crop and time of year (e.g., Changnon, 1971; Schiesser, 1990; Botzen et al., 2010). NOAA's *Storm Data* database does not include hail events at sizes less than 19 mm, so hail climatologies and risk assessments do not or cannot survey small hail events or risks (Ortega et al., 2009; Allen et al., 2019). Even small hail, from 3 mm, can physically damage young leaves and buds and mature, thus soft, fruit and wheat heads (e.g., Katz & Garcia, 1981; Kretchman et al., 1989; Sánchez et al., 1996; Bani et al., 2019; Yue et al., 2019). As such, convective storms that may not meet the typical criteria for severe storms yet produce large quantities of small hail have received increased attention in recent years (Kalina et al., 2016; Kumjian & Lebo, 2016; Kumjian et al., 2019).

Hail occurrence has been reported on every continent except Antarctica, with the majority of reports and largest historical datasets clustered around denser population centers in North America and Europe (Allen et al., 2019). Aside from dedicated field campaigns and hail pad networks in various regions, most hail reports are known to contain errors due to hail size referencing (e.g., pea, golf ball, egg, etc.) as well as melting that occurs between the time of hailfall and when the observation was made (Witt et al., 1998b; Blair & Leighton, 2012; Ortega et al., 2016).Satellite-based hail detection methodologies based on microwave scattering depressions and/or space borne radar measurements are useful for building climatologies of hail and assessing hail risk on a global scale (e.g., Cecil, 2009; Cecil & Blankenship, 2012; Ferraro et al., 2015; Mroz et al., 2017; Ni et al., 2017; Bang & Cecil, 2019; Le & Chandrasekar, 2021). In general, these types of climatologies (Fig. 1) reveal that hail (or its signature) is most prominent over the U.S. Great Plains region, southeast South America and central Africa, with hail also evident over Central America, the Tibetan Plateau region, South and Southeast Asia, Australia and Europe (Cecil & Blankenship, 2012; Bang & Cecil, 2019). It should be noted that the hail signature over central Africa is likely due to hail aloft that melts prior to reaching the surface (Bang 2023, personal communication).

The aim of this review paper is to provide an overview of hail and the representation of hail in numerical models (primarily cloud models).Although ground observations and remote sensing hail detection algorithms (HDAs) provide critical and insightful observations on hail, they are typically limited to widely spaced snap shots in time, which are on the order of 5 min for NEXRAD (Crum et al., 1993) to several hours or even daily for satellites (Hou et al., 2014; Skofronick-Jackson et al., 2017). Such observations do not and cannot capture the full range of processes associated with hail initiation, growth, sedimentation and melting and their sensitivity deep within convective clouds. As such, it is essential that accurate hail models continue to be developed and refined, including detailed parameterizations of hail processes capable of accurately representing the full range of hail processes across all environments.

Section 2 describes observed characteristics of hail and hailstorms, Sect. 3 details key factors affecting hail formation and growth processes within deep convection, and Sect. 4 covers details of various hail modeling schemes used within numerical cloud models. A general summary and future directions for modeling hail are presented in Sect. 5.

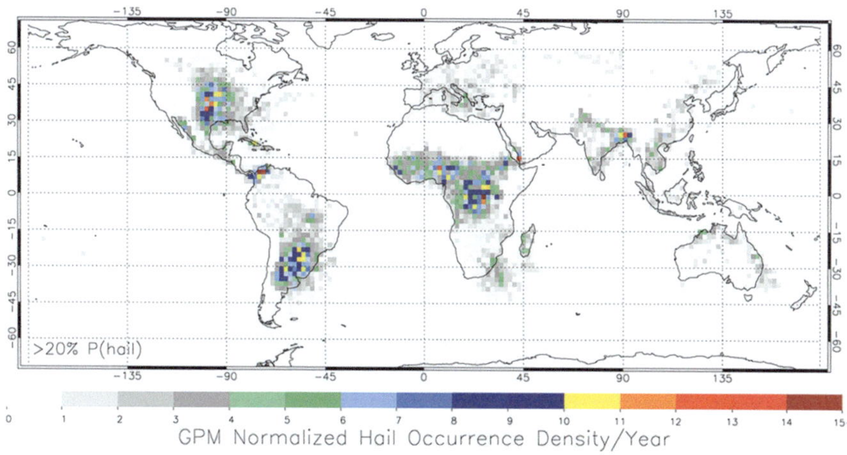

Fig. 1 Global climatology of satellite-estimated hail occurrence in the GPM domain. Figure adopted from Bang and Cecil (2019, their Fig. 7.)

2 Observations of Hail and Hailstorms

Hail generated by deep moist convection in nature varies over a wide range of sizes, with 'equivalent' diameters ranging from approximately 0.5 mm (Huschke, 1959) up to 13–17 cm and larger in extreme cases (Klimowski et al., 1998; Scharfenberg et al., 2005). The largest hailstones observed in the US include the 3 September 1970 Coffeyville, KS hailstone (Fig. 2a), which had a mass of 766 g and a maximum circumference of 44 cm (Roos, 1972), the 22 June 2003 Aurora, NE hailstone with a record maximum circumference of 47.6 cm and a diameter of 17.78 cm (Guyer & Ewald, 2004), and the recent 23 July 2010 Vivian, SD hailstone (Fig. 2b) that measured 20.3 cm in diameter, 47.3 cm in circumference, and weighed about 880 g (report from NWS Aberdeen, SD).And while individual small hailstones may seem insignificant, great quantities of small stones can create hazardous conditions for people and transportation (Fig. 2c).

The shapes of hailstones also vary widely in nature, with the most common shape resembling an ellipsoid. Browning and Beimers (1967) analyzed hailstones from Oklahoma storms and found that large hailstones almost always exhibit some degree of oblateness, which tends to increase with continued growth of the hailstone. A study by Knight (1986) also found increasing oblateness with increasing sizes for hailstones obtained from different geographical regions. English (1973) observed hailstone shapes ranging from conical to oblate spheroids, with conical shapes exhibited only by the smaller stones, and Carte and Kidder (1966) reported on apple-shaped hailstones that fell in South Africa. Other examples of the high degree of variability of hailstone shapes come from Knight and Knight (1970c) who observed two distinctly different symmetries (oblate spheroids and flattened, prolate stones) for comparably-sized large hailstones that fell from the same storm in a single location, as well as

Fig. 2 **a** Hailstone from Coffeyville, KS storm (after Knight & Knight, 1971), **b** Picture of the most massive hailstone ever recorded in the US, which fell from a severe hailstorm in Vivian, SD on 23 July 2010 (photo courtesy of NWS Aberdeen, SD), and **c** Small hail accumulation on roadway in Fort Collins, CO on June 15, 2009 (Photo courtesy of Adrian Loftus)

Knight et al. (2008) who reported on an unusual case in which large disk-shaped hailstones from a hailstorm over Boulder, CO were observed. Large hailstones often exhibit lobed structures (Fig. 2a, b) that initially arise from surface irregularities during the early growth stages (Browning, 1966; Knight & Knight, 1970a). These initial surface protrusions grow faster than the surrounding hailstone surface and become more marked with continued hail growth due to increased collection efficiency and heat transfer of the projections themselves (Browning, 1966; Bailey & Macklin, 1968). Browning (1966) also points out that the lobed structure of large hailstones indicates that the hail grew by primarily accreting small cloud droplets as opposed to large cloud droplets or raindrops. In addition, lobes that nearly cover the entire surface indicate the hailstone tumbled while growing, whereas lobes situated on only a portion of the surface reveal the hailstone fell with a preferred orientation while growing (Browning, 1966; Knight & Knight, 1970a).

The density of hailstones is often less than that of pure ice (0.917 g cm^{-3}) and tends to vary radially from core to surface as a result of trapped air bubbles within the concentric ice layers, although most measurements indicate bulk density values between 0.8 and 0.9 g cm^{-3} (Mossop & Kidder, 1961; Prodi, 1970; Pruppacher & Klett, 1980). Hail particles can have rather low densities during their initial growth from embryos (Farley, 1987b; Knight et al., 2008), and density typically increases rapidly as the hailstones grow through accretion of liquid water, though there are rare cases of low-density hail observed at the ground. For example, Knight and Heymsfield (1983) examined hailstones with unusually low densities (0.31 to 0.61 g cm^{-3} at sizes ranging from 0.63 to 1.54 cm) from a hailstorm that occurred over Boulder, CO on 4 March 1982. Farley (1987b) argues that low-density hail embryos experience increased rates of accretion and diffusional growth in addition to reduced terminal velocities, which can maintain the particle in favorable growth regions, and these combined factors can lead to larger hail.

There is also a great deal of variability in hail distributions within and among hail-producing storms (Marwitz, 1972a, 1972b; Changnon, 1973; English, 1973; Federer & Waldvogel, 1975; Nelson & Young, 1979; Ziegler et al., 1983; Miller et al., 1988, 1990; Musil et al., 1991, Knight et al., 2008). Marwitz (1972b) and Changnon (1973) noted the existence of small-scale 'hailstreaks' within surface hail swaths associated with individual thunderstorms and observed considerable variability of hail sizes and numbers within the hail streaks themselves. Federer and Waldvogel (1975) examined hail distributions of a multicell hailstorm in Switzerland and reported that large hail preceded small hail at the surface as well as maxima in number concentrations and median diameters that were offset from the center of the hail swath. From observations of hailstorms in South Africa, Carte and Kidder (1966) found that large diameter hailstones at the surface tended to occur in conjunction with smaller diameter stones (i.e., broad size spectra), whereas hailfalls comprising only small stones tended to have nearly monodisperse size spectra. Based on near-simultaneous aircraft and ground observations, Auer and Marwitz (1972) described hail size distributions as being very narrow (essentially monodisperse) and consisting of relatively few but large stones near organized updrafts as well as on the ground below the updrafts. Cheng and English (1983) and Cheng et al. (1985) investigated relationships between the parameters of fitted hail size distributions and storm characteristics, such as cloud base height and maximum water mass flux and found significant differences in the distribution parameters for individual storms.

It is notable that measurements of hail sizes and concentrations within deep convective clouds are few owing to the inherent danger to aircraft. Most in-cloud observations of hail were made by a now-retired armored T-28 aircraft during field campaigns in the central Plains of the US such as the National Hail Research Experiment (NHRE) in the 1970s (Musil et al., 1973; Sand, 1976), the Cooperative Convective Precipitation Experiment (CCOPE) in the early 1980s (Knight, 1982), the Verifications of the Origins of Rotation in Tornadoes Experiment (VORTEX; Rasmussen et al., 1994) in 1995, and the 2003 Joint Polarization Experiment (JPOLE; Ryzhkov et al., 2005) and Thunderstorm Electrification and Lightning Experiment (TELEX; MacGorman et al., 2008). Although these in-cloud hail observations are far fewer than surface observations, they have provided critical information of hail size distributions aloft and are still being used to develop and refine PSD and hail microphysical parameterizations for cloud models (e.g., Field et al., 2019).

Doppler radar reflectivity values exceeding roughly 50 to 55 dBZ are usually an indicator of hail (Chisholm, 1968; Mason, 1971; Foote & Wade, 1982), and values exceeding 65 dBZ have been observed in storms with large hail (Miller et al., 1988, 1990; Zrnić et al., 1993; Scharfenberg et al., 2005). Previous studies have also noted strong correlations between the regions of greatest reflectivity values and large hail observed at the ground (Browning et al., 1968; Mason, 1971; Foote & Wade, 1982; Aydin et al., 1986; Miller et al., 1988, 1990). Spatial reflectivity patterns such as WERs, BWERs and vaulted structures (e.g., Fig. 3) resulting from local minima in precipitation-sized particle concentrations have been found to be associated with particularly intense updrafts and are typical features of many hail-producing supercell and multicell storms (Browning & Ludlam, 1962; Marwitz, 1972a, 1972b, 1972c; Chisholm, 1973; Browning & Foote, 1976; Chalon et al., 1976; Foote & Wade, 1982; Heymsfield & Musil, 1982; Knight, 1984; Musil et al., 1986; Nelson, 1987; Musil et al., 1991). Several HDAs incorporate the vertically integrated reflectivity from ground-based radars to estimate hail occurrence and hail size. For example, higher values of vertically integrated liquid (VIL) density (Amburn & Wolf, 1997) tend to indicate a higher probability of a storm producing larger hail, although studies have also found this HDA exhibits low skill for estimating hail sizes in an operational setting (Edwards & Thompson, 1998; Blair et al., 2011). The Maximum Expected Size of Hail, or MESH, algorithm (Witt et al., 1998a) utilizes information on heights of reflectivity values (exceeding at least 40 dBZ) along with subfreezing temperature levels to compute a Severe Hail Index (SHI) from which MESH is then found through an empirically derived power law relationship.

Polarimetric radars, which can emit and receive linearly polarized horizontal (Z_H) and vertical (Z_V) waves, are able to provide information on types of hydrometeors as well as estimates of particle sizes in different regions of deep moist convection (Bringi et al., 1986; Höller et al., 1994; Hubbert et al., 1998; Tessendorf et al., 2005). There are several polarimetric variables that are typically used in conjunction with the standard reflectivity factor to identify hail in convection. Differential reflectivity (*ZDR*), computed from the ratio of received powers Z_H and Z_V, is related to the axis ratio and size of hydrometeors and can be used to distinguish regions of large oblate drops ($ZDR \geq +1$) as well as regions of wet, large, and/or tumbling hail ($-2 \leq ZDR \leq +1$) (Bringi et al., 1986; Höller et al., 1994; Straka et al., 2000). In their study of storms that produce large amounts of small hailstones in the U.S., Kumjian et al. (2019) noted such storms tend to exhibit unusually large differential attenuation (attenuation of Z_V relative to Z_H) signatures that are manifested as negative ZDR (swaths) regions extending downrange of the hail core. The authors also noted that storms producing large amounts of small hail seem to exhibit large values (often exceeding 10°km^{-1}) for specific differential phase shift (K_{DP}), which is proportional to number concentrations of nonspherical particles in a sampling volume, likely due

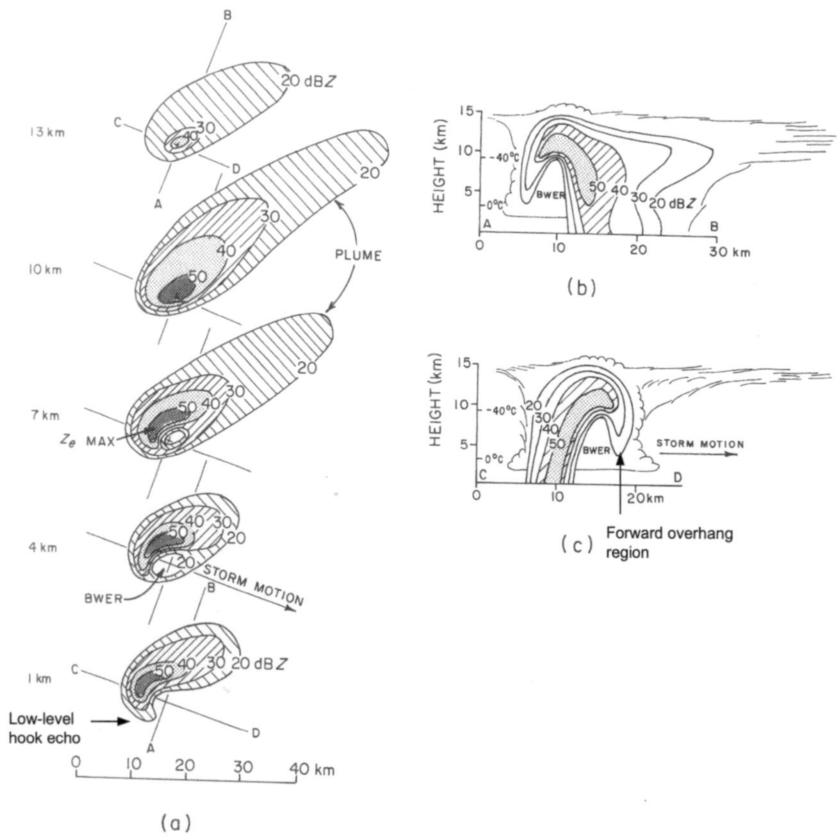

Fig. 3 Schematic illustrating the variation of radar reflectivity patterns with height in supercell thunderstorms observed in Alberta, Canada. Horizontal sections of reflectivity (dBZ) at various altitudes are shown in **a**. Vertical sections are shown in **b** and **c**. Salient features are labeled and cloud boundaries are sketched (Adapted from Chisholm & Renick, 1972)

to small melting hailstones. The linear depolarization ratio (*LDR*) is a function of the ratio of the cross-polar to the copolar powers received and is sensitive to hydrometeor shape, thermodynamic phase and canting (Straka et al., 2000). *LDR* has been shown to be a good indicator of hail above the melting level (Bringi et al., 1986; Höller et al., 1994) as well as large wet hail below the melting level (Carey & Rutledge, 1998; Hubbert et al., 1998), with increasing *LDR* values typically corresponding to larger hail sizes. The correlation coefficient (ρ_{HV}) between Z_H and Z_V is also useful to determine regions of rain, hail and rain mixed with hail (Hubbert et al., 1998; Straka et al., 2000; Zrnić et al., 1993). ρ_{HV} decreases from a value of unity due to increasing hail size, broadening of the hail size spectra, wetting of hail, and mixing of hail with liquid drops of various sizes (Balakrishnan & Zrnić, 1990). The diagnostic hail differential reflectivity (*HDR*) parameter (Aydin et al., 1986), originally formulated to distinguish hail from rain, can also be used to infer hail size (Kennedy & Detwiler, 2003; Depue et al., 2007). *HDR* is computed by subtracting a prescribed function of *ZDR* from the horizontally polarized reflectivity (Z_H), and values equal to or greater than 21 dB often signifies the presence of large diameter ($D \geq 19$ mm) hail (Depue et al., 2007).

Hail detection by ground-based radars provides a fairly accurate measure of hail within deep convective cores as well as near the surface (e.g., Cintineo et al., 2012; Dolan et al., 2013; Kumjian et al., 2019), yet many land regions still have inadequate or nonexistent coverage, with virtually no coverage over the oceans. The near global coverage of satellites makes them an attractive alternative for hail detection, and various methods have established empirical thresholds for passive and active sensor measurements as proxies for hail identification. HDAs using passive microwave radiometer measurements exploit brightness temperature (Tb) depressions resulting from scattering by ice particles within convection of upwelling radiation from the Earth's surface. In general, larger Tb depressions tend to be associated with greater hail probabilities. The 37 and 19 GHz microwave channels tend to be most successful at detecting hail (Cecil, 2009; Bang & Cecil, 2019), although HDAs utilizing higher frequency channels (e.g., 85, 150–170, 183 GHz) have also shown skill (Ferraro et al., 2015; Laviola et al., 2020; Mroz et al., 2017). Notably, most passive HDAs focus on identifying large hail. Notable drawbacks with using passive microwave are large footprints that can dampen Tb depressions, particularly at lower frequencies, hail occurrence in tropical regions tends to be overestimated, and the fact that different combinations of layers of ice scatterers of different sizes, and different depths can produce similar microwave Tb values (Cecil & Blankenship, 2012; Bang & Cecil, 2019).

Spaceborne radars with the Tropical Rainfall Measurement Mission (TRMM) and GPM have the unique ability to profile hydrometeors and can observe the large reflectivity values typically associated with hail above the freezing level (FL). Zipser et al. (2006) noted that the 40-dBZ height loosely correlates with updraft speeds capable of supporting growing hailstones in the mid-troposphere. Indeed, studies have found that the heights of radar reflectivity values exceeding 40 dBZ above the FL are good indicators of large hail aloft (Cecil, 2009; Mroz et al., 2017; Ni et al., 2017; Le & Chandrasekar, 2021; Seiki, 2021). The Dual-Frequency Precipitation Radar (DPR) aboard the GPM core satellite provides independent measurements within a volume at Ku- (13.6 GHz) and Ka-bands (35.5 GHz). Because Ka-band experiences greater attenuation than Ku-band for large ice particles and supercooled liquid, differences between these radar reflectivity measurements have been exploited in several HDAs (Mroz et al., 2017, 2018; Le & Chandrasekar, 2021). For example, an experimental HDA product in GPM V6 (Le & Chandrasekar, 2021) utilizes the maximum Ku-band reflectivity value along with the dual-frequency ratio (DFR) and storm top height information to flag the presence of hail or graupel in a column, though not necessarily falling at the surface. It should be noted that non-uniform beam filling and multiple scattering effects introduce additional uncertainty into these HDAs (Mroz et al., 2018).

3 Factors Affecting Hail Processes

3.1 Updraft Characteristics and Temperature Dependence

Convective updrafts must be strong enough to support hailstones, with larger hail typically associated with stronger updrafts (Ludlam, 1958; English, 1973; Browning & Foote, 1976; Musil et al., 1991). Nelson (1983) and Foote (1984) demonstrated that broad moderate updrafts (20 to 40 m s^{-1}) are more likely to produce large hail ($D > 2$ cm) than narrower moderate or even strong (>~35 m s^{-1}) updrafts owing to the longer residence time of growing hailstones in favorable growth regions. An updraft that tilts with height is another factor that can affect hail growth both directly and indirectly. The horizontal component of velocity within a tilted updraft allows for hail particles to be carried across the updraft and places a limit on the amount of time in which these particles can reside in favored growth regions as well as size sorting of hailstones such that the largest particles fall close to the updraft edge while smaller particles fall further away from the updraft (Browning, 1977). The degree of tilt may also govern whether or not particles falling out of an updraft at upper levels re-enter the updraft at lower levels and experience further growth (Browning, 1977). A tilted updraft also prevents much of the precipitation from accumulating within the updraft (e.g., precipitation loading, which would decrease the buoyancy of the updraft and cause it to decay) and allows for spatial separation between the updraft and the precipitation-induced downdraft thereby leading to a longer-lived storm in general (Browning & Ludlam, 1962; Browning, 1977). Hail growth occurs only at temperatures between 0 and −40 °C, with the majority of growth favored at temperatures between −10 and −25 °C (Foote, 1984; Nelson, 1983).

3.2 Hailstone Embryos

The sizes, numbers, and regions of millimeter-sized particles serving as 'hail embryos' can impact hail size distributions throughout deep convection. The term hail embryo is referred to here as a small diameter (~0.1 to ~5 mm) ice particle that serves as a precursor to the formation of a hailstone. The characteristics of hailstone embryos vary depending on geographical region and are largely dependent on cloud base temperature, with colder (warmer) cloud base temperatures tending to favor graupel (frozen raindrops) as the dominant hail embryo type (Knight, 1981). Additionally, Foote (1985) pointed out that collision-coalescence processes are typically too slow in continental thunderstorms to account for large raindrops that could subsequently freeze and serve as hail embryos. Detailed observations of hailstones from High Plains thunderstorms by Knight and Knight (1970b, 1979) and Knight et al. (1974) found that hail embryos are typically graupel particles, and to a lesser extent frozen supercooled raindrops.

Mossop and Kidder (1961) and Knight and Knight (1970b) initially suggested that a necessary step for the formation of large hail is the formation of hail embryos in one set of conditions followed by the 'injection' of these embryos into another set of conditions. Larger embryos in an updraft typically experience faster growth rates than smaller ones owing to their larger surface areas, though larger embryos don't necessarily lead to larger hail (Brimelow et al., 2002; Foote, 1984; Musil, 1970). In fact, List et al. (1968) suggested that growth of embryos to large hail sizes was more probable when the number of hail embryos is small owing to reduced competition for the liquid water available for growth. Paluch (1978) showed that more numerous embryos leads to increased competition for available water, which in supercells could limit hail sizes in one part of the storm while promoting hail growth in another region, depending on the initial locations of the embryos in relation to the updraft. Studies by Dye et al. (1983), Nelson (1983), Xu (1983), and Miller et al., (1983) demonstrated that the source regions of hail embryos dictate, to a degree, the trajectories that hail particles follow in a storm, and thus the ensuing growth rates encountered along those trajectories, ultimately affecting the evolution of hail size spectra.

Studies of a multicell thunderstorm over northeast Colorado by Heymsfield et al. (1980) and Heymsfield (1983) found that hail embryos first formed in developing 'feeder' cells (also referred to as flanking towers) adjacent to the main updraft from aggregates and graupel. These embryos were then advected by the environmental flow (as they sedimentated) into the primary updraft (Fig. 4) where they grew rapidly into hail particles. Similar findings were described in studies of multicellular hailstorms in Oklahoma by Ziegler et al. (1983), in Alberta by Cheng and Rogers (1988), and in Colorado by Brandes et al. (1995). Frozen drop embryos were also noted to originate from melted ice particles that were transported into the main updraft via low-level inflow (Heymsfield et al., 1980). In contrast, Kennedy and Detwiler (2003) examined hail origins in a NE CO multicell storm using in situ aircraft and polarimetric radar data and found that hail embryos were generated primarily via a recycling process of graupel particles from the forward portions of the storm rather than from feeder cells. Browning (1977) suggested an additional concept for hail origination in multicell storms in which hail embryos first develop in the cumulus stage of flanking towers, termed 'daughter clouds', adjacent to a mature stage cell. The mature cell eventually begins to dissipate while the daughter cell continues growing and reaches the mature stage, at which point the hail embryos experience a water-rich environment and grow rapidly to hail size. Thus, unlike the feeder cell concept in which hail embryos are physically transported from developing cells to an adjacent mature cell where they encounter pre-existing favorable growth conditions, hail embryos in a daughter cloud remain within the cell and find themselves in an increasingly favorable environment for hail growth as the cloud develops into a mature cell. Both of these concepts are based on the idea that hail embryos are most favored to grow into hailstones when they are incorporated into a developing updraft since their larger fall velocities will allow them to remain within the updraft core and take advantage of high liquid water content values (Heymsfield, 1983).

Numerous ideas exist regarding the sources of hail embryos in supercells, some of which include re-circulation of particles into the updraft from the 'embryo curtain'

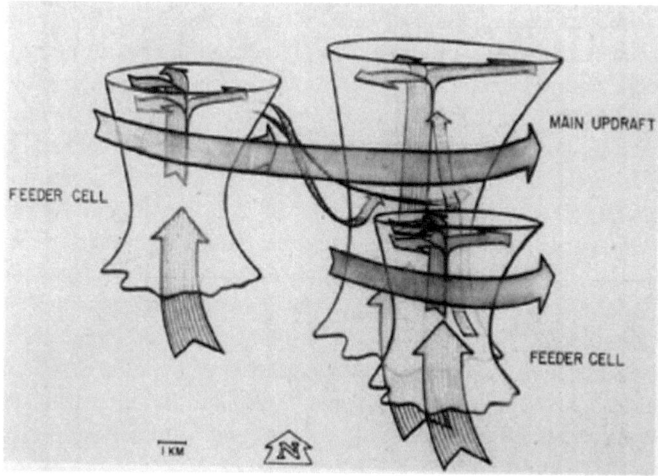

Fig. 4 Schematic representation of the hail formation process via feeder cells for a multicellular hailstorm. Arrows: no shading, environmental winds relative to the storm; vertical shading: updraft air; stippled shading: dominant hail-producing trajectories (From Heymsfield et al., 1980)

(also referred to in the literature as the forward overhang region, see Fig. 3c) encompassing the primary updraft periphery (Browning & Foote, 1976; Farley & Orville, 1986; Miller et al., 1988), and growing cells along the supercell flanks (Miller & Fankhauser, 1983; Krauss & Marwitz, 1984). Miller et al. (1990) showed a variety of embryo regions for a single supercell storm that included growing graupel particles in addition to drops from melted graupel along the updraft edges, shedding of drops from hail undergoing wet growth above the freezing level, shedding from melting hail below the freezing level, and even graupel ingested from cumulus congestus that merged with the supercell.

3.3 *Liquid Water Content*

Because hailstones grow primarily by accreting liquid water, which subsequently freezes (Ludlam, 1958; Schumann, 1938), the distribution of liquid water content (LWC) impacts accretion rates, and thus growth rates of hail, and can determine whether hailstones with surface temperatures near 0 °C undergo wet or dry growth (Browning, 1963; Bailey & Macklin, 1968; Lesins & List, 1986; Rasmussen & Heymsfield, 1987b; Cotton & Anthes, 1989). Of course, air temperature is another controlling factor for whether or not hailstones may experience wet growth, and as air temperatures decrease, higher values of LWC are required in order for wet growth to commence for a hailstone of a given size (Bailey & Macklin, 1968; Rasmussen & Heymsfield, 1987b) as shown in Fig. 5. Dry growth typically occurs in environments characterized by low LWC values and/or relatively cold temperatures (with respect to the hailstone surface temperature) such that heat transfer to the environment via conduction and evaporation is efficient. This efficient cooling of the hailstone surface allows all collected supercooled water to freeze, often trapping air bubbles within the ice structure that give the hail particle an opaque appearance. In environments with high LWC and/or relatively warm temperatures, wet growth of hail can occur in which all of the collected water by a hailstone is not immediately frozen; the hailstone surface becomes wet with a temperature very near 0 °C, and the ice structure assumes a clear appearance. The existence of liquid water on the hailstone surface alters the particle's thermal conductivity, and liquid may evaporate from the hailstone surface even under saturated conditions. The liquid layer can also alter the hailstone's drag characteristics, thereby impacting its fall velocity, and can appreciably affect its electromagnetic scattering properties, an important consideration for radar detection of hail. Lastly, some of the unfrozen collected water during wet growth is shed from the hailstone surface (Ludlam, 1958) and can be a significant source of surface rainfall (Rasmussen and Heymsfield, 1987b; List, 2010).

English (1973) noted that hailstones within time-varying updrafts tend to alternate between wet and dry growth regimes more frequently than those in steady-state updrafts and could help explain the alternating layers of clear and opaque ice commonly observed in hailstones. LWC has also been shown to affect shedding rates of liquid drops from large hail below the melting level, with increasing LWC generally leading to increased amounts of shedding (Rasmussen and Heymsfield, 1987a). Miller et al. (1988) computed hail trajectories in a supercell and reported that variability in LWC primarily affected the number rather than the sizes of hailstones, though a similar study by Miller et al. (1990) showed that substantial increases (decreases) in LWC led to an increase (decrease) in maximum hail size. List (1963) and Lesins and List (1986) describe an additional growth regime for hail, termed 'spongy growth', in which some of the collected unfrozen water is trapped in an ice matrix on the hailstone surface rather than being shed thereby allowing for greater

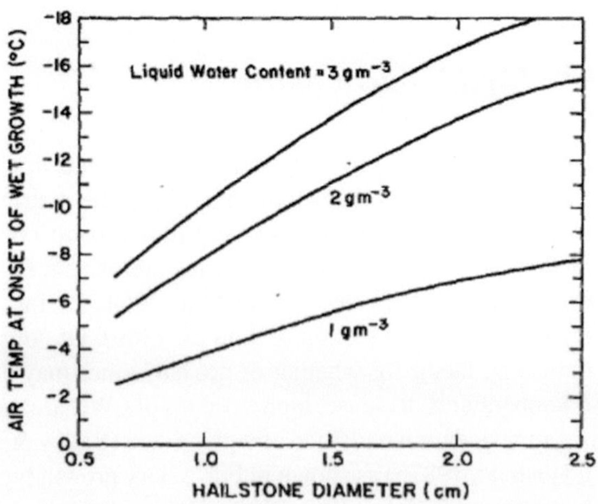

Fig. 5 Liquid water content required to produce wet growth on various hailstone sizes as a function of air temperature. Hailstones are assumed to have a density of 0.91 g cm^{-3}, and to collect cloud droplets with unit collection efficiency (From Rasmussen and Heymsfield, 1987b)

hailstone growth rates than suggested by the theoretical Schumann-Ludlam Limit (SLL). This growth regime is essentially a subcategory of the wet growth regime and therefore has similar dependencies on LWC and air temperature. Spongy growth was also investigated in wind tunnel experiments of icing by Macklin (1961) who found that spongy growth of ice occurred owing to the simultaneous collection of liquid droplets and ice crystals. Mason (1971) stated that spongy growth rates could increase by the additional collection of ice crystals and snowflakes. Analyses of the internal structures of freshly fallen hailstones by Browning et al. (1968) determined that many of the hail embryos grew spongy initially. However, based on examinations of numerous hailstones over several decades, Knight and Knight (2005) believe that spongy growth beyond the initial growth stages of natural hailstones is rather rare and typically only accounts for a minor fraction of hail wet growth, a conclusion also reached by Carte (1966) nearly 40 years prior.

3.4 Hail Fall Speeds and Surface Features

The shapes and sizes of hailstones largely dictate the terminal velocities and thus the accretion rates of hailstones. Browning (1963) claimed that oblate hailstones have lower critical LWC values than similar spherical hailstones and thus grow at a slower rate. However, calculations on hailstone growth rates by English (1973) showed that 'flatter' hailstones grow faster via accretion than do more spherical hailstones of similar size due to the increased sweep-out volume. The terminal velocity (V_{th}) of a hailstone (Eq. 1) is a function of its density ρ_h, its diameter D_h (equivalent spherical diameter), the drag coefficient C_D, and air density ρ_a (Charlton & List, 1972; Knight & Knight, 2001),

$$V_{th} = \left(\frac{4g\rho_h D_h}{3C_D\rho_a} \right)^{0.5}. \quad (1)$$

Observational and experimental measurements of hail terminal velocities have shown that C_D varies as a function of hail shape and to a lesser extent hail surface characteristics (Macklin & Ludlam, 1961; Charlton & List, 1972; Roos & Carte, 1973; Matson & Huggins, 1980; Knight & Heymsfield, 1983; Heymsfield & Wright, 2014). The value of C_D for a smooth sphere is roughly 0.45, and increasing oblateness tends to increase the value of C_D (Macklin & Ludlam, 1961; Matson & Huggins, 1980) and thus lead to slower fall speeds. Based on measurements of free falling artificial hailstones of various sizes and shapes, Macklin and Ludlam (1961) found values of C_D ranging from about 0.45 to about 0.8 and suggested a value of $C_D =$ 0.6 for diameters greater than 1 cm, whereas a value of $C_D = 0.8$ was implied in the empirically-derived hailstone terminal velocity formula of Roos and Carte (1973) for large ($3 > D > 12$ cm) hailstones. Matson and Huggins (1980) measured terminal velocities of primarily small (D of about 5 to 25 mm), naturally-falling hailstones

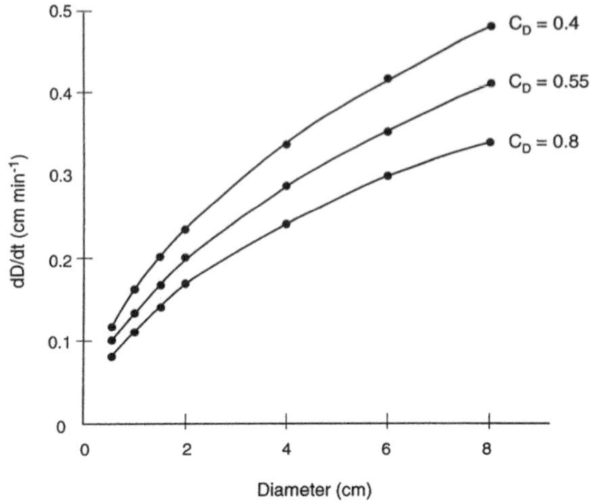

Fig. 6 Growth rates of hail for different drag coefficient values assuming air density at 500 mb and −20 °C (From Knight & Knight, 2001)

and reported C_D values ranging from 0.65 to 1.3 with mean values around 0.87 to 0.92. Thus, it is apparent that drag coefficients for natural hail are larger than that for smooth spheres having the same mass; surface features such as dimples and lobes further impact values of C_D (Heymsfield and Wright, 2014). The effect of C_D on hail growth rates is clearly evident in Fig. 6, which shows that, all else being equal, increasing values of C_D (increasing oblateness) generally correspond to slower growth rates.

Rough surfaces and protuberances (lobes) lead to increased heat transfer away from hailstones and more rapid freezing of collected water (Browning, 1966; Bailey & Macklin, 1968; Knight & Knight, 1970c). Browning (1966) also noted that surface lobes lead to increased drag (slower fall speeds) as well as increase the critical water content value at which hailstones transition to the wet growth regime. Surface lobes may also increase collision efficiencies between hailstones and cloud droplets, particularly for hail undergoing wet growth (Bailey & Macklin, 1968; Knight & Knight, 1970a), although Macklin and Bailey (1966, 1968) found that hail-cloud collision efficiencies depend more on hail size and tend to decrease with increasing hailstone diameter. This latter result is somewhat compensated by the fact that larger hailstones have greater terminal velocities and cross-sectional areas than smaller particles and can therefore collect more droplets per unit time, which ultimately leads to increasing growth rates for increasing hailstone size (Johnson, 1987; Xu, 1983). The ventilation factor, defined as the ratio of water mass flux to or from a moving versus a motionless hydrometeor (Pruppacher & Klett, 1980) depends on the size and surface roughness of the hailstone, with larger ventilation factors, and hence enhanced heat transfers between a hailstone and its environment, for increasing size and surface roughness (Bailey & Macklin, 1968). In spite of the fact that hailstone attributes affect growth rates of hail, the inherent difficulty and excessive computational cost of attempting to encompass the wide variety and complexity of hailstone features in microphysical models for use in three-dimensional simulations has forced many modelers to simply assume hailstones can be represented as smooth spheres.

3.5 Melting of Hail

Melting rates of hail depend on hail size, air temperature, relative humidity (RH), and the melting level height. Melting is faster at warmer air temperatures, higher RH values and for smaller hailstone sizes, which have a lower mass-to-area ratio, all of which lead to increased heat transfer from the environment to the hailstone surface (Pruppacher & Klett, 1980; Rasmussen et al., 1984). Lower RH values tend to slow melting rates due to reduced heat transfer resulting from competition between evaporative cooling and conductive heating at the hailstone surface (Foote, 1984; Rasmussen & Heymsfield, 1987a; Srivastava, 1987). In addition, Rasmussen and Heymsfield (1987b) found that shedding commences at a higher altitude for larger environmental RH values. The height of the melting level (ML) also governs both the amount and size of hail arriving at the ground as it determines the amount of time a falling hailstone spends in air with temperatures greater than 0 °C, and thus the amount of melting that can occur prior to reaching the surface (Foote, 1984; Rasmussen & Heymsfield, 1987a). Melting seems to increase oblateness of hailstones as well (Macklin, 1964; Browning & Beimers, 1967). Browning and Beimers (1967) attribute increased oblateness during melting to previous spongy ice growth at the top and bottom of the hailstone, with the less dense spongy ice melting faster than the denser ice on the sides of the stone. Macklin (1964) argued that differences in heat transfer rates to the melting hailstone over its surface accounted for the increases in oblateness, with greater transfer rates existing at the upstream and downstream of the stone and lower rates along the hailstone's sides.

4 Modeling of Hail

The majority of the precipitation produced within mid-latitude deep convection during the warm season is from frozen particles that often melt to rain prior to reaching the surface (Braham, 1964; Mason, 1971; Wisner et al., 1972; Dye et al., 1974; Knight et al., 1974; Farley & Orville, 1986; Knight & Knight, 2001; Khain et al., 2004). The dominant mechanism of precipitation formation in mid-latitude deep convective clouds is diffusional growth of ice crystals and subsequent riming to larger particle sizes (Knight, 1982). Even in deep convection over the tropics and sub-tropics, in which precipitation initially develops from warm rain processes, ice processes, once begun, can dominate the precipitation formation mechanism (Atlas et al., 2004; Smith et al., 1999). List (2010) asserts that collection and subsequent shedding of accreted liquid water from growing hailstones could be a major factor for transforming cloud droplets to raindrops in deep convection. The importance of hail in rainfall production was also noted by Rasmussen and Heymsfield (1987c), and there is evidence that melting hail may contribute significantly to intense rainfall and flash flooding events (Brimelow et al., 2017; Mahoney et al., 2012).

Ice processes can have significant impacts on the dynamics, thermodynamics, precipitation distributions, and longevity of deep convective clouds, particularly for severe convective storms (Johnson et al., 1993; van den Heever, 2001). Studies by Jewett et al. (1990) and Johnson et al. (1993) revealed increases in total surface precipitation, maximum vertical velocities, and perturbation potential temperatures aloft for supercell simulations in which ice processes were included compared to similar simulations that did not include ice microphysics. Comparisons between the 'no-ice' and 'ice' simulations of Johnson et al. (1993) also revealed that the inclusion of ice processes led to increased storm longevity owing to decreased downdraft strength, greater spatial separation of the downdraft from the updraft, and a 'warmer' low-level cold pool, factors which allowed the storm's gust front to propagate with the storm rather than propagating away from the storm as in the no-ice case. Representing rimed ice with hail-like versus graupel-like characteristics in models has been shown to significantly impact storm structure and precipitation associated with deep convection (e.g., McCumber et al., 1991; Gilmore et al., 2004a; Cohen & McCaul, 2006; Morrison & Milbrandt, 2011; Bryan & Morrison, 2012; Van Weverberg, 2013; Adams-Selin et al., 2013). Some simulations have also shown that greater numbers of small hail sizes can promote stronger low-level downdrafts due to increased cooling associated with greater melting and evaporation rates (Gilmore et al., 2004b; Srivastava, 1987) as well as deeper, colder, and faster propagating low-level cold pools (van den Heever & Cotton, 2004). It is clear from these studies that correct representation of microphysical processes associated with ice hydrometeors, particularly hail, in numerical simulations of deep convection is crucial for accurate prediction of precipitation and overall convective evolution.

Hail is highly dependent on convective updrafts. These updrafts not only serve as the primary source of the super-cooled cloud water needed for hail growth, but their size, shape, and intensity also impact the amount of residence time that hail and hail embryos spend in the hail growth region (Nelson, 1983, 1987; Ziegler et al., 1983; Dennis & Kumjian, 2017; Kumjian & Lombardo, 2020). Any deviations from the realistic simulation of these features will therefore adversely affect hail modeling regardless of how hail and its processes are parameterized. Thus, to properly simulate hail growth and hail properties, it is essential that convective updrafts are sufficiently resolved in cloud-scale models. Numerous studies have documented the substantial sensitivity of CRM simulations, including their cloud and dynamical properties and structures, to grid resolution, most notably the horizontal (Petch & Gray, 2001; Petch et al., 2002; Bryan et al., 2003; Grabowski et al., 2006; Lang et al., 2007). Bryan et al. (2003) conducted a series of squall line simulations over a span of grid resolutions ranging from 1 km down to 125 m. They found variations in precipitation amount, system phase speed, cloud depth, static stability, convective cell size, and convective overturning in response to increasing resolution and concluded that deep, moist convection may be on the order of only 1–2 km in scale in some environments, requiring grid spacing to be on the order of 100 m. Bryan and Morrison (2012) performed a series of sensitivity simulations, including horizontal resolutions from 4 km down to 250 m, for an observed midlatitude squall line. They found larger grid spacing resulted in slower development, more precipitation, and higher cloud tops, which are all consistent with larger convective cells that do not entrain midlevel air. Lebo and Morrison (2015) looked at the sensitivity of an idealized squall line to a variety of horizontal and vertical grid spacing. They identified two different regimes depending on the horizontal grid size. Notably, the convective core area changed substantially for horizontal grid sizes greater than 250 m but less so below that spacing. Tracer analysis further revealed that entrainment and detrainment were restricted for horizontal grid spacings greater than or equal to 500 m.

4.1 Microphysical Parameterizations

Representation of cloud and precipitation processes through the use of microphysical models is an integral part of all numerical cloud models, and there is a plethora of schemes of varying degrees of sophistication in use today. Microphysical schemes vary in the number of predicted hydrometeor species and types of microphysical processes represented and are divided into two classes: bulk microphysics and (spectral) bin microphysics (SBM), which are described in more detail in the following sections. The simplest microphysical schemes used in cloud models represent only warm rain processes following Kessler (1969), who developed parameterizations for condensation, evaporation, coalescence of cloud droplets into raindrops, collection of cloud particles by rain, and sedimentation in order to predict the mass mixing ratios of cloud and rain hydrometeors.

Many of the schemes following the Kessler paradigm have strived to include ice processes as well as more complete representation of the interactions between microphysics and cloud dynamics. Early microphysical models that incorporated simple ice processes include single-class ice schemes that predicted mass mixing ratios of cloud ice (Ogura & Takahashi, 1971) or hail (Wisner et al., 1972), and the two-class ice (cloud ice and hail) scheme of Bennets and Rawlins (1981), all of which parameterized the additional processes of sublimation, accretion of liquid water by ice particles, freezing and melting. Other more complex two-class ice models are those of Orville and Kopp (1977), which included wet and dry growth of hail as well as shedding of liquid from hail, Cotton et al. (1982), which modeled primary nucleation of ice crystals and allowed for large ice particles to be classified as either graupel or hail, and Koenig and Murray (1976), the latter being unique among these schemes at the time in that it predicted number concentrations of ice particles in addition to ice hydrometeor mass. Lin et al. (1983) developed a complex three-class ice scheme (commonly referred to as the LFO scheme) that predicted the mixing ratios of cloud ice, snow, and hail. A variation of the LFO scheme by Rutledge and Hobbs (1984) assigned graupel as the large ice category rather than hail. Later more advanced schemes include the four-class ice scheme (ice crystals, snow, graupel, and hail) of Ferrier (1994), the five-class ice scheme (pristine ice crystals, snow, aggregates, graupel, and hail) presented in Cotton et al. (2003), and even a ten-class ice scheme (two ice crystal habits, rimed cloud ice, snow, three graupel categories of differing densities, frozen drops, small hail, and large hail) developed by Straka and Mansell (2005). Both the number and type of predicted hydrometeor categories or species has an enormous impact on the microphysical and dynamical evolution of simulated deep moist convection. For example, Ferrier et al. (1995) found that predicting graupel and hail as separate categories produced more realistic storm characteristics for both tropical and mid-latitude convection.

4.1.1 Bulk Schemes with Hail

Bulk microphysical schemes employ analytical distribution functions to represent the hydrometeor distributions and usually predict one or two characteristics of the distribution such as mass mixing ratio (related to the 3rd moment) and/or particle concentration (related to the 0th moment). The two most commonly used distribution functions are the exponential distribution (e.g., Marshall & Palmer, 1948; Gilmore et al., 2004a, b) and the gamma distribution (e.g., Walko et al., 1995), although lognormal distribution functions may work as well (Feingold & Levin, 1986). A typical three-parameter gamma type distribution is given by Eq. (2), where D is the particle diameter, D_n is the characteristic diameter ($D_n = 1/\lambda$, where λ is called the slope parameter in different mathematical expressions for gamma distributions (e.g., Ferrier, 1994; Milbrandt & Yau, 2005a)), and Γ is the complete gamma function. The number density distribution for any hydrometeor of diameter D is given by Eq. (3), and in general, any moment P of the distribution can be computed using Eq. (4).The shape parameter (ν) is generally a fixed parameter and controls the relative amounts of smaller versus larger hydrometeors. A value of $\nu = 1$ corresponds to the commonly used exponential or Marshall-Palmer distribution, and larger ν values correspond to increasingly more narrow size distributions with the peak of the distribution curve approaching the mean diameter (i.e., the 1st moment) of the distribution.

$$f_{gam}(D\frac{1}{\Gamma(\nu)}\left(\frac{D}{D_n}\right)^{\nu-1}\frac{1}{D_n}exp\left(-\frac{D}{D_n}\right) \tag{2}$$

$$n(D) = N_t f_{gam}(D) \tag{3}$$

$$M(P) = \int_0^{\infty} f_{gam}(D)dD = D_n^P\frac{\Gamma(\nu+P)}{\Gamma(\nu)} \tag{4}$$

Observations of hail from thunderstorms at various locations suggest that exponential-type distributions adequately describe hail size spectra for some situations (Federer & Waldvogel, 1975; Musil et al., 1976; Cheng et al., 1985), although gamma distributions work better for others (Wang et al., 1987; Wong et al., 1988; Ziegler et al., 1983). The gamma distribution is advantageous for hail as it is capable of representing peaks often observed in the size spectra (Wong et al., 1988; Ziegler et al., 1983), whereas the exponential distribution inherently assumes that the smallest particles are most numerous and has been noted to poorly represent distributions at smaller diameters (Gilmore et al., 2004b; Straka & Mansell, 2005). Dessens and Fraile (1994) provided theoretical evidence that a collection of hailstones observed to conform to an exponential-type at the ground more correctly correspond to a gamma-type distribution while in free fall. However, Field et al. (2019) used in-cloud T-28 measurements (c.f., Sect. 2) to deduce that hail PSDs may exhibit more exponential-like distributions aloft, and more gamma-like distributions at the surface when melting processes are considered.

The single-moment (1M) Goddard 4ICE (cloud ice, snow, graupel and hail) scheme (Lang et al., 2014) was developed for the Goddard Cumulus Ensemble model (GCE, Tao & Simpson, 1993; Tao et al., 2003) to simulate intense convection. Built upon previous improvements to the GCE's 3ICE Rutledge and Hobbs graupel scheme (Lang et al., 2007, 2011) that reduced excessive graupel amounts and 40-dBZ echo penetrations aloft including new snow and graupel size mapping schemes to improve the distribution of radar echoes in the ice region, the 4ICE scheme adds hail processes from Lin et al. (1983), but includes many modifications and refinements as well as additional processes. The unrealistic buildup of graupel and hail in 4ICE are reduced by omitting their dry collection of ice and snow (Lang et al., 2007, 2014), resulting in more snow advected over a larger area, enhancing the often underestimated stratiform region. Riming and deposition effects on hail are included to treat ice growth more realistically at upper levels. The scheme includes parameterizations for rime splintering, immersion freezing, contact nucleation, and a SBM-based rain evaporation correction (Li et al., 2009) that reduces excessive evaporation, propagation and up-shear tilted convective cores. Snow and graupel size-mapping schemes, first introduced in Lang et al. (2011), wherein the characteristic sizes ($1/\lambda$, where λ is the slope parameter) of the exponential snow and graupel size distributions are predetermined as functions of both particle mixing ratio and air temperature with smaller sizes at colder temperatures and lower mixing ratios, have been steadily refined (Lang et al., 2014), including both an increased snow aggregation signature and a snow breakup effect (Tao et al., 2016), to further improve agreement with observed radar PDFs in the ice region. A simple hail size mapping scheme (Tao et al., 2016), which similarly assigns hail sizes ($1/\lambda$) as a function of both air temperature and particle mixing ratio, eliminates the need to select a hail intercept a priori and produced peak reflectivity profiles dominated by hail in good agreement with NEXRAD observations over a several-hour period for an intense squall line observed over Oklahoma during the Midlatitude Continental Convective Clouds Experiment (MC3E, Petersen & Jensen, 2012) field campaign.

Lang et al. (2014) showed that without the inclusion of a hail category, it was not possible for the 3ICE graupel schemes to produce the intense echoes observed in the 20 May 2011 MC3E squall line (Fig. 7). Nor could the graupel schemes capture the generally monotonically decreasing peak reflectivities with height due to the relatively slow graupel fall speeds, which resulted in excessive amounts of graupel being lofted into upper levels (Fig. 7b–d). Lang et al. (2014) also showed that even for

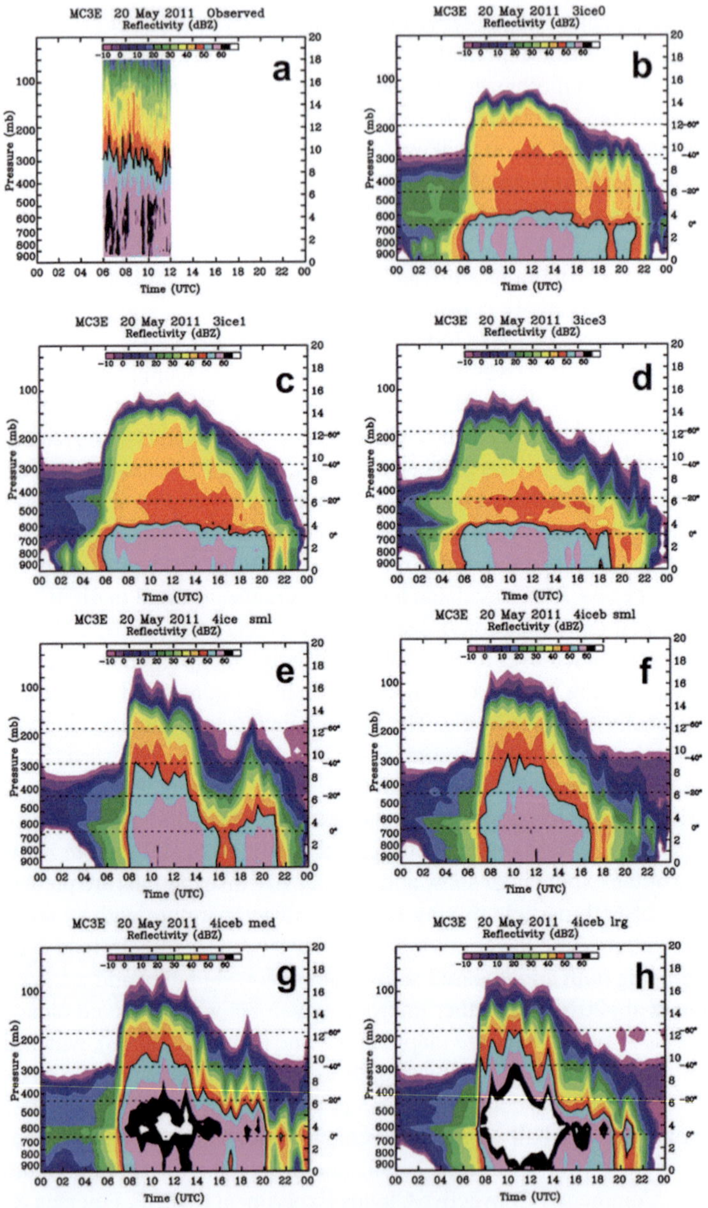

Fig. 7 Time–height cross sections of maximum radar reflectivity for the 20 May 2011 MC3E case **a** observed by NEXRAD radar and simulated using the **b** original 3ICE graupel, **c** level-1 improved 3ICE graupel (Lang et al., 2007), **d** level-3 improved 3ICE graupel (Lang et al., 2011), **e** new 4ICE with smaller hail, **f** new 4ICE with smaller hail and bin rain evaporation correction, **g** new 4ICE with moderate hail and bin rain evaporation correction, and **h** new 4ICE with larger hail and bin rain evaporation correction Goddard microphysics scheme. Right axes are heights (km), while horizontal dashed lines show the level of indicated environmental temperatures (°C). Times range from 0000 UTC 20 May to 0000 UTC 21 May 2011. Adapted from Lang et al., (2014, their Fig. 3)

moderate intensity convection, the inclusion of a frozen drops/hail category having higher fall speeds is necessary to prevent excessive graupel buildup aloft, resulting in unrealistic vertical gradients of peak reflectivity values above the freezing level.

The benefits of a double-moment (2 M) over a 1 M scheme include improvements in predicted surface precipitation and overall storm evolution, along with better representations of microphysical processes (Ferrier, 1994; Meyers et al., 1997; Reisner et al., 1998; Straka & Mansell, 2005; Milbrandt & Yau, 2006b; Mansell, 2008; Morrison et al., 2009; Thompson & Eidhammer, 2014). Bulk 2 M schemes have become more popular within cloud and mesoscale models, and there are multiple 2 M schemes currently available in the Weather Research and Forecasting (WRF) Model (Skamarock et al., 2008) for predicting hail. The Milbrandt and Yau (MY) (Milbrandt & Yau, 2005a) and NSSL (Mansell et al., 2010) bulk 2M microphysics are two such schemes that predict multiple liquid and ice categories, including separate graupel and hail categories, and are suited for simulating intense convection. Of note is that the NSSL scheme also predicts bulk density of graupel, which may improve prediction of hail embryo formation, as well as CCN thereby allowing for investigations of aerosol impacts on hail. The Morrison 2M scheme (Morrison et al., 2009) is another widely-used 2M scheme, although the user must specify whether the rimed ice category is either graupel or hail. Morrison and Milbrandt (2011) compared the Morrison and MY 2M schemes in simulations of an idealized supercell and reported distinctly different storm characteristics based on whether the schemes predicted either only graupel or only hail.

The Colorado State University (CSU) Regional Atmospheric Modeling System (RAMS) 2 M microphysics scheme (Meyers et al., 1997; Saleeby & Cotton, 2004) uses a three-parameter gamma PSD for three species of liquid (cloud and drizzle droplets, and rain) and five species of ice (small and large vapor grown crystals, aggregates, graupel, and hail). This scheme includes a bulk aerosol module (Saleeby & van den Heever, 2013) that explicitly simulates multiple aerosol types and includes cloud droplet activation, wet/dry deposition, and aerosol regeneration upon hydrometeor evaporation. Ice nucleation follows either Meyers et al. (1992) or DeMott et al. (2010) based on user specification and includes contact freezing, deposition-condensation freezing, and immersion freezing. Bin-emulating paradigms that incorporate pre-computed look-up-tables spanning a wide range of distribution parameters are used for collision-coalescence, hydrometeor sedimentation (Feingold et al., 1998), riming (Saleeby & Cotton, 2008), melting, and shedding (Meyers et al., 1997), the latter of which can occur owing to melting as well as during wet growth. An option to use size-dependent coefficients in fall speed relationships is also available, an important consideration for the wide range of hail fall speeds which are also influenced by particle density and surface characteristics (Mitchell, 1996). Unique to the RAMS microphysics, bulk liquid water fraction (LWF) values for graupel and hail are diagnosed through prediction of bulk particle internal energies (Walko et al., 1995) and are incorporated into category transition logic as well as melting processes.

Milbrandt and Yau (2005a, b) developed a triple-moment (3M) bulk scheme that additionally predicts hydrometeor radar reflectivity factors to obtain fully prognostic hydrometeor size distributions. In simulations of a severe hailstorm, the 3M scheme produced reflectivity, precipitation patterns, and hail sizes that resembled observations much more closely than simulations using 1M or 2M schemes (Milbrandt & Yau, 2006a, b). A separate investigation for tornadic supercells by Dawson et al. (2010) reached similar conclusions.

The RAMS triple-moment hail scheme (3MHAIL, Loftus et al., 2014) builds upon the RAMS 2 M scheme and additionally prognoses the spectral width parameter (ν, see Eq. 3.1) of the hail PSD by predicting the sixth moment (related to the radar reflectivity factor). 3MHAIL greatly expands the range of predicted hail sizes (diameters up to ~15 cm) beyond RAMS 2M (diameters up to ~4 cm) and uses particle density-based partitioning logic following Milbrandt and Yau (2005b) for hail initiation resulting from liquid-ice interactions and riming in order to better represent hail embryo formation. Improvements in representing hail sedimentation and melting processes allow for narrowing of the PSD and prevent excessive size sorting and artificial increases in mean hail sizes that can occur in many 2M microphysics schemes (e.g., Mansell et al., 2020). Whereas RAMS 2M assumes hailstone number losses are proportional to mass loss during the melting process, and all hail completely melts when LWF exceeds 0.95, 3MHAIL explicitly considers complete melting of the smallest hailstones based on the hail bulk LWF. Figure 8 compares time-height contour plots of melting for an ideal hail shaft comprised of falling small hailstones initially located above the 0 °C height using the RAMS 2M and 3MHAIL microphysics parameterizations. Three distinct melting zones are evident in the 3MHAIL case (Fig. 8b), with zone 1 (~2.3–3.5 km) associated with melting of the smallest hailstones ($D \leq 5$ mm). Within zone2 (~1 to 2.3 km), higher RH values reduce competition between evaporative cooling and convective heating at the hailstone surface, leading to increased environmental heat transfer and greater melting magnitudes. Zone 3 (0 to ~ 1 km) includes melting of all remaining hail to rain at grid points with bulk hail LWF >0.95. Conversely, the 2M case exhibits only a single contiguous region of complete melting below ~2.3 km (Fig. 8a), with melting absent above this height owing to bulk hail LWF values that are below 0.95. In addition, the 3MHAIL scheme realistically simulates shifts from broad hail size spectra aloft (small ν values) to narrow spectra (larger ν values) with time within and below the melting layer (Fig. 8c).The ability to fully predict the hail PSDs with 3MHAIL and similar 3M schemes has been shown to *vastly* improve the representation of hail formation, growth, sedimentation, and melting processes, as well as surface precipitation characteristics and overall convective evolution, relative to lower-order moment schemes (Milbrandt & Yau, 2005a, 2006a, b; Loftus et al., 2014; Loftus & Cotton, 2014). Figure 9 shows sample results from GCE simulations of an observed large hail-producing supercell from the Severe Thunderstorm Electrification and Precipitation Study (STEPS; Lang et al., 2004; Tessendorf et al., 2005) using 1M, 2M and 3M bulk schemes. A clear improvement is evident in the model's ability to predict realistic hail properties (e.g., narrow hail shafts and large hail reaching the surface) progressing from the 1M-4ICE to RAMS2M and 3MHAIL microphysics schemes.

Recently, some bulk microphysical schemes have shifted away from the traditional delineation imposed by predefined ice categories with fixed properties and instead employ prediction of evolving particle properties such as density (e.g.

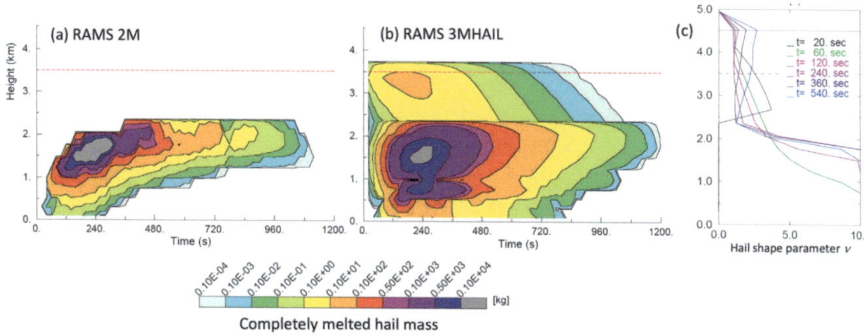

Fig. 8 Time-height plots of total mass (kg) of completely melted hailstones averaged over one minute at each vertical level for **a** RAMS 2M, and **b** RAMS 3MHAIL microphysics schemes. Panel **c** shows vertical profiles of ν through hail shaft core for 3MHAIL at selected times. Red dashed line at 3.5 km in all panels denotes approximate 0 °C isotherm height (Adapted from Loftus et al., 2014)

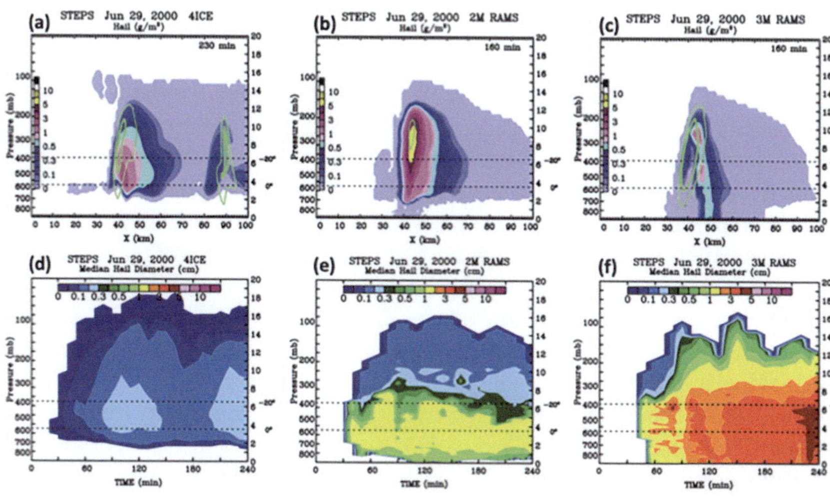

Fig. 9 Top row: vertical cross sections through hail cores of hail mass content (shaded) and vertical velocity (contours of 10 and 20 m s^{-1}) of GCE-simulated supercells using the 1M-4ICE, RAMS 2M, and 3MHAIL microphysics schemes. Bottom row: Time-height series of domain maximum hail median volume diameter for the simulations. Average heights of 0° and −20 °C are indicated by dashed lines in all panels

Mansell et al., 2010; Milbrandt & Morrison, 2013; Morrison & Milbrandt, 2015) or shape characteristics (Harrington et al., 2013; Jensen et al., 2017). The predicted particle properties (P3) scheme of Morrison and Milbrandt (2015) predicts rimed mass and volume in addition to mass mixing ratio and number concentration of a single ice category thereby allowing ice particle properties to evolve naturally through processes such as vapor deposition, aggregation, and riming. The P3 scheme also allows for multiple ice categories (Morrison & Milbrandt, 2016), as well as inclusion of bulk LWF for mixed-phase hydrometeors (Cholette et al., 2019), although the limited studies to date with these P3 scheme versions have not focused on hailstorms. Another recently developed microphysical parameterization is the 3M National Taiwan University (NTU) scheme (Tsai & Chen, 2020). Unlike predicting the 6th moment of the PSD as in the MY 3M and RAMS 3MHAIL schemes, this scheme instead predicts the 2nd moment of the PSD which is related to cross-sectional area. The NTU scheme combines the traditional bulk scheme aspects of predefined ice categories (e.g., separate graupel and hail species), as well also predicting adaptive growth habit for pristine ice and snow aggregates along with apparent density of rimed particles (graupel). Hailstones are initiated from graupel when the growth condition reaches the Shumann–Ludlam wet growth limit. Idealized simulations of a squall line in two dimensions showed the scheme produces results similar to using other 2M schemes; however, the ability of the NTU scheme to reasonably predict hail has not yet been fully tested and remains to be determined.

4.1.2 Bin Schemes

Bin or spectral microphysical schemes are formulated to predict the evolution of hydrometeors of discrete sizes. These schemes typically involve representation of the particle spectra with several tens of size (or mass) bins, and the evolution of the size distribution is explicitly calculated. Bin microphysical models are advantageous over bulk models in representing processes such as nucleation, collision-coalescence, and sedimentation of particles, though these advantages are offset by the considerable computing cost involved with predicting the evolution of individual size (mass) bins. These schemes are also useful to evaluate the components of the more efficient bulk microphysical schemes, particularly sedimentation processes (Milbrandt & Yau, 2005a; Mansell, 2010).

As with bulk schemes, bin microphysical models vary in complexity from treating only warm rain processes (e.g., Clark, 1973; Kogan, 1991) to inclusion of multiple ice species (e.g., Takahashi, 1976; Reisin et al., 1996; Khain et al., 2004). Simple bin models typically incorporate only one size distribution function to describe both cloud ice and frozen precipitation (e.g., Hall, 1980; Ovtchinnikov and Kogan, 2000), for which the smallest particles are interpreted as ice crystals and the larger particles are considered to be graupel and/or hail, whereas more complex schemes contain several size distribution functions for different types of ice particles (e.g., Takahashi, 1976; Reisin et al., 1996; Khain et al., 2004; Lynn et al., 2005). Size-dependent particle characteristics such as particle density, crystal habits and shape factors for ice species that affect collection rates and fall speeds are also capable of being represented in bin microphysical schemes (Takahashi, 1976; Hall, 1980; Farley & Orville, 1986; Chen & Lamb, 1994; Khain et al., 2004). For example, Farley (1987a) was able to simulate the initial low-density growth stages of hail through the incorporation of bin microphysics and found that enhanced hail growth and increased number concentrations of large hail resulted when allowing for variable hail density compared to using a fixed density. In addition, some spectral models (Farley & Orville, 1986; Chen & Lamb, 1994; Khain et al., 2004) include heat budget considerations to predict basic thermodynamic properties of ice particle bins, such as surface temperature and liquid water mass, which are especially important for hail growth processes. Simulation of the wide range of hail sizes remains a challenge for bin schemes, with one of the widely used bin schemes in the WRF model, the Hebrew University of Jerusalem, Israel SBM (Khain et al., 2004, 2010) only able to generate hail up to 1 cm diameter given the constraint on the number of mass bins predicted.

Although bin schemes often represent hydrometeor distributions and many microphysical processes more realistically than bulk schemes, their use in 3D cloud-resolving and mesoscale models is often not feasible owing to their enormous computational cost. That being the case, methods have been designed for bulk schemes to emulate a bin model for certain microphysical processes such as droplet nucleation, stochastic collection, melting and sedimentation (Feingold et al. 1998; Feingold & Heymsfield, 1992; Cotton et al., 2003; Saleeby & Cotton, 2004). For example, *bin-emulating* methods for collection, melting and sedimentation in the RAMS 3MHAIL scheme involve the use of separate bin models to create look-up tables spanning a wide range of possible distributions (Loftus et al., 2014). Another option is to employ a hybrid bin/bulk model such as Farley and Orville (1986), Johnson et al. (1993), and Guo and Huang (2002) in which the graupel/hail species is predicted using a bin approach and the remaining species are predicted using bulk microphysics.

4.2 Other Hail Prediction Methods

In operational settings, hail size has traditionally been predicted from forecast environmental parameters such as lapse rate, wind shear, temperature, height of the 0 °C wet-bulb temperature level, and convective available potential energy (CAPE), which have been formulated into indices like the Significant Hail Parameter (SHiP; https://www.spc.noaa.gov/help/ship.html). The HAILCAST 1D coupled cloud and hail model, developed by Poolman (1992)and improved upon by Brimelow et al. (2002), diagnoses the maximum expected hail diameter at the surface using a profile of predicted atmospheric temperature and moisture, and assumes a steady-state updraft. While these methods have shown reasonable accuracy, Adams-Selin et al. (2019) found that HAILCAST prognoses some hail-size ranges better than others depending on how it is configured, and Herman et al. (2018) noted that SHiP often performs poorly in high shear/low CAPE environments. Adams-Selin and Zeigler (2016) modified and integrated HAILCAST into the Advanced Research version of WRF (WRF-ARW; Skamarock et al., 2008) to provide hail size information at the surface with fairly good results (within 0.5 in of MESH hail size more than 60% of time when WRF successfully predicts convection). While some of the basic hail processes such as riming and melting/shedding are included in this model, the predicted hail sizes are sensitive to the assumed initial sizes and injection heights of hail embryos, and no feedbacks between hail and the parent storm are considered.

4.3 Hail Model Validation

It is critical to establish the ability of the microphysics to produce realistic hail fields and characteristics in relation to storm dynamical structures as well as their proper time evolution. The simplest evaluation involves comparisons of model-generated hail to surface observations of hail sizes and amounts, although such observations are not always available and provide little to no information on storm structure or convective-scale processes. More robust verification can be achieved via systematic and comprehensive evaluation against radar observations not depleted by significant attenuation. Comparisons of model results with satellite observations are another way to evaluate microphysical representation on a larger scale (e.g., Matsui et al., 2014).

Ground-based radars in particular offer a comprehensive view of convective systems, including their cores. This allows for statistical-type evaluations such as Cumulative Frequency with Altitude Diagrams (CFADs, Yuter & Houze, 1995) of model PDFs and modes at varying heights (e.g., Lang et al., 2014) as well as their time-varying attributes (Nicholls et al., 2017) to identify model biases. Instantaneous, detailed cross-sections are equally important for identifying unrealistic model structures that may result from microphysical representations (e.g., Lang et al., 2014; Tao et al., 2016). Polarimetric radar quantities allow further evaluation of microphysics via comparisons of simulated and observed hail-related polarimetric signatures (c.f. Sect. 2). The variability of polarimetric quantities can be combined into a hydrometeor classification algorithm (HID) able to identify regions of different bulk hydrometeors in storms as well as hydrometeor frequencies by altitude (Matsui et al., 2019, 2020). Most HID algorithms have a hail category (e.g., Dolan et al., 2013); some include both small and large hail categories (e.g., Vivekanandan et al., 1999; Tessendorf et al., 2005), offering a way to track the development of hail cores, making polarimetric radar a key aspect of validating hail microphysics schemes and models. These methods are often used to target specific related microphysical processes or assumptions for improvement.

Instrument simulators utilizing forward models serve to interface and integrate remote sensing measurements with model simulations, thereby facilitating the direct validation and intercomparison between model-predicted geophysical parameters and observations (Matsui et al., 2014). Meteorological and geophysical parameters predicted by the model are used as inputs to a forward model to drive the individual instrument simulators for calculating radiance, reflectivity and backscattered signals. There are several multi-instrument simulators available that consider hail in addition to other hydrometeors and include the Goddard Satellite Data Simulator Unit (G-SDSU, Masunaga et al., 2010; Matsui et al., 2009, 2013), and the Atmospheric Radiative Transfer Simulator (ARTS, Buehler et al., 2005; Eriksson et al., 2011). These simulators enable testing of the models' ability to simulate physical and meteorological processes in specified spatial and temporal domains by comparing model-derived synthetic radiance fields and reflectivity/backscatter profiles to those measured by satellite, surface, and/or airborne platforms. The POLArimetric Radar

Retrieval and Instrument Simulator (POLARRIS, Matsui et al., 2019) was developed to marry model output and radar observations in a common framework. POLARRIS includes both a forward model (POLARRIS-F) integrated into G-SDSU to maintain physical consistency with model microphysics and calculate polarimetric radar observations (Z_h, Z_{dr}, K_{dp}, ρ_{hv}) at precipitation radar wavelengths (X, C and S-band) and an inverse (retrieval and diagnostic) component (iPOLARRIS) based on CSU radar retrievals (e.g., Dolan & Rutledge, 2009; Dolan et al., 2013). POLARRIS was designed to be flexible so that varying assumptions and microphysics schemes can be utilized in POLARRIS-F and different retrievals (e.g., HID algorithms) adopted into iPOLARRIS. Figure 10 shows an example from MC3E of POLARRIS simulated data (left) using NASA Unified-WRF (NU-WRF; Peters-Lidard et al., 2015) output and ground-based C-band radar observations (right). POLARRIS captures well the main reflectivity and HID features of the observed storm; differences between the simulations and observations stem from NU-WRF forecast errors in timing and location.

One of the longstanding challenges for creating synthetic observations of hailstorms with instrument simulators deals with the electromagnetic scattering properties of large particles, especially when the particles are mixed-phase such as melting hail or hail under growing wet growth (Bringi & Seliga, 1977). Natural hailstones often exhibit irregular shapes and can have lobed structures (Knight & Knight, 1970a, c; Knight, 1986), although they are generally assumed to be spheroids in order to compute scattering properties via the Rayleigh approximation (e.g., Waldvogel et al., 1978; Jung et al., 2008) or T-matrix methods (Barber & Yeh, 1975; Aydin & Zhao, 1990; Balakrishnan & Zrnić, 1990; Depue et al., 2007; Loftus & Cotton, 2014). It is important to note that, even at S-band, hailstones larger than about 2 cm diameter are outside of the Rayleigh approximation (Kumjian, 2013), which assumes the particle size is much smaller than the wavelength. Discrete Dipole Approximation (DDA) methods can compute the scattering properties of arbitrary shapes and are thus better suited for hailstones. For example, Jiang et al. (2019) used the Amsterdam DDA

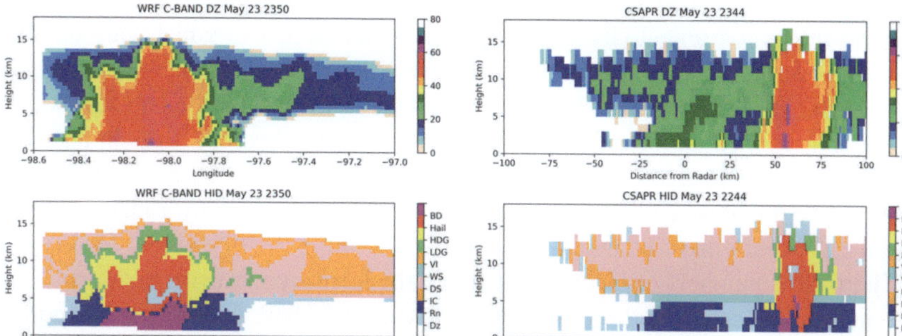

Fig. 10 An example of POLARRIS applied to a NU-WRF 1 M-4ICE simulation (left) and CSAPR C-band radar (right) for reflectivity (top) and HID (bottom) from the MC3E case (23 May 2011, ~23 UTC)

(ADDA) code (Yurkin & Hoekstra, 2011) to calculate scattering properties from digital renderings of actual hailstones and noted significant differences in backscattered reflectivity and polarimetric quantities compared to T-matrix computations using spheroids, even for small (2–3 cm) hailstones. The authors further showed that addition of thin liquid coatings on the synthetic hailstones led to enhancement of all simulated radar quantities.

5 Summary and Future Directions

Hail poses a significant threat to structures, vehicles, aircraft, livestock, and crops on an annual basis, especially in the U.S., and has been reported on every continent except Antarctica (Allen et al., 2019). There have been numerous observations of and observational studies on hail. Many of these are focused on surface observations, which have shown that hail can range in size from small mm-sized stones to as large as the famous Coffeyville, KS hailstone, which had a maximum circumference of 44 cm (Roos, 1972). Hailstones have been observed to have a variety of shapes, ranging from ellipsoid for the most common to more oblate for larger stones, which are also more likely to become highly asymmetric and have appendages. Examination of individual hailstones can also provide information on their density, which is often less than pure ice and can vary radially, revealing the different modes of growth. Hail size distributions are observed to vary widely from very broad to nearly monodisperse and to vary from storm to storm and even within the same hailstorm. In situ measurements of hail in deep convection are highly valuable and extremely rare due to the danger involved and were largely obtained from a single specialized T-28 aircraft.

Numerous studies, however, have exploited Doppler radar to safely but remotely study the characteristics of hail-producing storms. Various reflectivity thresholds have been established for the presence of hail and large hail. Specific radar signatures, namely WERs and BWERs, that result from the presence of intense updrafts are also typically associated with many hail-producing supercell and multicell storms. Column-integrated reflectivity values as well as the height to which various dBZ intensities penetrate can also be used to infer the presence of hail. Polarimetric radar variables such as ZDR, KDP, and LDR can provide additional information on the shape, orientation, and phase of remotely-sensed hydrometeors in addition to intensity, including hail. Satellite-based HDAs have also been developed using both passive microwave scattering signatures and active space-borne radar returns to assess hail occurrence on a global scale.

There are a numbers of factors involved in the formation, growth, sedimentation and melting of hail, including the number and type of hail embryos, the opportunity for growth in convective updrafts, the temperature profile, and availability of supercooled liquid water. Strong updrafts are needed to support hail, but the residence time hail and hail embryos have in those updrafts has a major impact on overall hail growth with broader updrafts and reduced horizontal flow relative to the updrafts resulting in larger hail. Hailstone embryos typically originate from graupel for colder cloud base temperatures and frozen drops from warmer cloud base temperatures. The number and size of these embryos affects subsequent hail growth with larger embryos growing faster (though not necessarily leading to large hail), while more numerous embryos result in increased competition for super cooled water and thus smaller hail. The majority of hail growth occurs between −10 and −25 °C. As air temperatures decrease, higher LWCs are required for wet growth for a given hailstone size. Dry growth typically occurs with lower LWCs and/or colder temperatures relative to the hailstone surface temperature. Hailstone terminal velocity depends on the hail density, diameter, drag coefficient C_D, and air density. Values of C_D have been found to range from about 0.45 for a smooth sphere to over 1.0 with increasing oblateness, leading to slower fall speeds. Hail melting depends on hail size, air temperature, RH, and melting level height with faster rates occurring for warmer air temperatures, higher RH values and smaller hailstone sizes. Higher RH values also lead to increased shedding.

The representation of hail and hail processes in atmospheric modeling must be parameterized as tracking individual hailstones is well beyond current computing capabilities, but it is also essential as hail processes can affect downdrafts and cold pools and hence overall storm evolution in addition to the radar and scattering signatures. There are a variety of cloud microphysics schemes, ranging from simpler 1M bulk to more detailed SBM schemes. Bulk schemes have assumptions on how the particles for a given species are distributed, including hail, with the most common for precipitation-sized particles being inverse exponential (Marshall-Palmer, Lin et al., 1983; Rutledge and Hobbs, 1984) or gamma (Saleeby & Cotton, 2004) with the latter having the advantage of being able to represent peaks, which are often observed in hail size spectra.

Bulk schemes that include both graupel and hail categories, even when using 1M, like the Goddard 4ICE scheme (Lang et al., 2014; Tao et al., 2016), have been shown to produce more realistic reflectivity signatures, including peak values and gradients near and above the freezing level, than when not including hail. Bulk schemes employing two (Milbrandt & Yau, 2005a; Mansell et al., 2010) or more moments have the potential to outperform 1M schemes (Ferrier, 1994; Morrison et al., 2009), especially in terms of sedimentation and melting when it comes to hail. By prognosing the spectral width, 3M schemes (Milbrandt & Yau, 2006a, b; Loftus et al., 2014) have been shown to produce more realistic reflectivity, precipitation patterns, and hail sizes, greatly expand the range of predicted hail sizes and prevent excessive size sorting and artificial increases in mean hail sizes relative to 2M schemes, which traditionally prognose just the mass and number concentrations. Several other innovations have been adopted into bulk schemes such as allowing particle density (Morrison & Milbrandt, 2015) or shape to vary rather than using pre-defined categories or predicting a different 3rd moment such as particle cross sectional area (Tsai & Chen, 2020).

Bin schemes predict the evolution of hydrometeor distributions using discrete size (or mass) bins and can more explicitly represent processes such as nucleation, collision-coalescence, and particle sedimentation but are very expensive. Simple bin schemes use one size distribution to describe both cloud ice and frozen precipitation

(Ovtchinnikov & Kogan, 2000) with small particles interpreted as ice crystals and large particles as graupel and/or hail while more complex schemes use several size distribution functions for different types of ice particles (Khain et al., 2004). Size-dependent ice particle characteristics such as density, habit and shape that affect collection rates and fall speeds as well as particle heat budget considerations that are important for hail growth can also be represented in bin schemes. Although SBM are costly, bulk schemes can be designed to emulate and thereby take advantage of bin schemes for certain microphysical processes such as droplet nucleation, stochastic collection, melting and sedimentation.

Another approach to hail modeling is ingest either forecast or simulated atmospheric profiles of temperature, moisture and shear into a 1D hail model such as SHiP or HAILCAST (Poolman, 1992) for the purposes of diagnosing the maximum expected hail diameter at the surface. These types of models can be useful in a forecast-type setting to assess hail threat but typically make assumptions regarding updrafts and are non-interactive with no feedback to the convection.

Validation is a critical component of hail modeling, so it is essential that microphysics schemes are systematically evaluated in their ability to produce realistic hail fields and characteristics as well as their time evolution relative to the storm dynamics. Comparing simulated and observed surface hail is one simple method, but such observations, though useful, are limited and provide little to no information on storm structure or convective-scale processes. The most comprehensive approach is via systematic, statistical comparisons against radar observations, including polarimetric, not depleted by significant attenuation, which can sample entire convective systems, including convective cores. These comparisons are often in the form of CFADs (or stacked PDFs) or HID analysis. Instrument simulators such as the G-SDSU (Matsui et al., 2009) are important for facilitating the direct comparison between simulated hydrometeors and observed instrument radiance, reflectivity and backscattered signals. Instantaneous, detailed cross-sections are also important for identifying unrealistic model structures that may not be apparent in the statistical comparisons as are time-varying comparisons. Comparisons against satellite observations are another way to evaluate microphysical parameterizations on a larger scale but cannot match the detail of ground-based radars and are susceptible to beam-filling issues. It is also important to note that other factors, namely grid resolution, can have a substantial impact on convective simulations in addition to the microphysics parameterization.

The representation of microphysical processes in numerical models continues to advance. One such advancement entails the use of a Smoothed Particle Hydrodynamics (SPH) model. The Snow Meshless Lagrangian Technique (SnowMeLT) model (Pelissier et al., 2023) applies physical laws governing heat transfer along with meltwater dynamics to simulate melting of 3Dsynthetic snowflakes at extremely high resolution (5–50 μm) with unprecedented realism. Work is underway to expand and transform the SnowMeLT model into a virtual cloud chamber capable of simulating both growth and melting processes for a variety of frozen particle types, including graupel and hail. High-resolution (0.008 cm) digital modeled hailstones from 3D scans of actual hailstones in the field (e.g., Giammanco et al., 2017) are available for helping advance this work. Such an enhanced virtual cloud chamber could allow for creation of a large database of hail growth and melting processes for developing, validating, and refining hail microphysical parameterizations in models. This database could also be used for simulating critically needed single-scattering properties of growing and melting hail for remote sensing applications (e.g. Jiang et al., 2019).

Understanding how hail events may change in response to expected climate change is vital given the enormous economic risk all hail poses to agriculture and property. Climate modeling studies examining this issue have primarily used projections of changing large-scale atmospheric indicators of severe convection as proxies for predicting trends in hail events. However, the coarse resolution of global and regional climate models (GCMs and RCMs, respectively) prevents simulation of convection itself, let alone the myriad of hail microphysical characteristics that are critical for determining potential impacts at the ground. Higher resolution convection-permitting models (CPMs) are better suited in this regard and can be applied to investigate hail trends in future climate scenarios; however, crucial details regarding hail size spectra remain largely unresolved leading to uncertainty in surface impacts.

Reviews by Allen et al. (2019) and Mahoney (2020) report that work on hail trends in both present and future climates assesses aggregate hail risk from GCM-generated environments that are statistically downscaled (e.g., Púčik et al., 2015; Tang et al., 2019), partially dynamically downscaled through offline hail models (e.g., Brimelow et al., 2017), or fully downscaled through convection-permitting simulations (e.g., Trapp et al., 2019). These environmental studies found no firm relationship between environmental CAPE and hail size nor a consensus in large hail trends. Increases in freezing level heights in future climate studies have led to general agreement that there will be a reduction in small hail. Projected increases in days with high CAPE environments that may favor severe weather (e.g., Prein et al., 2017; Rasmussen et al., 2017) may not actually result in more severe convective storms (Trapp et al., 2019) or large hail (Kumjian et al., 2019).

The expense of continuous long-running regional and global convection-permitting simulations has mostly limited the running of these simulations. One way around the expense of continuous simulations is continuous daily restarting. Hoogewind et al. (2017) performed a continuous daily restart of WRF at 4-km over the U.S. for 12-months forced by an RCP 8.5 climate scenario perturbation and 3 ice-class microphysics. Trapp et al. (2019) used the output to drive an offline hail growth model. This approach provides insight into potential changes in hail risk at the upper end of the hail size distribution over a broad area. From the storm-scale perspective, there are two important drawbacks. The daily-stepping approach does not accurately portray the continuity of development of long-lived (>12 h) convective systems (Giorgi, 2019), such as "plowable" hailstorms that produce large accumulations of small hail (Kalina et al., 2016; Kumjian et al., 2019). Offline hail growth models are affected by assumptions of steady updrafts and simple relationships between CAPE, vertical shear, and hail growth that overestimate hail frequency and size (Mahoney, 2020). Furthermore, such models cannot account for dynamical feedbacks between microphysical and storm-scale to mesoscale processes that can impact convective morphology and longevity (Allen et al., 2019). Because hail sizes and number densities can vary through the lifecycles of propagating systems, statistical analyses of populations of hailstorms from continuous regional, seasonal simulations will likely be required to understand how characteristics such as system sizes and hail size distributions might change in a future climate, possibly presenting new threats to agriculture, transportation, and wildlife.

References

Adams-Selin, R., & Ziegler, C. (2016). Forecasting hail using a one-dimensional hail growth model within WRF. *Monthly Weather Review, 144*, 4919–4939. https://doi.org/10.1175/MWR-D-16-0027.1

Adams-Selin, R. D., van den Heever, S. C., & Johnson, R. H. (2013). Impact of graupel parameterization schemes on idealized bow echo simulations. *Monthly Weather Review, 141*, 1241–1262. https://doi.org/10.1175/MWR-D-12-00064.1

Adams-Selin, R. D., Clark, A. J., Melick, C. J., Dembek, S. R., Jirak, I. L., & Ziegler, C. L. (2019). Evolution of WRF-HAILCAST during the 2014–16 NOAA/hazardous weather testbed spring forecasting experiments. *Weather and Forecasting, 34*, 61–79. https://doi.org/10.1175/WAF-D-18-0024.1

Allen, J. T., et al. (2019). Understanding hail in the earth system. *Reviews of Geophysics, 58*, e2019RG000665. https://doi.org/10.1029/2019rg000665.

Amburn, S. A., & Wolf, P. L. (1997). VIL density as a hail indicator. *Weather Forecasting, 12*, 473–478. https://doi.org/10.1175/1520-0434(1997)012%3c0473:VDAAHI%3e2.0.CO;2

Atlas, D., Ulbrich, C. W., & Williams, C. R. (2004). Physical origin of a wet microburst: Observations and theory. *Journal of Atmospheric Science, 61*, 1186–1196.

Auer, A. H., & Marwitz, J. D. (1972). Hail in the vicinity of organized updrafts. *Journal of Applied Meteorology, 11*, 748–752.

Aydin, K., & Zhao, Y. (1990). A computational study of polarimetric radar observables in hail. *IEEE Transactions on Geoscience and Remote Sensing, 28*, 412–422.

Aydin, K., Seliga, T. A., & Balaji, V. (1986). Remote sensing of hail with a dual linear polarization radar. *Journal of Climate Applied Meteorology, 25*, 1475–1484.

Bailey, I. H., & Macklin, W. C. (1968). Heat transfer from artificial hailstones. *Quarterly Journal of Royal Meteorological Society, 94*, 93–98.

Balakrishnan, N., & Zrnić, D. S. (1990). Use of polarization to characterize precipitation and discriminate large hail. *Journal of Atmospheric Science, 47*, 1525–1540.

Bang, S. D., & Cecil, D. J. (2019). Constructing a multifrequency passive microwave hail retrieval and climatology in the GPM domain. *Journal of Applied Meteorology and Climatology, 58*, 1889–1904. https://doi.org/10.1175/JAMC-D-19-0042.1

Bani, P., Grecchi, I., Ahmed, S., Ficuciello, V., Calamari, L., Tabaglio, V., & Minuti, A. (2019). Effects of defoliation on whole-plant maize characteristics as forage and energy crop. *Grass and Forage Science, 74*, 65–77. https://doi.org/10.1111/gfs.12397

Barber, P., & Yeh, C. (1975). Scattering of electromagnetic waves by arbitrarily shaped dielectric bodies. *Applied Optics, 14*, 2864–2872.

Bennetts, G. M., & Rawlins, F. (1981). Parameterization of the ice-phase in a model of midlatitude cumulonimbus convection and its influence on the simulation of cloud development. *Quarterly Journal of Royal Meteorological Society, 107*, 477–502.

Blair, S. F., Deroche, D. R., Boustead, J. M., Leighton, J. W., Barjenbruch, B. L., & Gargan, W. P. (2011). A radar-based assessment of the detectability of giant hail. *Electronic Journal of Severe Storms Meteorology, 6*(7).

Blair, S. F., & Leighton, J. W. (2012). Creating high-resolution hail datasets using social media and post-storm ground surveys. *Electronics Journal of Operational Meteorology, 13*(3), 32–45.
Botzen, W. J. W., Bouwer, L. M., & van den Bergh, J. C. J. M. (2010). Climate change and hailstorm damage: Empirical evidence and implications for agriculture and insurance. *Resource and Energy Economics, 32*, 341–362. https://doi.org/10.1016/j.reseneeco.2009.10.004
Braham, R. R., Jr. (1964). What is the role of ice in summer rain-showers? *Journal of Atmospheric Science, 21*, 640–645.
Brandes, E. A., Vivekanandan, J., Tuttle, J. D., & Kessinger, C. J. (1995). A study of thunderstorm microphysics with multiparameter radar and aircraft observations. *Monthly Weather Review, 123*, 3129–3143. https://doi.org/10.1175/1520-0493(1995)123%3c3129:ASOTMW%3e2.0.CO;2
Brimelow, J. C., Reuter, G. W., & Poolman, E. R. (2002). Modeling maximum hail size in Alberta thunderstorms. *Weather Forecasting, 17*, 1048–1062.
Brimelow, J. C., Burrows, W. R., & Hanesiak, J. M. (2017). The changing hail threat over North America in response to anthropogenic climate change. *Natural Climate Change, 7*, 516–522.
Bringi, V. N., & Seliga, T. A. (1977). Scattering from non-spherical hydrometeors. *Annals of Telecommunications, 32*, 392–397.
Bringi, V. N., Vivekanandan, J., & Tuttle, J. D. (1986). Multiparameter radar measurements in Colorado Convective Storms. Part II: Hail detection studies. *Journal of Atmospheric Science, 43*, 2564–2577.
Brown, T. M., Pogorzelski, W. H., & Giammanco, I. M. (2015). Evaluating hail damage using property insurance claims data. *Weather Climate Society, 7*, 197–210. https://doi.org/10.1175/wcas-d-15-0011.1
Browning, K. A. (1963). The growth of large hail within a steady up draught. *Quarterly Journal of Royal Meteorological Society, 89*, 491–506.
Browning, K. A. (1966). The lobe structure of giant hailstones. *Quarterly Journal of Royal Meteorological Society, 92*, 1–14.
Browning, K. A., & Beimers, J. G. D. (1967). The oblateness of large hailstones. *Journal of Applied Meteorology and Climatology, 6*, 1075–1081. https://doi.org/10.1175/1520-0450(1967)006%3c1075:TOOLH%3e2.0.CO;2
Browning, K. A., & Foote, G. B. (1976). Airflow and hail growth in supercell storms and some implications for hail suppression. *Quarterly Journal of Royal Meteorological Society, 102*, 499–533. https://doi.org/10.1002/qj.49710243303
Browning, K. A., & Ludlam, F. H. (1962). Airflow in convective storms. *Quarterly Journal of Royal Meteorological Society, 88*, 117–135.
Browning, K. A., Hallett, J., Harrold, T. W., & Johnson, D. (1968). The collection and analysis of freshly fallen hailstones. *Journal of Applied Meteorology and Climatology, 7*, 603–612. https://doi.org/10.1175/1520-0450(1968)007%3c0603:TCAAOF%3e2.0.CO;2
Browning, K. A. (1977). Hail: A review of hail science and hail suppression. Part I: Hail physics. In G. B. Foote & C. A. Knight (Eds.), *The Structure and Mechanisms of Hailstorms*. Meteorological Monographs (Vol. 38, pp. 1–39). American Meteorological Society.
Bryan, G. H., & Morrison, H. (2012). Sensitivity of a simulated squall line to horizontal resolution and parameterization of microphysics. *Monthly Weather Review, 140*, 202–225. https://doi.org/10.1175/MWR-D-11-00046.1
Bryan, G. H., Wyngaard, J. C., & Fritsch, J. M. (2003). Resolution requirements for the simulation of deep moist convection. *Monthly Weather Review, 131*, 2394–2416. https://doi.org/10.1175/1520-0493(2003)131%3c2394:RRFTSO%3e2.0.CO;2
Buehler, S. A., Eriksson, P., Kuhn, T., von Engeln, A., & Verdes, C. (2005). ARTS, the atmospheric radiative transfer simulator. *Journal of Quantitative Spectroscopy & Radiative Transfer, 91*(1), 65–93. https://doi.org/10.1016/j.jqsrt.2004.05.051
Carey, L. D., & Rutledge, S. A. (1998). Electrical and multiparameter radar observations of a severe hailstorm. *Journal of Geophysical Research, 103*, 13979–14000.
Carte, A. E. (1966). Features of Transvaal hailstorms. *Quarterly Journal of Royal Meteorological Society, 92*, 290–296.

Carte, A. E., & Kidder, R. E. (1966). Transvaal hailstones. *Quarterly Journal of Royal Meteorological Society, 92*, 382–391.

Cecil, D. J. (2009). Passive microwave brightness temperatures as proxies for hailstorms. *Journal of Applied Meteorology and Climatology, 48*, 1281–1286. https://doi.org/10.1175/2009JAMC2125.1

Cecil, D. J., & Blankenship, C. B. (2012). Toward a global climatology of severe hailstorms as estimated by satellite passive microwave imagers. *Journal of Climate, 25*, 687–703. https://doi.org/10.1175/JCLI-D-11-00130.1

Chalon, J. P., Fankhauser, J. C., & Eccles, P. J. (1976). Structure of an evolving hailstorm, Part I: General characteristics and cellular structure. *Monthly Weather Review, 104*, 564–575.

Changnon, S. A., Jr. (1971). Hailfall characteristics related to crop damage. *Journal of Applied Meteorology, 10*, 270–274.

Changnon, S. A., Jr. (1973). Hail sensing and small scale variability of windblown hail. *Journal of Weather Modification, 5*, 30–42.

Charlton, R. B., & List, R. (1972). Hail size distributions and accumulation zones. *Journal of Atmospheric Science, 29*, 1182–1193.

Chen, J., & Lamb, D. (1994). Simulation of cloud microphysical and chemical processes using a multicomponent framework. Part I: Description of the microphysical model. *Journal of Atmospheric Science, 51*, 2613–2630.

Cheng, L., & English, M. (1983). A relationship between hailstone concentration and size. *Journal of Atmospheric Science, 40*, 204–213.

Cheng, L., & Rogers, D. C. (1988). Hailfalls and hailstorm feeder clouds—An Alberta case study. *Journal of Atmospheric Science, 45*, 3533–3545. https://doi.org/10.1175/1520-0469(1988)045%3c3533:HAHFCA%3e2.0.CO;2

Cheng, L., English, M., & Wong, R. (1985). Hailstone size distributions and their relationship to storm thermodynamics. *Journal of Climate Applied Meteorology, 24*, 1059–1067.

Chisholm, A. J. (1968). Observations by 10-cm radar of an Alberta hailstorm in a sheared environment. In *13th Radar Meteorological Conference* (pp. 82–87). Montreal, American Meteorological Society.

Chisholm, A. J. (1973) *Alberta Hailstorms. Part I: Radar Case Studies and Airflow Models.* Meteorological Monographs (Vol. 36, pp. 1–36). American Meteorological Society.

Chisholm, A. J. & Renick, J. H. (1972). *The kinematics of multicell and supercell Alberta hailstorms.* Alberta Hail Studies, Research Council of Alberta Hail Studies Report No. 72-2, 24-31.

Cholette, M., Morrison, H., Milbrandt, J. A., & Thériault, J. M. (2019). Parameterization of the bulk liquid fraction on mixed-phase particles in the predicted particle properties (P3) scheme: Description and idealized simulations. *Journal of Atmospheric Science, 76*, 561–582. https://doi.org/10.1175/JAS-D-18-0278.1

Cintineo, J. L., Smith, T. M., Lakshmanan, V., Brooks, H. E., & Ortega, K. L. (2012). An objective high-resolution hail climatology of the contiguous United States. *Weather Forecasting, 27*, 1235–1248. https://doi.org/10.1175/waf-d-11-00151.1

Clark, T. L. (1973). Numerical modeling of the dynamics and microphysics of warm cumulus convection. *Journal of Atmospheric Science, 30*, 857–878.

Cohen, C., & McCaul, E. W., Jr. (2006). The sensitivity of simulated convective storms to variations in prescribed single-moment microphysics parameters that describe particle distributions, sizes, and numbers. *Monthly Weather Review, 134*, 2547–2565.

Cotton, W. R., Stephens, M. A., Nehkorn, T., & Tripoli, G. J. (1982). The Colorado State University three-dimensional cloud/mesoscale model—1982 Part II: An ice phase parameterization. *Journal De Recherches Atmospheriques, 16*, 295–320.

Cotton, W. R., Pielke, R. A., Sr., Walko, R. L., Liston, G. E., Tremback, C. J., Jiang, H., McAnnelly, R. L., Harrington, J. Y., Nicholls, M. E., Carrio, G. G., & McFadden, J. P. (2003). RAMS 2001: Current status and future directions. *Meteorological Atmospheric Physics, 82*, 5–29.

Cotton, W. R. & Anthes, R. A. (1989). *Storm and Cloud Dynamics* (883pp.). Academic.

Crum, T. D., Alberty, R. L., & Burgess, D. W. (1993). Recording, archiving, and using WSR-88D data. *Bulletin of the American Meteorological Society, 74*, 645–654. https://doi.org/10.1175/1520-0477(1993)074%3c0645:RAAUWD%3e2.0.CO;2

Dawson, D. T., II., Xue, M., Milbrandt, J. A., & Yau, M. K. (2010). Comparison of evaporation and cold pool development between single-moment and multimoment bulk microphysics schemes in idealized simulations of tornadic thunderstorms. *Monthly Weather Review, 138*, 1152–1171.

DeMott, P. J., et al. (2010). Predicting global atmospheric ice nuclei distributions and their impacts on climate. *Proceedings of the National Academy of Sciences of the United States of America, 107*(25), 11217–11222. https://doi.org/10.1073/pnas.0910818107

Dennis, E. J. & Kumjian, M. R. (2017). The impact of vertical wind shear on hail growth in simulated supercells. *Journal of Atmospheric Science, 74*, 641–663. https://doi.org/10.1175/JAS-D-16-0066.1.

Depue, T. K., Kennedy, P. C., & Rutledge, S. A. (2007). Performance of the hail differential reflectivity (HDR) polarimetric radar hail indicator. *Journal of Applied Meteorology and Climatology, 46*, 1290–1301.

Dessens, J., & Fraile, R. (1994). Hailstone size distributions in southwestern France. *Atmospheric Research, 33*, 57–73.

Dolan, B., & Rutledge, S. A. (2009). A theory-based hydrometeor identification algorithm for X-band polarimetric radars. *Journal of Oceanic Atmospheric Technology, 46*, 1196–1213. https://doi.org/10.1175/2009JTECHA1208.1

Dolan, B., Rutledge, S. A., Lim, S., Chandrasekar, V., & Thurai, M. (2013). A robust C-band hydrometeor identification algorithm and application to a long term polarimetric radar data set. *Journal of Applied Meteorology, 52*, 2162–2186.

Dye, J. E., Knight, C. A., Toutenhoofd, V., & Cannon, T. W. (1974). The mechanism of precipitation formation in northeastern Colorado cumulus III. Coordinated microphysical and radar observations and summary. *Journal of Atmospheric Science, 31*, 2152–2159.

Dye, J. E., Martner, B. E., & Miller, L. J. (1983). Dynamical-microphysical evolution of a convective storm in a weakly-sheared environment. Part I: Microphysical observations and interpretation. *Journal of Atmospheric Science, 40*, 2083–2096.

Edwards, R., & Thompson, R. L. (1998). Nationwide comparisons of hail size with WSR-88D vertically integrated liquid water and derived thermodynamic sounding data. *Weather Forecasting, 13*, 277–285. https://doi.org/10.1175/1520-0434(1998)013%3c0277:NCOHSW%3e2.0.CO;2

English, M. (1973). *Alberta Hailstorms. Part II: Growth of Large Hail in the Storm.* Alberta Hailstorms, Meteorological Monographs (Vol. 36, pp. 37–98). American Meteorological Society.

Eriksson, P., Buehler, S. A., Davis, C. P., Emde, C., & Lemke, O. (2011). ARTS, the atmospheric radiative transfer simulator, version 2. *Journal of Quantitative Spectroscopy and Radiative Transfer*. https://doi.org/10.1016/j.jqsrt.2011.03.001.

Farley, R. D. (1987a). Numerical modeling of hailstorms and hailstone growth. Part II: The role of low-density riming growth in hail production. *Journal of Climate Applied Meteorology, 26*, 234–254.

Farley, R. D. (1987b). Numerical modeling of hailstorms and hailstone growth. Part III: Simulation of an Alberta hailstorm—Natural and seeded cases. *Journal of Climate Applied Meteorology, 26*, 789–812.

Farley, R. D., & Orville, H. D. (1986). Numerical modeling of hailstorms and hailstone growth. Part I: Preliminary model verification and sensitivity tests. *Journal of Climate Applied Meteorology, 25*, 2014–2035.

Federer, B., & Waldvogel, A. (1975). Hail and raindrop size distributions from a Swiss multicell storm. *Journal of Applied Meteorology, 14*, 91–97.

Feingold, G., & Heymsfield, A. J. (1992). Parameterizations of condensational growth of droplets for use in general circulation models. *Journal of Atmospheric Science, 49*, 2325–2342. https://doi.org/10.1175/1520-0450(2004)043%3c0182:ALMAPN%3e2.0.CO;2

Feingold, G., & Levin, Z. (1986). The lognormal fit to raindrop spectra from frontal convective clouds in Israel. *Journal of Climate Applied Meteorology, 25*, 1346–1363.

Feingold, G., Walko, R. L., Stevens, B., & Cotton, W. R. (1998). Simulations of marine stratocumulus using a new microphysical parameterization scheme. *Atmospheric Research, 47–48*, 505–528. https://doi.org/10.1016/S0169-8095(98)00058-1

Ferraro, R., Beauchamp, J., Cecil, D., & Heymsfield, G. (2015). A prototype hail detection algorithm and hail climatology developed with the Advanced Microwave Sounding Unit (AMSU). *Atmospheric Research, 163*, 24–35. https://doi.org/10.1016/j.atmosres.2014.08.010

Ferrier, B. S. (1994). A double-moment multiple-phase four-class bulk ice scheme. Part I: Description. *Journal of Atmospheric Science, 51*, 249–280.

Ferrier, B. S., Tao, W., & Simpson, J. (1995). A double-moment multiple-phase four-class bulk ice scheme. Part II: Simulations of convective storms in different large-scale environments and comparisons with other bulk parameterizations. *Journal of Atmospheric Science, 52*, 1001–1033.

Field, P. R., Heymsfield, A. J., Detwiler, A. G., & Wilkinson, J. M. (2019). Normalized hail particle size distributions from the T-28 storm-penetrating aircraft. *Journal of Applied Meteorology and Climatology, 58*(2), 231–245. https://doi.org/10.1175/JAMC-D-18-0118.1

Foote, G. B. (1984). A study of hail growth using observed storm conditions. *Journal of Climate Applied Meteorology, 23*, 84–101.

Foote, G. B. (1985). Aspects of cumulonimbus classification relevant to the hail problem. *Journal De Recherches Atmospheriques, 19*, 61–74.

Foote, G. B., & Wade, C. G. (1982). Case study of a hailstorm in Colorado. Part I: Radar echo structure and evolution. *Journal of Atmospheric Science, 39*, 2828–2846.

Giammanco, I. M., Maiden, B. R., Estes, H. E., & Brown-Giammanco, T. M. (2017). Using 3d laser scanning technology to create digital models of hailstones. *Bulletin of the American Meteorological Society, 98*, 1341–1347. https://doi.org/10.1175/BAMS-D-15-00314.1

Gilmore, M. S., Straka, J. M., & Rasmussen, E. N. (2004a). Precipitation and evolution sensitivity in simulated deep convective storms: Comparisons between liquid-only and simple ice and liquid phase microphysics. *Monthly Weather Review, 132*, 1897–1916. https://doi.org/10.1175/1520-0493(2004)132%3c1897:PAESIS%3e2.0.CO;2

Gilmore, M. S., Straka, J. M., & Rasmussen, E. N. (2004b). Precipitation uncertainty due to variations in precipitation particle parameters within a simple microphysics scheme. *Monthly Weather Review, 132*, 2610–2627. https://doi.org/10.1175/MWR2810.1

Giorgi, F. (2019). Thirty years of regional climate modeling: Where are we and where are we going next? *Journal of Geophysical Research Atmospheres, 124*, 5696–5723. https://doi.org/10.1029/2018jd030094

Grabowski, W. W., et al. (2006). Daytime convective development over land: A model intercomparison based on LBA observations. *Quarterly Journal of Royal Meteorological Society, 132*, 317–344.

Guo, X., & Huang, M. (2002). Hail formation and growth in a 3D cloud model with hail-bin microphysics. *Atmospheric Research, 63*, 59–99.

Guyer. J. L. & Ewald, R. (2004). Record hail event—Examination of the Aurora, Nebraska supercell of 22 June 2003. In *22nd Conference on Severe Local Storms*, Hyannis, MA, Paper 11B.1.

Insurance Information Institute. (2020). *Hail: The hidden risk, an analysis of property exposure to damaging hail in 2017* (16pp.). https://www.iii.org/fact-statistic/facts-statistics-hail.

Hall, W. D. (1980). A detailed microphysical model within a two-dimensional dynamic framework: Model description and preliminary results. *Journal of Atmospheric Science, 37*, 2486–2507.

Harrington, J. Y., Sulia, K., & Morrison, H. (2013). A method for adaptive habit prediction in bulk microphysical models. Part I: Theoretical development. *Journal of Atmospheric Science, 70*, 349–364. https://doi.org/10.1175/JAS-D-12-040.1

van den Heever, S. (2001). *The impact of several hail parameters on simulated supercell storms* (265pp.) (Ph.D. dissertation, Department of Atmospheric Science, Colorado State University).

Herman, G. R., Nielsen, E. R., & Schumacher, R. S. (2018). Probabilistic verification of Storm Prediction Center convective outlooks. *Weather Forecasting, 33*, 161–184. https://doi.org/10.1175/WAF-D-17-0104.1

Heymsfield, A. J. (1983). Case study of a hali storm in Colorado. Part IV: Graupel and hail growth mechanisms deduced through particle trajectory calculations. *Journal of Atmospheric Science, 40*, 1482–1509. https://doi.org/10.1175/1520-0469(1983)040%3c1482:CSOAHI%3e2.0.CO;2

Heymsfield, A. J., & Musil, D. J. (1982). Case study of a hailstorm in Colorado. Part II: Particle growth processes at mid-levels deduced from in-situ measurements. *Journal of Atmospheric Science, 39*, 2487–2866.

Heymsfield, A. J., & Wright, R. (2014). Graupel and hail terminal velocities: Does a "supercritical" Reynolds number apply? *Journal of Atmospheric Science, 71*, 3392–3403. https://doi.org/10.1175/JAS-D-14-0034.1

Heymsfield, A. J., Jameson, A. R., & Frank, H. W. (1980). Hail growth mechanisms in a Colorado storm: Part II: Hail formation processes. *Journal of Atmospheric Science, 37*, 1779–1807. https://doi.org/10.1175/1520-0469(1980)037%3c1779:HGMIAC%3e2.0.CO;2

Höller, H., Bringi, V. N., Hubbert, J., Hagen, M., & Meischner, P. F. (1994). Life cycle and precipitation formation in a hybrid-type hailstorm revealed by polarimetric and Doppler radar measurements. *Journal of Atmospheric Science, 51*, 2500–2522.

Hoogewind, K. A., Baldwin, M. E., & Trapp, R. J. (2017). The impact of climate change on hazardous convective weather in the United States: Insight from high-resolution dynamical downscaling. *Journal of Climate, 30*, 10081–10100. https://doi.org/10.1175/jcli-d-16-0885.1

Hou, A. Y., et al. (2014). The global precipitation measurement mission. *Bulletin of the American Meteorological Society, 95*, 701–722. https://doi.org/10.1175/BAMS-D-13-00164.1

Hubbert, J., Bringi, V. N., Carey, L. D., & Bolen, S. (1998). CSU-CHILL polarimetric radar measurements from a sever hail storm in eastern Colorado. *Journal of Applied Meteorology, 37*, 749–775.

Huschke, R. E. (1959). *Glossary of Meteorology* (638pp.). American Meteorological Society.

Jensen, A. A., Harrington, J. Y., Morrison, H., & Milbrandt, J. A. (2017). Predicting ice shape evolution in a bulk microphysics model. *Journal of Atmospheric Science, 74*, 2081–2104. https://doi.org/10.1175/JAS-D-16-0350.1

Jewett, B. F., Wilhelmson, R. B., Straka, J. M., & Wicker, L. J. (1990). Impact of ice parameterization on the low-level structure of modeled supercell thunderstorms. In *16th conference on severe local storms* (pp. 275–280). Kananaskis Park, Alberta, American Meteorological Society.

Jiang, Z., Kumjian, M. R., Schrom, R. S., Giammanco, I., Brown-Giammanco, T., Estes, H., Maiden, R., & Heymsfield, A. J. (2019). Comparisons of electromagnetic scattering properties of real hailstones and spheroids. *Journal of Applied Meteorology and Climatology, 58*, 93–112. https://doi.org/10.1175/JAMC-D-17-0344.1

Johnson, D. B. (1987). On the relative efficiency of coalescence and riming. *Journal of Atmospheric Science, 44*, 1671–1680.

Johnson, D. E., Wang, P. K., & Straka, J. M. (1993). Numerical simulations of the 2 August 1981 CCOPE supercell storm with and without ice microphysics. *Journal of Applied Meteorology, 32*, 745–759.

Jung, Y., Zhang, G., & Xue, M. (2008). Assimilation of simulated polarimetric radar data for a convective storm using the ensemble Kalman filter. Part I: Observation operators for reflectivity and polarimetric variables. *Monthly Weather Review, 136*, 2228–2245. https://doi.org/10.1175/2007MWR2083.1

Kalina, E. A., Friedrich, K., Motta, B. C., Deierling, W., Stano, G. T., & Rydell, N. N. (2016). Colorado plowable hailstorms: Synoptic weather, radar, and lightning characteristics. *Weather Forecasting, 31*, 663–693. https://doi.org/10.1175/WAF-D-15-0037.1

Katz, R. W., & Garcia, R. R. (1981). Statistical relationships between hailfall and damage to wheat. *Agricultural Meteorology, 24*, 29–43. https://doi.org/10.1016/0002-1571(81)90031-5

Kennedy, P. C., & Detwiler, A. G. (2003). A case study of the origin of hail in a multicell thunderstorm using in situ aircraft and polarimetric radar data. *Journal of Applied Meteorology, 42*, 1679–1690.

Kessler, E., III. (1969). *On the Distribution and Continuity of Water Substance in Atmospheric Circulation.* Meteorological Monographs (Vol. 32, 84pp.). American Meteorological Society

Khain, A., Pokrovsky, A., Pinsky, M., Seifert, A., & Phillips, V. (2004). Simulation of effects of atmospheric aerosols on deep turbulent convective clouds using a spectral microphysics mixed-phase cumulus cloud model. Part I: Model description and possible applications. *Journal of Atmospheric Science, 61*, 2963–2982.

Khain, A., Lynn, B., & Dudhia, J. (2010). Aerosol effects on intensity of land falling hurricanes as seen from simulations with the WRF model with spectral bin microphysics. *Journal of Atmospheric Science, 67*, 365–384. https://doi.org/10.1175/2009JAS3210.1

Klimowski, B. A., Hjelmfelt, M. R., Bunkers, M. J., Sedlacek, D., & Johnson, L. R. (1998). Hailstorm damage observed from the GOES-8 satellite: The 5–6 July 1996 Butte-Meade storm. *Monthly Weather Review, 126*, 831–834.

Knight, N. C. (1981). The climatology of hailstone embryos. *Journal of Applied Meteorology, 20*, 750–755.

Knight, C. A. (1982). The Cooperative Convective Precipitation Experiment (CCOPE), 18 May- 7 August 1981. *Bulletin of the American Meteorological Society, 63*, 386–398.

Knight, C. A. (1984). Radar and other observations of two vaulted storms in northeastern Colorado. *Journal of Atmospheric Science, 41*, 258–271.

Knight, N. C. (1986). Hailstone shape factor and its relation to radar interpretation of hail. *Journal of Applied Meteorology and Climatology, 25*, 1956–1958. https://doi.org/10.1175/1520-0450(1986)025%3c1956:HSFAIR%3e2.0.CO;2

Knight, C. A., & Heymsfield, A. (1983a). Measurement and interpretation of hailstone density and terminal velocity. *Journal of Atmospheric Science, 40*, 1510–1516. https://doi.org/10.1175/1520-0469(1983)040%3c1510:MAIOHD%3e2.0.CO;2

Knight, N. C., & Heymsfield, A. J. (1983b). Measurement and interpretation of hailstone density and terminal velocity. *Journal of Atmospheric Science, 40*, 1510–1516. https://doi.org/10.1175/1520-0469(1983)040%3c1510:MAIOHD%3e2.0.CO;2

Knight, C. A., & Knight, N. C. (1970a). Lobe structures of hail. *Journal of Atmospheric Science, 27*, 667–671.

Knight, C. A., & Knight, N. C. (1970b). Hailstone embryos. *Journal of Atmospheric Science, 27*, 659–666.

Knight, C. A., & Knight, N. C. (1970c). The falling behavior of hailstones. *Journal of Atmospheric Science, 27*, 672–681.

Knight, C. A., & Knight, N. C. (1979). Results of a randomized hail suppression experiment in northeast Colorado. Part V. Hailstone embryo types. *Journal of Applied Meteorology, 18*, 1583–1588.

Knight, C. A., & Knight, N. C. (2005). Very large hailstones from Aurora, Nebraska. *Bulletin of the American Meteorological Society, 86*, 1773–1781.

Knight, C. A., Knight, N. C., Dye, J. E., & Toutenhood, V. (1974). The mechanism of precipitation formation in northeastern Colorado cumulus. I. Observations of the precipitation itself. *Journal of Atmospheric Science, 31*, 2142–2147.

Knight, C. A., Schlatter, P. T., & Schlatter, T. W. (2008). An unusual hailstorm on 24 June 2006 in Boulder, Colorado. Part II: Low-density growth of hail. *Monthly Weather Review, 136*, 2833–2848.

Knight, C. A. & Knight, N. C. (1971). *Bulletin of the American Meteorological Society, 52.* February issue cover photograph.

Knight, C. A. & Knight, N. C. (2001). *Hailstorms. Severe Convective Storms.* Meteorological Monographs (Vol. 50, pp. 223–249). American Meteorological Society.

Koenig, L. R., & Murray, F. W. (1976). Ice-bearing cumulus cloud evolution: Numerical simulation and general comparison against observations. *Journal of Applied Meteorology, 15*, 747–762.

Kogan, Y. (1991). The simulation of a convective cloud in a 3D model with explicit microphysics. Part I: Model description and sensitivity experiments. *Journal of Atmospheric Science, 48*, 1160–1189.

Krauss, T. W., & Marwitz, J. D. (1984). Precipitation processes within an Alberta supercell hailstorm. *Journal of Atmospheric Science, 41*, 1025–1034.
Kretchman, D., Jameson, M., & Willer, C. (1989). Effects of simulated hail on processing tomatoes (55pp.). Department of Horticulture, The Ohio State University, Ohio Agricultural Research and Development Center. https://kb.osu.edu/handle/1811/56210.
Kumjian, M. R. (2013). Principles and applications of dual-polarization weather radar. Part III: Artifacts. *Journal of Operational Meteorology, 1*(21), 265–274. https://doi.org/10.15191/nwajom.2013.0121
Kumjian, M. R., & Lombardo, K. (2020). A hail growth trajectory model for exploring the environmental controls on hail size: Model physics and idealized tests. *Journal of Atmospheric Science, 77*, 2765–2791. https://doi.org/10.1175/JAS-D-20-0016.1
Kumjian, M. J., Lebo, Z. J., & Ward, A. (2019). Large accumulations of small hail. *Journal of Applied Meteorology and Climatology, 58*, 341–364. https://doi.org/10.1175/JAMC-D-18-0073.1
Kumjian, M. J. & Lebo, Z. J. (2016). Large accumulations of small hail. In *28th Conference on Severe Local Storms*. Portland, OR, American Meteorological Society, 8A.4. https://ams.confex.com/ams/28SLS/webprogram/Paper301237.html.
Lang, T. J., Miller, L. J., Weisman, M., Rutledge, S. A., Barker, L. J., III., Bringi, V. N., Chandrasekar, V., Detwiler, A., Doesken, N., Helsdon, J., Knight, C., Krehbiel, P., Lyons, W. A., MacGorman, D., Rasmussen, E., Rison, W., Rust, W. D., & Thomas, R. J. (2004). The severe thunderstorm electrification and precipitation study. *Bulletin of the American Meteorological Society, 85*, 1107–1126. https://doi.org/10.1175/BAMS-85-8-1107
Lang, S., Tao, W., Simpson, J., Cifelli, R., Rutledge, S., Olson, W., & Halverson, J. (2007). Improving simulations of convective systems from TRMM LBA: Easterly and westerly regimes. *Journal of Atmospheric Science, 64*, 1141–1164. https://doi.org/10.1175/JAS3879.1
Lang, S. E., Tao, W., Zeng, X., & Li, Y. (2011). Reducing the biases in simulated radar reflectivities from a bulk microphysics scheme: Tropical convective systems. *Journal of Atmospheric Science, 68*, 2306–2320. https://doi.org/10.1175/JAS-D-10-05000.1
Lang, S. E., Tao, W., Chern, J., Wu, D., & Li, X. (2014). Benefits of a fourth ice class in the simulated radar reflectivities of convective systems using a bulk microphysics scheme. *Journal of Atmospheric Science, 71*, 3583–3612. https://doi.org/10.1175/JAS-D-13-0330.1
Laviola, S., Monte, G., Levizzani, V., Ferraro, R. R., & Beauchamp, J. (2020). A new method for hail detection from the GPM constellation: A prospect for a global hailstorm climatology. *Remote Sensing, 12*(21), 3553. https://doi.org/10.3390/rs12213553
Le, M., & Chandrasekar, V. (2021). Graupel and hail identification algorithm for the Dual-frequency Precipitation Radar (DPR) on the GPM core satellite. *Journal of Meteorological Society of Japan*. https://doi.org/10.2151/jmsj.2021-003
Lebo, Z. J., & Morrison, H. (2015). Effects of horizontal and vertical grid spacing on mixing in simulated squall lines and implications for convective strength and structure. *Monthly Weather Review, 143*, 4355–4375. https://doi.org/10.1175/MWR-D-15-0154.1
Lesins, G. B., & List, R. (1986). Sponginess and drop shedding of gyrating hailstones in a pressure-controlled icing wind tunnel. *Journal of Atmospheric Science, 43*, 2813–2825.
Li, X., Tao, W.-K., Khain, A. P., Simpson, J., & Johnson, D. E. (2009). Sensitivity of a cloud-resolving model to bulk and explicit bin microphysical schemes. Part II: Cloud microphysics and storm dynamics interactions. *Journal of Atmospheric Science, 66*, 22–40. https://doi.org/10.1175/2008JAS2647.1
Lin, Y.-H., Farley, R. D., & Orville, H. D. (1983). Bulk parameterization of the snow field in a cloud model. *Journal of Climate Applied Meteorology, 22*, 1065–1092. https://doi.org/10.1175/1520-0450(1983)022%3c1065:BPOTSF%3e2.0.CO;2
List, R. (1963). General heat and mass exchange of spherical hailstones. *Journal of Atmospheric Science, 20*, 189–197.
List, R., Charlton, R. B., & Buttuls, P. I. (1968). Mechanisms of hailstones in a one-dimensional steady-state model cloud. *Journal of Atmospheric Science, 25*, 1061–1074.

List, R. (2010) Hailstones, the main source of rain. In *13th Conference on Cloud Physics/13th Conference on Atmospheric Radiation*. Portland, OR, American Meteorological Society, 12.4.

Loftus, A. M., & Cotton, W. R. (2014). A triple-moment hail bulk microphysics scheme. Part II: Verification and comparison with two-moment bulk microphysics. *Atmospheric Research, 150*, 97–128. https://doi.org/10.1016/j.atmosres.2014.07.016

Loftus, A. M., Cotton, W. R., & Carrió, G. G. (2014). A triple-moment hail bulk microphysics scheme. Part I: Description and initial evaluation. *Atmospheric Research, 149*, 35–57. https://doi.org/10.1016/j.atmosres2014.05.013

Ludlam, F. H. (1958). The hail problem. *Nubila, 1*, 12–95.

Lynn, B. H., Khain, A. P., Dudhia, J., Rosenfeld, D., Pokrovsky, A., & Seifert, A. (2005). Spectral (bin) microphysics coupled with a mesoscale model (MM5). Part I: Model description and first results. *Monthly Weather Review, 133*, 44–58.

MacGorman, D. R., Rust, W. D., Schuur, T. J., Biggerstaff, M. I., Straka, J. M., Ziegler, C. L., Mansell, E. R., Bruning, E. C., Kuhlman, K. M., Lund, N. R., Biermann, N. S., Payne, C., Carey, L. D., Krehbiel, P. R., Rison, W., Eack, K. B., & Beasley, W. H. (2008). TELEX the thunderstorm electrification and lightning experiment. *Bulletin of the American Meteorological Society, 89*, 997–1014. https://doi.org/10.1175/2007BAMS2352.1

Macklin, W. C. (1961). Accretion in mixed clouds. *Quarterly Journal of Royal Meteorological Society, 87*, 413–424.

Macklin, W. C. (1964). Factors affecting the heat transfer from hailstones. *Quarterly Journal of Royal Meteorological Society, 90*, 84–90. https://doi.org/10.1002/qj.49709038309

Macklin, W. C., & Bailey, I. H. (1966). On the critical liquid water concentrations of large hailstones. *Quarterly Journal of Royal Meteorological Society, 92*, 297–300.

Macklin, W. C., & Bailey, I. H. (1968). The collection efficiencies of hailstones. *Quarterly Journal of Royal Meteorological Society, 94*, 393–396.

Macklin, W. C., & Ludlam, F. H. (1961). The fall speeds of hailstones. *Quarterly Journal of Royal Meteorological Society, 87*, 72–81. https://doi.org/10.1002/qj.49708737109

Mahoney, K. (2020). Extreme hail storms and climate change: Foretelling the future in tiny, turbulent crystal balls? *Bulletin of the American Meteorological Society, 101*, S17–S22. https://doi.org/10.1175/bams-d-19-0233.1

Mahoney, K., Alexander, M. A., Thompson, G., Barsugli, J. J., & Scott, J. D. (2012). Changes in hail and flood risk in high-resolution simulations over Colorado Mountains. *Natural Climate Change, 2*, 125–131.

Mansell, E. R. (2008). EnKF analysis and forecast predictability of a tornadic supercell storm. In *24th Conference on Severe Local Storms* (p. 5.2). American Meteorological Society, Savannah, GA.

Mansell, E. R. (2010). On sedimentation and advection in multimoment bulk microphysics. *Journal of Atmospheric Science, 67*, 3084–3094.

Mansell, E. R., Ziegler, C. L., & Bruning, E. C. (2010). Simulated electrification of a small thunderstorm with two-moment bulk microphysics. *Journal of Atmospheric Science, 67*, 171–194.

Mansell, E. R., Dawson, D. T., II., & Straka, J. M. (2020). Bin-emulating hail melting in three-moment microphysics. *Journal of Atmospheric Science, 77*, 3361–3385.

Marshall, J. S., & Palmer, W. M. (1948). The distribution of raindrops with size. *Journal of Meteorology, 5*, 165–166.

Marwitz, J. D. (1972a). The structure and motion of severe hailstorms. Part I: Supercell storms. *Journal of Applied Meteorology, 11*, 166–179.

Marwitz, J. D. (1972b). The structure and motion of severe hailstorms. Part II: Multi-cell storms. *Journal of Applied Meteorology, 11*, 180–188.

Marwitz, J. D. (1972c). The structure and motion of severe hailstorms. Part III: Severely sheared storms. *Journal of Applied Meteorology, 11*, 189–201.

Mason, B. J. (1971). *The Physics of Clouds* (2nd edn., 671pp.). Oxford University Press.

Masunaga, H., et al. (2010). Satellite data simulator unit: A multisensor, multispectral satellite simulator package. *Bulletin of the American Meteorological Society, 91*, 1625–1632.

Matson, R. J., & Huggins, A. W. (1980). The direct measurement of the sizes, shapes and kinematics of falling hailstones. *Journal of Atmospheric Science, 37*, 1107–1125. https://doi.org/10.1175/1520-0469(1980)037%3c1107:TDMOTS%3e2.0.CO;2

Matsui, T., Zeng, X., Tao, W.-K., Masunaga, H., Olson, W. S., & Lang, S. (2009). Evaluation of long-term cloud-resolving model simulations using satellite radiance observations and multifrequency satellite simulators. *Journal of the Seismological Society of Japan, 26*, 1261–1274. https://doi.org/10.1175/2008JTECHA1168.1

Matsui, T., et al. (2013). GPM satellite simulator over ground validation sites. *Bulletin of the American Meteorological Society, 94*, 1653–1660. https://doi.org/10.1175/BAMS-D-12-00160.1

Matsui, T., Santanello, J., Shi, J. J., Tao, W.-K., Wu, D., Peters-Lidard, C., Kemp, E., Chin, M., Starr, D., Sekiguchi, M., & Aires, F. (2014). Introducing multisensory satellite radiance-based evaluation for regional earth system modeling. *Journal of Geophysical Research Atmospheres, 119*, 8450–8475. https://doi.org/10.1002/2013JD021424

Matsui, T., Dolan, B., Rutledge, S. A., Tao, W.-K., Iguchi, T., Barnum, J., & Lang, S. E. (2019). POLARRIS: APOLArimetric radar retrieval and instrument simulator. *Journal of Geophysical Research Atmospheres, 124*, 4634–4657. https://doi.org/10.1029/2018JD028317

Matsui, T., Dolan, B., Iguchi, T., Rutledge, S. A., Tao, W.-K., & Lang, S. (2020). Polarimetric radar characteristics of simulated and observed intense convective cores for a midlatitude continental and tropical maritime environment. *Journal of Hydrometeorology, 21*, 501–517.

McCumber, M., Tao, W.-K., Simpson, J., Penc, R., & Soong, S.-T. (1991). Comparison of ice-phase microphysical parameterization schemes using numerical simulations of tropical convection. *Journal of Applied Meteorology, 30*, 985–1004.

Meyers, M. P., DeMott, P. J., & Cotton, W. R. (1992). New primary ice nucleation parameterizations in an explicit cloud model. *Journal of Applied Meteorology, 31*, 708–721. https://doi.org/10.1175/1520-0450(1992)031%3c0708:NPINPI%3e2.0.CO;2

Meyers, M. P., Walko, R. L., Harrington, J. Y., & Cotton, W. R. (1997). New RAMS cloud microphysics parameterization. II: The two-moment scheme. *Atmospheric Research, 45*, 3–39. https://doi.org/10.1016/S0169-8095(97)00018-5

Milbrandt, J. A., & Morrison, H. (2013). Predicting graupel density in a bulk microphysics scheme. *Journal of Atmospheric Science, 70*, 410–429. https://doi.org/10.1175/JAS-D-12-0204.1

Milbrandt, J. A., & Morrison, H. (2016). Parameterization of cloud microphysics based on the prediction of bulk ice particle properties. Part III: Introduction of multiple free categories. *Journal of Atmospheric Science, 73*, 975–995. https://doi.org/10.1175/JAS-D-15-0204.1

Milbrandt, J. A., & Yau, M. K. (2005a). A multimoment bulk microphysics parameterization. Part I: Analysis of the role of the spectral shape parameter. *Journal of Atmospheric Science, 62*, 3051–3064.

Milbrandt, J. A., & Yau, M. K. (2005b). A multimoment bulk microphysics parameterization. Part II: A proposed three-moment closure and scheme description. *Journal of Atmospheric Science, 62*, 3065–3081.

Milbrandt, J. A., & Yau, M. K. (2006a). A multimoment bulk microphysics parameterization. Part III: Control simulation of a hailstorm. *Journal of Atmospheric Science, 63*, 3077–3090.

Milbrandt, J. A., & Yau, M. K. (2006b). A multimoment bulk microphysics parameterization. Part IV: Sensitivity experiments. *Journal of Atmospheric Science, 63*, 3137–3159.

Miller, L. J., & Fankhauser, J. C. (1983). Radar echo structure, air motion, and hail formation in a large stationary multicellular thunderstorm. *Journal of Atmospheric Science, 40*, 2399–2418.

Miller, L. J., Dye, J. E., & Martner, B. E. (1983). Dynamical-microphysical evolution of a convective storm in a weakly-sheared environment. Part II: Airflow and precipitation trajectories from Doppler radar observations. *Journal of Atmospheric Science, 40*, 2097–2109.

Miller, L. J., Tuttle, J. D., & Knight, C. A. (1988). Airflow and hail growth in a severe northern high plains supercell. *Journal of Atmospheric Science, 45*, 736–762.

Miller, L. J., Tuttle, J. D., & Foote, G. B. (1990). Precipitation production in a large Montana hailstorm: Airflow and particle growth trajectories. *Journal of Atmospheric Science, 47*, 1619–1646.

Mitchell, D. L. (1996). Use of mass- and area-dimensional power laws for determining precipitation particle terminal velocities. *Journal of Atmospheric Science, 53*, 1710–1723. https://doi.org/10.1175/1520-0469(1996)053%3c1710:UOMAAD%3e2.0.CO;2

Morrison, H., & Milbrandt, J. (2011). Comparison of two-moment bulk microphysics schemes in idealized supercell thunderstorm simulations. *Monthly Weather Review, 139*, 1103–1130.

Morrison, H., & Milbrandt, J. A. (2015). Parameterization of cloud microphysics based on the prediction of bulk ice particle properties. Part I: Scheme description and idealized tests. *Journal of Atmospheric Science, 72*(1), 287–311. https://doi.org/10.1175/JAS-D-14-0065.1

Morrison, H., Thompson, G., & Tatarskii, V. (2009). Impact of cloud microphysics on the development of trailing stratiform precipitation in a simulated squallline: Comparison of one-and two-moment schemes. *Monthly Weather Review, 137*, 991–1007.

Mossop, S. C., & Kidder, R. E. (1961). Hailstorm at Johannesburg on 9th November, 1959, II. Structure of hailstones. *Nubila, 4*, 71–86.

Mroz, K., Battaglia, A., Lang, T. J., Cecil, D. J., Tanelli, S., & Tridon, F. (2017). Hail-detection algorithm for the GPM core satellite sensors. *Journal of Applied Meteorology and Climatology, 56*, 1939–1957. https://doi.org/10.1175/JAMC-D-16-0368.1

Mroz, K., Battaglia, A., Lang, T. J., Tanelli, S., & Sacco, G. F. (2018). Global precipitation measuring dual-frequency precipitation radar observations of hailstorm vertical structure: Current capabilities and drawbacks. *Journal of Applied Meteorology and Climatology, 57*, 2161–2178. https://doi.org/10.1175/JAMC-D-18-0020.1

Musil, D. J. (1970). Computer modeling of hailstone growth in feeder clouds. *Journal of Atmospheric Science, 27*, 474–482.

Musil, D. J., Sand, W. R., & Schleusener, R. A. (1973). Analysis of data from T-28 Aircraft penetrations of a Colorado hailstorm. *Journal of Applied Meteorology and Climatology, 12*, 1364–1370. https://doi.org/10.1175/1520-0450(1973)012%3c1364:AODFTA%3e2.0.CO;2

Musil, D. J., May, E. L., Smith, P. L., & Sand, W. R. (1976). Structure of an evolving hailstorm Part IV: Internal structure from penetrating aircraft. *Monthly Weather Review, 104*, 596–602.

Musil, D. J., Heymsfield, A. J., & Smith, P. L. (1986). Microphysical characteristics of a well-developed weak echo region in a High Plains supercell thunderstorm. *Journal of Climate Applied Meteorology, 25*, 1037–1051.

Musil, D. J., Christopher, S. A., Deola, R. A., & Smith, P. L. (1991). Some interior observations of southeastern Montana hailstorms. *Journal of Applied Meteorology, 30*, 1596–1612.

Nelson, S. P. (1983). The influence of storm flow structure on hail growth. *Journal of Atmospheric Science, 40*, 1965–1983.

Nelson, S. P. (1987). The hybrid multicellular-supercellular storm- an efficient hail producer. Part II: General characteristics and implications for hail growth. *Journal of Atmospheric Science, 44*, 2060–2073.

Nelson, S. P., & Young, S. K. (1979). Characteristics of Oklahoma hailfalls and hailstorms. *Journal of Applied Meteorology, 18*, 339–347.

Ni, X., Liu, C., Cecil, D. J., & Zhang, Q. (2017). On the detection of hail using satellite passive microwave radiometers and precipitation radar. *Journal of Applied Meteorology and Climatology, 56*, 2693–2709. https://doi.org/10.1175/JAMC-D-17-0065.1

Nicholls, S. D., Decker, S. G., Tao, W.-K., Lang, S. E., Shi, J. J., & Mohr, K. I. (2017). Influence of bulk microphysics schemes upon Weather Research and Forecasting (WRF) version 3.6.1 nor'easter simulations. *Geoscientific Model Development, 10*, 1033–1049. https://doi.org/10.5194/gmd-10-1033-2017

Ogura, Y., & Takahashi, T. (1971). Numerical simulation of the life cycle of a thunderstorm cell. *Monthly Weather Review, 99*, 895–911.

Ortega, K. L., Smith, T. M., Manross, K. L., Kolodziej, A. G., Scharfenberg, K. A., Witt, A., & Gourley, J. J. (2009). The severe hazards analysis and verification experiment. *Bulletin of the American Meteorological Society, 90*, 1519–1530. https://doi.org/10.1175/2009BAMS2815.1

Ortega, K. L., Krause, J. M., & Ryzhkov, A. V. (2016). Polarimetric radar characteristics of melting hail. Part III: Validation of the algorithm for hail size discrimination. *Journal of Applied Meteorological Climatol*ogy, 160203133521006. https://doi.org/10.1175/JAMC-D-15-0203.1.

Orville, H. D., & Kopp, F. J. (1977). Numerical simulation of the life history of a hailstorm. *Journal of Atmospheric Science, 34*, 1596–1618.

Ovtchinnikov, M., & Kogan, Y. L. (2000). An investigation of ice production mechanisms using a 3D cloud model with detailed microphysics. Part I: Model description. *Journal of Atmospheric Science, 57*, 2989–3003.

Paluch, I. R. (1978). Size sorting of hail in a three-dimensional updraft and implications for hail suppression. *Journal of Applied Meteorology, 9*, 763–777.

Pelissier, C., Olson, W., Kuo, K.-S., Loftus, A., Schrom, R., & Adams, I. (2023). A physically based, meshless Lagrangian approach to simulate melting precipitation. *Journal of Atmospheric Science, 80*, 353–373. https://doi.org/10.1175/JAS-D-22-0150.1

Petch, J. C., & Gray, M. E. B. (2001). Sensitivity studies using a cloud-resolving model simulation of the tropical west Pacific. *Quarterly Journal of Royal Meteorological Society, 127*, 2287–2306.

Petch, J. C., Brown, A. R., & Gray, M. E. B. (2002). The impact of horizontal resolution on the simulations of convective development over land. *Quarterly Journal of Royal Meteorological Society, 128*, 2031–2044.

Petersen, W. A. & Jensen, M. (2012). The NASA-GPM and DOE-ARM Midlatitude Continental Convective Clouds Experiment (MC3E), Earth Observer, 24, 12–18. http://pmm.nasa.gov/sites/default/files/document_files/Earth_Observer_Jan_2012_MC3E.pdf.

Peters-Lidard, C. D., et al. (2015). Integrated modeling of aerosol, cloud, precipitation and land processes at satellite-resolved scales. *Environmental Modelling & Software, 67*, 149–159. https://doi.org/10.1016/j.envsoft.2015.01.007

Poolman, E. R. (1992). *Die voorspelling van haelkorrelgroei in Suid-Afrika (The forecasting of hail growth in South Africa)* (113pp.) (M.S. thesis, Faculty of Engineering, University of Pretoria.

Prein, A. F., Liu, C., Ikeda, K., Bullock, R., Rasmussen, R. M., Holland, G. J., & Clark, M. (2017). Simulating North American mesoscale convective systems with a convection-permitting climate model. *Climate Dynamics, 55*, 95–110. https://doi.org/10.1007/s00382-017-3993-2

Prodi, F. (1970). Measurements of local density in artificial and natural hailstones. *Journal of Applied Meteorology and Climatology, 9*, 903–910. https://doi.org/10.1175/1520-0450(1970)009%3c0903:MOLDIA%3e2.0.CO;2

Pruppacher, H. R. & Klett, J. D. (1980). *Microphysics of Clouds and Precipitation* (714pp.). D. Reidel.

Púčik, T., Groenemeijer, P., Rýva, D., & Kolář, M. (2015). Proximity soundings of severe and nonsevere thunderstorms in Central Europe. *Monthly Weather Review, 143*, 4805–4821. https://doi.org/10.1175/mwr-d-15-0104.1

Rasmussen, R. M., & Heymsfield, A. J. (1987a). Melting and shedding of Graupel and hail. Part II: Sensitivity study. *Journal of Atmospheric Science, 44*, 2764–2782.

Rasmussen, R. M., & Heymsfield, A. J. (1987b). Melting and shedding of graupel and hail. Part III: Investigation of the role of shed drops as hail embryos in the August 1 CCOPE severe storm. *Journal of Atmospheric Science, 44*, 2783–2803.

Rasmussen, R. M. & Heymsfield, A. J. (1987c). Melting and shedding of graupel and hail. Part III: Investigation of the role of shed drops as hail embryos in the August 1 CCOPE severe storm. *Journal of Atmospheric Science, 44*, 2783–2803. https://doi.org/10.1175/1520-0469(1987)044<2783:MASOGA>2.0.CO;2

Rasmussen, R. M., Levizzani, V., & Pruppacher, H. R. (1984). A wind tunnel and theoretical study on the melting behavior of atmospheric ice particles: III. Experiment and theory for spherical ice particles of radius > 500 μm. *Journal of Atmospheric Science, 41*, 381–388.

Rasmussen, E. N., Straka, J. M., Davies-Jones, R. P., Doswell, C. A., III., Carr, F. H., Eilts, M. D., & MacGorman, D. R. (1994). The verifications of the origins of rotation in tornadoes experiment: VORTEX. *Bulletin of the American Meteorological Society, 75*, 997–1006. https://doi.org/10.1175/1520-0477(1994)075%3c0995:VOTOOR%3e2.0.CO;2

Rasmussen, K. L., Prein, A. F., Rasmussen, R. M., Ikeda, K., & Liu, C. (2017). Changes in the convective population and thermodynamic environments in convection-permitting regional climate simulations over the United States. *Climate Dynamics, 55*, 383–408. https://doi.org/10.1007/s00382-017-4000-7

Reisin, T., Levin, Z., & Tzvion, S. (1996). Rain production in convective clouds as simulated in an axisymmetric model with detailed microphysics. Part I: Description of the model. *Journal of Atmospheric Science, 53*, 497–519.

Reisner, J., Rasmussen, R. M., & Bruintjes, R. T. (1998). Explicit forecasting of supercooled liquid water in winter storm using the MM5 mesoscale model. *Quarterly Journal of Royal Meteorological Society, 124*, 1071–1107.

Roos, D. v.d. S. (1972). A giant hailstone from Kansas in free fall. *Journal of Applied Meteorology, 11*, 1008–1011.

Roos, D. S., & Carte, A. E. (1973). The falling behavior of oblate and spiky hailstones. *Journal of the Atmospheric Research, 7*, 39–52.

Rutledge, S. A., & Hobbs, P. V. (1984). The mesoscale and microscale structure and organization of clouds and precipitation in midlatitude cyclones. Part XII: A diagnostic modeling study of precipitation development in narrow cold-frontal rainbands. *Journal of Atmospheric Science, 41*, 2949–2972.

Ryzhkov, A. V., Schuur, T. J., Burgess, D. W., Heinselman, P. L., Giangrande, S. E., & Zrnić, D. S. (2005). The Joint Polarization Experiment: Polarimetric rainfall measurements and hydrometeor classification. *Bulletin of the American Meteorological Society, 86*, 809–824. https://doi.org/10.1175/BAMS-86-6-809

Saleeby, S. M., & Cotton, W. R. (2004). A large-droplet mode and prognostic number concentration of cloud droplets in the Colorado State University Regional Atmospheric Modeling System (RAMS). Part I: Module descriptions and supercell test simulations. *Journal of Applied Meteorology, 43*, 182–195. https://doi.org/10.1175/1520-0450(2004)043%3c0182:ALMAPN%3e2.0.CO;2

Saleeby, S. M., & Cotton, W. R. (2008). A binned approach to cloud-droplet riming implemented in a bulk microphysics model. *Journal of Applied Meteorology and Climatology, 47*, 694–703. https://doi.org/10.1175/2007JAMC1664.1

Saleeby, S. M., & van den Heever, S. C. (2013). Developments in the CSU-RAMS aerosol model: Emissions, nucleation, regeneration, deposition, and radiation. *Journal of Applied Meteorology and Climatology, 52*, 2601–2622. https://doi.org/10.1175/JAMC-D-12-0312.1

Sánchez, J. L., Fraile, R., de la Madrid, J. L., de la Fuente, M. T., Rodríguez, P., & Castro, A. (1996). Crop damage: The hail size factor. *Journal of Applied Meteorology, 35*, 1535–1541. https://doi.org/10.1175/1520-0450

Sand, W. R. (1976). Observations in hailstorms using the T-28 aircraft system. *Journal of Applied Meteorology and Climatology, 15*, 641–650. https://doi.org/10.1175/1520-0450(1976)015%3c0641:OIHUTT%3e2.0.CO;2

Scharfenberg, K. A., Miller, D. J., Schuur, T. J., Schlatter, P. T., Giangrande, S. E., Melnikov, V. M., Burgess, D. W., Andra, D. L., Jr., Foster, M. P., & Krause, J. M. (2005). The joint polarization experiment: Polarimetric radar in forecasting and warning decision making. *Weather Forecasting, 20*, 775–788.

Schiesser, H. H. (1990). Hailfall: The relationship between radar measurements and crop damage. *Atmospheric Research, 25*, 559–582. https://doi.org/10.1016/0169-8095(90)90038-E

Schumann, T. E. W. (1938). The theory of hailstone formation. *Quarterly Journal of Royal Meteorological Society, 64*, 3–21.

Seiki, T. (2021). Near-global three-dimensional hail signals detected by using GPM-DPR observations. *Journal of Meteorological Society of Japan, 99*, 379–402. https://doi.org/10.2151/jmsj.2021-018

Skamarock, W. C., Klemp, J. B., Dudhia, J., Gill, D. O., Barker, D. M., Duda, M. G., Huang, X.-Y., Wang, W., & Powers, J. G. (2008). *A description of the advanced research WRF version 3* (123pp.). NCAR technical note, NCAR/TN–475+STR. http://www2.mmm.ucar.edu/wrf/users/docs/arw_v3.pdf.

Skofronick-Jackson, G., et al. (2017). The Global Precipitation Measurement (GPM) mission for science and society. *Bulletin of the American Meteorological Society, 98*, 1679–1695. https://doi.org/10.1175/BAMS-D-15-00306.1

Smith, P. L., Musil, D. J., Detwiler, A. G., & Ramachandran, R. (1999). Observations of mixed-phase precipitation within a CaPE thunderstorm. *Journal of Applied Meteorology, 38*, 145–155.

Srivastava, R. C. (1987). A model of intense downdrafts driven by melting and evaporation of precipitation. *Journal of Atmospheric Science, 44*, 1752–1773.

Straka, J. M., & Mansell, E. R. (2005). A bulk microphysics parameterization with multiple ice precipitation categories. *Journal of Applied Meteorology, 44*, 445–466.

Straka, J. M., Zrnić, D. S., & Ryzhkov, A. V. (2000). Bulk hydrometeor classification and quantification using polarimetric radar data: Synthesis of relations. *Journal of Applied Meteorology, 39*, 1341–1372.

Takahashi, T. (1976). Hail in an axisymmetric cloud model. *Journal of Atmospheric Science, 33*, 1579–1601.

Tang, B. H., Gensini, V. A., & Homeyer, C. R. (2019). Trends in United States large hail environments and observations. *NPJ Climate Atmospheric Sciences, 2*, 45. https://doi.org/10.1038/s41612-019-0103-7

Tao, W.-K., & Simpson, J. (1993). The Goddard cumulus ensemble model. Part I: Model description. *Terrestrial Atmospheric and Oceanic Sciences, 4*, 19–54. https://doi.org/10.3319/TAO.1993.4.1.35(A)

Tao, W.-K., et al. (2003). Microphysics, radiation and surface processes in the Goddard Cumulus Ensemble (GEC) model. *Meteorology and Atmospheric Physics, 82*, 97–137.

Tao, W.-K., Wu, D., Lang, S., Chern, J.-D., Peters-Lidard, C., Fridlind, A., & Matsui, T. (2016). High-resolution NU-WRF simulations of a deep convective-precipitation system duringMC3E: Further improvements and comparisons between Goddard microphysics schemes and observations. *Journal of Geophysical Research Atmospheres, 121*, 1278–1305. https://doi.org/10.1002/2015JD023986

Tessendorf, S. A., Miller, L. J., Wiens, K. C., & Rutledge, S. A. (2005). The 29 June 2000 supercell observed during STEPS. Part I: Kinematics and microphysics. *Journal of Atmospheric Science, 62*, 4127–4150.

Thompson, G., & Eidhammer, T. (2014). A study of aerosol impacts on clouds and precipitation development in a large winter cyclone. *Journal of Atmospheric Science, 74*, 3636–3658. https://doi.org/10.1175/JAS-D-13-0305.1

Trapp, R. J., Hoogewind, K. A., & Lasher-Trapp, S. (2019). Future changes in hail occurrence in the United States determined through convection-permitting dynamical downscaling. *Journal of Climate, 32*, 5493–5509. https://doi.org/10.1175/jcli-d-18-0740.1

Tsai, T. C., & Chen, J. P. (2020). Multimoment ice bulk microphysics scheme with consideration for particle shape and apparent density. Part I: Methodology and idealized simulation. *Journal of Atmospheric Science, 77*, 1821–1850. https://doi.org/10.1175/JAS-D-19-0125.1

van den Heever, S., & Cotton, W. R. (2004). The impact of hail size on simulated storms. *Journal of Atmospheric Science, 61*, 1596–1609.

van Weverberg, K. (2013). Impact of environmental instability on convective precipitation uncertainty associated with the nature of the rimed ice species in a bulk microphysics scheme. *Monthly Weather Review, 141*, 2841–2849. https://doi.org/10.1175/MWR-D-13-00036.1

Vivekanandan, J., Zrnic, D. S., Ellis, S. M., Oye, R., Ryzhkov, A. V., & Straka, J. (1999). Cloud microphysics retrieval using S-band dual-polarization radar measurements. *Bulletin of the American Meteorological Society, 80*(3), 381–388.

Waldvogel, A., Federer, B., Schmid, W., & Mezeix, J. F. (1978). The kinetic energy of hailfalls. Part 2: Radar and hailpads. *Journal of Applied Meteorology, 17*, 1680–1693.

Walko, R., Cotton, W. R., Meyers, M. P., & Harrington, J. Y. (1995). New RAMS cloud microphysics parameterization. Part I: The single-moment scheme. *Atmospheric Research, 38*, 29–62.

Wang, P. K., Greenwald, T. J., & Wang, J. (1987). A three parameter representation of the shape and size distributions of hailstones—A case study. *Journal of Atmospheric Science, 44*, 1062–1070.

Wisner, C., Orville, H. D., & Meyers, C. (1972). A numerical model of a hail-bearing cloud. *Journal of Atmospheric Science, 29*, 1160–1181.

Witt, A., Eilts, M. D., Stumpf, G. J., Johnson, J. T., Mitchell, E. D. W., & Thomas, K. W. (1998a). An enhanced hail detection algorithm for the WSR-88D. *Weather Forecasting, 13*, 286–303. https://doi.org/10.1175/1520-0434(1998)013%3c0286:AEHDAF%3e2.0.CO;2

Witt, A., Eilts, M. D., Stumpf, G. J., Johnson, J. T., Mitchell, E. D. W., & Thomas, K. W. (1998b). Evaluating the performance of WSR-88D severe storm detection algorithms. *Weather Forecasting, 13*, 513–518. https://doi.org/10.1175/1520-0434(1998)013%3c0513:ETPOWS%3e2.0.CO;2

Wong, R. K. W., Chidambaram, N., Cheng, L., & English, M. (1988). The sampling variations of hail size distributions. *Journal of Applied Meteorology, 27*, 254–260.

Xu, J.-L. (1983). Hail growth in a three-dimensional cloud model. *Journal of Atmospheric Science, 40*, 185–203. https://doi.org/10.1175/1520-0469(1983)040%3c0185:HGIATD%3e2.0.CO;2

Yue, Y., Zhou, L., Zhu, A. X., & Ye, X. (2019). Vulnerability of cotton subjected to hail damage. *PLoS ONE, 14*, e0210787. https://doi.org/10.1371/journal.pone.0210787

Yurkin, M. A., & Hoekstra, A. G. (2011). The discrete-dipole approximation code ADDA: Capabilities and known limitations. *Journal of Quantitative Spectroscopy & Radiative Transfer, 112*, 2234–2247. https://doi.org/10.1016/j.jqsrt.2011.01.031

Yuter, S. E., & Houze, R. A., Jr. (1995). Three-dimensional kinematic and microphysical evolution of Florida cumulonimbus. Part II: Frequency distributions of vertical velocity, reflectivity, and differential reflectivity. *Monthly Weather Review, 123*, 1941–1963. https://doi.org/10.1175/1520-0493(1995)123%3c1941:TDKAME%3e2.0.CO;2

Ziegler, C. L., Ray, P. S., & Knight, N. C. (1983). Hail growth in an Oklahoma multicell storm. *Journal of Atmospheric Science, 40*, 1768–1791.

Zipser, E., Liu, C., Cecil, D., Nesbitt, S. W., & Yorty, S. (2006). Where are the most intense thunderstorms on Earth? *Bulletin of the American Meteorological Society, 87*, 1057–1071. https://doi.org/10.1175/BAMS-87-8-1057

Zrnić, D. S., Bringi, V. N., Balakrishnan, N., Aydin, K., Chandrasekar, V., & Hubbert, J. (1993). Polarimetric measurements in a severe hailstorm. *Monthly Weather Review, 121*, 2223–2238.

Observation and Simulation of Severe Storms Over China

Donghai Wang, Zhilin Zeng, Enguang Li, Qianhui Ma, Yuran Cao, Jia Li, and Ting Fan

Abstract Convective storms in China are often organized by multiple convective cells, and most of which develop to be a mesoscale convective system (MCS) that is characterized by an organized regions of deep convection within horizontal hundreds of kilometers and over a time span exceeding 6 h. Severe convective storm, generally as an intense convective cell, occurs within MCS, and is often further associated with hazardous weather, including extreme rainfall, severe winds, severe hail and tornadoes, some of which can be destructive and deadly. Given a large fraction of studies in China focusing on the climatological MCS and individual extreme events that are associated with severe convective storms, the present chapter aims to introduce the spatiotemporal character of the MCS in China, with special attention given to the characteristics, structures, and microphysical features of the severe convective storms that produce extreme rainfall, tornadoes, and severe hail based upon observation and simulation for some notorious cases occurred in the period of recent 10 years. This comprehensive analysis of convective storms in China provides valuable insights into their microphysical processes, tornado characteristics, hail storm dynamics, and overall organization patterns, contributing to improved weather forecasting, disaster preparedness, and mitigation efforts in the country.

D. Wang (✉) · E. Li · Q. Ma · Y. Cao · J. Li · T. Fan
Guangdong Province Key Laboratory for Climate Change and Natural Disaster Studies, Key Laboratory of Tropical Atmosphere-Ocean System, School of Atmospheric Sciences, Ministry of Education, Sun Yat-Sen University, Zhuhai, China
e-mail: wangdh7@mail.sysu.edu.cn

Z. Zeng (✉)
Guangdong Meteorological Observatory, and China Meteorological Administration Tornado Key Laboratory, Guangdong Meteorological Service, Guangzhou, China
e-mail: zengzhlin@mail3.sysu.edu.cn

D. Wang · J. Li · T. Fan
Southern Marine Science and Engineering Guangdong Laboratory (Zhuhai), Zhuhai, China

D. Wang
National Observation and Research Station of Coastal Ecological Environments in Macao, Macao Environmental Research Institute, Macau University of Science and Technology, Macao, China

S. Das and W-K. Tao (eds.), *Severe Storms*,
https://doi.org/10.1007/978-981-97-7075-5_11

Keywords Mesoscale convective system · Severe convective storm · Extreme rainfall · Tornadoes · Hail · Meso-to-micro scale characteristics

1 Introduction

The space-borne radar-based mesoscale convective system (MCS) climatology shows the maximum population over low-elevation regions, a marked decrease over the foothills, and the minimum frequency over the Tibetan Plateau (Chen et al., 2022). IR-based studies reported three MCS maxima in China, occurring over South China, Yangtze River valley (central-eastern China), and Tibetan Plateau. Over South China, MCS frequently develop along either cold front in April or quasi-stationarity front during the mei-yu season in May–June. MCS over the Yangtze River valley are often associated with active mei-yu fronts during June–July. Some MCS originate over the second-step terrain just east of the Tibetan Plateau and propagate eastward into the plains or coastal regions in eastern China.

Seasonal variations of MCS frequency in East Asia are closely related to the progression and retreat of the summer monsoon. MCS occurs frequently in southwestern China before the onset of the South China Sea summer monsoon, followed by over southern China after the onset of the South China Sea summer monsoon. As the monsoonal flows progress over inland China, MCS tends to develop along the high-altitude foothills over northern China, and subsequently decays rapidly with the South China Sea summer monsoon receding.

For a further study of the Chinese MCS, a detailed classification was made regarding histomorphology. Linear-mode MCS are classified into eight types according to the organizational mode, system-relative cell movement, and general features of the system. These eight types include trailing stratiform (TS) MCS, back-building/quasi-stationary (BB) MCS, trailing line/adjoining stratiform (TL/AS) MCS, leading stratiform (LS) MCS, parallel stratiform (PS) MCS, broken line (BL) MCS, embedded line (EL) MCS, and long line (LL) MCS. The first six types match the types previously proposed by studies abroad, with similarities in organizational mode, cell regeneration, and system movement. The last two types, particularly LL, have never been identified before. These two newly identified types of linear-mode MCS, which occur during the mei-yu season over the Yangtze River valley, are named according to their modal patterns. MCS are divided into linear and nonlinear modes, with the linear-mode MCS further classified into eight subtypes as described above. Nonlinear (NL) MCS feature unorganized strong convective echoes embedded within large areas of stratiform echoes. NL MCS can move either rapidly or slowly and are sometimes even quasi-stationary. Among the linear-mode MCS, TS, LS, PS, and BL MCS move rapidly, while TL/AS, BB, EB, and LL MCS are either slow-moving or quasi-stationary. Precipitation in fast-moving MCS is often light during early stages but heavy during mature stages. Heavy precipitation areas can therefore be roughly determined based on the direction and speed of MCS movement. Areas of heavy precipitation in quasi-stationary MCS can be estimated according to radar echoes. These estimates are particularly beneficial for forecasts of extreme heavy precipitation. The structures of NL MCS are more complicated than those of linear MCS, and present great challenges to realistic forecasting of heavy precipitation. The number and mean duration of various types of linear-mode MCSs observed over the Yangtze River valley during June–July between 2010 and 2012 show that TS is the dominant archetype among linear-mode MCSs in this region. BB is the second most common type of linear-mode MCS, followed by TL/AS, LS, PS, BL, LL, and EL. The percentage distribution of MCS archetypes shows that nonlinear MCS account for 48.5% of the total cases, while linear MCS account for 51.5% of the total. More than one fifth of the linear-mode MCS are TS MCS, and almost one fifth are BB MCS. The fractions of other linear-mode MCS archetypes range from 3.1 to 14.4%.

The most important concern for MCS remains the precipitation characteristics (Li et al., 2020). Over northern China, severe winds over the plains are frequently produced by squall lines, whereas they are more connected to isolated convective clusters over the mountains (Yang & Sun, 2018). Linear and nonlinear MCS account for 17 and 83% of the MCS in China, respectively. Linear MCS have much stronger convective intensity and heavier precipitation than nonlinear MCS. Though broad stratiform MCS have the weakest convection, they produce the maximum rain rate and the largest amount of heavy rainfall among nonlinear MCS. Vertical radar reflectivity profiles suggest that both ice-based and warm-rain processes play an important role in the precipitation processes of linear MCS over China. Intense precipitation exhibited the highest frequency in each morphology, especially in EL and TS systems (Chen et al., 2022).

The overall spatial distributions pattern of MCS precipitation amount is quite consistent with the frequency of MCS occurrence. Two MCS precipitation centers along the Nanling Mountains and Wuyi Mountains can be found in spring, and MCS precipitation contributes to over 45.0% to the total rainfall over some regions of south-eastern China. In summer, MCS precipitation extends to northward and covers a wider area over eastern China, compared with that in spring. Meanwhile, another MCS rainfall center shows up over the eastern periphery of the Tibetan Plateau. These results indicate that the long-lived and intense MCS play an important role in the water cycle over eastern China during warm seasons. In autumn, the MCS precipitation weakens and gradually disappears over southeastern China, the middle-to-lower reaches of the Yangtze River Basin, and the Yellow River basin. The MCS precipitation center over eastern Tibetan Plateau persists and contributes around 20–30% within some areas of the eastern periphery of the Tibetan Plateau, but the magnitude also shows a significant decrease compared with that in summer.

MCS causes many kinds of severe weather, which is highly disastrous and frequent across China. For instances, an extreme rainfall event occurred over Zhengzhou, a city in central China on 20 July 2021, leading to devastating floods within the metropolitan areas, leaving 292 death, 47 lost, and a direct economic loss of 114,269 billion CNY (Yin et al., 2021). A super cellular tornado hit Funing Jiangsu from about 1410 to 1500 local standard time (LST) on 23 June 2016, causing 98 fatalities and 846 injuries. It was the deadliest tornado in the past 40 years in China (Meng et al., 2018). In addition, hail is one of the severe-storm-producing hazards in China (Li et al., 2018), which is the second most frequent catastrophes following floods (Guan et al., 2014).

In this chapter, we delve into a comprehensive analysis of convective storms in China, covering five distinct parts. Firstly, we explore the Climatology of MCS in the country, providing insights into their spatial and seasonal variations, as well as their organization patterns. The second part investigates the microphysical processes, structures, and characteristics that give rise to extreme precipitation events. We then proceed to discuss the structure, vertical rotation characteristics, and evolutionary process of tornadoes in the third part. In the fourth part, we delve into the climatology, structure, and cloud microphysical characteristics of hail storms. Lastly, we conclude with a summary, synthesizing the key findings and implications derived from our comprehensive examination of convective storms in China.

2 Physical Process, Structure, and Characteristics of Severe Storms Producing Extreme Rainfall

One of the most frequent hazards that the severe storms produce in China is heavy and extreme rainfall, which sometimes leads to considerable losses of life and property. There are two types of heavy and extreme rainfall in the warm season in China according to the dominant dynamic forcing origin: one closely associated with synoptic weather systems, such as cold fronts and their circulations, tropical cyclones, and intertropical convergence zone; the other often occurring in the warm sector, a few hundred kilometers away from the subtropical weather systems (generally termed the warm-sector heavy rainfall). The former type of heavy rainfall occurs often across a widespread of China, while the latter takes place frequently over South China with a smaller horizontal span but a larger precipitation amount compared to the former. Two notorious cases of representative for the two types of extreme rainfall are illustrated in this section to display the characteristics of the severe storms associated with heavy and extreme rainfall.

An extremely heavy rainfall event occurred in Zhengzhou, central China, on 20 July 2021 and produced an hourly rainfall of 201.9 mm, which was the highest record over mainland China. The large-scale environmental conditions in this event involved the West Pacific Subtropical High (WPSH), Typhoon In-fa (2021) in the west Pacific, and a long-lived low-pressure vortex near Zhengzhou city. A large amount of warm moist air was transported from the ocean by the easterlies-to-southeasterlies due to Typhoon In-Fa (2021) and the WPSH, the latter of which was abnormally displaced northward. Due to blocking by the mountainous terrain to the north and a low-level mesoscale vortex west of Zhengzhou, there was notable convergence of low-level momentum and water vapor flux in Zhengzhou and surrounding regions, yielding very favorable conditions for high precipitable water and heavy rainfall. A warm-sector heavy rainfall occurred in Guangzhou, on 7 May 2017, with an observed rain record of 382.6 mm within 3 h, causing severe local flooding. Unlike the extreme rainfall event in Zhengzhou on 20 July 2021, this event occurs without strong large-scale forcing such as mid-level strong troughs, low-level jets, and cold front. Overall, this event occurred under the setting of little quasi-geostrophic forcing.

2.1 *Mesoscale Characteristics and Storm Evolution*

The overall evolution of the extreme-rain-producing storm occurring in Zhengzhou was the early stages and the later stagnation in Zhengzhou, while the formation of extreme rainfall in Guangzhou was closely related to the evolution into high-precipitation-supercell-like structure of the storm producing extreme rain rates. Figure 1 shows the extreme-rainfall-producing storm evolving into supercell-like structures. Observations of surface winds and radar reflectivity presents unambiguously some supercell-like structures, including the quasi-stationary mesoscale outflow boundary (MOB), inflow notch (Fig. 1c, d), and MOB-related flanking line (Fig. 1d). However, these supercell-like structures lasted for a very short time (less than $\sim$15 min) and dissipated after 0524 LST (Fig. 1f). Observations during this stage show that the storm intensified, and then evolved into supercell-like structures, which normally suggested the increasing updrafts within the storm, based upon $R = Ewq$ (Doswell et al., 1996), where R is the rainfall rate, E is the precipitation efficiency, q is the water vapor mixing ratio of the rising air, and w is the ascent rate. Therefore, the rain-rate enhancement was closely related to the formation of supercell-like storm structures, in which the increasing w (Doswell et al., 1996; Moller et al., 1994) may play an important role.

2.2 *Storm Structures*

The quasi-stagnation and structure of the storm of the heavy rain occurred in Guangzhou on 7 May 2017 are somewhat similar to those occurred in Zhengzhou 20 July 2021. Cross sections of radar reflectivity (only contoured above 40 dBZ) and radar radial velocities for the storm are shown in Fig. 2. The storm in a cross section presented multiple centroids (i.e., multiple maximum reflectivity cores) at the early minute rain-rate enhancement episode (Fig. 2a), and it gradually evolved into a single centroid, with a higher reflectivity factor (52 dBZ). From the view of radar radial velocities, the cold pool–related outbound velocity (toward the Guangzhou radar) manifested as dramatic changes including becoming deeper and stronger and was accompanied by its boundaries elevating, steepening, and surging ahead (not shown), all of which followed the storm intensification. The top boundary of the outbound velocity vertically extended to the maximum altitude of about 3 km at 0518 LST (Fig. 2d). Therefore, storm intensification was characterized by its progressively organized structure in conjunction with the elevated and increased outbound velocities.

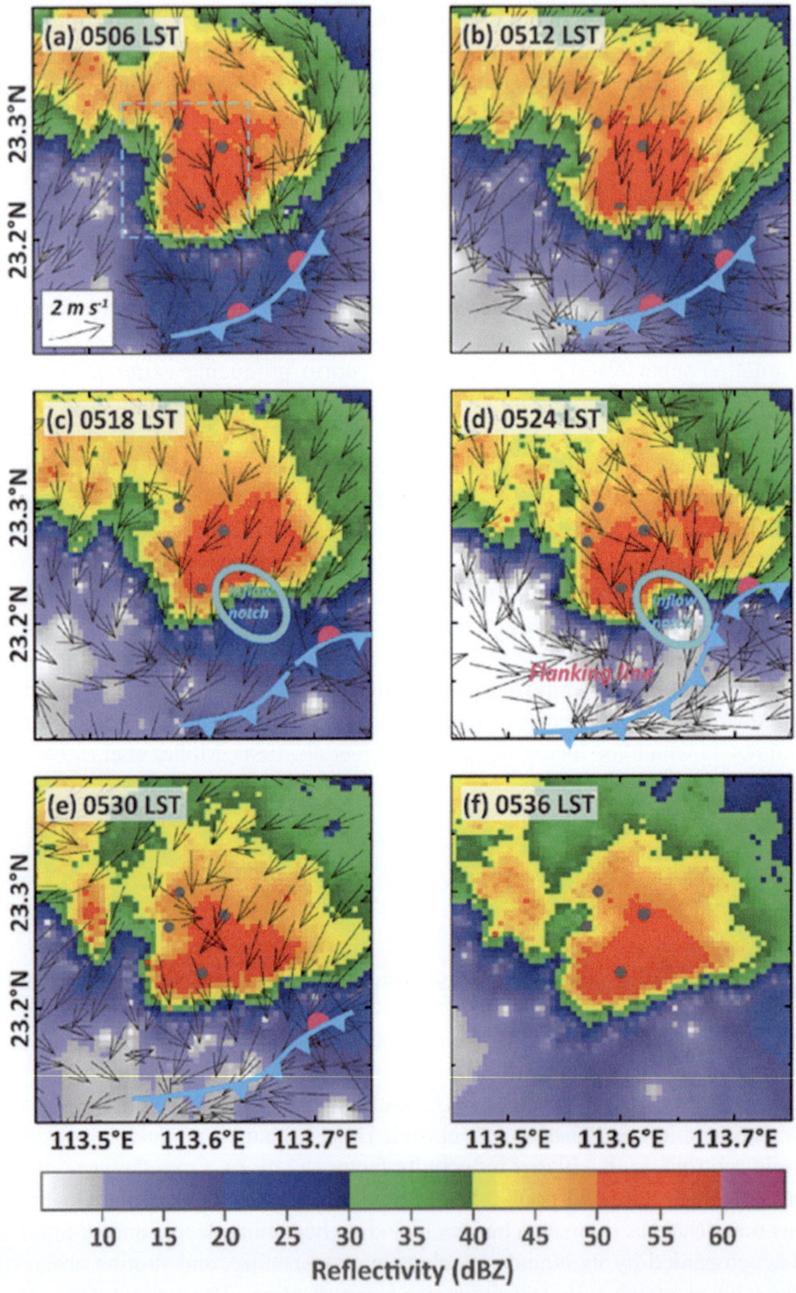

Fig. 1 Radar reflectivity (dBZ) views of the storm at 1.5° elevation at selected times. Gridded observed surface winds (arrows) are overlaid onto the reflectivity at corresponding times. Four rain gauges (baby blue dots) in all panels are given as location references for the storm. The MOB is denoted by the quasi-stationary frontal symbol in **a**–**e**. Cyan ellipses in **c** and **d** denote the locations of inflow notch, while flanking line is marked in **d**. Cited from Zeng and Wang (2022)

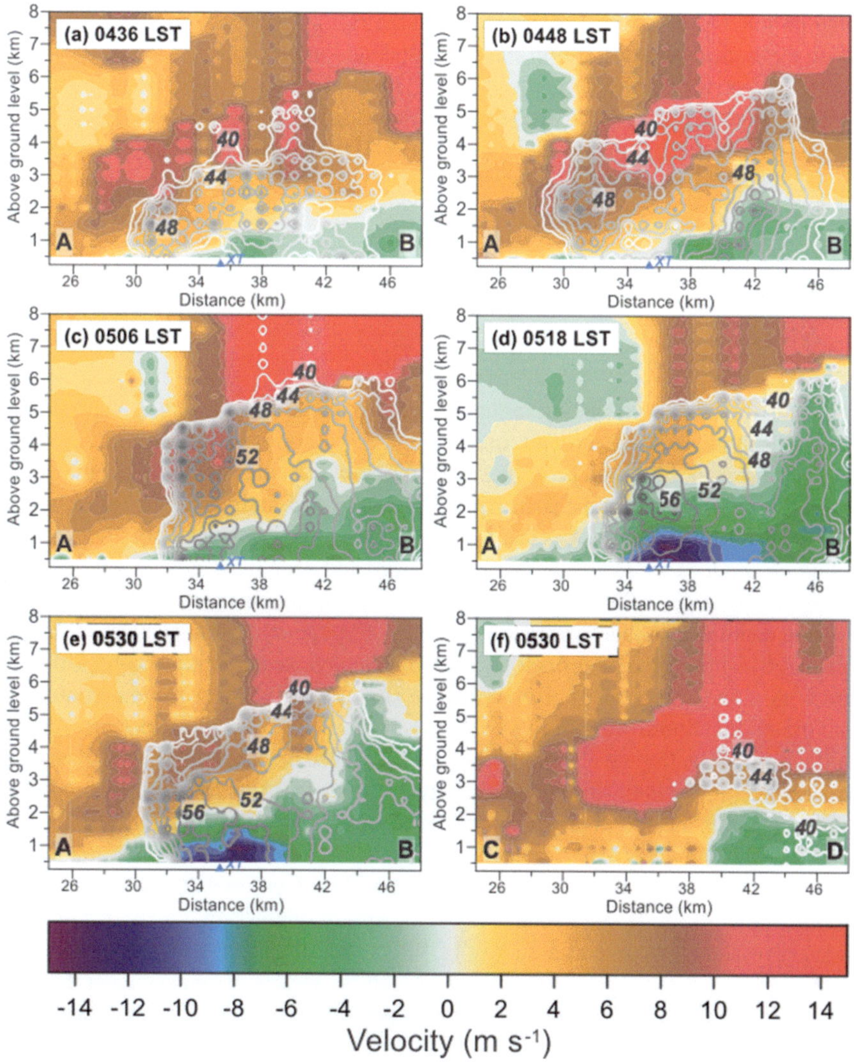

Fig. 2 **a–e** Vertical cross sections of radar reflectivity (dBZ; contoured above 40 dBZ) and radial velocity (shaded; m·s^{-1}) along the A–B line. **f** As in **a–e**, but cross section along the C–D line at 0530 LST 07 May 2017. The contours indicating reflectivity exceeding 40 dBZ are presented as gradient colors from white to black with 2 dBZ intervals. Abscissas in each panel denote the distance to Guangzhou radar. Cited from Zeng and Wang (2022)

2.3 Microphysical Processes

Large-scale precipitation efficiency (LSPE) and cloud-microphysical precipitation efficiency (CMPE) are defined as the ratio of surface rain rate to the sum of surface evaporation and moisture convergence, and the ratio of surface rain rate to the sum of condensation and deposition rates of supersaturated vapor, respectively. Water vapor flux convergence/divergence (QWVF) had a significant impact that influenced LPSE. To understand the influence of QWVF on LSPE, QWVF can be further decomposed into the water vapor convergence (QGRADV) and the water vapor advection (VGRADQ).

3 Structures, Characteristics, and Evolutions of Tornadic Storm

In China, the storms producing tornadoes tend to occur in large plains and typhoon periphery areas with flat ground, uniform underlying surface, and abundant water vapor (Xue et al., 2016). The highest frequency of tornadic storm in China occurs in July, with a secondary peak in April. There are two main types of tornadoes in China: super cellular tornadoes and tropical cyclone (TC) spawned tornadoes. Super cellular tornadoes are associated with prolonged and deep mesocyclone activity. For example, the super cellular tornado occurred in Funing, Jiangsu on 23 June 2016 was the deadliest tornado recorded in the past 40 years in China, causing 98 fatalities and 846 injuries. The tornado produced a damage swath that was 34.5 km long and with a maximum width of 4.1 km. The situ surface weather station recorded a maximum gust speed of 34.6 m s^{-1} (Meng et al., 2018). TC tornadoes are associated with shallow low-level vortex flow on a smaller scale, usually characterized by a smaller size, lower intensity, and less severe damage. TC tornadoes often occur in the outer rainbands located in the right-front or the northeast quadrant relative to the TC track within a 200–400-km annulus. These tornadoes often reside in miniature supercells in an environment characterized by a high low-level moisture content, a low lifting condensation level, a small surface dewpoint depression, small or moderate convective available potential energy (CAPE), and enhanced low-level shear due to increased surface friction from water to land, as well as boundary layer convergence (Zhao et al., 2017). These miniature supercells exhibit hook echo/appendages and tornado vortex signatures (TVSs) in Doppler radar observations.

In this section, we aim to introduce the two types of tornadoes via some representative case studies, including how the environmental conditions exhibit in these events, evolution characteristics and structures of the tornadic storms.

3.1 *Environmental Conditions for Tornadic Storms*

A super cellular tornado occurred in Funing, Jiangsu Province, China, from about 1410 to 1500 local standard time 23 June 2016, The Funing tornado occurred in a typical convection-producing synoptic environment. At 500 hPa, there was a shallow trough extending from quasi-stationary cold vortex. At lower levels, a distinct 700-hPa trough to the northwest and 850-hPa jets to the southwest that provided strong moisture convergence over Jiangsu. In central China, a distinct northeast-southwest oriented surface low was observed below these low-level jets. Specifically, there was a northwest-oriented surface warm tongue around Funing before the tornadoes formed.

The Funing tornado was spawned by a supercell that developed at the southern end of a northwest–southeast-oriented linear MCS (denoted by B in Fig. 3a), which developed in front of a much longer quasi-linear MCS (denoted by A in Fig. 3a). MCS-A was accompanied by a cold front and had a trailing stratiform precipitation region, while MCS-B formed in the warm sector and had a leading stratiform precipitation region from ~1100 LST (e.g., Fig. 3). The tornadic supercell storm (denoted by C in Fig. 3b) initiated to the northwest at about 0912 LST. It then moved to the northeast with MCS-B, intensified, and produced the Funing tornado from ~1410 to 1500 LST (Fig. 3c, d).

The Funing tornado was spawned by a supercell that developed at the southern end of a northwest–southeast-oriented linear MCS, which developed in front of a much longer quasi-linear MCS. A MCS was accompanied by a cold front and had a trailing stratiform precipitation region, while another MCS formed in the warm sector and

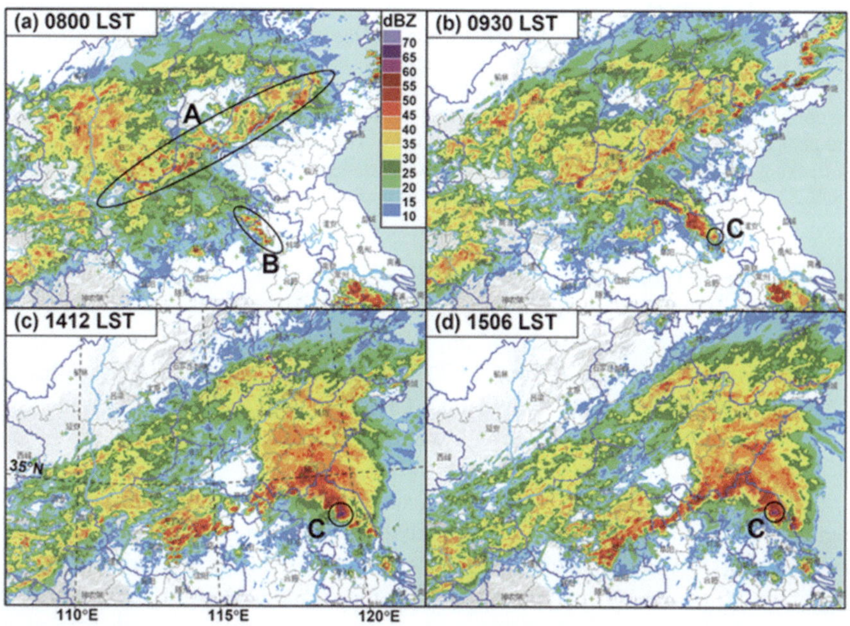

Fig. 3 Mosaic of composite radar reflectivity (dBZ) at **a** 0800, **b** 0930, **c** 1412, and **d** 1506 LST 23 Jun 2016. Capital letters A and B denote the southwest–northeast- and northwest-southeast-oriented convective lines, respectively. The small black circle and capital letter C in **b–d** denote the tornadic supercell. Cited from Meng et al. (2018)

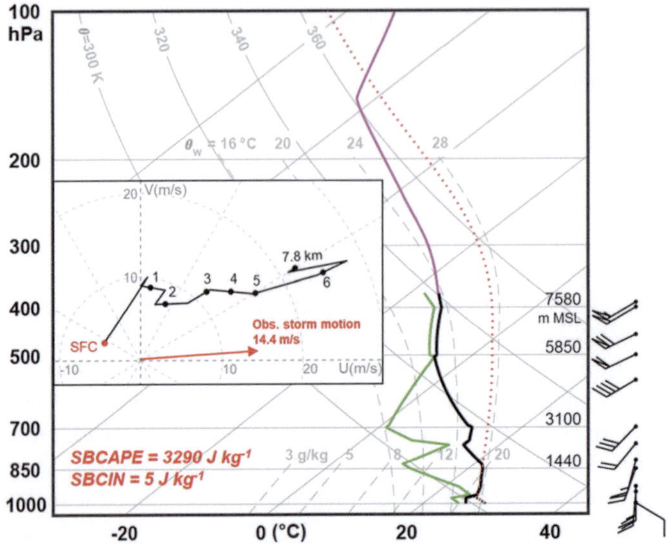

Fig. 4 Skew T–log*p* and hodograph (inset) diagrams near the tornadic storm at 1400 LST 23 Jun 2016. The ambient temperature and dewpoint are represented by the solid black and green lines, respectively. To obtain a more representative storm environment, the sounding was modified by assuming a well-mixed boundary layer using the mean surface values at four stations. The modified ambient temperature is plotted as a dotted black line, and the parcel that ascends undiluted from the surface is shown by the dotted red line. The upper-level ambient temperature, which was missing in the radiosonde observation, was filled with values obtained from the NCEP FNL analysis at the corresponding sounding location at 1400 LST 23 Jun 2016 (magenta). Half barbs, full barbs, and pennants denote 2, 4, and 20 m s^{-1}, respectively. The observed storm motion was obtained by tracking the centroid of the radial velocity maxima couplet of a 2.4° mesocyclone signature (at about 3 km AGL) from 1405 to 1433 LST. Cited from Meng et al. (2018)

had a leading stratiform precipitation region from ~ 1100 LST. The tornadic supercell storm initiated in northern Anhui Province at about 0912 LST. It then moved to the northeast with MCS, intensified, and produced the Funing tornado.

A near-storm sounding at 1400 LST 23 June 2016 showed the surface-based CAPE reaching 3290 J kg^{-1} with a small CIN of 5 J kg^{-1} (Fig. 4). The sounding also showed a 0–6 km AGL layer bulk wind difference of 27.2 m s^{-1} and a strong directional shear in the lowest 1 km of the atmosphere (Fig. 4).

Typhoon Mujigae was categorized as a super typhoon by the Chinese Meteorological Administration near the time of landfall in Guangdong province at 1400 LST 4 October 2015, which produced an EF3 tornado in Foshan city, the northeast quadrant of the TC center. When Typhoon Mujigae made landfall around 1.5 h before the genesis of the Foshan tornado, it was quite strong with the maximum wind speed of greater than 50 m s^{-1}, which was favorable for tornado genesis according to McCaul (1991; showing a mean intensity of 47.1 m s^{-1} for hurricanes a result of the increase of friction, the 0–6-km vertical wind shear increased from the sea toward inland in

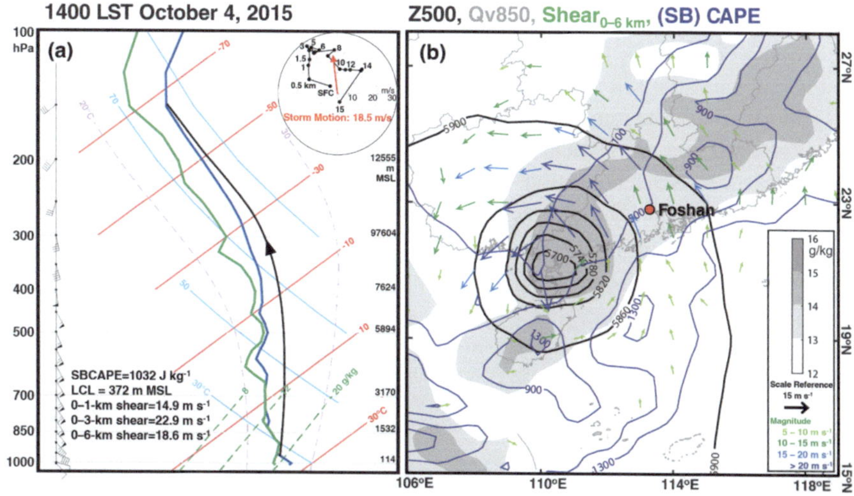

Fig. 5 **a** Skew T–log$p$ diagram showing the radiosonde observation in Hong Kong, China, at 1400 LST 4 Oct 2015. The path of the surface-based parcel is shown by the thick black line. The hodograph diagram is plotted in the upper-right corner and shows the storm-motion vector (red arrow). Numbers along the lines in the hodograph denote the altitude above the ground (km). Also shown on the left side of **a** are the horizontal wind barbs at some levels (half barb = 2 m s^{-1}, full barb = 4 m s^{-1}, and pennant = 20 m s^{-1}). **b** Geopotential height at 500 hPa (black contours; gpm), water vapor mixing ratio at 850 hPa (shaded; J kg^{-1}), surface-based CAPE (blue contours; J kg^{-1}), and 0–6-km vertical wind shear [vectors; the scale reference and magnitudes (reflected by different colors) of the vectors are shown in the bottom-right corner; m s^{-1}] at 1400 LST 4 Oct 2015 based on NCEP FNL. The location of Foshan is denoted by the red dot. Cited from Bai et al. (2017)

the northeast quadrant of the typhoon center (Fig. 5b).The environmental parameters derived from the near-storm radiosonde (Fig. 5a) at 1400 LST 4 October 2015.

3.2 *Evolution of the Tornadic Storms*

On 23 June 2016, the trend in the storm updraft strength agreed well with the narrowing and expansion of the size of the WER signature during the Funing tornado time period. A well-defined WER signature was observed starting at 1351 LST. The horizontal extent of this WER signature decreased after the genesis of the Funing tornado and reached a minimum at 1437 LST, which was approximately the midpoint of the tornado's life-span (Fig. 6b–f). After 1437 LST, the horizontal and vertical spatial extents of this WER signature began to expand again (Fig. 6d–f), which suggests updraft restrengthening. This restrengthening was likely associated with the occlusion process leading to the second tornado near Sheyang. Consequently, the demise of the Funing tornado was probably a result of a mismatch between the horizontal motions of the tornado vortex and the main storm-scale updraft of the parent supercell. Such a horizontal motion mismatch was likely because of the occlusion process in the tornado vortex due to the rear-flank gust front that wrapped cyclonically ahead of the tornado.

On 4 Oct 2015, the rainbands of Typhoon Mujigae that produced tornadoes were well organized in a narrow band shape and convectively active (e.g., Fig. 7a). Gaps between convective cells within these rainbands gradually increased as the cells approached the land. The convective cell that generated the Foshan tornado was initiated near the seashore, then intensified (in terms of composite radar reflectivity) immediately after it moved over land near Macau at 1418 LST, and then moved

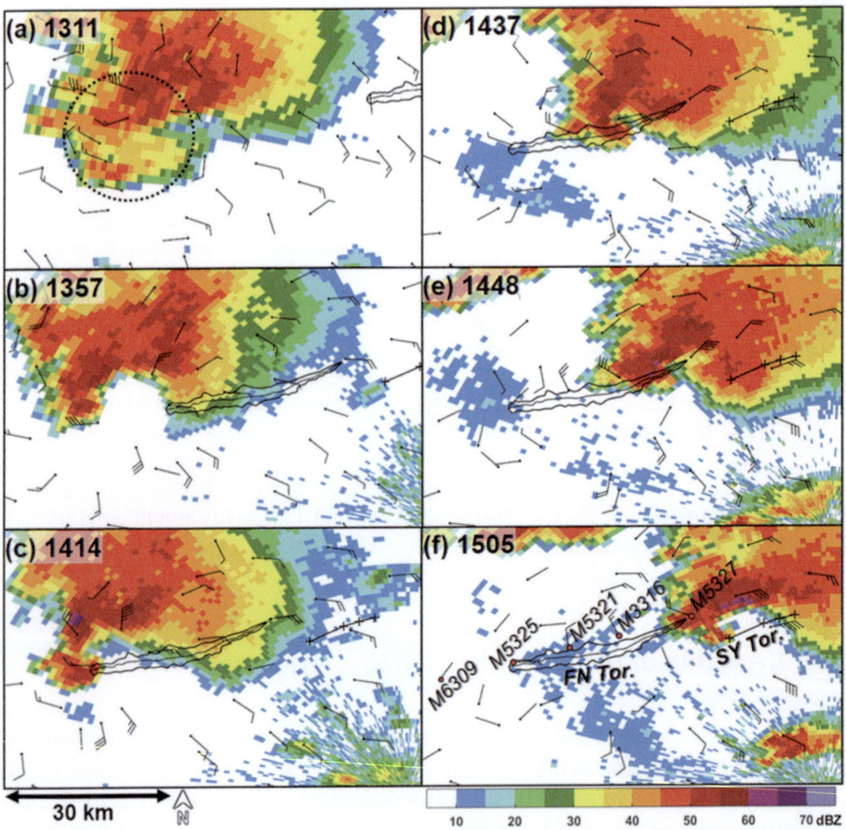

Fig. 6 Base reflectivity (shaded; dBZ) at the 0.5° elevation angle from the Yancheng radar at different times (LST) on 23 Jun 2016. Also plotted are surface gusts (half barb, 1 m s^{-1}; full barb, 2m s^{-1}; pennant, 10m s^{-1}) at the same time. The EF0 isopleth of the Funing tornado damage swath and the analyzed tornado track are both given by a black curve. The path of 0.5° tornado vortex signatures (TVSs) (black crosses) of the Sheyang tornado (SY Tor.) are also denoted. The dotted circle in **a** denotes the area where there are convergent and rotational wind patterns, as described in the text. The locations of surface weather stations (red dots) are given in **f**. The north direction and distance length scale are given at the bottom. Cited from Meng et al. (2018)

to the northwest along the rainband. A large gap sustained between this cell and its neighboring cell to the southeast until the dissipation of the tornado. This cell eventually developed into a supercell at around 1442 LST, when a mesocyclone was first identified. Details of the mesocyclone will be presented in the next section. The Foshan supercell had a well-organized hook echo (Fig. 7b–g). The tornado damage swath was located near the tip of the hook echo. Anomalously large reflectivity (greater than 64.5 dBZ) at the 0.5° elevation angle was observed almost right over the tornado track at 1536 and 1542 LST (Fig. 7f, g). The hook echo was most probably caused by the debris raised from the ground, since the maximum reflectivity occurred at the lowest level and decreased with height. The Foshan tornadic minisupercell was characterized by a low radar echo top, which was similar to the minisupercells that were observed in TC environments in the United States.

3.3 Vertical Structures and Rotational Features

During the time period of super cellular tornado in Funing on 23 June 2016, the WER signature was bounded above 4 km radar level (using 40-dBZ contours) at 1414 LST, whereas at 1437 LST, the vertical spatial extent of WER reached minimum and the height of the bounded WER dropped to below 3 km. The mesocyclone of the supercell was first identified at ~1340 LST when the rotational velocity V_{rot} [determined by the two-dimensional maximum V_{max} and minimum V_{min} radial velocity couplet (with a diameter of 2–10 km); $V_{rot} = |V_{max} - V_{min}|/2$] started to meet the mesocyclone threshold of 15 m s^{-1} and have a vertical extent greater than 3 km (from 1.1 to 5.4 km AGL) (diamonds in Fig. 8e). At this time, a 0.5° TVS was also detected. The gate-to-gate azimuthal radial velocity difference (DV) of a manually identified TVS in this study was required to be no less than 20 m s^{-1}. After ~1354 LST, the mesocyclone intensified and extended upward.

The mesocyclone signature evidently intensified, especially above 5 km AGL, from ~1400 LST (Fig. 8e). Six minutes later, at 1406 LST, the mesocyclone reached its highest measurable altitude of approximately 8 km AGL (Fig. 8e). A quite strong 0.5° TVS (DV = 59 m s^{-1}) was identified at ~1414 LST when apparent damage started to be observed on the ground. Almost simultaneously, station M5325 recorded an apparent enhancement of wind gusts. The TVS reached its peak intensity at 1437 LST, with a gate-to-gate azimuthal radial velocity difference of 84.5 m s^{-1} (Fig. 8e). The 0.5° peak rotational velocity (only for velocity maxima with cyclonic azimuthal shear within 10 km) also reached its maxima of 42.25 m s^{-1} (the same radial velocity pixels as TVS) at this time, which is close to the median of 0.5° peak V_{rot} associated with EF4 + tornadoes in the U.S. climatology (Smith et al., 2015). The TVS extended from the surface all the way up to about 12 km AGL.

The diameter of the mesocyclone is determined by the horizontal distance between the centers of the couplet of maximum and minimum radial velocities. The mesocyclone reached its maximum rotational velocity of 42.2 m s^{-1} at the 0.5° elevation angle at 1426 LST (Fig. 8e). After 1431 LST, the diameter of the mesocyclone

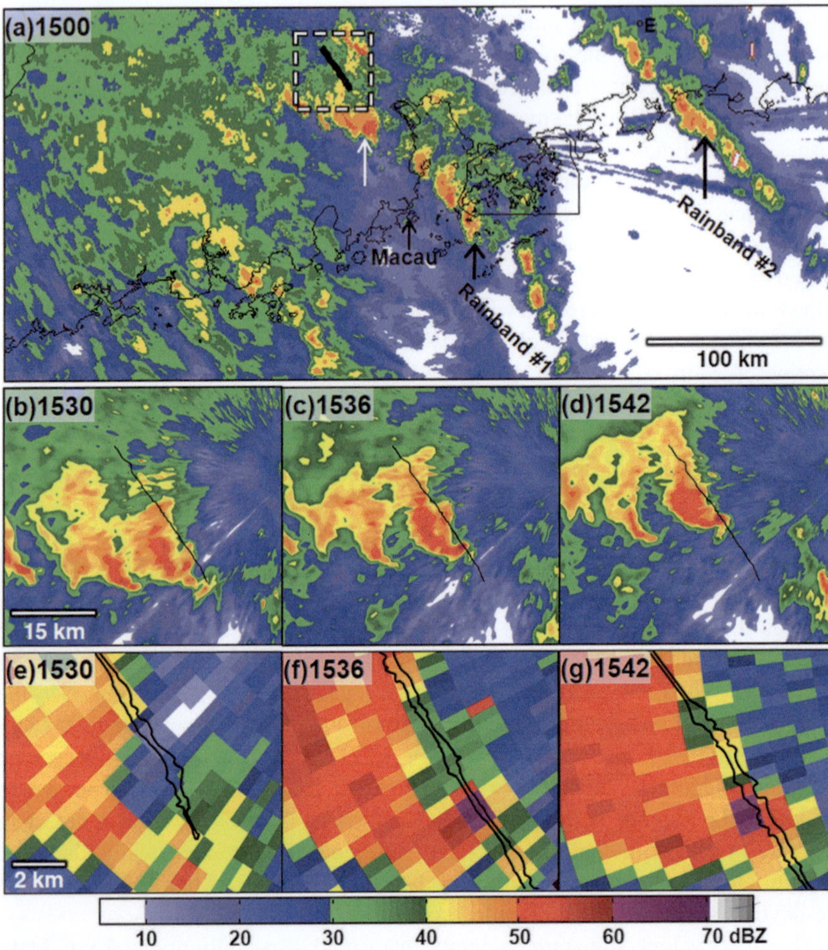

Fig. 7 **a** Composite radar reflectivity (shaded; dBZ) at 1500 LST 4 Oct 2015. The white arrow indicates the Foshan tornadic supercell. Base reflectivity (shaded; dBZ) at the 0.5° elevation angle in the dashed box in **a** are given at **b** 1530, **c** 1536, and **d** 1542 LST. The black line in **b**–**d** represents the Foshan tornado track. **e**–**g** Base reflectivity at the 0.5° elevation angle enlarged around the hook echo with EF0 isopleth (black) at **e** 1530, **f** 1536, and **g** 1542 LST. Distance scale for each row is given in **a**, **b**, and **e**. All the base reflectivity data were from the Guangzhou radar. Cited from Bai et al. (2017)

contracted to less than 2 km at the 0.5° elevation angle and thus could not be considered as a mesocyclone, although the rotation was still strong. During 1434–1440 LST, new mesocyclone signatures were identified and then moved to the east, became more collocated with the hook echo, and later produced the Sheyang tornado (Meng et al., 2018).

The Foshan tornado was produced by the landfalling Typhoon Mujigae on 4 Oct 2015. The mesocyclone was first detected at 1442 LST with a vertical extent

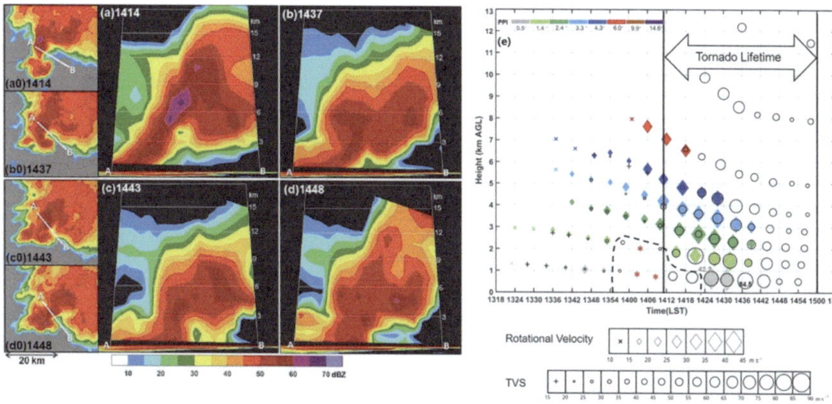

Fig. 8 **a–d** Vertical cross sections of radar reflectivity along the A–B lines as shown in the (a0)–(d0) 0.5° radar reflectivity, respectively, at different times on 23 Jun 2016. **e** Time–height evolution of the maximum gate-to-gate radial velocity difference (DV) of the TVS (circles; m s^{-1}) and the rotational velocity of the mesocyclone (diamonds with different colors denoting different elevation angles; m s^{-1}) manually identified from the Yancheng radar on 23 Jun 2016. The mesocyclone in the area circled by the dashed line could not be accurately identified because of obscuration by range-folded data. Similarly, red asterisks represent that the TVS could not be identified because of obscuration by range-folded data. Cited from Meng et al. (2018)

of approximately 2 km from 2 to 4 km AGL (Fig. 9d), lasting about 80 min. The maximum depth of the mesocyclone was mainly around 3 km AGL during the tornado life cycle. The mesocyclone diameter ranged from 1.5 to 6.1 km at all elevation angles (Fig. 9e). It experienced a sharp decrease near the tornado touchdown time (1520–1530 LST). In the volume scan at around 1536 LST, the diameter at the lowest level contracted to about 2.4 km. The intensity of the mesocyclone peaked in this volume scan with the maximum rotational velocity [$V_R = (|V_{SRVin}| + |V_{SRVout}|)/2$] of 34 m s^{-1} (Fig. 9f) and the rotational shear vorticity of 0.058 s^{-1} (Fig. 9d) obtained at the lowest level. The maximum 0.5° peak rotational velocity in this EF3 Foshan tornado was close to the median of rotational velocities associated with EF3 tornadoes in the United States (Smith et al., 2015). The most severe surface damage happened when the 0.5° mesocyclone was already in the weakening stage (Fig. 9d, f).

The maximum mesocyclone azimuthal shear [$(|V_{SRVin}| + |V_{SRVout}|)$/diameter; the magnitude of azimuthal shear is half of the rotational shear vorticity in Fig. 9d] in a volume scan experienced a sharp increase after the tornado touchdown and peaked at around 1544 LST at the 2.4° elevation angle with a value of 0.034 s^{-1}. The TVS was first detected between 1 and 2.2 km AGL about 10 min later than the time when the mesocyclone was first detected. It then developed both downward and upward. The strongest TVS occurred at 1536:55 LST at the same volume scan and elevation angle as those of the strongest mesocyclone (Fig. 9g, f). The strength (DV) of the strongest TVS reached 60 m s^{-1} at ~420 m AGL. Similar to the sequence of the strongest mesocyclone and surface damage, the most severe damage was observed about 4

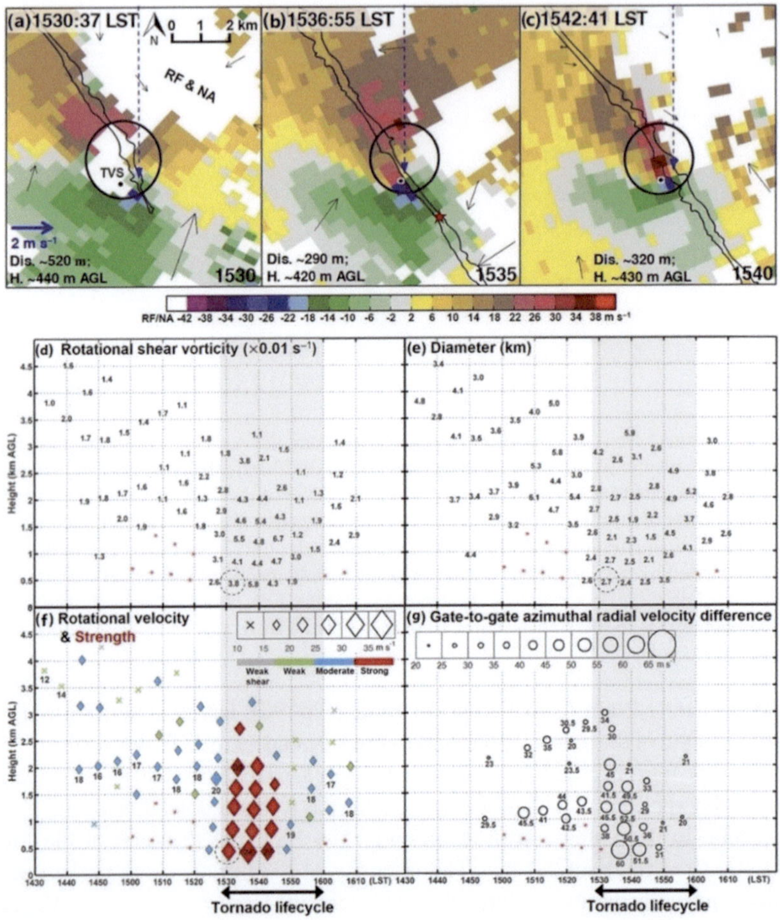

Fig. 9 **a–c** Radar radial velocity (shaded; m s^{-1}). Locations of the damage structures are indicated (dashed blue arrows). The times in **a–c** are the recorded receiving time of the TVS radar beams. Mesocyclones with the estimated diameter (black circles) and EF0 isopleth of the tornado (black contours) are indicated. Mesocyclone in **a** was estimated using the recognizable radial velocity pixels that satisfied the criterion, while the TVS location is roughly determined according to the morphology of the anomalously large radial velocity in the next volume scan [**b**]. The red star in **b** represents the estimated location of the tornado at 1536:55 LST. Vector difference of 10-m wind (black arrows) at the given time (LST; which is indicated by the number in the bottom-right corner in each panel) from that of 5 min before. Time-height diagrams of **d** rotational shear vorticity ($\times 0.01$ s^{-1}), **e** diameter (km), **f** storm-relative rotational velocity (m s^{-1}) of the 2D velocity-couplet signature of the Foshan tornadic mesocyclone, and **g** maximum gate-to-gate azimuthal DV (over an azimuthal distance of one beamwidth) greater than 20 m s^{-1} within the mesocyclone. In (f), the maximum values of storm-relative rotational velocity within a volume scan time are labeled by numbers, and the strength of rotation (color shading) is categorized using the Radar Operations Center mesocyclone strength nomogram. Heights where the **d–f** mesocyclone and **g** DV were undetectable (red asterisk) because of the range-folded value are indicated. Values in dashed circles in **d–f** were derived using the recognizable radial velocity pixels. Tornado life cycle is shaded in gray and denoted by the two-way arrow at the bottom for reference. Cited from Bai et al. (2017)

min after the peak strength of the low-level TVS. At 1542:41 LST, the 0.5° TVS (Fig. 9c) was corresponded to the DV decreasing to 51.5 m s^{-1} at the 0.5° elevation angle (Fig. 9g). The TVS dramatically weakened at the lowest level at 1548:58 LST and vanished at about 1600 LST. All the identified TVSs were below 3 km AGL during the entire event.

The collocation of the condensation funnel and the TVS location at the 0.5° elevation angle was estimated based on the video information together with the data from the Guangzhou radar. The identified 0.5° TVSs were found to be consistently located on the left side of the tornado centerline relative to the tornado motion (e.g., Fig. 9a–c). The mesocyclone center, however, was not always located on the same side as the corresponding TVS (e.g., Fig. 9b). The horizontal along-track displacement between the 0.5° TVS and the condensation funnel at the ground experienced an increase before the tornado reached its peak strength (Bai et al., 2017).

3.4 Comparison with Storm Producing Damaging Winds

In addition to tornadoes that can produce destructive winds, downbursts (or quasi-straight-line winds) can also cause serious damaging winds. There are differences in the direction of scattered debris between a tornado and a downburst (Fig. 10). The most significant features of the damage pattern caused by a tornado are a narrow damage swath and convergent surface debris (Fig. 10a), while straight-line winds tend to cause short and broad fan-shaped diverging surface debris patterns (Fig. 10b; Meng et al. 2014).

One of notorious damaging wind events in China was the Oriental Star that capsized in the Yangtze River within Jianli, Hubei Province on 1 June, 2015. A total of 442 lives were taken, among 454 passengers and crewmembers. The cruise ship capsized when it encountered strong winds of at least 31 m s^{-1} near the apex

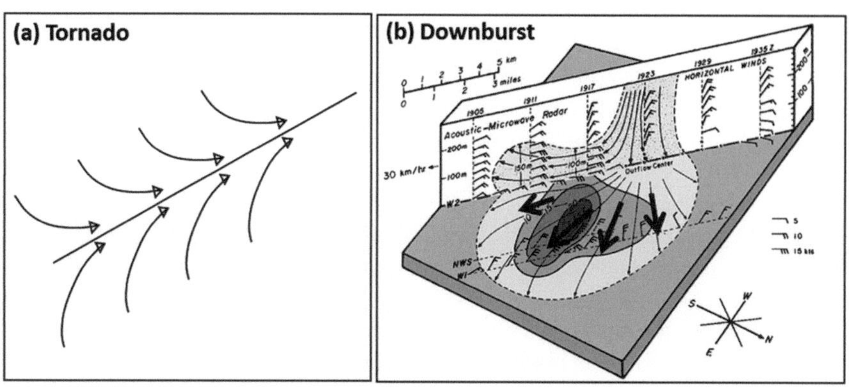

Fig. 10 Schematics of surface wind in a (**a**) tornado and a (**b**) downburst. Cited from Meng and Yao (2014)

of a bow echo embedded in a squall line. Such strong winds were likely caused by microburst straight-line wind and/or embedded small vortices. No adequate observational evidence of a tornado or gustnado was found in this event. According to the definition of a tornado by the Glossary of Meteorology (http://glossary.ametsoc.org/wiki/), three necessary conditions need to be satisfied for a vortex to be considered as a tornado: (1) rotating winds on the ground; (2) connection with a cloud-base rotation; and (3) location under a cumuliform cloud. From the damage point of view, a tornado usually causes a narrow damage swath with a stormscale curved or convergent debris. Gustnadoes are often mistaken for tornadoes due to difficulties in assessing condition (2) or ignorance of conditions (2) and/or (3). A small vortex that satisfies condition (1) and (3) but not (2) should be considered a gustnado if it forms right along the gust front. In this event, conditions (1) and (3) were met around the shipwreck location, but whether or not condition (2) was met remains unknown. Considering that the areas with convergent or curved trees were quite small and localized, without showing a narrow swath with a swathscale convergent or curved debris, and they did not appear along the gust front, they more likely resulted from small vortices forming on the flanks of surging outflow currents or microbursts, rather than tornadoes or gustnadoes (Meng et al., 2016).

4 Climatologies, Structures, Microphysical Characteristics for Hail Storm

This section aims to introduce the climatologies of hail storms in China, including the variation and frequency, followed by the hail storm structures and microphysical characteristics through some case studies.

4.1 Observed Climatologies of Hail Storms

Hail can occur almost everywhere, and the hail season has a diverse geographical distribution in China. Hail day frequency in northern China is typically higher than that in southern China. The annual climatology map very clearly shows the hail frequency (HF, that is mean hail days) features from 1980 to 2015, with greater HF values in mountainous areas than in plains areas (shown in Fig. 11a). The highest HF values were concentrated in the center of the Qinghai-Tibet Plateau (QT), and the second highest values occurred in northern mountain area. The highest HF value during 1980–2015 was 37.6 per year at Nagqu station in QT. HF exhibited a sharp gradient at the edge of the plain areas, near the boundary with the mountainous areas.

Li et al. (2016) found a continuous significant decrease in the annual station HF from the early 1980s for the whole of China (Fig. 11). However, the long-term variations of annual HF were different among the four separate regions. Figure 11c shows

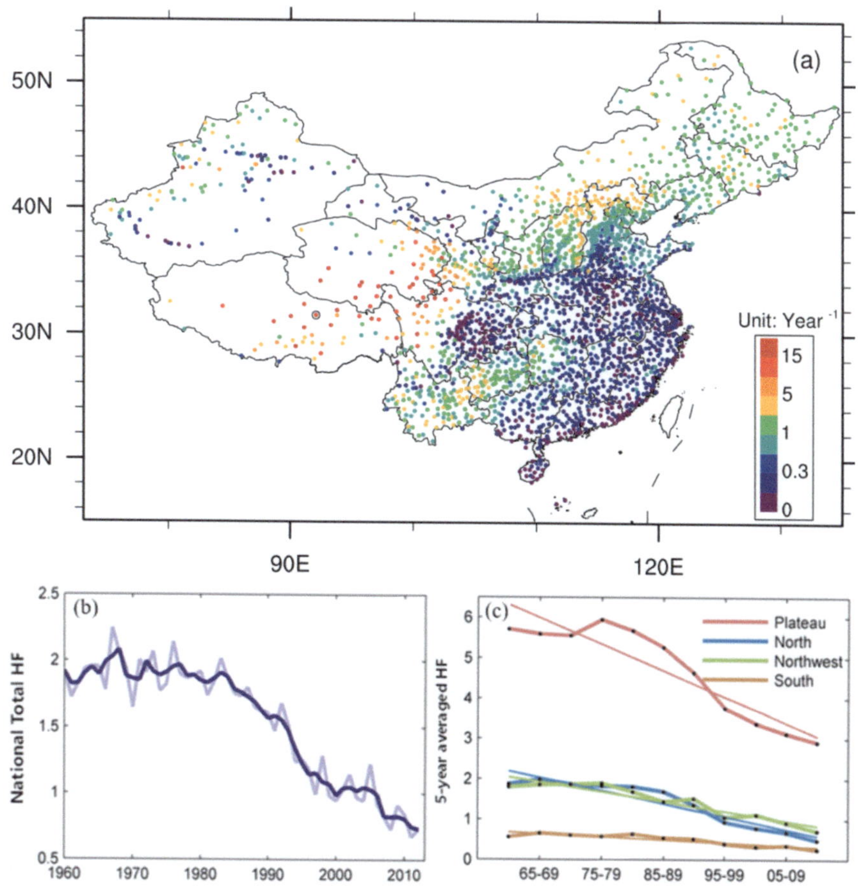

Fig. 11 **a** Distribution of annual mean HF at the stations from 1980 to 2015. Nagqu station (55,299; 31.58N, 92.18E; hollow circle) recorded the highest value (37.6 times per year) during this period in QT. **b** HF of China during 1960–2012 (light blue) and its moving average (dark blue). **c** The 5-yr average HF and trend line of four regions (red represents the plateau, blue represents the north, green represents the northwest, and yellow represents the south). Cited from Li et al. (2016)

that all four regions had experienced a significant decrease over several decades. The trend was 20.3251 (5 yr)$^{-1}$ for the plateau, 20.1592 (5 yr)$^{-1}$ for the north, 20.1194 (5 yr)$^{-1}$ for the northwest, and 20.0371 (5 yr)$^{-1}$ for the south. Compared to the other three regions, the decrease in the HF in the south accounted for the smallest share of the total decrease in HF. The annual HF in northern China remained constant from the 1960s to the 1980s but then started to decrease in the early 1990s. The annual HF in northwestern China also decreased, but with considerable interannual variability. Unlike in the north and northwest, the annual HF in the plateau increased slightly in the early 1970s, peaked in the late 1970s, and began to decline in the 1980s. The annual HF in the plateau declined 10 years earlier than in northern China. When considering Fig. 11b and c together, the decline in the annual HF over China after the 1980s was primarily due to a decrease in the north, northwest, and plateau regions (Li et al., 2016).

The trend of annual hail days for each station calculated from 1960 to 2012 showed that 37.8% of stations had a significant decreasing trend and 49.1% stations displayed a nonsignificant decreasing trend. In addition, 12% of stations displayed an increasing but nonstatistically significant trend. Compared with the results of a study of hail in China from 1960 to 2005 (Xie et al., 2008), more stations displayed a decreasing trend in annual hail days and there was almost no station that displayed a significant upward trend in the longer-term dataset.

With regard to the spatial distribution, most stations that had a decreasing trend in annual hail days were located in northern China. The maximum in the downward trend of annual hail days was located in the plateau area. The majority of stations in southern China displayed no trend, except for several provinces in southwestern China. Stations located in the plateau and mountainous areas tended to have a downward trend in annual hail days. In general, the trend in annual hail days reduced from northern to southern China and from western to eastern China, which was strongly related to the unique terrain configurations across the country.

The spatial distribution and trend of different hail diameters were different. The mean maximum hail diameter (MHD) increased from west to east and from north to south with obvious regional differences. The station mean MHD records were all less than 20 mm in QT. Most station mean MHD records were below 20 mm and only a small group exceeded 20 mm, in northern mountain area, northwest area, and southern mountainous area. However, most station mean MHD records equal to or exceeding 20 mm were found in southern plain area and northern plain area where the HF pattern was nearly the opposite of that of topographic height. In general, the mean MHD increased from plateaus to mountainous areas and increased further to the plain areas, with a wider range of the mean MHD from higher to lower elevation areas.

Similarly, the maximum value of MHD at each station during 1980–2015 also increased from west to east and from north to south. No station recorded a maximum MHD larger than 100 mm in the central area of the QT during the study period. Stations with maximum MHD equal to or greater than 100 mm were mainly located in southern plain area and northern plain area, distributed irregularly along their edges, near the border with the mountainous areas.

4.2 Vertical Structures for the Hail Storm: Two Cases Over Eastern China

A long-lasting hail-producing MCS occurred over Jiangsu Province in eastern China on 28 April 2015 under the influence of a cutoff low. According to the severe weather reports from the Chinese Meteorological Administration, this MCS was a prolific producer of chicken egg-sized hailstones (3–5 cm in diameter), intense lightning, and damaging surface winds (>23 m s^{-1}). This MCS occurred under a deep, semi-permanent upper-level East Asian trough (EAT), which could be traced back to 26 April 2015. Convection first formed at the northwestern border of Jiangsu Province. The storms organized into an MCS and moved southeastward, sweeping through much of the Jiangsu Province and eventually evolved into a bow echo. This MCS was proceeded by another MCS that developed on the evening of 27 April over eastern China and moved eastward into the East China Sea before the hail-producing MCS developed.

A three-stage conceptual model is proposed in Fig. 12, summarizing the initiation and organization of hail-producing MCS. Before the development of the main over-land MCS, precipitation from a proceeding MCS generates a cold pool off the eastern China coast over the Eastern China Sea (Fig. 12a).

Horizontal convergence and localized upward motion produced at the gust front by the outflow of the rearward/westward spreading cold pool act to tilt and stretch

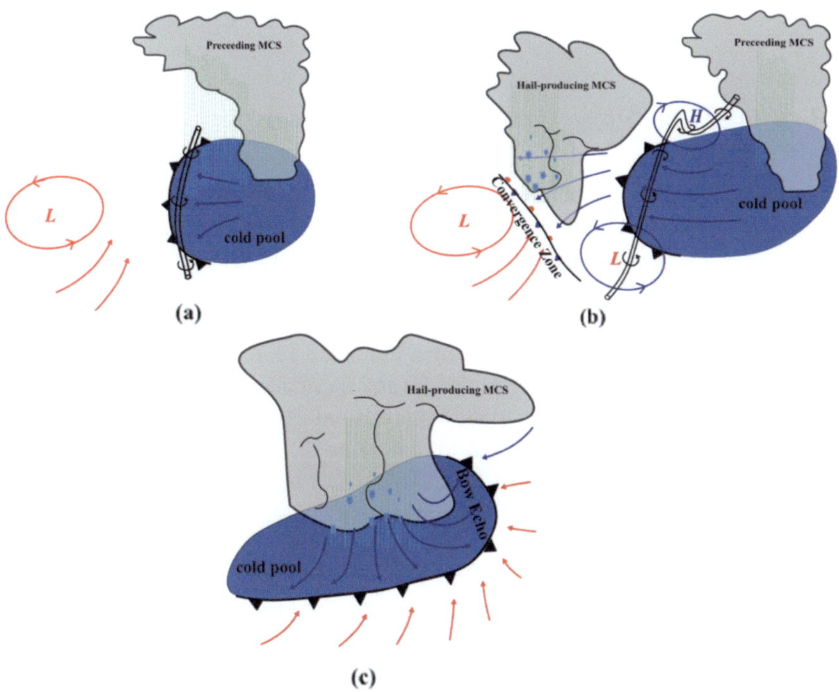

Fig. 12 a–c Conceptual model for the initiation and organization of a long-lasting multicellular hail-producing MCS following a preceding MCS to its east. A vortex tube is drawn in **a** and **b** whose sense of rotation is indicated by arrowed rings around the tube. Surface vortex circulations are drawn as red and blue arrowed circuits, and cold pool is shaded blue. Gust front is drawn like traditional cold front. Representative flows are indicated by curved arrows of different colors. Cited from Luo et al. (2020)

baroclinically horizontal vorticity, creating a couplet of mesoscale vortices on the south and north side of the cold pool (Fig. 12b). Combined with the background cyclonic circulation, the southern cyclonic vortex becomes the most prominent low-level feature over the East China Sea. At this time, another mesoscale low-level vortex exists over land that is associated with a small cutoff low at 500 hPa. Between the southwesterly flows on the southeast side of the overland vortex and the northeasterly winds on the northwest side of the over-ocean vortex is a strong convergence zone along which initial convective cells are triggered and subsequently intensified. On the southwest side of the convergence zone are moist southwesterly flows with moderately large CAPE, providing fuel to the organization of a long-lasting hail-producing MCS (Fig. 12b).

After the initiation, the storms organize into a northwest-southeast line and move southeastward. A cold pool is established underneath the MCS, which aided by cold advection from the East China Sea. The cold pool accelerates the southward and southeastward propagation of the MCS, establishing a west–east orientation along the gust front and later the development of eastward protruding bow at the east end of the convective line (Fig. 12c).

4.3 Cloud and Microphysical Characteristics

The hail microphysical budgets associated with cloud microphysics were examined with three schemes in eyewall and spiral-rainband areas during Typhoon Fitow (2013), that are the National Severe Storms Laboratory (NSSL) (Mansell et al., 2010) scheme, sensitivity test sets the terminal velocity of both graupel and hail in the NSSL (NSSLTest) scheme, and sensitivity test sets the terminal velocity of graupel in the NSSL (NSSLTestG) scheme. Figure 13 depicts vertical cross-sections of graupel melt and the accretion of cloud water by raindrops, and the mixing ratios of hail along the black line passing through the rainfall center in the spiral rainband area. These cross-sections were used to determine the cause for the higher hourly rainfall in the spiral rainband area in NSSLTestG compared to NSSL at 1400 and 1500 UTC on October 6, 2013. The figure shows the presence of hail in the spiral rainband where precipitation had been enhanced from NSSL to NSSLTestG and hail had decreased from NSSL to NSSLTestG accordingly near 29°N (28.5°N) at 1400 UTC (1500 UTC). The maximum mixing ratios of hail in NSSL was more than 100 mg·kg^{-1}, and that in NSSLTestG was about 100 mg·kg^{-1} with the area of graupel reduced.

Hail is produced from the accretion of raindrops by hail (qhlacr), accretion of cloud water by hail (qhlacw), accretion of snow by hail (qhlacs), and graupel converted to hail (qhlcnh). Hail will compete with graupel for cloud water, raindrops, and snow, and the decrease in graupel is beneficial to the growth of graupel and raindrops in regions where hail occurs (Fig. 14a, c). Thus, decreasing the terminal velocities of graupel in NSSLTestG substantially promotes graupel formation and constrains hail formation. This further hinders hail formation, as hail is mainly formed in the

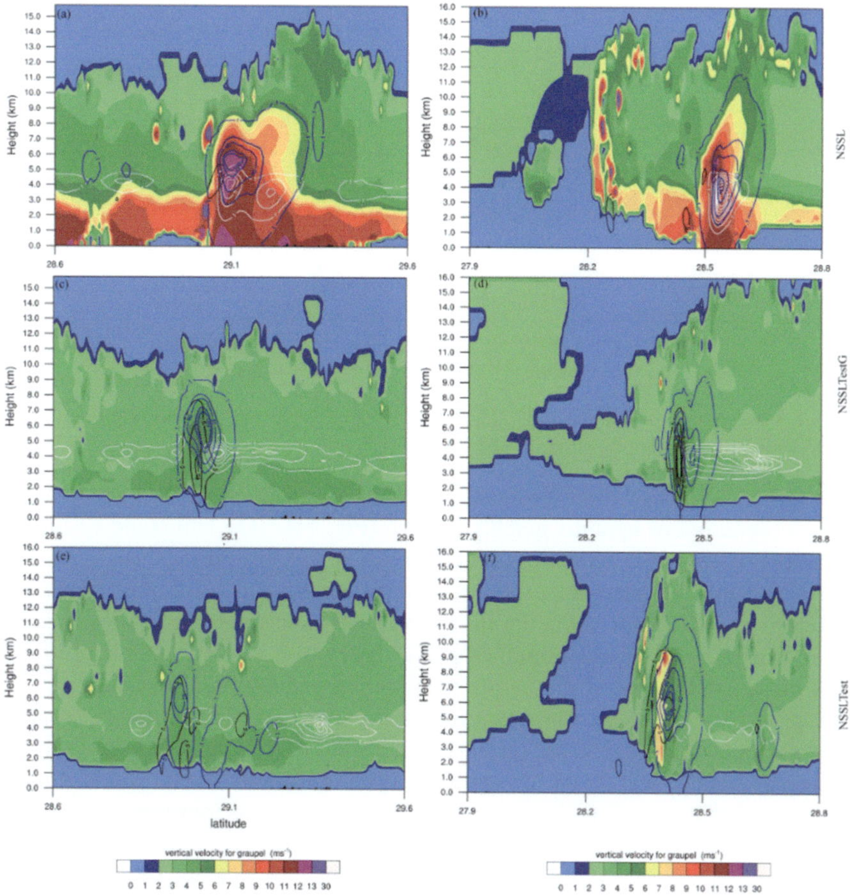

Fig. 13 The vertical cross-section of graupel melt (white contour lines; qhmlr) and accretion of cloud water by raindrop (black contour lines; qracw), and mixing ratios of hail (blue contour lines) along the black line passing through the center of the rainfall intensity in the spiral rainband for the (**a**, **b**) NSSL; (**c**, **d**) NSSLTestG; and (**e**, **f**) NSSLTest schemes at (**a**, **c**, **e**) 1400 UTC, and (**b**, **d**, **f**) 1500 UTC on 6 October 2013. The shaded contour lines denote the terminal velocity of graupel, and the white lines denote the graupel melt. The contour levels for the mixing ratios of hail are respectively 0.1, 20, 40, 80, 100, 200 mg·kg^{-1}·s^{-1}. The contour intervals for graupel melt in NSSL, NSSLTest, and NSSLTestG is 2.0 mg·kg^{-1}·s^{-1}. The contour interval for qracw is 10.0 g·kg^{-1} s^{-1}. Cited from Xu et al. (2021)

updrafts in spiral rainband. However, this scenario differs when both of the terminal velocities of graupel and hail are reduced in the NSSLTest scheme. The situation is more complex when reducing both the terminal velocities of graupel and hail. The simulation in NSSLTest can inhibit hail formation and promote graupel formation when graupel advection out of the updrafts occurs rapidly (4e), the melting of graupel can also inhibit updrafts, leading to decreased rainfall; In contrast, the simulation in NSSLTest can also promote hail formation by consuming graupel when graupel did

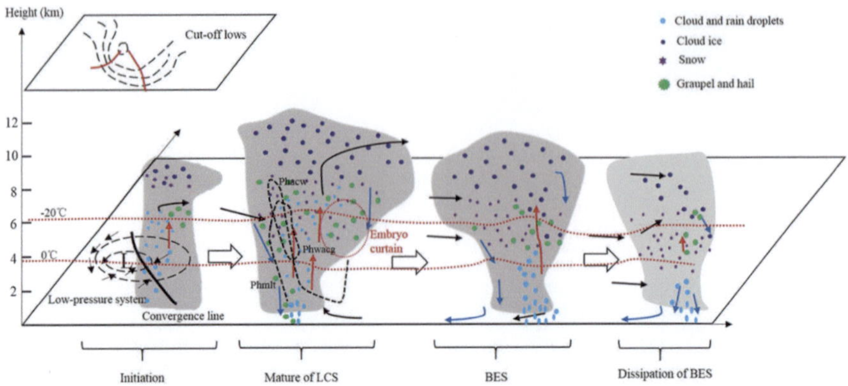

Fig. 14 Conceptual model of the hail-producing convective system that occurred in eastern China on 28 April 2015. Cited from Yin et al. (2021)

not advection out of updrafts quickly enough (Fig. 14f). The hail could be further promoted by accretion raindrops and cloud water, and this process was accompanied with increased latent heat release, and this enhanced upward movements, the enhanced updrafts further favor the hail formation by consuming raindrops and cloud water, leading to decreased rainfall (Xu et al. 2021).

A conceptual model of the hail-producing convective system that occurred on 28 April 2015 in eastern China is presented in Fig. 14. The convective system occurred under the synoptic background of cut-off lows (COLs) in the mid-to-upper troposphere and a mesoscale low-pressure system at lower levels, which formed unstable stratification. Convection triggered near the surface convergence line, and then organized into a northwest-southeast-oriented linear convective system.

In the mature period of the LCS, there were three well developed storms. Taking a strong radar reflectivity in the storm as an example, the low-level inflows in front of the storm ascended strongly in the main updraft region, with a maximum vertical velocity of approximately 28 m s^{-1}, and then flowed out at the upper troposphere. There was a middle-level inflow at the back edge of the storm and it sank significantly around the updraft region. The content of supercooled water above the level of −20 °C was sufficient at this moment, and hail particles were formed primarily through the accretion of liquid water between the heights of 4 and 9 km. Small hail particles with diameters less than 1 mm were located around the perimeter of the main updrafts, where the vertical velocity was not very strong. Large hail particles, meanwhile, with diameters greater than 4 mm, were distributed at the left edge of the main updrafts. Based on a hail growth model, the trajectories of hail particles were calculated. The particles originated from the middle levels (at about 4–5 km) on the right side (east side) of the updraft area, and were transformed westward and downward to the lower levels (at about 1 km). The particles continued to ascend along the left edge of the updraft zone and experienced one or more up-down recycles and fell down to the ground on the left (west) side of the updrafts.

When the storm evolved into a bow echo system, the intensity of the updraft was slightly weaker than that in the LCS stage, and snow and rain water were mainly generated in this stage. During the dissipation period of the bow echo system, the lower levels of the storm were dominated by the downdrafts (Yin et al. 2021).

5 Conclusions

This chapter presents a comprehensive analysis of convective storms in China. The introduction sets the foundation for the subsequent discussions, focusing on the Climatology of MCS across the country. It explores the spatial and seasonal variations of MCS and their organization patterns, providing valuable insights into the prevalence and characteristics of these convective systems in different regions of China. Section 2 reviews microphysical processes, structures and characteristics of convective storms that contribute to extreme precipitation events associated with convective storms. By investigating the mechanisms behind heavy rainfall. We shed light on the factors driving intense rainfalls and their potential impacts on local weather and hydrological systems. Section 3 reviews tornado-producing storms, including the structure, vertical rotation characteristics, and evolutionary processes. We provided crucial knowledge on the occurrence and behavior of tornadoes in China, emphasizing the understanding of the conditions conducive for tornado formation and their potential hazards to communities. Section 4 reviews a comprehensive analysis of hail storms, including the climatology, structure, and microphysical characteristics.

In conclusion, this chapter offers a comprehensive overview for the convective storms in China, covering various aspects such as their climatology, microphysical processes, tornado characteristics, and hail storm dynamics. The insights gained from the present chapter will aid in improving weather forecasting and preparedness for severe weather events, contributing to better disaster management and mitigation efforts in communities. In addition, to understand the patterns and characteristics of convective storms is vital for researchers and meteorologists to anticipate and respond effectively to their potential impacts, thus enhancing the resilience of communities in the face of extreme weather events in China.

References

Bai, L., Meng, Z., Huang, L., Yan, L., Li, Z., Mai, X., Huang, Y., Yao, D., & Wang, X. (2017). An integrated damage, visual, and radar analysis of the 2015 Foshan, Guangdong, EF3 Tornado in China produced by the landfalling Typhoon Mujigae (2015). *Bulletin of the American Meteorological Society, 98*, 2619–2640.

Chen, H., Xu, W., Liu, N., Sun, J., & Fu, J. (2022). Climatologies of mesoscale convective systems over China observed by spaceborne radars. *Monthly Weather Review, 150*, 2697–2717.

Doswell, C. A., III., Brooks, H. E., & Maddox, R. A. (1996). Flash flood forecasting: An ingredients-based methodology. *Weather and Forecasting, 11*, 560–581.

Guan, Y., Zheng, F., Zhang, P., & Qin, C. (2014). Spatial and temporal changes of meteorological disasters in China during 1950–2013. *Natural Hazards, 75*, 2607–2623.

Li, M., Zhang, Q., & Zhang, F. (2016). Hail day frequency trends and associated atmospheric circulation patterns over China during 1960–2012. *Journal of Climate, 29*, 7027–7044.

Li, M., Zhang, Q., Zhang, F., Zhang, D.-L., Sun, J., & Zhang, Q. (2018). A statistical analysis of hail events and their environmental conditions in China during 2008–15. *Journal of Applied Meteorology and Climatology, 57*, 2817–2833.

Li, P., Moseley, C., Prein, A. F., Chen, H., Li, J., Furtado, K., & Zhou, T. (2020). Mesoscale convective system precipitation characteristics over East Asia. Part I: Regional differences and seasonal variations. *Journal of Climate, 33*, 9271–9286.

Luo, L., Xue, M., & Zhu, K. (2020). The initiation and organization of a severe hail-producing mesoscale convective system in East China: A numerical study. *Journal of Geophysical Research: Atmospheres, 125*.

Mansell, E. R., Ziegler, C. L., & Bruning, E. C. (2010). Simulated electrification of a small thunderstorm with two-moment bulk microphysics. *Journal of the Atmospheric Science, 67*, 24.

McCaul, E. W., Jr. (1991). Buoyancy and shear characteristics of Hurricane-Tornado environments. *Monthly Weather Review, 119*, 1954–1978.

Meng, Z., & Yao, D. (2014). Damage survey, radar, and environment analyses on the first-ever documented Tornado in Beijing during the heavy rainfall event of 21 July 2012. *Weather and Forecasting, 29*, 702–724.

Meng, Z., Yao, D., Bai, L., Zheng, Y., Xue, M., Zhang, X., Zhao, K., Tian, F., & Wang, M. (2016). Wind estimation around the shipwreck of Oriental Star based on field damage surveys and radar observations. *Science Bulletin, 61*, 330–337.

Meng, Z., Bai, L., Zhang, M., Wu, Z., Li, Z., Pu, M., Zheng, Y., Wang, X., Yao, D., Xue, M., Zhao, K., Li, Z., Peng, S., & Li, L. (2018). The deadliest Tornado (EF4) in the past 40 years in China. *Weather and Forecasting, 33*, 693–713.

Moller, A., Doswell, C., III., Foster, M., & Woodall, G. (1994). The operational recognition of supercell thunderstorm environments and storm structures. *Weather and Forecasting, 9*, 327–347.

Smith, B. T., Thompson, R. L., Dean, A. R., & Marsh, P. T. (2015). Diagnosing the conditional probability of Tornado damage rating using environmental and radar attributes. *Weather and Forecasting, 30*, 914–932.

Xie, B., Zhang, Q., & Wang, Y. (2008). Trends in hail in China during 1960–2005. *Geophysical Research Letters, 35*.

Xu, H., Zhang, D., & Li, X. (2021). The impacts of microphysics and terminal velocities of graupel/hail on the rainfall of typhoon fitow (2013) as seen from the WRF model simulations with several microphysics schemes. *Journal of Geophysical Research: Atmospheres, 126*.

Xue, M., Zhao, K., Wang, M. J., Li, Z. H., & Zheng, Y. G. (2016). Recent significant tornadoes in China. *Advances in Atmospheric Sciences, 33*, 1209–1217.

Yang, X., & Sun, J. (2018). Organizational modes of severe wind-producing convective systems over North China. *Advances in Atmospheric Sciences, 35*, 540–549.

Yin, L., Ping, F., Xu, H., & Chen, B. (2021). Numerical simulation and the underlying mechanism of a severe hail-producing convective system in East China. *Journal of Geophysical Research: Atmospheres, 126*.

Zeng, Z., & Wang, D. (2022). On the local rain-rate extreme associated with a mesovortex over South China: Observational structures, characteristics, and evolution. *Monthly Weather Review, 150*, 1075–1096.

Zhao, K., Wang, M., Xue, M., Fu, P., Yang, Z., Chen, X., Zhang, Y., Lee, W., Zhang, F., Lin, Q., & Li, Z. (2017). Doppler radar analysis of a tornadic miniature supercell during the landfall of typhoon Mujigae (2015) in South China. *Bulletin of the American Meteorological Society, 98*, 1821–1831.

Wind Characteristics Associated with Turbulence in the Atmospheric Boundary Layer During Severe Storms

Jaya Singh

Abstract Understanding wind properties is very important in the atmospheric boundary layer (ABL) during extreme weather such as monsoons, typhoons, thunderstorms, and so on for the safety of human life and property. The study of wind flow and turbulence in the atmospheric boundary layer is not only a direct interest of meteorological science researchers but also for wind engineering researchers depending on different scales i.e., mesoscale and microscale. This chapter focuses on wind characteristics associated with the turbulence in the atmospheric boundary layer during severe storms and will discuss key aspects for both the research communities. Extreme weather that may bring strong wind and turbulence structure inside the ABL are discussed in detail. Chapter contents are divided into four sections: Sect. 1 Introduction, Sect. 2 Wind Speed during a severe storm, Sect. 3 Mean Wind Speed in the Atmospheric Boundary Layer, Sect. 4 Turbulence Characteristics of Wind Speed. Geostrophic wind and gradient wind, mean wind speed in the ABL, the discrepancy in mean wind speed during severe storms like typhoons, downbursts, tornados, and its turbulent structure inside the ABL, variation of wind speed concerning height above the ground surface, including a description of turbulence intensity, the integral scale of turbulence, power spectrum, and their variations with wind speed, are discussed in detail. Along with this, Typhoon Kai-Tak is mentioned as an illustration of a strong wind event to express the wind characteristics of strong winds.

Keywords Atmospheric motion · Severe storm · Wind speed · Turbulence · Atmospheric boundary layer

---

J. Singh (✉)
Department of Mathematics & Statistics, University of Wyoming, Laramie, WY 82071, USA
e-mail: jayasingh@ucar.edu; jsingh7@uwyo.edu

National Center for Atmospheric Research, Boulder, CO 80305, USA

S. Das and W-K. Tao (eds.), *Severe Storms*,
https://doi.org/10.1007/978-981-97-7075-5_12

1 Introduction

The atmospheric mechanism can be described in the quantities of wind velocity (i.e., horizontal and vertical wind), pressure, temperature, density, and moisture. The characteristics of these quantities are supervised by six equations: (i) the equation of state, (ii) the first law of thermodynamics, (iii) the equation of quantity of mass, (iv) the equation of quantity of moisture, (v) the horizontal equation of motion, and (vi) the vertical equations of motion. After the collection of required data, these equations can be combined to provide a significant illustration of atmospheric conditions. The computed values of the six variables attained by assimilation can then be used as an initial condition for further assimilation step. This consecutive estimation process is the basis of numerical weather prediction techniques that came into being following the increased availability of field measurements, satellites, and the development of modern electronic computers.

The description of the atmospheric motions can be seen as overlapping interconnected flows illustrated by a different range of scales approximately from one millimeter to thousands of kilometers. To evaluate such flows, it is suitable to distribute them according to their horizontal scale. In meteorology, three main groups of atmospheric motion scales are frequently defined: microscale, mesoscale, and synoptic scale. According to the description of (Humphreys, 1940; Byers, 1944; Haltiner & Martin, 1957; Matveev TT67-51,380; Miller, 1971; Neiburger et al., 1973; Goldstein, 1950; Fiedler & Panofsky, 1970),the synoptic scale involves flows with characteristics dimensions exceeding 500 km or so and time scales for two days or more. The microscale involves flows with characteristics dimensions of less than 20 km or so and a time scale of less than one hour. The mesoscale is defined by the dimensions and periods between those characteristics of the microscale and synoptic scale. Wind or Airflow is fundamentally caused by the temperature gradient due to variable solar heating of the earth's surface. It is initiated in a more immediate sense by density difference or pressure gradient between points of equal elevation. Temperature effect i.e., buoyancy effect, and mechanical effect i.e., velocity shear effect are the sources of turbulence production of atmospheric flow. The difference in solar radiation with time constructs the atmospheric boundary layer at different meteorological conditions, i.e., stable, unstable, and neutral conditions. In the atmosphere, a conversion in temperature i.e., an increment in temperature with height near the ground, can lead to air pollution problems such as smog being captured close to the ground surface. The formation of strong winds is closely related to global atmospheric circulations, and also related to smaller-scale temperature differences. However, in the case of strong wind, the fairly high frequency of the appearance of turbulence near the ground surface is one of the criteria that makes the atmospheric boundary layer different from the rest of the atmosphere.

2 Wind Speed During Severe Storms

2.1 Wind Hydrodynamics

Horizontal friction force slows down the moving air parcel on the surface of the earth and forms an atmospheric boundary layer with a vertical velocity profile through the turbulent mixing effect. In the atmosphere, this is the wind regime which is of direct interest to study for meteorological and wind engineering researchers. In the case of neutrally stratified flows, the range of the atmospheric boundary layers varies from a few hundred meters to several kilometers, depending on the wind intensity, roughness of the terrain, and angle of latitude (Simiu & Scanlan, 1986). In a synoptic boundary layer, the wind speed increases with height, and its magnitude at the top of the boundary layer is referred to as the gradient speed. However, outside the boundary layer, the wind flows generally with the gradient speed along the isobars. Inside the atmospheric boundary layer, the forces acting on the wind consist of friction force, pressure gradient force, Coriolis force, and centrifugal force.

The friction force reacts to prevent the movement of an air parcel in the atmosphere. Its direction is opposite to the movement of the air parcel and perpendicular to the Coriolis force.

The pressure gradient force is generated because of the pressure difference in the atmosphere, which performs from the high-pressure area to the low-pressure area. It is a force that accelerates the air parcel due to pressure difference present in the atmosphere (a force per unit mass) and results in wind.

Coriolis force acts upon a moving body in an independently rotating system caused by the earth's rotation. In the northern hemisphere, the Coriolis force acts to the right of the observer facing the flow direction, which results in an anti-clockwise rotation of wind around a low-pressure area.

Centrifugal force is an outward force, which is experienced by a moving object in a circular path away from the center of rotation. Its direction is away from the axis of rotation, and parallel to the axis of rotation.

The forces that are described above co-exist and govern the wind flow movement, depending on the contributions of the weather situation, wind climate, and height.

2.2 Geostrophic Wind and Gradient Wind

The pressure gradient force, the Coriolis force, and the centrifugal force are dominant forces above the surface of the earth where airflow can be assumed in friction-free conditions. Geostrophic wind is a wind that isparallel to isobars. In other words, the geostrophic wind is the resultant wind from the balance between the Coriolis force and the pressure gradient force. Hence, the geostrophic wind speed U_G is defined by the given equation (Cao, 2013)

$$-\frac{dp}{dn} + \rho_a U_G f_c = 0 \tag{2.2}$$

where ρ_a is the air density, f_c is the Coriolis parameter, and it is defined by f_c= 2Ω sinϕ = 1.454×10^{-4} sinϕ (rad/s), Ω is the earth's rotational speed, and ϕ is the latitude. In the northern hemisphere, the Coriolis force rediverts the winds toward the right. An evident force can be experienced by a body moving relative to the rotating earth. The wind blows from the poles toward the equator and turns to the right in the Northern Hemisphere and the left in the Southern Hemisphere because of this Coriolis effect.

The centrifugal forces are generated when the wind flow develops curvature, for example near the center of a tropical cyclone, which is equal to $\rho_a U_g^2/r$, where r is the radius of the curvature of isobars. During a cyclone when a low-pressure area is created, the centrifugal force poses in the same direction as the Coriolis force i.e., opposite to the pressure gradient force. In the Northern Hemisphere, this result is in the anti-clockwise rotation of wind around a low-pressure area. In this situation, the gradient wind speed U_g is defined by the given equation—

$$-\frac{dp}{dr} + \rho_a U_g f_c + \rho_a \frac{U_g^2}{r} = 0 \tag{2.3}$$

When the wind flow curves are in an anti-clockwise direction around a low-pressure area in this situation the gradient force U_g is smaller than the geostrophic wind speed U_G.However, in another case, it is larger than the geostrophic wind speed when the wind curves are clockwise around a high-pressure area.

Wind speed and the pressure gradient force are high within a tornado or cyclone eye where the radius of curvature of the isobar is small however, in this process the Coriolis force is small compared with the centrifugal force and thus can be neglected. Hence, this wind is balanced between the pressure gradient force and the centrifugal force, and it rotates around a low-pressure area. Such wind is known as a cyclo-strophic wind, whose velocity is defined by the given equation—

$$\frac{1}{\rho}\frac{\partial p}{\partial n} = \frac{v^2}{r} \tag{2.4}$$

2.3 Wind Flow Categories

There are three main categories of wind flow or air i.e., mean wind, waves, and turbulence shown in Fig. 1. Each category can happen separately or in the presence of others, and all these exist in the atmospheric boundary layer, where moisture, heat, momentum, and pollutants are dominated horizontally by the mean wind however, in the vertical by turbulence.

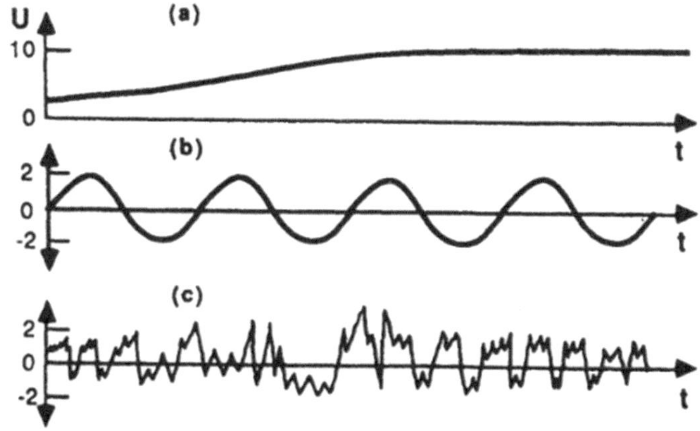

Fig. 1 **a** mean wind speed, **b** waves, and **c** turbulence, where U is the wind component in the x direction (Stull, 1988)

The advection process or rapid horizontal transport is caused because of the mean wind. In the atmospheric boundary layer, the horizontal mean winds of the order of 2–10 m/s are commonly present. Due to friction, mean wind speed slows down near the ground of the earth's surface. Despite that, vertical mean winds are considered much smaller, and it is usually of the order of millimeters to centimeters per second.

At night-time, the atmospheric boundary layer is very stable because of limited transport of heat, humidity, and other scalars for example pollutants thus waves can be observed in the atmosphere. However, these waves are effective at transporting momentum and energy and can be generated locally by mean-wind shear, and by mean flow over obstructions. Waves can also be generated from some distant source or event, for example, a thunderstorm or an explosion.

Turbulence which has a very high frequency is one of the most important wind categories that make the atmospheric boundary layer different from the rest of the atmosphere. Turbulence is also initially found in convective clouds, and near the jet stream where strong wind shear can create clear air turbulence. Sometimes turbulence can propagate due to atmospheric waves that may enhance the wind shear in localized regions. So, it can be noticed that wave phenomena can be associated with the turbulent transport of heat and pollutants.

An ordinary practice to study turbulence or waves is to divide meteorological variables such as temperature and wind into two parts i.e., (i) mean part, and (ii) perturbation part. The mean part reflects the effect of the mean temperature and mean wind, while the perturbation part represents either the wave effect or the turbulence effect that is overlapped on the mean wind (Stull, 1988).

2.4 *Strong Wind Associated with Severe Storms*

Many severe storms might result in strong winds such as monsoons, hurricanes/typhoons/cyclones, thunderstorms (downbursts), tornados, and so on.

2.4.1 Monsoon

Monsoons are seasonal winds, considered as the large-scale winds that form cells of the general circulation and develop around thermally produced continental in high, and low in the summer season. The effect of monsoons can be observed mostly in Asia as they have a substantial influence on the seasonal changes of weather. The period of monsoon and its influenced area may be longer than any other severe storm like a hurricane (typhoon/cyclone), thunderstorm, tornado, and so on. They are characterized by four parts: warm, rainy summer monsoon, cold, and dry winter monsoon.

2.4.2 Hurricane/Cyclone/Typhoon

Hurricanes also known as tropical cyclone is intense storms that generate all their energy from the latent heat released by the condensation process of water vapor, and originate over the oceans, mainly in summer and early autumn. Depending on a different definition, it is known as a typhoon in the East, a cyclone in Australia, and the Indian Ocean, and a hurricane in America. For instance, it is observed that a typhoon whose maximum mean wind speed near the center is greater than 17 m/s, and its diameter is between 100 and 2,000 km, usually of the order of several hundred kilometers. Also, the depth of the atmosphere involved is about 10–12 km. It can be noticed that the entire rotating body of a typhoon is like a compact disc from the ratio of the diameter to its thickness. The shape of the "eye" can be observed in the center of the storm, and the air inside is relatively dry and light, which arises gradually near the perimeter of the eye wall. A vortex is generated in which warm, moist air is convective at high altitudes, and forms tall convective clouds. Also, spiral rain band clouds can be often seen on the outside of the vortex (Fig. 2).

2.4.3 Thunderstorm

The formation of tall convective clouds in the atmosphere produced by the upward motion of warm, moist air is a necessary condition for the occurrence of thunderstorms. Such motion can be started by thermal instability on the mountain slopes or of a front. There are three classifications of thunderstorms i.e., thermally convective, orographic, and frontal. As the air moves upward, it cools, condenses, and forms

Fig. 2 Satellite image of Typhoon Kai-tak from August 2012 on NASA's Earth Observatory website (courtesy NASA)

cumulonimbus clouds. Many severe kinds of weather are associated with thunderstorms for their development and formation like downburst winds, heavy hailstorms, flash flooding, tornadoes, and waterspouts. The squally wind can be observed when the cold downdraft spreads over the ground in the form of a wall jet. The life cycle of thunderstorms associated with strong downdrafts is called the mature stage and usually lasts from 5 to 30 min (Mize & Cox, 1968). A schematic diagram of a thunderstorm cell in its three stages has been shown in Fig. 3. During the thunderstorm process, the sharp wind speed increases called as first gust, and it is associated with the transit of the discontinuity zone between the cold downdraft and the surrounding wind.

2.4.4 Tornado

The most powerful and destructive winds are contained during Tornado and consist of a vortex of air, typically of the order of 300 m in diameter, and moves concerning the ground with a speed of the order of 300–100 km/h in a path, approximately 15 km long (Duchene-Marullaz, 1972). Sometimes, the maximum tangential speed of tornados has been reported to be of the order of 350 km/h or may be higher. Tornadoes can be seen as funnel-shaped, as shown in Fig. 4. Its tangential speed is probably higher at the funnel edge and drops down towards the center and with increasing distance outside the funnel.

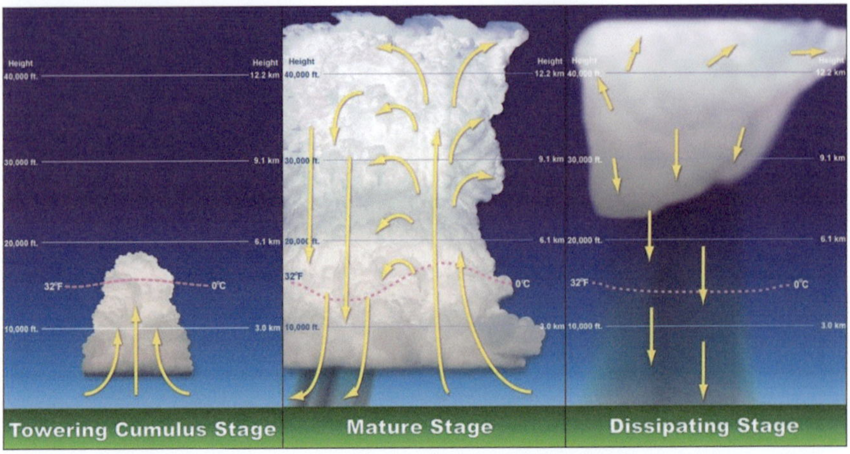

Fig. 3 Three stages of a thunderstorm (i) towering cumulus, (ii) mature, and (iii) dissipating (courtesy NOAA)

Fig. 4 Tornado funnel (courtesy NOAA)

Tornadoes have been reported frequently in the United States, and much less frequently in Australia, Western Europe, India, and Japan. In Japan, tornadoes are known as "Tatsumakis". The typical diameter for tatsumakisis observed in the order of 50 m. Its forward speeds are the order of 40–50 km/h, which are generally towards the northeast direction, and its maximum tangential speed is probably 200 km/h (Lieblein, 1953).

2.5 Atmospheric Boundary Layer

A friction-free wind cannot be observed near the earth's surface because of friction force due to the ground surface obstacles. In the atmospheric boundary layer, this friction force is perpendicular to the Coriolis force, and these two forces are balanced with the pressure gradient force. Thus, it can be seen that the wind direction cannot be parallel to the isobars, and shall turn to the left in the northern hemisphere. These friction force effects become more significant at a lower altitude, and the inclination angle of the wind direction from the isobar becomes larger near the ground surface. As the wind speed is reduced near the ground, wind direction also tends to shift anti-clockwise, and the tip of the velocity vector traces a spiral, which is called the Ekman spiral.

In the atmospheric boundary layer, the centrifugal force is usually small, and the wind flow is balanced among the pressure gradient force, Coriolis force, and the friction force, hence—

$$2\Omega V sin\varphi = -\frac{\partial P}{\partial x} + \frac{1}{\rho}\frac{\partial \tau_x}{\partial z} \tag{2.5}$$

$$2\Omega U sin\varphi = -\frac{\partial P}{\partial y} + \frac{1}{\rho}\frac{\partial \tau_y}{\partial z} \tag{2.6}$$

where, $\tau_x = -\rho\overline{u\prime w\prime}$ and $\tau_y = -\rho\overline{v\prime w\prime}$ are the Reynold stresses.

Considering the geostrophic wind (U_G, V_G) can be written in the following forms—

$$U_G = -\frac{1}{\rho f_c}\frac{\partial P}{\partial y} \tag{2.7}$$

$$V_G = \frac{1}{\rho f_c}\frac{\partial P}{\partial x} \tag{2.8}$$

and thus, the following equation can be obtained—

$$-f_c(V - V_G) = \frac{1}{\rho}\frac{\partial \tau_x}{\partial z} \tag{2.9}$$

$$f_c(U - U_G) = \frac{1}{\rho}\frac{\partial \tau_y}{\partial z} \tag{2.10}$$

Here Reynolds stress, $\tau = -\rho\overline{uw} = \rho\varepsilon\frac{\partial U}{\partial z}$ (where ε = Kinetic Eddy Viscosity), that is related toReynolds stress with the mean velocity profile, and the mean velocity can be estimated.

3 Mean Wind Speed in the Atmospheric Boundary Layer

As we discussed previously, friction force on the Earth's surface applies to the moving air parcel and generates turbulent mixing throughout the region, which later on forms the Atmospheric Boundary Layer. In the case of neutrally stratified flows, the depth of the boundary layer varies from a few hundred meters to thousands of kilometers, depending upon the roughness of the terrain, angle of latitude, and wind intensity. In the boundary layer, the wind speed increases with height, and its magnitude at the top of the boundary layer is generally known as the gradient speed. However, in the free atmosphere, the wind speed is approximate with the gradient speed along the isobars.

3.1 Mean Wind Speed in General ABL

The study of wind flow in the atmospheric boundary layer is not only a direct interest of meteorological science researchers but also to wind engineering researchers. Several studies have been done based on theoretical, and experimental data that include a description of mean wind profiles, different wind speeds in different roughness regimes, and the structure of atmospheric turbulence. It is considered that in strong winds associated with turbulence, mechanical turbulence dominates the heat convection, so the turbulent mixing tends to generate neutral stratification. It is just the same as in the shallow layer of incompressible fluid mixing tends to produce an isothermal state. In general, to calculate the mean wind speed profile in strong wind conditions, mathematical models i.e., the Logarithmic Law and the Power Law models are used, and descriptions are given below—

(i) The Logarithmic Law

According to the Logarithmic Law, the wind speed profile is defined as—

$$U = \frac{u_*}{k} \ln\left(\frac{z - d}{z_0}\right) \tag{3.1}$$

where z is the height above the surface, z_0 is the roughness length, U is the mean wind speed, u_* is the friction velocity, k is the Karman constant (~0.4), and d is the zero-plane displacement. The value of z_0 i.e., the roughness length is varied and depends upon the roughness ground. For this equation, generally uniform roughness condition is required. The Logarithmic Law of wind speed profile is generally considered at the lowest i.e., 100 m of the atmospheric boundary layer.

(ii) The Power Law

The wind speed profile according to the Power Law is defined as—

Table 1 Parameters for mean wind speed profile (Cao, 2013)

Terrain	Roughness length Z0 (m)	Power law index α	Zero-plane displacement d(m)
Sea, mudflats, snow cover flat land etc	0.000–0.003	0.1–0.13	0
Open flat countryside, fields with crops, fences and few trees etc (meteorological standard condition)	0.003–0.2	0.14–0.2	0.1
Dense woodland, domestic housing, suburban area	0.2–1	0.2–0.25	5
City	1–2	0.25–0.3	10
Large city center	2–4	0.3–0.5	20

$$U = U_r \left(\frac{z - d}{z_r} \right)^{\alpha} \tag{3.2}$$

where, U_r is the wind speed at the reference height z_r, α is the power law index, z is the height above the surface, d is the zero-plane displacement, and U is the mean wind speed. It is considered that from 100 m to the top of the atmospheric boundary layer wind speed profile of the power law is more accurate. The roughness length, power law index, and zero-plane displacement data have been shown in Table 1 depending on the different terrain categories.

3.2 Mean Wind Speed in Tornado

Tornados are often accompanied either by a cold front or a tropical cyclone (during the development of depression) and produce the strongest winds, whose peak gust value can be more than 100 m/s. It contains the most powerful of all winds. During the typhoon process, warm air flows into the front of the typhoon, and the atmosphere becomes very unstable. It is noticed that a developed cumulonimbus cloud can be the cause of a severe tornado or downburst, and it is generally difficult to identify damage. The Fujita scale from F0 to F5 is the representation of tornado damage, and it is seen that approximately 90% of all tornadoes are categorized as F3 or less (Letchford & Kwok, 2004). The estimated mean wind speed corresponding to the Fujita Scales is described in Table 2. Several practices have been done, and are still going on in understanding the vorticity source responsible for tornado genesis in the wind tunnel laboratory, the Computational Fluid Dynamics model, or the Weather Research and Forecasting model. A recent numerical study has been done of a “Tsukuba supercell tornado” that occurred on May 6th, 2012, and hit Tsukuba city, Japan based on F3 supercell to understand its vorticity sources using the Weather Research and Forecasting model incorporated with the Large Eddy Simulation (Tao & Tamura, 2020).

Table 2 Fujita scale for tornadoes (Letchford & Kwok, 2004)

Fujita scale	Damage description	Approximate wind speed (m/s)
F0	Light	18–32
F1	Moderate	33–50
F2	Considerable	51–70
F3	Severe	71–91
F4	Devastating	92–115
F5	Incredible	116–141

3.3 *Mean Wind Speed in Downburst*

In the case of a downburst, it is difficult to define an appropriate mean wind speed. However, using several techniques like filtering/conditional averaging can be an approach to estimate the time-varying mean wind speed describing the wind flow, and turbulence component because of convective and mechanical turbulence present in the atmospheric boundary layer. In Australia and USA, there are some instrument measurements using radars, and anemometer towers to describe a clear vision of the structure of a downburst. A basic structure of a downburst is a low-level jet where a maximum wind speed is reached at between 50 and 100 m, over which the mean wind speed is gradually reduced to the wind flow (Letchford & Kwok, 2004). Numerical experiments represented that the transient wind speed can be simulated best by a rolling vortex generated by the pulsed flow instead of the steady wall jet impacting the ground (Letchford et al., 2002; Mason et al., 2003). Several studies have been conducted to estimate the wind speed in a severe downburst, and the maximum wind speed after hitting the surface spreads out in all directions at approximately 240 km/h, as strong as a tornado and often producing damage. A mathematical model has been proposed to estimate the mean wind speed profiles in a steady wall jet, in which the horizontal wind speed component is described as (Oseguera & Bowles, 1988)—

$$U = \left(\frac{\lambda R^2}{2r}\right)\left\{1 - exp\left[-(r/R)^2\right]\right\}\left[exp(-z/z^*) - exp(-z/\varepsilon)\right] \tag{3.3}$$

where R is characteristic of the radius of the jet shaft, r is the radial coordinate from the center of the wall jet, z is the height above the ground, z^* is the characteristic height out of the boundary layer, λ is the scaling factor (per second). Figure 5 is an example of a wind speed profile in a downburst at the radius of maximum wind speed (Holmes, 2001).

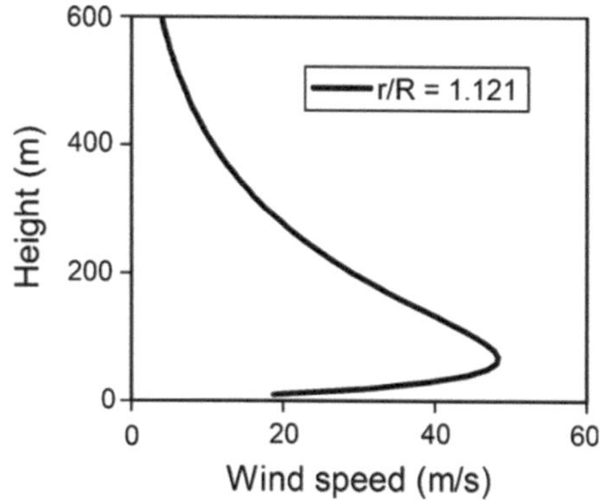

Fig. 5 Mean wind speed profile in a downburst (Holmes, 2001)

3.4 *Mean Wind Speed in Typhoon*

As mentioned previously, Typhoon Kai-Tak will be discussed here as an illustration of a strong wind event to express the wind characteristics of strong winds. The Weather Research and Forecasting (WRF) model has been widely utilized to simulate weather or air conditions because its computing core, software environment, and physical interface fit well with boundary layer simulation. The advantages of modular-based design and flexibility make WRF applicable to many scientific extension trials such as downscaling the nesting grids to a Large Eddy Simulation (LES) model level. We consider that the WRF model will be applied over a wide range of scientific and practical issues including typhoon wind speed warning at the local community level because it can be run on a moderate-level server and offers effective and feasible operation. Numerical simulation of the Kai-Tak typhoon has been performed using the Advanced Research WRF (ARW) model, version 3.7.1 (Skamarock & Klemp, 2008), which is based on fully compressible non-hydrostatic equations with a hydrostatic option. A time-split 2nd- or 3rd-order Runge–Kutta integration scheme is available for time advancing, in which an intermediate small step contributes to the stabilization of high-frequency turbulence.

Typhoon Kai-Tak (No. 1213) landed at Zhanjiang City, China with a track shown in Fig. 6a. The wind speeds of Kai-Tak were recorded by the China Meteorological Administration Public Meteorological Service Center with several propeller-type anemometers installed on a meteorological tower at heights 100, 70, 50, and 30 m above the ground at Donghai Island in Fig. 6b. The data obtained were 10-min-mean wind speeds. The meteorological tower was about 6.5 km north of the typhoon track. Nested multi-scale WRF model simulations are applied to reproduce the typhoon track and wind speed. In Fig. 6c, the typhoon simulation consists of 6 nested domains d01, d02, d03, d04, d05, and d06, each at the center of its parent domain. The nested WRF simulation of the present study adopts different turbulence treatments in two groups, using a PBL scheme in the first three domains (d01, d02, d03) and the LES model in the three inner domains (d04, d05, d06). The precise calculation of computational domain settings and physical schemes are given in the reference (Singh et al., 2020).

Figure 6d, the simulated typhoon track obtained by the first three domains (i.e., d01, d02, and d03), and compared with the observation data. The simulated tracks describe north-westerly movement during the entire integration period of the typhoon process. The non-nudge track shows that the simulated typhoon track deviated much further north compared to the observation track. The nudge track produces a much better position than the typhoon track and significantly reduces the deviation concerning its direction compared with non-nudge results. This suggests that the vortex has been assimilated well into the given initial condition.

Figure 7, illustrates the time series of 10-min mean wind velocities at heights of 30 m in Fig. 7a and 100 m in Fig. 7b. The wind speed history obtained in the simulation reproduces the tendency of velocity change with time at both heights, which shows that applying multi-scale nesting is beneficial in calculating the actual wind speed. However, significant differences can be found when the typhoon's eye is passing. When the typhoon eye passes, the simulated velocity plunges from approximately 23 to 8 m/s and from 28 to 9 m/s, followed by sharp velocity increases at heights of 30 and 100 m, respectively. This symbolic M-shaped variation tendency of mean velocity is not presented by field measurement. The main target of this study is the generation of micro-scale turbulence, so no further attempts are made to adjust the typhoon simulation.

Figure 8, presents the vertical profiles of mean velocity at two wind speed levels at different times, in which both wind velocity (a) at low wind speed and (b) at high wind speed profiles are comparable with observation data. In addition, it is shown that both vertical profiles of mean velocity at low and high wind speed conditions obey the logarithmic law within the height region where the mean velocity increases monotonically with height. In addition, the boundary layer height near the typhoon eye is about 400 m, which is lower than that before the typhoon eye arrived.

4 Turbulence Characteristics of Wind Speed

Turbulence is one of the small-scale features compared to the other scales of meteorological motions. Figure 9 represents the typical time and scale order of magnitude for micro and mesoscales (Orlanski, 1975). The turbulence which has a spatial scale smaller than 3 km, and in time scale it is shorter than 1 h or 10 min is noteworthy as microscale. Such small-scale features study is part of micrometeorology, and this microscale process involves the atmospheric boundary layer. Turbulence is so transient that the deterministic description and forecasting for an individual eddy are very difficult (Stull, 1988). Some stochastic methods can be explored to study the effect of the eddies. Turbulence characteristics are often associated with surface-layer phenomena near the earth's ground surface approximately 100 m in the atmospheric boundary layer.

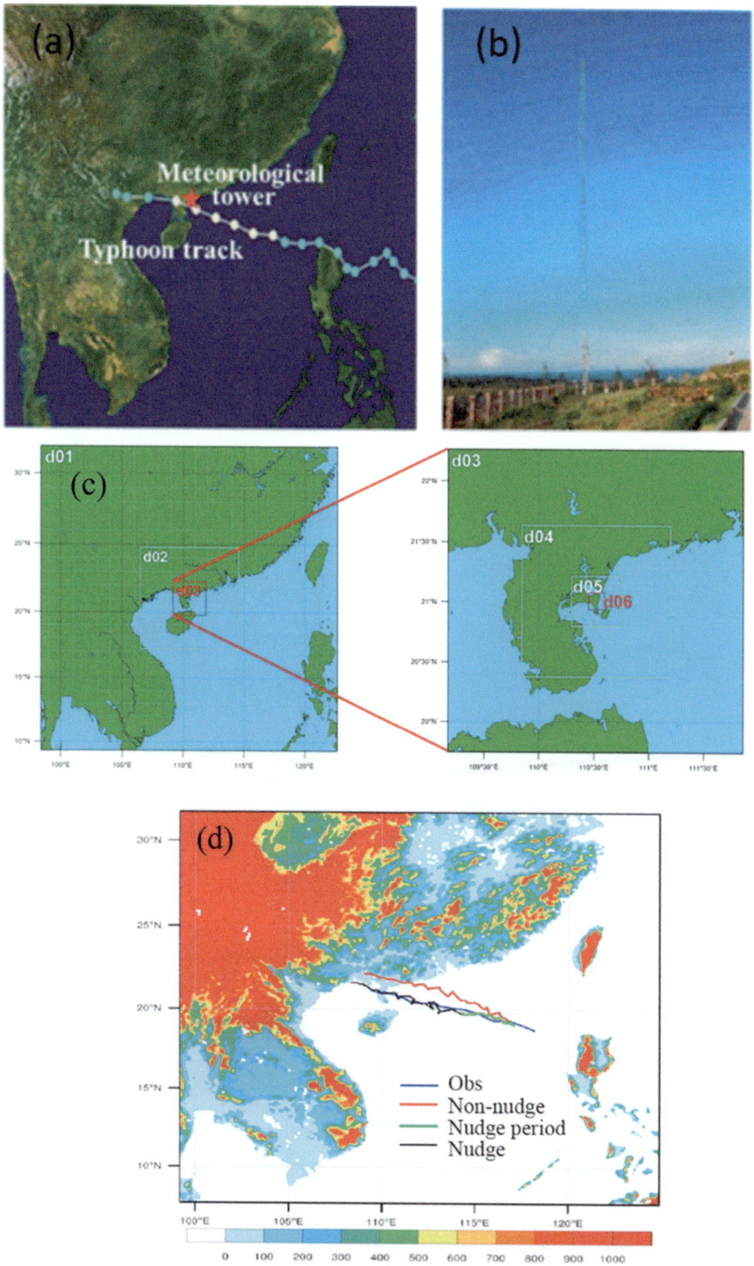

Fig. 6 **a** Track of Typhoon Kai-Tak, **b** Wind speed measurement at Donghai Island (Singh et al., 2020). **c** Layout of the nested domain (d01, d02, d03 with a PBL scheme; d04, d05, d06 with an LES model (Singh et al., 2020). **d** Kai-Tak Typhoon track obtained by first three domains (Singh et al., 2020)

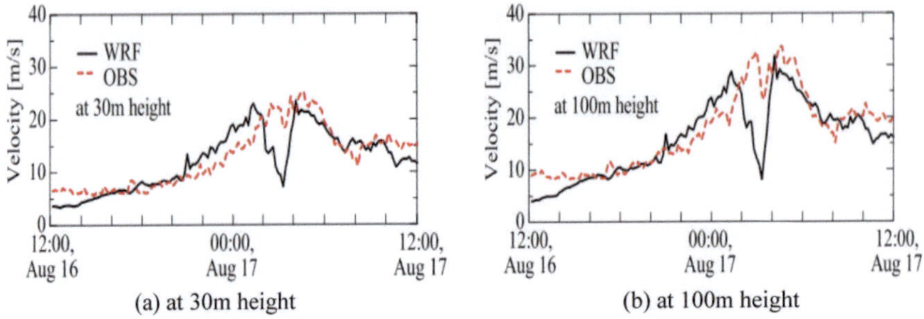

Fig. 7 Comparison of time-series of wind speed between simulation, and observation (Singh et al., 2020)

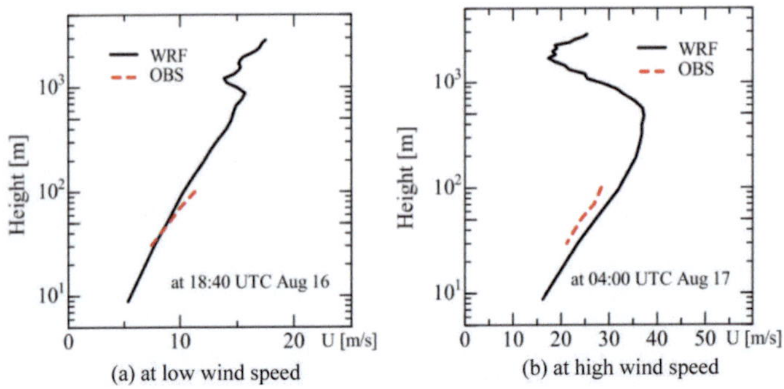

Fig. 8 Comparison of wind speed profile between simulation, and observation (Singh et al., 2020)

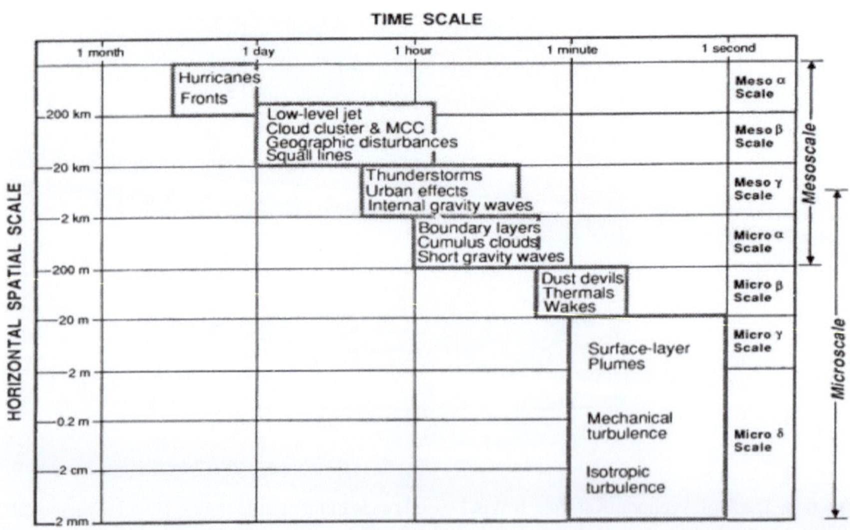

Fig. 9 Typical time and space orders-of-magnitude for micro and mesoscales (Orlanski, 1975)

4.1 Turbulence Intensity

Turbulence intensity is the simplest descriptor of atmospheric turbulence. Let, u(z) denote the wind fluctuations parallel to the direction of the mean wind speed in a turbulent flow passing a point with height z then the longitudinal turbulence is defined as (Simiu & Scanlan, 1986)—

$$I(z) = \frac{\overline{u^2(z)}^{1/2}}{U(z)} \tag{4.1}$$

where, U(z) is the mean wind speed at height z, and $\overline{u^2(z)}^{1/2}$ is the root mean square value of u. Similarly, lateral and vertical turbulence intensity can be defined.

A lot of studies have been done on turbulence intensity data evolution in the typhoon process (Zhao et al., 2019). The profiles of turbulence intensity have been discussed with the applicability of the cell perturbation method in the study of "Numerical generation of inflow turbulence by cell perturbation technique in WRF simulation" for both low and high wind speed stages (Singh et al., 2020). Figure 10a, b present the variations of turbulence intensity profiles with perturbation amplitude obtained from three perturbation cell sizes in the low and high wind speed stages, respectively. It is clearly shown that the turbulence intensity increases with perturbation amplitude as well as cell size. These results provide us with practical references for setting an appropriate combination of perturbation amplitude and cell size to achieve the specified turbulence statistics.

4.2 Turbulence Scale

In the mean wind speed, the velocity fluctuations pass a point can be transported by a superposition of ideal eddies. Each eddy can be defined at that point as a periodic fluctuation with circular frequency $\omega = 2\pi n$, where n is the frequency. An eddy wavelength can be defined as $\lambda = U/n$, where U = wind speed, and the eddy wave number, $k = 2\pi/\lambda$. The wavelength is the measure of eddy size.

The turbulence scale is defined from the average size of turbulent eddies of the wind flow. There are nine turbulence scales corresponding to the three-dimensional eddies, associated with longitudinal, transverse, and vertical components of the turbulent velocity u, v, and w. For instance, L_u^x, L_u^y, and L_u^z are respectively measures of the average longitudinal, transverse, and vertical size of eddies associated with the longitudinal velocity fluctuations (here, x is the direction of the mean wind U, and u is the longitudinal fluctuations). Mathematical expression of L_u^x can be defined as (Simiu & Scanlan, 1986)—

$$L_u^x = \frac{1}{\overline{u^2}} \int_0^\infty R_{u_1 u_2}(x) dx \tag{4.2}$$

A

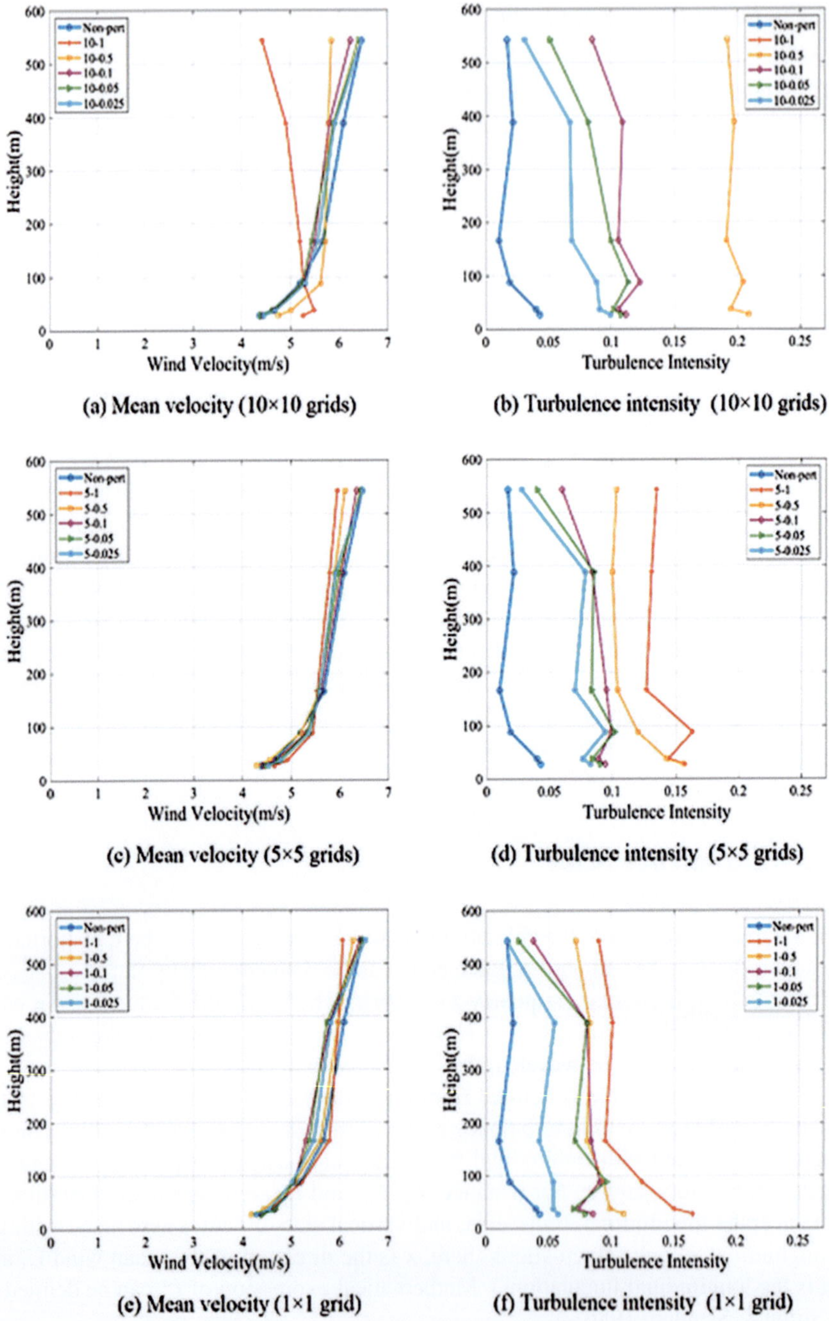

(a) Mean velocity (10×10 grids)

(b) Turbulence intensity (10×10 grids)

(c) Mean velocity (5×5 grids)

(d) Turbulence intensity (5×5 grids)

(e) Mean velocity (1×1 grid)

(f) Turbulence intensity (1×1 grid)

Fig. 10 **A** Variations of mean velocity, and turbulence intensity profiles with perturbation amplitude, and cell size at low wind speed stage (Singh et al., 2020). **B** Variations of mean velocity, and turbulence intensity profiles with perturbation amplitude, and cell size at high wind speed stage (Singh et al., 2020)

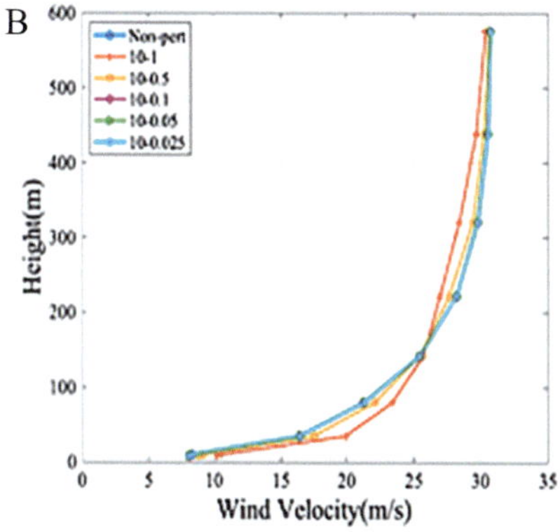

(a) Mean velocity (10×10 grids)

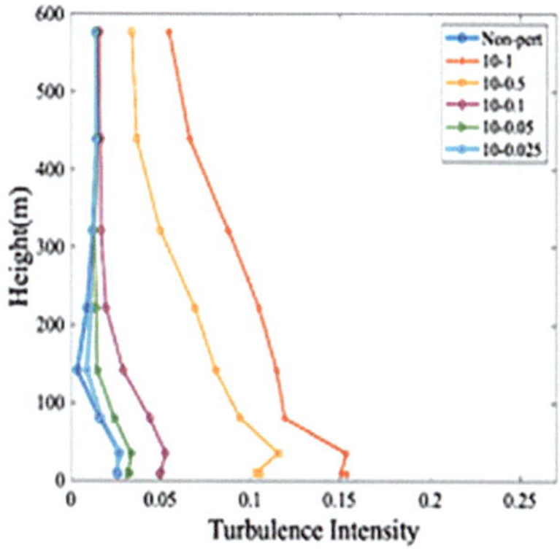

(b) Turbulence intensity (10×10 grids)

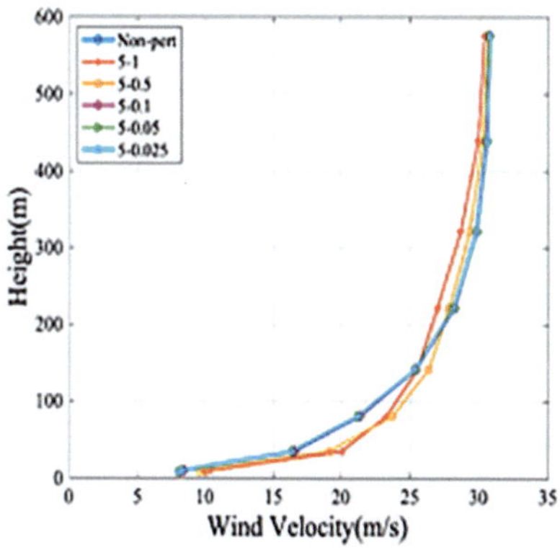

(c) Mean velocity (5×5 grids)

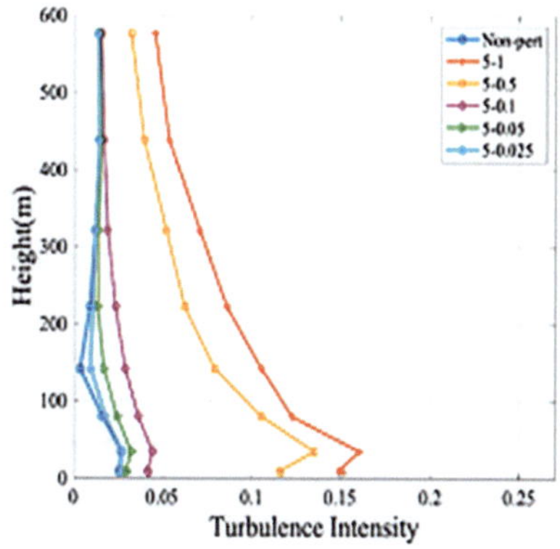

(d) Turbulence intensity (5×5 grids)

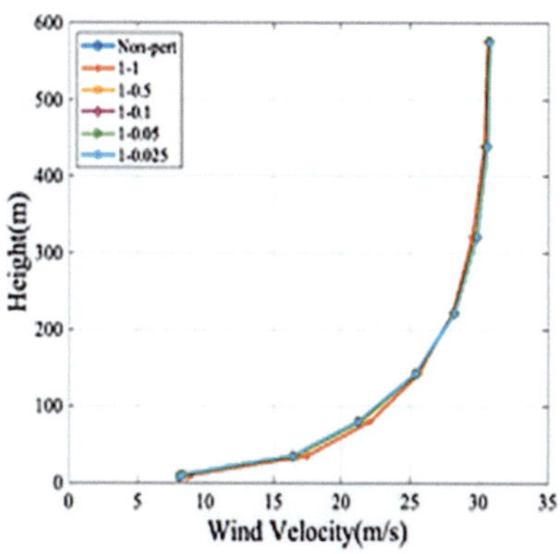

(e) Mean velocity (1×1 grid)

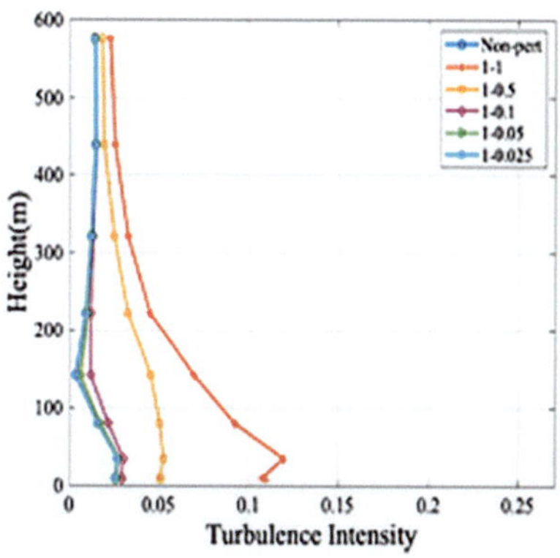

(f) Turbulence intensity (1×1 grid)

Fig. 10 (continued)

where, $R_{u_1u_2}(x)$ is the cross-covariance function of the longitudinal velocity components, $u_1 = u(x_1, y_1, z_1, t)$, and $u_2 = u(x_{1\pm}x, y_1, z_1, t)$ here t = time, $\overline{u^2}$ is the root mean square value of u_1 (and u_2). In the case of horizontally homogenous flow, L_u^x is independent of x_1, and y_1. Similarly, other turbulence integral scales can be defined.

4.3 Power Spectrum of Micro-meteorological Turbulence

In micro-meteorological turbulence, wind speed can be represented by the phenomena in which different sizes of eddies are transported by the mean flow both in time and in space. Large-size eddies present in long-period fluctuations known as low-frequency. However, small eddies present in short-period fluctuations known as high-frequency. A power spectral density function is a mathematical equation that describes the distribution of turbulence energy with frequency. It is also known as the "power spectrum".

In the case of micro-meteorological turbulence, the power spectrum schematically consists of three parts: production range, inertial range, and dissipation range. In the production range, the turbulence is generated as large eddies as a part of low frequency due to instabilities of the mean wind speed. In the inertial range, the large eddies break and transfer their momentum to the small eddies, then this sequence will go on to smaller ones, and so on. In the dissipation range, as the eddies become so small a part of high-frequency viscosity becomes predominant, and their energy is dissipated in terms of heat. Many mathematical expressions have been derived to express the power spectrum of micro-meteorological wind speed turbulence. A general mathematical expression of the power spectrum is given by the Fichtl and McVehill model (Cao, 2013)—

$$\frac{fS_u(f)}{\sigma_u^2} = \frac{4f^*}{(1 + \alpha f^{*\beta})^{\frac{5}{3\beta}}} \tag{4.3}$$

where,

$$f^* = \frac{fL_{ux}}{U}, \alpha = 1.5\frac{4^\beta}{b^\beta}, b = \frac{1.5^{\frac{1}{\beta}}\beta\Gamma(\frac{5}{3\beta})}{\Gamma(\frac{1}{\beta})\Gamma(\frac{2}{3\beta})}$$

And $\Gamma(-)$ is the Gamma function. It can be seen that,

$$S_u(f) \propto f^{(-\frac{5}{3})} \tag{4.4}$$

when f $\to \infty$. It is known as Kolmogorov's hypothesis of local isotropy in the inertial sub-range where the power spectrum follows a power of 5/3 law.

In the case of the Karman-type power spectrum, the value of $\beta = 2$. The Karman-type power spectrum is expressed as follows—

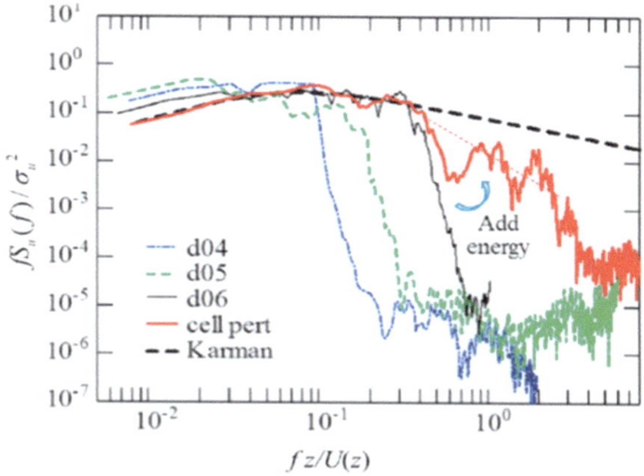

Fig. 11 Comparison of power spectra obtained in domains d04, d05, and d06 by LES with cell perturbation technique in d06 (Singh et al., 2020)

$$\frac{fS_u(f)}{\sigma_u^2} = \frac{4(\frac{fL_{ux}}{U})}{[1 + 70.8(\frac{fL_{ux}}{U})^2]^{\frac{5}{6}}} \tag{4.5}$$

Many researchers adopt the Karman-type spectrum as the expression of the wind spectrum. The power spectra can be well described by Karman-type spectra.

Power spectrum analysis is carried out to investigate the spectral energy distribution of wind speed fluctuation obtained by LES without and with the cell perturbation technique. The velocity data in the last 10 min of the one-hour data are utilized for spectrum analysis. Figure 11 compares the power spectra of wind speed fluctuations obtained at the center points of domains d04, d05, and d06 by LES without cell perturbation and the power spectra at the center point of domain d06 by the cell perturbation technique (Singh et al., 2020).

It is shown clearly in Fig. 11 that the power spectra obtained in domains d04, d05, and d06 by LES agree well with the Karman-type power spectrum in the lower frequency range but decay rapidly and deviate significantly from the Karman-type power spectrum in the higher frequency range. Meanwhile, it can be seen that the frequency number from which the power spectrum starts to deviate from the Karman type, referred to later as drop-down frequency, shifts to higher frequency numbers from d04 to d05 and d06. The increase in drop-down frequency is achieved by the increase in grid resolution in the simulation. The rapid decay of turbulence implies that the turbulence model (LES) itself is ineffective for improving high-frequency wind speed fluctuation energy. Therefore, in the present study, we just display the capacity and applicability of the cell perturbation technique in simulating inflow turbulence by setting different perturbation parameters. Another important result that can be seen in Fig. 11 is that the drop-down frequency for the cell perturbation case does not change compared to the non-perturbation case, indicating a different mechanism of turbulence development for cell perturbation. The increase in micro-scale turbulence is achieved by artificially adding micro-scale energy. For description and calculation setting, please refer to (Singh et al., 2020).

5 Summary

Understanding atmospheric wind characteristics associated with the turbulence in the atmospheric boundary layer during severe storms has an important role in stabilizing and strengthening human life, and man-made structures like wind energy assessment, disaster reduction from severe meteorological conditions such as typhoons, air pollution wind, and so on. The combination of the local atmospheric boundary layer with the heterogeneous surface has great complexity in a realistic wind environment and strongly affects man-made structures, which is the curiosity for both the research communities i.e., meteorological science researchers, and wind engineering researchers. Computational fluid dynamics (CFD), and field observation methods are widely utilized for studying wind characteristics, and turbulence associated with severe storms. However, it is still a matter of concern and complexity to examine the flow conditions influenced by the meteorological phenomenon, and hence an alternative approach of numerical simulation has been taken place. In meteorological science, the Weather Research and Forecasting (WRF) mesoscale meteorological model has been widely used in the fields of atmospheric science, meteorology, operational weather forecasting, and research purposes. Apart from this, the connection between mesoscale model WRF and microscale model large eddy simulation (LES) for wind characteristics associated with turbulence has been considered an effective approach for wind analysis influenced by meteorological conditions in the atmospheric boundary layer near the surface. This combined strength of WRF and LES has played a key role in overcoming the difficulties of inflow boundary conditions from the realistic atmospheric wind to understand the key characteristics of strong winds and their turbulence structure inside the ABL. For a better understanding of wind characteristics associated with the turbulence structure in the ABL during severe storms, the Typhoon Kai-Tak case study has been described in detail. Results based on the simulations using the WRF and LES model, the microscale wind speed fluctuations range frommacro-meteorological peak and micro-meteorological peak, in the power spectrum of natural wind speed, and its frequency range called the spectral gap close to the ground (van der Hoven, 1957), indicating that the micro-scale turbulence energy is not necessarily transferred from the macro-scale turbulence. It was recently shown by Larsen et al. (2016) that the spectral gap may be deep or shallow depending on the relative strengths of the two processes and the depth of the gap decreases with height. The wind-induced responses of typical structure during severe storms are depending on the flow-structure interaction, and on the dynamic features of the structure of concern. To achieve this, it is still a matter of concern that up to what frequency should be simulated? Thus, we could achieve timely and accurate early warnings of wind disaster risk.

Acknowledgements I would like to thank Dr. Amitava Bandopadhyay, Director, NAM S&T Centre, New Delhi for inviting me to write this article. Also, I would like to pay my sincere thanks to Prof. Someshwar Das, for motivating me with this manuscript. Some of the results presented in this paper are taken from the thesis during my Ph.D. work at Tongji University, Shanghai, China, and would like to thank my Ph.D. supervisor, Prof. Shuyang Cao, for his constant motivation, and academic support throughout this study.

References

Byers, H. R. (1944). *General Meteorology*. McGraw Hill.

Cao, S. (2013). *Strong Wind and their Characteristics, Advanced Structural Wind Engineering* (pp. 1–25). Springer.

Duchene-Marullaz, P. (1972). *Etude des vitessesmaximalesannuelles du vent* (pp. 1–20). Cahiers du Centre Scientifique et Technique du Bafiment, No. l3l, Cahier' 1118-1127.

Fiedler, F., & Panofsky, H. A. (1970). Atmospheric scales and spectral gaps. *Bulletin of the American Meteorological Society, 51*, 1114–1119.

Goldstein, H. (1950). *Classical Mechanics*. Addison Wisley.

Haltiner, G. J., & Martin, F. L. (1957). *Dynamical and Physical Meteorology*. McGraw- Hill.

Holmes, J. D. (2001). *Wind Loading of Structures* (p. 51). Spon Press, JDH, Consulting.

Humphreys, W. J. (1940). *Physics of the Air*. McGraw-Hill. (reprint Dover I964).

Larsen, X. G., Larsen, S. E., & Petersen, E. L. (2016). Full-scale spectrum of boundary-layer winds. *Boundary-Layer Meteorology, 159*, 349–371.

Letchford, C. W., & Kwok, K. (2004). Simulation and flow measurement of atmospheric boundary layer flows. In *State-of-the-art wind tunnel modeling and data analysis techniques for infrastructure and civil engineering application*. Croucher Advanced Study Institute.

Letchford, C. W., Mans, C., & Chay, M. T. (2002). Thunderstorms, their importance in wind engineering-a case for the next generation wind tunnel. *Journal of Wind Engineering and Industrial Aerodynamics, 90*, 1415–1433.

Lieblein, J. (1953). *A new method of analysing extreme value data*. National Bureau of Standards Report No. 2190, Washington.

Mason, M., Letchford, C. W., James, D. L. (2003). Pulsed jet simulation of a thunderstorm downburst. In *Proceedings of 11ICWE* (pp. 2249–2256). Lubbock.

L.T. Matveev, Physics of the Atmosphere, TT67-51380, U.S. Department of commerce, National Technical Information service, Springfield.

Miller, A. (1971). *Meteorology*. Charles E, Merrill.

Mize, J. H., & Cox, J. G. (1968). *Essentials of Simulation*. Prentice-Hall.

Neiburger, M., Edinger, J. G., & Bonner, W. D. (1973). *Understanding the Atmospheric Environment*. Freeman.

Orlanski, I. (1975). A rational subdivision of scales for atmospheric processes. *Bulletin of the American Meteorological Society, 56*, 529–530.

Oseguera, R. M., & Bowles, R. L. (1988). *A simple analytic 3-dimensional downburst model based on boundary layer stagnation flow*. NASA Technical Memorandum 100632. National Aeronautics and Space Administration.

Simiu, E., & Scanlan, R. (1986). *Wind Effects on Structures* (2nd ed., pp. 5–90). Wiley.

Singh, J., et al. (2020). Numerical generation of inflow turbulence by cell perturbation technique in WRF simulation. *Journal of Wind Engineering & Industrial Aerodynamics*. https://doi.org/10.1016/j.jweia.2020.104395

Skamarock, W. C., & Klemp, J. B. (2008). A time-split non-hydrostatic atmospheric model for weather research and forecasting applications. *Journal of Computational Physics, 227*(7), 3465–3485.

Stull, R. B. (1988). *An Introduction to Boundary Layer Meteorology* (1st ed., pp. 3–25). Dordrecht.

Tao, T., & Tamura, T. (2020). Numerical study of the 6 May 2012 Tsukuba Supercell Tornado: Vorticity sources responsible for tornadogenesis. *Monthly Weather Review, 148*, 1205–1228.

Van der Hoven, I. (1957). Power spectrum of horizontal wind speed in the frequency range from 0.0007 to 900cycles per hour. *Journal of Meteorology, 14*, 160–164.
Zhao, L., Cui, W., & Ge, Y. J. (2019). Measurement, modeling and simulation of wind turbulence in typhoon outer region. *Journal of Wind Engineering and Industrial Aerodynamics, 195*, 104021.

Dynamics of Rainstorms Over Southern Himalayan Slopes and Foothills and Their Simulation by Numerical Models

Hatsuki Fujinami, Shiori Sugimoto, and Kenichi Ueno

Abstract In boreal summer, the Asian monsoon circulation carries abundant water vapor from the Indian Ocean to the lower–middle troposphere in the Himalayan region. This creates conditions favorable for some of the heaviest orographically induced precipitation in the world over the steep and topographically complex southern slopes of the Himalayas. Here, we introduce summer precipitation variability, including extreme events over the Himalayas, and its dynamics based on studies using in situ observations, spaceborne precipitation radar, and cloud-resolving numerical models. Specifically, we focus on the diurnal cycle, low pressure systems (LPSs), intraseasonal oscillations (ISOs), and their interplay, which represent an important background environment for extreme precipitation events in the Himalayas. In the diurnal cycle of precipitation, twice-daily precipitation maxima corresponding to daytime and nighttime peaks occur across higher elevations, while a single peak appears at night over lower elevations. The daytime peak is the result of anabatic flow driven by surface heating of the slopes, whereas the nighttime peak is caused by acceleration of low-level monsoon flows (i.e., the nocturnal jet) from the Gangetic Plain toward the Himalayas. Additionally, LPSs such as monsoon lows and monsoon depressions, together with mid-latitude troughs and their interaction can modulate strong inflows of moisture toward the Himalayan slopes, occasionally leading to extreme precipitation events when the flow with abundant water vapor is forced to ascend over the Himalayan slopes. The locations of the LPSs and mid-latitude troughs are strongly controlled by ISOs that act as large-scale environmental conditions for the genesis and development of the LPSs and troughs. Thus, improved understanding of the controlling multiscale processes, including the diurnal cycle,

H. Fujinami (✉)
Institute for Space-Earth Environmental Research, Nagoya University, Nagoya, Japan
e-mail: hatsuki@nagoya-u.jp

S. Sugimoto
Japan Agency for Marine-Earth Science and Technology, Yokohama, Japan

K. Ueno
Institute of Life and Environmental Sciences, University of Tsukuba, Tsukuba, Japan

S. Das and W-K. Tao (eds.), *Severe Storms*,
https://doi.org/10.1007/978-981-97-7075-5_13

is essential for better prediction of extreme precipitation events and mitigation of related disasters over the Himalayas.

Keywords The Himalayas · Summer monsoon · Precipitation · Diurnal cycle · Monsoon lows · Intraseasonal oscillations · Extreme precipitation events

1 Introduction

In boreal summer (June–September), the Asian monsoon circulation system becomes established, driving low- and mid-level southerly winds toward the Himalayas (Fig. 1). These winds can carry abundant water vapor from ocean areas such as the Bay of Bengal (BoB) and the Arabian Sea, creating conditions favorable for some of the heaviest orographically induced precipitation in the world over the southern Himalayan slopes. Additionally, glaciers exist in the higher elevations of the Himalayas. Thus, the hydrological cycle in the Himalayas incorporates gaseous, liquid, and solid states of water within the narrow and steep zone of the southern slopes extending ~100 km latitudinally, which is indeed the epitome of the global water cycle. The cycle maintains the headwaters of major rivers such as the Indus, Ganges, and Brahmaputra, and provides water resources for South Asia's large population (Immerzeel et al., 2010). The mountain hydrological cycle is vulnerable and subject to the effects of recent global climate change. In the central–eastern Himalayas, most of the total annual precipitation falls in summer (Bohlinger and Sorteberg, 2018; Bookhagen, 2010; Salerno et al., 2015; Shea et al., 2015), which makes summer precipitation a crucial part of the regional hydroclimatology. Most of the extreme precipitation events occur in this season, and they often cause natural hazards such as flash floods and landslides (Bookhagen, 2010; Bohlinger and Sorteberg, 2018). The southern region of the Himalayas with its steep slopes and complex terrain is prone to extreme rainfall events in summer. Despite the occurrence of higher extreme rainfall amounts in low-elevation regions such as the Gangetic Plain, the occurrence of extreme events is approximately twice as high in the Himalayas in comparison with that in surrounding regions such as the Gangetic Plain and the Tibetan Plateau (Bookhagen, 2010).

The Himalayas feature extremely deep valleys with depth of 1000–3000 m, and abrupt ridges that are more than 6000 m in height. Most meteorological observatories at high elevation are located in the bottom of valleys. The topographic features, remoteness, and limitation of power supply over higher elevations of the Himalayas hamper the use of state-of-the-art ground observation systems such as ground precipitation radar that can observe the three-dimensional structure of precipitation systems continuously. Therefore, in situ observations by rain gauges remain vital for observing and monitoring the amount/variability of regional precipitation. In Nepal, there is a dense network of rain gauges in the Himalayas, which is maintained by the Department of Hydrology and Meteorology of Nepal. Most of the rain gauges are located at elevations of <3000 m above sea level (asl) (Shrestha et al., 2000).

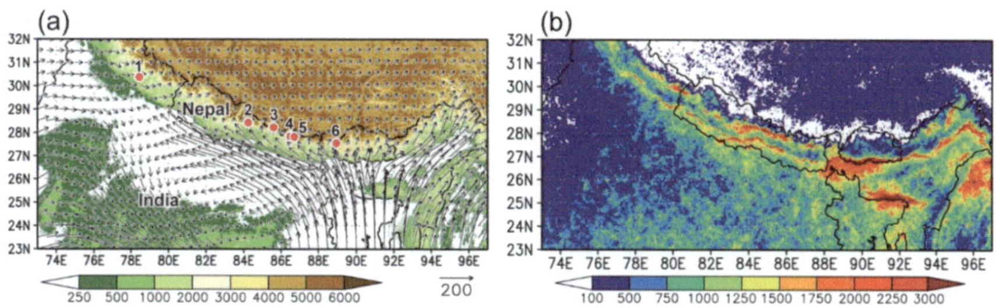

Fig. 1 **a** Topography (shading; m) and climatological mean (1998–2014) vertically integrated water vapor flux (vectors; kg m^{-1} s^{-1}) in summer (June–September) based on ERA5 data for the Himalayas. Red circles with numbers indicate approximate locations of river basins or valleys where an extreme precipitation event occurred (point 1) or in situ field observations were performed (points 2–6), which appear in this chapter. 1: Uttarakhand, 2: Marsyandi River Basin, 3: Langtang Valley, 4: Rolwaling Valley, 5: Khumbu, 6: Yadong Valley. **b** Climatology (1998–2014) of summer (June–September) surface rainfall (shading: mm) observed by TRMM-PR

Therefore, meteorological observations acquired at elevations of >3000 m asl in the Himalayas remain extremely scarce. Some in situ observation projects have provided valuable and important results in relation to meteorological elements at higher elevations, e.g., >3000 m asl. In contrast, long-term observations from spaceborne precipitation radar such as the Tropical Rainfall Measuring Mission (TRMM)-Precipitation Radar (PR) (1998–2014) and the Global Precipitation Measurement (GPM)-Dual-frequency Precipitation Radar (DPR) (2014–present) have provided new insights with regard to climatological features such as the topographic dependency of precipitation, diurnal variation of rainfall properties such as convective and stratiform rainfall, and the three-dimensional structure of rainfall systems around the Himalayas, together with snapshots of rainfall systems (e.g., Fujinami et al., 2017, 2021; Hirose et al., 2008, 2017; Houze et al., 2007; Medina et al., 2010; Shrestha et al., 2012). Unfortunately, the TRMM-PR and the GPM-DPR are unable to capture a single rainfall system continuously owing to their non-sun-synchronous orbits (Nakamura, 2021). Recently, some satellite-derived precipitation data with high spatiotemporal resolution, e.g., the Integrated Multi-satellitE Retrievals for GPM (IMERG) and Global Satellite Mapping of Precipitation (GSMaP), have become available for time series analysis and near real-time monitoring of precipitation, including extreme precipitation events globally (Kubota et al., 2020; Tan et al., 2019). Additionally, the hourly ERA5 (the fifth generation ECMWF atmospheric reanalysis) dataset on a $0.25° \times 0.25°$ grid represents high-quality global data of atmospheric circulation and land surface parameters such as surface heat fluxes (Hersbach et al., 2020; Martens et al., 2020). Such data have proven very useful in examining land–atmosphere interactions in terms of explaining diurnal cycles. The high spatiotemporal resolution of this dataset also allows examination of the detailed features of the diurnal cycles of meteorological variables in the region of the Himalayas more easily than would be possible using only 6-h temporal resolution and coarse-grid reanalysis data. A high-resolution cloud-resolving regional model is a powerful tool that can compensate for the lack of ground-based observations in the Himalayas. However, it is not easy to simulate the properties of precipitation over the Himalayas realistically because such precipitation is caused by various interactions between complex local mountain–valley circulations of various horizontal scale over the slopes and the large-scale atmospheric circulation with multiscale spatiotemporal variability (Ueno & Nakileza, 2023).

In this chapter, we review precipitation variability/dynamics in the Himalayas in summer, including extreme precipitation analyzed by in situ observations, spaceborne precipitation radar, satellite-derived precipitation, high spatiotemporal atmospheric reanalysis data such as ERA5, and cloud-resolving regional climate models. Specifically, we focus on the diurnal precipitation cycle over the Himalayas as a manifestation of land–atmosphere interaction, low pressure systems (LPSs) such as monsoon lows that generate over and around the BoB, intraseasonal oscillations (ISOs) such as the quasi-biweekly oscillation (QBW) and boreal summer intraseasonal oscillation (BSISO), and their interplay on precipitation over the Himalayas in the summer monsoon season. The multiscale interaction over the complex topography of the Himalayas can lead to frequent occurrence of extreme precipitation events.

2 Climatological Features of Summer Precipitation

Features of seasonal variation in precipitation differ from east to west and from lower to higher elevation in the Himalayas (Kansakar et al., 2004). The Himalayan region of central–eastern Nepal receives ~80% of its total annual precipitation in summer (June–September) (Bookhagen, 2010; Sharma et al., 2020; Shrestha, 2000; Shrestha et al., 2000). The highest rainfall is observed in the foothills of both the southern slopes of the Meghalaya Plateau and the Bhutan Himalayas where large moisture flux convergence occurs (Fig. 1). The amount of extreme rainfall, defined as the 90th percentile of summer precipitation, is high (daily amount: >50 mm) in both regions (Bookhagen, 2010). The two regions also have the largest QBW signal (i.e., fluctuation with period of ~14 days) in precipitation over the Asian monsoon region, as well as distinct nocturnal precipitation (Fujinami et al., 2014, 2017; Murata et al., 2017), suggesting that multiscale interaction among the diurnal cycle attributable to land surface processes, the QBW, and topography causes the heavy precpitation. The topography of the Himalayas also controls the spatial distribution of the amount of precipitation (Bookhagen and Burbank, 2006; Ichiyanagi et al., 2007; Kansakar et al., 2004; Shrestha et al., 2012). There are two zones of high rainfall at different elevation over the southern slopes of the Nepal Himalayas, which are oriented along the direction of strike. The two peaks are also observed in the Himalayas in India but not in Bhutan (Fig. 1b). The two zones are associated with the two-step topography that comprises the Sub-Himalayas at 500–1000 m asl and the Lesser Himalayas at ~2000 m asl (Bookhagen and Burbank, 2006; Nakamura, 2021; Shrestha et al., 2012). The peak over the Sub-Himalayas represents strong convective-type rainfall, whereas the peak over the Lesser Himalayas corresponds to frequent but less intense stratiform-type rainfall. Additionally, in the eastern Himalayas, precipitation systems become wider and more stratiform, whereas in the western Himalayas, precipitation falls mainly from small-scale but highly convective systems (Houze et al., 2007; Romatschke and Houze, 2011). The contrasting features of the precipitation systems are associated with the characteristics of synoptic-scale moisture transport (e.g., Fig. 1a) and the conditions of the land surface over which the low-level monsoon passes (Medina et al., 2010). In comparison with that of surrounding regions, the frequency of occurrence of extreme precipitation events is higher in the Himalayas. Moreover, higher-elevation areas in the Himalayas are characterized by an even greater number of extreme precipitation events during a single season, which is significantly decoupled from the rainfall pattern (e.g., Fig. 1), because even areas with overall lower amounts of rainfall have similar numbers of extreme rainfall events (Bookhagen, 2010).

3 Diurnal Cycle

Associated with seasonal migration of the large-scale atmospheric circulation around the Himalayas and the Tibetan Plateau, weakening of the mountain–valley circulation occurs when the upper-level westerlies weaken over the Himalayas owing to the northward shift of the subtropical jet from south of the Himalayas a few weeks before monsoon onset in the middle of June (Ueno et al., 2008). Then, the onset of monsoon rainfall characterized by nighttime precipitation is initiated with the occurrence and migration of LPSs such as monsoon lows/depressions in the northern part of the BoB (Lang and Barros, 2002; Ueno et al., 2008). The diurnal cycle is the building block that characterizes the hydrological cycle in mountainous regions during summer following onset of the monsoon.

First, we review the diurnal cycle signals detected by in situ observations in the Nepal Himalayas. The first full year of Himalayan meteorological observations, conducted at Lhajung (4420 m asl) in the glacierized catchment in the upper-Khumbu area of the eastern Nepal Himalayas (Fig. 1a, point 5), were obtained manually in the 1970s as part of a Japanese scientific expedition known as the Glaciological Expedition of Nepal (GEN). Analysis of the observations revealed prevalence of daytime (nighttime) precipitation on ridges (in valley bottoms) in summer (Ageta, 1976; Yasunari and Inoue, 1978), and the importance of nocturnal precipitation on the glacier mass balance was raised (Higuchi, 1977). Ohata et al. (1981) found that daytime anabatic winds (i.e., upslope winds) at Lhajung continued throughout the night in summer instead of being replaced by katabatic winds (i.e., downslope winds), indicating the prevalence of monsoon flows over the Himalayas. Clouds that tend to be initiated above the ridges by these anabatic winds in daytime, subsequently spread throughout the valley, and the cloud base gradually lowers to ground level in the early afternoon. The wind continues during the night accompanying drizzle and rainfall (Ohata et. al., 1981). From the mid-1990s to the early 2000s, multiyear in situ observations by an automatic weather station (AWS) were conducted at Shangboche (3833 m) in the Khumbu area as a part of the Cryosphere Research in the Himalayas (CREH) program, in which assessment of the interannual variabilities of mountain weather or validation of satellite measurements were conducted (Ueno et al., 2001; Ueno and Aryal, 2008; Yamamoto et al., 2011). The Pyramid Meteorological Station (5035 m asl) also started continuous monitoring of mountain weather at the foot of Mt. Everest under the Ev-K^2-CNR Project (Bollasina et al., 2002). During 2002–2004, five AWSs (including the Shangboche and Pyramid stations) located at elevation of 2660–5035 m asl were categorized as reference sites of the Coordinated Enhanced Observation Period Project (CEOP) (Koike, 2004), and the seasonal progress of the variability of diurnal precipitation in relation to local circulations was examined consistently in the elevation range of 2000–5000 m asl (Ueno et al., 2008). The diurnal precipitation cycles showed a single nighttime peak at ~23:00 Local Time (LT at 90°E) at lower elevations, i.e., below the Shangboche station, and a double peak at ~15:00 and ~23:00 LT at stations at higher elevation. In the surface wind system, upslope winds were found to continue until 03:00 LT at stations at higher elevation, similar to the findings of Ohata et al. (1981), while at lower elevations, upslope winds also appeared but with weak downslope winds also occurring sporadically during night. In the central Nepal Himalayas, in situ observations were conducted using a network of 20 meteorological stations at elevations of 528–4435 m asl in the Marsyandi River valley during 1999–2001 (Barros et al., 2000; Barros and Lang, 2003; Lang and Barros, 2002) (Fig. 1a, point 2). Analysis of these observations also revealed that the diurnal cycle of summer precipitation has two peaks that correspond to daytime and nighttime at higher elevations above 2000 m asl, while stations at lower elevations (i.e., below 1500 m asl) showed a single peak of precipitation at night. Surface winds also presented a pattern of daytime upvalley winds and nocturnal downvalley flow at lower elevations, whereas stations at higher elevations did not show such a reversal, similar to the observations in the Khumbu area. Radiosonde observations revealed that the wind reversal in the lower-elevation areas appears from the surface to ~1000 m above ground level. In regions other than the Nepal Himalayas, a high-elevation (2800–4500 m asl) rain gauge network observed precipitation during 2016–2018 in the Yadong Valley, which is located between Sikkim, India, and Bhutan (Ouyang et al., 2020) (Fig. 1a, point 6). Analysis of the observations revealed a clear change in the diurnal cycle of precipitation depending on surface elevation. At stations lower than 3000 m asl, precipitation only occurred at night, whereas the precipitation presented a double-peak characteristic in the afternoon and at night at higher-elevation stations. The afternoon peak was found to become more prominent with increasing elevation.

Precipitation at higher elevation controls the glacier mass balance. In situ measurements by AWSs located on and around Yala glacier in Langtang Valley, central Nepal Himalayas (Fig. 1a, point 3), revealed that the diurnal cycle of summer precipitation has two peaks, corresponding to daytime and nighttime (Shea et al., 2015). A similar feature was observed in the observations from an AWS located close to the terminus of Trakarding glacier at 4806 m asl in Rolwaling Valley, eastern Nepal Himalayas (Fujinami et al., 2021) (Fig. 1a, point 4). Thus, the double-peak feature of the diurnal cycle of precipitation is probably also dominant in high-elevation glacierized areas. Figure 2a shows the two peaks in summer precipitation that occur at around 14:00–16:00 and 00:00–02:00 LT, as observed by a tipping-bucket rain gauge with 0.1-mm resolution at the site of the AWS in Rolwaling Valley, together with other relevant meteorological variables. Rolwaling Valley is aligned in the west-northwest to east-southeast direction around the AWS, with elevation increasing toward the east-southeast (Fig. 1b of Fujinami et al., 2021). Downward shortwave radiation warms the surface after sunrise, causing the surface air temperature to rise (Fig. 2b). The warmed slope surface induces a west-northwesterly or northwesterly anabatic flow in Rolwaling Valley during 12:00–17:00 LT, with peak wind speed that exceeds 2.5 m s^{-1} at 13:00–14:00 LT (Fig. 2c). This anabatic flow carries moist air along the valley, which cools adiabatically as it travels up the slope, promoting condensation. High relative humidity of around 95% and weak southerlies (wind speed of <1 m s^{-1}) were observed at around the time of the nighttime precipitation peak.

Long-term observations from the TRMM-PR that elucidate the spatial distribution of the diurnal cycle of rainfall across the Nepal Himalayas (Fig. 1b), can extend the results from the point-measurement scale to the synoptic scale. Figure 3 shows the spatial distribution of the pattern of the diurnal cycle of rain frequency in summer, classified by cluster analysis. Rain frequency is defined as the number of rainfall events as a percentage of the total number of observations from a grid of TRMM-PR observations. We constructed the TRMM-PR diurnal climatology to apply the cluster analysis. One striking feature is the band of high rain frequency with peaks at 16:00 and 00:00 LT, consistent with the peaks in the results based on in situ observations over higher elevations, which are classified as cluster 7 (Fig. 3a and h). Over lower elevations below 500 m asl, the diurnal cycle has a single peak, which occurs at 06:00

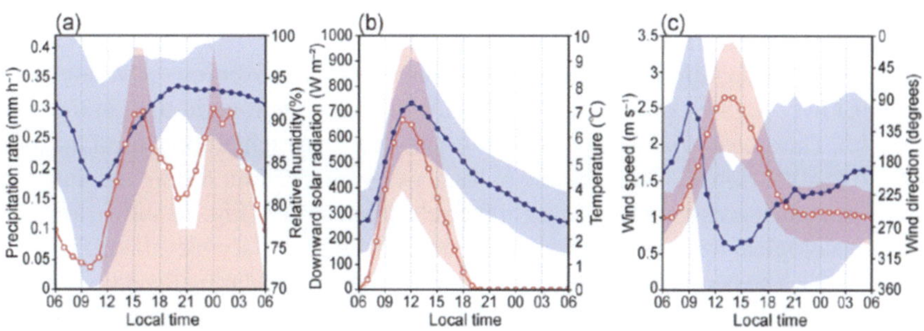

Fig. 2 Three-year (2016, 2017, and 2018) mean diurnal cycles of meteorological variables measured during summer (June–September) at the Rolwaling automatic weather station: **a** precipitation rate (mm h^{-1}; red line with open circles), determined using a tipping-bucket rain gauge with 0.1-mm resolution, and relative humidity (%; blue line with closed circles); **b** downward shortwave radiation (W m^{-2}; red line with open circles) and surface air temperature (°C; blue line with closed circles), and **c** wind speed (m s^{-1}; red line with open circles) and wind direction (degrees from North (0 = northerlies), blue line with closed circles). Shading for precipitation indicates the range between the lower (25th percentile) and upper (75th percentile) quartiles, and ±1 standard deviation for the other variables (Fujinami et al., 2021)

LT (clusters 4 and 6, Fig. 3a, e, and g). Rain frequency for most grid cells over areas with elevation of >4000 m asl in the Nepal Himalayas is categorized into clusters 2 and 3 (Fig. 3c and d), including the grid cells that are closest to the AWSs on Trakarding Glacier and Yala Glacier. Rain frequency for these cells is <5%, which is much lower than the value of ~50% recorded at the AWSs (Fujinami et al., 2021). The minimum rainfall rate that the TRMM-PR sensor can detect is ~0.7 mm h^{-1}, which is higher than the rate for many rainfall events recorded at the AWSs (Fig. 2a). Surface clutter in PR observations associated with the complex deep valley system of the Himalayas also affects the rainfall measurements for high-elevation areas (Hirose et al., 2017). Therefore, it appears difficult to observe light rainfall and the contrasting features of the diurnal cycles between ridges and valleys in regional-scale topography from TRMM-PR observations, as shown by Ageta (1976) and Yasunari and Inoue (1978).

Rainfall properties can provide information that is useful for understanding the mechanisms that drive a precipitation system. Figure 4 shows latitude–time sections of the properties of rainfall revealed in TRMM-PR data around Rolwaling Valley. Rainfall occurs to the south of 27.2°N, where the surface elevation is <500 m asl, until 10:00 LT. The unconditional rain rate (i.e., the mean rainfall rate irrespective of whether or not it was raining) and the rain frequency are both minimum during 09:00–11:00 LT over the slopes. After 12:00 LT, rainfall occurs widely over slopes that are between 500 and 4000 m asl (from 27°N to 28°N), with maxima at around 27.3°N and 27.6°N (Fig. 4a). Rainfall starts earlier in the day at lower elevations than at higher elevations. At the peak at around 2000 m asl, the rainfall rate decreases at around 18:00 LT, increases again thereafter, and has a maximum at around 23:00–00:00 LT. This clear nighttime enhancement of rainfall at around 2000 m asl extends from 27.5°N to 28.0°N. The rainfall continues until 03:00 LT, after which the area of high precipitation shifts rapidly to the south of 27.2°N. The diurnal pattern of rainfall rate is similar to that of rainfall frequency (Fig. 4b). The conditional rain rate (i.e., the mean rainfall rate during rain events) is higher over low-elevation areas than that over high-elevation areas, and tends to be higher at night than in daytime (Fig. 4c). The area in which convective rain accounts for a high proportion of the total rainfall (>50%) spreads from lower- to higher-elevation terrain during the day, and stratiform rain increases at night over higher-elevation terrain (Fig. 4d and e), forming the climatological features of rainfall type that depend on surface elevation (Shrestha et al., 2012). The height of the rainfall system is <6500 m asl during the day, whereas it exceeds 7000 m asl at night (Fig. 4f). The height difference of the rainfall system between daytime and nighttime is consistent with the results determined for cloud based on IR and visible cloud imagery from a geostationary satellite (Kurosaki and Kimura, 2002). The difference in rainfall properties between daytime and nighttime reflects the driving mechanism of precipitation.

Daytime rainfall is caused by upslope air flow (anabatic slope flow) that is driven thermally by surface heating during the day (Barros and Lang, 2003; Bhatt and Nakamura, 2006; Fujinami et al., 2021; Ouyang et al., 2020; Ueno et al., 2008; Yang et al., 2018). Upslope flows move moist air to higher altitudes as the ground elevation increases, and condensation therefore occurs over the slopes. Daytime surface

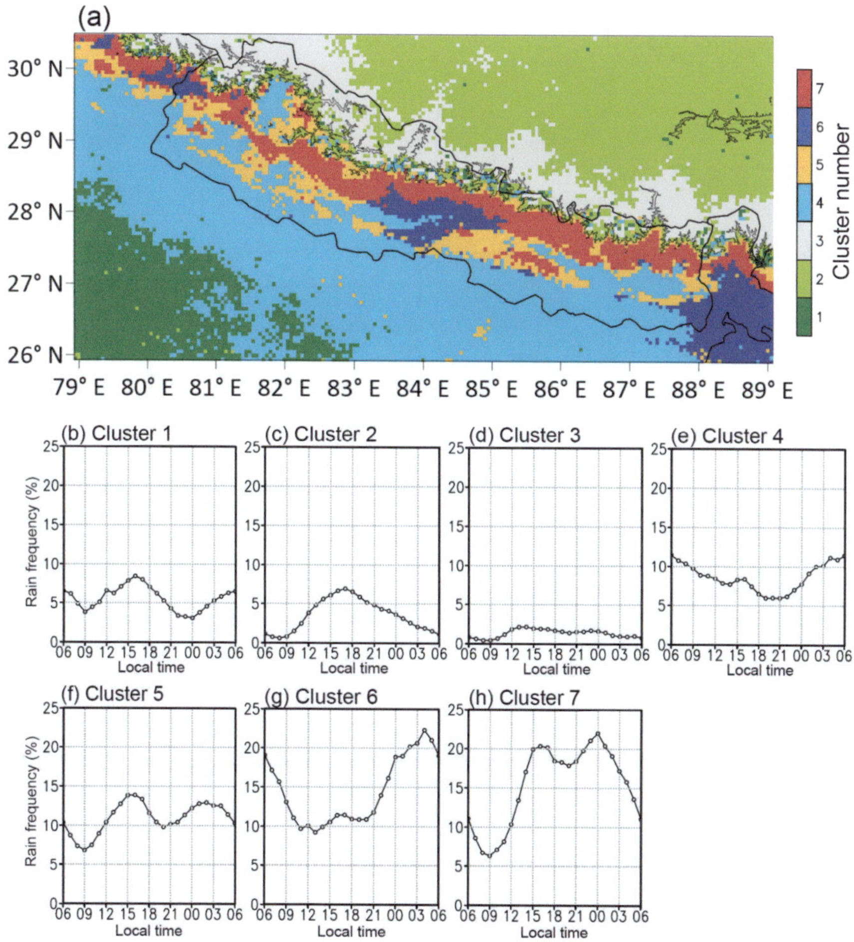

Fig. 3 **a** Spatial distribution of the seven clusters of the climatological mean (1998–2014) diurnal cycle in rain frequency in summer, calculated from TRMM PR data using cluster analysis. Thin solid lines are topographic contours at 4000 m. Lower panels: the mean diurnal cycle of precipitation frequency in **b** cluster 1, **c** cluster 2, **d** cluster 3, **e** cluster 4, **f** cluster 5, **g** cluster 6, and **h** cluster 7 (Fujinami et al., 2021)

heating causes greater instability in the near-surface atmosphere, which suggests that convective-type rainfall dominates precipitation over the slopes during the day (Fig. 4d). In contrast, there are several mechanisms proposed for the occurrence of nocturnal rainfall. Barros and Lang (2003) proposed a possible mechanism based on intensive in situ observations performed in 2001. At night, the speed of the upslope wind becomes weak near the surface, even reversing direction in the valley at lower elevations (Fig. 16 of Barros and Lang 2003). This process enhances convergence of the large-scale monsoon flow at the foot of mountains, leading to upward motion that can initiate convection over lower elevations. This convergence is reduced during

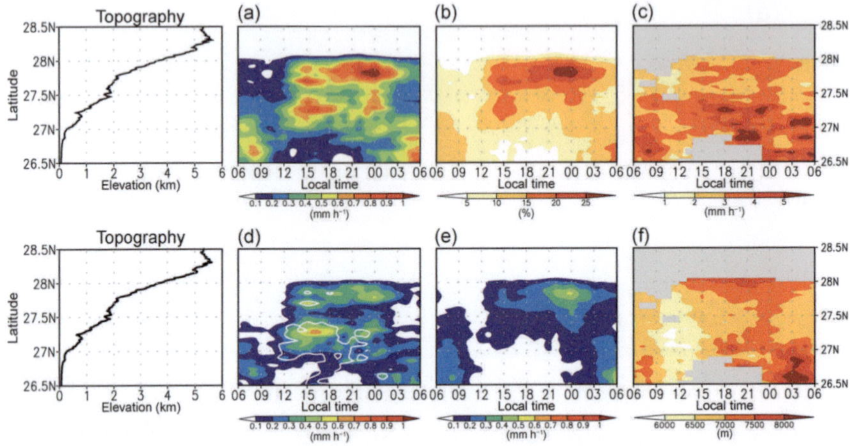

Fig. 4 Left panels: average surface elevation between 85.5°E and 86.5°E (km asl). Latitude–time (LT at 90°E) cross sections of climatological mean diurnal cycles in rainfall properties from TRMM-PR between 85.5°E and 86.5°E in summer (June–September) for 17 years (1998–2014): **a** unconditional rain rate (mm h^{-1}); **b** rainfall frequency (%); **c** conditional rain rate (mm h^{-1}); **d** convective-type rainfall (shading; mm h^{-1}), white contours show where convective-type rain accounts for 50% of the total rainfall; **e** stratiform-type rainfall (mm h^{-1}); and **f** storm height above sea level (shading; m) (Fujinami et al., 2021)

the daytime because the thermally forced upslope and upvalley flow reduces the spatial gradient in wind velocity. On the basis of atmospheric reanalysis data and the behavior of convective systems elicited from TRMM-PR data, Bhatt and Nakamura (2006) suggested that downslope flows (gravity currents) driven by nighttime rainfall at higher elevations enhance nocturnal moist convection over the Himalayas and that this convection is enhanced by cloud-top radiative cooling (Fig. 12 of Bhatt and Nakamura 2006).

In both studies, the large-scale low-level monsoon inflow toward the Himalayan slopes is presumed broadly constant during both day and night. However, the large-scale low-level monsoon flow over the Gangetic Plain windward of the Himalayas does exhibit strong diurnal variation, as do the winds over the slopes of the Himalayas (Chen, 2020; Fujinami et al., 2017). The wind speed at 950–925 hPa over the Gangetic Plain to the south of the Himalayas is generally a minimum in the afternoon and reaches a maximum over the plain at night, and such a diurnal cycle in the wind system is maintained by subcontinental-scale land–atmosphere interaction (Fujinami et al., 2021; 2022a). The vertically well-mixed layer associated with daytime heating of the land surface decelerates the low-level monsoon flow because of vertical turbulent mixing within the mixed layer, which acts as a frictional force on the low-level horizontal wind. In contrast, surface cooling after sunset generates a stable surface layer during the night. Thus, the wind speed accelerates at night over the plain because of reduction in the frictional force and inertial oscillation of the winds (Blackadar, 1957; Stensrud, 1996). The vertical structure shows a clear low-level jet (LLJ) structure between 950 and 925 hPa (i.e., a nocturnal LLJ). It is well known that the LLJ enhances the orographically induced nocturnal rainfall maximum over the slopes of the Meghalaya Plateau, which represents one of the areas in the world with the heaviest rainfall south of the Himalayas (Fig. 1b) (Fujinami et al., 2017; Kataoka and Satomura, 2005; Sato, 2013; Terao et al., 2006). The incoming moisture flux toward the Himalayas, which is controlled by the low-level winds, strengthens twice daily, creating maxima in the southerly component of the flux at 12:00 and 00:00 LT from the foothills to the southern slopes of the Himalayas (Fig. 5). The timing of the strengthening of the moisture inflow toward the southern slopes corresponds closely to the twice-daily rainfall maxima observed at the AWS in Rolwaling Valley and evidenced in the TRMM-PR data (Fig. 4). The area over which the southerlies are enhanced extends from the foothills to higher-elevation areas of the Himalayas during the day, and extends up to the Gangetic Plain at night. During the day, near-surface upslope flows are evident in the Himalayas, while at night, the near-surface upslope flows weaken and the southerly airflow toward the Himalayas is enhanced at 950–900 hPa (500–700 m asl) over the foothills. As the flows move toward the high-elevation Himalayas, they become subject to gentle orographic lifting, which enhances stratiform-type rainfall (Fig. 4e). Thus, large-scale moisture flux convergence occurs over terrain stretching from the foothills to higher elevations in the Himalayas. Fujinami et al. (2022a) revealed that low-level southerlies are greatly enhanced at night parallel to the coastline over the western BoB, which then flow across the Gangetic Plain, enhancing moisture transport toward land and producing nocturnal precipitation in South Asia. This process suggests that land surface effects of the Indian subcontinent, including both their dynamic and thermodynamic roles, can also be major influencing factors regarding the nocturnal enhancement of monsoon flow over the Gangetic Plain. Thus, the diurnal cycle of large-scale low-level monsoon flow and its large-scale convergence can be a major factor in the initiation of nocturnal rainfall over the Himalayas. Additionally, weakening of the upslope flow over the Himalayas can help enhance moisture convergence (Barros and Lang, 2003). Once moist convection occurs over the slopes, gravity currents associated with the convection might interact with the incoming large-scale monsoon flow and enhance convection further (Bhatt and Nakamura, 2006).

Reasonable simulation in precipitation and local-scale circulation related to deep valley systems is not easy when using coarse-resolution models. Actually, even a numerical experiment with 5-km horizontal resolution cannot well represent the nocturnal precipitation maximum over the Himalayan slopes and foothills during the mature monsoon season (e.g., Sugimoto and Takahashi, 2016). Recent numerical experiments with cloud-resolving models (i.e., $\leq$2 km horizontal resolution), e.g., the Weather Research and Forecasting (WRF) model, improved the simulation of the diurnal cycle in precipitation as well as near-surface temperature in comparison with the results of coarse-resolution experiments (Collier and Immerzeel, 2015; Karki et al., 2017). In 2-km-mesh experiments, surface friction along Himalayan valleys increases in association with the complex topography, which decelerates the surface wind and moisture fluxes and reduces the wet bias over the Tibetan Plateau (Lin et al., 2018). Sugimoto et al. (2021) also reproduced the observed diurnal cycles of

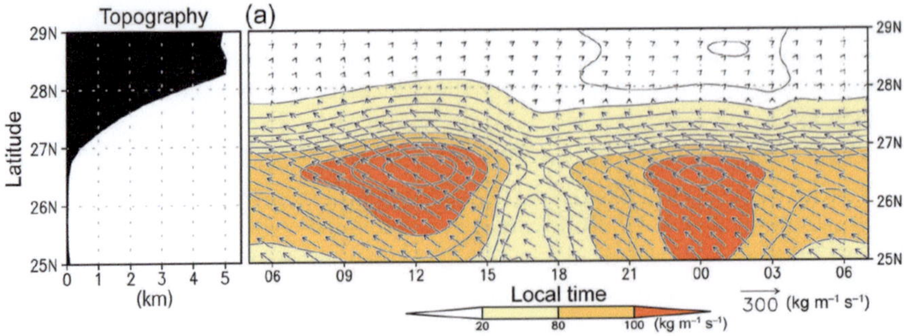

Fig. 5 **a** Latitude–time (LT at 90°E) cross section of climatological mean diurnal cycle of vertically integrated water vapor flux (vectors; kg m^{-1} s^{-1}) and the meridional component of water vapor flux (shading; kg m^{-1} s^{-1}) averaged over 85.5°–86.5°E. The diurnal cycle was constructed by averaging ERA5 reanalysis data for each hour during summer (June–September) for 17 years (1998–2014). Left panel shows the average elevation for 85.5–86.5°E (km asl) from ERA5 data (Fujinami et al., 2021)

precipitation such as the two precipitation peaks in the afternoon and at midnight over higher elevations and a single nocturnal peak at lower elevations over the Himalayas during summer using the WRF model. Then, they investigated the physical process inducing nocturnal precipitation over the Himalayas. They found that precipitation that occurred over mountain ridges in the Himalayas during the afternoon was associated with a coupled system between large-scale monsoon flows and local-scale upslope winds, which expanded horizontally toward lower-elevation areas through the night. During nighttime, the downslope wind was caused by radiative cooling at the surface and was intensified by evaporative cooling by hydrometeors in the near-surface layer (Fig. 6). Consequently, convergence between the downslope wind and the synoptic-scale monsoon flow with nocturnal acceleration promoted nocturnal precipitation over the Himalayas and to the south, as well as the moisture convergence driven by orography and synoptic-scale circulation patterns.

Such a cloud-resolving model that can reasonably simulate the precipitation process over the Himalayas was applied to understand the sensitivities of meteorological variables on physical schemes and land surface conditions. For example, Orr et al. (2017) conducted four sensitivity experiments using different microphysics schemes in a cloud-resolving model, and suggested that the cold-rain process of hydrometeors is a key factor in accurate simulation of precipitation. A 1-km sensitivity experiment by Potter et al. (2018) suggested that the strength of the daytime upslope wind is controlled by the pressure gradient associated with the presence or absence of glaciers. According to in-situ observation, precipitation over the Himalayan slopes is enhanced near glacier fronts because of convergence between cold-dry outflows over glaciers and warm-moist monsoon flows (i.e., glacier–air interaction), as suggested by in Lin et al. (2021). They conducted a non-glacier experiment and proposed the

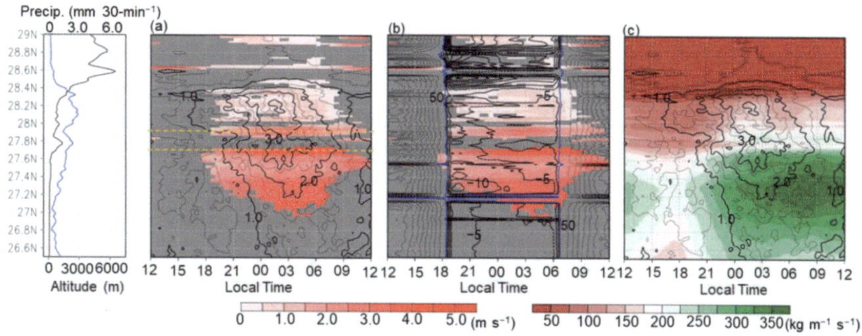

Fig. 6 Composites of time series for **a** wind speed with a northerly component at 10 m above the surface (shading) and vertically integrated hydrometeors (contours with 0.5-mm intervals) along 84°E for nocturnal precipitation events over the western slope area (83°–85°E) of the Nepal Himalayas. **b** As in **a** but for net radiation at the surface (contours). The contour intervals in **b** are 50 and 5 W m^{-2} for positive (upward) and negative (downward) values, respectively. **c** As in **a**, but for magnitude of vertically integrated moisture flux (shading). Left panel shows topography (black) and precipitation at 03:00 LT (blue) along 84°E. In **a** and **b**, wind speed with a southerly component is shaded gray. Blue line in **b** indicates 0 W m^{-2}. Dashed yellow lines in **a** indicate latitudes of 27.7°N and 27.9°N. The values were simulated by WRF with 2 km resolution during the mature monsoon seasons between 2003 and 2010 (Sugimoto et al., 2021)

idea that glacier retreat increases precipitation and snow accumulation in higher-elevation regions of the Himalayas because of glacier–air interaction. Thus, a cloud-resolving model can be an important tool for improved understanding of the precipitation process from regional to synoptic scales, including the response of precipitation to global climate change.

4 Multiscale Process Leading to Heavy Precipitation

LPSs, generated over and around the BoB, contribute a large fraction of the total summer monsoon rainfall over South Asia (Hurley and Boos, 2015; Krishnamurthy and Ajayamohan, 2010). Humid and convectively unstable environments with high cyclonic vorticity and high sea surface temperature favor LPS genesis over the northwestern BoB, despite relatively strong vertical wind shear due to the vertical structure of the large-scale Asian monsoon circulation (Diaz and Boos, 2019, 2021; Ditchek et al., 2016; Fujinami et al., 2020; Samanta et al., 2018). Typically, LPSs form over the head of the BoB and move northwestward over the Indian subcontinent, with a life cycle of 3–6 days and horizontal scale of ~1000–2000 km (Godbole, 1977; Sikka, 1977). The LPSs that occur around the Gangetic Plain over the northeastern Indian subcontinent tend to have smaller horizontal scale, i.e., ~600 km (Hatsuzuka et al., 2014). The LPSs around India are categorized into lows, depressions, deep depressions, and cyclonic storms on the basis of their intensity (Hurley and Boos, 2015; Saha et al., 1981). Extreme precipitation and catastrophic floods in South Asia are often associated with LPSs (Bohlinger and Sorteberg, 2018; Bohlinger et al., 2019; Dimri et al., 2017; Houze et al., 2011; Joseph et al., 2015; Webster et al., 2011). Active moist convection processes, including mesoscale convective systems, which occur frequently within the LPSs, help maintain humid conditions inside the LPSs from the surface to the middle troposphere (Fujinami et al., 2017, 2020; Hunt et al., 2016a, 2016b). Additionally, the vertical structure of the LPSs reaches the upper troposphere. Therefore, the LPSs can enhance the moist southerlies/southeasterlies moving toward the Himalayas on the eastern side of the LSPs from the lower to middle troposphere. Without LPSs, the transport of moisture toward the Himalayas is realized only by the low-level monsoon flow (typically <1500-m thick) in the lower troposphere and anabatic upslope winds over the Himalayan slopes (Dong et al., 2017; Fujinami et al., 2017). Thus, LPSs play an important role in the transport of water vapor toward the high mountain areas such as the Himalayas and the Tibetan Plateau throughout the troposphere, affecting precipitation processes in these regions (Dong et al., 2016, 2017; Fujinami and Yasunari, 2004; Lang and Barros, 2002; Ueno et al., 2008). Therefore, the onset of monsoon precipitation tends to correspond to the first occurrence of the LPSs over the northern BoB. Precipitation systems within the LPSs and orographic precipitation resulting from interactions between the horizontal moist winds from the LPSs and the Himalayan slopes promote precipitation over the Himalayas. The LPSs also influence the diurnal cycle of precipitation over the Himalayas. Nocturnal precipitation over Himalayan slopes is enhanced by the LPSs because the stronger than usual southeasterlies from the LPSs and the diurnal variation enhance nocturnal convergence (Hunt et al., 2022; Sugimoto et al., 2021). Actually, low-level winds exhibit a remarkable diurnal cycle even under the influence of the LPSs and ISOs (Fujinami et al., 2017). Land surface effects of the Indian subcontinent can regulate the wind speed over the Gangetic Plain diurnally (Fujinami et al., 2021; 2022a).

Statistical analysis of extreme precipitation events in the Himalayas in summer and the synoptic-scale atmospheric circulation revealed that LPSs and the mid-latitude trough are major factors in causing extreme precipitation (Bohlinger et al., 2017; Houze et al., 2017; Vellore et al., 2016). Figure 7 presents a sketch of the process leading to extreme precipitation in Nepal as conceived by Bohlinger et al. (2017). The mid-level trough from the mid-latitudes could force an LPS to change direction and ultimately lead it to Nepal where extreme precipitation occurs. The relationship is applicable to the western Himalayas in India, west of Nepal. Houze et al. (2017) provided detailed analysis of the storm that caused a devastating flood in June 2013 in Uttarakhand (India) in the western Himalayas (Fig. 1a, point 1). An eastward-propagating upper-level trough in the Asian westerly jet extended abnormally far southward and reached the Himalayas. The southern end of the trough merged with an LPS moving westward across India. The merged system produced a persistent moist low-level flow oriented normal to the Himalayas that advected a large amount of water vapor into the Uttarakhand region. The orographic lifting of the flow over the Himalayan barrier produced heavy continuous rain over the region for 2–3 days. The precipitation in this case was largely stratiform in nature, although embedded convection of moderate depth occurred along the foothills. Vellore et al. (2016) described the

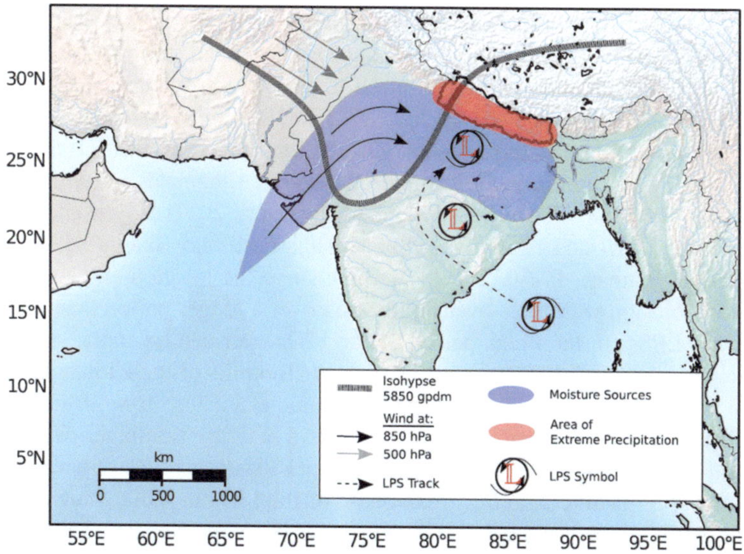

Fig. 7 Conceptual sketch of processes leading to extreme precipitation in Nepal (Bohlinger et al., 2017)

common large-scale circulation pattern connecting the mid-latitudes and the South Asian monsoon region, similar to Bohlinger et al. (2017). They also identified key elements of the monsoon–mid-latitude interaction, e.g., mid-latitude Rossby wave breaking, west–northwest propagation of monsoon LPSs from the BoB across the Indian subcontinent, eddy shedding of the Tibetan anticyclone, ageostrophic motions and transverse circulation across the Himalayas, and strong moist convection over the Himalayan foothills.

Quasi-stationary Rossby waves often appear and propagate eastward along the Asian subtropical jet in summer because the Asian subtropical jet acts as a waveguide for Rossby waves (Ambrizzi et al., 1995; Enomoto et al., 2003; Terao, 1998). On rare occasions, the amplitude of the waves increases intensively around the Tibetan Plateau, resulting in Rossby wave breaking and deepening of a trough around the Himalayas. Rossby wave propagation and westward shedding of the Tibetan anticyclone often vary on a quasi-biweekly time scale (i.e., the QBW) (Fujinami and Yasunari, 2004, 2009; Fujinami et al., 2014; Wang and Duan, 2015). The QBW mode is strongly linked with convective activity over the Tibetan Plateau and its surrounding regions including the Himalayas (Yasunari, 1976). The active phase of convective activity over the southern Tibetan Plateau is induced by a similar atmospheric circulation that causes the extreme precipitation over Nepal (Fig. 7 in the current chapter and Fig. 15 of Fujinami and Yasunari 2004). In short, the ISOs of the Asian subtropical jet and the Tibetan anticyclone are important in creating the environment for monsoon–extratropical interactions that cause extreme precipitation events.

Similar to the ISOs in the mid-latitudes, monsoon precipitation over Asia does not remain constant within the summer monsoon season but exhibits clear ISO with high and low precipitation phases, referred to as active and break phases, respectively. The submonthly scale (10–20 days) ISO or QBW and a 30–60-day oscillation called the Boreal Summer Intraseasonal Oscillation (hereafter, BSISO) are the two main modes of ISO in the Asian summer monsoon (Annamalai and Slingo, 2001; Goswami and Mohan, 2001; Hatsuzuka and Fujinami, 2017; Hoyos and Webster, 2007; Krishnamurti and Ardanuy, 1980; Krishnamurti and Bhalme, 1976; Wang et al., 2006; Yasunari, 1979). The ISOs can change the location and speed of the low-level monsoon westerlies over South Asia over a longer period than that associated with LPSs. In the break period of the ISOs over central India, strong low-level monsoon westerlies move northward to the foothills of the Himalayas, which provide conditions favorable for bringing abundant moisture toward the foothills of the Himalayas and enhancing local convective activity. Bohlinger et al. (2019) described a multiscale process associated with quasi-stationary Rossby waves along the Asian subtropical jet and the break phase of the ISOs in India, which led to an extreme precipitation event and devastating floods in central Nepal on July 19, 2007. The mid-latitude trough related to Rossby wave propagation was located to the north of the Himalayas, providing large-scale ascent over the Himalayas that prepared an environment favorable for the development of deep convection. The flow was orographically lifted by the topography of the foothills of the Himalayas, releasing convective instability and triggering the extreme precipitation event. Additionally, the low-level flow in the break period of the ISOs supplied an unusually warm and moist inflow toward the Himalayas, which caused a wide intense convective system.

The understanding of the process underlying the occurrence of heavy rainfall and mesoscale convective systems at higher elevations in the Himalayas remains inadequate. Therefore, we need to accumulate more case studies of various types of extreme precipitation event. Next, based mainly on the results of Hirata et al. (2023), we discuss a multiscale process within an LPS, a mesoscale convective system over the Himalayan slopes, and the ISOs over South Asia that led to heavy precipitation observed by the rain gauge network in Rolwaling Valley on July 8, 2019, under the international Himalaya Precipitation Study (HiPRECS) (Fujinami et al., 2021, 2022a; Hirata et al., 2023; Sunako et al., 2019). In this event, a deep mid-latitude trough was not observed around the Himalayas. Six rain gauge stations at elevations of ~1000 to ~4200 m asl form the precipitation observation network in Rolwaling Valley (Hirata et al., 2023). Heavy precipitation (112 mm day^{-1}) was observed on July 8, 2019 at Dongang station (2790 m asl) in Rolwaling Valley. At the other stations, precipitation tended to increase on the same day but did not exceed 100 mm day^{-1}. Most of the precipitation (81 mm) at Dongang occurred during 19:00–23:00 LT. Figure 8 shows the spatial distributions of precipitation and vertically integrated moisture flux at 21:00 LT. A monsoon LPS was present to the south of the Nepal Himalayas with active precipitation areas. A large southeasterly moisture flux appeared in the eastern–northern part of the LPS from the central to eastern Nepal Himalayas. The LPS lifted moisture from the lower troposphere and then transported it laterally into the central–eastern Nepal Himalayas within the middle troposphere, increasing the water vapor content over Rolwaling Valley (Hirata et al., 2023). A zonally elongated rainband developed along ~28°N over the central–eastern Nepal Himalayas where the moisture flux moved over the slopes. At 18:00 LT, a zonally elongated cloud/precipitation system was observed on the southern side of Dongang (Fig. 9b). The cloud system comprised local deep convective areas with horizontal scale of ~10–20 km (<232 K) and surrounding stratiform areas (>232 K). The cloud system moved northward from 19:00 LT and passed over Dongang at 21:00 LT. Thus, the heavy rainfall event was caused by the passage over Dongang station of the deep convective area inside the zonally elongated cloud/precipitation system. Numerical simulation using the Cloud-Resolving Storm Simulator (CReSS) (Tsuboki & Sakakibara, 2002) with 2-km horizontal resolution showed that surface heat fluxes from the slope surface prevailed under the mid-tropospheric (~500 hPa) southeasterly flow associated with the LPS around a mountain ridge on the upwind side of Dongang until 19:00 LT, enhancing convective instability (defined as $\partial\theta e/\partial z < 0$) over the slopes. Forced orographic lifting led to the release of convective instability, which triggered the development of a mesoscale precipitation system with a zonally elongated distribution of precipitation/cloud over the Himalayas (Hirata et al., 2023). The behavior of the simulated precipitation system resembled that of the observed cloud system passing over Dongang during the intense precipitation event. An area of strong convection (scale: ~20 km) was successfully simulated near Dongang station in the zonally elongated precipitation band (Fig. 10). However, the simulated precipitation rate was underestimated at the grid closest to Dongang station, which might reflect different locations of the intense precipitation area between the simulation and the observed data. The maximum precipitation at Dongang occurred at 21:00 LT, which was slightly earlier than the peak phase of nocturnal precipitation in Rolwaling Valley. Strong influence by the synoptic-scale environmental flow associated with the LPS might have modulated the mean diurnal cycle of precipitation seen in Fig. 2a.

The LPS that caused heavy rainfall over Rolwaling Valley was generated under the influence of two different ISOs, i.e., the BSISO and QBW. These two ISO modes have different spatiotemporal structures. The BSISO exhibits northward propagation of convection and circulation anomalies from the equatorial Indian Ocean (e.g., Hoyos and Webster, 2007; Kemball-Cook and Wang, 2001; Wang et al., 2006; Yasunari, 1979), whereas the QBW is characterized by westward propagation of convection and circulation anomalies from the western Pacific (e.g., Chen and Chen, 1993; Fujinami et al., 2014; Kikuchi and Wang, 2009; Krishnamurti and Ardanuy, 1980). The development of LPSs often occurs under a favorable large-scale environment such as a humid lower troposphere with low-level strong horizontal cyclonic wind shear and weak vertical wind shear (Kikuchi, 2021; Kikuchi and Wang, 2010; Yanase et al., 2012). The QBW and BSISO can modulate the large-scale environment substantially and largely determine the behavior of the LPSs (Goswami et al., 2003; Hatsuzuka and Fujinami, 2017; Krishnamurthy and Ajayamohan, 2010). Hatsuzuka and Fujinami (2017) proposed a process for initiation of the LPS over the Gangetic Plain around Bangladesh through combination of the BSISO and QBW modes (Fig. 11). A cyclonic anomaly of the BSISO moves northward from the equator to the BoB. The cyclonic anomaly enhances a westerly flow over the BoB, resulting in enhancement of

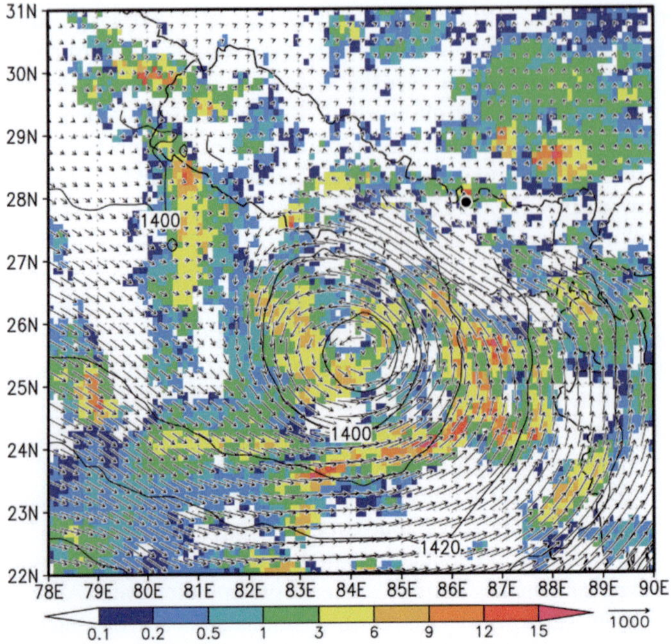

Fig. 8 Precipitation (shading; mm h^{-1}) from IMERG, 850-hPa geopotential height (contours), and vertically integrated water vapor flux from the ERA5 dataset at 21:00 LT at 90°E on July 8, 2019. The black circle indicates the location of Dongang in Rolwaling Valley (Fujinami et al., 2022b)

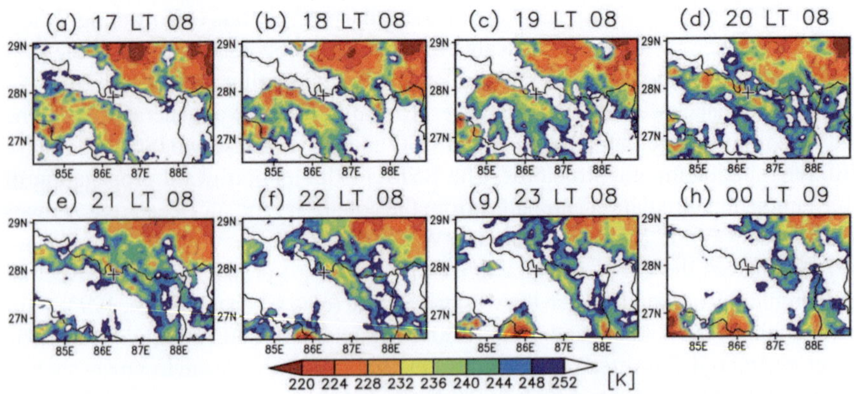

Fig. 9 **a–h** Maps of infrared brightness temperature derived from Meteosat from 17:00 LT on July 8, 2019 to 00:00 LT on July 9, 2019. Black cross indicates the location of Dongang (Hirata et al., 2023)

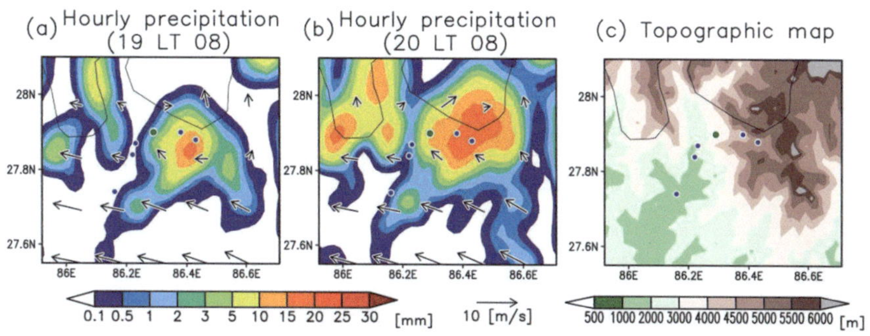

Fig. 10 Maps of hourly precipitation at the surface (shading; mm) and horizontal wind vector at 400 hPa (arrows; m s^{-1}) from numerical simulation at **a** 19:00 LT and **b** 20:00 LT on July 8, 2019. **c** Map of topography used in the numerical simulation. Closed circles indicate the locations of rain gauge stations in Rolwaling Valley. Dongang station is represented by the largest circle (Hirata et al., 2023)

the meridional cyclonic shear of zonal winds over the northern BoB and Bangladesh. The westward propagating anticyclonic anomaly of the QBW from the western North Pacific also contributes to enhancement of the strong cyclonic shear. Additionally, the two ISO modes contribute to the increase in mid-tropospheric humidity around the region of LPS genesis. Thus, the two ISOs provide environmental conditions favorable for LPS genesis.

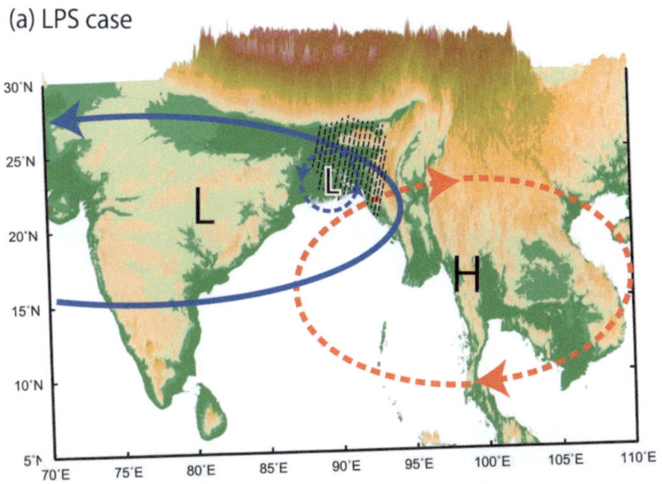

Fig. 11 Schematic of the relationship between the BSISO mode (solid arrows) and QBW mode (dashed arrows) when LPSs are present around Bangladesh. Symbols H and L represent the centers of anticyclonic (orange arrows) and cyclonic (blue arrows) anomalies associated with the ISO modes, respectively (Hatsuzuka and Fujinami, 2017)

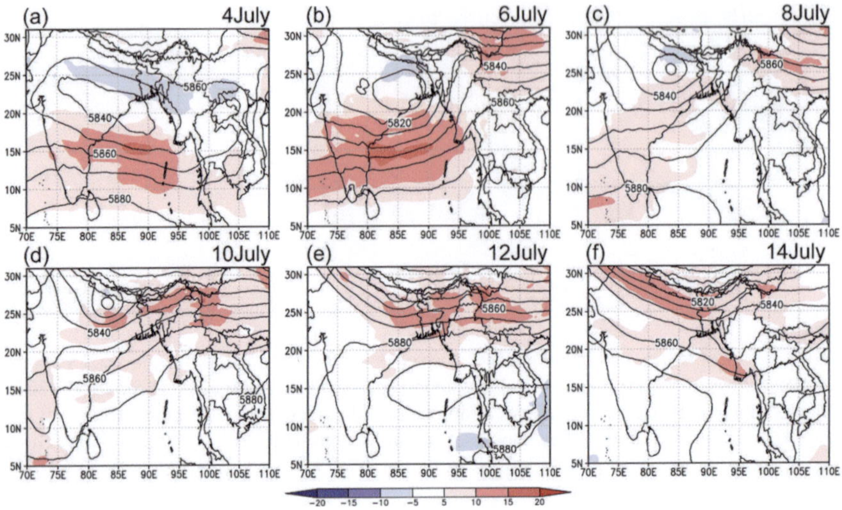

Fig. 12 Geopotential height (contours) and zonal wind component (shading) at 500 hPa from the ERA5 dataset from **a** July 4 to **f** July 14, 2019. Contour interval for geopotential height is 10 gpm (Fujinami et al., 2022b)

The behavior of the LPS observed on July 8, 2019 to the south of the Himalayas (Fig. 8) was similar to that of the process proposed by Hatsuzuka and Fujinami (2017). Figure 12 presents the time sequence of geopotential height at 500 hPa during July 4–14, 2019 to show the spatiotemporal relationship among the LPS, BSISO, and QBW for this case (Fujinami et al., 2022b). The lower–middle troposphere showed a spatial structure in terms of geopotential height similar to that at 500 hPa. The center of the monsoon trough was located over the Indian subcontinent and the BoB from 15° to 20°N. A local maximum of westerlies appeared along 15°N, while easterlies occurred around the Gangetic Plain (Fig. 12a). Northward propagation of the strong westerlies until July 6 was associated with the BSISO mode. The strong westerlies expanded northward to the head of the BoB, creating a zone of strong cyclonic shear to its north. Simultaneously, an anticyclonic circulation related to the QBW mode emerged over the Indochina Peninsula and moved westward from July 6, resulting in strengthening of the cyclonic shear associated with the BSISO mode. The anticyclone reached the BoB on July 12. July 6 corresponded to the day of LPS genesis (Hirata et al., 2023). The closed low pressure area to the south of the Himalayas on July 8 (Fig. 12c) corresponded to the LPS that caused heavy rainfall in Rolwaling Valley (Fig. 8). The LPS enhanced the easterlies in the mid-troposphere over the Himalayas during July 6–10. The northward shift of the westerlies by the ISOs pushed the LPS northward closer to the Himalayas during July 8–10. Finally, the LPS dissipated after July 12 around the foothills of the Himalayas. The strong westerlies that blew around the Himalayas came from the mid-latitudes with a drier air mass along the western periphery of the Tibetan Plateau (Fig. 12f). The northward shift of the strong monsoon westerlies around the Himalayas, which corresponded to the break phase

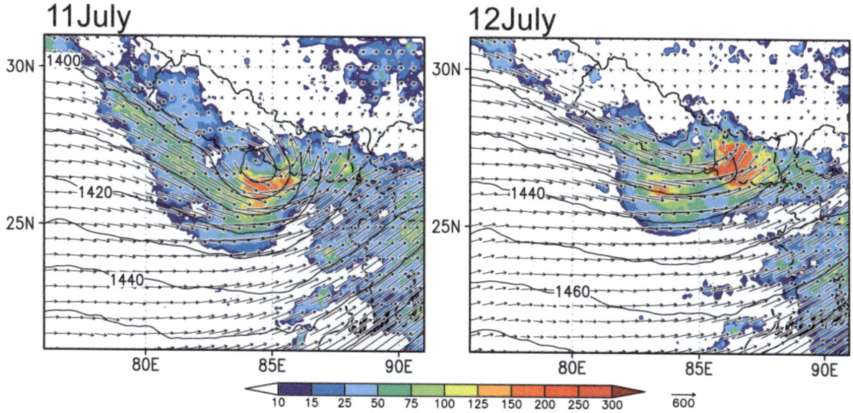

Fig. 13 Precipitation (shading; mm day^{-1}) from IMERG, geopotential height (contours; gpm) at 850 hPa, and vertically integrated water vapor flux (vectors; kg m^{-1} s^{-1}) from the ERA5 dataset on **left** July 11 and **right** July 12, 2019. Contour interval for geopotential height is 10 gpm

of the ISO in central India, and the LPS concentrated the flux of incoming moisture and its convergence over the slopes of the central–eastern Nepal Himalayas, leading to the heavy rainfall event and flooding on July 12 (Fig. 13). On July 12, the daily precipitation reached was >200 mm over eastern Nepal, where abundant moisture flux oriented normal to the Himalayan slopes advected large amounts of water vapor.

5 Concluding Remarks

In this chapter, we reviewed the variability of summer precipitation in the Himalayas, including extreme events, and its dynamics based on studies using in situ observations, spaceborne precipitation radar, high spatiotemporal atmospheric reanalysis data, and cloud-resolving numerical models. In particular, we focused on the diurnal cycle of precipitation, and the LPSs, ISOs, and their interplay that can be important in creating an environment favorable for extreme precipitation events in the Himalayas.

In the diurnal cycle of precipitation, twice-daily precipitation maxima corresponding to daytime and nighttime peaks occur across higher elevations of the Nepal Himalayas, while a single peak appears at lower elevations at nighttime. The two maxima are the result of different land surface effects. The daytime peak results from upslope flows driven by surface heating of the slopes, whereas the nighttime peak is caused by acceleration of low-level monsoon flows toward the Himalayas, known as nocturnal LLJs over the Gangetic Plain. The amplitude of the peaks is affected by synoptic-scale disturbances such as LPSs and ISOs. Therefore, the diurnal cycle does not always show the two distinct peaks each day. However, precipitation tends to occur frequently in the afternoon and at nighttime at higher elevations. Understanding how interactions between the diurnal cycle and synoptic-scale disturbances might affect the occurrence of an extreme precipitation event warrants further investigation. To unravel the dynamics, we might need cloud-resolving simulations with horizontal resolution better than 1 km capable of resolving the regional topography and its interaction with the large-scale atmospheric circulation, as well as advanced in situ observations such as those realized by ground precipitation radar. Additionally, further observational and numerical studies are required to better understand how precipitation systems, which are strongly influenced by the land surface effects of both the Himalayan slopes and the Indian subcontinent, maintain the Himalayan glaciers at higher elevations.

Common features of the atmospheric circulation that lead to heavy precipitation events in the Himalayas are abundant moisture transport toward the Himalayan slopes and its uplift. Day-to-day fluctuations of the moisture inflow are actually affected by the synoptic-scale atmospheric circulation in the mid-latitudes and in the tropics in summer. Specifically, LPSs such as monsoon lows, monsoon depressions, and mid-latitude troughs and their interaction create an environment that concentrates moist strong winds onto the Himalayan slopes and enhances forced uplift of the moist air, leading to heavy rainfall. The locations of the LPSs and the mid-latitude troughs are strongly controlled by ISOs on larger spatiotemporal scales that create an environment suitable for the genesis and development of LPSs and troughs. To better understand and predict heavy precipitation events, we need to improve understanding of the controlling multiscale processes over the complex terrain of the Himalayas.

Acknowledgements This study was supported by JSPS KAKENHI (grant number JP18KK0098 and JP23KK0064) and from JSPS and SNSF under the Joint Research Projects (JRPs grant number 20191503). Portions of this study were conducted as part of a collaborative research program of the Institute for Space-Earth Environmental Research, Nagoya University. We thank James Buxton MSc, from Edanz (https://jp.edanz.com/ac), for editing a draft of this manuscript.

References

Ageta, Y. (1976). Characteristics of precipitation during monsoon season in Khumbu Himal Glaciological Expedition to Nepal. *Seppyo, 38*(Special issue), 84–88.

Ambrizzi, T., Hoskins, B. J., & Hsu, H. H. (1995). Rossby-wave propagation and teleconnection patterns in the austral winter. *Journal of the Atmospheric Sciences, 52*, 3661–3672.

Annamalai, H., & Slingo, J. M. (2001). Active/break cycles: Diagnosis of the intraseasonal variability of the Asian Summer Monsoon. *Climate Dynamics, 18*, 85–102.

Barros, A. P., Joshi, M., Putkonen, J., & Burbank, D. W. (2000). A study of the 1999 monsoon rainfall in a mountainous region in central Nepal using TRMM products and rain gauge observations. *Geophysical Research Letters, 27*, 3683–3686.

Barros, A. P., & Lang, T. J. (2003). Monitoring the monsoon in the Himalayas: Observations in central Nepal, June 2001. *Monthly Weather Review, 131*, 1408–1427.

Bhatt, B. C., & Nakamura, K. (2006). A climatological–dynamical analysis associated with precipitation around the southern part of the Himalayas. *Journal of Geophysical Research-Atmospheres, 111*, D02115.

Blackadar, A. K. (1957). Boundary layer wind maxima and their significance for the growth of nocturnal inversions. *Bulletin of the American Meteorological Society, 38*, 283–290.

Bohlinger, P., & Sorteberg, A. (2018). A comprehensive view on trends in extreme precipitation in Nepal and their spatial distribution. *International Journal of Climatology, 38*, 1833–1845.

Bohlinger, P., Sorteberg, A., Liu, C. H., Rasmussen, R., Sodemann, H., & Ogawa, F. (2019). Multi-scale characteristics of an extreme precipitation event over Nepal. *Quarterly Journal of the Royal Meteorological Society, 145*, 179–196.

Bohlinger, P., Sorteberg, A., & Sodemann, H. (2017). Synoptic conditions and moisture sources actuating extreme precipitation in Nepal. *Journal of Geophysical Research-Atmospheres, 122*, 12653–12671.

Bollasina, M., Bertolani, L., & Tartari, G. (2002). Meteorological observations in the Khumbu Valley, Nepal Himalayas, 1994–1999. *Bulletin of Glaciological Research, 19*, 1–11.

Bookhagen, B. (2010). Appearance of extreme monsoonal rainfall events and their impact on erosion in the Himalaya. *Geomatics, Natural Hazards and Risk, 1*, 37–50.

Bookhagen, B., & Burbank, D. W. (2006). Topography, relief, and TRMM-derived rainfall variations along the Himalaya. *Geophysical Research Letters, 33*, L08405.

Chen, G. (2020). Diurnal cycle of the Asian summer monsoon: Air pump of the second kind. *Journal of Climate, 33*(5), 1747–1775.

Chen, T. C., & Chen, J. M. (1993). The 10–20-day mode of the 1979 Indian monsoon—Its relation with the time-variation of monsoon rainfall. *Monthly Weather Review, 121*, 2465–2482.

Collier, E., & Immerzeel, W. W. (2015). High-resolution modeling of atmospheric dynamics in the Nepalese Himalaya. *Journal of Geophysical Research-Atmospheres, 120*, 9882–9896.

Diaz, M., & Boos, W. R. (2019). Barotropic growth of monsoon depressions. *Quarterly Journal of the Royal Meteorological Society, 145*, 824–844.

Diaz, M., & Boos, W. R. (2021). The influence of surface heat fluxes on the growth of idealized monsoon depressions. *Journal of the Atmospheric Sciences, 78*, 2013–2027.

Dimri, A. P., Chevuturi, A., Niyogi, D., Thayyen, R. J., Ray, K., Tripathi, S. N., Pandey, A. K., & Mohanty, U. C. (2017). Cloudbursts in Indian Himalayas: A review. *Earth-Science Reviews, 168*, 1–23.

Ditchek, S. D., Boos, W. R., Camargo, S. J., & Tippett, M. K. (2016). A genesis index for monsoon disturbances. *Journal of Climate, 29*, 5189–5203.

Dong, W., et al. (2016). Summer rainfall over the southwestern Tibetan Plateau controlled by deep convection over the Indian subcontinent. *Nature Communications, 7*, 10925.

Dong, W., Lin, Y. L., Wright, J. S., Xie, Y. Y., Xu, F. H., Xu, W. Q., & Wang, Y. (2017). Indian monsoon low-pressure systems feed up-and-over moisture transport to the southwestern Tibetan Plateau. *Journal of Geophysical Research-Atmospheres, 122*, 12140–12151.

Enomoto, T., Hoskins, B. J., & Matsuda, Y. (2003). The formation mechanism of the Bonin high in August. *Quarterly Journal of the Royal Meteorological Society, 129*, 157–178.

Fujinami, H., & Yasunari, T. (2004). Submonthly variability of convection and circulation over and around the Tibetan Plateau during the boreal summer. *Journal of the Meteorological Society of Japan, 82*, 1545–1564.

Fujinami, H., & Yasunari, T. (2009). The effects of mid-latitude waves over and around the Tibetan Plateau on submonthly variability of the East Asian summer monsoon. *Monthly Weather Review, 137*, 2286–2304.

Fujinami, H., Fujita, K., Takahashi, N., Sato, T., Kanamori, H., Sunako, S., & Kayastha, R. B. (2021). Twice-daily monsoon precipitation maxima in the Himalayas driven by land surface effects. *Journal of Geophysical Research-Atmospheres, 126*, e2020JD034255.

Fujinami, H., Fujita, K., Takahashi, N., Sato, T., Hirata, H., Kanamori, H., Sato, Y., Gurung, T. R., Kayastha, R. B., & Shrestha, M. L. (2022b, March 22–26). Precipitation and its variability in the high-elevation area of the Nepal Himalayas. In *Seventh WMO international workshop on monsoons (IWM-7)*, India (online) (Invited talk).

Fujinami, H., Hirata, H., Kato, M., & Tsuboki, K. (2020). Mesoscale precipitation systems and their role in the rapid development of a monsoon depression over the Bay of Bengal. *Quarterly Journal of the Royal Meteorological Society, 146*, 267–283.

Fujinami, H., Sato, T., Kanamori, H., & Murata, F. (2017). Contrasting features of monsoon precipitation around the Meghalaya Plateau under westerly and easterly regimes. *Journal of Geophysical Research-Atmospheres, 122*, 9591–9610.

Fujinami, H., Sato, T., Kanamori, H., & Kato, M. (2022a). Nocturnal southerly moist surge parallel to the coastline over the western Bay of Bengal. *Geophysical Research Letters, 49*, e2022GL100174.

Fujinami, H., Yasunari, T., & Morimoto, A. (2014). Dynamics of distinct intraseasonal oscillation in summer monsoon rainfall over the Meghalaya-Bangladesh-western Myanmar region: Covariability between the tropics and mid-latitudes. *Climate Dynamics, 43*, 2147–2166.

Godbole, R. V. (1977). The composite structure of the monsoon depression. *Tellus, 29*, 25–40.

Goswami, B. N., & Mohan, R. S. A. (2001). Intraseasonal oscillations and interannual variability of the Indian summer monsoon. *Journal of Hydrometeorology, 14*, 1180–1198.

Goswami, B. N., Ajayamohan, R. S., Xavier, P. K., & Sengupta, D. (2003). Clustering of synoptic activity by Indian summer monsoon intraseasonal oscillations. *Geophysical Research Letters, 30*, 1431.

Hatsuzuka, D., & Fujinami, H. (2017). Effects of the South Asian monsoon intraseasonal modes on genesis of low pressure systems over Bangladesh. *Journal of Climate, 30*, 2481–2499.

Hatsuzuka, D., Yasunari, T., & Fujinami, H. (2014). Characteristics of low pressure systems associated with intraseasonal oscillation of rainfall over Bangladesh during boreal summer. *Monthly Weather Review, 142*(12), 4758–4774.

Hersbach, H., Bell, B., Berrisford, P., Hirahara, S., Horanyi, A., Munoz-Sabater, J., et al. (2020). The ERA5 global reanalysis. *Quarterly Journal of the Royal Meteorological Society, 146*, 1999–2049.

Higuchi K. (1977). Effect of nocturnal precipitation on the mass balance of the Rikha Samba glacier, Hidden Valley, Nepal. *Seppyo, 39*(Special issue), 84–88.

Hirata, H., Fujinami, H., Kanamori, H., Sato, Y., Kato, M., Kayastha, R. B., Shrestha, M. L., & Fujita, K. (2023). Multiscale processes leading to heavy precipitation in the eastern Nepal Himalayas. *Journal of Hydrometeorology, 24*, 641–658.

Hirose, M., Oki, R., Shimizu, S., Kachi, M., & Higashiuwatoko, T. (2008). Fine-scale diurnal rainfall statistics refined from eight years of TRMM PR data. *Journal of Applied Meteorology and Climatology, 47*, 544–561.

Hirose, M., Takayabu, Y. N., Hamada, A., Shige, S., & Yamamoto, M. K. (2017). Spatial contrast of geographically induced rainfall observed by TRMM PR. *Journal of Climate, 30*, 4165–4184.

Houze, R. A., Rasmussen, K. L., Medina, S., Brodzik, S. R., & Romatschke, U. (2011). Anomalous atmospheric events leading to the summer 2010 floods in Pakistan. *Bulletin of the American Meteorological Society, 92*, 291–298.

Houze, R. A., McMurdie, L. A., Rasmussen, K. L., Kumar, A., & Chaplin, M. M. (2017). Multiscale aspects of the storm producing the June 2013 flooding in Uttarakhand, India. *Monthly Weather Review, 145*, 4447–4466.

Houze, R. A., Wilton, D. C., & Smull, B. F. (2007). Monsoon convection in the Himalayan region as seen by the TRMM Precipitation Radar. *Quarterly Journal of the Royal Meteorological Society, 133*, 1389–1411.

Hoyos, C. D., & Webster, P. J. (2007). The role of intraseasonal variability in the nature of Asian monsoon precipitation. *Journal of Climate, 20*, 4402–4424.

Hunt, K. M. R., Turner, A. G., Inness, P. M., Parker, D. E., & Levine, R. C. (2016a). On the structure and dynamics of Indian monsoon depressions. *Monthly Weather Review, 144*, 3391–3416.

Hunt, K. M. R., Turner, A. G., & Parker, D. E. (2016b). The spatiotemporal structure of precipitation in Indian monsoon depressions. *Quarterly Journal of the Royal Meteorological Society, 142*, 3195–3210.

Hunt, K. M. R., Turner, A. G., & Schiemann, R. K. H. (2022). Katabatic and convective processes drive two preferred peaks in the precipitation diurnal cycle over the Central Himalaya. *Quarterly Journal of the Royal Meteorological Society, 148*, 1731–1751.

Hurley, J. V., & Boos, W. R. (2015). A global climatology of monsoon low-pressure systems. *Quarterly Journal of the Royal Meteorological Society, 141*, 1049–1064.

Ichiyanagi, K., Yamanaka, M. D., Muraji, Y., & Vaidya, B. K. (2007). Precipitation in Nepal between 1987 and 1996. *International Journal of Climatology, 27*, 1753–1762.

Immerzeel, W. W., van Beek, L. P. H., & Bierkens, M. F. P. (2010). Climate change will affect the Asian water towers. *Science, 328*, 1382–1385.

Joseph, S., Sahai, A. K., Sharmila, S., Abhilash, S., Borah, N., Chattopadhyay, R., Pillai, P. A., Rajeevan, M., & Kumar, A. (2015). North Indian heavy rainfall event during June 2013: Diagnostics and extended range prediction. *Climate Dynamics, 44*, 2049–2065.

Kansakar, S. R., Hannah, D. M., Gerrard, J., & Rees, G. (2004). Spatial pattern in the precipitation regime of Nepal. *International Journal of Climatology, 24*, 1645–1659.

Karki, R., ul Hasson, S., Gerlitz, L., Schickhoff, U., Scholten, T., & Böhner, J. (2017). Quantifying the added value of convection-permitting climate simulations in complex terrain: A systematic evaluation of WRF over the Himalayas. *Earth System Dynamics, 8*, 507–528.

Kataoka, A., & Satomura, T. (2005). Numerical simulation on the diurnal variation of precipitation over northeastern Bangladesh: A case study of an active period 14–21 June 1995. *SOLA, 1*, 205–208.

Kemball-Cook, S., & Wang, B. (2001). Equatorial waves and air–sea interaction in the Boreal summer intraseasonal oscillation. *Journal of Climate, 14*, 2923–2942.

Kikuchi, K. (2021). The Boreal Summer Intraseasonal Oscillation (BSISO): A review. *Journal of the Meteorological Society of Japan, 99*, 933–972.

Kikuchi, K., & Wang, B. (2009). Global perspective of the Quasi-Biweekly Oscillation. *Journal of Climate, 22*, 1340–1359.

Kikuchi, K., & Wang, B. (2010). Formation of tropical cyclones in the northern Indian Ocean associated with two types of tropical intraseasonal oscillation modes. *Journal of the Meteorological Society of Japan, 88*, 475–496.

Koike, T. (2004). The Coordinated Enhanced Observing Period: An initial step for integrated global water cycle observation. *WMO Bulletin, 53*, 115–121.

Krishnamurthy, V., & Ajayamohan, R. S. (2010). Composite structure of monsoon low pressure systems and its relation to Indian rainfall. *Journal of Climate, 23*, 4285–4305.

Krishnamurti, T. N., & Ardanuy, P. (1980). 10 to 20-day westward propagating mode and breaks in the monsoons. *Tellus, 32*, 15–26.

Krishnamurti, T. N., & Bhalme, H. N. (1976). Oscillations of a monsoon system. 1. Observational Aspects. *Journal of the Atmospheric Sciences, 33*, 1937–1954.

Kubota, T. et al. (2020). Global satellite mapping of precipitation (GSMaP) products in the GPM Era. In V. Levizzani, C. Kidd, D. B. Kirschbaum, C. D. Kummerow, K. Nakamura, & F. J. Turk (Eds.), *Satellite Precipitation Measurement. Advances in Global Change Research* (Vol. 67). Springer.

Kurosaki, Y., & Kimura, F. (2002). Relationship between topography and daytime cloud activity around the Tibetan Plateau. *Journal of the Meteorological Society of Japan, 80*, 1339–1355.

Lang, T. J., & Barros, A. P. (2002). An investigation of the onsets of the 1999 and 2000 monsoons in central Nepal. *Monthly Weather Review, 130*, 1299–1316.

Lin, C., Chen, D., Yang, K., & Ou, T. (2018). Impact of model resolution on simulating the water vapor transport through the central Himalayas: Implication for models' wet bias over the Tibetan Plateau. *Climate Dynamics, 51*, 3195–3207.

Lin, C., Yang, K., Chen, D., Guyennon, N., Balestrini, R., Yang, X., Acharya, A., Ou, T., Yao, T., Tartari, G., et al. (2021). Summer afternoon precipitation associated with wind convergence near the Himalayan glacier fronts. *Atmospheric Research, 259*, 105658.

Martens, B., Schumacher, D. L., Wouters, H., Munoz-Sabater, J., Verhoest, N. E. C., & Miralles, D. G. (2020). Evaluating the land–surface energy partitioning in ERA5. *Geoscientific Model Development, 13*(9), 4159–4181.

Medina, S., Houze, R. A., Kumar, A., & Niyogi, D. (2010). Summer monsoon convection in the Himalayan region: Terrain and land cover effects. *Quarterly Journal of the Royal Meteorological Society, 136*, 593–616.

Murata, F., Terao, T., Fujinami, H., Hayashi, T., Asada, H., Matsumoto, J., & Syiemlieh, H. J. (2017). Dominant synoptic disturbance in the extreme rainfall at Cherrapunji, Northeast India, based on 104 years of rainfall data (1902–2005). *Journal of Climate, 30*, 8237–8251.

Nakamura, K. (2021). Progress from TRMM to GPM. *Journal of the Meteorological Society of Japan, 99*, 697–729.

Ohata, T., Higuchi, K., & Ikegami, K. (1981). Mountain–valley wind system in the Khumbu Himal, East Nepal. *Journal of the Meteorological Society of Japan, 59*, 753–762.

Orr, A., Listowski, C., Couttet, M., Collier, E., Immerzeel, W., Deb, P., & Bannister, D. (2017). Sensitivity of simulated summer monsoonal precipitation in Langtang Valley, Himalaya, to cloud microphysics schemes in WRF. *Journal of Geophysical Research-Atmospheres, 122*, 6298–6318.

Ouyang, L., Yang, K., Lu, H., Chen, Y., Zhu, L., Zhou, X., & Wang, Y. (2020). Ground-based observations reveal unique valley precipitation patterns in the central Himalaya. *Journal of Geophysical Research-Atmospheres, 125*, e2019JD031502.

Potter, E. R., Orr, A., Willis, I. C., Bannister, D., & Salerno, F. (2018). Dynamical drivers of the local wind regime in a Himalayan valley. *Journal of Geophysical Research-Atmospheres, 123*, 13, 186–113, 202.

Romatschke, U., & Houze, R. A. (2011). Characteristics of precipitating convective systems in the South Asian monsoon. *Journal of Hydrometeorology, 12*, 3–26.

Saha, K., Sanders, F., & Shukla, J. (1981). Westward propagating predecessors of monsoon depressions. *Monthly Weather Review, 109*, 330–343.

Salerno, F., et al. (2015). Weak precipitation, warm winters and springs impact glaciers of south slopes of Mt. Everest (central Himalaya) in the last 2 decades (1994–2013). *The Cryosphere, 9*, 1229–1247.

Samanta, D., Hameed, S. N., Jin, D., Thilakan, V., Ganai, M., Rao, S. A., & Deshpande, M. (2018). Impact of a narrow coastal Bay of Bengal sea surface temperature front on an Indian summer monsoon simulation. *Scientific Reports, 8*, 17694.

Sato, T. (2013). Mechanism of orographic precipitation around the Meghalaya Plateau associated with intraseasonal oscillation and the diurnal cycle. *Monthly Weather Review, 141*, 2451–2466.

Sharma, S., Hamal, K., Khadka, N., & Joshi, B. B. (2020). Dominant pattern of year-to-year variability of summer precipitation in Nepal during 1987–2015. *Theoretical and Applied Climatology, 142*, 1071–1084.

Shea, J. M., Wagnon, P., Immerzeel, W. W., Biron, R., Brun, F., & Pellicciotti, F. (2015). A comparative high-altitude meteorological analysis from three catchments in the Nepalese Himalaya. *International Journal of Water Resources Development, 31*, 174–200.

Shrestha, A. B., Wake, C. P., Dibb, J. E., & Mayewski, P. A. (2000). Precipitation fluctuations in the Nepal Himalaya and its vicinity and relationship with some large scale climatological parameters. *International Journal of Climatology, 20*, 317–327.

Shrestha, D., Singh, P., & Nakamura, K. (2012). Spatiotemporal variation of rainfall over the central Himalayan region revealed by TRMM Precipitation Radar. *Journal of Geophysical Research-Atmospheres, 117*(D22), D22106.

Shrestha, M. L. (2000). Interannual variation of summer monsoon rainfall over Nepal and its relation to the Southern Oscillation Index. *Meteorology and Atmospheric Physics, 75*, 21–28.

Sikka, D. R. (1977). Some aspects of the life history, structure and movement of monsoon depressions. *Pure and Applied Geophysics, 115*, 1501–1529.

Stensrud, D. J. (1996). Importance of low-level jets to climate: A review. *Journal of Climate, 9*, 1698–1711.

Sugimoto, S., & Takahashi, H. G. (2016). Effect of spatial resolution and cumulus parameterization on simulated precipitation over South Asia. *SOLA, 12A*, 7–12.

Sugimoto, S., Ueno, K., Fujinami, H., Nasuno, T., Sato, T., & Takahashi, H. G. (2021). Cloud-resolving-model simulations of nocturnal precipitation over the Himalayan slopes and foothills. *Journal of Hydrometeorology, 22*, 3171–3188.

Sunako, S., Fujita, K., Sakai, A., & Kayastha, R. B. (2019). Mass balance of Trambau Glacier, Rolwaling region, Nepal Himalaya: In situ observations, long-term reconstruction and mass-balance sensitivity. *Journal of Glaciology, 65*, 605–616.

Tan, J., Huffman, G. J., Bolvin, D. T., & Nelkin, E. J. (2019). Diurnal cycle of IMERG V06 precipitation. *Geophysical Research Letters, 46*, 13584–13592.

Terao, T. (1998). Barotropic disturbances on intraseasonal time scales observed in the mid-latitudes over the Eurasian continent during the northern summer. *Journal of the Meteorological Society of Japan, 76*, 419–436.

Terao, T., Islam, M. N., Hayashi, T., & Oka, T. (2006). Nocturnal jet and its effects on early morning rainfall peak over northeastern Bangladesh during the summer monsoon season. *Geophysical Research Letters, 33*, L18806.

Tsuboki, K., & Sakakibara, A. (2002). Large-scale parallel computing of cloud resolving storm simulator. In H. P. Zima et al. (Eds.), *High Performance Computing* (pp. 243–259). Springer.

Ueno, K., & Aryal, R. (2008). Impact of tropical convective activity on monthly temperature variability during non-monsoon season in the Nepal Himalayas. *Journal of Geophysical Research, 113*, D18112.

Ueno, K., & Nakileza, B. (2023). Atmospheric envelopes and glacial retreat. In F. O. Sarmiento (Ed.), *Montology Palimpsest. Montology* (Vol. 1). Springer.

Ueno, K., et al. (2001). Meteorological observations during 1994–2000 at the automatic weather station (GEN-AWS) in Khumbu region, Nepal Himalayas. *Bulletin of Glaciological Research, 18*, 23–30.

Ueno, K., Toyotsu, K., Bertolani, L., & Tartari, G. (2008). Stepwise onset of monsoon weather observed in the Nepal Himalaya. *Monthly Weather Review, 136*, 2507–2522.

Vellore, R. K., Kaplan, M. L., Krishnan, R., Lewis, J. M., Sabade, S., Deshpande, N., Singh, B. B., Madhura, R. K., & Rama Rao, M. V. S. (2016). Monsoon–extratropical circulation interactions in Himalayan extreme rainfall. *Climate Dynamics, 46*, 3517–3546.

Wang, B., Webster, P., Kikuchi, K., Yasunari, T., & Qi, Y. J. (2006). Boreal summer quasi-monthly oscillation in the global tropics. *Climate Dynamics, 27*, 661–675.

Wang, M., & Duan, A. (2015). Quasi-biweekly oscillation over the Tibetan Plateau and its link with the Asian summer monsoon. *Journal of Climate, 28*, 4921–4940.

Webster, P. J., Toma, V. E., & Kim, H. M. (2011). Were the 2010 Pakistan floods predictable? *Geophysical Research Letters, 38*, L04806.

Yamamoto, M., Ueno, K., & Nakamura, K. (2011). Comparison of satellite precipitation products with rain gauge data for the Khumb Region, Nepal Himalayas. *Journal of the Meteorological Society of Japan, 89*, 597–610.

Yanase, W., Satoh, M., Taniguchi, H., & Fujinami, H. (2012). Seasonal and intraseasonal modulation of tropical cyclogenesis environment over the Bay of Bengal during the extended summer monsoon. *Journal of Climate, 25*, 2914–2930.

Yang, K., Guyennon, N., Ouyang, L., Tian, L. D., Tartari, G., & Salerno, F. (2018). Impact of summer monsoon on the elevation-dependence of meteorological variables in the south of central Himalaya. *International Journal of Climatology, 38*, 1748–1759.

Yasunari, T. (1976). Spectral analysis of monsoonal precipitation in the Nepal Himalaya. *Seppyo, 38*(Special issue), 59–65.

Yasunari, T. (1979). Cloudiness fluctuations associated with the Northern Hemisphere summer monsoon. *Journal of the Meteorological Society of Japan, 57*, 227–242.

Yasunari, T., & Inoue, J. (1978). Characteristics of monsoonal precipitation around peaks and ridges in Shorong and Khumbu Himal. *Journal of the Japanese Society of Snow and Ice, 40*, 26–32.

Physical and Dynamical Characteristics of Severe Storms: Bangladesh Perspective

T. Rashid, D. A. Quadir, F. Akter, S. D. Shuvo, M. Islam, M. N. Ahasan, and M. A. Rahman

Abstract The chapter describes the physical and dynamical characteristics of severe storms in the perspective of Bangladesh. An event on 30 March 2018 has been selected as an extreme event that had occurred in the north-eastern part of Bangladesh which damaged crop and houses, and at least six people were killed in different regions of Bangladesh due to severe hailstorms. The high-quality data from European Centre for Medium Range Weather Forecasts (ECMWF) Reanalysis (ERA5) sites were used. The satellite imagery from INSAT of 30 March 2018 were also used. The analysis has shown that lowering of the pressure occurs over Bangladesh and adjacent areas of India within the pre-existing east–west oriented low-pressure trough. The strong warm south-westerly low-level winds pump a plentiful amount of moisture from the head Bay of Bengal in to the land areas of Bangladesh. The cool and dry airmass enters the area from the northern mountain and forms a discontinuity with dryline interface. The surface temperature distribution shows the thermal discontinuity and vector winds indicate the airmass discontinuity. These conditions initiate the convective activities to the north of Bangladesh as early as 0000 UTC, which is clearly seen in the satellite imagery. The area of occurrence of the severe storms are characterized by the high value of low-level relative vorticity in the order of > $150 \times 10^{-5}\ s^{-1}$, high value of low level convergence in the order of > $40 \times 10^{-5}\ s^{-1}$, high value of vertical velocity of wind higher than 12 m/s, high value of CAPE higher than 3000 J/kg and low value of CINE (<50 J/kg), high value of cloud-water mixing ratio in the order of > 30 g/kg and very high value of ice-water mixing ratio higher than 80 g/kg and low-level reflectivity > 50 dBZ with vertical extents upto 150 hPa level with a core at 500 hPa level. The above physical and dynamical characteristics of severe storms are found for the thunderstorm case and may be considered as the prerequisites for genesis of the severe storms over Bangladesh.

Keywords Severe storm · WRF model · Physical and dynamical characteristics · Instability

---

T. Rashid (✉) · D. A. Quadir · F. Akter · S. D. Shuvo · M. Islam · M. N. Ahasan · M. A. Rahman
Department of Meteorology, University of Dhaka, Dhaka, Bangladesh
e-mail: towhida_rashid@yahoo.com

S. Das and W-K. Tao (eds.), *Severe Storms*,
https://doi.org/10.1007/978-981-97-7075-5_14

1 Introduction

Bangladesh is situated within the latitudes 20°34′–26°38′ N and longitudes 88°01′–92°41′ E with the foothills of the Himalayas to its north, northeast and the eastern side of Bangladesh with the Bay of Bengal to the south. Thunderstorms are characterized by stormy winds (60–150 km/h), which sometimes become extremely severe and turn into tornadoes. The country experiences a good number of thunderstorms (herein-after refer as storms) mainly occurring in the pre-monsoon period (March–May) which have great destructive potential and often gain the strength of a tornado (wind speed > 300 km/hour) accompanied by high magnitude squalls, gusty winds, torrential rains, lightning, and hails. They cause extensive damage to crops, infrastructures, and lives and livelihoods in the areas across which they traverse.

Thunderstorms occur almost everywhere in the world due to atmospheric instability and strong moisture convergence causing severe convections. More than 1600–2400 thunderstorms occur at any moment around the world (Ahrens, 2013). The size of thunderstorms in Bangladesh varies from a few kilometres to a few hundreds of kilometres, usually with a life span of less than an hour (Saha & Quadir, 2016). These authors mentioned that multi-cell thunderstorms developed due to organized intense convection may have life spans of several hours and may travel over hundreds of kilometres.

Thunderstorms are very common in Bangladesh and its adjacent areas of India such as Gangetic West Bengal, Assam, Orissa, Jharkhand, and some parts of Bihar. The places of formation of the thunderstorms pass over Bangladesh may either be located within Bangladesh or further north over Sub-Himalayan West Bengal, Sikkim, Assam valley, the plateau of Meghalaya, or to the west of West Bengal, and even over the plateau of Jharkhand in India—a good number of those systems then move to Bangladesh (Fig. 1). Thunderstorms occur predominantly in the pre-monsoon months, especially during March–May; however, they are also found to occur in other seasons. The pre-monsoon (March–May) thunderstorms are also designated as Nor'westers in this region, so called because the storms usually move towards Bangladesh from a northwesterly direction. These are also locally known as 'Kalbaishakhi' in Bengali because of the heavy damage to life and property inflicted by them during the month of Baishakh (from mid-April to mid-May), the first month of the Bengali calendar.

The local storms may be categorized as severe storms if it is associated with hail, thunder, lightning, high gusty winds, etc. (Houze, 1981; Chevuturi et al., 2015). And these severe storms occur with strong vertical wind shear, which is an ideal condition for severe convection (Orville & Kopp, 1977; Chevuturi et al., 2015). A number of favourable physical and thermodynamic conditions have to combine together for the formation of thunderstorms which may then turn into severe storms.

Doswell (1992) and later McNulty (1995) pointed out that—the three most important factors for the occurrence of severe storms include—intense instability, a sufficiently deep humid layer in the lower and middle troposphere, and an updraft to initiate convection. Warm air rising and cold air sinking (convection) play a key

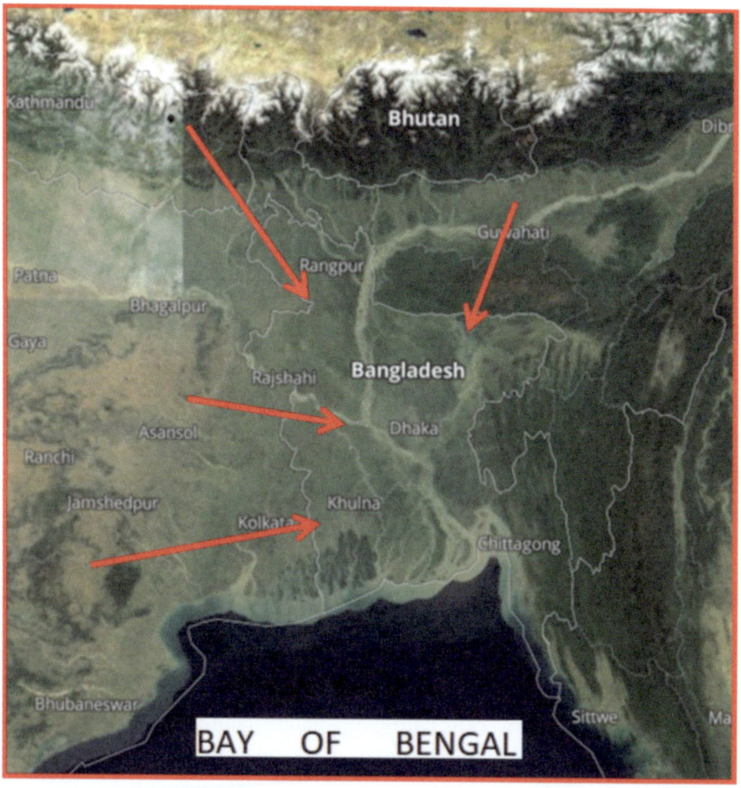

Fig. 1 Map of Bangladesh and surrounding areas showing the directions along which the thunderstorms cells move toward Bangladesh from its neighbouring territories. The arrows show the direction

role in the formation of severe thunderstorms. Instability and moisture conditions are commonly used to determine the strength of a storm's hydrodynamic conditions (rate of rain release, formation of hydrometeors, etc.) while wind shear determines the storm's movement and life span (Weisman & Rotunno, 2000, 2004; Weisman et al., 1988; Chaudhari et al., 2010). A strong wind shear, by its own, is able to produce supercells and make storms as well as to continue the process of updraft and downdraft separation. Eventually, the rotating updraft falls into the storm with the help of directional shear (Das et al., 2015). The maximum updraft speed measured by radars for a typical cumulonimbus over Europe by Schmidt et al. (2012) was about 10.5 m/s. But, Uma and Rao (2009) found the strongest updrafts in deep convection with a value of 15–20 m/s as shown by RADAR data of India. Stratiform rain areas are found behind the initial line while squall lines bow out because of the formation of the high-pressure system (Cotton & Anthes, 1989; Davis et al., 2004; Houze, 1993). During the occurrence of severe storms, strong wind is observed which is a threat to aviation; the highest number of aviation hazards are caused by thunderstorms (Litta et al., 2012).

The results of the model-simulated values are always lower than the satellite-derived hydrometeor profiles. The value of wind shear containing 6–8 m/s is generally situated in the lower troposphere. Solenoidal field and warm advection are also found where geopotential contours cut across isothermal flow over Bangladesh at 300 hPa (Das et al., 2015). According to these authors, the strong low-level heating and cooperative positive vorticity in the wind field support the initial rise and growth of the thermals. Cumulonimbus systems characterized by deep convective core echo structures occur almost exclusively over land, where extreme buoyancy is produced through sensible heat flux from heated land surfaces. Studies of convective regimes over the northern Indian Ocean adjoining Sri Lanka and Maldives have been carried out under special field experiments (Webster et al., 2002).

Brooks et al. (2003) demonstrated that the threat of significant severe convective weather increases with increasing Convective Available Potential Energy (CAPE). Karmakar and Alam (2006) concluded on the fact that almost 94% of Nor'westers (pre-monsoon thunderstorms) over Bangladesh occur when the Severe Weather Index (SWI) is greater than or equal to 100. Lopez and Bauer (2007) mentioned that low level temperature and relative humidity are useful in forecasting the likelihood of the occurrence of a thunderstorm. According to Zipser et al. (2006) the greater convective updraft speed indicates more intense storms that carry more ice into the anvils and produce more lightning.

The thunderstorms mainly originate over heated land masses with an adequate supply of moisture. In India and Bangladesh, these thunderstorms reach severity when the relatively cooler and dryer continental air mass meets warm and moist air from the ocean (SAARC STORM, 2005). The advection of cool air in the middle troposphere over the warm and moist air in the lower level enhances the probability of thunderstorm formation. Ahasan et al. (2014) in a modelling exercise with the WRF model have shown a clear convergence in the lower troposphere and divergence over the upper troposphere at 200 mb. Further to note that vertical wind shear ranging from 12 to 15 m/s between 1000 and 500 hPa favours the development of super-cell thunderstorms, some of which were found to be associated with tornadoes (Ahasan et al., 2014; SAARC STORM, 2005).

Atmospheric instability indices of various types as observed in early hours of the day are helpful to forecasting the thunderstorms 3–4 h ahead. Karmakar and Alam (2005) have shown that a range of the Showalter Index (SI) within the values between −3 and 3 corresponds to a high frequency of thunderstorm activities over the region of eastern India and Bangladesh. A decrease of SI values is closely connected with the sharp increase of the probability of convection. It was shown that the modified Showalter Index is a suitable predictor for the convective activity of this region. The Total Total Index (TTI) between 40–45 and K-Index (KI) ranging from 35 to 40 are favourable for thunderstorm formation (Jayakrishnan & Babu, 2014). Karmakar and Quadir (2014) have shown that the thunderstorms over Bangladesh and surrounding territories of India are formed if the moisture content is high in the morning time with a mixing ratio > 16–22.4 g/kg (RH > 85%). These authors have also shown that the high vertical gradient of the equivalent potential temperature (EPT) at 0.00UTC in the lower troposphere is an indicator for thunderstorm development in the late hours of the corresponding day. EPT decreases rapidly with height up to the 500hPalevel and then it increases, indicating the probable occurrence of thunderstorms. It may be noted that the formulation of EPT considers mixing ratio as an element; as a result, the value of EPT, as well as its vertical gradient, play an important role in moist convection. Owing to the short temporal and spatial scale of thunderstorms, forecasting the precise time and location of occurrences of thunderstorms remains as a challenging task.

Tyagi et al. (2011) investigated the pre-monsoon convective activities over Kolkata (22.52°N, 88.37°E) (close to Bangladesh) during STORM field phase 2006–2008 using 12 UTC radiosonde data and thermodynamic indices. The study suggests that the indices with highest skill for thunderstorm prediction are KI, DCI, SWEAT, DEW, HI, RH, LI, TTI, while the prediction efficiency is poor for CAPE, BRN, and BI. Observed values of these indices also reveal that scattered, multi-cellular thunderstorms are possible over these areas during pre-monsoon months. Ferdaus et al. (2021) analysed the thunderstorm predictability on the basis of some of the indices as described above and found that most of the indices sharply increase a few hours (1.5–6 h) before the storms.

The present chapter provides an introductory section (Sect. 1) which provides the background information and a brief review of the physics and dynamics of the thunderstorms of Bangladesh and adjoining territories of India, Climatological Characteristics of Thunderstorms (Sect. 2), Physical and Dynamical Characteristics of Thunderstorm: Bangladesh Perspectives, which has a few important subsections describing some aspects of thunderstorm, such as satellite observations, pressure or geopotential fields, winds, vorticity, divergence, vertical wind shears and thermodynamic parameters and atmospheric instability indices obtained from observations and NWP modelling for a limited cases (Sect. 3) and a Conclusion (Sect. 4), which summarises findings with a list of references at the end.

2 Climatology of Thunderstorms

The climatology of thunderstorms was produced by Saha and Quadir (2016) using the data for the period 1976–2010. Around 143 thunderstorms formed in Bangladesh during the period of 35 years. The monthly distribution shows that the highest number of thunderstorms occurs in the month of May and the next highest in June (Fig. 2). The monthly distribution of the country average thunderstorm frequency shows that the highest number corresponds to the month of May and the next highest to June and September with frequency of 27.1, 22.9 and 20.3 respectively. Thunderstorm frequency is low in the peak monsoon months of July and Augusts (around 15). The seasonal distribution of the thunderstorms over Bangladesh is shown in Table 1 respectively. The months of November–December and January rarely have

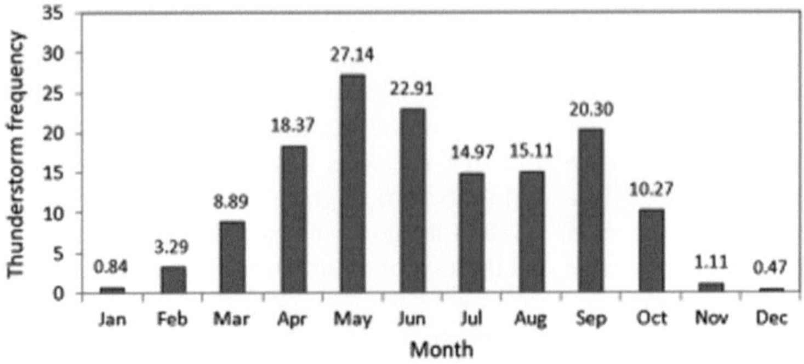

Fig. 2 Monthly distribution of thunderstorms over Bangladesh (Saha & Quadir, 2016)

Table 1 Seasonal and annual climatology of thunderstorms (Saha & Quadir, 2016)

Season	Average thunderstorm frequency
Winter	4.6
Pre-monsoon	54.41
Monsoon	73.29
Post-monsoon	11.38
Annual	143.68

any thunderstorm. Thunderstorms are associated with rainfall and hail in the pre-monsoon months. Nearly 20% of the Bangladesh total annual rainfall comes in the pre-monsoon season, most of which is associated with thunderstorm activity. The distribution of hourly frequency of Bangladesh thunderstorms shows that tornado occurrences are more frequent in the afternoon and early part of the night (Yamane et al., 2010 and Bhattacharya et al., 2013) (Figs. 3, 4 and 5).

This has been used by many researchers to understand regional trends, frequency, spatial distribution, and severity of storms in different parts of the world (Kunz et al., 2009; Berthet et al., 2011; Piani et al., 2005; García-Ortega et al., 2014; Cao, 2008; Brooks & Dotzek, 2008; Xie et al., 2008; Leslie et al., 2008). Most of this research indicate increase in the storm frequency and intensity in the recent years. Some detail study on thunderstorm climatology has found that there is a increasing trend of severe storm events over Bangladesh (Saha & Quadir, 2016) in a 35 years of data from 1976 to 2010 with strong peaks in the early 1990s. It is seen that the thunderstorm frequency slowly decreases after the Nineties' (Sabrina et al., 2020). Saha and Quadir (2016) have further shown that the month of May in the years with high thunderstorm frequency have negative anomalies of geopotential heights in the lower troposphere and positive anomalies in the higher troposphere. The specific humidity distribution shows that the lower troposphere is characterised by high positive anomaly. The v-component of the wind has positive anomaly, which helps the moisture supply over the land areas helping the convective activities. It was evident that the time series of thunderstorm frequency of May and the corresponding

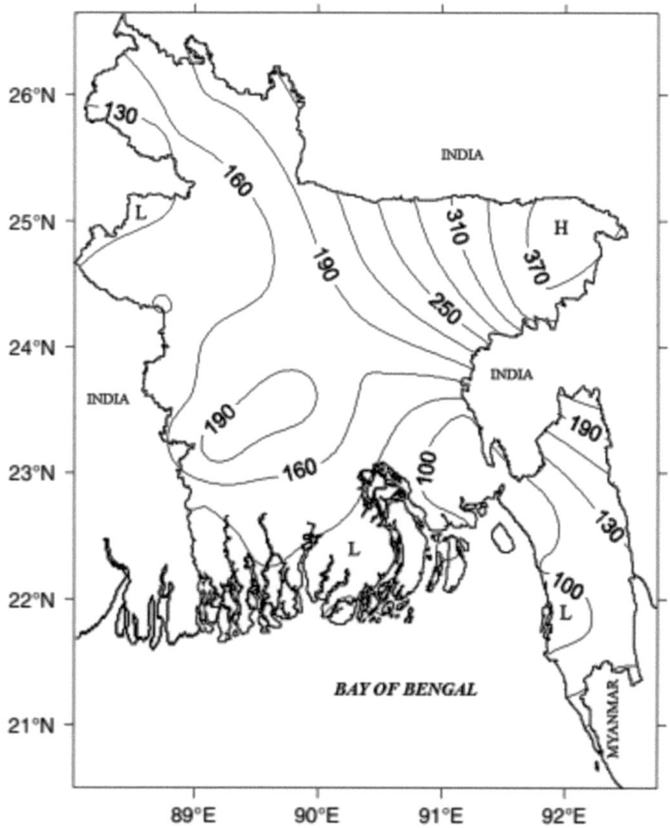

Fig. 3 The geographical distribution of annual thunderstorm frequency in Bangladesh using data from 29 stations in Bangladesh from 1976 to 2010

Vertical Wind Shear (VWS) between 850 and 500 hPa levels shows weak correlation with correlation coefficient (CC) of 0.27 which is statistically significant with $p < 0.05$ as per the student's t-test. These findings are agreeable with those by Ahasan et al. (2014). Yamane et al. (2010) have shown that low level VWS acts as an important positive factor for formation and organization of super cell thunderstorms. CAPE and strong low-level wind shear increases the likelihood of increasing frequency. The links of the thunderstorm activity with atmospheric instabilities are discussed in the following section mainly through NWP modelling of a single thunderstorm case, analysis of ERA5 data and synthesizing the results of earlier studies on the dynamic and thermodynamic characteristics of Bangladesh thunderstorms.

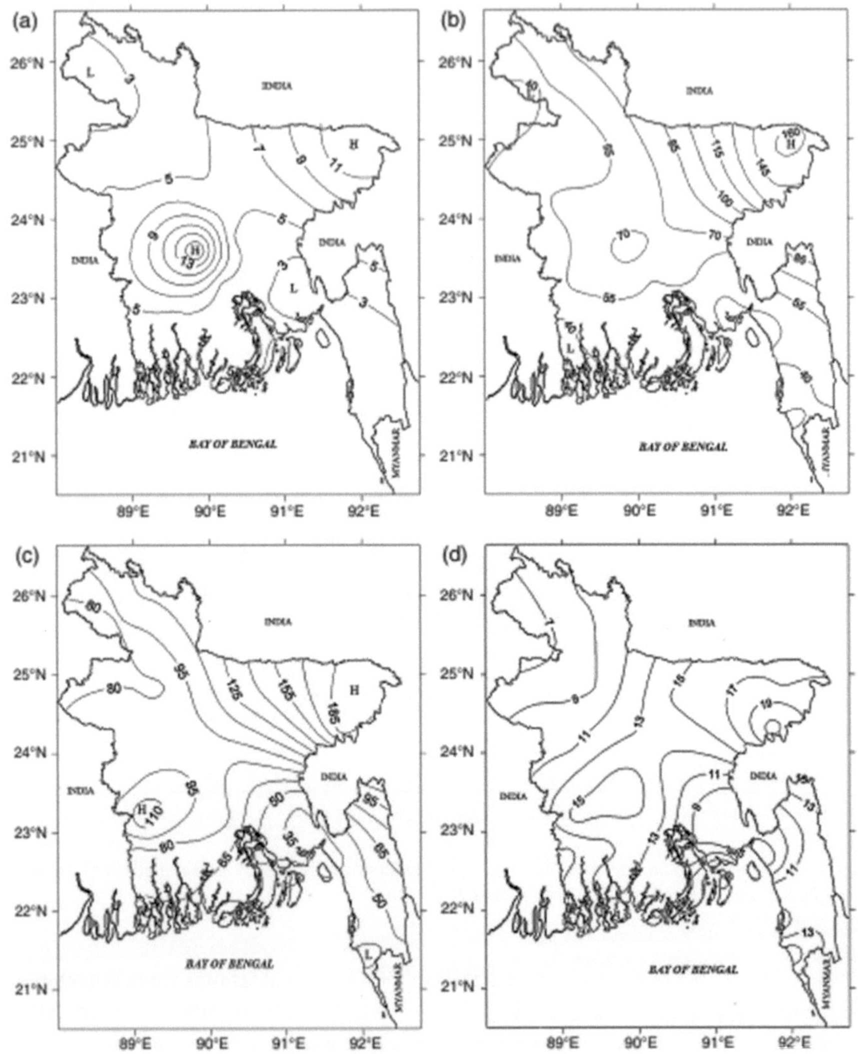

Fig. 4 Same as Fig. 3, but for the seasons—winter (**a**), pre-monsoon (**b**), monsoon (**c**) and post-monsoon (**d**)

3 Physical and Dynamical Characteristics of Thunderstorm: Bangladesh Perspectives

The study on the physical and dynamical characteristics of the severe storms in the perspective of Bangladesh has been performed using an event of severe thunderstorms which lashed over the north-western, northcentral, central and east-central part of Bangladesh on 30 March 2018 and discussed in the following sections.

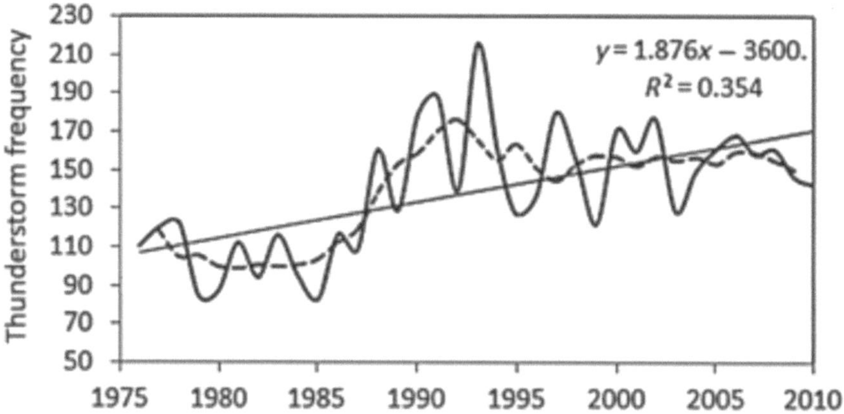

Fig. 5 Monthly frequency distribution time series of thunderstorms over Bangladesh (Saha & Quadir, 2016)

3.1 Characteristics of Thunderstorms as Seen from Satellites

The selected case of the thunderstorms in Bangladesh was investigated using the INSAT3D imageries and their development and movement have been monitored throughout their lifespan. The severe thunderstorm associated with rain, hail, and gusty winds occurred at 0000 UTC of 30 March 2018 over India adjacent to the border of north-western Bangladesh. The system then grew further by around 05:00 UTC and moved over Bangladesh and covered the areas of Rangpur, Gaibandha, Mymensingh, and nearby districts. At 1100 UTC the system intensified further over the north and north-central parts of Bangladesh. As reported in the newspaper, the severe thunderstorms damaged crop and houses, and at least six people were killed in different regions of Bangladesh. The INSAT-3D meteorological satellite cloud images from 0000 to 1400 UTC on 30 March 2018 at an interval of 1 h demonstrate the development and movement of well-defined multiple cells of convective systems (Fig. 6). There were 3 cells at 00 UTC and 5 cells at 03 UTC, which later merged and then 3 distinct cells were identified. These cells continued up to 08 UTC. Two of the three cells have the width of about 100 km each and the third one had the width of about 50 km. The highest intensity of the thunderstorm cells was observed at 1100 UTC (Ahasan et al., 2022).

3.2 Physical Characteristics

This section has been constructed using the ERA-5 5 km resolution gridded dataset of the ECMWF and analysis was made of the meteorological variables at low level

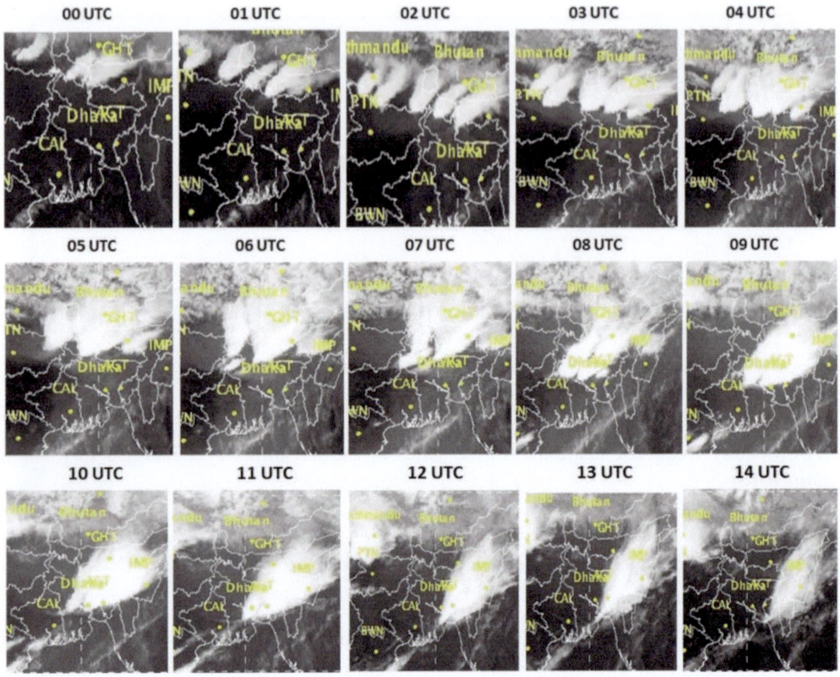

Fig. 6 The INSAT-3D meteorological satellite cloud images during 0000 UTC to 1400 UTC of 30 March 2018. The present study was conducted on the thunderstorm case shown in the image

(1000 hPa) and higher troposphere (200 hPa) for the area within 20–30°lat and 88–98°lon. The results are produced and discussed in the following subsections.

3.2.1 Mean Sea Level Pressure

Mean Sea Level Pressure (SLP) is one of the key variables of the atmosphere which modulated the conditions for a severe storm formation and intensification. The recognition of the role of SLP is investigated using the climatology of MSLP field (1975–2022) over the study region (Fig. 7a), which shows that the Gangetic plains including the West Bengal and Bangladesh is influenced by an east–west oriented low-pressure trough with lowest mean SLP of around 1009. This ensures that a trough of low pressure climatologically dominates over this region during this period. However, the pressure field varies over time and space. As already shown in the satellite imagery that a multi-cell thunderstorm developed on 30 March 2018. The distribution of ERA5 mean sea level pressure (hPa) valid for 0900 UTC (2-h earlier than the mature stage of the storm) of 30 March 2018 shows that the sea level pressure of the pre-monsoon trough has lowered to a value of around 1003 hPa in the core region of the trough, which extended towards West Bengal and Bangladesh when pressure was found to be 1005–1007 hPa over Bangladesh. The norther part of the trough experienced relatively strong south-to-north gradient. The pressure anomaly in that date and time indicated strong negative values, the axis of which elongated from west to east and passed over West Bengal and Bangladesh. This situation plays an important role for low level convergence along with advection of moisture from the Bay of Bengal and creates a congenial atmosphere for the convections to begin with ultimate genesis and intensification of severe storms over Bangladesh. Such a find was supported by Prosad (2006).

A few other literatures dealing the thunderstorm investigations have found that the sea level pressure dropped to the values as low as 996–998 hPa for some other cases of thunderstorms (Ahasan et al., 2014; Ahasan & Debsarma, 2015; Karmakar et al., 2017). These authors depicted that intense micro-lows may be formed within the trough, which activated local severe storms.

Quadir et al. (2020) have shown that a number of core regions of lowest pressure lower than 1004 hPa form over Uttar Pradesh in India. These low-pressure areas

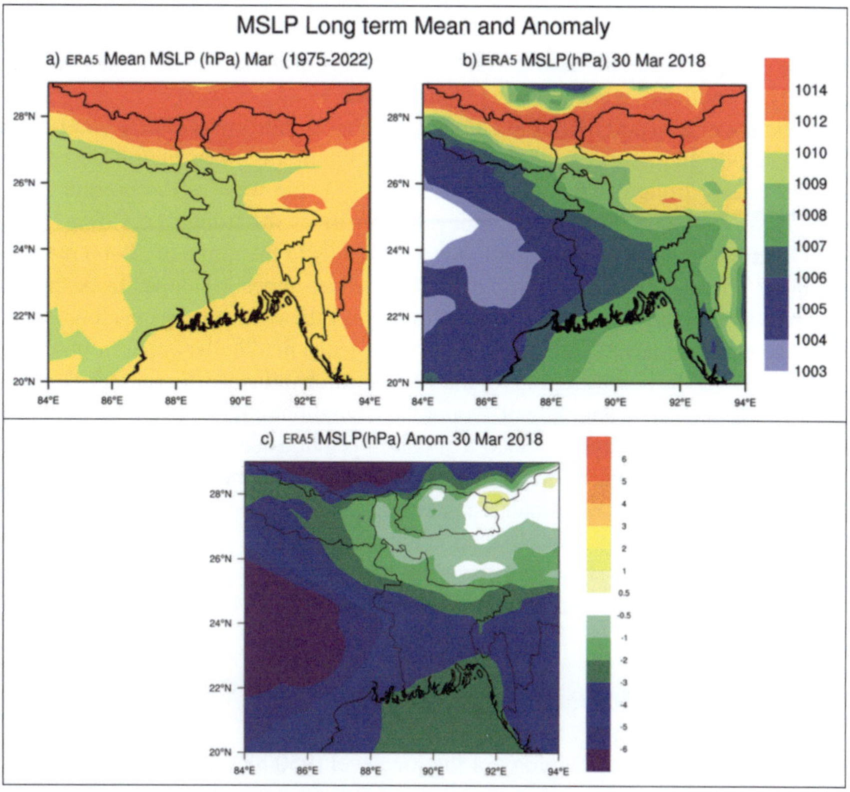

Fig. 7 Spatial distribution of ERA5 Mean Sea Level Pressure: **a** long-term climatology for March (1975–2022), **b** SLP valid for 0900 UTC of 30 March 2018 and **c** anomaly of SLPof 30 March 2018 with reference to the climatology

advance to the east and intensify to severe thunderstorms over Bangladesh in the later part of the day.

3.2.2 Low- and Upper-Level Geopotentials and Winds

Spatial distribution of low level (1000 hPa) geopotential height and vector wind climatology and an analysis of the above field for 30 March are shown in Fig. 8a, b, The anomaly of these fields are shown in Fig. 8c. It reveals from the Fig. 8a that the strong south-westerly winds blow from the Bay of Bengal towards the Bangladesh which carries plentiful of moisture to Bangladesh and adjoining territories of India. In addition, a trough in low level winds is observed over the sub-Himalayan West Bengal of India. The trough and wind intensity are high on March 30, 2018. A strong west-north-westerly wind of about 20 m/s is found over the areas of the occurrence of the storms (Fig. 8b). This carries very dry and cool mountainous wind which meet the very wet and warm air from the Bay of Bengal forming a dry line. Usually, thunderstorm forms over the southern side of this dry line. The anomaly of Geopotential Height and wind fields exhibit a congenial atmospheric condition for formation of the severe thunder storms with lowering of the height fields and positive anomaly of southerly winds blowing from the sea.

In the upper air (200 hPa), the climatology shows uniform westerly winds with the magnitude increasing toward north (20–40 m/s) (Fig. 9a). Figure 9b reveals that a westerly trough is passing over and lying to the west of the present study area. Thus, the study area lies in anticyclonic part of the trough providing upper air divergence and strong west-south-westerly winds in the range of 40–60 m/s (Jet Streams). The presence of subtropical jet stream at upper the upper tropospheric level provides a mechanism for strong vertical wind shear, favouring development of severe convections (Quadir et al., 2020). A divergence field in the upper troposphere is also seen over the nearby regions of the storms which indicates the vertical extent of the storm. The negative anomaly of geopotential height fieldin200 hPa level indicates the presence of strong trough. Such a situation provides congenial atmosphere for thunderstorm formation. A several studies expressed the similar opinion on the low level wind convergence over small areas, strong low level winds from Bay of Bengal carrying moisture to inlands in association with the upper the tropospheric Jet Strems and strong vertical wind shear that play roles in the thunderstorm formation (Ahasan et al., 2014, Ahasan & Debsarma, 2015; Karmakar et al., 2017; Quadir et al., 2020).

3.2.3 Low Level (950 hPa) Relative Vorticity

Spatial distribution of climatology of the low level (950 hPa) is shown in Fig. 10a. The relative vorticity is analysed for a case of thunderstorms that occurred at 1100 UTC of 30 March 2018. Climatology of the relative vorticity in the pre-monsoon

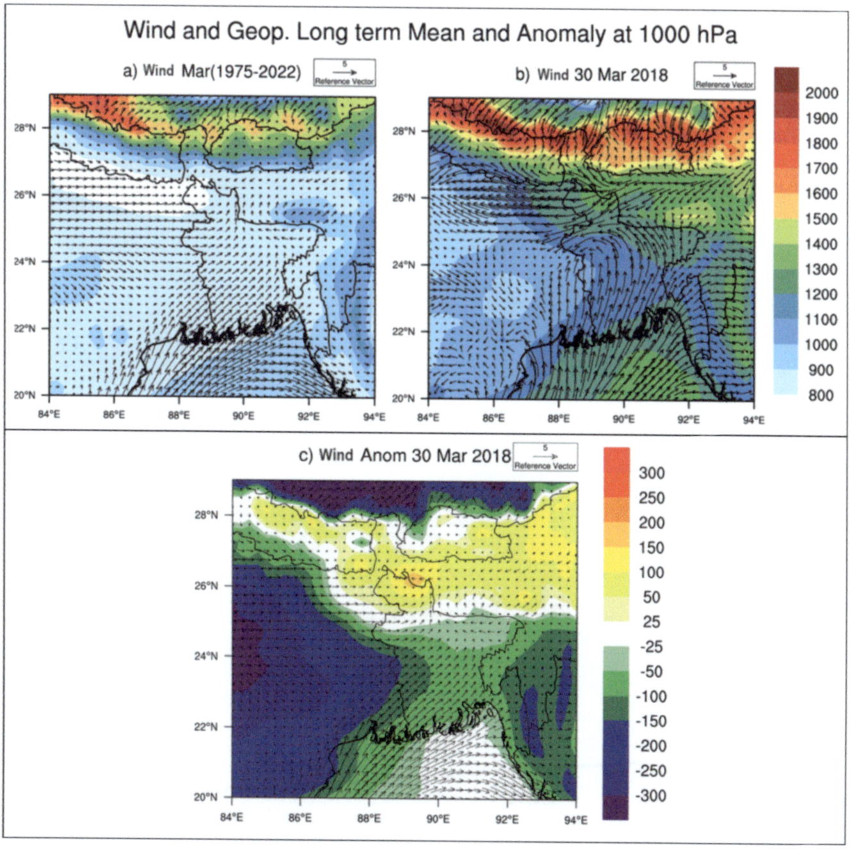

Fig. 8 Spatial distribution of ERA5 Geopotential (m2/s2) at 1000 hPa **a** long term mean values for March (1975–2022) **b** values valid for 0900 UTC of 30 March 2018 **c** anomalies between long term mean and 30 March 2018

months exhibits the presence of a east-west oriented narrow strip of weak relative vorticity with values (10–70) × 10^{-5} s^{-1}which extends up the northern part of Bangladesh. For a thunderstorm day (30 March. 2018 in this analysis), it is seen that the positive vorticity field is found to concentrate in a more southerly position oriented in narrow strips along east–west direction with very high values ranging from (80–150) × 10^{-5} s^{-1}. High positive values of vorticity at the lower troposphere indicate the outset of vigorous convective activities over the respective areas. A very interesting feature of the vorticity fields is their orientation in the east-west elongated stripes with alternating positive and negative vorticity signifying ascending and descending motion within the convective cells. The vorticity anomaly field exhibits the positive values of vorticity is oriented to the south of the negative vorticity each parameters having their orientation along west–east (Fig. 10c), which implies that a regional scale restructuring of the vorticity fields takes place during the thunderstorm

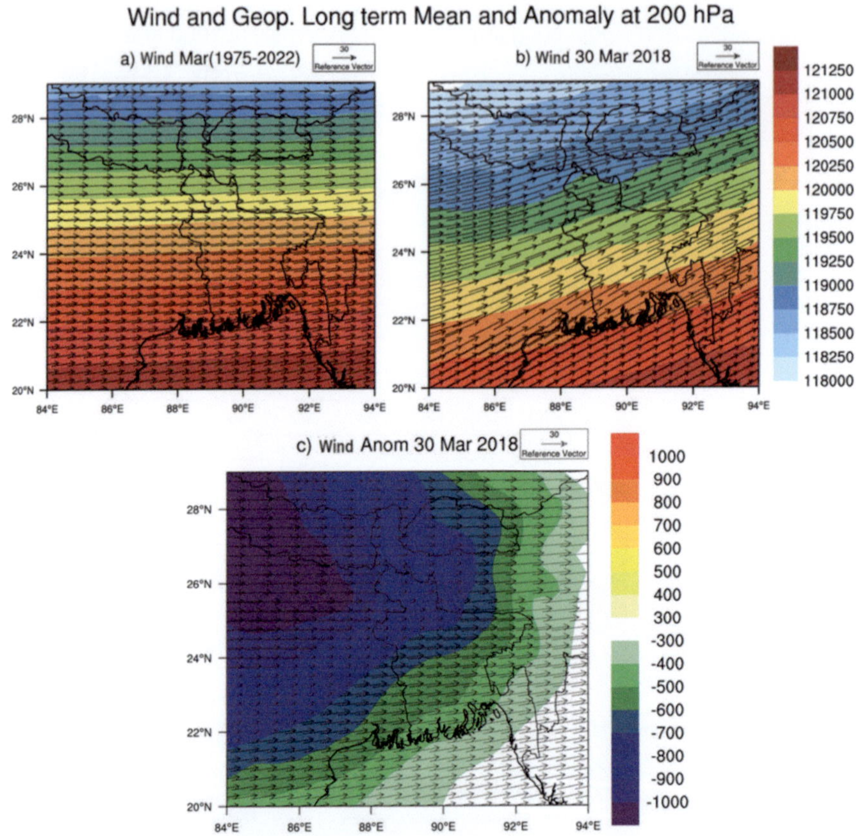

Fig. 9 Spatial distribution of ERA5 Wind and Geopotential (m2/s2) at 200 hPa **a** long term mean values for March (1975–2022) **b** values valid for 0900 UTC of 30 March 2018 **c** anomalies between long term mean and 30 March 2018. The geopotential is shaded

formation. The authors Ahasan et al. (2014, 15) and Karmakar et al. (2017) in their modelling studies of the Bangladesh thunderstorm found high relative vorticity in the thunderstorm areas.

3.2.4 Low-Level (850 hPa) Divergence Field

Spatial distribution of pre-monsoon climatology of 850 hPa using ERA-5 data, model derived divergence of the same level valid at 1100 UTC of 30 March 2018 and corresponding anomaly of divergence are presented in Fig. 11a–c. The ERA-5 climatology indicates very low convergence over the northern Bangladesh and adjoining areas to the west. The distribution of divergence for 1100 UTC of 30 March 2018 shows that the divergence at 950 hPa level that the area of occurrence of thunderstorms is

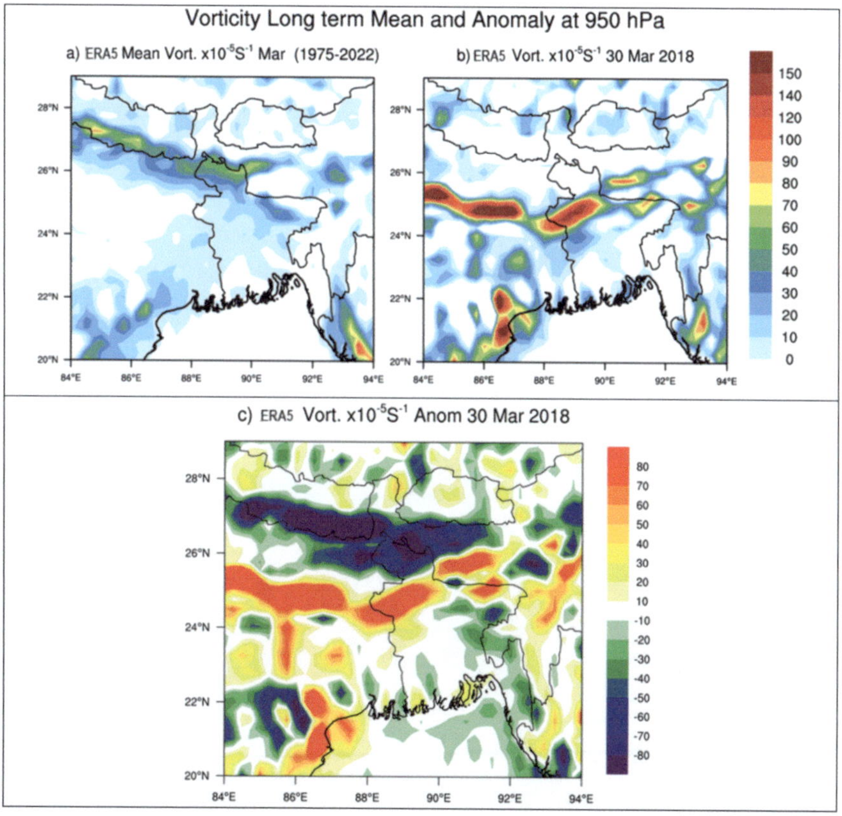

Fig. 10 Spatial distribution of ERA5 Vorticity ($\times 10^{-5}\mathrm{S}^{-1}$) at 950 hPa **a** long term mean values for March (1975–2022) **b** values valid for 0900 UTC of 30 March 2018 **c** anomalies between long term mean and 30 March 2018

characterized by high value of divergence in the order of -20 to -40 ($\times 10\ \mathrm{s}^{-1}$). Presence of low-level convergence and high level of divergence is one of the favourable atmospheric conditions for the occurrence of severe thunderstorms. It indicates the vertical extent of the storms with strong updraft over the tropospheric column.

Karmakar et al. (2017) observed similar characteristics in case of the distribution of divergence for the same level. At 1430 UTC, the thunderstorm area is covered by dominant convergence over the place of occurrence and is surrounded by divergence and the same is found to dominate at 1730 UTC which matches with the interpretation of the horizontal wind distribution and evolution of the vertical wind fields. The convergence is found to be of the order of -10 to $-20 * 10^{-5}\ \mathrm{s}^{-1}$ at 1730 UTC.

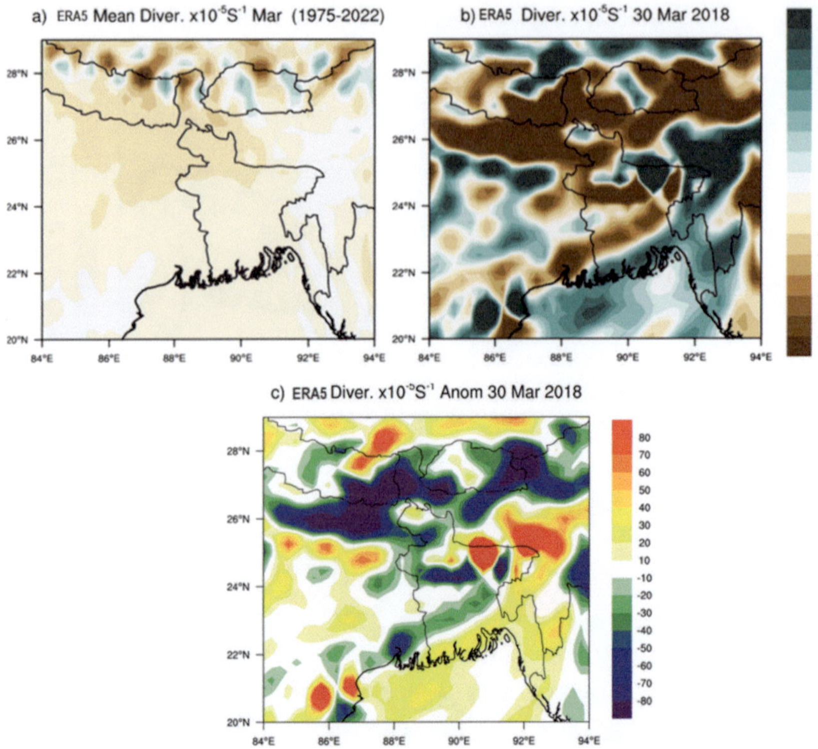

Fig. 11 Spatial distribution of ERA5 divergence ($\times 10^{-5}s^{-1}$) at 950 hPa **a** long term mean values for March (1975–2022), **b** values valid for 0900 UTC of 30 March 2018 and **c** anomalies between long term mean and 30 March 2018

3.3 Thermodynamic Characteristics

3.3.1 Horizontal (950 hPa Level) and Vertical Profile of Relative Humidity

The climatology of the horizontal distribution of relative vorticity at 950 hPa level for the month of March using ERA5 data is presented in Fig. 12a. It is found that Bangladesh and adjacent areas have relative humidity of around 45–60%. The model simulated horizontal distribution of RH% at 950 hPa for 30 March 2018 (Fig. 12b) shows that the southern and northern part of Bangladesh is covered by with highly humid air (70–85%). Over Napal a few core areas of high humidity is identified with very high humidity which matches in position and shape with the convective cells as seen in the satellite cloud imagery during the time of early development of 4–5 convective cells (see Fig. 6). To the north of this region, a dry belt appears providing a dryline of cool air. The low-level wind field in Fig. 8 suggests that south and south-westerly warm winds continuously pumping moisture from the Bay of

Bengal towards inland of Bangladesh. These moist and warm winds from the south as well as cool and dry winds from the north meet together and forms a dry line which helps to the development of convective activities leading to the formation of thunderstorms.

As already mentioned, that the incursion of cool and dry air from north as well as moist and warm air from the south (Bay of Bengal) cause the thunderstorm cells to develop. The region to the west is dry and constitute a dryline feature at the interface of the low-humid areas. It is known that such a moisture discontinuity is some times good for the convections to initiate thunderstorms development. The anomaly of the humidity field confirms the conditions as discussed earlier.

Vertical profile of model simulated relative humidity (%) along 24.98N latitude for 1100 UTC of 30 March 2018 is presented in Fig. 13a, it is observed that the RH of 70–90% reaches the 200 hPa level. In the distribution of RH% from ERA5 data, it is found that high humidity extends up to the level 300 hPa, with very high in the

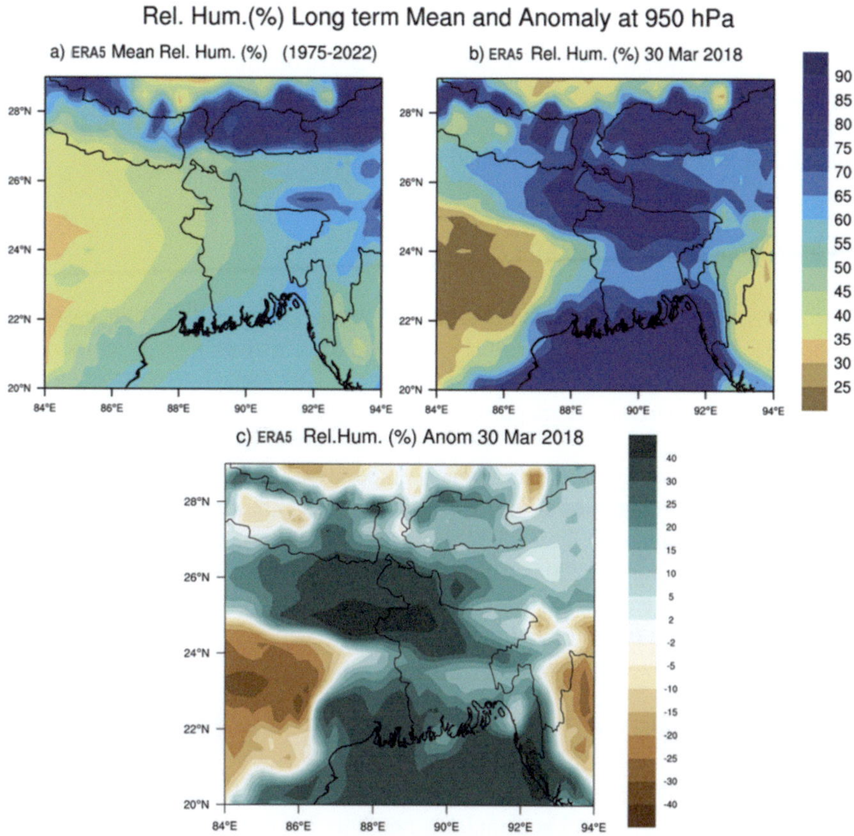

Fig. 12 Spatial distribution of model simulated horizontal profile of relative humidity (%) with vector winds at 850 hPa level valid at 1100 UTC of 30 March 2018

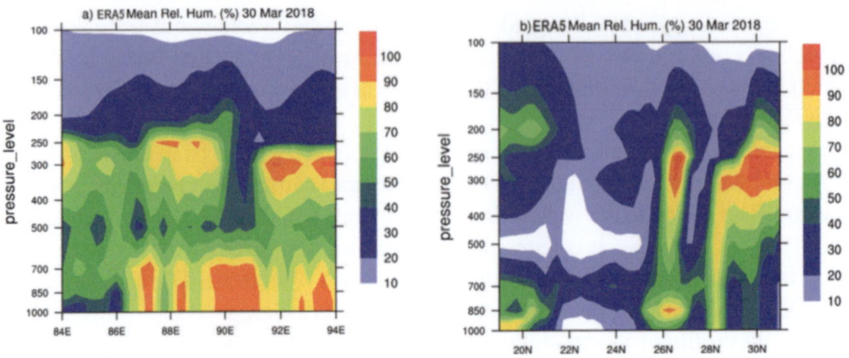

Fig. 13 ERA5 vertical profile along 27° latitude during 08:00 UTC of 30 March for 0800 UTC, 2018 of relative humidity (%) **a** along East–west direction across 24.98° N latitude **b** along North–South direction across 89°E longitude

lower atmosphere and upper atmosphere while it is rather low in the mid-troposphere. Consulting other recent works, it is seen that the above distribution of horizontal and vertical distribution of moisture is more or less common, but there are variations due to the variation of the geographical location of the thunderstorm activity (Ahasan et al., 2014; Karmakar et al., 2017; Paul et al., 2019).

3.3.2 Temperature at 950 hPa

Spatial distribution of ERA5 climatology of temperature at 950 hPa has been produced in Fig. 14a, which depicts that a relatively dry airmass with high temperature extends from west /north-west towards Bangladesh. The temperature distribution at 0900 UTC of 30 March 2018 are presented during the severe storm in Fig. 14b shows that the airmass of low temperature (16 °C) prevailed in the northern Bangladesh. The lower temperature to the north and higher temperature to the south of this location produced a strong temperature contrast. This situation is more clearly visible in the temperature anomaly, which indicated a temperature drop of 4 °C from the normal. (Fig. 14c). The contrast area matches with the convergence of surface wind and high humidity where the convections were initiated in early morning of the local time as was demonstrated in the satellite imagery (Fig. 6).

3.3.3 Precipitation

The climatology of the precipitation for the month of March is shown in Fig. 15a. Spatial distribution of ERA5 accumulated precipitation from 0900 to1400 UTC of the mature stage of the storm on 30 March 2018 are presented in Fig. 15b. It is observed that the long term mean of precipitation is not prominent on the northern

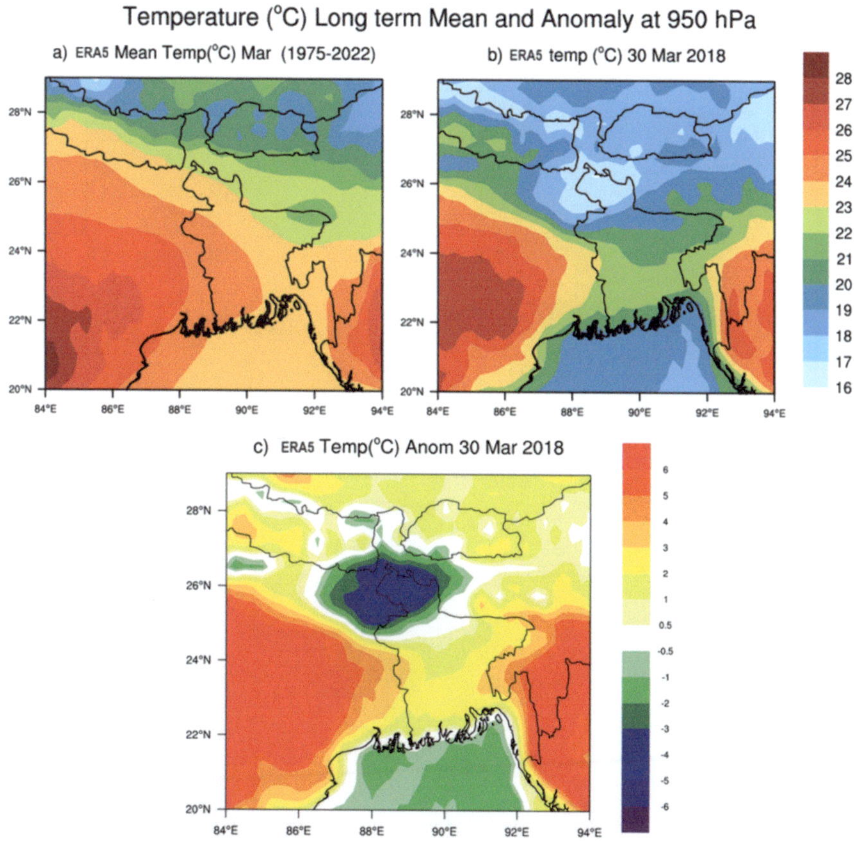

Fig. 14 ERA5 temperature at 950 hPa of **a** mean of month March (1975–2022) **b** accumulated at 09:00 UTC of 30 March, 2018 **c** anomaly of 30 March, 2018

most regions of Bangladesh (Fig. 15a) but during the severe storm of 30 March 2018, 10–26 mm of accumulated precipitation has been observed which varied over space (Fig. 15b). The precipitation anomalies shows more or less similar pattern of precipitation distribution in the northern regions as depicted in Fig. 15c.

3.3.4 Reflectivity

Spatial distribution of horizontal profile of low level (950 hPa) reflectivity (dBZ) at 1200 UTC of 30 March 2018 presented in Fig. 16a. Reflectivity indicates the structure and intensity of the storms. From the Fig. 16a, it is observed that the horizontal profile of the model-simulated reflectivity indicates the clear signatures of the storm at 1100 UTC of 30 March over the north-central areas of the country when severe storms had

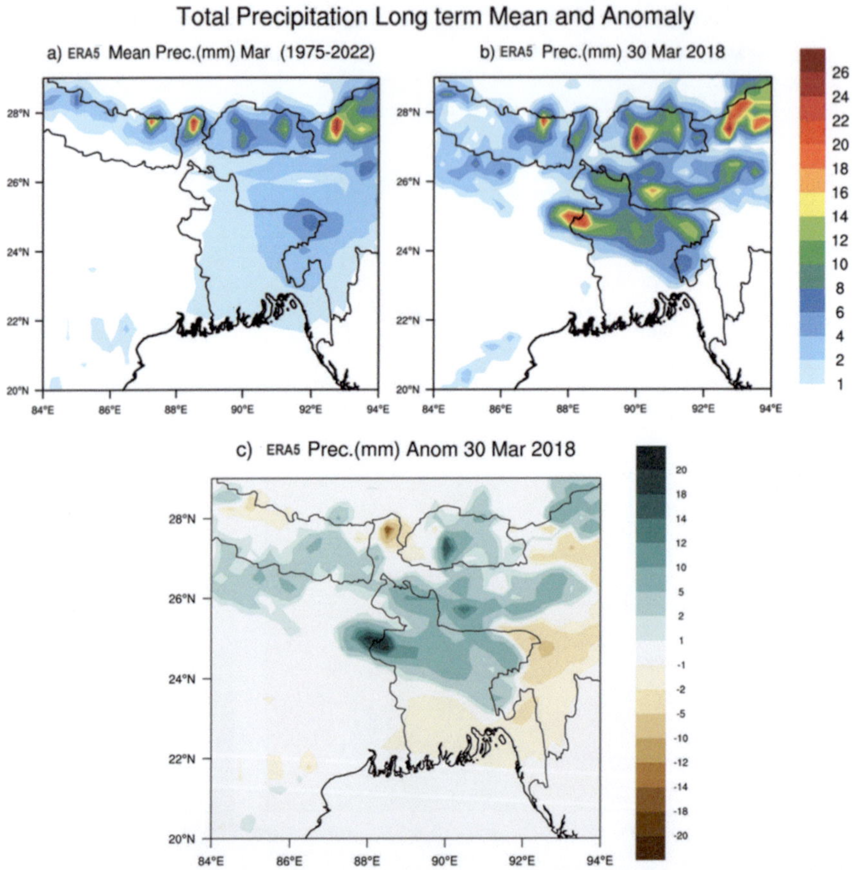

Fig. 15 ERA5 precipitation of **a** mean of month March (1975–2022) **b** accumulated during 06:00–20:00 UTC of 30 March, 2018 **c** anomaly of 30 March, 2018

occurred are justified with the cloud images of the same time reference. The model simulated reflectivity is found to be in the order of >0 dBZ.

Spatial distribution of vertical profile of reflectivity (dBz) along 90.58E longitude at 1200 UTC of 30 March 2018 are presented in Fig. 16b. It is observed that the vertical profile of the reflectivity at 1100 UTC of 30 March 2018 along 90.58 longitude extends up to 150 hPa. Vertical reflectivity is found to be high in the order of 45–50 dBZ.

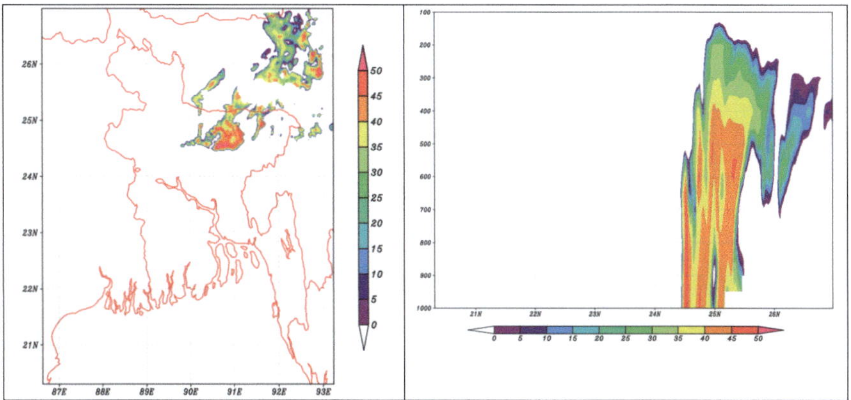

(a) Horizontal profile of reflectivity (dBZ) | (b) Vertical profile of reflectivity (dBz))

Fig. 16 Spatial distribution of horizontal profile of low level (950 hPa) reflectivity (dBZ) at 1200 UTC of 30 March 2018 (**a**) and vertical profile of reflectivity (dBz) along 90.58E longitude at 1200 UTC of 30 March 2018 (**b**)

3.3.5 Convective Available Potential Energy (CAPE) and Convective Inhibition Energy (CINE)

The investigation of Convective Available Potential Energy (CAPE) at 1100 UTC of 30 March 2018 shows that the area of the occurrence of the severe storms are characterized by high value of CAPE in the order of >1800 J/kg (Fig. 17). The Convective Inhibition Energy (CINE) at 1100 UTC of 30 March 2018 exhibits low values (<50 J/kg) in the area of the occurrence of the severe storms (Fig. 18). The anomaly fields have clearly shown the presence of high positive anomaly of CAPE and negative anomaly for CINE over the thunderstorm and adjoining areas.

3.3.6 Cloud Water and Ice Water Mixing Ratio

Spatial distribution of model simulated cloud-water mixing ratio at 0800 UTC of 30 March 2018 are shown in Fig. 19a and ice-water mixing ratio in Fig. 19b. The mixing ratios vary in the atmosphere from nearly zero in the cold and dry air to as much as 30 g/kg in warm and moist air. From the figures, it reveals that the area of occurrence of the selected severe storm are characterized by the high value of cloud-water mixing ratio with values higher than 37 g/kg in the lower atmosphere (Fig. 19a), while the ice-water mixing ratio had the values higher than150g/kg specially in the upper atmosphere (Fig. 19b). Such conditions are considered as characteristic thermodynamic features of the thunderstorms, which provides the indications of high probability of hail formation.

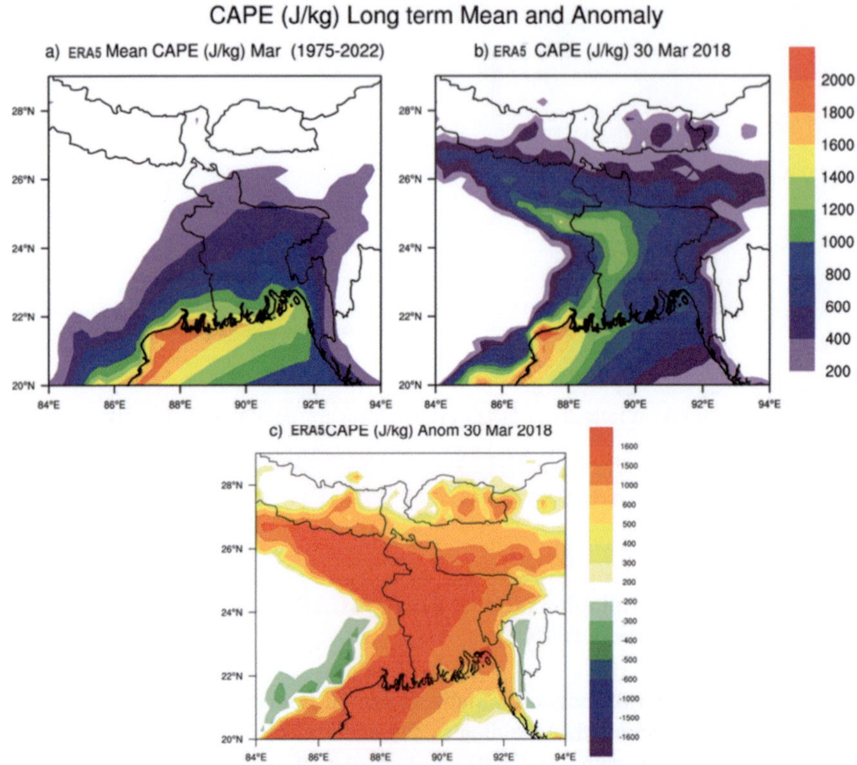

Fig. 17 ERA5 Convective Available Potential Energy (CAPE) of **a** mean of month March (1975–2022) **b** during 0700 UTC of 30 March, 2018 **c** anomaly of 30 March, 2018

The vertical time section of the cloud-water mixing ratio and ice-water mixing ratio for the thunderstorm for the 30 March shows that the former extends from 870 to 300 hPa level in the vertical extent and the latter extends from 500 to 120 hPa. The high density of ice-water ratio supported the strong hail activity over the northwestern Bangladesh in around 0500–0600 UTC. Karmakar et al. (2017) found that the ice-water mixing ratio higher than 140 g/kg extended from 580 to 250 hPa for a hail producing thunderstorm (Table 2).

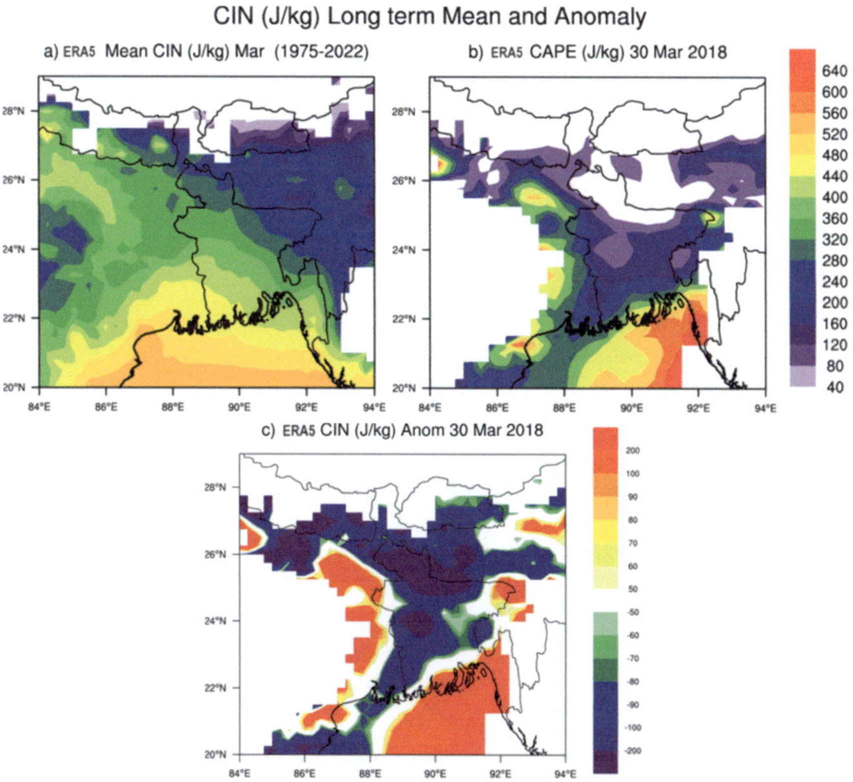

Fig. 18 ERA5 Convective Inhibition Energy (CINE) of **a** mean of month March (1975–2022) **b** during 0700 UTC of 30 March, 2018 **c** anomaly of 30 March, 2018

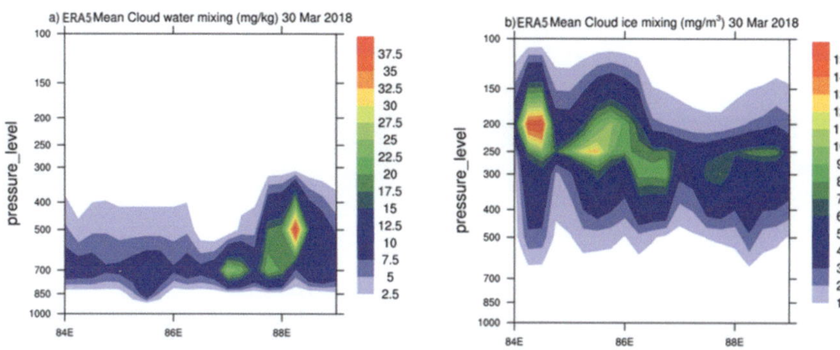

Fig. 19 ERA5 vertical profile along 27°latitude during 08:00 UTC of 30 March, 2018 of **a** cloud water mixing ratio **b** cloud ice mixing ratio

Table 2 Synthesized results of all studied parameters

Sl	Parameters	Descriptive features		Range of values
		ERA5 data on 30 March 2018	Other references	
1	The convective cells in the satellite Imagery	4–5 convective cells appeared to the north in the southern plane and hills over Nepal near the Bangladesh border. The imagery shows hourly development of the convective systems and movement to Bangladesh	There are innumerable examples of thunderstorm formation and forecasting using satellite data	Up to100 km
2	Mean sea level pressure	The climatology of pressure field of March indicates the existence of an east–west oriented trough with lowest pressure of around 1009 hPa	No citation	1009 hPa
3	Sea level pressure on 30 March 2918	The MSLP of 30 March, 2018 shows the lowest pressure of around 1005 hPa over Bangladesh	Prasad (2006) found the lowest pressure at around 1003–1004 hPa; Ahasan and Debsarma (2014), Ahasan et al. (2015): deeper pressure field of 996–998 hPa Karmakar et al. (2017): 1004 hPa with cells of micro lows	996–1004 hP
2	Low-level winds with geopotential height	~15 ^{5} ms^{-1} south-westerly	20 m/s Ahasan and Debsarma (2014), Ahasan et al. (2015) Karmakar et al. (2017) referred to medium to high and very high of around 40–50 m/s	15–50 m s^{-1}
3	Upper-level winds with geopotential height	~40–60^{5} ms^{-1} south-westerly A westerly trough was observed over the western part of the study region	No citation	40–60 ms^{-1}

(continued)

Table 2 (continued)

Sl	Parameters	Descriptive features		Range of values
		ERA5 data on 30 March 2018	Other references	
4	Low level (850 hPa) relative vorticity	Strong low level relative vorticity > 150×10^{-5} s^{-1}	Quadir et al. (2020) 10–20 s^{-1}, Ahasan and Debsarma (2014), Ahasan et al. (2015): 20–30 $\times 10^{-5}$ s^{-1} Bhuiya et al. (2022): 30–40 $\times 10^{-5}$ s^{-1} Karmakar et al. (2017): 5 April 2015: 150×10^{-5} s^{-1} at 950 hPa	10–150 $\times$ 10^{-5} s^{-1}
5	Low-level convergence (950 hPa) convergence	>-10 to -20×10^{-5} s^{-1}	Karmakar et al. (2017), 5 April 2015: -10 to $-20 \times$ 10^{-5} s^{-1} at 200 hPa	-10 to $-20 \times$ 10^{-5} s^{-1}
6	Vertical velocity (w) of winds	6 m s^{-1}	Karmakar et al. (2017): 5 April 2015: 0.2–7 m s^{-1}	0.2–7 m s^{-1}
7	Horizontal (850 hPa level) profile of relative humidity	>90%	Karmakar et al. (2017): 5 April 2015: > 85% at 950 hPa	>85%
8	Vertical profile of relative humidity	Extended up to 200 hPa	Karmakar et al. (2017): above 200 hPa	Above 200 hPa
9	Reflectivity	10–50 dBZ	Karmakar et al. (2017), 5 April 2015: 60 dBZ	10–60 dBZ
10	Convective available potential energy (CAPE)	>3800 J/kg	Karmakar et al. (2017), 5 April 2015: > 3000 J/K	1000–4000 J/kg

(continued)

Table 2 (continued)

Sl	Parameters	Descriptive features		Range of values
		ERA5 data on 30 March 2018	Other references	
11	Convective inhibition energy (CINE)	<50 J/kg	<50 J/kg	<30 J/kg
12	Cloud water mixing ratio	>30 g kg^{-1} in the lower atmosphere and vertically extends upto around 400–500 hPa	Karmakar et al. (2017), 5 April 2015: 114 g kg^{-1} up to 600 hPa	30–114 g kg^{-1} Upper range: 600–400 hPa
13	Ice water mixing ratio	>80 g kg^{-1} vertically extends upto 150 hPa	140 g kg^{-1} extends from 580 to 250 hPa	Ice-water mixing ratio 89–140 g kg^{-1} Upper range: up to 150 hPa level
14	Precipitation	26 mm	Karmakar et al. (2017): 26 mm	Around 26 mm

4 Conclusion

Based on the results and discussion on the physical and dynamical characteristics of thunderstorms of Bangladesh, the following major conclusion are drawn. The results for the selected thunderstorm of 30 March 2018 have been used.

1. The satellite imagery provided indications of the development of the convective cells as early as 0000 UTC of 30 March 2018, with clear 5 cells which moved to northwestern Bangladesh at around 0500 UTC. Thus, satellite infrared data are very useful for predicting thunderstorms.
2. In the pre-monsoon season of the Gangetic plain is climatologically dominated by a west to east oriented trough of low pressure that extends up to Bangladesh. This semipermanent trough suffers day to day variation, and the thunderstorms are formed on the dates of low-pressure anomalies. In the distribution of the MSLP of 30 March 2018, it reveals that the pressure reaches as low as 1005 hPa and multicell thunderstorm was found to develop over the northern Bangladesh.
3. The development of the thunderstorm was supported by the pumping of the moisture by the low level strong south-westerly wind. A wind flow of around 15 ms^{-1} was observed at 850 hPa level which converged with the relatively cool and dry wind from the mountainside in the north Bangladesh and adjacent areas to its north and west forming a line of discontinuity. Such a condition supported the multicell convection in the early hours (0000 UTS) of the day and thunderstorms were formed with string wind associated with hails. The chronological development of the thunderstorm is evident in the satellite imagery at different hours of the day.

4. In the 200 hPa wind and geopotential height fields, a subtropical western disturbance was recognized to the west of the study region on 30 March 2018. The thunderstorm area was found to be situated in the divergent part of the disturbance providing an anticyclonic circulation at top of the troposphere. This along with south-westerly Jetstream (40–60 ms^{-1}) promotes the convective systems to intensify to thunderstorm.
5. The high value of low-level relative vorticity in the order of (–60 to $-150) \times 10^{-5}\ s^{-1}$ is seen to be oriented in east–west strips over the vast area centring the area of occurrence of the storms on 30 March 2018.
6. The area of occurrence of the severe storms are characterized by high values of convergence. For the case of 30 March 2018, the convergence was found to be in the range $(-10$ to $-40) \times 10^{-5}\ s^{-1}$.
7. The areas of the occurrence of the severe storm of 30 March 2018 occupied with high values of vertical velocity of wind (>12 ms^{-1}) as compared to the neighbourhoods which allows the updraft of the storm to sustain for a longer period of time and provides the necessary favourable condition for the development of severe storm like tornadoes along with hailstorms and rotating air to be associated with the storm into the updraft.
8. The area of the occurrence of severe storms on 30 March 2018 was characterized by high relative humidity in the order of >90% due to south-westerly low-level winds. The vertical distribution of relative humidity shows that high relative humidity higher than 90% reached above 200 hPa level.
9. The area of the occurrence of severe storms are characterized by reflectivity higher than 50 dBZ for the case 30 March 2018. In addition, the vertical distribution of the reflectivity is found to extend up to 200 hPa or above.
10. High value of CAPE in the order of >3000 J/ kg and low value of CINE with values lower than 50 J/ kg were observed over the area of the occurrence of the severe storms.
11. The selected severe storm was characterized by the high value of cloud-water mixing ratio higher than 30 g/kg in the lower atmosphere and ice-water mixing ration higher than 80 g/kg occupies between 600 and 100 hPa.

The above concluding remarks highlight the major physical and dynamical characteristics of the thunderstorms which provide knowledge on essential conditions of the thunderstorm formation and will foster the local level thunderstorm forecasting techniques.

Acknowledgements The authors would like to thank to National Center for Atmospheric Research (NCAR) for providing the model code. The authors would also like to thanks to USGS for providing topography and land-use data and NCEP for providing initial and lateral boundary condition data to run the model. Thanks are due to the Director of the Bangladesh Meteorological Department (BMD) for providing the necessary data and information.

References

Ahasan, M., Rashid, T., Quadir, D., & Sultana, S. (2022). Simulation of a hailstorm event over Bangladesh using WRF model. *The Dhaka University Journal of Earth and Environmental Sciences, 10*(3), 123–130. https://doi.org/10.3329/dujees.v10i3.59078

Ahasan, M. N., & Debsarma, S. K. (2015). Impact of data assimilation in simulation of thunderstorm (squall line) event over Bangladesh using WRF model, during SAARC–STORM pilot field experiment 2011. *Natural Hazards, 75*(2), 1009–1022.

Ahasan, M. N., Quadir, D. A., Khan, K. A., & Haque, M. S. (2014). Simulation of a thunderstorm event over Bangladesh using WRF-ARW model. *Journal of Mechanical Engineering, 44*(2), 124–131.

Ahrens, D. C. (2013). *Meteorology Today. Cengage Learning*. Brooks/Cole.

Brooks, H. E., & Dotzek, N. (2008). The spatial distribution of severe convective storms and an analysis of their secular changes. *Climate Extremes and Society, 35*, 53.

Brooks, H. E., Lee, J. W., & Craven, J. P. (2003). The spatial distribution of severe thunderstorm and tornado environments from global reanalysis data. *Atmospheric Research, 67*, 73–94.

Berthet, C., Dessens, J., & Sánchez, J. L. (2011). Regional and yearly variations of hail frequency and intensity in France. *Atmospheric Research, 100*(4), 391–400.

Bhattacharya, S., Chattopadhyay, D., & Mukhopadhyay, A. (2013). Changing dimensions of food security in a globalized world: A review of the perspectives for environment, economy and health Int. *Research Journal of Environmental Sciences, 2*, 67–73.

Cao, Z. (2008). Severe hail frequency over Ontario, Canada: Recent trend and variability. *Geophysical Research Letters, 35*(14).

Chaudhari, H. S., Sawaisarje, G. K., Ranalkar, M. R., & Sen, P. N. (2010). Thunderstorms over a tropical Indian station, Minicoy: Role of vertical wind shear. *Journal of Earth System Science, 119*(5), 603–615.

Chevuturi, A., Dimri, A. P., Das, S., Kumar, A., & Niyogi, D. (2015). Numerical simulation of an intense precipitation event over Rudraprayag in the central Himalayas during 13–14 September 2012. *Journal of Earth System Science, 124*, 1545–1561.

Cotton, W., & Anthes, R. (1989). *Storm and Cloud Dynamics*. International Geophysics Series (Vol. 44). Academic Press, Inc.

Davis, C., Atkins, N., Bartels, D., Bosart, L., Coniglio, M., Bryan, G., Cotton, W., Dowell, D., Jewett, B., Johns, R., & Jorgensen, D. (2004). The bow echo and MCV experiment: Observations and opportunities: Observations and opportunities. *Bulletin of the American Meteorological Society, 85*(8), 1075–1094.

Das, S., Das, M. K., Sarkar, A., & Rahman, M. M. (2015). Composite characteristics of Nor'westers based on observations and simulations. *Atmospheric Research, 158*, 158–178.

Doswell, C. A., III. (1992). Forecaster workstation design: Concepts and issues. *Weather and Forecasting, 7*(2), 398–407.

Ferdaus, J., Quadir, D., Alam, M., Panda, S., Das, S., Ahasan, M., Rabbani, K., & Shuvo, S. (2021). Prediction of thunderstorms based on atmospheric instability indices over Bangladesh using WRF-ARW model. *JALAWAAYU, 1*(2). https://doi.org/10.3126/jalawaayu.v1i2.41008.

García-Ortega, E., Hermida, L., Hierro, R., Merino, A., Gascón, E., Fernández-González, S., Sánchez, J. L., & López, L. (2014). Anomalies, trends and variability in atmospheric fields related to hailstorms in north-eastern Spain. *International Journal of Climatology, 34*(11), 3251–3263.

Houze, R. A., Jr. (1981). Structures of atmospheric precipitation systems–A global survey. *Radio Science, 16*, 671–689 (Invited paper).

Houze, R. A., Jr. (1993). *Cloud Physics*.

Jayakrishnan, R. P., & Babu, A. C. (2014). Assessment of convective activity using stability indices as inferred from radiosonde and MODIS data. *Atmospheric and Climate Sciences, 4*, 122–130.

Karmakar, S., & Alam, M. (2005). On the sensible, latent heat energy and potential energy of the troposphere over Dhaka before the occurrence of nor'westers in Bangladesh during the pre-monsoon season. *Mausam, 56*(3), 671–680.

Karmakar, S., & Alam, M. M. (2006). Instability of the troposphere associated with thunderstorms/nor'westers over Bangladesh during the pre-monsoon season. *MAUSAM, 57*(4), 629–638.

Karmakar, S., & Quadir, D. A. (2014). Study on the potential temperatures of the troposphere associated with local severe storms and their distribution over Bangladesh and neibourhood during the pre-monsoon season. *Journal of Engineering Sciences, 05*(1), 13–30.

Karmakar, S., Quadir, D. A., & Das, M. K. (2017). Numerical simulation of physical and dynamical characteristics associated with the severe thunderstorm on April 5, 2015 at Kushtia and Jhenaidah. *Natural Hazards, 86*(3), 1127–1146.

Kunz, M., Sander, J., & Kottmeier, C. (2009). Recent trends of thunderstorm and hailstorm frequency and their relation to atmospheric characteristics in southwest Germany. *International Journal of Climatology: A Journal of the Royal Meteorological Society, 29*(15), 2283–2297.

Leslie, L. M., Buckley, B. W., & Leplastrier, M. (2008). The operational impact of QuikSCAT winds in Perth, Australia: Examples and limitations. *Weather and Forecasting, 23*(1), 183–193.

Lopez, P., & Bauer, P. (2007). "1D+ 4DVAR" assimilation of NCEP stage-IV radar and gauge hourly precipitation data at ECMWF. *Monthly Weather Review, 135*(7), 2506–2524.

Litta, A. J., Mohanty, U. C., Das, S., & Idicula, S. M. (2012). Numerical simulation of severe local storms over east India using WRF-NMM mesoscale model. *Atmospheric Research, 116*, 161–184.

McNulty, R. P. (1995). Severe and convective weather: A central region forecasting challenge. *Weather and Forecasting, 10*(2), 187–202.

Orville, H. D., & Kopp, F. J. (1977). Numerical simulation of the life history of a hailstorm. *Journal of the Atmospheric Sciences, 34*(10), 1596–1618.

Piani, F., Crisci, A., De Chiara, G., Maracchi, G., & Meneguzzo, F. (2005). Recent trends and climatic perspectives of hailstorms frequency and intensity in Tuscany and Central Italy. *Natural Hazards and Earth System Sciences, 5*(2), 217–224.

Paul, P., Imran, A., Islam, M. J., Kabir, A., Jaman, S., & Syed, I. M. (2019). Study of pre-monsoon thunderstorms and associated thermodynamic features over Bangladesh using WRF-ARW model. *Dhaka University Journal of Science, 67*(2), 151–156.

Prasad, K. (2006). Environment and synoptic conditions associated with Nor'westers and Tornadoes in Bangladesh: An appraisal based on Numerical Weather Prediction (NWP) guidance products, SAARC Meteorological Research Centre, Scientific Report.

Quadir, D. A., Rashid, T., Akhter, S., & Ahasan, M. N. (2020). Investigation of meteorological conditions favourable for formation of thunderstorm occurred over Bangladesh through simulation by WRF-ARW model. *Dhaka University Journal of Science, 68*(1), 71–77.

SAARC STORM. (2005). The SAARC STORM: A Coordinated Field Experiment on Severe Thunderstorm Observations and Regional Modeling over the South Asian Region. *Bulletin of the American Meteorological Society, 95*(4), 603–617.

Sabrina, S. S., Rashid, T., Das, S., Panda, S. K., & Mallik, M. A. K. (2020). Prediction of hailstorm during pre-monsoon period over Bangladesh using NWP model. In *International Conference on Earth and Environmental Sciences & Technology for Sustainable Development (ICEEST 2020).* Dhaka, Bangladesh.

Schmidt, A., Carslaw, K. S., Mann, G. W., Rap, A., Pringle, K. J., Spracklen, D. V., Wilson, M., & Forster, P. M. (2012). Importance of tropospheric volcanic aerosol for indirect radiative forcing of climate. *Atmospheric Chemistry and Physics, 12*, 7321–7339.

Saha, T. R., & Quadir, D. A. (2016). Variability and trends of annual and seasonal thunderstorm frequency over Bangladesh. *International Journal of Climatology, 36*(14), 4651–4666. https://doi.org/10.1002/joc.4663

Tyagi, B., Naresh Krishna, V., & Satyanarayana, A.N.V. (2011). Study of thermodynamic indices in forecasting pre-monsoon thunderstorms over Kolkata during STORM pilot phase 2006–2008. *Natural Hazards, 56*, 681–698. https://doi.org/10.1007/s11069-010-9582-x.

Uma, K. N., & Rao, T. N. (2009). Characteristics of vertical velocity cores in different convective systems observed over Gadanki, India. *Monthly Weather Review, 137*(3), 954–975.

Weisman, M. L., & Rotunno, R. (2000). The use of vertical wind shear versus helicity in interpreting supercell dynamics. *Journal of the Atmospheric Sciences, 57*(9), 1452–1472.

Weisman, M. L., & Rotunno, R. (2004). "A theory for strong long-lived squall lines" revisited. *Journal of the Atmospheric Sciences, 61*(4), 361–382.

Weisman, M. L., Klemp, J. B., & Rotunno, R. (1988). Structure and evolution of numerically simulated squall lines. *Journal of Atmospheric Sciences, 45*(14), 1990–2013.

Webster, P. J., Bradley, E. F., Fairall, C. W., Godfrey, J. S., Hacker, P., Houze, R. A., Jr., Lukas, R., Serra, Y., Hummon, J. M., Lawrence, T. D. M., & Russell, C. A. (2002). The JASMINE pilot study. *Bulletin of the American Meteorological Society, 83*(11), 1603–1630.

Xie, Z. T., Coceal, O., & Castro, I. P. (2008). Large-eddy simulation of flows over random urban-like obstacles. *Boundary-Layer Meteorology, 129*, 1–23.

Yamane, Y., Hayashi, T., Dewan, A. M., & Akter F. (2010). *Severe local convective storms in Bangladesh: Part 1. Climatology. Atmospheric Research, 95*(4), 400–406.

Zipser, E. J., Cecil, D. J., Liu, C., Nesbitt, S. W., & Yorty, D. P. (2006). Where are the most intense thunderstorms on Earth? *Bulletin of the American Meteorological Society, 87*(8), 1057–1072.

Assimilation of Surface Mesonet Observations and Ground-Based Radar and Lidar Data for Improved Prediction of Mesoscale Convective Systems

Zhaoxia Pu and Linbo Wei

Abstract Mesoscale convective systems (MCSs) are organized convective weather systems that cover an area of several hundred kilometers and last for several hours or more. To improve the observation and prediction of MCSs, a dense network of observing systems is required. The surface Mesonet provides better coverage of surface observations than conventional meteorological networks. By assimilating surface meteorological Mesonet observations into numerical weather prediction (NWP) models, more accurate initial conditions can be obtained. Ground-based radar and lidar data offer valuable information about the structure and evolution of MCSs. Radar data, such as reflectivity and Doppler velocity, provide insights into precipitation patterns, storm intensity, and wind fields within the storms. Lidar data, which uses laser beams to measure atmospheric properties, can provide vertical wind profiles. Assimilating radar and lidar data into NWP models can enhance the representation of convective processes within MCSs, resulting in improved predictions of their track, intensity, and timing. Data assimilation involves incorporating these observational data into numerical models using variational or ensemble-based assimilation methods. These methods aim to find the best fit between the observations and the model's simulated state, effectively blending the observations with the model's background state. The assimilation process enables a more accurate representation of the initial atmospheric conditions and better captures the complex dynamics and microphysics associated with these convective systems, ultimately enhancing the skill and reliability of MCS forecasts. This chapter utilizes MCS cases over the Southern Great Plains of the United States to illustrate the significance of assimilating surface Mesonet meteorological observations, radar reflectivity and radial velocity, and lidar profiling data into the mesoscale Weather Research and Forecasting (WRF) model.

Z. Pu (✉)
Department of Atmospheric Sciences, University of Utah, Salt Lake City, UT, USA
e-mail: Zhaoxia.Pu@utah.edu

L. Wei
College of Atmospheric Sciences, Lanzhou University, Lanzhou, China
e-mail: weilb@lzu.edu.cn

S. Das and W-K. Tao (eds.), *Severe Storms*,
https://doi.org/10.1007/978-981-97-7075-5_15

Keywords Data assimilation · Mesoscale convective systems · Surface observations · Radar · Lidar · Numerical simulations

1 Introduction

Severe weather prediction has become a prominent focus in atmospheric science due to its significant social and economic impacts. Many severe weather systems are associated with mesoscale convective systems (MCSs), making accurate prediction of MCSs crucial for accurate severe weather forecasting. The prediction of MCSs by regional and global models relies on accurate initial conditions, physical parameterizations, and an understanding of the processes governing convective cloud life cycles.

Among these factors, the accuracy of initial conditions plays a pivotal role in determining the accuracy of numerical simulations and predictions of MCSs and severe storms. Specifically, the forecast accuracy of mesoscale and object-oriented models relies heavily on accurately predicting the initiation of convection, commonly known as convective initiation. Previous studies, such as one by Crook (1996), have shown that convective initiation is sensitive to low-level vertical temperature and moisture gradients. Even a small change in these gradients can make a significant difference in the simulated storm intensity. These critical variations are within the typical range of observational boundary layer variability, posing a challenge for accurately forecasting convective rainfall initiation and intensity. In this context, Fabry (2006) demonstrated the relationship between the radar-derived near-surface refractivity field and convection initiation.

Understanding the lifting mechanisms at boundaries and their interaction with the ambient wind shear profile is also crucial for storm initiation and maintenance. Early numerical simulations by Droegemeier and Wilhelmson (1987), Rotunno et al. (1988), and Parsons (1992) suggested that storm intensity, structure, and lifetime are influenced by the environmental wind profile and the strength of the cold pool. Additionally, moisture content, convergence strength, and shear values affect storm initiation and development. The presence of shear above the boundary layer can inhibit convective development.

The interaction between the land surface and atmospheric boundary layer processes is another important aspect. Studies (e.g., Holt et al., 2006) have shown that changes in surface characteristics affect surface fluxes of sensible and latent heat, which in turn modify air temperature and moisture content in the lower boundary layer. These changes propagate upward through the boundary layer, impacting boundary layer growth, convective available potential energy (CAPE), and convective initiation, contributing to frontogenetic forcing, boundary layer clouds, and precipitation timing and amount.

Surface cold pools, characterized by evaporatively cooled downdraft air spreading horizontally under precipitating clouds, play a vital role in maintaining and organizing multicell thunderstorms and deep convection systems (Engerer et al., 2008).

The boundary between a cold pool and its environment is called the outflow boundary. It is a primary mechanism for the maintenance and organization of multicell thunderstorms and deep convective systems (Thorpe et al., 1982; Rotunno et al., 1988; Szeto & Cho, 1994). For instance, the leading edge of the outflow boundary is termed the gust front. Many investigators (e.g., Wakimoto, 1982; Droegemeier & Wilhelmson, 1987; Mueller & Carbone, 1987) have indicated that gust fronts play an important role in thunderstorm dynamics. The speed of a gust front is governed by the density difference between the cold pool and its environment, as well as the cold pool depth (Wakimoto, 1982). The gust front maintains a thunderstorm by forcing warm, moist air upward (Purdom, 1976; Wilson & Schreiber, 1986), or by triggering new convective cells via low-level convergence (Engerer et al., 2008). One study has shown that the lifting of air along mesoscale-sized cold pool boundaries can produce moist unstable layers of more than 100 hPa in depth that influence the internal structure of convective systems (Bryan & Fritsch, 2000). In addition, the interaction between the surface outflow and the low-level shear produces much deeper and less inhibited lifting than would be possible without the low-level shear, making it easier for new cells to form and grow as old cells decay. Subsequent interaction between the cold outflow and the ambient shear can initiate new convection and promote long-lasting, organized convective systems (Rotunno et al., 1988; Thorpe et al., 1982).

Despite the importance of cold pools in the initiation and evolution of MCSs, the representation of cold pools presents a challenge in numerical models. Studies by Bryan et al. (2005) and Trier et al. (2006) suggest that observed cold pools are often deeper than simulated ones. If the models fail to represent cold pool characteristics correctly, then their ability to correctly predict convective MCS evolution is likely reduced. In addition, to identify a cold pool, the near-surface temperature should be accurately simulated. However, as shown in previous studies, numerical models often have problems predicting near-surface atmospheric conditions such as surface 2-m temperature (e.g., Liu et al., 2008; Stensrud et al., 2009; Zhang et al., 2013). This inaccurate surface temperature prediction/simulation is associated with the misrepresentation of soil moisture conditions, land use, and atmospheric boundary layer conditions (e.g., Massey et al., 2014). Due to these problems, numerical weather models often fail to represent cold pools accurately, thus preventing further studies that could lead to understanding and predicting the interaction between cold pools and MCSs.

Recognizing the importance of accurate atmospheric conditions in simulating mesoscale convective systems, field campaigns have been conducted to obtain comprehensive observations to improve our understanding and prediction of mesoscale convective systems over both land and ocean. Notable field experiments include the International H2O Project (IHOP_2002; Weckwerth et al., 2004), VORTEX2 (Wurman et al., 2012), and Plains Elevated Convection at Night (PECAN; Geerts et al., 2017) over the US Southern Great Plains (SGP). These field programs have proven that intense observing periods (IOPs) with comprehensive instrument sampling of atmospheric PBL structure can greatly enhance our understanding of PBL processes related to mesoscale convective systems (Blumberg et al., 2017;

Degelia et al., 2019). However, it is still of great importance to evaluate the impacts of data from ground-based and routine networks on numerical simulations and predictions of MCSs.

This chapter introduces data assimilation concepts and demonstrates the impacts of data assimilation on numerical simulations of MCSs using case studies over the SGP. The specific observations utilized in the chapter include surface Mesonet observations, radar observations, and lidar profiling observations.

2 Data Assimilation Methods and a Mesoscale Model

Data assimilation plays a crucial role in incorporating available observations into numerical models to estimate accurate initial conditions (Kalnay, 2003). In the context of numerical weather prediction and simulation, observations are obtained from various sources such as weather stations, ships, buoys, and satellite instruments. However, these conventional observations often have a coarse spatial resolution, which limits their effectiveness in predicting severe mesoscale storms. Therefore, a dense observation network like the surface Mesonet becomes necessary to capture the finer details. Additionally, radar measurements are essential due to their ability to sample local convective and environmental conditions more frequently. Ground-based lidar is a valuable tool that can sample aerosols in the atmosphere and provide information about the vertical wind profile. The assimilation of these observations into numerical models can improve the accuracy of mesoscale severe weather prediction.

The primary objective of data assimilation is to determine the best possible atmospheric state by combining observations and short-range forecasts. It involves a sequential time-stepping procedure where a previous model forecast is compared with newly received observations. The model state is then adjusted to match the observations, a new forecast is initiated, and this process is repeated. The step of updating the model state based on observations is known as analysis, while the short model forecast used in this step is referred to as the background. Different types of data assimilation systems are used in research and operational environments, each employing different algorithms to handle the background and observational information.

a. *A 3-dimensional variational (3D-VAR) data assimilation method*

In the 3D-VAR (three-dimensional variational) approach, a cost function is defined that is proportional to the square of the distance between the analysis and both the background and the observations. This cost function is minimized to obtain the analysis.

The cost function, denoted as J, can be expressed as follows:

$$J = 1/2\left\{\left[y^o - H(\mathbf{x})\right]^T R^{-1}\left[y^o - H(\mathbf{x})\right] + 1/2(\mathbf{x} - \mathbf{x}^b)^T B^{-1}\left(\mathbf{x} - \mathbf{x}^b\right)\right\} \quad (1)$$

In Eq. (1), the first term of the cost function measures the distance between the analysis (denoted as $\mathbf{x}$) and the observations (denoted as $\mathbf{y}^o$), scaled by the inverse of the observation error covariance matrix R^{-1}. The second term measures the distance between the analysis and the background (denoted as $\mathbf{x}^b$), scaled by the inverse of the background error covariance matrix (B^{-1}). The minimum of the cost function, denoted as $\mathbf{x}_a$, represents the analysis field. In the 3D-VAR approach, the minimization of Eq. (1) is performed directly, allowing for greater flexibility and simultaneous global use of the available data.

b. *A 4-dimensional variational (4D-VAR) data assimilation method*

An important extension of the 3D-VAR approach is the 4D-VAR (four-dimensional variational) method, which allows for observations distributed within a time interval (t_0, t_n). In 4D-VAR, the cost function is modified to incorporate the temporal dimension and includes terms measuring the distance to the background at the beginning of the interval and a summation over time.

The cost function in 4D-VAR can be represented as:

$$J[\boldsymbol{x}(t_0)] = \frac{1}{2}\left[\boldsymbol{x}(t_0) - \boldsymbol{x}^b(t_0)\right]^T \mathbf{B}_0^{-1}\left[\boldsymbol{x}(t_0) - \boldsymbol{x}^b(t_0)\right] + \frac{1}{2}\sum_{i=0}^{N}\left[H(\boldsymbol{x}_i) - y_i^0\right]^T \mathbf{R}_i^{-1}\left[H(\boldsymbol{x}_i) - y_i^0\right] \quad (2)$$

In Eq. (2), the first term measures the distance between the initial state of the model at time t_0 (denoted as $\mathbf{x}(t_0)$) and the background state ($\mathbf{x}_b$), scaled by the inverse of the background error covariance matrix (B^{-1}). The second term is a summation over time, where for each observation at time t_i (denoted as $\mathbf{y}_i$), the distance between the observation and the model-simulated observation ($H(\mathbf{x}(t_i))$) is measured and scaled by the inverse of the observation error covariance matrix (R^{-1}). The control variable, to which the cost function is minimized, is the initial state of the model within the time interval ($\mathbf{x}(t_0)$). The analysis at the end of the interval is given by the model integration from the solution $\mathbf{x}(t_n) = M_0\,[\mathbf{x}(t_0)]$, where M_0 represents the model operator. Therefore, the model acts as a strong constraint, ensuring that the analysis solution satisfies the model equations. In essence, 4D-VAR aims to find an initial condition ($\mathbf{x}(t_0)$) that leads to a forecast that best fits the observations within the assimilation interval (t_0 to t_n).

c. *Ensemble-based data assimilation (EnKF)*

Because of its ability to estimate the evolution of forecast uncertainty, known as the error of the day, ensemble forecasting has been applied in data assimilation. In an ensemble Kalman filter, an ensemble of K data assimilation cycles is carried out simultaneously. All the cycles assimilate the same real observations, but to maintain them as realistically independent, different sets of *random perturbations are added to the observations* assimilated in each member of the ensemble data assimilations. Another type of ensemble Kalman filter is known as the deterministic or square root filter. Here there is a single data assimilation and the new analysis perturbations are obtained from the background (forecast) ensemble perturbations using a square root algorithm.

After completing the ensemble of analyses at time t_{i-1}, and the K forecasts $x^f(t_i) = M^k\left[x^a(t_{i-1})\right]$, one can obtain an estimate of the forecast error covariance from the K forecasts $x_k^f(t_i)$, as

$$P^f \approx \frac{1}{K-1}\sum_{k=1}^{K}\left(x_k^f - \overline{x^f}\right)\left(x_k^f - \overline{x^f}\right)^T \tag{3}$$

where the overbar represents the ensemble average. Then the analysis is achieved as $x^a = x^f + K\left[y^0 - H\left(x^f\right)\right]$ with a Kalman gain matrix $K = P^f H^T\left(HP^T H^T + R\right)^{-1}$, where H is the observational operator and R is the observational error covariance.

d. *Hybrid data assimilation*

Because 3D-VAR and 4D-VAR use static background covariances (B term in Eq. (1) and B_0 in (2)) from the statistics of short-range forecasts, they do not account for the flow dependency of atmospheric behaviour. A hybrid technique that combines variational methods (such as 3D-VAR or 4D-VAR) and EnKF has proven to be very successful in recent studies (Hamill & Snyder, 2000). Typically, hybrid data assimilation uses variational data assimilation (3D-VAR or 4D-VAR) as the main framework, while adopting the flow-dependent error covariance term generated from the EnKF method. This integration of variational and ensemble-based techniques addresses the limitations of static background covariances in variational methods by capturing the evolving nature of atmospheric conditions. By combining the strengths of both approaches, hybrid data assimilation offers improved accuracy and reliability in estimating the state of the atmosphere.

e. *Mesoscale community Weather Research and Forecasting (WRF) model*

The Weather Research and Forecasting (WRF) model is a mesoscale community numerical weather prediction system designed to serve both operational forecasting and atmospheric research needs. It features multiple dynamic cores. The WRF model is widely used in research communities, particularly the advanced research version known as WRF ARW (or WRF; Skamarock et al., 2008, 2021). WRF ARW, developed by the National Center for Atmospheric Research (NCAR), is based on an Eulerian solver for fully compressible non-hydrostatic equations. It is cast in flux conservation form, utilizing a mass (hydrostatic pressure) vertical coordinate. The solver employs a third-order Runge–Kutta time integration scheme and a split-explicit second-order time integration scheme for the acoustic and gravity-wave modes. The fully conservative flux divergence integration utilizes 5th-order upwind-biased advection operations, with selectable run-time schemes ranging from second to sixth order. The WRF model incorporates various options for cumulus, microphysics, planetary boundary layer (PBL), and radiation physical processes.

In addition to WRF ARW, several methods (such as those described in Sects. 2.a–2.d) have been developed over the years as data assimilation options (see Skamarock et al., 2021; Huang et al., 2009). The U.S. National Centers for Environmental Prediction (NCEP) operational data assimilation system, known as the Gridpoint Statistical Interpolation (GSI) system (Wu et al., 2002), has also been adopted for WRF model data assimilation needs.

3 Results and Discussion

To demonstrate the impacts of regional data assimilation on numerical simulations and predictions of MCSs, several cases from the U.S. Southern Great Plains (SGP) were examined. The SGP region is known for its frequent and active severe weather systems during the warm season, making it an ideal location for studying and improving the prediction of these convective systems. By focusing on the SGP, we were able to assess the effectiveness of regional data assimilation techniques in capturing the unique characteristics and dynamics of MCSs over this region. The goal was to evaluate how the assimilation of various observational data sources, such as surface Mesonet observations, ground-based radar, and lidar data, could enhance the numerical simulation and prediction of severe weather events in the SGP.

a. *Impact of assimilating Oklahoma Mesonet surface data on numerical simulations of MCSs*

To improve the near-surface atmospheric conditions associated with the initiation, evolution, and dissipation of MCSs, it is crucial to accurately analyze near-surface atmospheric conditions such as temperature and moisture. However, numerical models often encounter difficulties in representing near-surface atmospheric conditions, including surface 2-m temperature (e.g., Zhang et al., 2013). Moreover, in previous years, assimilating surface 2-m temperature data presented a particular challenge in operational 3D-VAR, as neither the NCEP nor ECMWF global reanalysis assimilated surface 2-m temperature for a long time. To address the problem of surface data assimilation and identify an effective method for assimilating such data, Pu et al. (2013) conducted a study comparing the 3D-VAR and EnKF methods for surface data assimilation and short-range weather forecasting. The study found that the EnKF method could overcome early challenges in surface data assimilation, resulting in improved analysis of surface conditions and better short-range weather forecasts, mainly due to its flow-dependent background covariances.

To examine the impact of assimilating surface observations from the Oklahoma Mesonet (Fig. 1) on numerical simulations of MCSs, hourly surface Mesonet observations, including wind, temperature, and pressure, were assimilated into the WRF model from 12 to 15 UTC on May 19, 2011. The forecast was then extended until 12 UTC on May 21, 2011. Figure 2 demonstrates the significant impact of surface Mesonet observations on predicting a convective system during the period from 0000 to 1800 UTC on May 20 over Oklahoma. Specifically, without assimilating surface

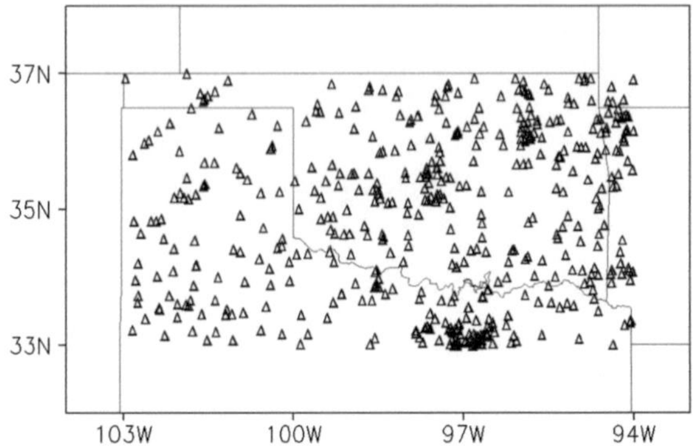

Fig. 1 Surface Mesonet stations over Oklahoma and near the Texas Panhandle (*source* https://mesonet.org)

Mesonet observations (control exp.), the WRF forecast failed to capture the development of the convective system. However, with the assimilation of surface Mesonet data, the model provided a better prediction of the initiation and evolution of the convective system, including a reasonable representation of rainbands.

Assimilating Oklahoma Mesonet data was also found to significantly improve the numerical simulation of convective properties. Not only were the time and location of convective initiation better represented, but there was also a notable enhancement in quantitative precipitation forecasting (QPF) within the numerical simulation. To quantify the precipitation forecasting skill, the equitable threat scores (ETS) were derived from a contingency table approach (e.g., Wilks, 1995) using the following equations:

$$\mathrm{ETS} = \frac{A}{A + B + C}$$

For a given threshold, A represents the number of grid points that the model forecast and observation exceed the threshold; B denotes the number of grid points that the model forecast exceeds the threshold, but the observation does not; and C is the number of grid points when the model forecast does not reach the threshold, but the observation exceeds it. Figure 3 shows the ratio of the (ETS) for 1-h accumulated precipitation between data assimilation (EXP) and CTRL experiments from 0100 UTC 20 May to 2200 UTC 20 May 2001. It indicates the quantitative precipitation forecasting (QPF) skill in the simulation is improved by data assimilation, as most ratios are greater than one.

b. *Improving cold pool representation*

In addition to their role in maintaining MCSs, cold pools generated by convective systems can be important in triggering secondary convective initiation. Accurate numerical simulations of cold pools are not only essential for representing the evolution of convective properties but can also influence the prediction of secondary convective initiation. With this in mind, and recognizing the positive impact of surface observations on numerical simulations of MCSs, we further evaluated the ability

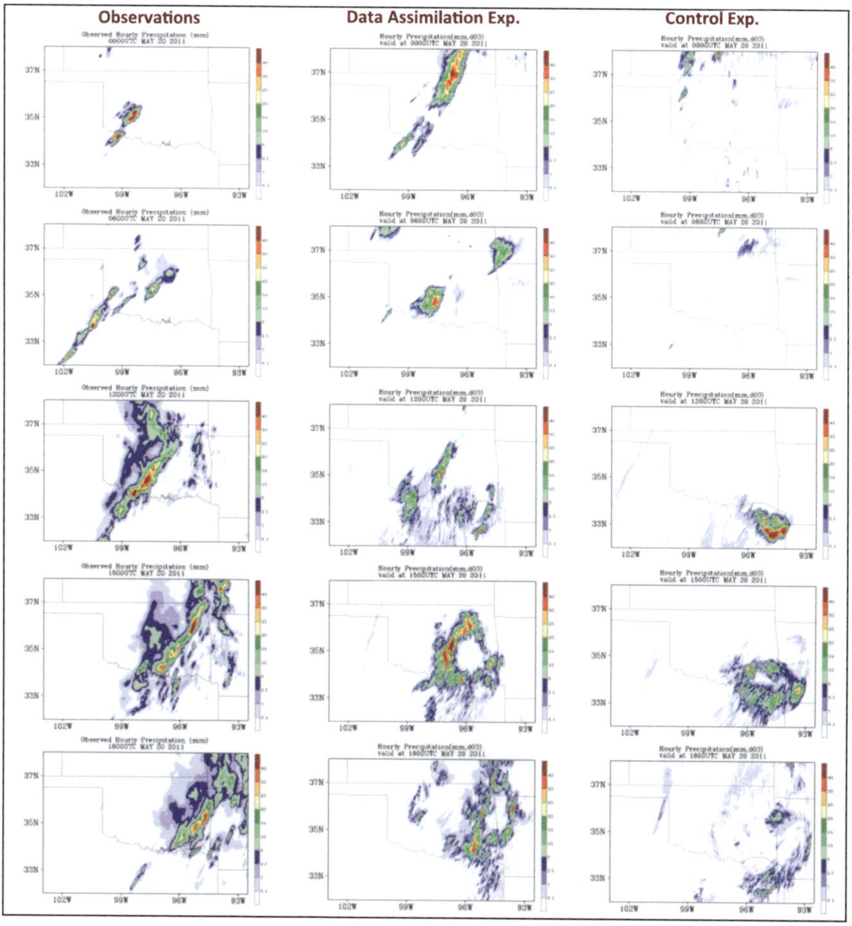

Fig. 2 Hourly precipitation (unit: mm) in model domain 3 (1.33 km horizontal resolution), valid at 00 UTC (1st row), 06 UTC (2nd row), 12 UTC (3rd row), 15 UTC (4th Row), and 18 UTC (5th Row) 20 May 2011. The observations (left column) compared with numerical simulation results with (middle column) and without (right column) assimilation of surface Mesonet observations

of the WRF model to represent cold pools in numerical simulations of convective systems, both with and without data assimilation.

We used the convective systems of May 20, 2011, as an example. Since numerical simulations of convective systems are sensitive to the choice of microphysics and planetary boundary layer (PBL) schemes in the WRF model, we examined the cold pool in several simulations with different configurations of WRF physical parameterizations (see Table 1 for details). Near-surface atmospheric conditions, such as 2-m temperature and surface pressure, from each numerical simulation, were compared with Oklahoma Mesonet observations.

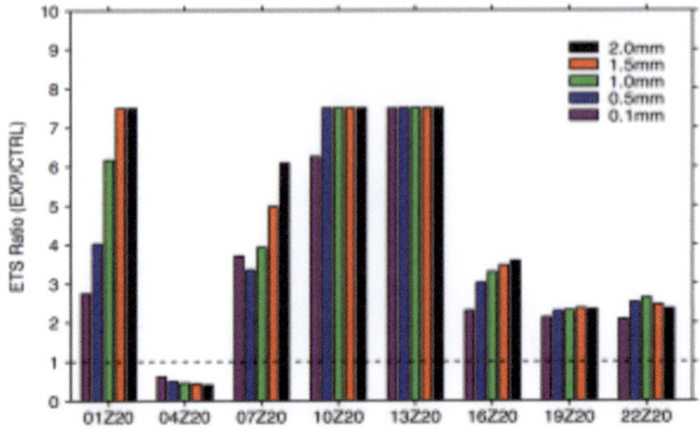

Fig. 3 Ratio of equitable threat scores (ETS) for 1-h accumulated precipitation between data assimilation (EXP) and CTRL experiments from 0100 UTC 20 May to 2200 UTC 20 May 2001. When the ratio is greater than 1, the quantitative precipitation forecasting (QPF) skill in the simulation is improved by data assimilation

Table 1 Configurations of WRF sensitivity experiments

Exp	Microphysics	Surface layer physics	PBL	Data assimilation
CTRL	LIN	Monin–Obukhov (Eta)	MYJ	N
WSM6-MYJ	WSM6	Monin–Obukhov (Eta)	MYJ	N
LIN-YSU	LIN	Monin–Obukhov (MM5)	YSU	N
EXP	LIN	Monin–Obukhov (Eta)	MYJ	Y

To effectively validate a cold pool in numerical simulations, the surface cold pool should first be identified in the observations. We utilized Oklahoma Mesonet observations over 16 years (1997–2012) to determine the climatic state on May 20 and the weekly average around May 20, 2011, for each station, to eliminate the impact of diurnal variation on identifying the cold pool in the observations. Then, the model-simulated temperature was compared with the observations to verify whether the model accurately simulated the cold pool. It was determined that a cold pool occurred around 0400–0600 UTC on May 20, 2011.

The numerical results indicate that the WRF simulation of surface temperature was sensitive to the choices of physical parameterizations (microphysics and PBL). However, none of these numerical simulations could predict the occurrence of a cold pool. Only when surface Mesonet data were assimilated within the initial few hours of the simulations was the WRF model able to generate a reasonable representation of the surface cold pool (refer to Figs. 4 and 5).

According to Weiss and Bluestein (2002), convective initiation is characterized by interactions among cold fronts, drylines, and cold pools (refer to the conceptual model in Fig. 6a). In our study, we observed that the convective initiation that took

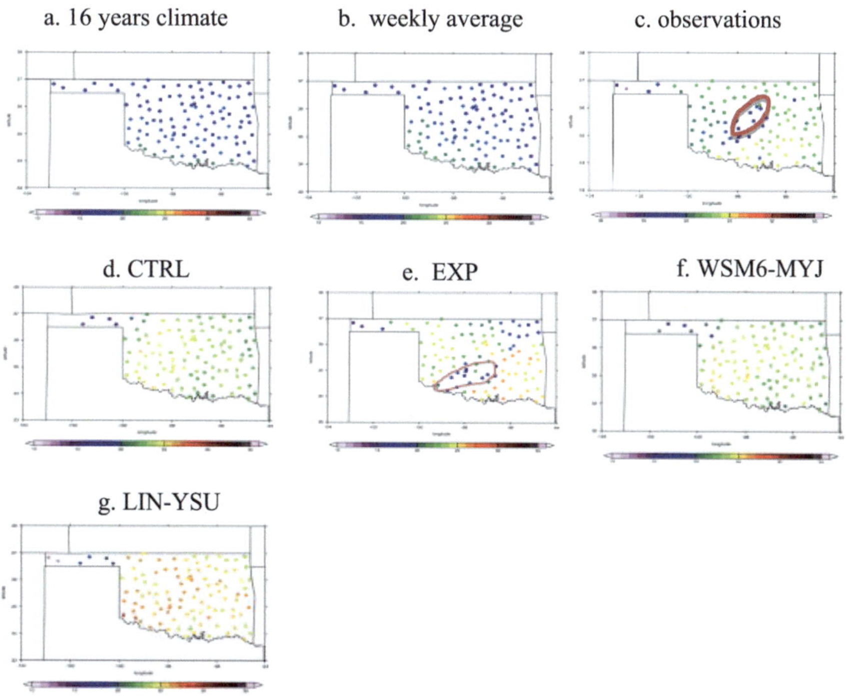

Fig. 4 Surface temperature at 0500 UTC 20 May 2011 over Oklahoma Mesonet stations from observations and various numerical simulations. The red circle denotes a surface cold pool

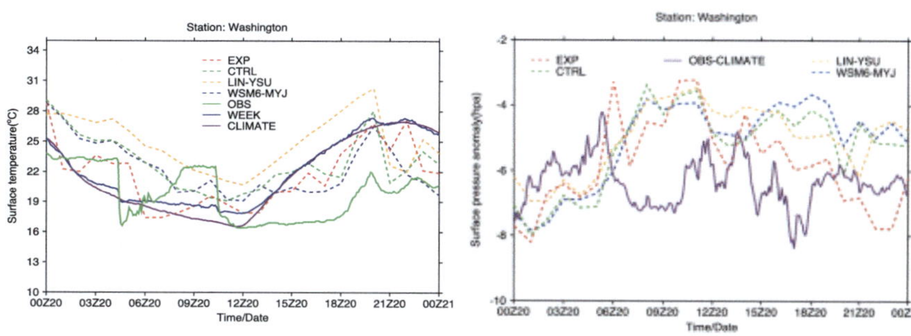

Fig. 5 The diurnal variation of surface temperature (left) and pressure (right) between 0000 UTC 20 May and 0000 UTC 21 May 2011 at Washington Station, OK. Observations show a cold pool occurring at 04–06 UTC 20 May. Surface temperature and pressure forecasts are sensitive to the choice of model physical parameterization scheme. However, only the experiment with data assimilation (EXP) captured the cold pool event with the drop in surface temperature and increase in surface pressure

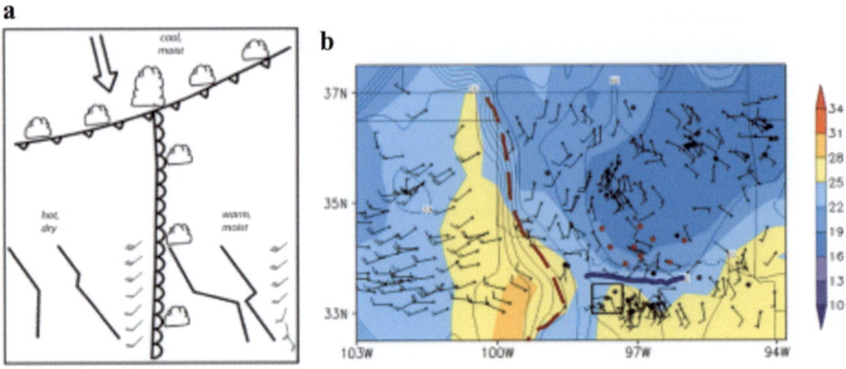

Fig. 6 **a** Conceptual model of cold front, dryline, and cold pool interactions (Weiss & Bluestein, 2002). **b** Surface weather map at 1800 UTC 20 May 2011: wind barbs, temperature (shaded), relative humidity (contours). Red dots denote the cold pool. The thick blue line marks the cold front. The purple dashed line represents the dryline. The small black box indicates the area of convective initiation

place at 1800 UTC on May 20, 2011, could be explained by this conceptual model (see Fig. 6b). Additionally, we found that only when data assimilation was applied did the WRF simulation capture the surface cold pool and accurately predict the occurrence of secondary convective initiation near the observed time.

c. *Impact of assimilating multi-time wind profiles on numerical simulations of MCSs and QPF*

Most routine measurements obtained from climate study facilities are collected from individual sites over long periods. While these single-station data are valuable for many studies, it can be challenging to obtain comprehensive 3-dimensional spatial structures of atmospheric boundary layers that include distinct features of deep convection. To address this issue, research has focused on exploring the capabilities of data assimilation techniques in retrieving 4-dimensional (both spatial and temporal) atmospheric boundary layer structures using multi-time profiling data from a single station, along with single-level surface Mesonet observations. In the context of this problem, wind profiling observations acquired from the Goddard Lidar Observatory for Winds (GLOW) during IHOP_2002 were assimilated into the WRF ARW model using its 4D-VAR system. The impact of this data assimilation on the numerical simulation of a squall line occurring between June 12 and 13, 2002, over the region spanning from the Kansas and Oklahoma border to the Texas Panhandle, was assessed. Data assimilation experiments presented in this section are detailed in Zhang and Pu (2011); we reinterpreted some results here to demonstrate the importance of assimilating boundary layer profiling observations in the numerical prediction of MCSs.

Figure 7 illustrates the composite radar reflectivity mosaic observed during a 6-h period at 2-h intervals, corresponding to the initiation, organization, intensification, and maturation stages of the convection during the IHOP_2002 field experiment.

Convection was observed to initiate around 2100 UTC on June 12, with isolated convection (groups A and B, as depicted in Fig. 7a) forming along the dryline from the Kansas and Oklahoma border to the Texas Panhandle (Fig. 7a). These groups gradually intensified along the dryline (Fig. 7b). Group A extended northeastward and reached its maximum intensity, while group B continuously organized and intensified. By 0100 UTC on June 13 (Fig. 7c), group A continued to move southeastward and extend northeastward, weakening and dissipating after 2 h (Fig. 7d). Group B, on the other hand, continued to intensify and extend eastward, eventually developing into a squall-line structure, with convection intensifying along the Kansas and Oklahoma border by 0100 UTC on June 13 (Fig. 7c). Subsequently, the squall line moved southeastward (Fig. 7d) and dissipated after 1100 UTC on June 13, 2002.

During the IHOP_2002 field experiment, which took place from May 13, 2002, to June 25, 2002, a total of more than 240 h of wind profiles were gathered using the ground-based GLOW Doppler wind lidar instrument. Following initial quality control and pre-processing steps, a data product was generated with a temporal resolution of 10 min and a vertical resolution of 100 m for altitudes below 3 km, while altitudes above 3 km had a vertical resolution of 200 m. These wind profiles provide detailed information on the wind structure from the surface up to approximately 7 km. Figure 8 displays the time series of GLOW wind profiles assimilated in our study, specifically from 1800 to 2100 UTC on June 12, 2002. These profiles encompass a wide range of altitudes and offer valuable insights into the vertical distribution of wind patterns during this time.

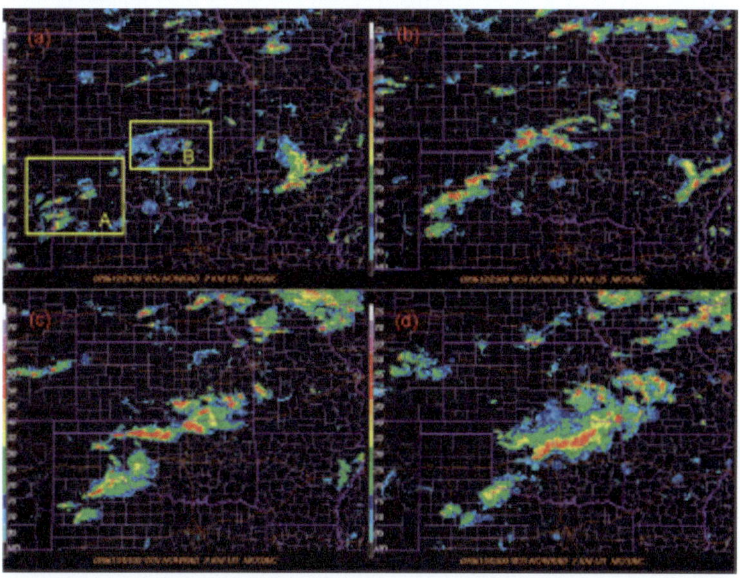

Fig. 7 Radar observed composite reflectivity mosaic at **a** 2100 UTC 12 June 2002, **b** 2300 UTC 12 June 2002, **c** 0100 UTC 13 June 2002, **d** 0300 UTC 13 June 2002 (The data come from the WSI IHOP_2002 2-km resolution sector mosaic reflectivity imagery dataset)

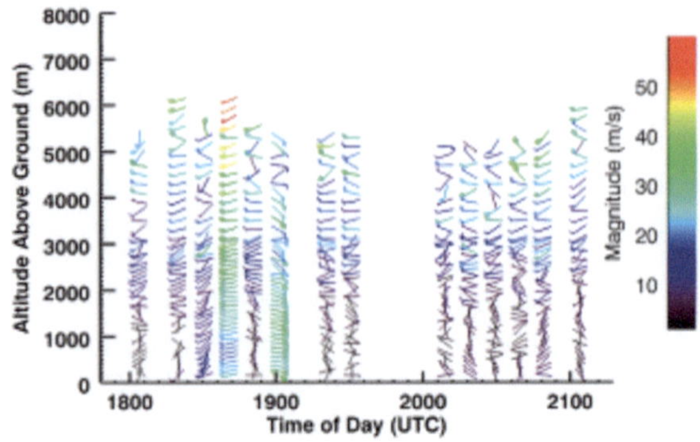

Fig. 8 The time series of GLOW wind profiles at the Homestead profiling site from 1800 to 2100 UTC June 2002. Different colors represent various wind speeds

Figure 9 shows the analysis increments as the differences between the control experiment (CTRL) and the 4D-VAR data assimilation experiment (4D-VAR) of wind *u*- and *v*-components, temperature, and water vapor mixing ratio at 850 hPa at the end of the data assimilation (2100 UTC 12 June 2002). Large differences were found in the regions along the dryline system. The wind fields were strengthened over most of the model domain, especially around the Texas Panhandle, while along the Texas and Oklahoma border, weaker winds were found in the 4D-VAR experiment. After data assimilation, large temperature increases (Fig. 9c) were produced in the northwest corner of Texas and along the Texas and Oklahoma border. In these regions, the temperature increased up to 3.0 K. A large moisture increase was found near the southwestern corner of Kansas and the western corner of the Oklahoma and Texas border, while a large moisture decrease appeared in northwest Oklahoma. Along the Kansas and Texas border, the water vapor mixing ratio increased up to 1.0 g kg^{-1} and increases up to 2.5 g kg^{-1} were found in the northwest corner of the Texas and Oklahoma border, while in northwest Oklahoma the water vapor mixing ratio decreased by 2.0 g kg^{-1}. Therefore the moisture gradient was reduced in the dryline region after data assimilation. It should be noted from the above results that although only wind profiles from a single station were assimilated into the WRF model by its 4D-VAR system, the analysis increments (Fig. 9) during the data assimilation showed a reasonable spatial distribution. The multi-time observations at a single station influenced the spatial distribution of the state variables. More importantly, even though only wind information was assimilated, the temperature and moisture fields were also adjusted.

To further examine the data assimilation impacts, Fig. 10a and b compare the horizontal wind fields at 850 hPa at 2100 UTC 12 June 2002 from CTRL (Fig. 10a) and 4D-VAR (Fig. 10b). It is apparent that major differences between CTRL and 4D-VAR occurred mainly in the region where convective initiation took place. Compared with the multi-radar reflectivity observations (Fig. 7a), the simulated locations of the convection in CTRL were farther to the northeast than the observed locations. The locations of the convection initiation in the 4D-VAR experiment were much closer to the observed convection initiation locations. The simulated horizontal and vertical wind fields in the 4D-VAR experiment were more realistic than those in the control

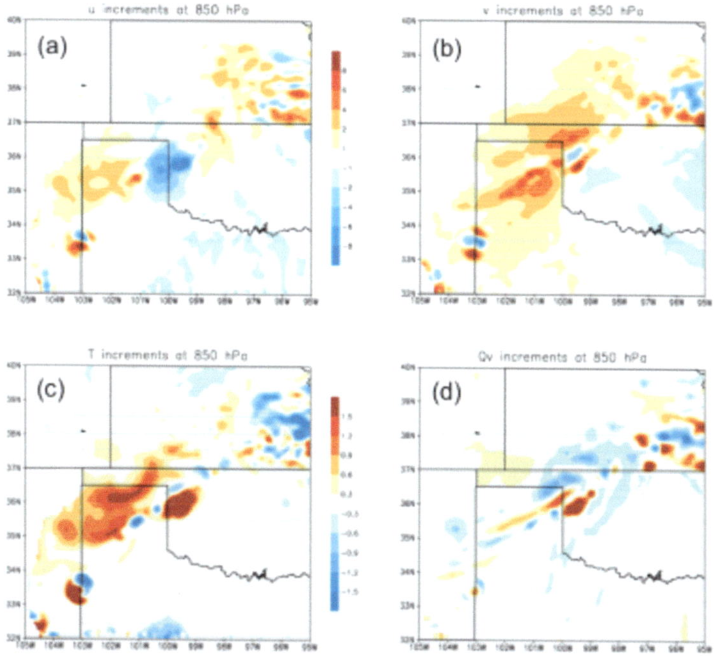

Fig. 9 Differences between the control experiment (CTRL, without assimilation of wind profiles) and 4D-VAR experiments for the **a** u-component (m s^{-1}), **b** v-component (m s^{-1}), **c** temperature (K), and **d** water vapor mixing ratio (g kg^{-1}) at 850 hPa at 2100 UTC 12 June 2002

experiment. The split of two convective cells was evident in the 4D-VAR experiment (Fig. 10d). As a consequence, the skill of quantitative precipitation forecasting was greatly enhanced (see Zhang & Pu, 2011 for details).

d. *Radar data assimilation and cloud initialization with radar observations*

It has been demonstrated that assimilating ground-based Doppler radar observations can improve forecasts and simulations of MCSs. Previous studies have indicated that the assimilation of radial velocity-derived winds and radar reflectivity can enhance numerical simulations and predictions of severe storms, as shown by Pu et al. (2009) and other references (e.g., Sun & Wang, 2013) for studies over the SGP.

Recent advancements and research examples utilizing various data assimilation methods, described in Sect. 2, have been extensively documented in numerous publications. In this chapter, we omit our sample studies but present an illustrative example of cloud initialization using radar observations.

Accurate numerical prediction of MCSs at a cloud-permitting scale poses a significant challenge. Many instances of poor MCS forecasts can be attributed to errors in the initial conditions, particularly the lack of cloud initialization in several data assimilation systems. Consequently, the representation of clouds in numerical models relies heavily on the cumulus and microphysical processes during the spin-up phase.

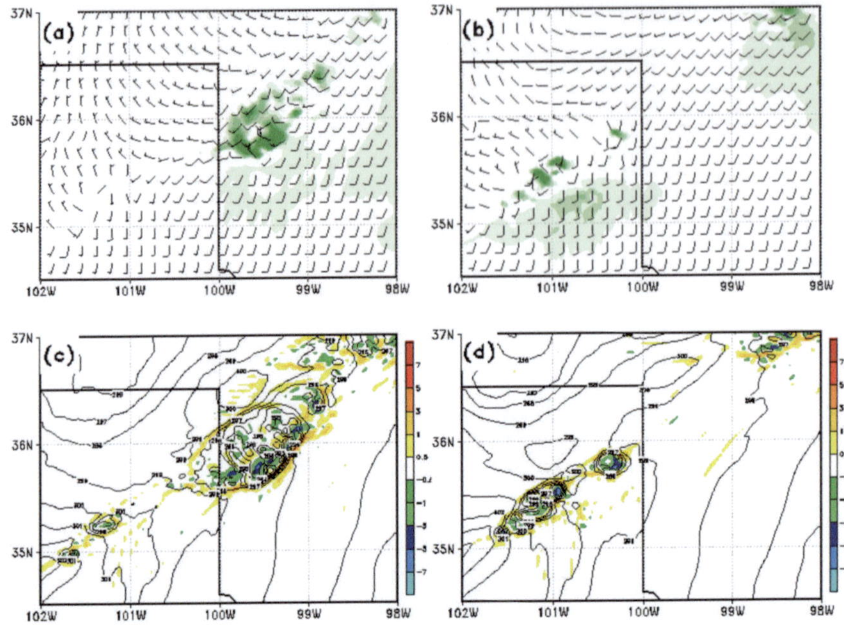

Fig. 10 The horizontal and vertical wind fields (m s^{-1}) and temperature (K) at 850 hPa at 2100 UTC 12 June 2002 from the control run (**a**, **c**) and data assimilation experiments (**b**, **d**). Panels **a** and **b** represent the horizontal wind field: the shaded area indicates the region with wind speed exceeding 8 m s^{-1}. Panels **c** and **d** represent the vertical velocity (shaded) and temperature fields (contour lines; contour interval is 0.5 K)

To investigate the impact of cloud initialization on numerical simulations of MCSs at a cloud-permitting scale, we utilized the NCEP GSI data assimilation system along with its cloud analysis package (e.g., Hu et al., 2008, 2018), which was modified for WRF model applications. We assimilated U.S. Next Generation Weather Radar (NEXRAD) observations into the WRF model. Initially, radar reflectivity observations were assimilated using the GSI cloud analysis package to derive the initial cloud analysis of hydrometeors (refer to Fig. 11). Subsequent forecasts indicated that cloud initialization reduced the model spin-up effects and improved the simulation of convective systems. Furthermore, radar radial velocities were also assimilated, in conjunction with cloud initialization, to obtain a realistic local dynamic field. The impacts of assimilating NEXRAD observations, particularly cloud initialization, were comprehensively evaluated.

For our study, we utilized an MCS case from May 2011. We initiated the cloud analysis and simulation at 06 UTC on May 24, 2014, when the MCS was in its mature stage. As shown in Fig. 12, cloud analysis provided a means to initialize cloud conditions at the start of the model simulation, significantly reducing the cloud spin-up period in the analysis and forecasts.

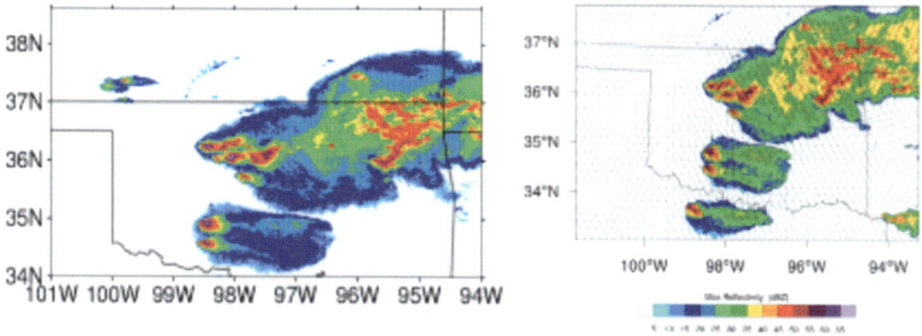

Fig. 11 Composite radar reflectivity (DBZ) from NEXRAD radar (left), and WRF model initial conditions with cloud analysis (right) at 06 UTC 24 May 2011, when the MCS was in its mature stage. Without cloud analysis, the model's initial reflectivity remained zero in the whole model domain. The figures reveal that cloud analysis successfully generated initial cloud conditions

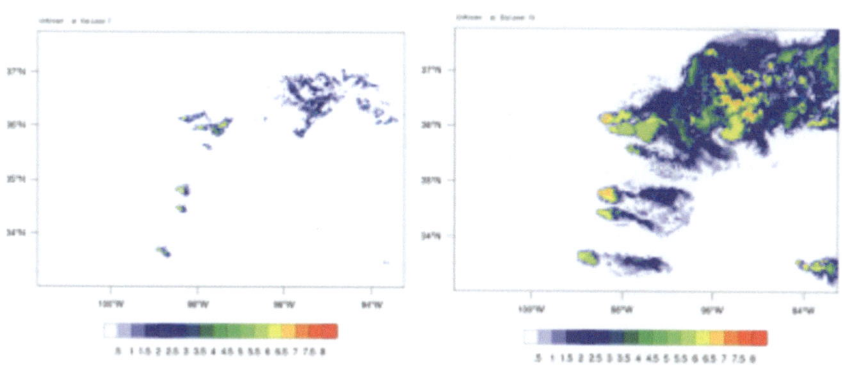

Fig. 12 The cloud rainwater mixing ratio (left) at sigma level 7 and snow mixing ratio (right) at sigma level 15, produced by the cloud analysis scheme for the WRF model at the initial time (06 UTC 24 May 2011)

4 Summary and Conclusions

Analysis of the selected cases in this chapter demonstrates that regional data assimilation has a significant impact on the accuracy and skill of numerical models in predicting MCSs. The assimilation of surface Mesonet observations, ground-based radar, and lidar data has been proven to enhance the prediction of MCSs by improving the representation of atmospheric conditions and dynamics.

Surface Mesonet observations provide crucial information about near-surface atmospheric conditions such as temperature, humidity, and wind. These observations play a vital role in initializing and validating numerical models, thereby improving the accuracy of simulations. The results presented in Sect. 3.2 indicate that assimilating surface observations holds promise as an effective approach to enhancing the accuracy of numerical simulations of cold pools within MCSs. With improved representation of cold pools, a better understanding of the interactions among cold fronts, drylines, and cold pools can be achieved.

The 4D-VAR method allows improvement in 4-dimensional atmospheric boundary layer structures by incorporating multi-time profiling data from a single station. Routine observations from single stations can provide valuable information regarding the 3-dimensional spatial structures of atmospheric boundary layers, including deep convection. Lidar data, particularly Doppler wind lidar observations, offer additional insights into vertical wind structure, aiding in the initialization and verification of the model's atmospheric boundary layer.

Ground-based radar data contribute to understanding the three-dimensional structure of MCSs, including cloud development, precipitation patterns, and wind fields. Assimilating radar reflectivity and radial velocity data into numerical models improves the models' representation of precipitation and storm-scale dynamics. The assimilation of radar data, combined with cloud analysis schemes, allows for more accurate specification of cloud properties, leading to improved analysis and simulation of cloud life cycles during MCSs. Furthermore, this approach provides a way to better understand cloud microphysical processes and their evolution and interaction with dynamics during MCSs.

Overall, the utilization of regional data assimilation techniques in the U.S. Southern Great Plains has demonstrated positive impacts on the numerical simulation and prediction of MCSs. By incorporating observational data specific to the region, researchers and forecasters can gain valuable insights into the complex dynamics of severe weather events, thereby improving the accuracy and reliability of forecasts in this high-impact weather region. The concepts and applications demonstrated in this chapter are not limited to the U.S. Southern Great Plains but can be applied to many other regions worldwide where severe weather occurs.

References

Blumberg, W. G., Wagner, T. J., Turner, D. D., & Correia, J., Jr. (2017). Quantifying the accuracy and uncertainty of diurnal thermodynamic profiles and convection indices derived from the atmospheric emitted radiance interferometer. *Journal of Applied Meteorology and Climatology, 56*, 2747–2766.

Bryan, G. H., Ahijevych, D., Davis, C., Trier, S., & Weisman, M. (2005). Observations of cold pool properties in mesoscale convective systems during BAMEX. In *11th Conference on Mesoscale Processes*, Albuquerque, NM, American Meteorological Society, CD-ROM, JP5J.12.

Bryan, G. H., & Fritsch, M. J. (2000). Moist absolute instability: The sixth static stability state. *Bulletin of the American Meteorological Society, 81*, 1207–1230.

Crook, N. A. (1996). Sensitivity of moist convection forced by boundary layer processes to low-level thermodynamic fields. *Monthly Weather Review, 124*, 1767–1785.

Degelia, S. K., Wang, X., & Stensrud, D. J. (2019). An evaluation of the impact of assimilating AERI retrievals, kinematic profilers, rawinsondes, and surface observations on a forecast of nocturnal convection initiation event during the PECAN field campaign. *Monthly Weather Review, 147*, 2739–2764.

Droegemeier, K. K., & Wilhelmson, R. B. (1987). Numerical simulation of thunderstorm outflow dynamics. Part I: Outflow sensitivity experiments and turbulence dynamics. *Journal of Atmospheric Science, 44*, 1180–1210.

Engerer, N. A., Stensrud, D. J., & Coniglio, M. C. (2008). Surface characteristics of observed cold pools. *Monthly Weather Review, 136*, 4839–4849.

Fabry, F. (2006). The spatial variability of moisture in the boundary layer and its effect onconvection initiation: Project-long characterization. *Monthly Weather Review, 134*, 79–91.

Geerts, B., et al. (2017). The 2015 plains elevated convection at night field project. *Bulletin of the American Meteorological Society, 98*, 767–786.

Hamill, T. M., & Snyder, C. (2000). A hybrid ensemble Kalman filter–3D variational analysis scheme. *Monthly Weather Review, 128*, 2905–2919.

Holt, T. R., Niyogi, D., Chen, F., Manning, K., LeMone, M. A., & Qureshi, A. (2006). Effect of land-atmosphere interactions on the IHOP 24–25 May 2002 convection case. *Monthly Weather Review, 134*, 113–133.

Hu, M., Weygandt, S., Benjamin, S., & Xue, M. (2008). Ongoing development and testing of generalized cloud analysis package within GSI for initializing Rapid Refresh Preprints. In *13th Conference on Aviation, Range and Aerospace Meteorology*, New Orleans, LA, American Meteorology Society, 7.4.

Hu, M., Ge, G., Zhou, C., Stark, D., Shao, H., Newman, K., Beck, J., & Zhang, X. (2018). *Grid-point Statistical Interpolation (GSI) user's guide version 3.7* (149pp.). Developmental Testbed Center. https://dtcenter.org/community-code/gridpoint-statistical-interpolation-gsi/documentation.

Huang, X. Y., et al. (2009). Four-dimensional variational data assimilation for WRF: Formulation and preliminary results. *Monthly Weather Review, 137*, 299–314.

Kalnay, E. (2003). *Atmospheric modeling* (p. 341). Cambridge University Press.

Liu, Y., et al. (2008). The operational mesogamma-scale analysis and forecast system of the U.S. army test and evaluation command. Part II: Interrange comparison of the accuracy of model analyses and forecasts. *Journal of the Applied Meteorology and Climatology, 47*, 1093–1104.

Massey, J. D., Steenburgh, W. J., Hoch, S. W., & Knievel, J. C. (2014). Sensitivity of near-surface temperature forecasts to soil properties over a sparsely vegetated dryland region. *Journal of the Applied Meteorology Climatology, 53*, 1976–1995.

Mueller, C. K., & Carbone, R. E. (1987). Dynamics of a thunderstorm outflow. *Journal of Atmospheric Science, 44*, 1879–1898.

Parsons, D. B. (1992). An explanation for intense frontal updrafts and narrow cold-frontal rainbands. *Journal of Atmospheric Science, 49*, 1810–1825.

Pu, Z., Li, X., & Sun, J. (2009). Impact of airborne Doppler radar data assimilation on the numerical simulation of intensity changes of Hurricane Dennis near a landfall. *Journal of Atmospheric Science, 66*, 3351–3365.

Pu, Z., Zhang, H., & Anderson, J. A. (2013). Ensemble Kalman filter assimilation of near surface observations over complex terrain: Comparison with 3DVAR for short-range forecasts. *Tellus, 65*, 19620.

Purdom, J. F. (1976). Some uses of high-resolution GOES imagery in the mesoscale forecasting of convection and its behavior. *Monthly Weather Review, 104*, 1474–1483.

Rotunno, R., Klemp, J. B., & Weisman, M. L. (1988). A theory for strong, long-lived squall lines. *Journal of Atmospheric Science, 45*, 463–485.

Skamarock, W. C., Klemp, J. B., Dudhia, J., Gill, D. O., Barker, D. M., Wang, W., & Powers, J. G. (2008). *A description of the advanced research WRF version 3*, NCAR Technical Note. NCAR/TN!468+STR, National Center for Atmospheric Research, Boulder, Colorado.

Skamarock, W. C., Klemp, J. B., Dudhia, J., Gill, D. O., Liu, Z., Berner, J., & Huang, X.-Y. (2021). A description of the advanced research WRF model version 4.3 (No. NCAR/TN-556+STR). https://doi.org/10.5065/1dfh-6p97.

Stensrud, D. J., Yussouf, N., Dowell, D. C., & Coniglio, M. C. (2009). Assimilating surface data into a mesoscale model ensemble: Cold pool analyses from spring 2007. *Atmospheric Research, 93*, 207–220.

Sun, J., & Wang, H. (2013). Radar data assimilation with WRF 4D-Var. Part II: Comparison with 3D-Var for a squall line over the U.S. Great Plains. *Monthly Weather Review, 141*, 2245–2264.

Szeto, K. K., & Cho, H. R. (1994). A numerical investigation of squall lines. Part II: The mechanics of evolution. *Journal of Atmospheric Science, 51*, 425–433.

Thorpe, A. J., Miller, M. J., & Moncrieff, M. W. (1982). Two-dimensional convection in non-constant shear: A model of mid-latitude squall lines. *Quarterly Journal of Royal Meteorological Society, 108*, 739–762.

Trier, S. B., Davis, C. A., Ahijevych, D. A., Weisman, M. L., & Bryan, G. H. (2006). Mechanisms supporting long-lived episodes of propagating nocturnal convection within a 7-day WRF model simulation. *Journal of Atmospheric Science, 63*, 2437–2461.

Wakimoto, R. M. (1982). The life cycle of thunderstorm gust fronts as viewed with Doppler radar and rawinsonde data. *Monthly Weather Review, 110*, 1060–1082.

Weckwerth, T. M., et al. (2004). An overview of the International H2O Project (IHOP_2002) and some preliminary highlights. *Bulletin of the American Meteorological Society, 85*, 253–278.

Weiss, C. C., & Bluestein, H. B. (2002). Airborne pseudo-dual Doppler analysis of a dryline-outflow boundary intersection. *Monthly Weather Review, 130*, 1207–1226.

Wilks, D. S. (1995). *Statistical Methods in Atmospheric Sciences* (467pp.). Academic Press.

Wilson, J. W., & Schreiber, W. E. (1986). Initiation of convective storms at radar-observed boundary layer convergent lines. *Monthly Weather Review, 114*, 2516–2536.

Wu, W.-S., Purser, R. J., & Parrish, D. F. (2002). Three-dimensional variational analysis with spatially inhomogeneous covariances. *Monthly Weather Review, 130*, 2905–2916.

Wurman, J., Dowell, D., Richardson, Y., Markowski, P., Rasmussen, E., Burgess, D., Wicker, L., & Bluestein, H. B. (2012). The second verification of the origins of rotation in Tornadoes experiment: VORTEX2. *Bulletin of the American Meteorological Society, 93*, 1147–1170.

Zhang, H., Pu, Z., & Zhang, X. (2013). Examination of errors in near-surface temperature and wind from WRF numerical simulations in regions of complex terrain. *Weather Forecasting, 28*, 893–914.

Zhang, L., & Pu, Z. (2011). Four-dimensional assimilation of multi-time wind profiles over a single station and numerical simulation of a mesoscale convective system observed during IHOP_ 2002. *Monthly Weather Review, 139*, 3369–3388.

Rapid Update Cycle Data Assimilation for Severe Storm Prediction

Ashish Routray, Devajyoti Dutta, Shivaji Singh Patel, John P. George, and V. S. Prasad

Abstract The National Centre for Medium Range Weather Forecasting (NCMRWF) has set up a 1 hrly updated 1.5 km High Resolution Rapid Refresh (HRRR) Data Assimilation (DA) system with regional NCUM-R forecast model that is configured for three specific domains over the Indian region. In this study, we considered the model domain configured over the northeast region. The main objective of the present study is to demonstrate the assimilation of DWR observations in the HRRR DA system for the simulation of thunderstorms over the northeast states of India. Two set-up numerical experiments viz CTL (without DA) and HRR (Assimilation of radial velocity and reflectivity observations along with surface, sonde and satellite observations in 1 hrly 4DVAR HRRR DA system) are carried out. The analysis innovation (O-A) is relatively less than the background innovation (O-B) for all variables, suggesting that the assimilation of observations is properly utilized in the assimilation cycle. The location and distribution of rainfall are better simulated in the HRR experiment than in the CTL simulation. But the CTL simulated a high amount of rainfall that slightly shifted northeastward corresponding the observations. The peak rainfall is well brought out in the HRR simulation with minimal time lag while compared with the CTL simulation. The CTL simulation shows a high amount of rainfall during the early hours which may be due to the model spin-up. The temporal evolution of the convective system is well captured by the HRR with a higher correlation than in the CTL experiment. The ETS at different rainfall thresholds clearly shows an improvement in the HRR simulation than the CTL simulation. The forecast skill (%) of HRR with respect to the CTL is more at higher thresholds of rainfall. It is deduced that the high intensity of precipitation is well simulated by the HRR experiment than in the CTL experiment. At the same time, frequency bias is also considerably less in the HRR simulation than the CTL simulation in all cases. The object-based verification of rainfall clearly explains that the HRR experiment simulated the location and pattern of rainfall well with minimal errors than in the CTL simulation. But the volume errors are slightly more in the HRR simulation than

A. Routray (✉) · D. Dutta · S. S. Patel · J. P. George · V. S. Prasad
National Centre for Medium Range Weather Forecasting (NCMRWF), Ministry of Earth Sciences (MoES), A-50, Sector-62, Noida, UP 201309, India
e-mail: ashish.routray@gov.in

S. Das and W-K. Tao (eds.), *Severe Storms*,
https://doi.org/10.1007/978-981-97-7075-5_16

the CTL. It is deduced that the HRR produces high intensity rainfall over the domain corresponding to the observation. Overall, the total errors are reduced in the HRR experiment while compared with the CTL. The present study results suggested that the hourly updated cyclic assimilation in the high-resolution modeling approach with the assimilation of DWR observations has a beneficial impact on such convective weather systems.

Keywords HRRR data assimilation · Innovations · DWR observations · Object based verification · Thunderstorms · Frequency bias

1 Introduction

The storm associated with strong winds (between 80 and 93 kmph) and/or, hail (13–25 mm in diameter) is called severe thunderstorm. The thunderstorms are mesoscale systems of vigorous localized convective activity associated with squall, thunder, lightning and heavy rain and hail. These are causes of heavy losses in agriculture, housing, property and loss of life. The thunderstorms occur from cumulonimbus cloud with high vertical extent due to heavy convection that lead disastrous weather over a particular region. Different types of thunderstorms occur like single cell thunderstorm has spatial extent few kilometers with life span less than an hour and multi cell thunderstorm developed from organized convective activity, having life span several hours and travel more than 100 km. The devastating nature of thunderstorm depends upon its horizontal and vertical extent. It is well known that the frequency of the severe thunderstorms is higher during the pre-monsoon season (April–May) over Indian landmass. Generally, the severe thunderstorms are produced during the pre-monsoon season over Gangetic West Bengal, Jharkhand, Odisha, Bihar, Assam and northeastern states (Fig. 1) due to the incursion of warm moist air from ocean overrides of hot dry continental air. In these regions, thunderstorms are also locally called the Nor'westers because the systems mainly travel from northwest to southeast direction. Therefore, the causality due to thunderstorm is highest in these regions as compared to other parts of India that affect the socio-economic structure of eastern and north-eastern part of India. Therefore, it is very important to improve the prediction of the thunderstorm well in advance.

To understand the genesis, evolution, life cycle, dynamics and physical mechanisms of the thunderstorms, various observational field experiments over India like severe thunderstorms-observations and regional modeling (STORM) have been carried out in the past to provide comprehensive observational and modeling studies (Mohanty et al., 2006, 2007; Das et al., 2014). The observations were collected during the STORM field experiment over the domain depicted in Fig. 1. In recent years, the high-resolution non-hydrostatic mesoscale models with advanced parameterization schemes (Weiss et al., 2006) have been capable of simulating the evolution, thermodynamic and physical mechanisms associated with the system (Lopez et al., 2007; Farely et al., 1992; Brooks & Wilhelmson, 1992). Many studies were carried

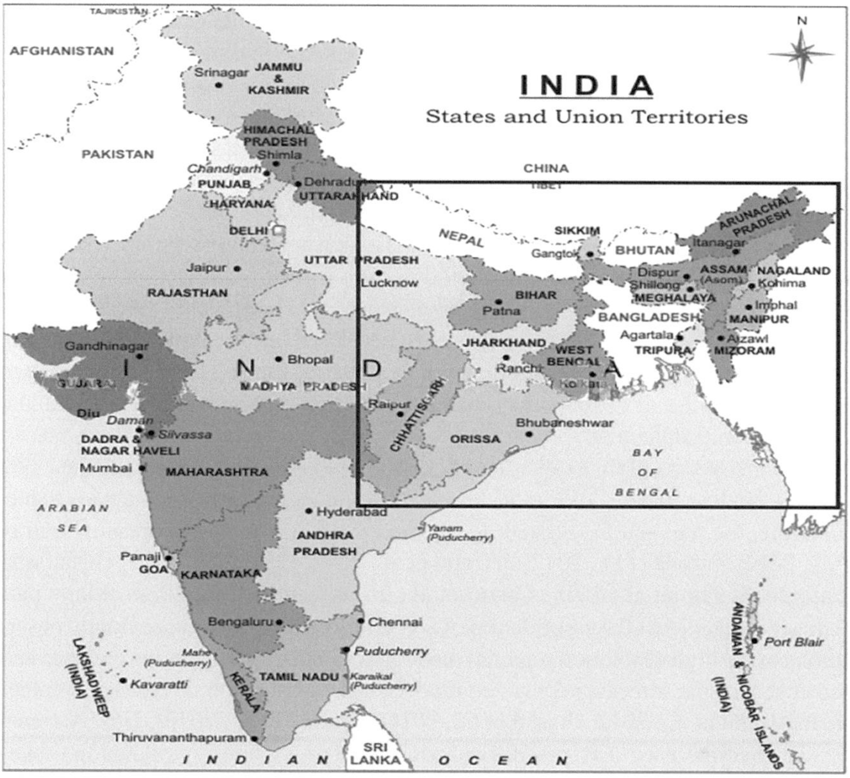

Fig. 1 STORM field experiment domain

out to simulate the location and intensity of the Nor'westers over Eastern & North Eastern India including Bangladesh through different mesoscale models and found mixed results (Das et al., 2006, 2016, and Tyagi et al., 2011, 2013; etc.). The performance of the mesoscale models significantly depends upon the quality of initial conditions, which come from the large-scale global model analyses. Those analyses have limitations viz. low resolution and poor representation of localized mesoscale features. Hence, data assimilation (DA) techniques that utilize the various observations collected from different platforms are essential components in the numerical model to further improve the initial conditions and enhance the forecast skill of the NWP models (Daley, 1991). The past decades have seen high-resolution regional models with three or four-dimensional variational assimilation techniques (3D/4DVAR) being used in the simulation of high-impact weather events (Kalnay, 2003; Routray et al., 2013, 2016; Dutta et al., 2019; etc.).

The high-impact convective weather events evolved with different microphysical and dynamical processes that operate at very small horizontal and time scales. The scale of eddies in the case of dust devils is around a few meters at the same time thunderstorms/hailstorms vary from the range of one to tens of km in clustered form.

Those extreme weather systems vertically extend from middle to high atmospheric levels about 10–15 km. To resolve their properties, the high-resolution mesoscale models must be configured with a horizontal resolution of about 1–2 km as well as the increase of frequency of assimilation cyclic. This demand can be achieved due to the advancement of high-performance computing as well as the availability of diverse observations in recent times for the prediction of extreme events accurately well in advance. Therefore, the convective scale DA remains an active area of research (Bannister et al., 2020; Routray et al., 2022). The high-resolution regional DA system with both variational and ensemble methods show a positive impact on the forecast of severe weather (Dutta et al., 2019; Gustafsson et al., 2018). Fabry and Sun (2010) studied the sensitivity of the WRF model to perturbed observations and found that after a few hours of model integration, the model atmosphere is mostly sensitive to the vertical profiles of wind, temperature and moisture in the lower level and middle level of the troposphere.

In recent years, high temporal and spatial resolution observational data like Doppler Weather Radar (DWR) and Automatic Weather Station (AWS) are available. Therefore, the regional DA system in convective scale is more beneficial (Routray et al., 2010; Kuroda et al., 2012; Ferreira et al., 2017; Lien et al., 2017; Gustafsson et al., 2018; Zhu et al., 2019; Carrió et al., 2019; Dutta et al., 2022). Many past studies suggested that the assimilation of DWR reflectivity and radial velocity observations in the high-resolution regional models has a positive impact on analyses and also enhances the forecast skill of the model for the simulation of extreme weather events (Dutta et al., 2022; Prasad et al., 2014; Routray et al., 2010). U.S. National Weather Service used a Rapid Refresh (RR) system of hourly updated assimilation cycle with 13 km grid spacing in a Hybrid Gridpoint Statistical Interpolation (GSI) analysis system (Benjamin et al., 2016). Pan et al. (2018) have tested a hybrid system with a ~13 km resolution RR analysis forecast system. Similarly, Pal et al. (2021) used RR DA with WRF-Hydro to study flash flood event and encouraging result is documented. Given the need for more accurate prediction of high-impact weather systems on a short-time scale, the High-Resolution Rapid Refresh (HRRR) an hourly-updated, storm-resolving model with 3-km horizontal grid spacing DA and forecast model infrastructure developed in NOAA by taking the advantage of latest weather observations (Benjamin et al., 2016; Dowell et al. 2022). The HRRR system guides swiftly evolving meso-convective weather phenomena (Kain et al., 2006; Smith et al., 2008; Clark et al. 2012). Routray et al., 2022 set up a 1hry HRRR 4DVAR DA system with 1.5 km resolution convection-permitting NCUM-R by assimilating conventional, satellite observations along with DWR radial velocity and reflectivity over the Indian region. The preliminary results of the HRRR system showed the beneficial impact of the assimilation of Indian DWR observations on the rainfall forecast of a very heavy precipitation event over Mumbai from 5 to 7 August 2020. The main objective of the present study is to demonstrate the assimilation of DWR observations in the HRRR DA system for the simulation of thunderstorms over the northeast states of India. The HRRR DA system is developed with high resolution (1.5 km) regional NCUM-R-4DVAR forecast/analysis system.

The following section illustrates the pre-processing of DWR observations. A brief description of each component of the forecast model and 4DVAR DA system is provided in Sect. 3. Section 4 describes the synoptic situation associated with the two thunderstorm cases used in this study. The detail of the numerical experiment considered and data used in this study is given in Sect. 5. The results and discussion are presented in Sect. 6 and the broad conclusion drawn from the current study is given in Sect. 7.

2 Pre-processing of DWR Reflectivity and Radial Velocity

India comprises 34 DWRs operating with different frequencies viz. 19-S band; 3-C band and 1-X band which are depicted in Fig. 2a. These DWRs are operated with pulse repetition frequency between 250 and 1200 Hz. The India Meteorological Department (IMD) has opted for two scanning strategies a long-range (400 km) scan with the lowest two elevation angles and a short-range (250 km) of 10 elevation angles. For assimilation purposes, only a short range of 10 elevation angle scans are considered. NCMRWF receives DWR observations from IMD on regular basis. IMD performed a few quality controls (QC) at the station level on raw DWR observations like the removal of ground and sea clutters through discrete filtering techniques (Sachidananda & Zrnic, 2000).

Typically, Fig. 2b depicts the amount of DWR observations received at NCMRWF from various DWR stations on a particular time valid at 00 UTC 7th June 2021 as an example. Once received partially QC DWR observations from IMD, various QC checks are applied to the DWR observations through an in-house developed stand-alone QC package (Dutta et al., 2019; Jyothi et al., 2017). The quality control procedures have been applied for DWR radial velocity and reflectivity separately. The detailed procedures are explained here. The DWR observations required further QC such as partial blockage from ground clutter, anomalous beam propagation effects and biological scatters (birds, insects etc.). The radar beam geometry is also connected with errors generated by the beam broadening effect causing an increase in errors with an increase in distance between the neighboring bins and the height of the DWR beam above ground also increases as the distance of the beam increases from the site. External signals from the Sun that interfere with radar signals visible as spikes pointing from the Sun to toward the radars site along with small-scale measurement noise termed as speckles can be removed effectively using geometrical methods that analyze the variability in the data along and across the DWR beam (Smalley et al., 2008).

Other types of error like attenuation in heavy rain cause error, anomalous propagation of the radar beam from non-meteorological (terrain) and meteorological due to an unusual distribution of temperature and humidity modified the refractive index. The generation of the Quality Index (QI) plays a vital role in the quality control of DWR data as it provides information about the quality of the DWR data quantitatively (Ośródka & Szturc, 2015; Jyothi et al., 2017; Michelson et al., 2021). The

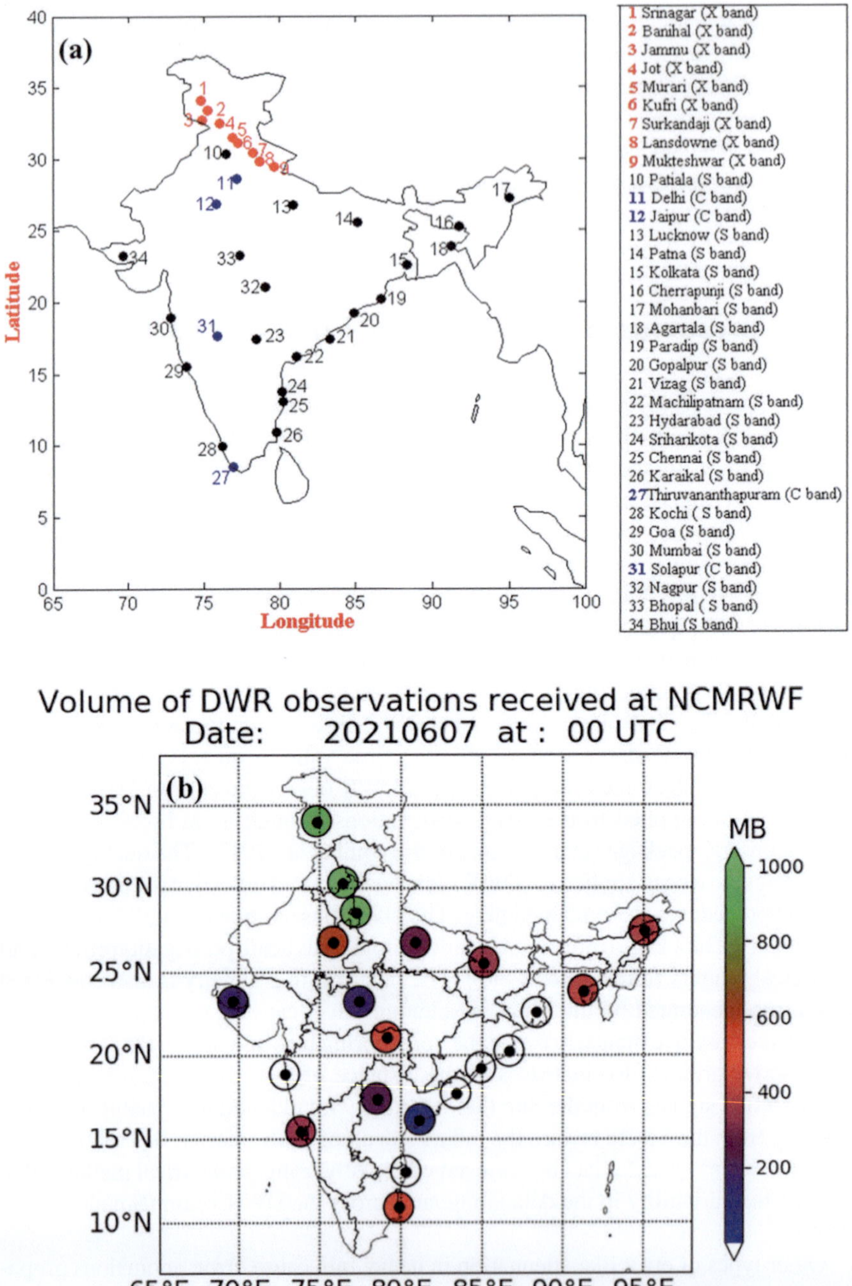

Fig. 2 **a** IMD DWRs network over India (*Source* IMD, New Delhi). **b** DWR observations received at NCMRWF valid at 00 UTC 07th June 2021

parameters considered for total average QI computation for DWR reflectivity are beam widening, attenuation, bright band and nonmeteorological echo. In addition to the above parameter, the standard deviation of radial velocity is considered to calculate the total average QI for DWR radial velocity. The average QI is calculated by taking into uncertainty caused by various factors in the measurement of DWR reflectivity and a simple equation of computation of QI is given below.

$$QI_{av} = \sum_{1}^{n} W(i)QI(i)$$

where W(i) is the coefficient of each weight assigned to QI; (i) of each factor. In the case of DWR reflectivity, the average QI above 0.5 is considered for assimilation and in the case of radial velocity the average QI above 0.8 is considered for assimilation. The accepted QI for assimilation is assigned quality flag 1 and the rejected QI is assigned a quality flag 7. Thereafter, some level of QC is also performed in the OPS module before assimilation. For the observations close to the radar, we removed the observations below 3 km as well as the user-specified maximum distance from the radar and time rage. The DWR observations are compared with the model's first guess, the radial velocity will discard if the model departure (O-B) is more than 10 m/s. The study by Stewart et al., 2008 suggests that the large departure of O-B causes the slow convergence in the assimilation cyclic. Here we applied a 3 $\times$ 3 Laplace filter to reject the pixels which are too vary from their neighbors. The observations are rejected when they are more than one standard deviation of the average of supper-obbing cell and also rejected if the supper observation cell has only one observation. The thinning of dense observations is taken care of by the OPS module. Here, we applied 8 km for the thinning of the DWR observations. Rihan et al. (2008) and Simonin et al. (2014) explained details about the QC, thinning and supper-obbing methods of the DWR observations. Here, we provide an example in Fig. 3 for processing the raw DWR observations from the Chennai radar station, model innovation (O-B) and super-observations that are used in the assimilation cyclic. The observations around 3 km (i.e. user specified) from the radar are discarded in the assimilation cyclic because the maximum clutters are available within this range. Here, we considered the observations available within100km radius from DWR location (Routray et al., 2021). The data which lie outside and time ranges is rejected. The raw DWR radial velocity is compared against the model first guess, if the O-B exceeds 10 m/s then discarded the data. It well known that the large values of O-B mainly increase the total error that causes the slow convergence at the time of assimilation (Stewart et al., 2008). The observation does not have the support of at least 3-valid neighbouring observations, it will consider as spurious data and rejected. At the same time, the pixels rejected when they are very different from the surrounding data by employing a 3 by 3 Laplace filter. The raw DWR observations more than one standard deviation of the mean super-obbing cell are discarded. If one observation is available within the super-obbing cell, it is not considered for assimilation. In this study we considered 8 km to thin the DWR observations. Dutta et al., (2019) described the details of

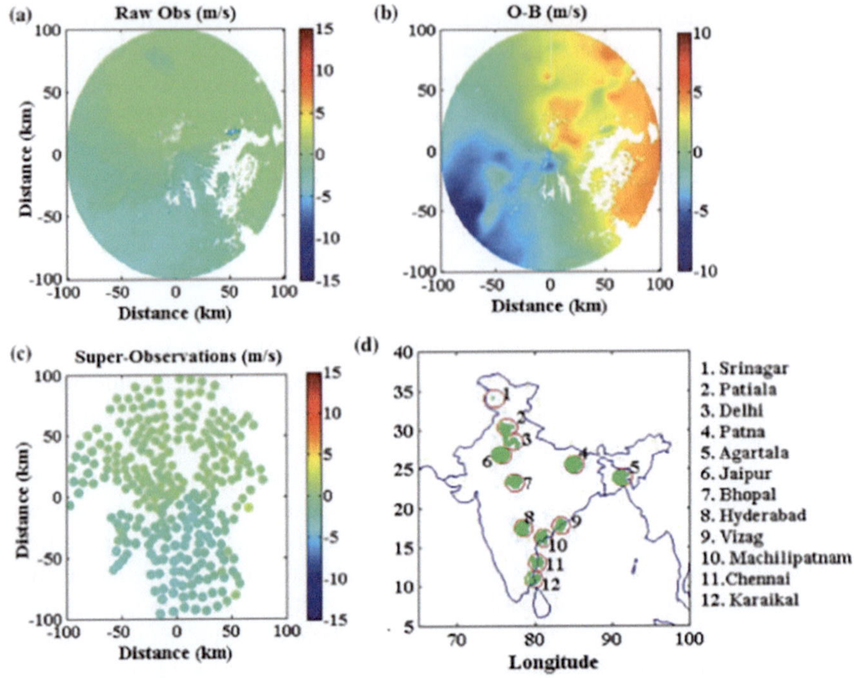

Fig. 3 **a** Raw DWR radial wind observations; **b** Observed minus background **c** super-observations and **d** DWR station wise assimilated observations as an example (Dutta et al., 2019)

the particular figures and super-obbing procedures (Rihan et al., 2008 and Simonin et al., 2014) applied for DWR observations. After the QC procedure as discussed above, the DWR radial velocity and reflectivity are packed in Grib2 format through a stand-alone software developed at NCMRWF which will be used for assimilation purposes.

3 Forecast and Assimilation Systems

The atmospheric forecast model and DA systems of NCUM-R used for HRRR are briefly described in this section.

3.1 Forecast Model

The National Centre for Medium Range Weather Forecasting (NCMRWF) has adopted the UK Met Office Unified Model (UM) for operational weather forecasting which is named as NCMRWF Unified Model (NCUM). The uniqueness of the model is for a seamless approach where a single model is used across both timescales from now-casting to centennial and spatial scales from convective to climate earth system modeling. The chief model parameters are the three components of wind (u, v, w); potential temperature; exner pressure; density as well as three moisture components like vapour, cloud water, and cloud ice. In the dynamical core of the model semi-Lagrangian advection and semi-implicit time stepping are used to solve compressible non-hydrostatic equations of motion (Davies et al., 2005). The model used a diagnostic cloud scheme Smith (1990) and the model microphysics based on Wilson and Ballard (1999) single-moment scheme with a prognostic and warm rain scheme based on Boutle et al. (2014). The details of physical parameterization schemes available within the framework of NCUM forecast model can be found in Davies et al. (2005) and Routray et al. (2016). The overall description of the model configuration and physical parameterization schemes are given in Table 1.

3.2 4DVAR Assimilation System

The major components of the NCUM-R DA system includes (i) Observation Processing System (OPS), which prepares the observations for the analysis system (ii) Four-dimensional variational data assimilation system (4DVAR), which produces the analysis increment and (iii) Regional version of NCMRWF Unified Model (NCUM-R), used for the preparation of short forecasts (known as "first guess" or "background") for the DA system. Analysed surface variables are interpolated from global or regional model analyses. They are snow, Sea Surface Temperature (SST), sea Ice and Soil Moisture (SM). These analyses are reconfigured (interpolated) over the regional model domain and used as surface boundary conditions for the NCUM-R forecast model. The 4DVAR DA system implemented at NCMRWF is an incremental four-dimensional scheme (Rawlins et al., 2007) with a single outer loop of a nonlinear model providing linearization states and background fields for the assimilation of various observations. The previous forecast (here 1 hrly) is considered as a background state to calculate the increments. The low-resolution (lower than the analyses system) linear Perturbation Forecast (PF) model is generally used in the 4DVAR DA system to produce time varying increments over the assimilation window (Ballard et al., 2016). The incremental analysis updates (IAU) based 4DVAR DA system is to find the perturbation δw by minimizing the penalty function of background, observation and weak constraint terms. The IAU based 4DVAR DA system is defined as follows:

$$
\begin{aligned}
J(\delta w) &= \frac{1}{2}(\delta w - \delta w^b)^T B^{-1}(\delta w - \delta w^b) \\
&+ \frac{1}{2}\sum_{i=1}^{n}(y - y^O)^T (E + F)^{-1}(y - y^O) + J_c
\end{aligned}
$$

Table 1 Model configuration and physical parameterization schemes used in the current study

Parameters	Regional NCUM (NCUM-R)
Governing equations	Compressible non-hydrostatic equations with terrain-following hybrid-height vertical coordinate
Horiz. resolution and points	~1.5 km (0.014*0.014); Points: 1000 × 1000
Vertical layers	L70 (terrain following hybrid-height; model top upto ~40 km)
Assimilation technique with assimilation cycle	1 hrly updated 4DVAR data assimilation algorithm
Model time step	60 s
Initial & Boundary Conditions (IC & BC)	IC: downscaling from regional 4 km model analysis to 1.5 km model domain in 1st cycle and previous cycle forecast from 1.5 km model used as first guess for next cycle BC: from regional NCUM that interpolated to 1.5 km domain & updated 01 hrly
Spatial discretization	Finite difference method
Convection process	Explicit convection
Time integration/advection	Semi-implicit, semi-Lagrangian, predictor corrector numerical scheme (Davies et al., 2005)
Radiation process	Spectral band radiation (Zhong et al., 2008)
Surface process	Joint UK Land Environment Simulator (JULES) land surface Scheme (Smith et al., 2007)
PBL process	First order non-local boundary layer scheme (Lock et al., 2000)
Microphysics	Single-moment Wilson and Ballard (1999), with a number of modifications including the use of prognostic rain, and a warm rain scheme based on Boutle et al. (2014)
Gravity wave drag	A new orographic drag scheme which contains better physical basis
Surface boundary condition	Surface Analysis + Climatology

where δw is the low-resolution analysis increment. The calculation in the inner loop is linearized about the current best 'guess' (background field) full-resolution state $\mathbf{x}^g$ and a single pass through the outer loop i.e. $\mathbf{x}^g = \mathbf{x}^b$, where $\mathbf{x}^b$ is the background model state. y is the observation vector. B, E and F represent the background error, observation error from instrumental and representative error covariance respectively. H is a non-linear observation operator that relates the model space to the observation space ($y = Hx$). Jc denotes a weak constraint term. The studies (Li et al., 2018; Rawlins et al., 2007; Routray et al., 2020) clearly described the IAU based 4DVAR

DA in the NCUM modeling system. The complete flow chart of the 4DVAR analysis system is given in Fig. 4a.

The observation Processing System (OPS) processes and prepares the quality controlled observations for the 4DVAR DA system. All the meteorological observations are quality controlled against background values and neighbouring observations. Based on the information available in the station list file, OPS assigns flags for observations whether that observation will be used or rejected. Also, it checks the probability of error (PGE-Probability Gross Error), consistency in data etc. The quality control procedures include checking for extreme values, performing buddy checking, etc. The OPS outputs are files containing quality controlled and processed

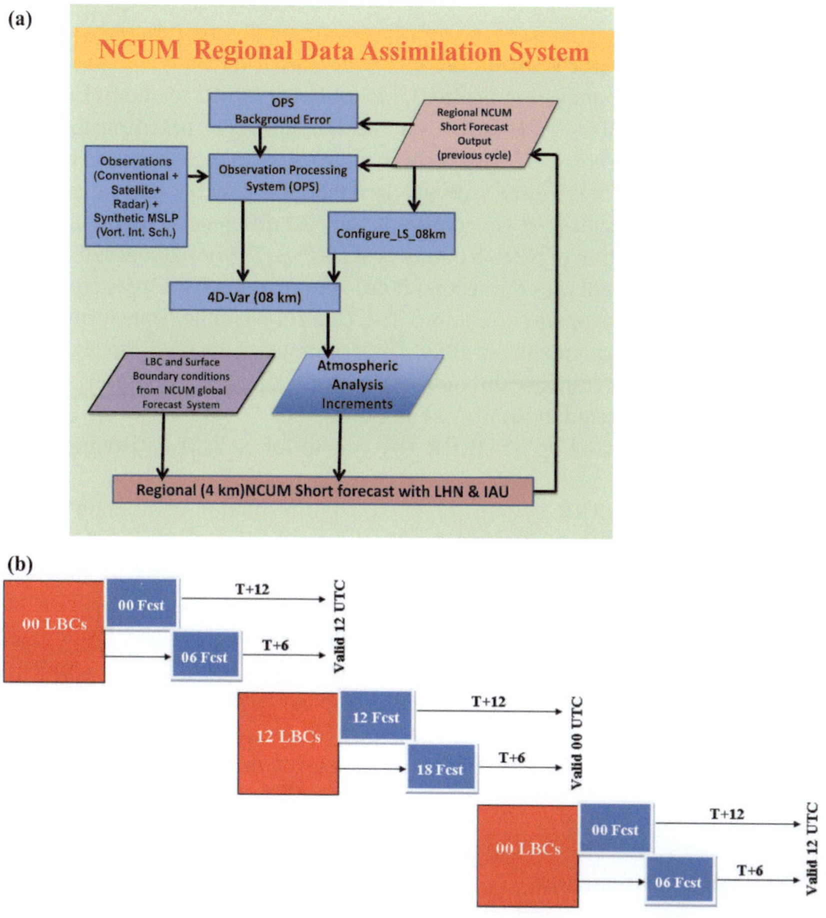

Fig. 4 **a** Schematic diagram of 4DVAR analysis system (Routray et al., 2020). **b** Procedure to generate forecast differences for NMC method (*Source McMillan,* 2015; http://code.metoffice.gov.uk)

observations (VarObs), the model forecast interpolated in time and position to match observation in the VarObs file (i.e. VarCx).

The observation thinning is carried out in both time and space, depending on the type. The satellite sounding data are thinned carefully to produce one retrieval for every desired (details are given in Table 2) square box, having first passed several standard quality-control checks using a 1D-VAR retrieval based on radiative transfer calculations through RTTOV software. The observation inputs (i.e. obstore format) for the OPS system are prepared by the "NCMRWF observation preprocessing system" using observations available at NCMRWF from various sources including observations available through IMD, GTS, ftp site (NOAA-NESDIS), ISRO (MOSDAC) etc. (Kumar et al., 2018).

DA system, which provides the error information of background fields. The quality of the assimilation largely depends on the representation of error covariance. Generally, the background error statistics are computed based on various methods such as observation-space (innovation method), model-space (NMC method), Ensemble method, etc. We calculated the BES over the Indian region based on the NMC method (Parrish & Derber, 1992), which derived from differences between different length forecasts valid for the same time and used the same later boundary conditions. Figure 4b clearly describes the generation of forecast difference used in the NMC method for the calculation of BES (Routray et al., 2019). The horizontal spreading in the NCUM-R DA system uses the second order auto-regressive (SOAR) correlation function to model horizontal correlation. The Control Variable Transform (CVT) module is used for the calculation of BES. The details about the CVT module, calculation of BES and their response through single-observation (pseudobs) test over the Indian region can be found in Routray et al. (2019).

The NCMRWF has set up an HRRR DA system for NCUM-R (Routray et al., 2022).

Hourly updating HRRR system for the 1.5 km resolution convection permitting NCUM-R has been configured for three specific domains over the Indian region (Fig. 5a). The DWR reflectivity observations from various locations are included in the HRRR data assimilation system. The regional 4DVAR HRRR DA system assimilates available conventional, satellite observations and DWR radial velocity and reflectivity with higher temporal frequency (10 min) in the assimilation cyclic. The lateral boundary conditions for the forecast model are obtained from the Indian domain 4 km resolution operational NCUM-R forecast model. The schematic diagram of the 1 hrly HRRR assimilation cycle is given in Fig. 5c.

4 Synoptic Conditions

In this study two thunderstorm cases are considered which occurred on 7th June 2021 (Case 1) and 20th April 2020 (Case 2) over the northeast region. The synoptic conditions associated with these two cases are briefly discussed below:

Table 2 Observation types, thinning and variables utilized in the 4DVAR DA system

Observation types	Observation description	Thinning (km)	Assimilated variables
SONDE	Radiosonde observations, upper-air wind profile from Pilot balloons, wind profilers	–	u, v, T, q
Surface	Surface observations from land and ocean (land SYNOP, SHIP, BUOY)	–	u, v, T, q, P_s
Aircraft/AMDAR	Upper-air temperature and winds from Aircraft	–	u, v and T
AHIClear	Advanced Himawari Imager radiances from Himawari-8	40	Direct radiances
AIRS	Atmospheric Infrared Sounder of MODIS	30	
ATMS	Advanced Technology Microwave Sounder in NPP satellite	40	
ATOVS	AMSU-A, AMSU-B/MHS, HIRS from NOAA-18 & 19, MetOp-A&B	40	
CrIs	Cross-track Infrared sensor observations in NPP satellite	40	
IASI	Infrared Atmospheric Sounding Interferometer from Metop-A&B	40	
IN3DSndr	INSAT-3D sounder radiances	40	
SEVIRIClear	Cloud clear observations from SEVIRI of METEOSAT 8 &11; 24 km thinning	40	
GPSRO	Global Positioning System Radio Occultation observations from various satellites (including MT-ROSA)	–	Bending angle
Satwind	Atmospheric Motion Vectors from various geostationary and polar orbiting satellites (including INSAT-3D)	80	u, v
Scatwind	Advanced scatterometer in MetOp-A&B, ScatSat-1,WindSat	80	u, v
RADWIND and RADREF	Radial wind observations from Indian 12-DWRs	8	Radial wind & reflectivity

Case 1

A cyclonic circulation was noticed over the Southeast Bay of Bengal between 5.8 km to 7.6 km above the mean sea level (MSL) on 7th June 2021. During the same period, a trough at 0.9 km above MSL is laid from East Uttar Pradesh up to West Bengal. Due to these synoptic systems' interaction, a favorable condition was developed for the formation of thunderstorm activity over Northeast India and adjoining West Bengal. Hence, fairly widespread thunderstorms along with rainfall occurred over this region. The three hourly INSAT-3D imageries (Fig. 6a) clearly depict the evolution of convective cloud bands over the region. From the figures, the brightness temperature (BT) gradually decreased along with time and found that the low values of BT were

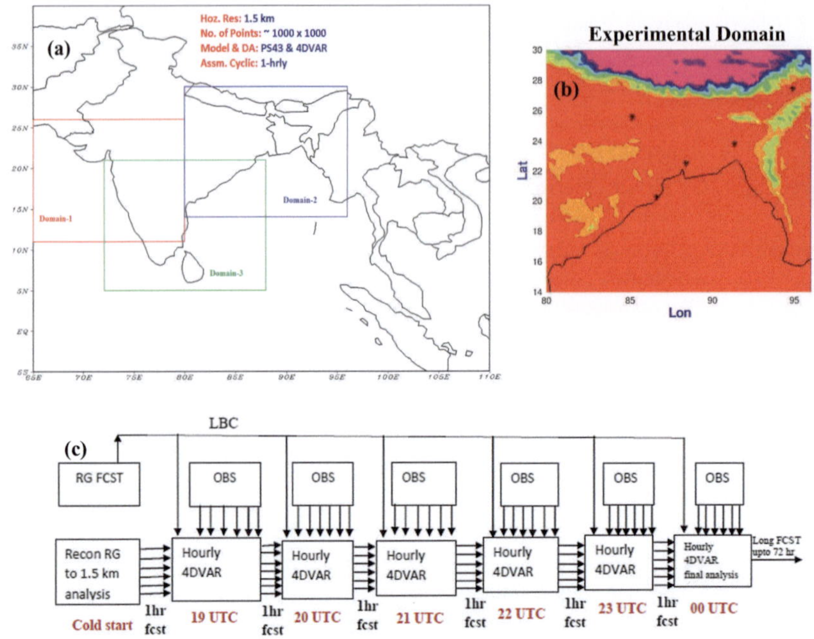

Fig. 5 **a** Different domains of the HRRR DA system configured for the Indian region; **b** Zoom of Domain-2 with terrain in shaded (*represent the location of DWRs) and **c** Schematic diagram of 1 hrly HRRR assimilation cycle

noticed over the northeast region from 9 to 15 UTC on 7th June 2021. Rainfall above 5 cm was observed in some parts of the northeast, especially over Tripura and more than 3 cm of rain occurred over some parts of West Bengal during 7th June 2021.

Case 2

A western disturbance (WD) was formed over the northwest Himalayas and adjoining regions on 19th April 2020 and it moved northeast direction. A cyclone circulation above 0.9 km above MSL appeared over the Bay of Bengal in favorable position that continuously pumped the moisture towards the northeast and adjoining part of eastern India and interacted with the WD environment. The presence of strong wind convergence in the lower level as well as the upper level features are favorable for thunderstorm activity over the northeast and adjoining parts of eastern India. The three hourly INSAT-3D imageries (Fig. 6b) show the organized convective cloud bands with low values of BT over the regions. The temporal variation of INSAT-3D imageries clearly show the convective system moved from west to east. As per the IMD report, the thunderstorms were accompanied by moderate to heavy rainfall at many stations such as Williamnagar (7 cm), Goalpara (7 cm), Kokrajhar (6 cm), Mokokchang (5 cm), Cherrapunji (5 cm) and Agartala (3 cm) on 20th April 2020.

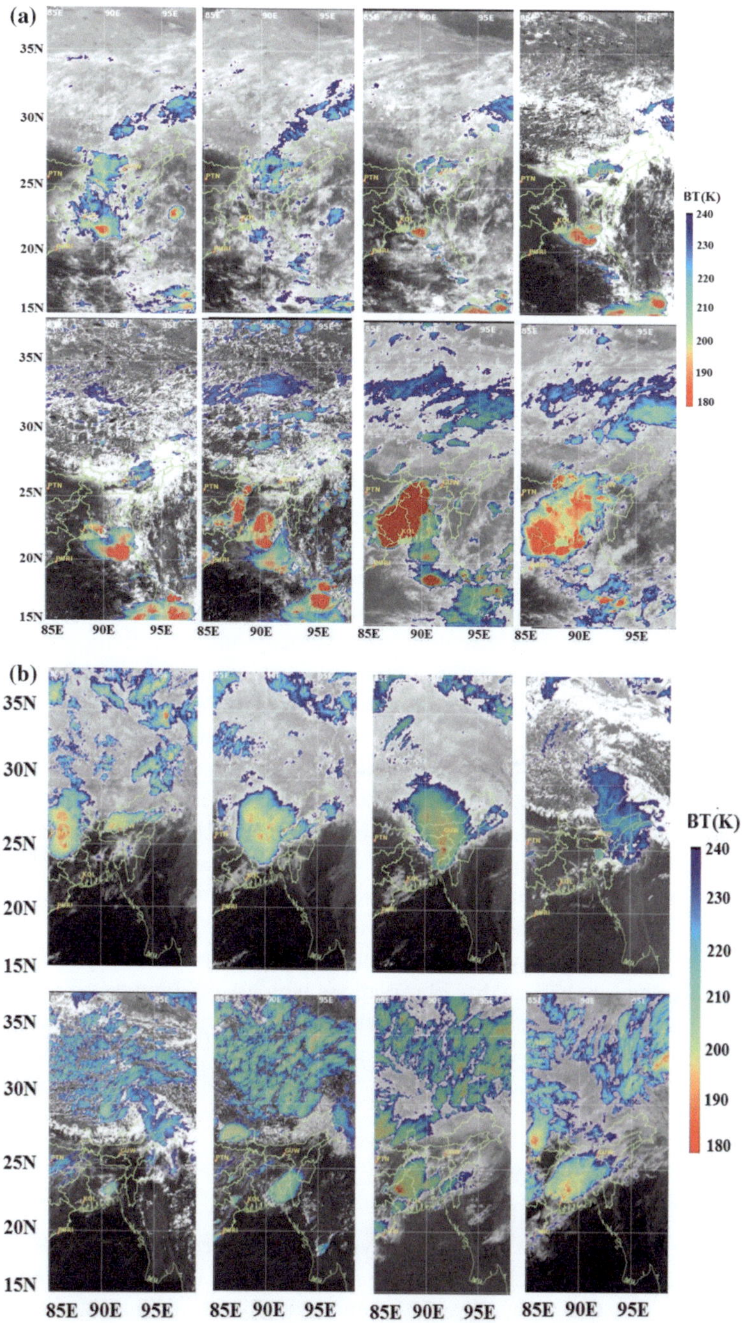

Fig. 6 Three hourly INSAT-3D imageries with brightness temperature (BT; °K; shade) valid for **a** Case 1 from 18 UTC 06th to 15 UTC 07th June 2021 and **b** Case 2 from 18UTC 19–15th UTC 20th April 2020

5 Numerical Experiments

To assess the beneficial impact of a 1 hrly updated HRRR DA system by simulating two convective events that occurred over the northeast region. The NCUM model is reconfigured to the regional domain (NCUM-R) over northeast India (14°–30° N and 80°–97° E) with a horizontal resolution of 1.5 km and 70 hybrid vertical levels extended up to height 38.5 km (Fig. 5b). Here, two sets of numerical experiments are carried out using the high-resolution NCUM-R forecast model with a single domain (Fig. 5b) of 1.5 km resolution and 70 hybrid vertical levels. The first set-up numerical experiment named as CTL is conducted without data assimilation. The model initial and boundary conditions are provided from the 4 km NCUM-R model. The second experiment i.e. HRR is conducted with an hourly assimilation cycle using surface, sonde, satellite, DWR radial velocity and reflectivity observations. The assimilation cycle is started 6 h before from the model integration in each case. The 1st hour forecast from the previous cycle is considered as the background for the next cycle. So, in this process, six assimilation cycles are performed to initialize the model before the model integration for 72 h forecast. The detail of the procedure of the hourly HRRR DA system is given in Fig. 5c. In the experimental domain (Fig. 5b), five DWR stations (Patna, Kolkata, Paradip, Agartala and Mohanbari) are available, however, only Agartala DWR station radial velocity and reflectivity are available for assimilation. Other DWR station observations are missing during the period. A typical volume scan of Agartala DWR consists of 10 elevation angles (0.197, 0.99, 1.99, 2.99, 4.49, 5.99, 8.99, 12.02, 16.01 and 21.01) with area coverage is 250 km in which up to 100 km data range is considered for assimilation. We used 10 min interval DWR observations with a cut-off time ± 30 min. Figure 7a represents the raw radar reflectivity from Agartala station received on 7th June 2021. Before the assimilation of DWR observations, various QC checks were applied that were already discussed in the previous section. Figure 7b shows the reflectivity data of Agartala DWR with the QI flag. Here, the value corresponds to one indicating the accepted value for assimilation i.e. good quality reflectivity and more than one reflectivity is considered as erroneous data i.e. bad quality observations which would not be considered for the assimilation. The number of observations utilized in the assimilation cycles is represented in Fig. 7c.

6 Results and Discussion

The effectiveness of the data assimilation system is evaluated through the comparison between the mean of O-B (observation minus first guess) i.e. background innovation as well as the mean of O–A (observation minus analysis) i.e. analysis innovations generally. The analysis and background innovations are calculated based on the observations utilized (Fig. 7c) during the assimilation cyclic. The root mean square errors (RMSE) of analysis/background innovations are given in Fig. 7d. It is seen

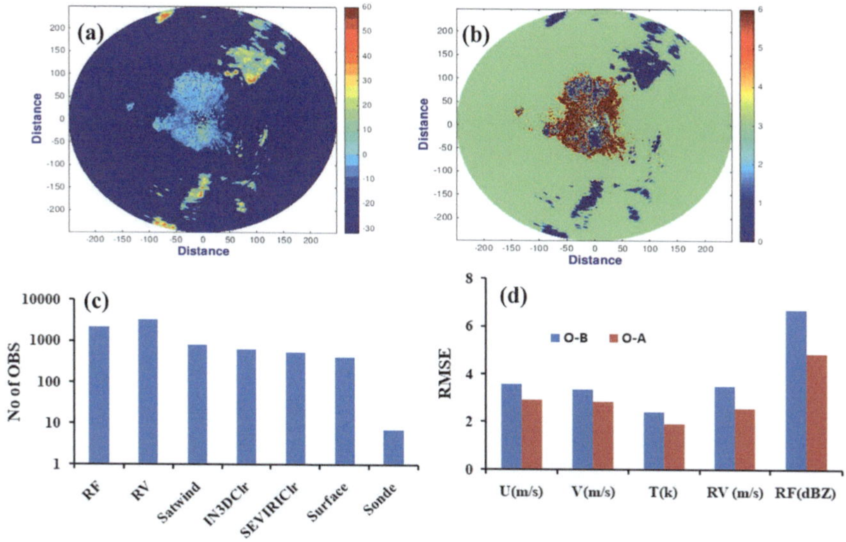

Fig. 7 **a** DWR reflectivity from Agartala; **b** DWR reflectivity QI plot after QC checks valid at 00:02:25 UTC of 07th June 2021. **c** Number of observations used in the assimilation cycle and **d** RMSE of O-B and O-A for Case 1

from the figure, RMSEs of various meteorological parameters are high in the O-B than the O-A. One thing observed from the Fig. 7d, the O-B and O-A errors obtained from the reflectivity (RF) are relatively more than the others but the O-A error is reduced reasonably well as compared to the O-B. The particular figure suggests that the efficacy of the DA system where the O-A significantly reduced than the O-B errors in all the variables. Hence, it is deduced that the observations are assimilated properly in the assimilation cycle and have a beneficial impact on the model analysis.

The model simulated 24 h accumulated rainfall valid at 03 UTC 8th June 2021(Case 1) obtained from the two numerical experiments i.e. CTL and HRR are compared with the Agartala DWR derived rainfall; Global Precipitation Measurement (GPM) and IMD-NCMRWF satellite rain gauge merge rainfall (INSRG) data (Mitra et al., 2003, 2009), which are shown in Fig. 8a–e respectively. The observed rainfall clearly shows widespread heavy rainfall of about 8–16 cm around the vicinity of Agartala DWR. From the CTL simulation (Fig. 8d), it is found that the simulated heavy rainfall patches are shifted more northeastwards and also not able to simulate the heavy rainfall over the eastern part of Agartala. However, it is noticed that the HRR (Fig. 8e) is able to reproduce the orientation, pattern and volume of rainfall in a better way than the CTL simulation. These features are reasonably well correlated with the observations. One can observe from Fig. 8e that areal coverage of light precipitation (2–4 cm) is slightly higher as compared to the others.

The temporal variation of area-averaged (24°–26° N; 91°–93.5° E) 1 hrly accumulated satellite estimated (GPM) rainfall and DWR derived rain-rate along with the model simulated precipitation for day 1 of Case 1 are demonstrated in Fig. 9a.

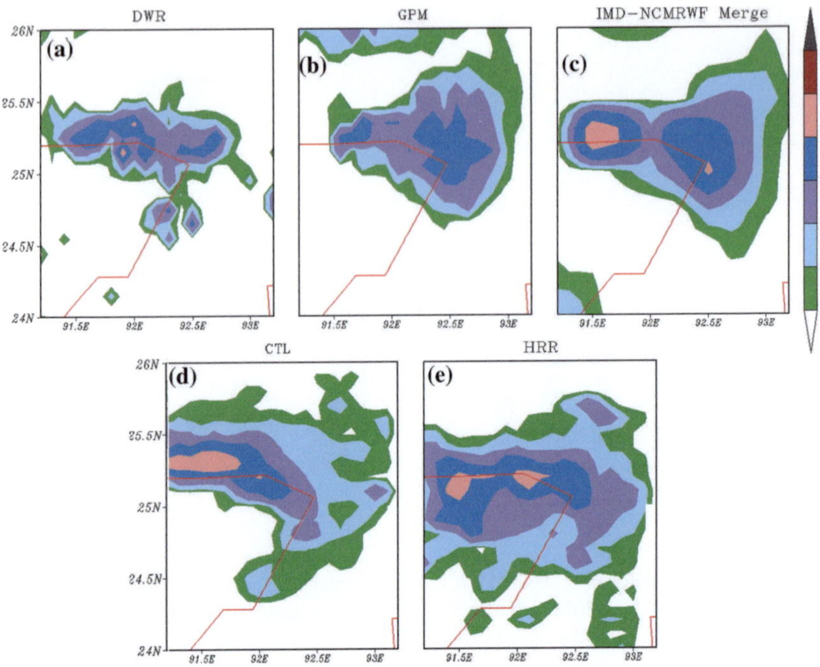

Fig. 8 24 h accumulated rainfall from **a** Agartala DWR; **b** GPM satellite; **c** merged satellite-rain gauge analysis (NCMRWF-IMD); **d** CTL and **e** HRR experiment valid at 03 UTC 8 June 2021 (day-1; Case 1)

It is noticed that the satellite estimated (GPM) rain-rate is underestimated the peak rainfall over the reason as compared to the DWR derived rain-rate. It is obvious that the satellite estimated rainfall is fully dependent upon the pass of the satellite over the region. The CTL simulated rainfall is showing a higher amount of rainfall trend throughout the forecast hours. However, the HRR simulated rainfall trend is reasonably well matched with the DWR derived rain-rate in all the forecast hours. The observations show two spells of maximum rainfall at 6 UTC and 13 UTC respectively. The 1st spell of maximum rainfall at 6 UTC is not captured by the CTL simulation. In the case of 2nd spell of maximum rainfall, the initiation of convection is showing early 3 h in CTL and 1 h in HRR simulations. However, the 1st spell of rainfall is captured by the HRR simulation with a time lag of 2 h as compared to the observation. It is also observed that the CTL experiment simulated the higher amount of rainfall during initial hours up to 9 h compared to the observations that may be caused by the model spin-up. The rainfall pattern is well adjusted in the HRR simulation after the assimilation of DWR reflectivity during the initial forecast hours (Prasad et al., 2014). From the figure (Fig. 9a), the quick decrease of rainfall pattern after 14 UTC is also well simulated by both experiments, but the HRR simulation trend is well comparable with the observed trend of rainfall. The correlation coefficient (CC) of the HRR experiment calculated with respect to the DWR derived rain-rate is relatively higher (0.68) than the CC from CTL (0.58). From the CC, it is seen that a 20% improvement is found in the HRR simulated rainfall than in the CTL experiment. It meant that the temporal evolution of thunderstorm is better captured in the HRR experiment than in the CTL experiment. Many recent studies also cited that the assimilation of high-resolution temporal/spatial observations through the HRRR DA system appreciably improved the forecast skill of model rainfall associated with the highly convective systems (Milan et al., 2019; Dillon et al., 2021; Routray et al., 2022).

The evaluation of the quality of the forecast using a statistical skill score is important to understand the uncertainty in the forecast. The model simulated precipitations are further evaluated using statistical skill score Equitable Threat Score (ETS) and bias. The ETS and bias are calculated with respect to the DWR derived rain-rate. For that, a contingency table of 2 × 2 is used by Mason, 2003. The ETS is defined as,

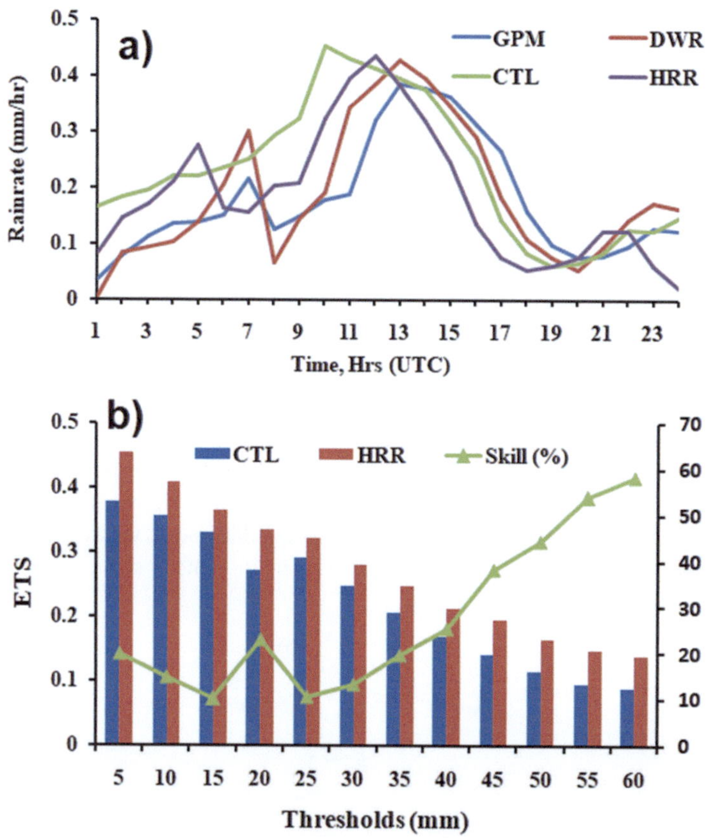

Fig. 9 **a** Area averaged (24°–26° N; 91°–93.5° E) temporal (1 hrly) variation of rainfall (mm/h) up to 24 h (00 UTC 7–8th June 2021) from GPM and DWR observations along with day 1 forecast from CTL and HRR simulations of Case 1. **b** ETS along with % of improvement of HRR over CTL (line in right hand *Y*-axis) of day 1 (24 h) for different thresholds of rainfall (mm)

$$ETS = \frac{a - a_r}{a + c + b - a_r}$$

where a: Total number of correct forecasts that occurred; b: Number of incorrect forecast occurrences; c: Number of the incorrect forecast of non-occurrences and a_r are hits random defined as $a_r = \frac{(a+b)(a+c)}{n}$, here n is the total number of forecasts i.e. (a + b + c + d) and d is the number of correct forecast of non-occurrence. The ETS score ranges from 0 to 1 with 1 representing perfect forecast. Figure 9b shows ETS with different thresholds of rainfall (mm) along with the % of improvement of HRR over the CTL experiment for Case 1. It is seen that the ETS values are higher in the lower threshold of rainfall and it is gradually decreased with the increase of the threshold values in both experiments. However, the ETS values are relatively high in the HRR experiment in all thresholds as compared to the CTL experiment. It is also seen that the gain of improvement is comparatively high in the higher threshold of rainfall (> 50 mm) in HRR with respect to the CTL. It is clearly deduced that the high intensity of rainfall is better simulated by the HRR experiment than by the CTL experiment. The % of improvement is relatively less in the case of a lower threshold of rainfall. The CTL simulation is reasonably well represented by the low-intensity rainfall than the high-intensity rainfall.

The frequency bias of precipitation is calculated with respect to the DWR derived rain-rate for both the cases at 10 mm rainfall threshold depicted in Fig. 10a and b respectively. In both cases, the rainfall is over predicted in the CTL and HRR experiments. However, the frequency bias is quite less in the HRR experiment when compared with the CTL throughout the forecast hours. It is also noticed that the frequency bias of rainfall is less in the early forecast hours and gradually increases with the increase of forecast time in both cases. It is suggested the high amount of rainfall is underestimated in both experiments, however, the frequency bias is relatively less in the HRR experiment.

To understand the impact of HRR simulations over CTL runs and quantify the systematic error in both experiments, the rainfall forecast from both simulations is further verified through object-oriented methods like the Contiguous Rain Area (CRA) method. The CRA method calculates the displacement, volume (intensity), size and shape (spatial pattern) for precipitation. The total mean square error (MSE)

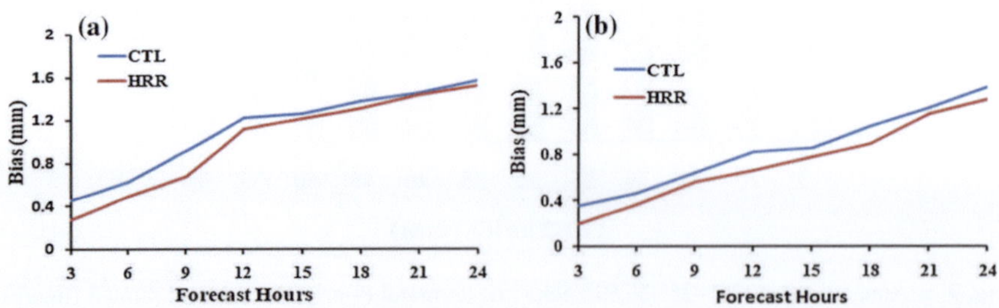

Fig. 10 Rainfall bias (3 hrly) with different forecast hours (up to 24 h) at 10 mm threshold for **a** Case 1 and **b** Case 2

is calculated from the errors due to displacement, volume and pattern components, it is presented in the equation form as given below:

$MSE_{total} = MSE_{displacement} + MSE_{volume} + MSE_{pattern}$; where, $MSE_{displacement}$ is calculated from total mean square error difference from before and after translation.

$MSE_{displacement} = MSE_{total}\text{-}MSE_{shifted}$; where $MSE_{shifted} = (Fx\text{-}X)^2$; here Fx is shifted forecast and X is CRA mean observed value after the shift. $MSE_{volume} = (F\text{-}X)^2$, here F is the CRA mean forecast values after the shift. The details about the CRA method can be found in the studies (Ebert & McBridge, 2000; Ebert & Gallus, 2009; Dutta et al., 2019; Routray et al., 2021). The % of individual error (volume; displacement and pattern); total MSE and vector displacement error are calculated for both experiments with respect to GPM rainfall with a 10 mm threshold from day-1 forecast for Case 1 and Case 2 depicted in Fig. 11a–f respectively. It is seen that the volume displacement errors (Fig. 11a and d) are relatively higher in the HRR experiment than in the CTL in both cases. It is suggested that the intensity of rainfall is slightly higher than predicted by the HRR simulation than the CTL simulation. At the same time, the displacement and pattern errors are reasonably less in the HRR as compared to the CTL simulation in all cases. In Fig. 11b and e, the overall errors (total MSE) are reduced in the HRR simulations while comparing with the CTL simulation. Similarly, the vector displacement errors (Fig. 11c and f) are considerably less in the HRR experiment than in the CTL simulation in both cases.

7 Summary and Conclusions

The current study assesses the impact of hourly updated HRRR DA system by assimilating DWR observations (radial velocity and reflectivity) in the prediction of two convective systems that occurred over the northeast region. For this purpose, two numerical experiments are conducted i.e. CTL without data assimilation and HRR with hourly update cycle DA system by assimilating the surface, sonde, satellite, radial velocity and reflectivity observations from Agartala DWR. The assimilation cycle is started 6 h before the model integration in each case. The main conclusions drawn from the present study are given below.

The RMSE of background innovation (O-B) of meteorological variables is relatively higher than the RMSE of analysis innovations (O-A). It is also observed that the innovations of reflectivity are higher than the other parameters. But the RMSE of O-A is comparatively less than of O-B in reflectivity. It is suggested the assimilation of observation is properly utilized in the assimilation cycle.

From the CTL simulation, it could not produce the location and distribution of heavy rainfall which is shifted northeastward corresponding to the observations. The orientation and intensity of rainfall are better simulated in the HRR simulation. These features are well correlated with the DWR derived rain and other observations. It is also noticed that the areal spread of rainfall is slightly higher in the HRR experiment than in the CTL and observations.

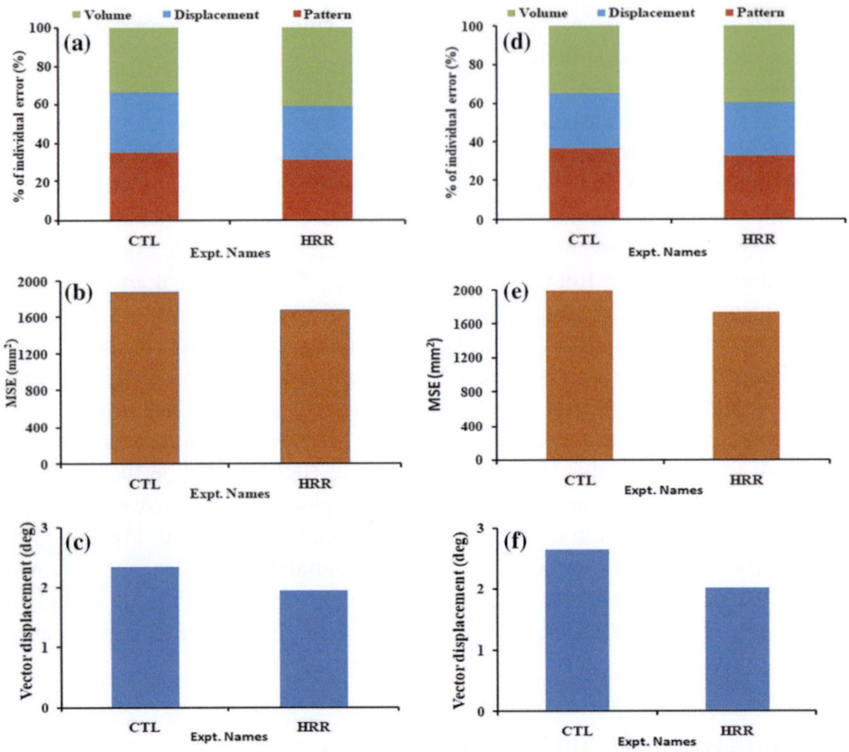

Fig. 11 CRA verification of precipitation for threshold 10 mm **a** % of contribution of volume, displacement and pattern errors; **b** total Mean Square Error (MSE) and **c** vector displacement error (deg.) for Case 1 valid at 00 U

The temporal variation of mean rainfall from the HRR is well correlated with the DWR derived rain-rate throughout forecast hours. The two intense rainfall spells were observed during the period which is well brought in the HRR simulations with 1–2 time lags as compared to the observations. However, in the CTL simulation, the initial intense rainfall is missing and 2nd maxima rainfall is simulated with a higher time lag ~3. Subsequently, the CTL experiment simulated a higher amount of rainfall during initial hours up to 9 h compared to the observations that were caused due to the model spin-up. The rainfall correlation is higher (0.68) in the HRR simulation than the CTL (0.58). It is deduced that the temporal evolution of the convective system is better captured by the HRR than the CTL experiment. Recent studies have also reported that the HRR DA system considerably improved the forecast skill of the model mainly rainfall forecast (Milan et al., 2019; Dillon et al., 2021; Routray et al., 2022).

The ETS at different thresholds shows improvement in HRR simulations in comparison with the CTL. In both simulations, the ETS values are reduced as the thresholds increase. However, it is seen that % of improvement of ETS at a higher threshold of rainfall is considerably high in the HRR with respect to the CTL. It is inferred that the high intensity of precipitation is well simulated by the HRR experiment than the CTL experiment. At the same time frequency bias is also considerably less in the HRR simulation than the CTL simulation in all cases.

The object-based verification of rainfall clearly shows that the overall errors (total MSEs) are reduced in the HRR experiment when compared with the CTL. But the volume errors are slightly more in the HRR simulation than in the CTL. It is deduced that the HRR produces high-intensity rainfall over the domain corresponding to the observation. But the location, pattern and vector displacement errors are reasonably less in the HRR as compared to the CTL simulation in all cases.

The present study is an initial assessment of the hourly cyclic HRRR DA system with a regional NCUM forecast model to further improve the prediction/simulation of high-impact weather phenomena. It suggested that the hourly updated cyclic assimilation in the high-resolution modeling approach with assimilation of DWR observations has a beneficial impact on such convective weather systems. To further support the present study, one can do additional by considering more diverse weather events with more quality control DWR observations and highly dense spatial and temporal observations.

Acknowledgements The authors thank IMD for providing the DWR observations, satellite imageries and observation data sets used in the present study. The authors acknowledged the GPM team, NASA for providing satellite-merged rainfall data to validate the model simulated results. We also thank the Head, NCMRWF, MoES for providing the necessary support for the completion of this piece of research work. Authors sincerely thank to anonymous reviewer and editor for their valuable comments/suggestion to further enhance quality of the chapter.

References

Ballard, S. P., Li, Z., Simonin, D., & Caron, J.-F. (2016). Performance of 4D-Var NWP based now casting of precipitation at the Met Office for summer 2012. *Quarterly Journal of the Royal Meteorological Society, 142*, 472–487. https://doi.org/10.1002/qj.2665

Bannister, R. N., Chipilski, H. G., & Martinez-Alvarado, O. (2020). Techniques and challenges in the assimilation of atmospheric water observations for numerical weather prediction towards convective scales. *Quarterly Journal of the Royal Meteorological Society, 146*, 1–48. https://doi.org/10.1002/qj.3652

Benjamin, S. G., Weygandt, S. S., Brown, J. M., Hu, M., Alexander, C. R., Smirnova, T. G., Olson, J. B., James, E. P., Dowell, D. C., Grell, G. A., Lin, H., Peckham, S. E., Smith, T. L., Moninger, W. R., Kenyon, J. S., & Manikin, G. S. (2016). A north American hourly assimilation and model forecast cycle: The rapid refresh. *Monthly Weather Review, 144*, 1669–1694. https://doi.org/10.1175/MWR-D-15-0242.1

Boutle, I. A., Eyre, J. E., & Lock, A. P. (2014). Seamless stratocumulus simulation across the turbulent gray zone. *Monthly Weather Review, 142*, 1655–1668. https://doi.org/10.1175/MWR-D-13-00229.1

Brooks, H. E., & Wilhelmson, R. B. (1992). Numerical simulation of a low-precipitation supercell thunderstorm. *Meteorology and Atmospheric Physics, 49*, 3–17.

Carrió, D., Homar, V., & Wheatley, D. (2019). Potential of an EnKF storm-scale data assimilation system over sparse observation regions with complex orography. *Atmospheric Research, 216*, 186–206. https://doi.org/10.1016/j.atmosres.2018.10.004

Clark, A. J., et al. (2012). An overview of the 2010 Hazardous Weather Testbed Experimental Forecast Program Spring Experiment. *Bull. Amer. Meteor. Soc., 93*, 55–74. https://doi.org/10.1175/BAMS-D-11-00040.1

Daley, R. (1991). *Atmospheric data analysis* (p. 457). Cambridge University Press.

Das, S., Abhilash, S., Das Gupta, M., George, J. P., Kalsi, S. R., Banerjee, S. K., Thampi, S. B., Pradhan, D., & Mohankumar, K. (2006). Assimilation of Doppler weather radar wind in a mesoscale model and their impact on simulation of thunderstorms and severe weather systems. Report no. NMRF/RR/01/2006, 120 pages, Published by NCMRWF, A-50, Sector-62, NOIDA, India 201307.

Das, S., Islam, M. N., & Das, M. K. (2016). Simulation of severe storms of tornadic intensity over Indo-Bangla region. *Mausam, 67*(2), 479–492.

Das, S., Mohanty, U. C., Tyagi, A., Sikka, D. R., Joseph, P. V., Rathore, L. S., Habib, A., Baidya, S. K., Sonam, K., & Sarkar, A. (2014). The SAARC STORM: A coordinated field experiment on severe thunderstorm observations and regional modeling over the South Asian Region. *Bulletin of the American Meteorological Society, 95*, 603–617.

Davies, T., Cullen, M. J. P., Malcolm, A. J., Mawson, M. H., Staniforth, A., White, A. A., et al. (2005). A new dynamical core for the Met Office's global and regional modeling of the atmosphere. *Quarterly Journal of the Royal Meteorological Society, 131*, 1759–1782.

Dillon, M. E., Maldonado, P., Corrales, P., Skabar, Y. G., Ruiz, J., Sacco, M., Cutraro, F., Mingari, L., Matsudo, C., Vidal, L., Rugna, M., Hobouchian, M. P., Salio, P., Nesbitt, S., Saulo, C., Kalnay, E., & Miyoshi, T. (2021). (2021) A rapid refresh ensemble based data assimilation and forecast system for the RELAMPAGO field campaign. *Atmospheric Research, 264*, 105858.

Dowell, D. C., et al. (2022). The High-Resolution Rapid Refresh (HRRR): An hourly updating convection-allowing forecast model. Part I: Motivation and system description. *Weather Forecasting, 37*, 1371–1395. https://doi.org/10.1175/WAF-D-21-0151.1

Dutta, D., Routray, A., Lodh, A., George, J. P., & Mitra, A. K. (2022). Impact of DWR radial wind on simulation of western disturbances using NCUM-R modeling system. *Meteorology and Atmospheric Physics, 134*, 81. https://doi.org/10.1007/s00703-022-00919-0

Dutta, D., Routray, A., Preveen Kumar, D., & George, J. P. (2019). Regional data assimilation with the NCMRWF Unified Model (NCUM): Impact of Doppler weather radar radial wind. *Pure and Applied Geophysics, 176*, 4575–4597.

Ebert, E. E., & McBride, J. L. (2000). Verification of precipitation in weather systems: Determination of systematic errors. *Journal of Hydrology, 239*, 179–202.

Ebert, E. E., & Gallus, W. A. (2009). Toward better understanding of the contiguous rain area (CRA) verification method for spatial verification. *Weather and Forecasting, 24*, 1401–1415.

Fabry, F., & Sun, J. (2010). For how long should what data be assimilated for the mesoscale forecasting of convection and why? Part 1: On the propagation of initial conditions errors and their implications for data assimilation. *Monthly Weather Review, 138*, 242–255.

Farely, R. D., Wang, S., & Orville, H. D. (1992). A comparison of 3D model results with observations for an isolated CCOPE thunderstorm. *Meteorology and Atmospheric Physics, 49*, 187–207.

Ferreira, R. C., Herdies, D. L., Vendrasco, É. P., Beneti, C. A. A., & Biscaro, T. S. (2017). Impact of radar data assimilation in a mesoscale convective system: A Case of Study. *Rev. Brasil. Meteorol., 32*, 447–458. https://doi.org/10.1590/0102-77863230011

Gustafsson, N., Janjić, T., Schraff, C., Leuenberger, D., Weissmann, M., Reich, H., Brousseau, P., & Montmerle, T. (2018). Survey of data assimilation methods for convective-scale numerical weather prediction at operational centres. *Quarterly Journal Royal Meteorological Society, 144*, 1218–1256. https://doi.org/10.1002/qj.3179

Jyothi, A. K., Dutta, D., Devarajan, P. K., George, J. P., & Rajagopal, E. N. (2017). Quality characterization of reflectivity and radial velocity observed by Indian Doppler weather radars. *Journal of Applied Remote Sensing, 11*(3), 036026. https://doi.org/10.1117/1.JRS.11.036026

Kain, J. S., Weiss, S. J., Levit, J. J., Baldwin, M. E., & Bright, D. R. (2006). Examination of convection-allowing configurations of the WRF Model for the prediction of severe convective weather: The SPC/NSSL Spring Program 2004. *Weather Forecasting, 21*, 167–181. https://doi.org/10.1175/WAF906.1

Kalnay, E. (2003). *Atmospheric modeling, data assimilation and predictability* (364 pp.). Cambridge University Press.

Kumar, S., Jayakumar, A., Bushair, M. T., Buddhi Prakash, J., George, G., Lodh, A., Indira Rani, S., Mohandas, S., George, J. P., & Rajagopal, E. N. (2018). Implementation of new high resolution NCUM analysis-forecast system in Mihir HPCS (p. 22) (NCMRWF Technical Report No. NMRF/TR/01/2018).

Kuroda, T., Fujita, T., Seko, H., & Saito, K. (2012). *Construction of mesoscale LETKF Data assimilation experiment system* (p. 3).

Li, Z., Ballard, S. P., & Simonin, D. (2018). Comparison of 3D-Var and 4D-Var data assimilation in an NWP-based system for precipitation now casting at the met office. *Quarterly Journal Royal Meteorological Society, 144*, 404–413. https://doi.org/10.1002/qj.3211

Lien, G. Y., Miyoshi, T., Nishizawa, S., Yoshida, R., Yashiro, H., Adachi, S. A., Yamaura, T., & Tomita, H. (2017). The near-real-time SCALE-LETKF system: A case of the September 2015 Kanto-Tohoku heavy rainfall. *SOLA, 13*, 1–6. https://doi.org/10.2151/sola.2017-001

Lock, A. P., Brown, A. R., Bush, M. R., Martin, G. M., & Smith, R. N. B. (2000). A new boundary layer mixing scheme. Part I: Scheme description and single-column model tests. *Monthly Weather Review, 128*, 3187–3199.

Lopez, L., Garcia-Ortega, E., & Sanchez, J. L. (2007). A short-term forecast model for hail. *Atmospheric Research, 83*, 176–184.

Mason, I. B. (2003). Binary events. In I. T. Jolliffe & D. B. Stephenson (Eds.), *Forecast verification—a practitioner's guide in atmospheric science* (240pp.). Wiley.

McMillan, M. (2015). SINGV DA User Guide, UK Met Office (33pp.). http://code.metoffice.gov.uk.

Michelson, D. B., Lewandowski, R., Szewczykowski, M., Beekhuis, H., Haase, G., Mammen, T., & Johnson, D. (2021). EUMETNET OPERA weather radar information model for implementation with the HDF5 file format, Version 2.4, EUMETNET OPERA Document.

Milan, M., Macpherson, B., Tubbs, R., Dow, G., Inverarity, G., Mittermaier, M., Halloran, G., Kelly, G., Li, D., Maycock, A., Payne, T., Piccolo, C., Stewart, L., & Wlasak, M. (2019). Hourly 4D-Var in the Met Office UKV operational forecast model. *Quarterly Journal Royal Meteorological Society, 2020*(146), 1281–1301. https://doi.org/10.1002/qj.3737

Mitra, A. K., Bohra, A. K., Rajeevan, M. N., & Krishnamurti, T. N. (2009). Daily Indian precipitation analysis formed from a merge of rain-gauge data with the TRMM TMPA satellite-derived rainfall estimates. *Journal of the Meteorological Society of Japan, 87A*, 265–279. https://doi.org/10.2151/jmsj.87A.265

Mitra, A. K., Das Gupta, M., Paliwal, R. K., & Singh, S. V. (2003). Observed daily large-scale rainfall patterns during BOBMEX-99. *Proceedings of Indian Academy of Sciences (earth & Planetary Science), 112*(2), 223–232.

Mohanty, U.C., et al. (2006). Weather summary during pilot experiment of severe thun-derstorms observations and regional modeling (STORM) programme, 177. India Department of Science and Technology Report.

Mohanty, U. C., et al. (2007). Weather summary during pilot experiment of severe thun-derstorms observations and regional modeling (STORM) programme, 179. India Department of Science and Technology Report.

Ośródka, K., & Szturc, J. (2015). Quality-based generation of weather radar Cartesian products. *Atmospheric Measurement Techniques, 8*, 2173–2181. https://doi.org/10.5194/amt-8-2173-2015

Pal, S., Dominguez, F., Dillon, M. E., Alvarez, J., Garcia, C. M., Nesbitt, S. W., & Gochis, D. (2021). Hydrometeorological observations and modeling of an extreme rainfall event using WRF and WRF-Hydro during the RELAMPAGO field campaign in Argentina. *Journal of Hydrometeorology, 22*, 331–351. https://doi.org/10.1175/JHM-D-20-0133.1

Pan, Y., Xue, M., Zhu, K., & Wang, M. (2018). A prototype regional GSI-based EnKF variational hybrid data assimilation system for the rapid refresh forecasting system: Dual-resolution implementation and testing results. *Advances in Atmospheric Sciences, 35*, 518–530. https://doi.org/10.1007/s00376-017-7108-0

Parrish, D. F., & Derber, J. C. (1992). The National Meteorological Center's spectral statistical-interpolation analysis system. *Monthly Weather Review, 120*, 1747–1763.

Prasad, S. K., Mohanty, U. C., Routray, A., Osuri Krishna, K., Ramakrishna, S. S. V. S., & Niyogi, D. (2014). Impact of Doppler weather radar data on thunderstorm simulation during STORM pilot phase—2009. *Natural Hazards, 74*, 1403–1427. https://doi.org/10.1007/s11069-014-1250-0

Rawlins, F. R., Ballard, S. P., Bovis, K. R., Clayton, A. M., Li, D., Inverarity, G. W., Lorenc, A. C., & Payne, T. J. (2007). The Met Office global four-dimensional data assimilation system. *Quarterly Journal Royal Meteorological Society, 133*, 347–362. https://doi.org/10.1002/qj.32

Rihan, F. A., Collier, C. G., Ballard, S. P., & Swarbrick, S. J. (2008). Assimilation of Doppler radial winds into a 3D-Var system: Errors and impact of radial velocities on the variational analysis and model forecasts. *Quarterly Journal Royal Meteorological Society, 134*, 1701–1716.

Routray, A., Dutta, D., Lodh, A., & George, J. P. (2021). Impact of the assimilation of DWR-derived precipitation rates through latent heat nudging on simulation of rainfall events over Indian region using NCUM-R. *Journal of Hydrology, 596*, 126072. https://doi.org/10.1016/j.jhydrol.2021.126072

Routray, A., Dutta, D., Amarjyothi, K., George, J. P., & Mitra, A. K. (2022, March). Radar reflectivity assimilation in a rapid refresh system for high resolution NCUM-R (pp. 14–16), UM Users Newsletter Issue 23. https://info.metoffice.gov.uk/1XRP-7M6U2-A7D10B6DB39886D7WZLD758770223FD64687D6/cr.aspx.

Routray, A., Lodh, A., Dutta, D., & George, J. P. (2020). Study of an Extremely Severe Cyclonic Storm "Fani" over Bay of Bengal using regional NCUM modeling system: A case study. *Journal of Hydrology, 590*, 125357. https://doi.org/10.1016/j.jhydrol.2020.125357

Routray, A., Mohanty, U. C., Rizvi, S. R. H., Niyogi, D., Krishna, K. O., & Pradhan, D. (2010). Impact of Doppler weather radar data on numerical forecast of Indian monsoon depressions. *Quarterly Journal Royal Meteorological Society*. https://doi.org/10.1002/qj.678

Routray, A., Mohanty, U. C., Osuri, K. K., & Kiran Prasad, S. (2013). Improvement of monsoon depressions forecast with assimilation of Indian DWR data using WRF-3DVAR analysis system. *Pure and Applied Geophysics, 170*, 2329–2350.

Routray, A., Mohanty, U. C., Osuri, K. K., Kar, S. C., & Niyogi, D. (2016). Impact of satellite radiance data on simulations of Bay of Bengal tropical cyclones using the WRF-3DVAR modeling system. *IEEE Transactions on Geoscience and Remote Sensing, 54*(4), 2285–2303.

Routray, A., Rajasree, V. P. M., Dutta, D., & George, J. P. (2019). *New background error statistics for regional NCUM 4DVAR data assimilation system* (p. 23) (NCMRWF Technical Report No. NMRF/TR/06/2019).

Sachidananda, M., & Zrnic, D. S. (2000). Clutter filtering and spectral moment estimation for Doppler weather radars using staggered pulse repetition time (PRT). *Journal of Atmospheric and Oceanic Technology, 17*, 323–331.

Simonin, D., Ballard, S. P., & Li, Z. (2014). Doppler radar radial wind assimilation using an hourly cycling 3D-Var with a 1.5 km resolution version of the Met Office Unified Model for nowcastings. *Quarterly Journal of the Royal Meteorological Society, 140*, 2298–2314.

Smalley, D. J., Mann, E., Ivaldi, C. F., & Bennett, B. J. (2008). Improving weather RADAR data quality for aviation weather needs. In *13th conference on Aviation, Range, and Aerospace Meteorology (ARAM)*, New Orleans, LA, American Meteorological Society.

Smith, D. M., Cusack, S., Colman, A. W., Folland, C. K., Harris, G. R., & Murphy, J. M. (2007). Improved surface temperature prediction for the coming decade from a global climate model. *Science, 317*, 769–799.

Smith, R. N. (1990). A scheme for predicting layer cloud and their water content in a general circulation model. *Quarterly Journal Royal Meteorological Society, 116*, 435–460. https://doi.org/10.1002/qj.49711649210

Smith, T. L., Benjamin, S. G., Brown, J. M., Weygandt, S., Smirnova, T., & Schwartz, B. (2008). Convection forecasts from the hourly updated, 3-km High-Resolution Rapid Refresh (HRRR) model. In *24th conference on severe local storms*, Savannah, GA, American Meteorological Society, 11.1. https://ams.confex.com/ams/pdfpapers/142055.pdf.

Stewart, L. M., Dance, S. L., & Nichols, N. K. (2008). Correlated observation errors in data assimilation. *International Journal for Numerical Methods in Fluids, 58*, 1521–1527.

Tyagi, B., Naresh Krishna, V., & Satyanarayana, A. N. V. (2011). Skill of thermodynamic indices for forecasting pre-monsoon thunderstorms over Kolkata during STORM pilot phase 2006–2008. *Natural Hazards, 56*, 681–698. https://doi.org/10.1007/s11069-010-9582-x

Tyagi, B., Naresh, S. A. N. V., & Krishna, V. (2013). Thermodynamical structure of atmosphere during pre-monsoon thunderstorm season over Kharagpur as revealed by STORM data. *Pure and Applied Geophysics, 170*, 675–687.

Weiss, S. J., Bright, D. R., Kain, J. S., Levit, J. J., Pyle, M. E., Janjic, Z. I., Ferrier, B. S., Du, J. (2006). Complementary use of short-range ensemble and 455 KM WRF–NMM model guidance for severe weather forecasting at the Storm Prediction Centre. In *23rd conference severe local storms*. ST Louis MO. American Meteorological Society, CD-ROM. 8.5.

Wilson, D. R., & Ballard, S. P. (1999). A microphysically based precipitation scheme for the UK Meteorological Office Unified Model. *Quarterly Journal Royal Meteorological Society, 125*, 1607–1636. https://doi.org/10.1002/qj.49712555707

Zhong, W., Osprey, S. M., Gray, L. J., & Haigh, J. D. (2008). Influence of the prescribed solar spectrum on calculations of atmospheric temperature. *Geophysical Research Letters, 35*, L22813. https://doi.org/10.1029/2008GL035993

Zhu, K., Xue, M., Pan, Y., Hu, M., Benjamin, S. G., Weygandt, S. S., & Lin, H. (2019). The impact of satellite radiance data assimilation within a frequently updated regional forecast system using a GSI-based ensemble Kalman filter. *Advances in Atmospheric Sciences, 36*, 1308–1326. https://doi.org/10.1007/s00376-019-9011-3

Ensemble Forecasting of Severe Storms Over Indian Region

Paromita Chakraborty, S. Kiran Prasad, and Someshwar Das

Abstract The geographical location, topography and synoptic scenario cause Indian subcontinent to be a land prone to frequent thunderstorm activities during the pre-monsoon, monsoon and post-monsoon seasons. Ensemble forecasting has proved to be a successful way of dealing with the inherent uncertainties of weather and climate forecasts. NEPS-G and GEFS are the two high-resolution (12 km) models operational in India for generating global medium-range ensemble forecasts. The convective scale regional model at NCMRWF is designed specifically for prediction of extreme events over Indian region. The severe weather events like thunderstorms are likely to be increasing in frequency over next few decades projected under the climate change scenario. Early warning in medium, short and nowcasting range is therefore essential for storm preparedness and risk management. Different ensemble forecast products are tailored based on the high resolution global and regional EPS for prediction of severe weather events like that of thunderstorms and associated rainfall and lightning activities that can be useful for early warning.

Keywords Ensemble forecasting · Thunderstorms · Numerical modeling · Convective scale · Probability · Severe weather

Abbreviations

AROME Application of Research to Operations at Mesoscale

P. Chakraborty (✉)
Centre for Climate Research Singapore, Meteorological Services Singapore, 36 Kim Chuan road, Singapore 537054, Singapore
e-mail: paromitaiitd2012@gmail.com; paromita_chakraborty@nea.gov.sg

P. Chakraborty · S. K. Prasad
National Centre for Medium Range Weather Forecasting (NCMRWF), Ministry of Earth Sciences, Government of India, A-50, Sector-62, Noida-201309, India

S. Das
South Asian Meteorological Association (SAMA), Pune, India

S. Das and W-K. Tao (eds.), *Severe Storms*,
https://doi.org/10.1007/978-981-97-7075-5_17

ARPEGE	Action de Recherche Petite Echelle Grande Echelle
3DVAR	Three-Dimensional Variational
ARW	Advanced Research WRF
AS	Arabian Sea
BIMSTEC	Bay of Bengal Initiative for Multi-Sectoral Technical and Economic Cooperation
BoB	Bay of Bengal
BPs	Boundary Condition Perturbations
BS	Brier Score
BSS	Brier Skill Score
BVs	Breeding Vectors
CAPE	Convective Available Potential Energy
CDO	Climate Data Operators
CIN	Convective Inhibition Energy
CNTL	Control
COSMO	Consortium for Small scale Modeling
CRPS	Continuous Ranked Probability Score
CS	Cyclonic Storm
CSI	Critical Success Index
DEM	Digital Elevation Model
ECMWF	European Centre for Medium-Range Weather Forecasts
ENDGame	Even Newer Dynamics for General atmospheric modeling of the environment
EPS	Ensemble Prediction System
EPSgram	Ensemble Meteogram
EFI	Extreme Forecast Index
ETKF	Ensemble Transform Kalman Filter
GEFS	Global Ensemble Forecast System
GPM	Global Precipitation Measurement
GrADS	Grid Analysis and Display System
GSI	Grid-point Statistical Interpolation
HEI	Hail Environmental Index
HRRR	High-Resolution Rapid Refresh
HSS	Heidke Skill Score
IITM	Indian Institute of Tropical Meteorology
IMERG	Integrated Multi-satellite Retrievals for GPM
INSAT	Indian National Satellite
IPs	Initial Condition Perturbations
J&K	Jammu and Kashmir
LAMEPS	Limited Area Ensemble Prediction System
Lat	Latitude
LAYRH	Layer Relative Humidity
LFC	Lightning Flash Count
Lon	Longitude
LPI	Lightning Potential Index

MOGREPS	Met Office Global and Regional Ensemble Prediction System
NCEP	National Centers for Environmental Prediction
NCMRWF	National Centre for Medium Range Weather forecasting
NCR	National Capital Region
NEPS	NCMRWF Ensemble Prediction System
NEPS-G	NCMRWF Ensemble Prediction System-Global
NEPS-R	NCMRWF Ensemble Prediction System-Regional
NWP	Numerical Weather Prediction
NWS	National Weather Service
PBL	Planetary Boundary Layer
PEACE	ARPEGE based ensemble system
PPs	Physics Perturbations
PSS	Peirce's Skill Score
RMSE	Root Mean Squared Error
ROC	Relative Operating Characteristic
RP	Random Parameters
RPS	Ranked Probability Score
RPSS	Ranked Probability Skill Score
SCP	Super Cell Composite Parameter
SCS	Severe Cyclonic Storm
SKEB	Stochastic Kinetic Energy Backscatter
SPR	Spread-error Ratio
SREF	Short Range Ensemble Forecast
SREPS	Short Range Ensemble Prediction System
SRTM	Shuttle Radar Topography Mission
SV	Singular Vector
Tmax	Maximum temperature
Tmin	Minimum temperature
UK	United Kingdom
UKMO	United Kingdom Met Office
UP	Uttar Pradesh
THORPEX	The Observing-system Research and Predictability Experiment
TIGGE	THORPEX Interactive Grand Global Ensemble
WRF	Weather Research and Forecasting
VSCS	Very Severe Cyclonic Storm

1 Introduction

1.1 Thunderstorms Over Indian Region

Over Indian region intense thunderstorms occur mainly during the pre-monsoon season, which is from March to May. During this time, the hot and humid air from the Bay of Bengal and the Arabian Sea converges over the region, leading to enhanced convective activity. These thunderstorms are usually associated with strong winds, heavy rainfall, lightning and sometimes hail. The topography of the region, including the Himalayas and the Deccan Plateau, helps to enhance convection by forcing air to rise and cool, leading to the formation of thunderstorms. Multi-cell thunderstorms (also known as squall lines) which develop due to organized intense convection may have a life span of several hours and spread over a few hundreds of kilometers. Severe thunderstorms generally travel from northwest to southeast direction over the eastern and northeastern parts of India and Bangladesh during the pre-monsoon season and hence are also called the Nor'westers (Tyagi et al., 2012; Nayak and Mandal, 2014; Chattopadhyay, et al., 2017; Sahu et al., 2020; Das, 2022; Sahu and Tyagi, 2022). At the beginning of the monsoon season (June–September) strong convection and availability of abundant moisture in the lower troposphere provide favorable conditions for thunderstorms activity over Indian region. The India Meteorological Department closely monitors the weather patterns during different seasons to issue severe weather warnings (Mohapatra et al., 2022) and protect the public from potential harm.

Post-monsoon thunderstorms in India are rarer and occur mainly in the southern and southeastern regions. These thunderstorms are less intense compared to summer thunderstorms, and are often accompanied by gusty winds and lightning (Sarkar et al., 2020; Tyagi, 2007).

In both pre- and post-monsoon seasons, the frequency and intensity of thunderstorms in India can be influenced by several factors, including the El Niño-Southern Oscillation, the Indian Ocean Dipole, and the Inter-Tropical Convergence Zone (Sinha et al., 2016, 2020). The frequency of thunderstorms is highest during the pre-monsoon season over the Indian subcontinent. Three ingredients necessary for a thunderstorm to occur are moisture, instability, and lifting. Another ingredient is wind shear which further enhances the intensity of the system to form severe thunderstorms (Das et al., 2014; Tyagi et al., 2011, 2012, 2013; Sen Roy et al., 2019).

Instability is a critical factor in the formation and development of thunderstorms. It refers to the condition where the atmosphere becomes more unstable as it rises, allowing air parcels to continue to rise on their own. This instability is often caused by warm, moist air at the surface being heated and rising, which results in the development of cumulus clouds. The greater the instability, the stronger the updrafts within the thunderstorm, which can lead to the growth of the storm and the development of severe weather such as strong winds, hail, and tornadoes.

Lifting and buoyancy play key roles in the formation and development of thunderstorms. Lifting is the process by which air rises due to forces such as converging winds, fronts, and terrain. When this lifting takes place in a moist and unstable environment, the moist air rises and cools, leading to the formation of cumulus clouds. Buoyancy, on the other hand, refers to the ability of a body to float in a fluid. If an air parcel is warmer and less dense than the surrounding air, it will rise due to buoyancy, creating an updraft. The combined effect of lifting and buoyancy can lead to the rapid development of thunderstorms and severe weather.

The amount of moisture present in the planetary boundary layer (Wyngaard, 1985) is referred to as the PBL moisture. Low convective inhibition (CIN) refers to the lack of atmospheric stability, allowing for easier upward movement of air and the development of thunderstorms. CAPE, or Convective Available Potential Energy, is a measure of the amount of energy available for thunderstorm development. When there is high PBL moisture, low CIN, and high CAPE, the atmospheric conditions are favorable for the development of thunderstorms.

The prediction of thunderstorms has been very challenging but became feasible with the advent of advanced numerical weather prediction systems. The prediction of thunderstorms with advanced deterministic and ensemble modeling systems have been reviewed in this study. The global deterministic and ensemble modeling systems are briefly described in the subsequent sections followed by different perturbation methods, importance of resolution and ensemble size, and then ensemble consensus and probabilistic forecast verification. Section 2 deals with the discussion on the different ensemble models being run and ensemble products generated operationally. Convective scale modeling used for severe weather prediction in India is discussed in Sect. 4. Convective scale EPS, Regional Ensemble Prediction system and probabilistic forecast methods are discussed in Sect. 5. Forecast performance of these methods are examined based on some case studies, by subjective intercomparison of corresponding weather maps, EPSGRAM, Skew-T ensemble profile. Various skill scores are computed, and the results are discussed for verification of convective weather prediction.

1.2 Deterministic Prediction System and Transition to Ensemble Methods

Historically, weather forecasters have focused on finding the most likely outcome for future weather, which was considered the most crucial aspect of forecasting. In deterministic numerical weather prediction (NWP), a single forecast is generated by integrating a weather model from the initial state of the atmosphere. The early approaches of NWP were strictly deterministic, but in the 1950s, Jule Charney and his research team made a groundbreaking 24-h forecast of the large-scale atmospheric flow (Charney et al., 1950). By the late 1950s, it was believed that forecasting beyond several days was possible with a single NWP model. However, it became clear that deterministic prediction had its limits due to uncertainties in initial conditions, model formulations, and the chaotic nature of the atmosphere. These uncertainties, including inaccuracies in initial state estimation and model representation of atmospheric processes, make it challenging to quantify the individual contributions of these errors. To address this issue, an ensemble prediction system (EPS) can be used to form a consistent probability distribution function that describes uncertainty in forecasts, which is crucial for a wide range of users. The combination of probability and determinism in forecasting is known as Ensemble Prediction and was first proposed by Eady (1951) as a solution to the strictly deterministic approach in NWP.

1.3 Global Ensemble Systems

The idea of ensemble forecasting was conceived in numerical weather prediction by Lorenz (1965) and Epstein (1969). Leith (1974) implemented the use of random perturbations in a perfect model environment, also known as Monte Carlo forecasting. Lewis (2005) provides a comprehensive review of the early research on ensemble forecasting and its predictability. The first operational ensemble prediction system was put into use by the European Centre for Medium-range Weather Forecast in the 1990s (Palmer et al., 1993; Molteni et al., 1996a, b), followed by the US National Centers for Environmental Prediction (Toth and Kalnay, 1993) and the Recherche en Prévision Numérique in Canada (Houtekamer et al., 1996). However, due to limitation of resources, the ensemble prediction systems often have to be run at a lower resolution because they require multiple forecast integrations.

1.4 Non-linearity, Ensemble Mean and Spread

An EPS typically consists of multiple perturbed forecasts, along with a control forecast. The control forecast is generated by simulating the model from the best estimated state of the atmosphere, as determined by the data assimilation system. The other ensemble forecasts are created by introducing perturbations or errors into the initial conditions and model physics. In the early stages of forecasting, errors tend to grow almost linearly over time. The set of ensemble forecasts begins as a symmetrical cloud of possible solutions centered around the control analysis, but it eventually transforms into an irregularly shaped cloud due to nonlinear processes that displace the control forecast from the center. Small-scale and more unstable phenomena reach the critical level of nonlinearity faster, such as in the case of convective precipitation where linearity may be lost in just a few hours, while large-scale features may retain linearity for several days. While the initial condition errors remain small, the forecast and actual conditions remain close in phase space (Buizza, 2002), the deterministic forecast can still exhibit good skill. Beyond the range of linear error growth, the deterministic forecast becomes less reliable. The ensemble mean, which is the average of all the ensemble forecasts, can still be considered as a single forecast representing the best estimate of future atmospheric conditions. It filters out the unpredictable components of the forecast by retaining only the components that are consistent among the ensemble members, through the nonlinear evolution of the perturbations.

One crucial aspect of ensemble forecasting is its ability to reflect the reliability of the forecast. The spread which represents the degree of disagreement between the ensemble members, is used to gauge confidence in the forecast. A small spread implies that there is a good level of agreement among the members, which provides a high degree of confidence in the forecast. The forecaster can then determine the importance of a single deterministic forecast, or the ensemble control based on the spread. A small spread in the ensemble gives confidence in the single forecast, while a larger spread implies that the uncertainty in the forecast must be considered. If the single model forecast deviates significantly from the ensemble mean compared to the spread, then the control forecast should not be given much weight. As the spread grows, it becomes less appropriate to solely rely on a single forecast, be it the high-resolution or control forecast of the ensemble, as the most likely scenario. In such cases, all the solutions in the ensemble must be considered when evaluating the likelihood of different forecast scenarios. According to Buizza (1997), a successful ensemble prediction system is expected to meet the following three criteria:

(i) The spread of the ensemble should be comparable to the error of the control forecast
(ii) A small spread should indicate a high level of skill in the control forecast
(iii) The verifying analysis should fall within the range represented by the ensemble forecasts.

Ensemble forecasting not only provides a way to quantify the uncertainty of the forecast but also make probabilistic quantification feasible. Traditional single forecast methods rely on error statistics over time to estimate forecast errors and assume that errors are constant for a given lead time. However, this is not always the case in real time. EPS systems were designed to calculate forecast uncertainty dynamically, considering the initial conditions, model limitations, and atmospheric evolution in different weather conditions. The use of EPS systems can provide the best estimate of forecast uncertainty.

1.5 Perturbation Methods

Ensemble forecasting methods used by various operational centers across the world mainly differ in the way initial condition perturbations are generated. A straightforward method to generate perturbations is to add random noise to the original analysis, which is known as the Monte Carlo method. However, this method fails to account for the "growing errors of the day" as the analysis errors tend to grow much quicker than the random initial perturbations (Hollingworth, 1980; Hoffman and Kalnay, 1983; Kalnay & Toth, 1996). To overcome this limitation, a new class of methods, called "Breeding" and "Singular Vector," was developed and implemented globally. These methods account for the growing errors in the initial perturbations. The

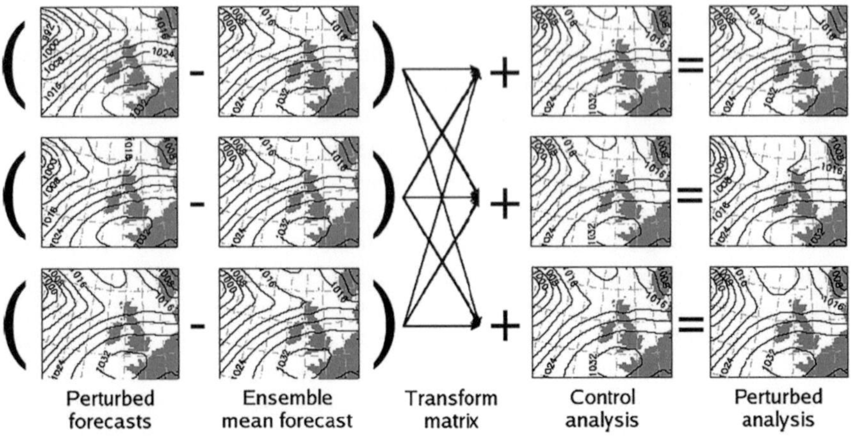

Fig. 1 Graphical representation of the ETKF method (*source* Bowler et al., 2008)

breeding vectors (BVs) (Toth and Kalnay, 1993) method is used to generate perturbations to create the initial condition at NCEP. ECMWF uses the singular vector (SV) approach (Buizza and Palmer, 1995; Molteni et al., 1996a, b). The Met Office in the UK generates initial perturbations using the Ensemble Transform Kalman Filter (ETKF) (Bishop et al., 2001) in its Global and Regional Ensemble Prediction System (MOGREPS). The ETKF is similar to the error breeding method, but with some differences, as described by Toth and Kalnay (1993). In ETKF (Fig. 1), the analysis perturbation of each member is a linear combination of the forecast perturbations, resulting in improved performance over the error breeding method (Wang et al., 2004).

1.6 Importance of Ensemble Size

The usefulness of ensemble forecasting is largely dependent on correctly representing the uncertainties in the initial conditions through perturbations. A limited number of ensemble members can result in sampling error and an underestimation of forecast error covariance by the ETKF, making ensemble size a crucial factor in determining the efficiency of the ensemble system. Despite the desire to increase ensemble size to improve the average skill of a probabilistic forecast, doing so also increases computational cost. Studies have shown benefits in increasing ensemble size, particularly over mid-latitudes (Buizza et al., 1997; Buizza and Palmer, 1998; Barkmeyer et al., 1999; Mullen and Buizza, 2002; Ma et al., 2012), but the extent of improvement varies based on the verification metric used. In the tropics, where forecast error is linked to more complex convective systems (Zhang and Krishnamurti, 1999; Kalnay, 2003), increasing ensemble size may also improve accuracy in estimating forecast error structure (Kay et al., 2013). The trade-off between prediction skill and computational cost involved for operational EPS is of interest for short to medium-range ensemble forecasts over the tropics. A recent study by Chakraborty et al. (2020) examined the impact of ensemble size on the performance of medium-range ensemble forecasts for tropical cyclones over the Bay of Bengal and found that in most cases, a larger ensemble size provided better uncertainty information. The study also found that for longer forecast lead times, an ensemble of larger size had less mean track forecast error compared to an ensemble of smaller size and the control member.

1.7 Importance of Resolution and Ensemble Size

The high spatial resolution of a NWP model allows for more accurate representation of the structure and intensity of weather systems, thereby reducing forecast errors for mesoscale phenomena (Roebber et al., 2004). An optimal ensemble size can be achieved by combining current time forecasts with lagged ensemble forecasts, where a larger ensemble size improves the estimation of forecast uncertainty. The ensemble size and model resolution are two key factors in determining the EPS's forecast performance. High resolution EPS enables a probabilistic forecast of extreme weather events with sufficient resolution to resolve small-scale features and makes it feasible to generate the location specific forecasts like ensemble Meteogram (EPSgram) and storm following Meteograms. Forecasts of extreme events, like temperature, wind speed, and precipitation, have benefited more from improvements in the forecast system than average weather forecasts (Magnusson et al., 2014; Durai and Roy Bhowmik, 2014). Probabilistic forecast products for extreme events, such as rainfall, temperature, and strong winds, have been evaluated and their performance varied by event type (Sharpe et al., 2018). Blenkinsop et al. (2018) studied extreme short-duration rainfall events, linking observations and models to understand the mechanisms associated with an extreme rainfall that can lead to flash flooding. Chakraborty et al. (2019) reported the performance of high-resolution EPS for extreme weather events like heatwaves, cold waves and cyclonic storms. Chakraborty et al. (2021) evaluated the probabilistic quantitative forecast of very heavy precipitation events by a high-resolution EPS in India to add confidence to the ensemble mean prediction.

1.8 Ensemble Spread RMSE Relationship

One of the key benefits of ensemble forecasting is that it allows for the expression of uncertainty in the predictions. An ensemble is considered to be highly accurate when all sources of forecast uncertainty are effectively sampled. EPS are designed to represent the full uncertainty range as realistic as possible. To evaluate the accuracy of EPS, the spread-skill relationship is frequently studied (Brown et al., 2013 and

MacLeod et al., 2018). Palmer et al. (2005) demonstrated that in an ideal ensemble, the mean spread should be equivalent to the root mean square error of the ensemble mean over a given period. The ensemble spread is expressed as:

$$spread = \sqrt{\frac{1}{M-1}\sum_{m=1}^{M}(f_m - \overline{f})^2} \tag{1}$$

where M is the ensemble size, f_m is the forecast value of the *m*th member and $\overline{f}$, the ensemble mean, is given by

$$\overline{f} = \frac{1}{M}\sum_{m=1}^{M} f_m \tag{2}$$

The spread-error ratio (SPR) is calculated as the square root of the ratio of the mean ensemble variance to the mean squared error of the ensemble mean in comparison to the verifying observations (Grimit and Mass 2007; Hopson 2014). SPR > 1 indicates over-dispersion (under-confidence), whereas SPR < 1 indicates under-dispersion (overconfidence) in the forecasts. The under-dispersion between ensemble members is often more pronounced when errors stem from unresolved atmospheric processes at sub-grid scales, which can lead to amplified uncertainties in derived forecast weather parameters (Kalnay, 2003). Factors contributing to prediction uncertainties include errors in initial conditions, model physics, and underlying assumptions. Under-dispersion is frequently seen in many EPS (such as those by Toth et al., 2006; Bowler and Mylne, 2009; Tennant et al., 2011; Chakraborty et al., 2021) due to their inability to fully capture all sources of uncertainty and their growth during the model integration time.

1.9 *Ensemble Consensus and Probabilistic Forecast Verification*

Evaluating the quality of an ensemble system often involves comparing the performance of the ensemble mean to that of a single control forecast (Chakraborty et al., 2023). Several deterministic scores have been widely used for decades to assess the performance of non-probabilistic forecasts, including the anomaly correlation coefficient (Murphy and Epstein, 1989; Jolliffe and Stephenson, 2003), RMSE (Fortin et al., 2014), Peirce's skill score or true skill statistic (Hanssen & Kuipers, 1965; Peirce, 1884), Threat score, or critical success index (Mesinger, 2008; Schaefer, 1990), and Heidke skill scores (Heidke, 1926).

Probabilistic quantitative forecasts provide a way to quantify forecast uncertainty, which is crucial for building confidence in using these forecasts when it comes to critical threshold values. With enough lead time, probabilistic forecasts allow people to take necessary actions, such as protecting property and evacuating the area, if necessary, which helps reduce the risk of property damage and fatalities resulting from heavy rains and subsequent flooding (Durai et al., 2015). To monitor the performance of a modeling system and provide useful feedback for improvement, forecast verification is essential. The accuracy of the forecasting system can be indicated by the correspondence between the forecast distribution and the distribution of observations or analyses. Numerous methods have been adapted from Murphy and Winkler (1992) and Murphy (1997) to check this correspondence, such as the double factorization of the joint distribution of forecasts and observations (Murphy and Winkler, 1987). To assess the quality of ensemble predictions, a suite of verification metrics can be used, including probabilistic scores (such as the Brier score, Brier skill score, ranked probability score, and ranked probability skill score), reliability diagrams, relative operating characteristic curves, and rank histograms (Hamill et al., 2000). The Brier score (Brier, 1950) measures the mean-squared error of the probability forecasts over the verification sample (binary) for a specific threshold of a specific variable. To verify the reliability and resolution (Murphy, 1973; Stanski et al., 1989) of an ensemble prediction, it's important to use a suite of verification measures and to understand the results considering the different focus of each score as different verification methods provide insights into different aspects of the ensemble forecast (Wei et al., 2008).

Ensemble forecasting aims to provide reliable probability forecasts, where the forecast probabilities align with the observed relative frequencies. In the medium-range forecasting domain, the focus is on the requirement that the ensemble spread reflects the uncertainty in the mean. In the seasonal forecasting domain, the emphasis is on the reliability of the ensemble forecasts, which should have the same climatological variance as the truth (Johnson and Bowler, 2009). Although a system that always forecasts the climatic probability of the atmospheric state is reliable, it is not useful for weather forecasting. That's why, resolution, another aspect of probabilistic forecasting, must be taken into consideration. Resolution is the extent to which the ensemble forecasts distinguish between different observed events. The Brier score, reliability diagram, and ROC curves are used to assess the probabilities of dichotomous events, while the CRPS is used to assess the probabilities of continuous variables treated as categorical forecasts (Chakraborty et al., 2020, 2021, 2023). The probabilistic skill (BSS and RPSS) of rainfall that exceeds specific thresholds of varying magnitudes is usually compared to a reference or long-term climatology.

2 The Global Ensemble Prediction Systems Operational in India

2.1 Global Ensemble Prediction System at NCMRWF (NEPS-G)

A global EPS with a resolution of 12 km, based on the Met Office Global and Regional Prediction Systems (Bowler et al., 2008; Walters et al., 2017), was put into operation at the National Centre for Medium Range Weather Forecasting (NCMRWF) (NEPS-G) in 2018 (Mamgain et al., 2018). The previous version of the EPS (Sarkar et al., 2016) had a resolution of 33 km (NEPS-G33) and ensemble size of 45 members (44 perturbed and 1 control). Both versions of the EPS use the ENDGame (Thuburn, 2016) dynamical core. The model physics is perturbed by Stochastic Kinetic Energy Backscatter (SKEB) (Tennant et al., 2011) and Random Parameters (RP) (Bright and Mullen, 2002; Sanchez et al., 2016) schemes. The RP method takes into account the uncertainties in the empirical parameters of physical parameterization, as well as simulating non-deterministic processes that are not captured by different parameterizations. However, the transfer of energy from small to large scales in the atmosphere through physical processes is hard to replicate in numerical weather prediction models, leading to a reduction in kinetic energy. To mitigate this, the SKEB method is applied in the Unified Model to replenish the lost kinetic energy (Tennant et al., 2011).

The analysis from the deterministic Unified model at NCMRWF (Kumar et al., 2018) is used as the initial condition for the control forecast. The perturbations in the initial conditions are produced by the ETKF (Bishop et al., 2001) four times daily (00, 06, 12, and 18 UTC) based on the previous 6-h short forecast for the variables u, v, θ, q, and exner pressure at all levels. These perturbations are added to the global deterministic analysis to create perturbed initial conditions. The control and 11 perturbed ensemble members run from the initial conditions at 00 UTC of the current day, while another set of 11 perturbed members, which run from the 12 UTC initial conditions of the previous day, are combined during post-processing to produce 23 members ($11 + 11 + 1$ control) ensemble forecast products (Chakraborty et al., 2019) with a 10-day lead time.

2.2 Global Ensemble Forecast System (GEFS) at IITM

The Global Ensemble Forecast System (GEFS) is a numerical weather prediction model consisting of 20 ensemble members and 1 control run that provides probabilistic forecasts for various weather variables (Ma et al., 2012; Zhou et al., 2022). It is developed by the National Centers for Environmental Prediction (NCEP) of the National Weather Service (NWS) in the United States. It is implemented with a resolution of approximately 12.5 km (T1534) by Indian Institute of Tropical Meteorology (IITM) and is run by the IMD (Mukhopadhyay et al., 2018, 2022). The model is initialized daily at 00 UTC with initial conditions from NCMRWF and produces 10-day forecasts at 3-h intervals. It is also used for severe weather forecasting over Indian region (Mukhopadhyay et al., 2022).

3 Short to Medium Range Ensemble Forecast Products Useful for Extreme Weather Events

This section illustrates the probabilistic predictions for extreme weather phenomena of different scales which would help forecasters interpret and make optimal use of the probabilistic medium-range ensemble forecast products for severe weather over the Indian region.

3.1 How Probabilistic Forecasts Are Generated at NCMRWF

NEPS-G forecast products are generated in grib2 format at 6 hourly intervals. All fields are bi-linearly interpolated to a 0.125° lat/lon grid. Some of the forecast products are also being generated at one hourly interval. Each post-processed file contains 252 hours ensemble forecasts of 11 members from 00 UTC of current day merged with 11 members from 12 UTC of previous day. The data are extracted using wgrib2 utilities and processed for computation of ensemble statistics using Climate Data Operators (CDO) (Schulzweida et al., 2006). The files in wgrib2 format are directly read in Grid Analysis and Display System (GrADS), for computation and plotting the ensemble statistics of some of the specific parameters. Various software packages, like python, R and GrADS are being used for graphical visualization of the products. The operational forecast products from NEPS-G are generated as per user needs at national as well as international (Conroy et al., 2023; Dunion et al., 2023) level.

To make the best use of the products, the timely generation of these high-resolution forecasts is necessary and the same became feasible after the commissioning of state-of-the-art Mihir high performance computing system at NCMRWF. The location specific probabilistic forecasts are issued all over India at city and district level in real time. Forecast products are also being developed for block level as well to meet future demands. The development of products largely depends on user demands and necessities of particular meteorological services. The NEPS-G forecast products can be used as guidance to forecasters and to provide direct input to help end-users such as regional and national authorities, to make the optimal decision for preventive action for severe weather events. Chakraborty et al. (2019) outlines various ensemble forecast products generated using NEPS-G for medium-range time scale.

3.2 *Postage Stamp Maps*

A set of small maps showing the scenarios in each individual ensemble member forecast helps the forecaster to assess the possible risks of extreme events. Postage stamp maps are intended to explain the spread in the forecast solutions in synoptic terms, in particular the extreme weather. Daily forecast (day-1 to day-10) of postage stamp maps are routinely prepared from the 6-hourly outputs containing control and ensemble forecast information of 22 members all valid for the same time.

3.2.1 Stamps with Winds at 850 hPa

Two very severe cyclonic storms (VSCS) simultaneously developed over Arabian Sea and Bay of Bengal in 2018 and was one of the rarest of rare events. The genesis of VSCS Titli over Bay of Bengal took place 45 hours after the genesis of VSCS Luban over Arabian Sea. Titli was the most destructive cyclonic storm to strike Indian coast during 2018 (IMD cyclone division report 2018). Figure 2 shows the distribution of wind streamlines associated with one cyclonic circulation persisting at BoB (Titli) and another one at AS (Luban). 21 out of 22 members have captured the wind intensity of Luban (in CS stage) between 30 and 40 m/s in day-5 forecast valid at 00 UTC 10th October 2018. This agrees well with IMD observed best track intensity of this system which was 33 m/s at 00 UTC 10th Oct 2018. The wind speed around vortex of Titli (in CS stage) was between 20–30 m/s for most of the members. 7th ensemble member predicted the wind speed beyond 40 m/s for TC Titli. The observed (IMD) wind intensity was 28 m/s for Titli in SCS stage.

3.2.2 Rainfall Stamps

Figure 3 depicts the postage stamp maps of day-7 NEPS forecast of 24 hours accumulated rainfall on 11th October 2018 associated with the cyclonic circulations seen in Fig. 2. This chart shows the very heavy to extremely heavy rainfall amount of 32–64 cm/day for TC Luban captured by 14 out of 22 members. The heavy to very heavy precipitation between 8 and 16 cm/day was predicted at Bay of Bengal for TC Titli by 16 out of 22 members, 7 days in advance.

3.3 *Departures*

Departure gives a measure of the extent to which ensembles over- or under-forecast the risks of anomalous or extreme weather events with respect to climatology. It also shows how well the forecasts can span the full climatological range. Daily rainfall departures are calculated with respect to model climatology since model has

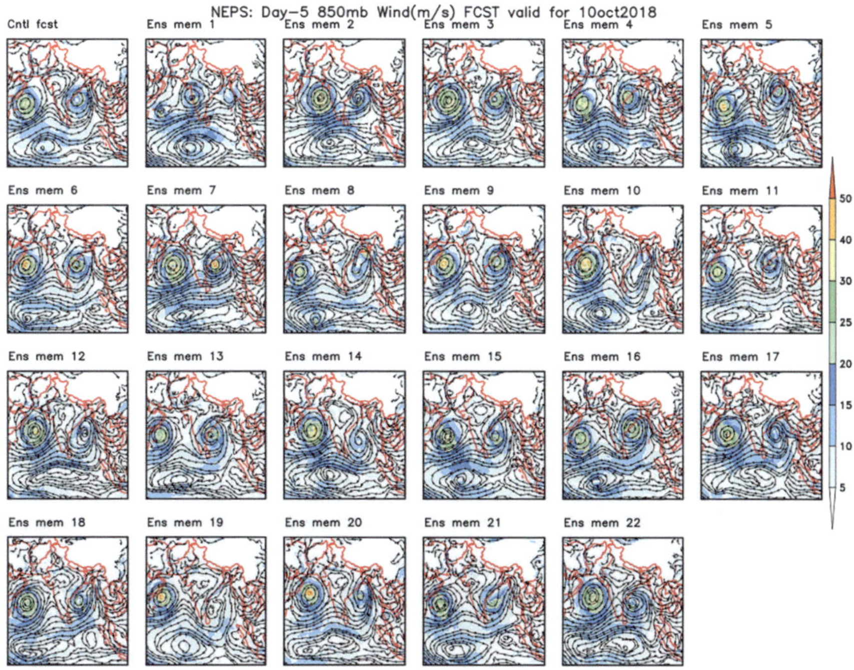

Fig. 2 Postage stamp maps of day-5 forecasts of winds (m/s) at 850 hPa for CNTL and each of 22 ensemble members of NEPS-G valid for 00 UTC 10th October 2018

generally underestimated the extreme values in past as compared to observed values. Therefore, the signature of the extremes can be captured better if departure in forecast is considered with respect to model mean values. Also, observations at resolution comparable to the model resolution are unavailable and downscaling or interpolation may lead to shift in location of extreme values. The climatological reference is the average values from MOGREPS data between 2007 and 2015, available in the TIGGE portal. The areas where the rainfall probabilities are more than 90th and 99th percentiles are also illustrated in these figures.

3.3.1 Rainfall Departure

It can be seen from Fig. 4a–c that rainfall more than 4 cm over normal is predicted at South Jammu and Kashmir, Himachal Pradesh, Punjab, west Uttar Pradesh and west interior Odisha on 23rd July 2018. The model has predicted little excess (1–2 cm) over normal for Jharkhand, Chhattisgarh and coastal Maharashtra. The predicted rainfall is 3–4 cm below normal in Bangladesh and North Bengal and that by 1–2 cm in Bihar, east UP and east Rajasthan. The contours in Fig. 4d and e represent the 90th and 99th percentile rainfall of MOGREPS climatology and the shaded areas depicts the excess in predicted rainfall over climatology. Extreme rainfall of 2.5–3 cm in excess

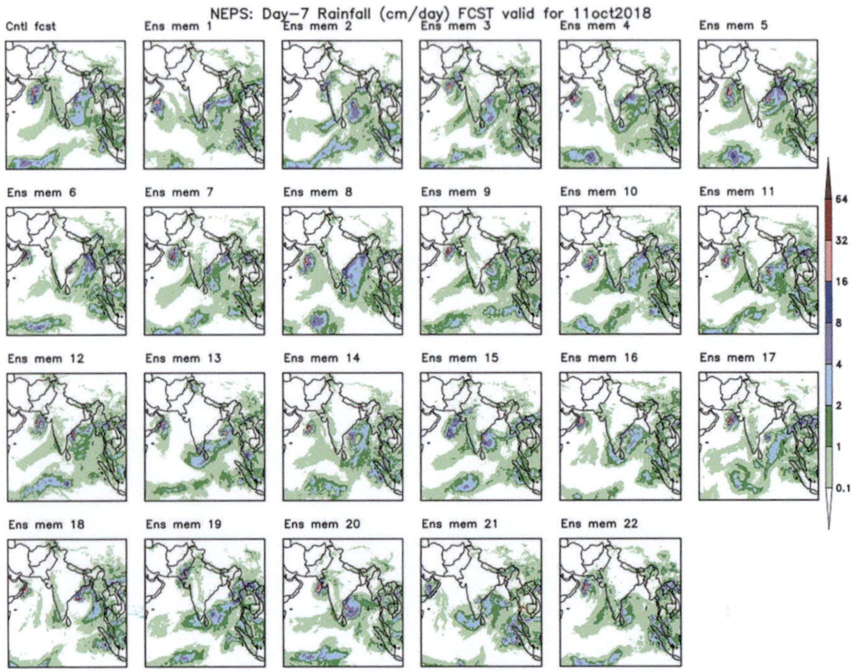

Fig. 3 Postage stamp maps of 24-h-accumulated day-7 rainfall (cm/day) forecasts for CNTL and each of 22 ensemble members of NEPS-G valid for 00 UTC 11th October 2018

over climatological 99th percentile over parts of South J&K, Himachal Pradesh and Punjab is predicted on this day. It is over 2.5 cm in excess over 90th percentile over Chhattisgarh and west interior Odisha.

3.3.2 Temperature Departure

Daily maximum surface temperature (Tmax) and minimum surface temperature (Tmin) departures are calculated for NEPS-G with respect to 2007–2013 mean values from MOGREPS data. Areas with probabilities of extreme cold (Tmin < 10th percentile) and extreme warm (Tmax > 90th percentile) temperatures are marked in shades of blue and red respectively. The departure for Tmin is shown below in Fig. 5.

Ensemble mean Tmin (Fig. 5a) predicted for 30th December 2018 was below −4 °C for northern J&K region, Sikkim and North–east India. This is same as or more than normal (Fig. 5b) for these regions and are depicted (red) in the Tmin departure (Fig. 5c). The Tmin between 0 and 3 °C over north central India is associated with a cold wave passing across these zones as depicted by the departure of −4 °C in Fig. 5c.

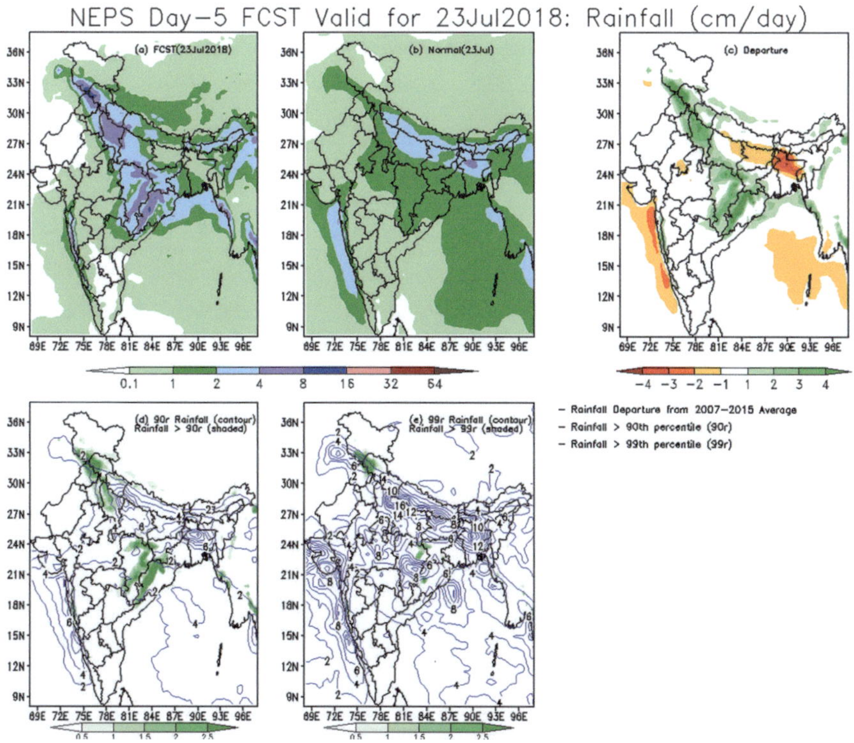

Fig. 4 **a** Day-5 forecast of ensemble mean precipitation **b** Normal (2007–2015 average) precipitation **c** Departure **d** Precipitation > 90th percentile **e** Precipitation > 99th percentile based on NEPS-G valid for 23rd July 2018

3.4 Ensemble Meteogram (EPSGRAM)

Ensemble meteogram or EPSgram is one of the most commonly used prognostic presentations of location specific forecasts from the ensemble prediction systems. The boxplots show the full range of distribution of the surface meteorological variables (temperature at 1.5 m, relative humidity at 1.5 m, wind speed at 10 m, rainfall and mean sea level pressure). The boxes show a range of 25–75% percentile values and whiskers show the range between minimum and maximum values. The red line joins median values. Model output variables are extracted from the forecast at six hourly intervals to summarize forecasts at one location.

Cyclonic storm Daye which formed over east-central BoB crossed south Odisha and north Andhra Pradesh coast close to Gopalpur (Odisha) as a cyclonic storm with a wind speed of 16–20 m/s gusting to 22 m/s during 1900–2000 UTC of 20th September. The instability associated with the system is depicted by the surface parameters in the EPSgram for Ganjam district based on initial conditions of 00 UTC 19th September 2018 shown in Fig. 6. The median dips down by 2 °C for

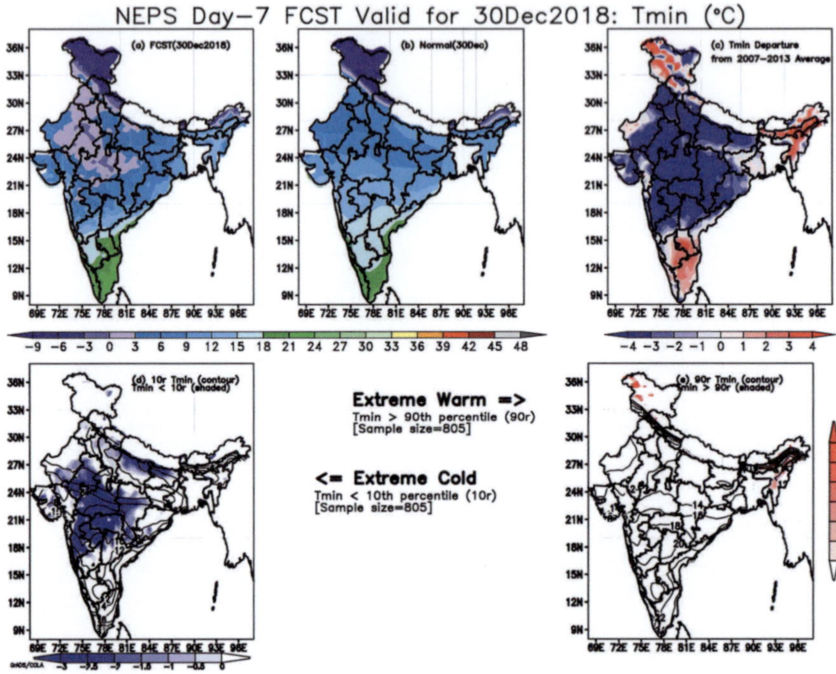

Fig. 5 **a** Day-7 forecast of ensemble mean minimum temperature (Tmin) **b** Normal (2007–2013 average) Tmin **c** Tmin Departure **d** Tmin < 10th percentile **e** Tmin > 90th percentile based on NEPS-G

temperature at 2 m within 24 hours along with a precipitation peak of 40 mm/6 h on 20th September. The surface wind intensifies to 16 m/s with a drop of 8 hPa in MSLP on 18 UTC of the same day. This is followed by gradual stability in the surface parameters over the following days.

3.5 *Ensemble Mean and Spread in Rainfall from NEPS-G33 Versus NEPS-G*

Very heavy rainfall of 120–200 mm was observed (Fig. 7a) in Integrated Multi-satellite Retrievals for Global Precipitation Measurement (IMERG GPM) over the Western Ghats on 10th June 2018. Day-3 forecast of the ensemble mean rainfall from NEPS-G33 (Fig. 7b) of 33 km horizontal resolution and NEPS-G (Fig. 7c) of 12 km resolution based on 00 UTC 7th June 2018 initial conditions, have been compared with respect to the rainfall observations. Although both the systems have captured the locations of heavy rainfall of 60–120 mm, the NEPS-G ensemble mean could predict the very heavy rainfall amounts between 120–200 mm day^{-1}. The spatial distribution of light (10–20 mm day^{-1}), moderate (20–60 mm day^{-1}), and heavy

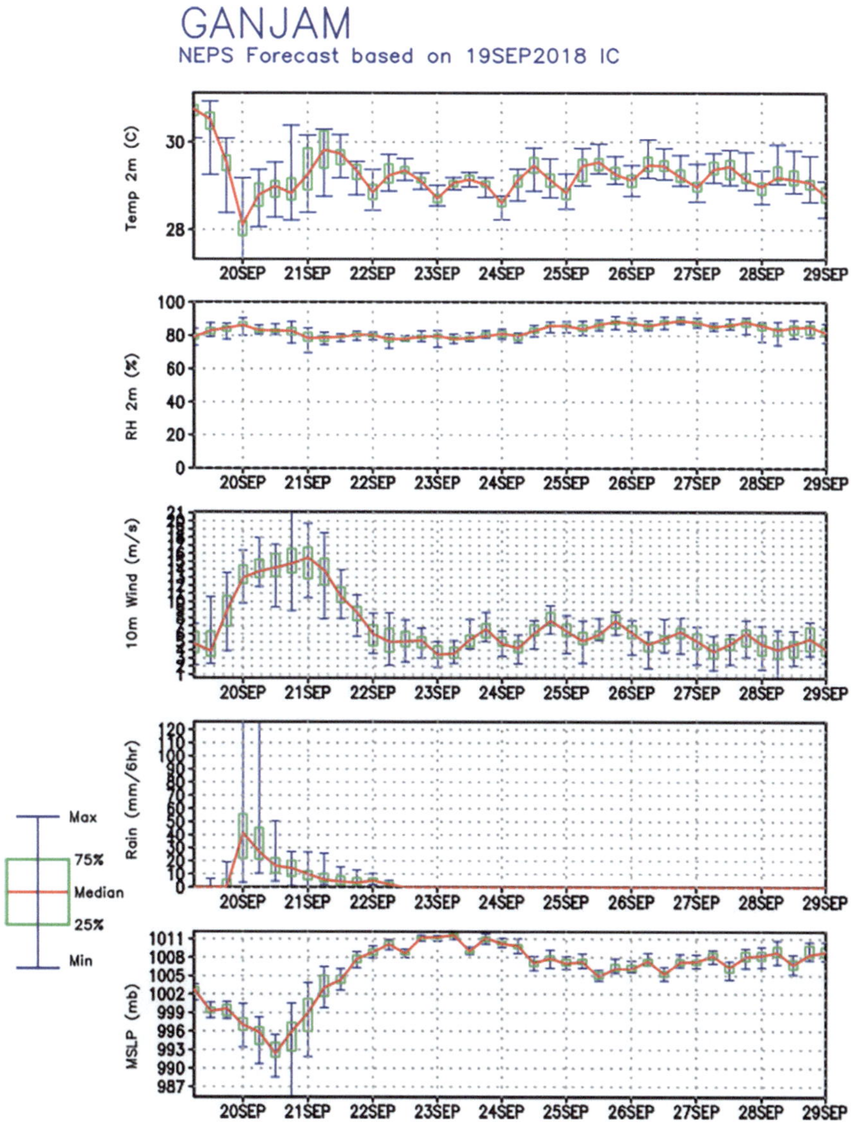

Fig. 6 EPSGRAM forecast of Temperature (°C) & Relative humidity at 2 m (%), 10 m wind (m/s), rainfall (mm/6 h) and MSLP (hPa) from NEPS-G at Ganjam based on 19th September 2018 initial condition

(60–120 mm day^{-1}) precipitation is over-predicted by both the ensemble systems. The over-prediction of light and moderate rainfall over oceanic regions is lesser in NEPS-G.

The ensemble spread is also higher in NEPS-G as compared to NEPS-G33 over places receiving very heavy (120–200 mm day^{-1}) and extremely heavy

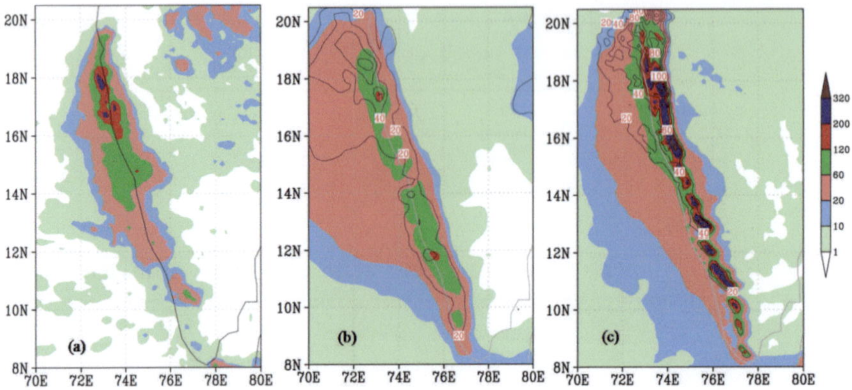

Fig. 7 **a** IMERG GPM rainfall (mm day^{-1}) for 10th June 2018 over the Western Ghats region; Ensemble Mean (shaded) and Spread (contours) in day-3 rainfall forecast valid for 10th June 2018 from **b** NEPS-G33 and **c** NEPS-G

(>200 mm day^{-1}) rainfall. Since NEPS-G33 under-predicted the rainfall amounts, the magnitude of ensemble spread is lesser. In contrast, NEPS-G over-predicted the rainfall amounts, and the ensemble spread is also higher. Larger values of the ensemble mean and spread in precipitation forecast of NEPS-G indicate the ability of the individual members to predict very heavy and extremely heavy precipitation but with a lot of disagreement between the forecast locations and magnitude. Buizza and Palmer (1998) also reported an increase in ensemble spread with the increase in resolution of ECMWF EPS.

3.6 *Spaghetti Diagram for NEPS-G33 Versus NEPS-G*

Spaghetti diagrams display the spatial distribution of isolines of one or two values for a field of interest from all the ensemble members to highlight probability distributions for a variable on a region. The distance between members gives a notion of uncertainty. Also, it gives an idea of the probability distribution of the forecast for the displayed contours. Figure 8 illustrates the contours of 150 mm precipitation predicted by 45 members (including control) from NEPS-G33 and that by 23 members (including control) of NEPS-G, each member forecast depicted in different colors. Contours near locations of heavy rainfall are much closely spaced in NEPS-G as compared to NEPS-G33, suggesting more agreement among the members on locations in the higher resolution EPS than that in the other EPS.

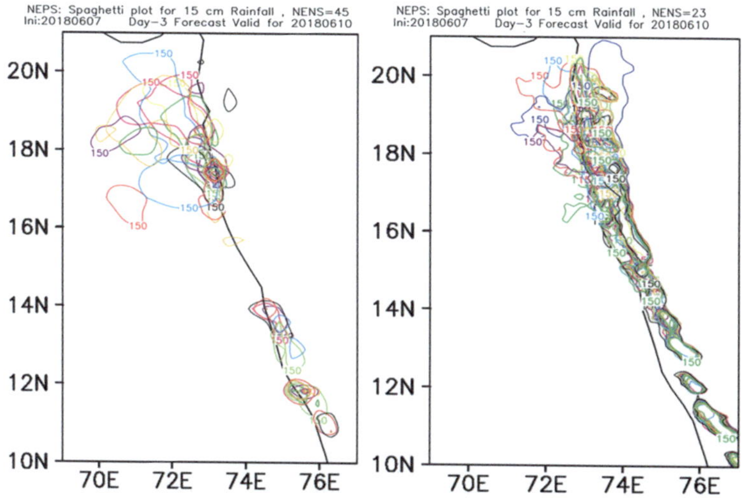

Fig. 8 Day-3 forecast of Spaghetti of 150 mm day^{-1} precipitation from **a** NEPS-G33 and **b** NEPS-G valid for 10th June 2018 over the Western Ghats

3.7 Skew-T Ensemble Profile

Skew-T plots can provide valuable information for forecasting thunderstorms, as they display the vertical profile of temperature, dew point, and wind in the atmosphere. By examining these parameters, meteorologists can identify the potential for instability, moisture, and wind shear, which are important ingredients for thunderstorm development. Skew-T plots can also be used to diagnose the type of thunderstorm that is most likely to form, such as a single-cell, multi-cell, or supercell thunderstorm, based on the atmospheric conditions present. However, it is important to note that other tools and data sources should also be used in conjunction with Skew-T plots for a comprehensive thunderstorm forecast.

Delhi and NCR experienced heavy rainfall accompanied with hailstorms on 7 February 2019. Figure 9 shows the Skew-T profiles tailored for ensemble prediction of thunderstorms. The instability developed at 00 UTC of 7th February 2019 is depicted by the thermodynamic vertical profile at New Delhi based on NEPS-G. Dry adiabats (grey lines), temperature lines (blue), mixing ratio lines (purple), moist adiabats (green), and wind barbs with direction at vertical levels are colored according to wind speed. The hodograph displays absolute and storm-relative helicity, with trace (green) and storm motion vector (red arrow). The thick lines of temperature (yellow), dew-point temperature (light blue) and parcel (red) sounding correspond to ensemble mean. The long-dashed lines represent ensemble maximum and minimum values. The dotted lines represent the 75th and 25th percentile values. The thermodynamic parameters inside the box on right correspond to ensemble mean values.

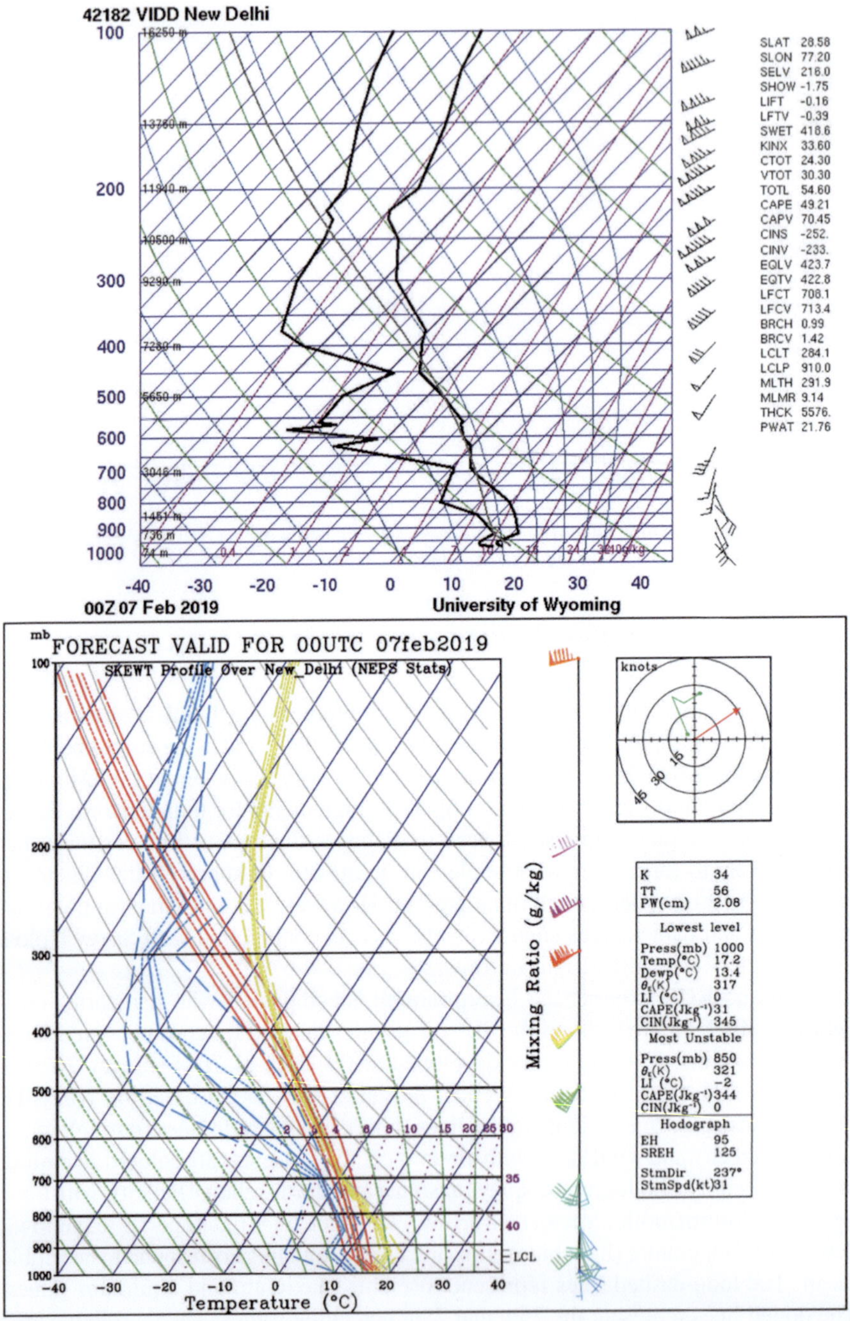

Fig. 9 **a** Observed sounding for New Delhi on 7th February 2019; **b** SkewT profile from NEPS-G at New Delhi on 7th February 2019

3.8 Short Range Forecasts for Thunderstorms at IITM

Some thunderstorm indices are used to predict the likelihood and severity of thunderstorms in GEFS IITM model which include Hail Environmental Index (HEI) (Lepore et al., 2018), Super Cell Composite Parameter (SCP) (Carbin et al., 2016), Layer Relative Humidity (LAYRH) (Rajeevan et al., 2012) and surface wind gust (GUST) shown in Fig. 10. The threshold for each index used for calculating the probability of occurrence among each ensemble member is HEI > 1; SCP > 3; LAYRH > 70; Surface wind gust > 25 knots. By utilizing the GEFS model output, these probability indices are developed for thunderstorm prediction over Indian Region (Mukhopadhyay et al., 2022). These indices take into account various atmospheric variables such as temperature, humidity, and wind, and provide information on the stability of the atmosphere and the likelihood of thunderstorm development. Figure 10a shows that there is more than 75% probability of occurrence of hail over Bangladesh, north-eastern states of India and south Karnataka on 14th May 2022. There is more than 75% probability of the SCP values greater than 3 (Fig. 10b) implying conducive environment for thunderstorm intensification over Bangladesh and north-eastern states. More than 75% probability of LAYRH > 70 in Fig. 10c depicts high moisture content of the atmosphere over south peninsular India and north-eastern states favorable for thunderstorm development over these regions. Figure 10d shows considerable surface wind gust >25 knots over Rajasthan and west Madhya Pradesh.

3.9 The Extreme Forecast Index (EFI) Precipitation

The Extreme Forecast Index (EFI), developed at ECMWF (Dutra et al., 2013; Lalaurette, 2003; Zsoter, 2006), measures the difference between the ensemble forecast distribution and model climate distribution. It enables detection of extreme weather events without defining thresholds in space or time and thus can be applied to observation-sparse areas. The EFI (Fig. 11) ranges from −1 to +1, with magnitudes between 0.5 and 0.8 suggesting unusual weather, and greater than 0.8 indicating very unusual weather.

3.9.1 Multi-model Ensemble Forecast Guidance by IMD for Severe Weather

The multi-model ensemble forecast guidance at IMD combines the outputs of multiple global and regional NWP models for severe storms associated with tropical cyclones (Kotal and Roy Bhowmik, 2011; Kotal, et al., 2014;, 2015; Kaur et al., 2020; Saranya Ganesh et al., 2019). In the case of a tropical cyclone, the track, intensity, and potential impact of the storm are predicted. The forecasts generated by each model are combined to produce a consensus forecast (Roy Bhowmik and

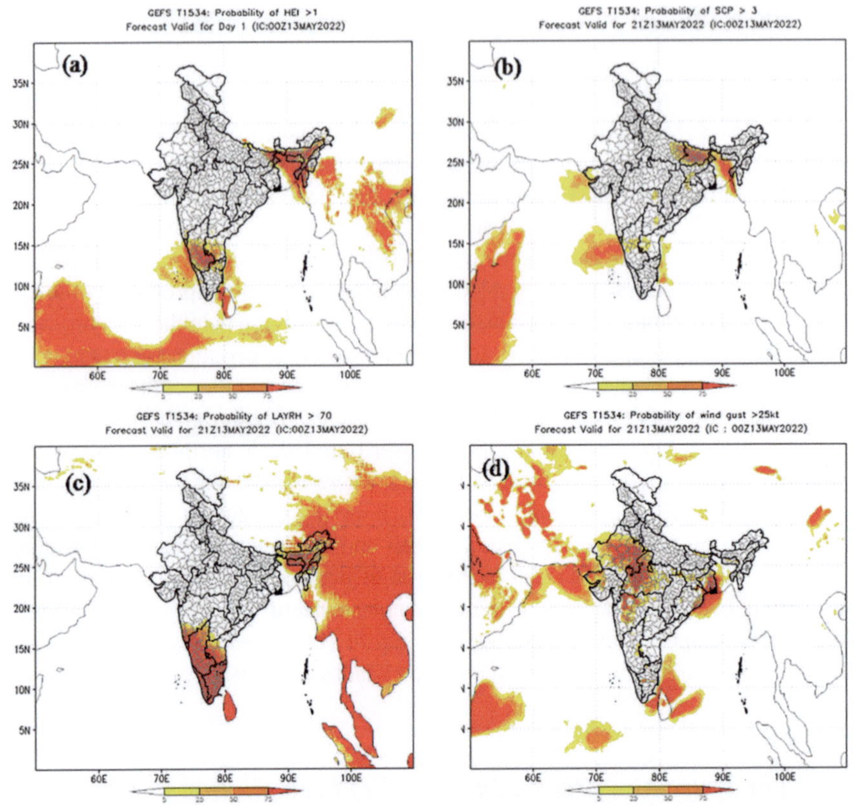

Fig. 10 GEFS Probabilistic forecast of **a** HEI > 1 (day-1) valid for 00 UTC 14 May 2022, **b** SCP > 3, **c** LAYRH > 70 and **d** GUST > 25 knots valid for 21 UTC 13th May 2022

Durai, 2010, 2012). This approach helps to provide a forecast of the potential impact of the severe weather event, including information on wind speeds, storm surge, and rainfall. Many operational agencies still employ consensus techniques to generate the official forecast paths, and the majority of them still depend on an un-weighted consensus of four to nine NWP models (Conroy et al., 2022). In addition to the multi-model ensemble approach, the IMD also uses satellite imagery, radar data, and other observational data to provide real-time updates on the development and movement of severe weather systems.

The IMD issues severe weather forecast guidance with four types of weather conditions associated with thunderstorms such as lightning, strong gusty winds and hail at different time scales: medium-range (1–5 days), short-range (24–72 h), and very short-range or now-cast scale (3 h). The forecasts closest to the events are most accurate and location specific. These location specific nowcasts, which are generated and transmitted at every 3-h intervals, utilize the latest observations of the events from the Indian and Global observatory and Meteorological satellites,

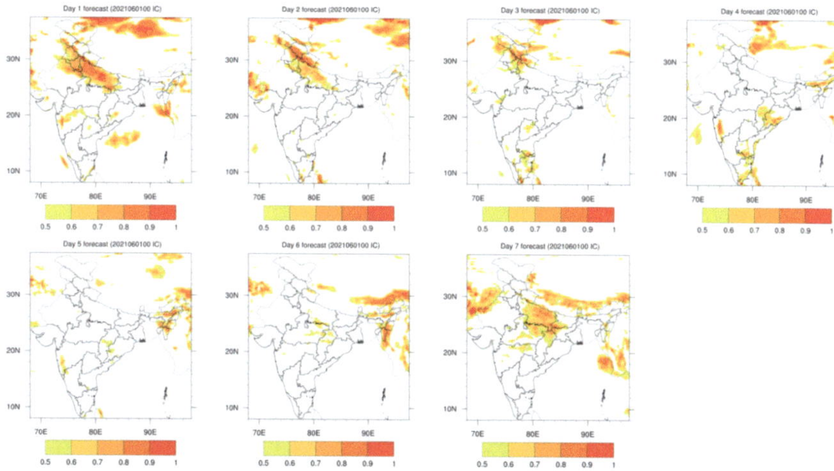

Fig. 11 Extreme Forecast Index (EFI) of precipitation from GEFS day-1 to day-7 forecasts based on 1st June 2021 00 UTC initial conditions

Doppler Weather Radar network and the data from the lightning detector network of the Indian Air Force and IITM. The use of NWP models and other observing tools has improved the ability of the IMD to accurately predict thunderstorms in India (Sen Roy et al., 2021). However, the prediction of severe weather is still a challenging task, as the atmospheric processes that lead to these events are highly complex and can be influenced by a variety of factors, including land-surface processes, topography, and the interaction of the atmosphere with the oceans.

4 Convective Scale Modeling Used for Severe Weather Prediction in India

One of the most challenging problems in weather forecasting is the prediction of extreme rainfall events associated with severe thunderstorms possibly leading to flash floods. Medium to long-range deterministic or ensemble NWP models still face difficulties in correctly predicting the intensity and location of events that can be triggered and amplified by very small-scale features (Ashrit et al., 2020). Forecasting such spurious events is further difficult at areas characterized by a complex, steep orography close to a coastline, where intense precipitation can be enhanced by specific topographic features (2017b; Abhilash et al., 2012; Chakravarty et al., 2022; Sarkar et al., 2017a). Sen Roy (2009) studied precipitation patterns in India for winter, dry-summer, and wet-summer monsoon seasons and reported rising trends in extremely heavy precipitation events, mostly in the high-elevation regions of the northwestern Himalaya as well as along the foothills of the Himalaya extending south into the Indo-Gangetic basin. Physical attribution of extreme precipitation changes is more challenging in the tropics and subtropics because of the role of large-scale vertical motions, for which very few direct observations exist (Tandon et al., 2018).

Convective scale models are particularly useful for predicting severe weather because they are able to simulate the processes that occur on scales of a few kilometers to several tens of kilometers, which is the typical scale of thunderstorms and other severe weather events in India (Srivastava et al., 2010, 2012). Rajeevan et al. (2012) developed a prognosis probabilistic model based on the Weather Research and Forecasting (WRF) for prediction of lightning over south-east India. Chen and Wang (2015) investigated thunderstorms prediction creating convective scale lagged ensemble forecasts using WRF model coupled with cloud resolving models taking initial and boundary conditions from NCEP Reanalysis. Sandeep et al. (2021) tested the McCaul et al. (2009) scheme in NCMRWF regional (4 km) unified model to predict the lightning flash count (LFC) based on the graupel water path amount and reported the LFC to be a useful tool for day-to-day prediction of thunderstorm activity and intensity. Mohan et al. (2021) evaluated different lightning parameterization schemes and Lightning Potential Index (LPI) in the Advanced Research version of the WRF (ARW) model to simulate lightning flash counts over Maharashtra, India and indicated a promising future for the operational forecast of lightning prediction. Umakanth et al. (2021) pronounced the importance of INSAT-3D and INSAT-3DR satellite data in WRF ARW model in the prediction of thunderstorms over Andhra Pradesh region. Halder et al. (2021) computed the LPI as a measure of the potential for charge generation and separation that leads to lightning flashes, using best suitable physics parameterization schemes in cloud resolving (1 km) WRF-ARW model with initial and boundary conditions from Global Forecast System (Mukhopadhyay et al., 2019). Vani et al. (2022) demonstrated the real-time operational lightning forecast based on storm parameters in the WRF-ARW Model with different microphysics schemes to be skillful during 2019 and 2020 pre-monsoon over India. Kumar et al. (2022) used an ensemble of different physics parameterization in WRF model and obtained probability of lightning threat coherent to IITM lightning data. Banik et al. (2022) obtained improved simulated lightning density for a few events after assimilation of ground-based lightning data of IITM into the specialized electrical physics-based WRF model tested at IMD.

The High-Resolution Rapid Refresh (HRRR) model forecasting system simulates the location and timing of the convective systems with better accuracy due to the frequent assimilation of available observations. It is an hourly updating weather forecasting model with a forecast duration of 18 hours. It uses WRF ARW model and assimilates conventional, satellite radiance and radar observations on an hourly basis using the Grid-point Statistical Interpolation (GSI) 3DVAR technique. It is a convection-allowing model ($\leq$3 km) with 45 vertical levels running operationally at NCMRWF on an experimental basis. It is useful for nowcasting of weather events such as lightning, thunderstorms, heavy rainfall, etc.

5 Short Range Ensemble Prediction of Severe Weather

The global ensemble systems essentially are intended to mainly focus on the uncertainties due to synoptic-scale baroclinic instability for a forecast lead time of 3–10 days. But the uncertainty on shorter time and length scales is still significant and leads to large uncertainty in finer weather details at a local scale. Addressing this issue is therefore necessary and has made to shift focus towards developing EPS to tackle the uncertainties on a short-range.

Throughout the recent decades, various short-range ensemble prediction systems have been developed using different approaches to generate perturbed initial conditions. These systems are run at various operational facilities around the world. One such system is NCEP's SREF, which uses two limited area models and boundary conditions from NCEP's medium-range global ensemble (Tracton et al., 1998; Stensrud et al., 1999). Another multi-model system, SREPS, developed by the Spanish Met Service, INM, runs with five regional models and boundary conditions from different NWP centers worldwide (Garcia-Moya et al., 2007). However, the main disadvantage of multi-model systems is that the ensemble members tend to differ systematically from each other, leading to different skills and a clustering of forecasts by model (Alhamed et al., 2002). This can result in the forecast probability distribution being dominated by the distribution of models or schemes rather than synoptically-dependent forecast uncertainty. These issues are not present in single-model-based ensemble prediction systems. Meteo-France has developed PEACE, a single-model-based system based on the ARPEGE model with 11 ensemble members. The Norwegian Meteorological Institute's EPS, LAMEPS (Frogner et al., 2006), covers Scandinavia with boundary conditions and initial condition perturbations from ECMWF-EPS.

5.1 Convective Scale EPS

Convection permitting forecasts are highly sensitive to uncertainties in initial conditions, boundary conditions, and physical processes due to the highly non-linear and short-lived nature of convective processes. Therefore, an ensemble approach is needed to account for uncertainty at a local scale. To address this, many operational forecasting centers have shifted from short-range EPS to convective scale EPS. The increase in computing power has also led to the development of convective scale ensemble systems worldwide. The convective scale ensemble at Meteo-France uses the 2.5 km AROME model, nested within the ARPEGE EPS, and runs with 12 members twice a day (Bouttier et al., 2012; Nuissier et al., 2012). The German Weather Service operates a 2.8 km ensemble (COSMO-DE-EPS), which is nested within a regional-scale ensemble and runs with 20 members 8 times a day for 21 h (Bouallégue et al., 2013).

The UKMO has developed MOGREPS-UK, a convective scale ensemble prediction system, to address the issue of uncertainty in short-term forecasts. The resolution of this system is 2.2 km and is nested inside the global EPS (MOGREPS-G). To effectively address the issue of uncertainty, MOGREPS considers three key requirements—appropriate method of generating initial condition perturbations, addressing uncertainties from approximations in model formulation, and handling model error within a single model environment (Bowler et al., 2008).

To enhance the accuracy of forecasts for high-impact weather events, many centers globally have implemented convective scale EPS (Roberts et al., 2022). These models can more accurately predict heavy precipitation events, but with a trade-off of uncertain location and structure. A convective scale EPS can assist forecasters in determining the potential for a weather event and its extent, thus reducing the uncertainty of the event's occurrence. The design of these ensembles has been influenced by the weather system being studied and the available computational resources, and various strategies being explored, such as increasing ensemble size, improving horizontal resolution, creating initial condition perturbations, incorporating stochastic physics, and implementing multi-model and time-lagged ensembles.

Hagelin et al. (2017) conducted a study using the UK ensemble and found that increasing the ensemble size was more effective than decreasing the horizontal resolution, particularly in the case of a small ensemble. Raynaud and Bouttier (2017), in their study of the AROME-France-EPS system, also found that a 34-member EPS with a 2.5 km resolution performed better than a 12-member EPS with a 1.5 km resolution. However, Schwartz et al. (2017) reported the opposite results in their study using the WRF model, finding that a 10-member EPS with 1 km grid spacing was better at predicting precipitation over central United States than a 30-member EPS with 3 km grid spacing. These conflicting results highlight that the best strategy depends on the relationship between the number of ensemble members and the EPS's horizontal resolution, the meteorology of the area, and the model used. Additionally, Raynaud and Bouttier (2017) recommended that post-processing techniques like neighborhood post-processing and time-lagging ensembles could lead to improved forecasts with a lower computational burden. Kühnlein et al. (2014) evaluated the sensitivity of initial condition perturbations (IPs), boundary condition perturbations (BPs), and physics perturbations (PPs) and found that IPs had a significant impact in the first 6 hours of the forecast, which diminished quickly as BPs and PPs took over. PPs continued to have an impact throughout the forecast as they correspond to tunable empirical values in the physics schemes.

The downscaling strategy is widely used by centers around the world to initially implement a convective scale ensemble system, primarily due to its simplicity. In this method, the driving ensemble provides the initial and boundary conditions, which are then interpolated to the domain of the convective scale ensemble. The high-resolution model in the convective scale ensemble then dynamically adds high-resolution information to the driving ensemble, allowing local scale features to develop. Additionally, it is desirable to have the IPs and BPs from the same driving ensemble to prevent conflict between independent IPs and BPs. However, this approach may not fully address the uncertainty in the initial conditions at the convective scale. Alternatively, a high-resolution regional analysis from a regional data assimilation system can be used where an ensemble data assimilation system is not available. In this strategy, perturbations from the driving ensemble are interpolated to the regional domain and added to the high-resolution analysis to create the perturbed initial conditions for the ensemble members (Tennant, 2015). Although this method has its own drawbacks, it is still advantageous as the high-resolution analysis contains better information at finer scales than the downscaled global analysis.

5.2 *NCMRWF Regional Ensemble Prediction System (NEPS-R)*

NCMRWF Regional Ensemble Prediction System (NEPS-R) (Kiran Prasad et al., 2019; Jayakumar et al., 2021; Sarkar et al., 2021) is based on the regional version of MOGREPS with 12 members (1 control + 11 perturbed members). While NEPS-G running at a resolution of 12 km addresses the uncertainty on a synoptic scale, NEPS-R runs at a convective scale resolution (4 km) and is aimed at addressing the uncertainty in finer weather information at a local scale. NEPS-R runs with the initial and boundary conditions from its global counterpart (NEPS-G). It operates with 80 vertical levels and produces forecasts up to 75 hours. It runs on UM version 11.1 covering Indian mainland, Himalayan and the BIMSTEC (Bay of Bengal Initiative for Multi-Sectoral Technical and Economic Cooperation) countries (62°E–106°E, 6°S–41°N), and with a horizontal grid resolution of 4 km, using 1200 grid points in the east–west and north–south using a time step of 120 seconds. Further, the model uncertainty is handled by RP scheme which takes care of uncertainty caused by the choice of tunable parameters in physical parameterization schemes and uncertainty due to sub-grid scale (not fully resolved) processes (McCabe et al., 2016). Realistic dust and soil moisture are taken from the global analysis instead of climatological values. SRTM (Shuttle Radar Topography Mission) and CartoSat digital elevation models (DEM) orography and Indian Space Research Organization (ISRO) Land use/ Land cover (LuLC) are used, which are developed in-house. The lightning parameterization in NEPS-R is based on McCaul et al. (2009), and is used to predict the lightning flash rate tuned to the Indian region based on the study of Sandeep et al. (2021).

5.3 *Probabilistic Forecasts Based on NEPS-R*

Many probabilistic forecast products are operationally generated every day, out of which probabilistic forecast of precipitation and lightning and stamp plots of precipitation are particularly of interest in thunderstorm and lightning prediction.

5.3.1 Probability of Exceedance of Precipitation Associated with Thunderstorms

Probability of occurrence of an event can be quantified in terms of exceedance of a threshold. It is the most consistent way to convey forecast uncertainty information. The event threshold often corresponds to the point at which the user has to take some action to mitigate for the potential damage due to a significant weather event.

Widespread lightning and rainfall occurred on 31st August 2022 over the Western Ghats and the Himalayan belt. The meteorological conditions at 00 UTC include a cyclonic circulation over South coastal Andhra Pradesh and neighborhood extending up to Lakshadweep Area and adjoining southeast Arabian Sea, is seen upto 5.8 km above mean sea-level, and a trough from southeast Bay of Bengal to the cyclonic circulation over south coastal Andhra Pradesh and neighborhood exists upto 3.1 km above sea-level. Further, a western disturbance lay as a trough in middle & upper-level westerlies running along Lon 87°E and north of Lat 22°N. Figure 12 illustrates the 24-hours-accumulated ensemble mean rainfall along with the probabilistic rainfall for thresholds >0.25 cm/day, >1.56 cm/day, >6.55 cm/day, >11.5 cm/day and 19.5 cm/day (thresholds are based on IMD criteria for rainfall amounts). Figure 12a gives the mean picture of rainfall from all the 12 members, which indicates that thunderstorm induced heavy to very heavy rainfall to the order of 4–16 cm is forecasted over the west coast of India, Nepal and the northeastern states of Assam, Arunachal Pradesh and Meghalaya. Light to moderate rainfall is predicted over Bihar, Sub-Himalayan West Bengal, north Bangladesh, Uttar Pradesh, Madhya Pradesh, Chhattisgarh, and Jharkhand and at isolated places over Odisha, Telangana, Andhra Pradesh, Gujarat, Jammu and Kashmir, Himachal Pradesh and Uttarakhand. The probabilistic forecast for different thresholds indicates more than 90% (areas displayed in red color) chance of rainfall of threshold >0.25 cm (Fig. 12b) and >1.56 cm (Fig. 12c) over the west coast of India, Nepal and the northeastern states of Assam, Arunachal Pradesh and Meghalaya, which is more likely to realize. The thresholds of > 11.5 cm (Fig. 12e) displays percentages less than 50, which is less likely to occur and the threshold of >19.5 cm (Fig. 12f) displays percentages less than 30 indicating a remote chance of extremely heavy rain to occur over the predicted regions.

5.3.2 Probability of Exceedance of Lightning

Figure 13 shows the 24-hours-accumulated ensemble mean lightning flashes (f/day) along with lightning flash thresholds >1/day, >10/day and >20/day. Figure 13a illustrates widespread lightning flashes forecasted over India except for the western and northern part of the country. The peak lightning is predicted over the same areas as the peak rainfall (Fig. 12) over the west coast of India and over a few locations in Nepal. Probabilistic lightning flashes of threshold >1 (Fig. 13b) displays a pattern similar to that of the ensemble mean with greater than 90% probability over areas of the west coast of India, Nepal, Chhattisgarh and Odisha which is more likely to occur. Further, the probability of >10 (Fig. 13c) is high (50–90%) at a few locations

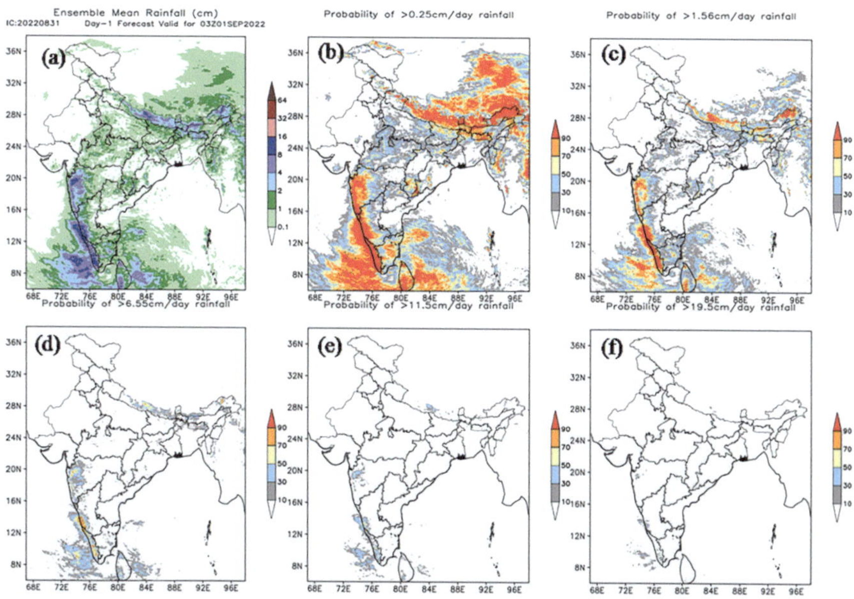

Fig. 12 **a** Ensemble mean rainfall (cm) forecast (day-1) and probabilistic quantitative precipitation forecast of NEPS-R for day-1, **b** >0.25 cm, **c** >1.56 cm, **d** >6.55 cm, **e** >11.5 cm and **f** >19.5 cm valid for 03 UTC 1st September 2022

over the west coast of India, likely to realize and the probability of >20 (Fig. 13d) is less likely (probability less than 50%) to occur at a few locations over the west coast of India.

5.3.3 Regional Ensemble Stamps

Figure 14 displays the day-1 rainfall forecast stamp plots for each the 12 ensemble members. The stamp plots indicate that the location of forecasted rainfall is more or less similar in all the members except for minor displacement in the peak rainfall, minor differences in intensity and spread of the rainfall, indicating good agreement among the members. The advantage of NEPS-R is that it minimizes the over prediction of precipitation by the control member (deterministic) and also minimizes the location error in the precipitation forecast due to the spread among the members. But the possibility of underpredicting the extreme precipitation by the ensemble mean remains as a drawback of the system. The underprediction of extreme precipitation in case of a cyclone is reported by Sarkar et al. (2022) where the extreme precipitation due to the TC-Amphan at the time of landfall is missed out by the ensemble mean but was predicted by the members with varying locations. In such cases, the stamp plots provide forecast information by each member which could be used by the forecaster to assess the possible risk scenario.

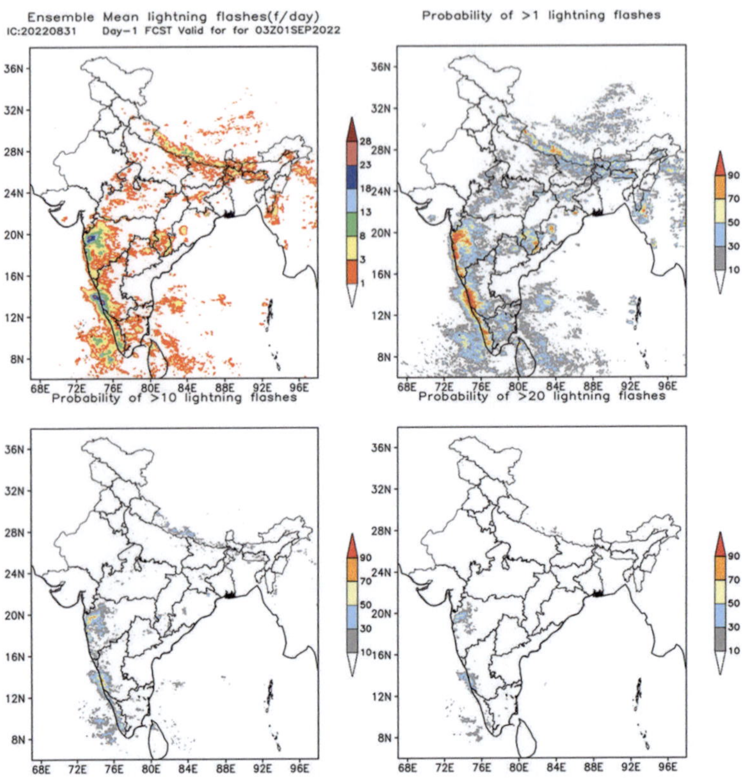

Fig. 13 **a** Ensemble mean lightning flashes (f/day) forecast (day-1) and probabilistic quantitative lightning flash forecast of NEPS-R for day-1, **b** >1 f/day, **c** >10 f/day, **d** >20 f/day valid for 03 UTC 1st September 2022

Figure 15 shows the day-1 rainfall forecast associated with widespread thunderstorm activity on 2nd May 2018 over north India along with the observation. The prevailing meteorological conditions at 00 UTC include a cyclonic circulation lying over Haryana and neighborhood in lower levels, inducing strong easterlies in the lower levels and aiding moisture incursion over the northern parts of the country including western Himalayan region. Further, a western disturbance lay as a trough in middle & upper-level westerlies running along Lon 68°E and north of Lat 32°N. It can be seen from the figure that ensemble member 1 and 2 have completely missed the event with little or no rainfall predicted by them. This indicates the possibility of the control member failing to predict the event with marginal changes in the initial conditions, which also could be witnessed in this case where the control and many other members have missed predicting the rainfall over north Madhya Pradesh and adjoining Uttar Pradesh. This rainfall patch over north Madhya Pradesh and adjoining Uttar Pradesh is predicted by ensemble member 10 and its forecast also is the best one close to the observed rainfall except for the over-predicted rainfall over a few locations. It is also observed that such scenarios of highly varying forecasts

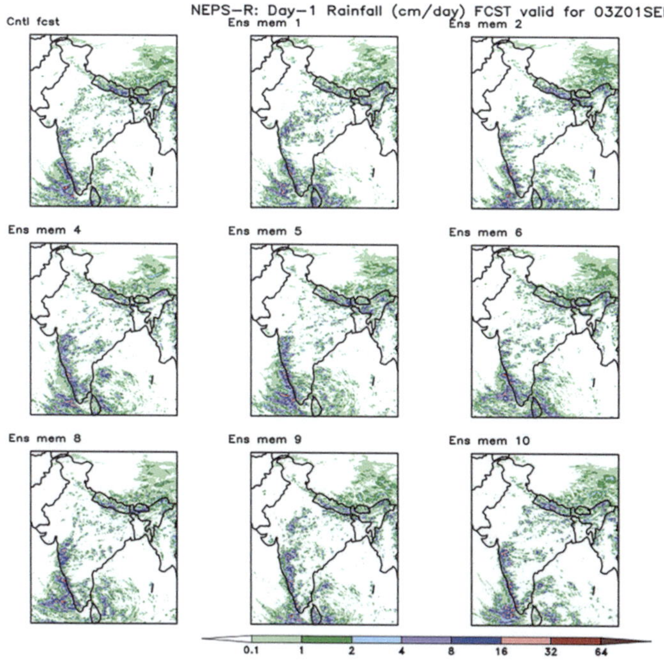

Fig. 14 The postage stamp maps of 24 hours accumulated rainfall forecast of day-1 for each of the 12 members valid for 03 UTC 1st September 2022

of the members are found during the pre-monsoon season during weak synoptic forcing. However, during the monsoon season, where strong synoptic forcing exists, the members are in good agreement with each other except for minor displacement in the peak rainfall, minor differences in intensity and spread of the rainfall.

6 Summary

This chapter outlines prediction of thunderstorms and associated severe weather using ensemble forecasting approach which is operational in India and methods which are being developed through ongoing research activities. Some important aspects of the ensemble prediction are discussed and the modeling infrastructure operational at the national centers for probabilistic prediction of severe weather are detailed. The forecast products tailored for severe weather prediction mainly thunderstorms over India are illustrated for some events at real time. Some of these products are generated on operational basis at the national centers while rests of them are still in development and testing stage. Representing the physical phenomena driven by convective processes over tropics is a well-known challenge in NWP systems. Moreover, the spurious nature of the severe weather events and their associated sudden

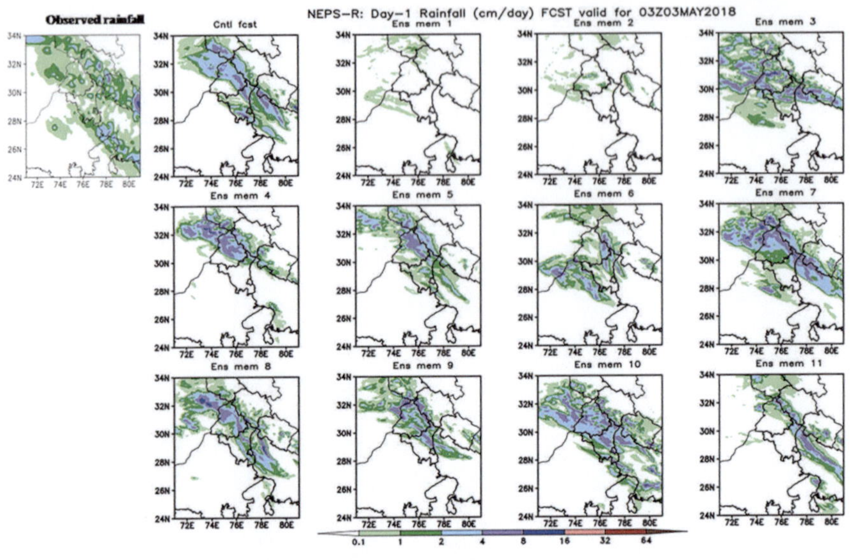

Fig. 15 Observed rainfall (cm) and postage stamp maps of 24 hours accumulated rainfall (cm) forecast of day-1 for each of the 12 members valid for 03 UTC 3rd May 2018

impact necessitate the development of convection scale models for better resolving the sub grid scale processes. The uncertainties underlying in the physics parameterizations of the deterministic convective models can be addressed and mapped using the ensemble systems at convection scale. This enables improving the representation of the physical features in short range forecasts and development of probabilistic high resolution forecast products tailored for severe weather. The short-range forecasts of thunderstorm indices, lightning probabilities and stamp shown can help assess the risk associated with the severe storms and plan for the disaster mitigation in time. The interpretations of the short to medium range probabilistic ensemble forecast products would help forecasters and researchers make optimal use of these products as per requirement and for planning further advancements of severe weather modeling over Indian region.

Acknowledgements The authors acknowledge the Ministry of Earth Sciences (MoES), Government of India, New Delhi for providing the resources required for this study. Authors thank Head, NCMRWF for motivation and encouragement in the study. The authors acknowledge the scientists at NCMRWF, Indian Institute of Tropical Meteorology (IITM), Pune and India Meteorological Department (IMD) for their valuable support, suggestions and feedback.

References

Abhilash, S., Sahai, A. K., Mohankumar, K., George, J. P., & Das, S. (2012). Assimilation of Doppler weather radar radial velocity and reflectivity observations in WRF-3DVAR system for short-range forecasting of convective storms. *Pure and Applied Geophysics, 169*, 2047–2070.

Alhamed, A., Lakshmivarahan, S., & Stensrud, D. J. (2002). Cluster analysis of multimodel ensemble data from SAMEX. *Monthly Weather Review, 130*, 226–256.

Ashrit, R., Sharma, K., Kumar, S., Dube, A., Karunasagar, S., Arulalan, T., Mamgain, A., Chakraborty, P., Kumar, S., Lodh, A., & Dutta, D. (2020). Prediction of the August 2018 heavy rainfall events over Kerala with high-resolution NWP models. *Meteorological Applications, 27*(2), e1906.

Banik, T., Pattanaik, D. R., Das, A. K., Pawar, S. D., & Kundu, S. S. (2022). A report on numerical weather prediction products for sectoral applications.

Barkmeyer, J., Buizza, R., & Palmer, T. N. (1999). 3D-Var Hessian singular vectors and their potential use in the ECMWF Ensemble Prediction System. *Quarterly Journal of the Royal Meteorological Society, 125*(558), 2333–2351.

Bishop, C. H., Etherton, B. J., & Majumdar, S. J. (2001). Adaptive sampling with the ensemble transform Kalman filter. Part I: Theoretical aspects. *Monthly Weather Review, 129*, 420–436.

Blenkinsop, S., Fowler, H. J., Barbero, R., Chan, S. C., Guerreiro, S. B., Kendon, E., Lenderink, G., Lewis, E., Li, X. F., Westra, S., & Alexander, L. (2018). The INTENSE project: Using observations and models to understand the past, present and future of sub-daily rainfall extremes. *Advances in Science and Research, 15*, 117–126.

Bouallégue, Z., Theis, S., & Gebhardt, C. (2013). Enhancing COSMO-DE ensemble forecasts by inexpensive techniques. *Meteorologische Zeitschrift., 22*, 49–59.

Bouttier, F., Vie, O., & Raynaud, L. (2012). Impact of stochastic physics in a convection-permitting ensemble. *Monthly Weather Review, 140*, 3706–3721.

Bowler, N. E., & Mylne, K. R. (2009). Ensemble transform Kalman filter perturbations for a regional ensemble prediction system. *Quarterly Journal of the Royal Meteorological Society: A Journal of the Atmospheric Sciences, Applied Meteorology and Physical Oceanography, 135*(640), 757–766.

Bowler, N. E., Arribas, A., Mylne, K. R., Robertson, K. B., & Beare, S. E. (2008). The MOGREPS shortrange ensemble prediction system. *Quarterly Journal of the Royal Meteorological Society, 134*, 703–722.

Brier, G. W. (1950). Verification of forecasts expressed in terms of probability. *Monthly Weather Review, 78*(1), 1–3.

Bright, D. R., & Mullen, S. L. (2002). Short-range ensemble forecasts of precipitation during the southwest monsoon. *Weather and Forecasting, 17*(5), 1080–1100.

Brown, B., Ebert, E., Fowler, T., Gilleland, E., Kucera, P., & Wilson, L. (2013). WWRP/WGNE joint working group on forecast verification research.

Buizza, R. (1997). Potential forecast skill of ensemble prediction and spread and skill distributions of the ECMWF ensemble prediction system. *Monthly Weather Review, 125*(1), 99–119.

Buizza, R., & Palmer, T. (1995). The singular vector structure of the atmospheric general circulation. *Journal of Atmospheric Science, 52*, 1434–1456.

Buizza, R., & Palmer, T. N. (1998). Impact of ensemble size on ensemble prediction. *Monthly Weather Review, 126*(9), 2503–2518.

Buizza, R., Gelaro, R., Molteni, F., & Palmer, T. N. (1997). The impact of increased resolution on predictability studies with singular vectors. *Quarterly Journal of the Royal Meteorological Society, 123*(540), 1007–1033.

Buizza, R. (2002). Chaos and weather prediction January 2000. European Centre for medium-range weather meteorological training course lecture series ECMWF.

Carbin, G. W., Tippett, M. K., Lillo, S. P., & Brooks, H. E. (2016). Visualizing long-range severe thunderstorm environment guidance from CFSv2. *Bulletin of the American Meteorological Society, 97*(6), 1021–1031.

Chakraborty, P., Sarkar, A., Kumar, S., George, J. P., Rajagopal, E. N., & Bhatla, R. (2020). Assessment of NCMRWF global ensemble system with differing ensemble populations for tropical cyclone prediction. *Atmospheric Research, 244*, 105077.

Chakraborty, P., Sarkar, A., Bhatla, R., & Singh, R. (2021). Assessing the skill of NCMRWF global ensemble prediction system in predicting Indian summer monsoon during 2018. *Atmospheric Research, 248*, 105255.

Chakraborty, P., Dube A., Singh, H., Arulalan, T., Prasad, S. K., Mamgain, A, Singh, A., Kumar, S., Sarkar, A., Ashrit, R, & Rajagopal, E. N. (2019). Generation of probabilistic forecast products from NCMRWF Ensemble Prediction System (NEPS). NMRF/TR/03/2019.

Chakravarty, K., Patil, R., Saha, S., Murugavel, P., Yadav, P., & Pandithurai, G. (2022). Unraveling the contrasting features of rainfall microphysics for the tropical cyclone Tauktae (2021) and a severe thunderstorm event as observed over the orographic region of Western Ghat.

Chakraborty, P., Dube, A., Sarkar, A., Mitra, A. K., Bhatla, R., & Singh, R. S. (2023). How much does a high-resolution global ensemble forecast improve upon deterministic prediction skill for the Indian summer monsoon?. *Meteorology and Atmospheric Physics, 135*(4), 33. https://doi.org/10.1007/s00703-023-00966-1

Charney, J. G., Fjortoft, R., & von Neumann, J. (1950). Numerical integration of the barotropic vorticity equation. *Tellus, 2*, 237–254.

Chattopadhyay, N., Devi, S. S., John, G., & Choudhari, V. R. (2017). Occurrence of hailstorms and strategies to minimize its effect on crops. *Mausam, 68*(1), 75–92.

Chen, Y.W., & Wang, J. (2015, July). The ensemble forecasting technique of the thunderstorm. In *2015 International Conference on Artificial Intelligence and Industrial Engineering* (pp. 163–165). Atlantis Press.

Conroy, A., Brammer, A., Chakraborty, P., Chen, G., Cowan, L., Dunion, J., Feng, X., Ford, M., Rivett, R., Sarkar, A., & Titley, H. (2022). Track forecast: Operational capability and new techniques (in combination with TC-PFP Stage 1, IWTC report (Vol. 10, 4.4, 26pp.).

Conroy, A., Titley, H., Rivett, R., Feng, X., Methven, J., Hodges, K., Brammer, A., Burton, A., Chakraborty, P., Chen, G., & Cowan, L. (2023). Track forecast: Operational capability and new techniques-Summary from the Tenth International Workshop on Tropical Cyclones (IWTC-10). *Tropical Cyclone Research and Review, 12*(1), 64–80.

Das, S., Mohanty, U. C., Tyagi, A., Sikka, D. R., Joseph, P. V., Rathore, L. S., Habib, A., Baidya, S. K., Sonam, K., & Sarkar, A. (2014). The SAARC STORM: A coordinated field experiment on severe thunderstorm observations and regional modeling over the South Asian Region. *Bulletin of the American Meteorological Society, 95*(4), 603–617.

Das, S. (2022). Challenges in predicting extreme weather events over the South Asian Region. In *Extreme Natural Events: Sustainable Solutions for Developing Countries* (pp. 51–106). Springer Nature Singapore.

Dunion, J. P., Davis, C., Titley, H., Greatrex, H., Yamaguchi, M., Methven, J., ... & Vigh, J. L. (2023). Recommendations for improved tropical cyclone formation and position probabilistic Forecast products. *Tropical Cyclone Research and Review, 12*(4), 241–258.

Durai, V. R., & Roy Bhowmik, S. K. (2014). Prediction of Indian summer monsoon in short to medium range time scale with high resolution global forecast system (GFS) T574 and T382. *Climate Dynamics, 42*, 1527–1551.

Durai, V. R., Roy Bhowmik, S. K., Rao, Y. R., & Bhardwaj, R. (2015). Verification of quantitative precipitation forecasts from operational ensemble prediction systems over India. *Mausam, 66*(3), 479–496.

Dutra, E., Diamantakis, M., Tsonevsky, I., Zsoter, E., Wetterhall, F., Stockdale, T., Richardson, D., & Pappenberger, F. (2013). The extreme forecast index at the seasonal scale. *Atmospheric Science Letters, 14*, 256–262. https://doi.org/10.1002/asl2.448

Eady, E. (1951). The quantitative theory of cyclone development. In T. Malone (Ed.), *Compendium of Meteorology* (pp. 464–469). American Meteorological Society.

Epstein, E. S. (1969). Stochastic dynamic prediction. *Tellus, 21*, 739–759.

Fortin, V., Abaza, M., Anctil, F., & Turcotte, R. (2014). Why should ensemble spread match the RMSE of the ensemble mean? *Journal of Hydrometeorology, 15*(4), 1708–1713.

Frogner, I.-L., Haakenstad, H., & Iversen, T. (2006). Limited-area ensemble predictions at the Norwegian Meteorological Institute. *Quarterly Journal of the Royal Meteorological Society, 132*, 2785–2808.

Garcia-Moya, J. A., Callado, A., Santos, C., Santos, D., & Simarro, J. (2007). Multi-model ensemble for short-range predictability. In *3rd International Verification Methods Workshop*. ECMWF. http://www.ecmwf.int/newsevents/meetings/workshops/2007/jwgv/workshop/presentations/indexposter.htm.

Grimit, E. P., & Mass, C. F. (2007). Measuring the ensemble spread–error relationship with a probabilistic approach: Stochastic ensemble results. *Monthly Weather Review, 135*(1), 203–221.

Hagelin, S., Son, J., Swinbank, R., McCabe, A., Roberts, N., & Tennant, W. (2017). The met office convective-scale ensemble, MOGREPS-UK. *Quarterly Journal of the Royal Meteorological Society, 143*, 2846–2861.

Halder, M., Kanase, R., Mukhopadhyay, P., Halder, S., Pawar, S. D., Domkawle, M., & Pandey, A. C. (2021). Latest approaches to thunderstorm/lightning and severe weather forecasting using high resolution numerical model.

Hamill, J., Haddad, J. M., & McDermott, W. J. (2000). Issues in quantifying variability from a dynamical systems perspective. *Journal of Applied Biomechanics, 16*(4), 407–418.

Hanssen, A. W. & Kuipers, W. J. (1965). *On the relationship between the frequency of rain and various meteorological parameters: with reference to the problem of objective forecasting*. NA.

Heidke, P. (1926). Measures of success and goodness of wind force forecasts by the gale-warning service. *Geografiska Annaler Series A Physical Geography, 8*, 301–349.

Hoffman, R. N., & Kalnay, E. (1983). Lagged average forecasting, an alternative to Monte Carlo forecasting. *Tellus, 35A*, 100–118.

Hollingsworth, A. (1980). *An experiment in Monte Carlo forecasting. Workshop on stochastic-dynamic forecasting*. ECMWF, Shinfield Park.

Hopson, T. M. (2014). Assessing the ensemble spread–error relationship. *Monthly Weather Review, 142*(3), 1125–1142.

Houtekamer, P. L., Lefaivre, L., Derome, J., Ritchie, H., & Mitchell, H. L. (1996). A system simulation approach to ensemble prediction. *Monthly Weather Review, 124*, 1225–1242.

Jayakumar, A., Mohandas, S., George, J. P., Duttta, D., Routray, A., Prasad, S. K., Sarkar, A., & Mitra, A. K. (2021). NCUM Regional Model Version 4 (NCUM-R: V4), NMRF/TR/03/2021.

Johnson, C., & Bowler, N. (2009). On the reliability and calibration of ensemble forecasts. *Monthly Weather Review, 137*(5), 1717–1720.

Jolliffe, I. T., & Stephenson, D. B. (Eds.). (2003). *Forecast Verification: A Practitioner's Guide in Atmospheric Science* (p. 240). Wiley.

Kalnay, E. (2003). *Atmospheric Modeling, Data Assimilation and Predictability*. Cambridge University Press.

Kalnay, E. & Toth, Z. (1996). The breeding method. In *Proceedings of the Seminar on Predictability*, held at ECMWF on 4–8 September 1995. (Available from ECMWF, Shinfield Park, Reading, Berkshire RG29AX, e.kooij@ecmwf.int)

Kaur, M., Krishna, R. P. M., Joseph, S., Dey, A., Mandal, R., Chattopadhyay, R., Sahai, A. K., Mukhopadhyay, P., & Abhilash, S. (2020). Dynamical downscaling of a multimodel ensemble prediction system: Application to tropical cyclones. *Atmospheric Science Letters, 21*(8), e971.

Kay, J. K., Kim, H. M., Park, Y. Y., & Son, J. (2013). Effect of doubling the ensemble size on the performance of ensemble prediction in the warm season using MOGREPS implemented at the KMA. *Advances in Atmospheric Sciences, 30*(5), 1287–1302.

Kotal, S. D., & Roy Bhowmik, S. K. (2011). A multimodel ensemble (MME) technique for cyclone track prediction over the North Indian Sea. *Geofizika, 28*(2), 275–291.

Kotal, S. D., Bhattacharya, S. K., & Roy Bhowmik, S. K. (2014). Development of NWP based objective cyclone prediction system (CPS) for North Indian Ocean tropical cyclones—evaluation of performance. *Tropical Cyclone Research and Review, 3*(3), 162–177.

Kotal, S. D., Bhattacharya, S., Roy Bhowmik, S. K., & Kundu, P. K. (2015). Verification of forecasts of IMD NWP based cyclone prediction system (CPS) for cyclones over the north Indian seas during 2013. *Mausam, 66*(3), 497–510.

Kühnlein, C., Keil, C., Craig, G. C., & Gebhardt, C. (2014). The impact of downscaled initial condition perturbations on convective-scale ensemble forecasts of precipitation. *Quarterly Journal of the Royal Meteorological Society, 140*, 1552–1562.

Kumar, A., Das, S., & Panda, S. K. (2022). Numerical simulation of a widespread lightning event over north India using an ensemble of WRF modeling configurations. *Journal of Atmospheric and Solar-Terrestrial Physics, 241*, 105984.

Kumar, S., Jayakumar, A., Bushair, M. T., Prakash J.B., George, G., Lodh, A., Rani, S. I., Mohandas, S., George, J. P., & Rajagopal, E. N. (2018). Implementation of new high resolution NCUM analysis-forecast system in Mihir HPCS.

Lalaurette, F. (2003). Early detection of abnormal weather conditions using a probabilistic extreme forecast index. *Quarterly Journal of Royal Meteorological Society, 129*, 3037–3057. https://doi.org/10.1256/qj.02.152

Leith, C. E. (1974). Theoretical skill of Monte Carlo forecasts. *Monthly Weather Review, 102*, 409–418.

Lepore, C., Tippett, M. K., & Allen, J. T. (2018). CFSv2 monthly forecasts of tornado and hail activity. *Weather and Forecasting, 33*(5), 1283–1297.

Lewis, J. M. (2005). Roots of ensemble forecasting. *Monthly Weather Review, 133*, 1865–1885.

Lorenz, E. N. (1965). A study of the predictability of a 28-variable atmospheric model. *Tellus, 17*, 321–333.

Ma, J., Zhu, Y., Wobus, R., & Wang, P. (2012). An effective configuration of ensemble size and horizontal resolution for the NCEP GEFS. *Advances in Atmospheric Sciences, 29*(4), 782–794.

MacLeod, D., O'Reilly, C., Palmer, T., & Weisheimer, A. (2018). Flow dependent ensemble spread in seasonal forecasts of the boreal winter extratropics. *Atmospheric Science Letters, 19*(5), e815.

Magnusson, L., Haiden, T., & Richardson, D. (2014). Verification of extreme weather events: Discrete predictands. European Centre for Medium-Range Weather Forecasts.

Mamgain A., Sarkar A., Dube A., Arulalan T., Chakraborty P., John. P. G., & ajagopal E. N. (2018). Implementation of very high resolution (12 km) global ensemble prediction system at NCMRWF and its initial validation, NMRF/TR/02/2018.

McCabe, A., Swinbank, R., Tennant, W., & Lock, A. (2016). Representing model uncertainty in the Met Office convection-permitting ensemble prediction system and its impact on fog forecasting. *Quarterly Journal of Royal Meteorological Society, 142*, 2897–2910.

McCaul, E. W., Goodman, S. J., LaCasse, K. M., & Cecil, D. J. (2009). Lightning threat using cloud-resolving model simulations. *Weather Forecasting, 24*, 709–729.

Mesinger, F. (2008). Bias adjusted precipitation threat scores. *Advances in Geosciences, 16*, 137–142.

Mohan, G. M., Vani, K. G., Hazra, A., Mallick, C., Chaudhari, H. S., Pokhrel, S., Pawar, S. D., Konwar, M., Saha, S. K., Das, S. K., & Deshpande, S. (2021). Evaluating different lightning parameterization schemes to simulate lightning flash counts over Maharashtra, India. *Atmospheric Research, 255*, 105532.

Mohapatra, M., Kumar, N., Sharma, M., Devi, S. S., Senroy, S., Devi, K. S., & Singh, K.K. (2022). Short to medium range weather forewarning system in India. In *Social and Economic Impact of Earth Sciences* (pp. 1–29). Springer Nature Singapore.

Molteni, F., Buizza, R., Palmer, T. N., & Petroliagis, T. (1996a). The ECMWF ensemble prediction system: Methodology and validation. *Quarterly Journal of the Royal Meteorological Society, 122*, 73–119.

Molteni, F., Buizza, R., Palmer, T. N., & Petroliagis, T. (1996b). The new ECMWF ensemble prediction system: Methodology and validation. *Quarterly Journal of Royal Meteorological Society, 122*, 73–119.

Mukhopadhyay, P., Prasad, V. S., Krishna, R. P. M., Deshpande, M., Ganai, M., Tirkey, S., Sarkar, S., Goswami, T., Johny, C. J., Roy, K., & Mahakur, M. (2019). Performance of a very high-resolution global forecast system model (GFS T1534) at 12.5 km over the Indian region during the 2016–2017 monsoon seasons. *Journal of Earth System Science, 128*, 1–18.

Mukhopadhyay, P., Krishna, R. P. M., Deshpande, M., Ganai, M., Tirkey, S., Goswami, T., Sarkar, S., Malviya, S., Kanase, R., Roy, K., & Prasad, V.S. (2018). High resolution (12.5 km) ensemble prediction system based on GEFS: Evaluation of extreme precipitation events over Indian region.

Mukhopadhyay, P., Goswami, T., Tirkey, S., Kanase, R.D., Phani Murali Krishna, R., & Deshpande, M. (2022). Severe weather events over the Indian region: Insights from ensemble prediction system. In *Social and Economic Impact of Earth Sciences* (pp. 49–59). Springer Nature Singapore.

Mullen, S. L., & Buizza, R. (2002). The impact of horizontal resolution and ensemble size on probabilistic forecasts of precipitation by the ECMWF ensemble prediction system. *Weather and Forecasting, 17*(2), 173–191.

Murphy, A. H. (1973). A new vector partition of the probability score. *Journal of Applied Meteorology, 12*(4), 595–600.

Murphy, A. H., & Epstein, E. S. (1989). Skill scores and correlation coefficients in model verification. *Monthly Weather Review, 117*(3), 572–582.

Murphy, A. H., & Winkler, R. L. (1987). A general framework for forecast verification. *Monthly Weather Review, 115*(7), 1330–1338.

Murphy, A. H., & Winkler, R. L. (1992). Diagnostic verification of probability forecasts. *International Journal of Forecasting, 7*(4), 435–455.

Murphy, A. H. (1997). Forecast" verification. In *Economic Value of Weather and Climate Forecasts* (p. 19).

Nayak, H. P., & Mandal, M. (2014). Analysis of stability parameters in relation to precipitation associated with pre-monsoon thunderstorms over Kolkata, India. *Journal of Earth System Science, 123*, 689–703.

Nuissier, O., Joly, B., Vie, B., & Ducrocq, V. (2012). Uncertainty of lateral boundary conditions in a convective-permitting ensemble: A strategy of selection for Mediterranean heavy precipitation event. *Natural Hazards, 12*, 2993–3011.

Palmer, T. N., Doblas-Reyes, F. J., Hagedorn, R., & Weisheimer, A. (2005). Probabilistic prediction of climate using multi-model ensembles: From basics to applications. *Philosophical Transactions of the Royal Society B: Biological Sciences, 360*(1463), 1991–1998.

Palmer, T. N., Molteni, F., Mureau, R., Buizza, R., Chapelet, P., & Tribbia, J. (1993). Ensemble prediction. In *Proceedings of the ECMWF Seminar on Validation of Models over Europe* (Vol. 1, pp. 21–66). ECMWF, Shinfield Park, Reading, UK.

Peirce, C. S. (1884). The numerical measure of the success of predictions. *Science, 93*, 453–454.

Prasad, S. K., Sarkar, A., Mamgain, A., & Rajagopal, E. N. (2019). *Implementation of NCMRWF regional ensemble prediction system (NEPS-R)*. NMRF/TR/09.

Rajeevan, M., Madhulatha, A., Rajasekhar, M., Bhate, J., Kesarkar, A., & Rao, B. A. (2012). Development of a perfect prognosis probabilistic model for prediction of lightning over south-east India. *Journal of Earth System Science, 121*, 355–371.

Raynaud, L., & Bouttier, F. (2017). The impact of horizontal resolution and ensemble size for convective-scale probabilistic forecasts. *Quarterly Journal of the Royal Meteorological Society, 143*, 3037–3047.

Roberts, B., Gallo, B. T., Jirak, I. L., & Clark, A. J. (2022). The High Resolution Ensemble Forecast (HREF) system: Applications and performance for forecasting convective storms. *Authorea Preprints*.

Roebber, P. J., Schultz, D. M., Colle, B. A., & Stensrud, D. J. (2004). Toward improved prediction: High-resolution and ensemble modeling systems in operations. *Weather and Forecasting, 19*(5), 936–949.

Roy, S. S., Mohapatra, M., Tyagi, A., & Roy Bhowmik, S. K. (2019). A review of Nowcasting of convective weather over the Indian region. *Mausam, 70*(3), 465–484.

Roy Bhowmik, S. K., & Durai, V. R. (2010). Application of multimodel ensemble techniques for real time district level rainfall forecasts in short range time scale over Indian region. *Meteorology and Atmospheric Physics, 106*, 19–35.

Roy Bhowmik, S. K., & Durai, V. R. (2012). Development of multimodel ensemble based district level medium range rainfall forecast system for Indian region. *Journal of Earth System Science, 121*, 273–285.

Sahu, R. K., & Tyagi, B. (2022). Spatial variation of thermodynamic indices over north-east India during pre-monsoon thunderstorm season. *Journal of Atmospheric and Solar-Terrestrial Physics, 232*, 105868.

Sahu, R. K., Dadich, J., Tyagi, B., & Vissa, N. K. (2020). Trends of thermodynamic indices thresholds over two tropical stations of north-east India during pre-monsoon thunderstorms. *Journal of Atmospheric and Solar-Terrestrial Physics, 211*, 105472.

Sanchez, C., Williams, K. D., & Collins, M. (2016). Improved stochastic physics schemes for global weather and climate models. *Quarterly Journal of the Royal Meteorological Society, 142*(694), 147–159.

Sandeep, A., Jayakumar, A., Sateesh, M., Mohandas, S., Prasad, V. S., & Rajagopal, E. N. (2021). Assessment of the efficacy of lightning forecast over India: A diagnostic study. *Pure and Applied Geophysics, 178*, 205–222.

Saranya Ganesh, S., Abhilash, S., Sahai, A. K., Joseph, S., Chattopadhyay, R., Mandal, R., Dey, A., & Phani, R. (2019). Genesis and track prediction of pre-monsoon cyclonic storms over North Indian Ocean in a multi-model ensemble framework. *Natural Hazards, 95*, 823–843.

Sarkar, A., Chakraborty, P., Singh, B. K., & Bhatla, R. (2017a). Simulation of Nor'Wester by using a nested mesoscale model. *International Journal, 4*(02), 64–73.

Sarkar, A., Dutta, D., Chakraborty, P., & Das, S. (2017b). Numerical diagnosis of situations causing heavy rainfall over the Western Himalayas. *Modeling Earth Systems and Environment, 3*, 515–531.

Sarkar, A., Dutta, D., Chakraborty, P., Dutta, S. K., Majumdar, S., Payra, S., & Bhatla, R. (2020). Influence of cumulus convection and cloud microphysics parameterizations on the prediction of western disturbances. *Meteorology and Atmospheric Physics, 132*, 413–426.

Sarkar, A., Kumar, S., Dube, A., Prasad, S. K., Mamgain, A., Chakraborty, P., Ashrit, R., & Mitra, A. K. (2021). Forecasting of tropical cyclone using global and regional ensemble prediction systems of NCMRWF: A review. *Mausam, 72*(1), 551–561.

Sarkar, A., Prasad, S. K., Mamgain, A., Dube, A., Chakraborty, P., Kumar, S., Karunasagar, S., Thota, M. S., Shanker, G., Ashrit, R., & Mitra, A. K. (2022). Probabilistic forecasting of Super Cyclone 'Amphan' using NCMRWF global and regional ensemble prediction systems. *Journal of Earth System Science, 131*(4), 260.

Sarkar, A., Chakraborty, P., John, P.G., & Rajagopal, E. N. (2016). Implementation of unified model based ensemble prediction system at NCMRWF (NEPS). NMRF/TR/02/2016.

Schaefer, J. T. (1990). The critical success index as an indicator of warning skill. *Weather and Forecasting, 5*(4), 570–575.

Schulzweida, U., Kornblueh, L., & Quast, R. (2006). CDO user guide. *Climate Data Operators, Version, 1*(6), 205–209.

Schwartz, C. S., Romine, G. S., Fossell, K. R., Sobash, R. A., & Weisman, M. L. (2017). Toward 1-km ensemble forecasts over large domains. *Bulletin of American Meteorological Society, 145*, 2943–2969.

Sen Roy, S. (2009). A spatial analysis of extreme hourly precipitation patterns in India. *International Journal of Climatology: A Journal of the Royal Meteorological Society, 29*(3), 345–355.

Sen Roy, S., Saha, S. B., Roy Bhowmik, S. K., & Kundu, P. K. (2019). Diurnal variability of convection over northwest Indian subcontinent observed by the Doppler weather radar data. *Meteorology and Atmospheric Physics, 131*, 1577–1604.

Sen Roy, S., Sharma, P., Sen, B., Sathi Devi, K., Sunitha Devi, S., Gopal, N. K., Kumar, N., Mishra, K., Katyar, S., Singh, S. P., & Balakrishnan, S. (2021). A new paradigm for short-range

forecasting of severe weather over the Indian region. *Meteorology and Atmospheric Physics, 133*, 989–1008.
Sharpe, M. A., Bysouth, C. E., & Stretton, R. L. (2018). How well do Met Office post-processed site-specific probabilistic forecasts predict relative-extreme events? *Meteorological Applications, 25*(1), 23–32.
Sinha, M., Yadav, R. K., & Chakraborty, P. (2016). Island modeling using unstructured grid during a tropical storm. *International Journal of Oceanography, 2016*(1), 5834572.
Sinha, M., Jha, S., & Chakraborty, P. (2020). Indian Ocean wind speed variability and global teleconnection patterns. *Oceanologia, 62*(2), 126–138.
Srivastava, K., Roy Bhowmik, S. K., Sen Roy, S., Thampi, S. B., & Reddy, Y. K. (2010). Simulation of high impact convective events over Indian region by ARPS model with assimilation of Doppler weather radar radial velocity and reflectivity. *Atmosfera, 23*(1), 53–73.
Srivastava, K., Lau, S. Y., Yeung, H. Y., Cheng, T. L., Bhardwaj, R., Kannan, A. M., Roy Bhowmik, S. K., & Singh, H. (2012). Use of SWIRLS nowcasting system for quantitative precipitation forecast using Indian DWR data. *Mausam, 63*(1), 1–16.
Stanski, H. R., Wilson, L. J., & Burrows, W. R. (1989). *Survey of common verification methods in meteorology* (114pp.). WMO World Weather Watch Technical Report 8.
Stensrud, D. J., Brooks, H. E., Du, J., Tracton, M. S., & Rogers, E. (1999). Using ensembles for short-range forecasting. *Monthly Weather Review, 127*, 433–446.
Tandon, N. F., Zhang, X., & Sobel, A. H. (2018). Understanding the dynamics of future changes in extreme precipitation intensity. *Geophysical Research Letters, 45*(6), 2870–2878.
Tennant, W. (2015). Improving initial condition perturbations for MOGREPS-UK. *Quarterly Journal of the Royal Meteorological Society, 141*, 2324–2336.
Tennant, W. J., Shutts, G. J., Arribas, A., & Thompson, S. A. (2011). Using a stochastic kinetic energy backscatter scheme to improve MOGREPS probabilistic forecast skill. *Monthly Weather Review, 139*(4), 1190–1206.
Thuburn, J. (2016). ENDGame: The new dynamical core of the met office weather and climate prediction model (pp. 27–33). UK success stories in Industrial Mathematics.
Toth, Z., & Kalnay, E. (1993). Ensemble forecasting at the NMC: The generation of perturbations. *Bulletin of American Meteorological Society, 74*, 2317–2330.
Toth, Z., Talagrand, O., & Zhu, Y. (2006). The attributes of forecast systems: A general framework for the evaluation and calibration of weather forecasts. *Predictability of Weather and Climate, 584*, 595.
Tracton, M. S., Du, J., Toth, Z., & Juang H. (1998). Short-range ensemble forecasting (SREF) at NCEP/EMC. In *12th Conference on Numerical Weather Prediction* (pp. 269–272). Phoenix, American Meteorological Society.
Tyagi, A. (2007). Thunderstorm climatology over Indian region. *Mausam, 58*(2), 189–212.
Tyagi, B., Naresh Krishna, V., & Satyanarayana, A. N. V. (2011). Study of thermodynamic indices in forecasting pre-monsoon thunderstorms over Kolkata during STORM pilot phase 2006–2008. *Natural Hazards, 56*, 681–698.
Tyagi, A., Sikka, D. R., Goyal, S., & Bhowmick, M. (2012). A satellite based study of pre-monsoon thunderstorms (Nor'westers) over eastern India and their organization into mesoscale convective complexes. *Mausam, 63*(1), 29–54.
Tyagi, B., Satyanarayana, A. N. V., & Vissa, N. K. (2013). Thermodynamical structure of atmosphere during pre-monsoon thunderstorm season over Kharagpur as revealed by STORM data. *Pure and Applied Geophysics, 170*, 675–687.
Umakanth, N., Satyanarayana, G. C., Naveena, N., Srinivas, D., & Rao, D. B. (2021). Statistical and dynamical based thunderstorm prediction over southeast India. *Journal of Earth System Science, 130*, 1–18.
Vani, K. G., Mohan, G. M., Hazra, A., Pawar, S. D., Pokhrel, S., Chaudhari, H. S., Konwar, M., Saha, S. K., Mallick, C., Das, S. K., & Deshpande, S. (2022). Evaluation and usefulness of lightning forecasts made with lightning parameterization schemes coupled with the WRF model. *Weather and Forecasting, 37*(5), 709–726.

Walters, D., Brooks, M., Boutle, I., Melvin, T., Stratton, R., Vosper, S., Wells, H., Williams, K., Wood, N., Allen, T., & Bushell, A. (2017). The met office unified model global atmosphere 6.0/6.1 and JULES global land 6.0/6.1 configurations. *Geoscientific Model Development*, *10*(4), 1487–1520.

Wang, X., Bishop, C. H., & Julier, S. J. (2004). Which is better, an ensemble of positive–negative pairs or a centered spherical simplex ensemble? *Monthly Weather Review, 132 (7)*, 1590–1605. https://doi.org/10.1175/1520-0493(2004)132<1590:WIBAEO>2.0.CO;2

Wei, M., Toth, Z., Wobus, R., & Zhu, Y. (2008). Initial perturbations based on the ensemble transform (ET) technique in the NCEP global operational forecast system. *Tellus A: Dynamic Meteorology and Oceanography, 60*(1), 62–79.

Wyngaard, J. C. (1985). Structure of the planetary boundary layer and implications for its modeling. *Journal of Applied Meteorology and Climatology*, *24*(11), 1131–1142.

Zhang, Z. H. A. N., & Krishnamurti, T. N. (1999). A perturbation method for hurricane ensemble predictions. *Monthly Weather Review, 127*(4), 447–469.

Zhou, X., Zhu, Y., Hou, D., Fu, B., Li, W., Guan, H., Sinsky, E., Kolczynski, W., Xue, X., Luo, Y., & Peng, J. (2022). The development of the NCEP global ensemble forecast system version 12. *Weather and Forecasting, 37*(6), 1069–1084.

Zsoter, E. (2006). *Recent developments in extreme weather forecasting* (pp. 8–17). ECMWF Newsletter, No. 107, ECMWF, Reading, United Kingdom.

Lightning Forecasting and Utilization of AI/ML in Early Warnings

Rituparna Sarkar and Parthasarathi Mukhopadhyay

Abstract In recent years, the application of Artificial Intelligence (AI) in different sectors of society has become one of the most common technologies. As AI becomes a useful tool, incorporating AI in weather forecasts has also started rising. Lightning is one of India's leading causes of casualty associated with natural disasters. An early warning system for lightning forecast is of utmost importance. However, the atmospheric processes associated with lightning are yet to be fully understood, making it challenging to simulate in a numerical weather prediction (NWP) model for operational purposes. Considering all these factors, in this chapter, we have discussed an AI-based deep learning approach for lightning prediction. This new AI-based lightning prediction system uses the Indian Meteorology Department (IMD) Weather Research and Forecasting (WRF) model at 9 km resolution as input which presently do not have any lightning product. The Indian Institute of Tropical Meteorology (IITM) Lightning Location Network (LLN) observation data is used as target vector to train the model. Due to the high imbalance of rare and severe events (RSEs) like lightning, we have used a novel approach of two-auto encoder-based classification (C_{2AE}) to train a deep-learning algorithm to correctly encode meteorological features associated with lightning and non-lightning incidents and, when the trained C_{2AE} model is tested on a new dataset, it is able to correctly identify lightning events with statistical significance.

R. Sarkar · P. Mukhopadhyay
Indian Institute of Tropical Meteorology, Ministry of Earth Sciences, Pune 411008, India
e-mail: rituparna.sarkar@tropmet.res.in

R. Sarkar
Department of Atmospheric and Space Sciences, Savitribai Phule Pune University, Pune 411007, India

Present Address:
P. Mukhopadhyay (✉)
Earth and Environmental Sciences, Indian Institute of Science Education and Research Berhampur, Berhampur, India
e-mail: parthasarathi64@gmail.com

S. Das and W-K. Tao (eds.), *Severe Storms*,
https://doi.org/10.1007/978-981-97-7075-5_18

Keywords Lightning · Autoencoders · Early warning system · Imbalanced classification · Deep learning · LLN · WRF · IMD

1 Introduction

Although atmospheric processes have always affected human civilization, our collective understanding of severe storms streamlined slightly over a century ago (Brooks et al., 2019). From the beginning of the twentieth century, meteorologists worldwide tried to understand the large-scale processes driving deep convective systems and the microphysical processes that happen inside these systems. One of the earliest such works were by Desai (1950) and Desai and Rao (1954), outlining and classifying the thunderstorms called 'Nor'westers'. In consecutive decades, research in understanding and forecasting thunderstorms would rapidly increase (Maddox, 1987).

Atmospheric instability leads to the formation of cumulonimbus clouds around the globe. The spatiotemporal scales of cumulonimbus clouds determine the events' severity and potential hazardous nature (Maddox, 1987). When such convections produce hail larger than 1 inch and wind gusts exceeding 50 kt, they are called "severe thunderstorms" (Ellis & Miller, 2016). If the convective process can sustain itself to trigger congestus and rainfall within an area of 100 km, it is called a *mesoscale convective system* (Houze, 2004). Globally, the tropics experience the highest number of thunderstorms, with the Indian subcontinent experiencing the most intense storms during the premonsoon season (March–April–May) (Cecil et al., 2015; Zipser et al., 2006).

Within a thunderstorm, the updraft of ice and water particles beyond freezing level and the downdraft of cloud droplets, hails and graupels lead to collisions and charge separation. The downward mixed-phase hydrometeors gain a negative charge between freezing level to −40 °C, and the upward-moving ice crystals become positively charged at the cirriform cloud top. With two oppositely charged regions, the cumulus cloud acts as a dipole (Dash & Wettlaufer, 2003; Gaskell & Illingworth, 1980; Helsdon et al., 2002; Illingworth, 1985; MacGorman et al., 2001; Reynolds et al., 1957; Saunders, 2008; Williams, 1985, 1988, 1989; Williams et al., 1991). As the potential difference between the positive and negative pole goes beyond a critical point, a discharge happens in the form of lightning flash where a bidirectional ionized channel propagates as a positive and negative leader away from the initiation point with a positive leader depositing 20–30 Coulumbs in the earth surface (Kasemir, 1960; Mansell et al., 2002; Mason, 1953; Williams & Montanyà, 2019). As the groundward leader reaches the surface, a channel of positive charge from the surface is created, leading to new step formation and subsequent return strokes (Zeng et al., 2016). Although cloud-to-ground lightning poses a significant threat, it is increasingly evident that intra-cloud lightning begins during the growth phase of the storm, indicating the importance of total lightning observation for improving the skill of early warning systems (MacGorman et al., 2011; Meng et al., 2019; Wu et al., 2016).

With the launch of Tropical Rainfall Measuring Mission (TRMM) Lightning Imaging Sensor (LIS) (Cecil et al., 2014), a significant amount of study has examined lightning climatology and its variation over different spatiotemporal scales. Due to high land–ocean contrast, it is evident that the northern hemisphere tropics experience the most intense storms and, as an extent, lightning, with the Indian subcontinent experiencing most lightning during the March–April–May months (Murugavel et al., 2012; Tinmaker et al., 2022; Williams, 2009; Zipser et al., 2006). As the convective events are driven by diurnal cycles, it is consistently evident that the diurnal cycle of lightning peaks during the afternoon except for lightning events from the orographically driven storms and land-sea breeze-influenced coastal storms (Sen Roy & Balling, 2013).

Due to a strong correlation between temperature and lightning, many studies have looked into lightning in a changing climate (Carey et al., 2003; Gilmore & Wicker, 2002; Smith et al., 2000; Williams et al., 2004, 2005, 2019; Williams, 2005) with recent publications showing the possibility of an increase in intense storms days over the subcontinent (Glazer et al., 2021). With the possibility of more lightning events and an increase in lightning-related deaths over the subcontinent in the past two decades (Cooper & Holle, 2019; Ray et al., 2021; Srivastava, 2022), the Indian government has taken many initiatives to improve lightning warning system with a significant increase in the skills of nowcasting across the country (Sen Roy et al., 2021).

Attempts to forecast lightning have been going on for decades (MacGorman et al., 2001; Price & Rind, 1993, 1994; Rawlins, 1982; Takahashi, 1987; Ziegler & Macgorman, 1994), which can be classified into two different categories:

1. Lightning forecast using proxy variables (Barthe et al., 2010; Lopez, 2016),
2. Lightning forecasts based on statistical techniques (Rajeevan et al., 2012; Shafer & Fuelberg, 2006).

There are also attempts to explicitly simulate lightning at the grid level (Mansell et al., 2002), while others focus on deriving indices as a proxy of lightning flash (Yair et al., 2010).

Due to the average lifetime of a storm being less than 30 min (Wilson et al., 1998) with the associated potential of hazard, nowcasting thunderstorms and lightning has been a priority for a long time. Historically, the nowcasting is based on the RADAR observation and climatological data of storms. In India, an automated decision tree-based algorithm using RADAR, satellite, ground-based and upper air observation and analysis of past 24 h forecast is used to give nowcast over 600 stations 1–3 h in advance (Roy et al., 2021; Sen Roy et al., 2021). Along with thunderstorm nowcasting, an early warning system for cloud-to-ground lightning has also greatly interested meteorologists. With the increase in real-time lightning monitoring (Bürgesser, 2017; Mudiar et al., 2018; Pawar et al., 2017; Utsav et al., 2019), early warnings of lightning using single-station sensor data are also attempted (Bennett, 2018; Mostajabi et al., 2019). Despite significant improvement in nowcasting, the operational very short-range forecasting has lag and difficulty in predicting severe storm events (Vani et al., 2022). In 2021, the National Disaster Management Authority (NDMA) formed a protocol for early warning dissemination of thunderstorms and lightning to minimize lightning casualty by emphasizing the importance of lightning warning one hour in advance (*A Protocol for Early Warning Dissemination on Thunderstorm and Lightning*, 2021). In their 2022 bulletin, WMO emphasizes the possibilities of using a Machine Learning (ML) based algorithm to develop an early warning system using existing climatology data, real-time observation and forecast data ("Vol. 71(1): Early Warning and Anticipatory Action," 2022).

It has been shown that a hybrid system combining deterministic numerical weather prediction (NWP) model and machine learning techniques can better predict atmospheric processes with less computational power (Pathak et al., 2018; Reichstein et al., 2019). Unsupervised deep learning techniques have been used to forecast different events associated with earth system processes (Hagos et al., 2020; Ou et al., 2019; Saha et al., 2016; Xayasouk et al., 2020). Chinese Academy of Meteorological Sciences has developed operational lightning nowcasting using a machine learning approach that combines RADAR, satellite, and ground-based observation data (Meng et al., 2019). Similar storm nowcasting is deployed for severe storm events over Great Planes, USA (Mecikalski et al., 2021). For thunderstorm and lightning forecasts with 24 h lead time, deterministic model output has been used as input for a machine learning-based prediction system with significant confidence (Collins & Tissot, 2016; Collins & Tissot, 2015; Hewage et al., 2020; Kamangir et al., 2020; Putro, 2020; Wang et al., 2014).

As we have discussed so far, there is insufficient understanding of the physical processes behind thunderstorms and their association with lightning. Yet lightning-related deaths have been on increasing trend in India. This, along with the possibility of an increase in intense storms and lightning in a changing climate, has made lightning an essential subject for policymakers. However, the existing operational forecasts for lightning using NWP models have limitations, for which many now focus on data-driven lightning forecasts. Although AI-based lightning prediction systems are tested and operational in many countries, India has yet to deploy its own.

Therefore, in this chapter, we aim to develop an AI-based lightning prediction model that uses the existing IMD WRF operational forecast as input with IITM LLN dataset in training the model. For this purpose, we have taken 9 km resolution WRF model outputs provided by IMD which presently does not have any lightning parameterization. Hence the new AI-based lightning prediction system proposed in this chapter will help provide a lightning forecast at 9 km resolution. In Data and Methodology (Sect. 2), we first gave a detail of IMD WRF data and IITM LLN data, data preprocessing, followed by model description, experiment setup and model verification techniques. In Results and Discussion (Sect. 3), we have a detailed discussion of the model and statistical evaluation. We then summarise our findings and further research in Conclusions (Sect. 4).

2 Data and Methodology

2.1 Data

As discussed in Sect. 1, to develop an early warning system using IMD WRF output, we have taken IMD WRF 9 km resolution output as our input. Followed by studies (Mukhopadhyay et al., 2010; Taraphdar et al., 2010) showing advanced research WRF (ARW) model capability in resolving monsoonal rainfall, IMD operationally adopted WRF at 9 km resolution since 2010, whose model dynamic and physics are described in detail by Das et al. (2016). WRF at this resolution is also used for sub-basin rainfall forecasts (Das et al., 2022). ARW WRFv3.4 is run twice daily (00 UTC and 12 UTC) with IITM GFS T1543 as the initial condition with 72 h of lead time. For this study, we have used daily 00 UTC forecast output at 24-h lead time.

We have used IITM LLN observed lightning flash density to train the deep learning model as our target vector. IITM LLN was initially operational in Maharashtra, India, as the Maharashtra Lightning Location Network (Pawar et al., 2017). Since 2018, it has expanded to cover India with over 82 active locations (Sarkar et al., 2022).

We have chosen a study region between 15.5°N–22.5°N and 72.5°E–81.5°E. Figure 1 details the study domain and the distribution of 9 km $\times$ 9 km grids of IMD WRF output. The locations of IITM LLN sensors are shown as blue circular dots in Fig. 2. To coincide with IMD WRF 9 km output with IITM LLN observation data, we chose 2019 and 2020 as our study period. IITM LLN data has been gridded over the study region with 9 km resolution with hourly intervals to calculate the total lightning flash density in the last hour (in a flash km^{-2} h^{-1}). Figure 3 shows the mean hourly lightning flash density (flash count 10^{-3} km^{-2} day^{-1}) between 2019 and 2020. The distribution of lightning flashes shows the influence of the north–south distribution of Western Ghat on deep convection and associated lightning.

Therefore, we have IMD WRF output as our input features and gridded IITM LLN observation as the target output, totaling 6,052,167 data points. As our goal is to develop a robust deep-learning-based algorithm that can classify lightning and non-lightning events, we have removed latitude, longitude and datetime information from the dataset. This forces the model to learn the information from the features associated with lightning and non-lightning events independent of space and time. In order to prepare IITM LLN dataset for training we have converted observed total lightning flash density into dichotomous "yes"/ "no" observation. This is achieved by converting greater than zero total lightning flash count as "yes" and zero total flash count as "no" for each data point. Figure 4 shows the distribution of lightning (red) and non-lightning (blue) events for every 1000 datapoint of our dataset. Out of 6,052,167 data points, only 12,130 data points are associated with lightning, which is only 0.20% of our total dataset. This identifies the highly imbalanced nature of lightning data, indicating lightning as a rare and severe event (RSE), making it harder to forecast.

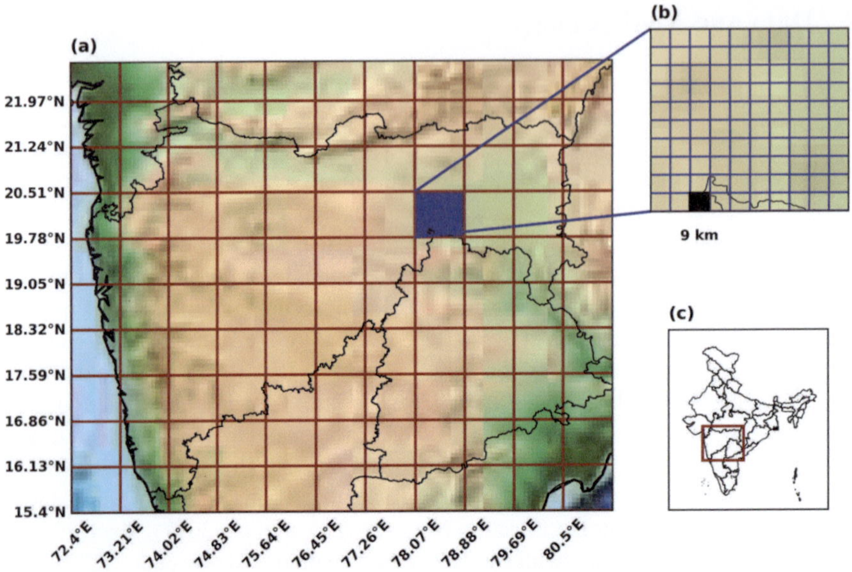

Fig. 1 **a** Domain selected for the present study (adapted from Pawar et al. (2017)). Each red box represents 100 grid domains as shown in **b** zoomed over blue-shaded box gridded with a 9 km × 9 km grid **c** shows domain location over India

Fig. 2 IITM LLN station locations shown with blue circles over study domain

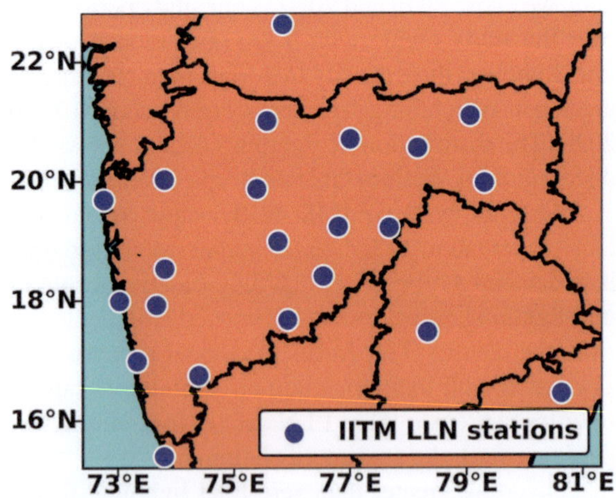

2.2 *Deep Learning Autoencoder-Based Lightning Prediction Model*

Autoencoders are unsupervised deep learning architecture consisting of an encoder and a decoder (Bengio, 2009). Autoencoders first encode the input layer onto a hidden layer, which is then reconstructed back to the input layer (Rumelhart et al., 1986).

Fig. 3 Mean hourly lightning flash density (flash count $10^{-3}\ km^{-2}\ h^{-1}$) between 2019 and 2020, obtained from IITM LLN and gridded for 9 km × 9 km resolution

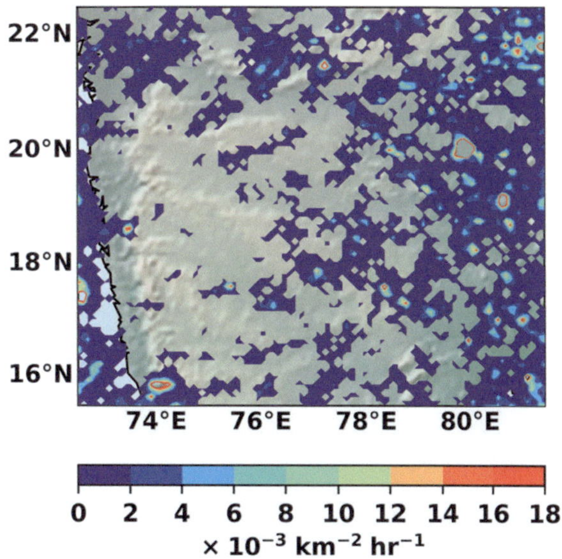

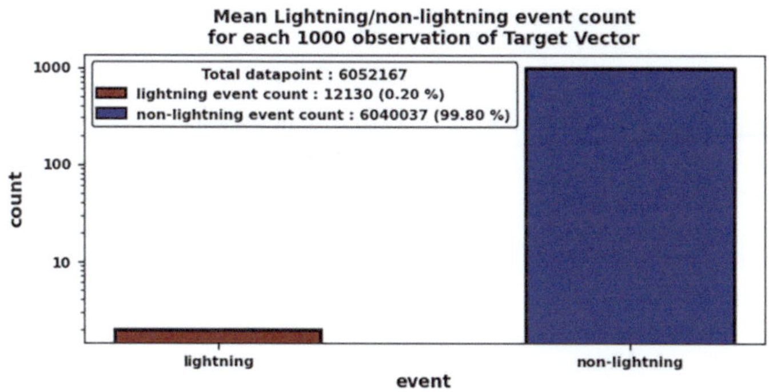

Fig. 4 Barplot showing number lightning (red) and non-lightning (blue) events per each 1000 datapoint of total 6,052,167 data points

Using multilayer perception, autoencoders attempt to minimize the error between the input and reconstructed input (Bourlard & Kamp, 1988). Since its inception, autoencoders have become the building block for unsupervised deep-learning neural networks (Kalchbrenner & Blunsom, 2013; Ranzato et al., 2007; Rifai et al., 2011; Schölkopf et al., 1998; Vincent et al., 2010). Figure 5 shows a schematic of an autoencoder with one hidden layer. The encoder maps input x to hidden layer y using weight w:

$$y = s(wx + b) \tag{1}$$

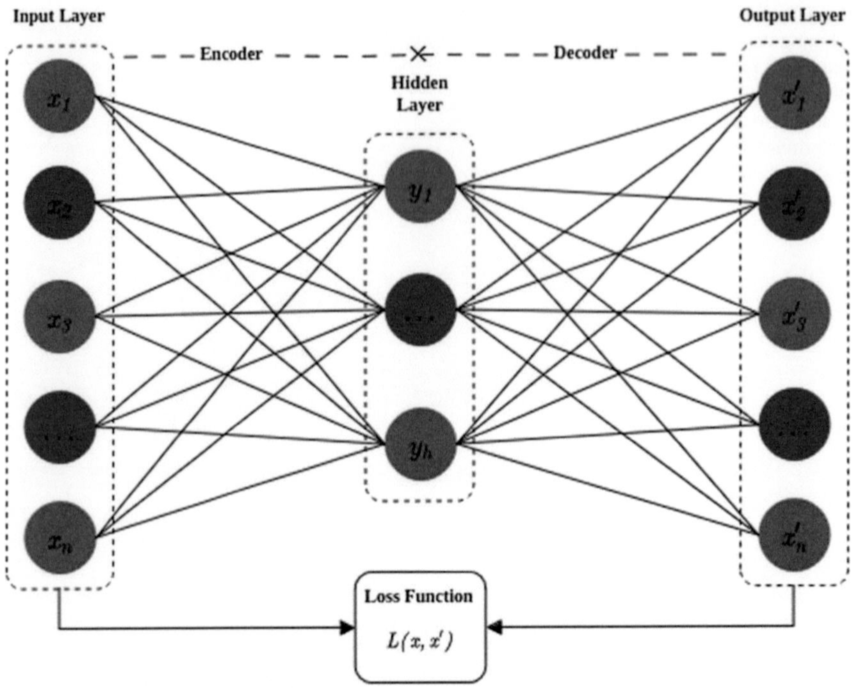

Fig. 5 Schematic of autoencoders in C_{2AE}, x, y and x' with input layers (x), hidden layer (y) and output layer x'. Loss function (L) is calculated between input and output layer

where b is the offset matrix and s non-linearity. The decoder then reprojects y back to the input plane as:

$$x\prime = s(w\prime y + b\prime) \tag{2}$$

where $w\prime$ and $b\prime$ are the weight matrix and offset parameter, respectively, for the decoder.

$$L = f(x, x\prime) \tag{3}$$

The reconstruction error can be expressed as:

With unsupervised training through multilayer perception, the autoencoder minimizes the L, also known as the loss function. If L is cross-entropy, then:

$$L = -\sum\nolimits_{k=1}^{n}\left[\frac{x_k}{\log x_k\prime} + \frac{1 - x_k}{\log 1 - x_k\prime}\right] \tag{4}$$

One major constraint for deep learning-based classification methods like autoencoders is the "curse of imbalance training set"(Kubát & Matwin, 1997), where the dataset used for training not being distributed equally among different classes. This leads to models only learning to correctly classify the majority class. As seen in Fig. 4, only 0.20% of the total dataset represents lightning events, making the dataset extremely skewed. To solve this issue, many studies have developed different preprocessing techniques such as random oversampling, random under sampling (Fernández et al., 2018; Kulkarni et al., 2020), Synthetic Minority Oversampling or SMOTE (Chawla et al., 2011). The main drawback of adopting any of these methods is either the removal of critical information (undersampling), overfitting (oversampling) or added noise (SMOTE) (Fernández et al., 2018). To address this, some studies address the imbalance at the algorithm level, such as one-class classification (Sotiropoulos et al., 2014), anomaly detection based on support vector (Schölkopf et al., 1998), interpretable anomaly detection such as local interpretable model-agnostic explanations (Ribeiro et al., 2016), Shapley additive explanations (Lundberg & Lee, 2017).

For our problem, we have taken reference from Tomescu et al. (2021), which uses a two-autoencoder-based classifier model (C_{2AE}). In this model, two separate autoencoders (AE_+ and AE_-), one of which (AE_+) is trained on the positive dataset (lightning), while the other (AE_-) is trained on the negative dataset (non-lightning). As autoencoders are built to learn the hidden features of any given input and reproduce the input as closely as possible, a trained AE_+ will have less error in reproducing lightning datasets while AE_- will have more significant errors. The error output will be reversed for both the autoencoder for non-lightning cases. After training both autoencoders, a new test dataset is introduced to the model. Unlike the training set, the test set consists of lightning (+) and non-lightning (−) events. As the autoencoders reconstruct the input separately, a new class c is calculated as given by:

$$c = \text{argmin}_{c \in \{+,-\}} L_c(x, x\prime) \tag{5}$$

where the loss functions,L_+ and L_- from AE_+ and AE_-, respectively. If $c = 1$, then we assign the event as lightning and if $c = 0$, then as non-lightning. Figure 6 details the dataset preprocessing and model setup, including unsupervised training using lightning and non-lightning training sets and supervised fine-tuning using the combined test set.

2.3 Experiment Set Up

From IMD WRF generated output, 26 parameters associated with convective, dynamics, and physical processes have been selected as input features or predictors. Table 1 shows the names of all the features selected, their long name, units, as well as processes these parameters are associated with. Since not every selected feature has the same dimension, features on separate pressure levels are considered as individual parameters. This is also important since each vertical level plays different roles in convective events. Therefore, the total number of features used as preliminary

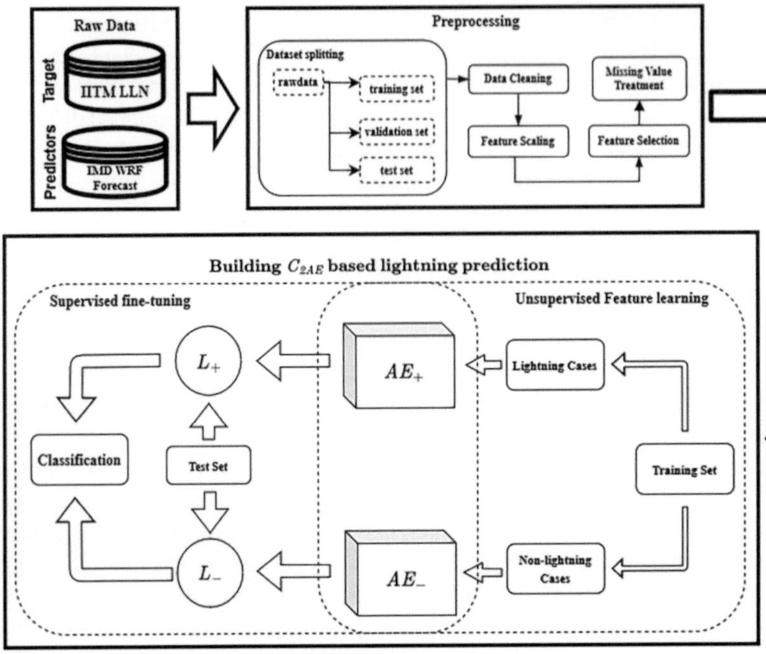

Fig. 6 Schematic showing data preprocessing, unsupervised feature learning and supervised fine-tuning of C_{2AE} with raw data consisting of IMD WRF Forecast as predictors and IITM LLN as target vector

input is 107, with lightning as our target vector. Since many of the parameters in our dataset can have hidden dependencies with each other, this can lead to overfitting of the deep learning model. To avoid this, we have calculated the Spearman rank correlation (Spearman, 1904) between the features to identify their co-dependency. In Fig. 7, we have plotted a dendrogram of all 107 parameters with Spearman Rank Correlation as a distance matrix on the x-axis. We have chosen 0.5 as our threshold, which gives us separate 4 clusters (pink, cyan, yellow and red). Black shows the cluster threshold. From this analysis, we have chosen 47 independent features from our original 107 parameters.

For model architecture, we need to build two separate autoencoders, where each autoencoder will take input features and reconstruct the information in the hidden layer, as seen in Fig. 5. Following Bourlard & Kamp (1988), we have done principal component analysis (PCA) for the selected 47 feature parameters to define the number of neurons in our hidden layers. The Fig. 8 scree plot shows that the first 20 components have 90% cumulative variance. Hence, we can retain 90% of information even when the feature dimension is reduced from 47 to 20. Therefore, we have selected one hidden layer with 20 neurons for our model.

Table 1 Predictors selected for C_{2AE}

Parameter name	Long name	Units	Processes associated with
	Surface		
ACPCPsfc	Surface convective precipitation	kg/m^2	Convection
CAPEsfc	Surface convective available potential energy	J/kg	Convection
CINsfc	Surface convective inhibition	J/kg	Convection
HGTclb	Cloud base height	gpm	Convection
HGTclt	Cloud top height	gpm	Convection
TMPclt	Cloud top temperature	K	Convection
GUSTsfc	Surface gust wind	s^{-1}	Dynamic
HGTsfc	Surface geopotential height	gpm	Dynamic
PRESsfc	Surface pressure	Pa	Dynamic
PRMSLmsl	mean sea level pressure	Pa	Dynamic
TMP2m	2-m height temperature	K	Dynamic
VWSHclm	Atmospheric column vertical speed shear	s^{-1}	Dynamic
APCPsfc	Surface total precipitation	kg/m^2	Physical
HGT0deg	Isotherm height	gpm	Physical
TCDCclm	Atmospheric column total cloud cover	$\%$	Physical
ULWRFclm	Atmospheric column upward long wave flux	W/m^2	Physical
PWATclm	Atmospheric column precipitable water	kg/m^2	Physical
	Pressure		
ABSVprs	Absolute vorticity	s^{-1}	Dynamic
POTprs	Potential temperature	K	Dynamic
TMPprs	Temperature	K	Dynamic
UGRDprs	u wind	s^{-1}	Dynamic
VGRDprs	v wind	s^{-1}	Dynamic
VVELprs	Pressure w wind	s^{-1}	Dynamic
DPTprs	Dew point temp	K	Physical
RHprs	Relative humidity	$\%$	Physical
SPFHprs	specific humidity	kg/kg	Physical

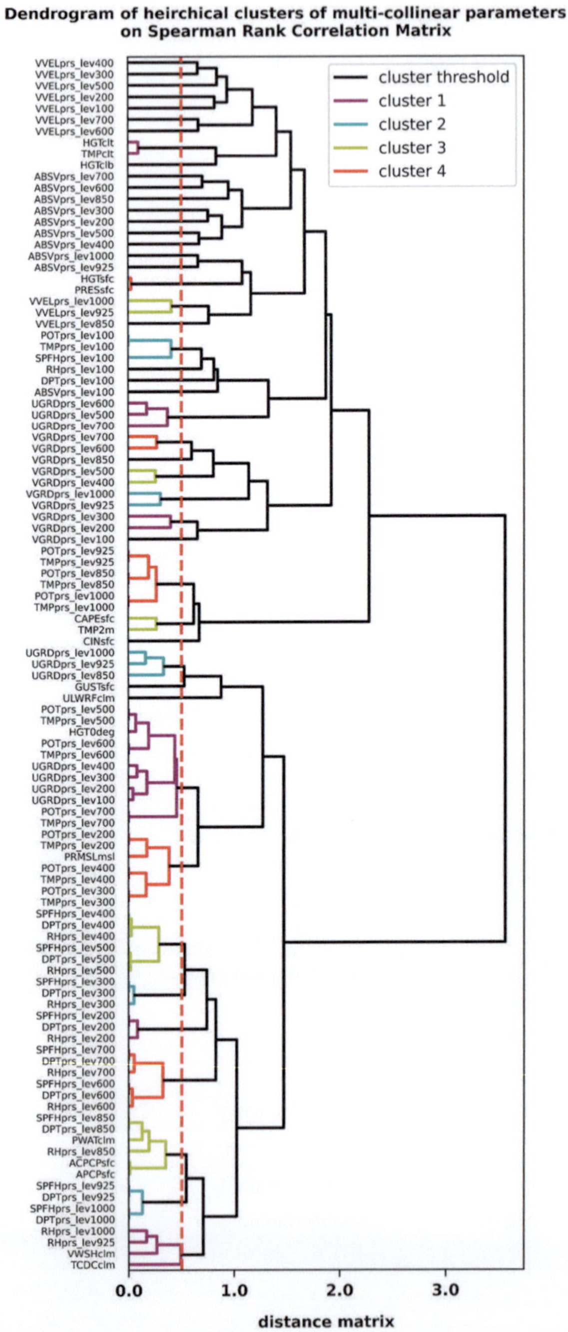

Fig. 7 Dendrogram plot showing hierarchical clusters of features based on Spearman Rank Correlation Matrix. Red dotted line shows the threshold for the distance matrix. From each cluster, one feature is selected for the model. Features on pressure level are represented by {feature name}$_{\mathrm{LEV}}$ {pressures level in hPa}

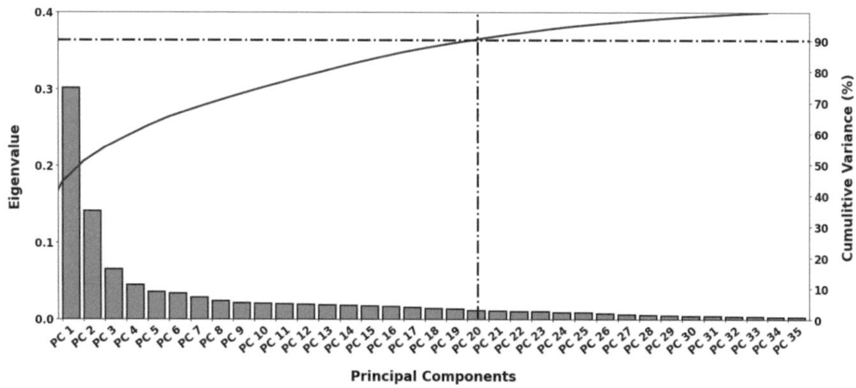

Fig. 8 Scree plot showing eigenvalues (left) of first 35 principal components and cumulative variance (%, right). The dotted line shows 90% cumulative variance achieved by the first 20 principal components. Scree plot showing eigenvalues (left) of first 35 principle

2.4 *Statistical Methods Used for Model Verification*

For the last two centuries, researchers have developed many different methods that can accurately verify the skill of a forecast (Wilks, 2006). One popular method of verifying deterministic forecasts is the contingency table, first developed by Pearson (1904). However, while verifying predictions of rare and severe events (RSEs), traditional skill scores derived from the contingency table can fail to correctly identify the actual strength of the forecast (Murphy, 1991). Similarly, the standard matrices used to verify deep learning-based neural networks can fail to correctly validate the model for the minority class if the dataset is highly imbalanced (Kulkarni et al., 2020).

Addressing these issues, first, we have investigated the degree of dataset imbalance. Based on the level of imbalance, we have then computed forecast skill scores developed specifically for RSEs. For analyzing model performance from a machine learning perspective, we have also examined matrices designed for imbalanced classification problems. We have also investigated prediction errors in terms of probability statements.

Table 2 shows a sample contingency table of a binary classification problem. Imbalance ratio (IR) is a measure of the degree of class imbalance (Fernández et al., 2018):

$$IR = \frac{b+d}{a+b} \tag{6}$$

Imbalance is a helpful tool for understanding the distribution of positive and negative classes. A similar score is used in forecast verification, known as base rate (Hogan & Mason, 2011):

Table 2 Sample contingency table

	Observation			
	Yes (+)	No (−)	Total	
Forecast				
Yes (+)	a Hits	b False alarms	$a + b$ Yes fcsts	Marginal of Forecast
No (−)	c Misses	d True negative	$c + d$ No fcsts	
Total	$a + c$ Yes obs	$b + d$ No obs	***N*** **Total fcsts**	
	Marginal of Observation			

$$p = \frac{a + c}{N} \tag{7}$$

p is a constant that gives us the probability of the event occurring.

For the verification of RSE, one of the important skill scores is bias or bias frequency:

$$bias = \frac{a + b}{a + c} \tag{8}$$

If $bias > 1$, the model is over-forecasting while the model is under-forecasting if $bias < 1$. For RSE, it is crucial to quantify the amount of over-forecasting to evaluate the quality of the forecast (Murphy, 1991).

One of the essential criteria for selecting appropriate skill scores for RSE forecast is to select skill scores that are less prone to "hedging". One such score is the *Odds Ratio* (*OR*):

$$OR = \frac{a \times d}{b \times c} \tag{9}$$

The *Odds Ratio* quantifies the probability of correctly forecasting positive (+) events for every false alarm (Stephenson, 2000).

Another difficulty of RSE forecast is that often, with sporadic events, the skill score tends to converge to zero as $p \to 0$. To resolve this issue, Stephenson et al. (2008) developed Extreme Dependency Score (*EDS*):

$$EDS = \frac{2\log\left(\frac{a+c}{N}\right)}{\log\left(\frac{a}{N}\right)} - 1 \tag{10}$$

EDS is independent of *bias* and, for extremely rare events, can be a measure of hit rate $\frac{a}{(a+c)}$ exponent.

One of the most popular matrices used for deep-learning-based classification models is *Accuracy*:

$$Accuracy = \frac{a + d}{N} \quad (11)$$

However, for the imbalanced dataset, accuracy fails to represent the skill of the positive forecast (Kulkarni et al., 2020). This can be handled through the use of *Balanced Accuracy*:

$$Balanced\ Accuracy = \frac{1}{2}\left(\frac{a}{a+c} + \frac{d}{b+d}\right) \quad (12)$$

Although it should be noted that for the forecast of RSEs such as lightning, *Balanced Accuracy* fails to evaluate the skill for positive class.

One way to identify how accurately the model can classify negative classes without being influenced by class imbalance is by computing *Specificity* (Japkowicz, 2013):

$$Specificity = \frac{d}{b + d} \quad (13)$$

Another popular metric used for the assessment of predicting correctly the events is F_β:

$$F_\beta = \left(1 + \beta^2\right) \times \frac{Recall \times Precision}{\beta^2 \times Recall + Precision} \quad (14)$$

where $Recall = \frac{a}{(a+c)}$ and $Precision = \frac{a}{(a+b)}$. β is a coefficient representing the importance of *Recall* concerning *Precision*, where *Precision* represents the accuracy of positive prediction, and *Recall* measures the true positive rate (Branco et al., 2017; Japkowicz, 2013; Kulkarni et al., 2020). For $\beta = 1$, *weighted* F_1 measures the weighted average of F_1 for both positive and negative forecasts while $macro - averaged$ F_1 gives the unweighted mean F_1 score (Ferri et al., 2009).

Geometric mean or $G - mean$, defined by Kubat et al., 1998 is independent of class imbalance and identifies the cost of every positive event forecast:

$$G - mean = \sqrt{Specificity \times Recall} \quad (15)$$

We have also investigated model fidelity in terms of probability along with binary classification matrices for RSE events and imbalanced deep learning problems. One such popular skill score is *Brier Score*, measuring the correlation between forecast and observation and the frequency distribution of forecast probability (Brier, 1950):

$$Brier\ Score = \frac{1}{N}\sum\nolimits_{i=0}^{N-1} (y_i - \mathrm{P}_i)^2 \quad (16)$$

where y_k is the forecast and P_i observation probability. For binary discrete classification *Brier Score* represents the mean squared error. Another popular score is the *Bayes Error Rate*, representing the minimum error of a model correctly identifying the class of the given event (Tumer & Ghosh, 2003). *Bayes Error Rate* is robust and can be used to verify different models for the same events. It is expressed as:

$$\frac{\mathrm{P}(\hat{y}=1)\mathrm{P}(y=1|\hat{y}=1)}{\mathrm{P}(y=1)} \tag{17}$$

where P(y) is the probability of event y.

3 Results and Discussion

IMD WRF at 9 km resolution presently does not have lightning products. Therefore in this chapter we intent to develop an AI-based lightning product with IMD WRF output for 9 km resolution as our input. To develop this NWP-ML hybrid lightning prediction system, we have taken IITM LLN observed flash density as our target vector. However, one of the challenges for accurately forecasting lightning is the high imbalance between lightning and non-lightning events, as seen in Fig. 4. Only 0.2% of the total dataset represents lightning events. To overcome this challenge, we have taken a novel approach and developed a two-autoencoder-based deep learning classification C_{2AE}, as detailed in Sect. 2 based on Tomescu et al. (2021). From IMD WRF output, initially, 107 predictors were chosen as input vectors based on existing literature. However, as atmospheric processes are non-linear, many of the predictors can have co-dependency among themselves, which can cause a deep learning model to overfit. To find independent parameters, we have derived Spearman rank correlation and a rank-correlation matrix, as shown in Fig. 7, and selected 47 parameters as input features for our model. For C_{2AE} architecture, we have assumed a linear encoding–decoding. Therefore, a PCA is carried out on the selected 47 features. As shown in Fig. 8, the first 20 principal components cumulatively contribute to 90% variance. Based on this finding, 20 neurons are used in the hidden layer. As AE_+ is trained on lightning training data and AE_- on non-lightning training data, therefore creating a significant difference in dataset size between training data for both the autoencoders. Hence, batch size for stochastic gradient descent, training epoch and learning rates are chosen separately. The model architecture and hyperparameters are given in Table 3. During training a deep learning algorithm, we aim to minimize the loss function L. We have plotted a learning curve for both autoencoders (Fig. 9). Here, we can see that with increasing training epoch, L is decreasing exponentially until the change in L is minuscule even as the training epoch continues.

After completing the unsupervised training for individual autoencoders, we introduced a new dataset comprising lightning and non-lightning events. After both the

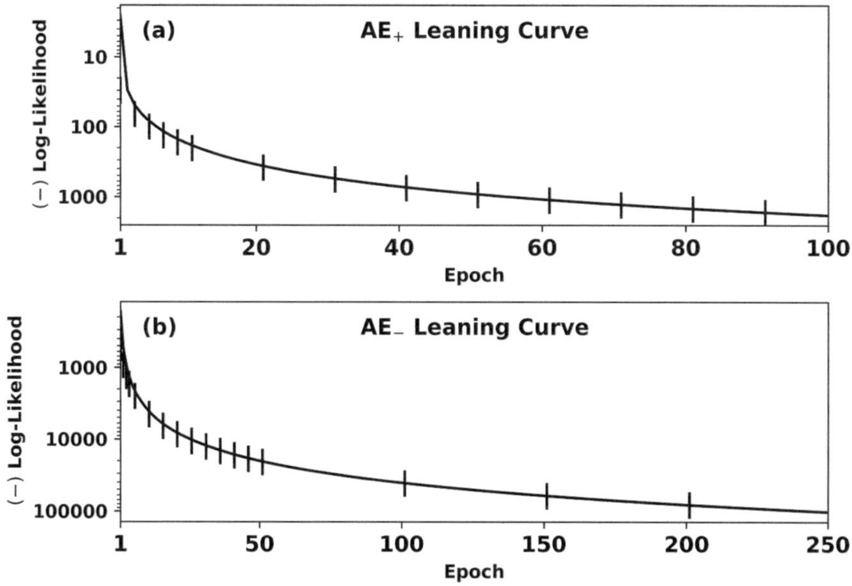

Fig. 9 Learning curve for (**a**) AE_+ and (**b**) AE_- showing negative log-likelihood as a function of training epoch

Table 3 Model architecture, hyperparameters for **a** AE_+ and **b** AE_-

(a) AE_+		(b) AE_-	
Hyperparameters	Values	Hyperparameters	Values
Input dimension	47	Input dimension	47
Number of hidden layer	1	Number of hidden layer	1
Hidden layer neurons	[20]	Hidden layer neurons	[20]
Added corruption	[0.0]	Added corruption	[0.0]
Batch size	32	Batch size	128
Epoch	100	Epoch	250
Learning rate	0.01	Learning rate	0.001

autoencoders try to recreate the test dataset, loss functions are derived, and prediction of lightning or non-lightning is given based on c value, where $c = 0$ denotes no lightning event and $c = 1$ as lightning event. As discussed in Sect. 2.1, observed dataset also are converted into "yes"/"no" categories where in each data point zero total lightning flash density is considered as "no" lighting event and total lightning flash density greater than zero as "yes" lightning event. Therefore a contingency table (Table 4) is derived based on the test data output, and different statistical skill scores are computated using the contingency table, as discussed in Sect. 2.4. The result is shown in Table 5.

Table 4 Contigency table for verification of categorical lightning events as forecasted by the model for the test dataset

		Observation		
		Yes	No	
Forecast	Yes	185 (Hits)	40965 (False alarm)	41150 (Yes fcst)
	No	3797 (Misses)	1941979 (True negative)	1945776 (No fcst)
		3982 (Yes Obs)	1982944 (No Obs)	1986926 (Total)

Table 5 Statistical skill score derived from Table 4, where a, b, c and d are hits, false alarms, misses and true negatives, respectively. Blue, yellow, pink and green colored rows represent data imbalance, scores used for rare and severe weather forecasts, metrics used for imbalanced categorical machine learning classifiers and probability scoring methods, respectively. Scores with statistically significant values are shown in bold

Scores	Formula	Range	Values
IR	$\frac{b+d}{a+c}$	For balanced data → **1**	**497**
Base rate (p)	$p = \frac{a+c}{N}$	For rare and severe events, **p → 0**	**0.2004%**
bias	$\frac{a+b}{a+c}$	under-forecasting < **1**; over-forecasting > **1**	10.33
OR	$\frac{a \times d}{b \times c}$	For Random Forecast → **1**	**2.31**
EDS	$\frac{\log \frac{a+c}{N}}{\log \frac{a}{N}} - 1$	**[−1, 1]; EDS > 0** if **p → 0** & forecast is not random	**0.34**
Accuracy	$\frac{a+d}{N}$	**[0, 1];** Perfect score → **1**	**0.9775**
Balanced Accuracy	$\frac{1}{2}\left(\frac{a}{a+c} + \frac{d}{b+d}\right)$	**[0, 1];** Perfect score → **1**	**0.51**
Specificity	$\frac{d}{b+d}$	**[0, 1];** Perfect score → **1**	**0.9793**
F1 weighted	$\left(\frac{a}{a+b}\right)\left(\frac{a+c}{N}\right)+\left(\frac{d}{c+d}\right)\left(\frac{b+d}{N}\right)$	**[0, 1];** Perfect score → **1**	**0.99**
F1 macro	$\frac{1}{2}\left(\frac{a}{a+b} + \frac{d}{c+d}\right)$	**[0, 1];** Perfect score → **1**	**0.5**
G Mean	$\sqrt{\left(\frac{a}{a+b}\right)\left(\frac{a}{a+c}\right)}$	**[0, 1];** Perfect score → **1**	**0.21**
Brier Score	$\frac{1}{N}\sum_{i=0}^{N-1}(y_i - P_i)^2$	**[0, 1];** Perfect score → **0**	**0.02**
Bayes Error Rate	$\frac{P(\hat{y} = 1)P(y = 1 \mid \hat{y} = 1)}{P(y = 1)}$	**[0, 1];** Perfect score → **0**	**0.05**

From the contingency table (Table 5), we can see that out of 1,986,926 test data points, only 3797 (Yes Obs in Table 4) or 0.2% lightning events were observed. This skewness in the dataset makes it difficult to determine the skill of a forecast from a forecasting and a machine learning perspective. For this reason, we have selected statistical methods that address the imbalance in the dataset, are resistant to hedging, are designed for weather forecasting and deep learning model classification problems, and give us a probabilistic error. In Table 5, the skill scores used for this study, their equations, range and values are shown, where a, b, c, d, and N are Hits, False Alarm, Misses, True Negative and Total, respectively, from Table 3.

IR, or the imbalance ratio in Table 5, shows a value of 497, indicating 497 non-lightning observations for each lightning event. The base rate (p) is 0.2004%, representing the percentage of lightning in the total dataset. *OR* and *EDS* were developed for weather events categorized as Rare and Severe Events with $p \rightarrow 0$, both of which show statistically significant values for our model. Apart from skills used in weather forecasting, we have also calculated different scores used in machine learning-based classifiers with an imbalance binary dataset (Table 5, pink) showing significant skill. Similarly, we have calculated the *Brier Score*, which shows 0.02, while the *Bayes Error Rate* shows 0.05, both of which calculate the probability of correct and incorrect forecast.

4 Conclusions

This chapter first discusses the existing literature on lightning early warning systems. As physical and dynamic lightning processes are still poorly understood and considering potential damages caused by lightning, there is a growing demand from scientists and policymakers to invest in AI-driven lightning early warning systems. However, as shown in Fig. 4, the lightning-to-non-lightning event ratio is highly skewed. Therefore, it is challenging for any data-driven model to learn features from lightning events in order to correctly classify a forecast between lightning and non-lightning. This "curse of imbalance dataset" can be dealt within the preprocessing stage of building an AI model. However these methods pose limitations by removing important information from the dataset or making the model overfit. Instead, in this chapter, we focused on the architecture of the AI model itself, which is built around exploiting the imbalanced nature of the dataset.

We have built a lightning warning system that can be used for operational purposes. As IMD produces 9 km resolution WRF forecasts every day twice (00 UTC and 12 UTC) with 72 h lead time, we have selected the 24-h lead time forecast (00 UTC) as our input. We have referenced Tomescu et al. (2021) for the architecture to build a novel, two-autoencoder-based deep-learning model (C_{2AE}) consisting of a positive autoencoder and a negative autoencoder. To train the model, we have used IITM LLN lightning location observation data. The positive autoencoder AE_+ is trained with lightning data, and the negative autoencoder AE_- is trained with non-lightning data. After training both autoencoders, AE_+ would classify lightning data with fewer errors while AE_- will have a larger error. Exploiting this, we introduce a decision function based on loss functions. The detailed architecture of this model is shown in Fig. 6.

When tested on an entirely new dataset consisting of both lightning and non-lightning events, various skill score analyses were performed to evaluate model performance. As shown in Table 5, even as the base rate tends to zero, this new hybrid model can perform well in terms of skills designed to measure the model fidelity for skewed events with statistically significant values. The probabilistic error measured using the *Bayes Error Rate* is 5%, showing that the novel AI-NWP hybrid model can be used as a warning system for lightning forecasts.

We would also like to add that one of the difficulties in forecasting RSEs has been high false alarms. Although C_{2AE} can give optimistic lightning forecasts, we are also aware of the high false alarms present in the current model. To reduce the false alarm in the future, we would like to use non-linear hidden layers (Bourlard & Kamp, 1988), which can help the autoencoders learn hidden features that may not be possible with a linear hidden layer. We would also like to add spatiotemporal information to our dataset (Ploton et al., 2020) to identify the clustered lightning active regions. However, for lightning, the cost of misses is higher than that of false alarms (Barnes et al., 2007).

Hence, a carefully constructed deep learning algorithm with a simple architecture can forecast lightning with sufficient skill for a 24-h lead time at 9 km resolution. Therefore, this study may be instrumental for the future of lightning early warning systems in India.

Acknowledgements Indian Institute of Tropical Meteorology (IITM) and Ministry of Earth Sciences (MoES), Government of India are responsible for the resources and financial support behind this study, and the authors cordially thank them for this. The authors acknowledge Dr. Mrutyunjay Mohapatra, DGM, Indian Meteorology Department (IMD) and Dr. Ananda Kumar Das, IMD Numerical Weather Prediction Division, for curating and providing WRF 9 km resolution output for our study. The authors also acknowledge Dr. Sunil D. Pawar, project director of Thunderstorm Dynamics, for maintaining and providing IITM LLN lightning observation data. Laboratoire d'Informatique des Systèmes d'Apprentissage (LISA), University of Montreal, developed the Theano python package, which has been integral to our study. All the codes are run on Pratyush High-Performance Computer at IITM. We also extend our gratitude to Sayanti Bardhan and Dr. Siddharth Kumar for their insights on the subject. Lastly, we acknowledge the editors and reviewers of this Monograph for allowing us to publish our study. We are also indebted to valuable comments from anonymous reviewers. This study is part of the Ph.D. thesis of Rituparna Sarkar, funded by Innovation in Science Pursuit for Inspired Research (INSPIRE), Department of Science and Technology, Government of India.

References

A protocol for early warning dissemination on thunderstorm and lightning. (2021). https://ndma.gov.in/Resources/Technical-Documents.

Barnes, L. R., Gruntfest, E. C., Hayden, M. H., Schultz, D. M., & Benight, C. (2007). False alarms and close calls: A conceptual model of warning accuracy. *Weather and Forecasting, 22*(5), 1140–1147. https://doi.org/10.1175/WAF1031.1

Barthe, C., Deierling, W., & Barth, M. C. (2010). Estimation of total lightning from various storm parameters: A cloud-resolving model study. *Journal of Geophysical Research Atmospheres, 115*(24), 1–17. https://doi.org/10.1029/2010JD014405

Bengio, Y. (2009). Learning deep architectures for AI. *Foundations and Trends® in Machine Learning, 2*(1), 1–127. https://doi.org/10.1561/2200000006

Bennett, A. J. (2018). Warning of imminent lightning using single-site meteorological observations. *Weather, 73*(6), 187–193. https://doi.org/10.1002/wea.2782

Bourlard, H., & Kamp, Y. (1988). Auto-association by multilayer perceptrons and singular value decomposition. *Biological Cybernetics, 59*(4–5), 291–294. https://doi.org/10.1007/BF00332918

Branco, P., Torgo, L., & Ribeiro, R. P. (2017). A survey of predictive modeling on imbalanced domains. *ACM Computing Surveys, 49*(2), 1–50. https://doi.org/10.1145/2907070, https://dl.acm.org/doi/10.1145/2907070

Brier, G. W. (1950). Verification of forecasts expressed in terms of probability. *Monthly Weather Review, 78*(1), 1–3. https://doi.org/10.1175/1520-0493(1950)078%3C0001:VOFEIT%3E2.0.CO;2

Brooks, H. E., Doswell, C. A., III., Zhang, X., Chernokulsky, A. M. A., Tochimoto, E., Hanstrum, B., De, E., Nascimento, L., Sills, D. M. L., Antonescu, B., & Barrett, B. (2019). A century of progress in severe convective storm research and forecasting. *Meteorological Monographs, 59*(1), 18.1-18.41. 10.1175/AMSMONOGRAPHS-D-18-0026.1

Bürgesser, R. E. (2017). Assessment of the World Wide Lightning Location Network (WWLLN) detection efficiency by comparison to the Lightning Imaging Sensor (LIS). *Quarterly Journal of the Royal Meteorological Society, 143*(708), 2809–2817. https://doi.org/10.1002/qj.3129

Carey, L. D., Rutledge, S. A., & Petersen, W. A. (2003). The relationship between severe storm reports and cloud-to-ground lightning polarity in the contiguous United States from 1989 to 1998. *Monthly Weather Review, 131*(7), 1211–1228. https://doi.org/10.1175/1520-0493(2003)131%3c1211:TRBSSR%3e2.0.CO;2

Cecil, D. J., Buechler, D. E., & Blakeslee, R. J. (2014). Gridded lightning climatology from TRMM-LIS and OTD: Dataset description. *Atmospheric Research, 135–136*, 404–414. https://doi.org/10.1016/j.atmosres.2012.06.028

Cecil, D. J., Buechler, D. E., & Blakeslee, R. J. (2015). TRMM LIS climatology of thunderstorm occurrence and conditional lightning flash rates. *Journal of Climate, 28*(16), 6536–6547. https://doi.org/10.1175/JCLI-D-15-0124.1

Chawla, N. V., Bowyer, K. W., Hall, L. O., & Kegelmeyer, W. P. (2011). SMOTE: Synthetic minority over-sampling technique. *Journal of Artificial Intelligence Research, 16*, 321–357. https://doi.org/10.1613/jair.953

Collins, W., & Tissot, P. (2015). An artificial neural network model to predict thunderstorms within 400 km2 South Texas domains. *Meteorological Applications, 22*(3), 650–665. https://doi.org/10.1002/met.1499

Collins, W. G. & Tissot, P. (2016). Thunderstorm predictions using artificial neural networks. In *Artificial Neural Networks—Models and Applications.* InTech. https://doi.org/10.5772/63542.

Cooper, M. A. & Holle, R. L. (2019). *Current Global Estimates of Lightning Fatalities and Injuries* (pp. 65–73). Springer. https://doi.org/10.1007/978-3-319-77563-0_6.

Das, A., Yadav, B., Charu, C., & Dhingra, J. (2022). Evaluation of performance of WRF (ARW) and GFS for quantitative precipitation forecast and its value addition over India river sub-basins during recent years. *MAUSAM*, *73*(2), 315–340. https://doi.org/10.54302/mausam.v73i2.5480.

Das, A. K., Kundu, P. K., Bhowmik, S. K. R., & Rathee, M. (2016). Verification of real-time WRF-ARW forecast in IMD during monsoon 2010. *MAUSAM*, *67*(2), 333–356. https://doi.org/10.54302/mausam.v67i2.1301.

Dash, J. G., & Wettlaufer, J. S. (2003). The surface physics of ice in thunderstorms. *Canadian Journal of Physics, 81*(1–2), 201–207. https://doi.org/10.1139/p03-011

Desai, B. N. (1950). Mechanism of Nor'westers of Bengal. *MAUSAM*, *1*(1), 74–76. https://doi.org/10.54302/mausam.v1i1.3958.

Desai, B. N., & Rao, Y. P. (1954). On the cold pools and their role in the development of Nor'westers over West Bengal and Eastern Pakistan. *Indian Journal of Meteorology and Geophysics, 5*(3), 243–248.

Early warning and anticipatory action. *Bulletin, 71(1)* (2022). http://public.wmo.int/en/resources/bulletin.

Ellis, A., & Miller, P. (2016). The emergence of lightning in severe thunderstorm prediction and the possible contributions from spatial science. *Geography Compass, 10*(5), 192–206. https://doi.org/10.1111/GEC3.12265

Fernández, A., García, S., Galar, M., Prati, R. C., Krawczyk, B., & Herrera, F. (2018). *Learning from Imbalanced Data Sets.* Springer International Publishing. https://doi.org/10.1007/978-3-319-98074-4

Ferri, C., Hernández-Orallo, J., & Modroiu, R. (2009). An experimental comparison of performance measures for classification. *Pattern Recognition Letters, 30*(1), 27–38. https://doi.org/10.1016/j.patrec.2008.08.010

Gaskell, W., & Illingworth, A. J. (1980). Charge transfer accompanying individual collisions between ice particles and its role in thunderstorm electrification. *Quarterly Journal of the Royal Meteorological Society, 106*(450), 841–854. https://doi.org/10.1002/qj.49710645013

Gilmore, M. S., & Wicker, L. J. (2002). Influences of the local environment on supercell cloud-to-ground lightning, radar characteristics, and severe weather on 2 June 1995. *Monthly Weather Review, 130*(10), 2349–2372. https://doi.org/10.1175/1520-0493(2002)130%3c2349:IOTLEO%3e2.0.CO;2

Glazer, R. H., Abraham Torres-Alavez, J., Coppola, E., Giorgi, F., Das, S., Ashfaq, M., & Sines, T. (2021). Projected changes to severe thunderstorm environments as a result of twenty-first century warming from RegCM CORDEX-CORE simulations. *Climate Dynamics, 57*, 1595–1613. https://doi.org/10.1007/s00382-020-05439-4

Hagos, S., Feng, Z., Plant, R. S., & Protat, A. (2020). A machine learning assisted development of a model for the populations of convective and stratiform clouds. *Journal of Advances in Modeling Earth Systems, 12*(3). https://doi.org/10.1029/2019MS001798.

Helsdon, J. H., Gattaleeradapan, S., Farley, R. D., & Christopher Waits, C. (2002). An examination of the convective charging hypothesis: Charge structure, electric fields, and Maxwell currents. *Journal of Geophysical Research Atmospheres, 107*(22), ACL 9-1–ACL 9-26. https://doi.org/10.1029/2001JD001495.

Hewage, P., Trovati, M., Pereira, E., & Behera, A. (2020). Deep learning-based effective fine-grained weather forecasting model. *Pattern Analysis and Applications, 24*, 343–366. https://doi.org/10.1007/s10044-020-00898-1

Hogan, R. J., & Mason, I. B. (2011). Deterministic forecasts of binary events. In *Forecast Verification* (pp. 31–59). Wiley. https://doi.org/10.1002/9781119960003.ch3.

Houze, R. A. (2004). Mesoscale convective systems. *Reviews of Geophysics, 42*(4), 1–43. https://doi.org/10.1029/2004RG000150

Illingworth, A. J. (1985). Charge separation in thunderstorms: Small scale processes. *Journal of Geophysical Research, 90*(D4), 6026–6032. https://doi.org/10.1029/JD090iD04p06026

Japkowicz, N. (2013). Assessment metrics for imbalanced learning. In *Imbalanced Learning* (pp. 187–206). John Wiley & Sons, Inc. https://doi.org/10.1002/9781118646106.ch8.

Kalchbrenner, N., & Blunsom, P. (2013). *Recurrent Continuous Translation Models.* Association for Computational Linguistics.

Kamangir, H., Collins, W., Tissot, P., & King, S. A. (2020). A deep-learning model to predict thunderstorms within 400 km ^{2} South Texas domains. *Meteorological Applications, 27*(2), 1–17. https://doi.org/10.1002/met.1905

Kasemir, H. W. (1960). A contribution to the electrostatic theory of a lightning discharge. *Journal of Geophysical Research, 65*(7), 1873–1878. https://doi.org/10.1029/jz065i007p01873

Kubat, M., Holte, R. C., & Matwin, S. (1998). Machine learning for the detection of oil spills in satellite radar images. *Machine Learning, 30*(2–3), 195–215. https://doi.org/10.1023/a:1007452223027

Kubát, M., & Matwin, S. (1997). Addressing the curse of imbalanced training sets: One-sided selection. In *International Conference on Machine Learning.*

Kulkarni, A., Chong, D., & Batarseh, F. A. (2020). Foundations of data imbalance and solutions for a data democracy. In *Data Democracy* (pp. 83–106). Elsevier. https://doi.org/10.1016/B978-0-12-818366-3.00005-8.

Lopez, P. (2016). A lightning parameterization for the ECMWF integrated forecasting system. *Monthly Weather Review, 144*(9), 3057–3075. https://doi.org/10.1175/MWR-D-16-0026.1

Lundberg, S., & Lee, S.-I. (2017). *A unified approach to interpreting model predictions*. http://arxiv.org/abs/1705.07874.

MacGorman, D. R., Apostolakopoulos, I. R., Lund, N. R., Demetriades, N. W. S., Murphy, M. J., & Krehbiel, P. R. (2011). The timing of cloud-to-ground lightning relative to total lightning activity. *Monthly Weather Review, 139*(12), 3871–3886. https://doi.org/10.1175/MWR-D-11-00047.1

MacGorman, D. R., Straka, J. M., & Ziegler, C. L. (2001). A lightning parameterization for numerical cloud models. *Journal of Applied Meteorology, 40*(3), 459–478. https://doi.org/10.1175/1520-0450(2001)040%3c0459:ALPFNC%3e2.0.CO;2

Maddox, R. A. (1987). Mesoscale and severe storm meteorology. *Reviews of Geophysics, 25*(3), 329. https://doi.org/10.1029/RG025i003p00329

Mansell, E. R., MacGorman, D. R., Ziegler, C. L., & Straka, J. M. (2002). Simulated three-dimensional branched lightning in a numerical thunderstorm model. *Journal of Geophysical Research: Atmospheres, 107*(9–10). https://doi.org/10.1029/2000jd000244.

Mason, B. J. (1953). On the generation of charge associated with Graupel formation in thunderstorms. *Quarterly Journal of the Royal Meteorological Society, 79*(342), 501–509. https://doi.org/10.1002/qj.49707934206

Mecikalski, J. R., Sandmæl, T. N., Murillo, E. M., Homeyer, C. R., Bedka, K. M., Apke, J. M., & Jewett, C. P. (2021). Random forest model to assess predictor importance and nowcast severe storms using high-resolution radar–GOES satellite-lightning observations. *Monthly Weather Review, 149*(6), 1725–1746. https://doi.org/10.1175/MWR-D-19-0274.1

Meng, Q., Yao, W., & Xu, L. (2019). Development of lightning nowcasting and warning technique and its application. *Advances in Meteorology, 2019*, 1–9. https://doi.org/10.1155/2019/2405936

Mostajabi, A., Finney, D. L., Rubinstein, M., & Rachidi, F. (2019). Nowcasting lightning occurrence from commonly available meteorological parameters using machine learning techniques. *NPJ Climate and Atmospheric Science, 2*(1), 1–15. https://doi.org/10.1038/s41612-019-0098-0

Mudiar, D., Pawar, S. D., Hazra, A., Konwar, M., Gopalakrishnan, V., Srivastava, M. K., & Goswami, B. N. (2018). Quantification of observed electrical effect on the raindrop size distribution in tropical clouds. *Journal of Geophysical Research: Atmospheres, 123*(9), 4527–4544. https://doi.org/10.1029/2017JD028205

Mukhopadhyay, P., Taraphdar, S., Goswami, B. N., & Krishnakumar, K. (2010). Indian summer monsoon precipitation climatology in a high-resolution regional climate model: Impacts of convective parameterization on systematic biases. *Weather and Forecasting, 25*(2), 369–387. https://journals.ametsoc.org/view/journals/wefo/25/2/2009waf2222320_1.xml

Murphy, A. H. (1991). Probabilities, odds, and forecasts of rare events. *Weather and Forecasting, 6*(2), 302–307. https://doi.org/10.1175/1520-0434(1991)006%3C0302:POAFOR%3E2.0.CO;2

Murugavel, P., Pawar, S. D., & Gopalakrishnan, V. (2012). Trends of convective available potential energy over the Indian region and its effect on rainfall. *International Journal of Climatology, 32*(9), 1362–1372. https://doi.org/10.1002/joc.2359

Ou, C., Yang, J., Du, Z., Zhang, X., & Zhu, D. (2019). Integrating cellular automata with unsupervised deep-learning algorithms: A case study of urban-sprawl simulation in the Jingjintang urban agglomeration, China. *Sustainability, 11*(9), 2464. https://doi.org/10.3390/su11092464

Pathak, J., Wikner, A., Fussell, R., Chandra, S., Hunt, B. R., Girvan, M., & Ott, E. (2018). Hybrid forecasting of chaotic processes: Using machine learning in conjunction with a knowledge-based model. *Chaos an: Interdisciplinary Journal of Nonlinear Science, 28*(4), 041101. https://doi.org/10.1063/1.5028373

Pawar, V. S., Domkawale, M. A., Pawar, S. D., Salvekar, P. S., & Pradeep Kumar, P. (2017). Inter annual variability of tropospheric NO 2 and tropospheric ozone over Maharashtra (India): The role of lightning. *Remote Sensing Letters, 8*(11), 1015–1024. https://doi.org/10.1080/2150704X.2017.1346398

Pearson, K. (1904). On the theory of contingency and its relation to association and normal correlation. In *Mathematical Contributions to the Theory of Evolution* (Vol. XIII, 1st edn., pp. 1–46). Dulua and Co.

Ploton, P., Mortier, F., Réjou-Méchain, M., Barbier, N., Picard, N., Rossi, V., Dormann, C., Cornu, G., Viennois, G., Bayol, N., Lyapustin, A., Gourlet-Fleury, S., & Pélissier, R. (2020). Spatial validation reveals poor predictive performance of large-scale ecological mapping models. *Nature Communications, 11*(1), 4540. https://doi.org/10.1038/s41467-020-18321-y

Price, C., & Rind, D. (1993). What determines the cloud-to-ground lightning fraction. *Geophysical Research Letters, 20*(6), 463–466.

Price, C., & Rind, D. (1994). Modeling global lightning distributions in a general circulation model. *Monthly Weather Review, 122*(8), 1930–1939. https://doi.org/10.1175/1520-0493(1994)122<1930:MGLDIA>2.0.CO;2.

Putro, W. S. (2020). Development of Sumatera early warning convective system (Sancis) for thunderstorm prediction model. *Geographica: Science & Education Journal, 1*(1), 1–7. https://doi.org/10.31327/gsej.v1i1.1072.

Rajeevan, M., Madhulatha, A., Rajasekhar, M., Bhate, J., Kesarkar, A., & Rao, B. V. A. (2012). Development of a perfect prognosis probabilistic model for prediction of lightning over southeast India. *Journal of Earth System Science, 121*(2), 355–371. https://doi.org/10.1007/s12040-012-0173-y

Ranzato, M. A., Boureau, Y.-L., & Lecun, Y. (2007). Sparse feature learning for deep belief networks. In *Advances in Neural Information Processing Systems* (Vol. 20).

Rawlins, F. (1982). A numerical study of thunderstorm electrification using a three dimensional model incorporating the ice phase. *Quarterly Journal of the Royal Meteorological Society, 108*(458), 779–800. https://doi.org/10.1002/qj.49710845804

Ray, K., Giri, R. K., Ray, S. S., Dimri, A. P., & Rajeevan, M. (2021). An assessment of long-term changes in mortalities due to extreme weather events in India: A study of 50 years' data, 1970–2019. *Weather and Climate Extremes, 32*, 100315. https://doi.org/10.1016/j.wace.2021.100315.

Reichstein, M., Camps-Valls, G., Stevens, B., Jung, M., Denzler, J., Carvalhais, N., & Prabhat, F. (2019). Deep learning and process understanding for data-driven Earth system science. *Nature, 566*(7743), 195–204. https://doi.org/10.1038/s41586-019-0912-1

Reynolds, S. E., Brook, M., & Gourley, M. F. (1957). Thunderstorm charge separation. *Journal of Meteorology, 14*, 426–436. https://doi.org/10.16309/j.cnki.issn.1007-1776.2003.03.004.

Ribeiro, M. T., Singh, S., & Guestrin, C. (2016). *Why Should I Trust You?: Explaining the Predictions of Any Classifier*. http://arxiv.org/abs/1602.04938.

Rifai, S., Vincent, P., Muller, X., Glorot, X., & Bengio, Y. (2011). *Contractive Auto-Encoders: Explicit Invariance During Feature Extraction.*

Roy, S. S., Mohapatra, M., Tyagi, A., & Bhowmik, S. K. R. (2021). A review of nowcasting of convective weather over the Indian region. *MAUSAM, 70*(3), 465–484. https://doi.org/10.54302/mausam.v70i3.227.

Rumelhart, D. E., Hinton, G. E., & Williams, R. J. (1986). Learning representations by back-propagating errors. *Nature, 323*(6088), 533–536. https://doi.org/10.1038/323533a0

Saha, M., Mitra, P., & Nanjundiah, R. S. (2016). Predictor discovery for early-late Indian Summer Monsoon using stacked autoencoder. *Procedia Computer Science, 80*, 565–576. https://doi.org/10.1016/j.procs.2016.05.337

Sarkar, R., Mukhopadhyay, P., Bechtold, P., Lopez, P., Pawar, S. D., & Chakravarty, K. (2022). Evaluation of ECMWF lightning flash forecast over Indian subcontinent during MAM 2020. *Atmosphere, 13*(9), 1520. https://doi.org/10.3390/atmos13091520

Saunders, C. (2008). Charge separation mechanisms in clouds. *Space Science Reviews, 137*(1–4), 335–353. https://doi.org/10.1007/s11214-008-9345-0

Schölkopf, B., Smola, A., & Müller, K.-R. (1998). Nonlinear component analysis as a kernel eigenvalue problem. *Neural Computation, 10*(5), 1299–1319. https://doi.org/10.1162/089976698300017467

Sen Roy, S., & Balling, R. C. (2013). Diurnal variations in summertime lightning activity in Tropical Asia. *International Journal of Climatology, 33*(7), 1830–1836. https://doi.org/10.1002/joc.3658

Sen Roy, S., Sharma, P., Sen, B., Sathi Devi, K., Sunitha Devi, S., Gopal, N. K., Kumar, N., Mishra, K., Katyar, S., Pratap Singh, S., Balakrishnan, S., Singh, C., Srivastava, K., Lotus, S., Paul, S., Singh, B., Gupta, J. P., Bandopadhyay, S., Das, G., … Mohapatra, M. (2021). A new paradigm for short-range forecasting of severe weather over the Indian region. *Meteorology and Atmospheric Physics, 133*(4), 989–1008. https://doi.org/10.1007/s00703-021-00788-z.

Shafer, P. E., & Fuelberg, H. E. (2006). A statistical procedure to forecast warm season lightning over portions of the Florida Peninsula. In *86th AMS Annual Meeting* (pp. 851–868).

Smith, S. B., LaDue, J. G., & MacGorman, D. R. (2000). The relationship between cloud-to-ground lightning polarity and surface equivalent potential temperature during three tornadic outbreaks. *Monthly Weather Review, 128*(9), 3320–3328. https://doi.org/10.1175/1520-0493(2000)128%3c3320:TRBCTG%3e2.0.CO;2

Sotiropoulos, D., Giannoulis, C., & Tsihrintzis, G. A. (2014). A comparative study of one-class classifiers in machine learning problems with extreme class imbalance. In *IISA 2014 - 5th International Conference on Information, Intelligence, Systems and Applications* (pp. 362–364). https://doi.org/10.1109/IISA.2014.6878723.

Spearman, C. (1904). The proof and measurement of association between two things. *The American Journal of Psychology, 15*(1), 72. https://doi.org/10.2307/1412159

Srivastava, S. (2022). *Annual lightning report 2021–2022: Executive summary*. http://cropc.org/lightning-report.php.

Stephenson, D. B. (2000). Use of the "odds ratio" for diagnosing forecast skill. *Weather and Forecasting, 15*(2), 221–232. https://doi.org/10.1175/1520-0434(2000)015%3C0221:UOTORF%3E2.0.CO;2.

Stephenson, D. B., Casati, B., Ferro, C. A. T., & Wilson, C. A. (2008). The extreme dependency score: a non-vanishing measure for forecasts of rare events. *Meteorological Applications, 15*, 41–50. https://doi.org/10.1002/met.53.

Takahashi, T. (1987). Determination of lightning origins in a thunderstorm model. *Journal of Meteorological Research Japan, 28*(44), 5241–5244.

Taraphdar, S., Mukhopadhyay, P., & Goswami, B. N. (2010). Predictability of Indian summer monsoon weather during active and break phases using a high resolution regional model. *Geophysical Research Letters, 37*(21), n/a-n/a. https://doi.org/10.1029/2010GL044969.

Tinmaker, M. I. R., Ghude, S. D., Dwivedi, A. K., Islam, S., Kulkarni, S. H., Khare, M., & Chate, D. M. (2022). Relationships among lightning, rainfall, and meteorological parameters over oceanic and land regions of India. *Meteorology and Atmospheric Physics, 134*(1), 5. https://doi.org/10.1007/s00703-021-00841-x

Tomescu, V.-I., Czibula, G., & Niţică, Ş. (2021). A study on using deep autoencoders for imbalanced binary classification. *Procedia Computer Science, 192*, 119–128. https://doi.org/10.1016/j.procs.2021.08.013

Tumer, K., & Ghosh, J. (2003). Bayes error rate estimation using classifier ensembles. *International Journal of Smart Engineering System Design, 5*(2), 95–109. https://doi.org/10.1080/10255810305042

Utsav, B., Deshpande, S. M., Das, S. K., Pandithurai, G., & Niyogi, D. (2019). Observed vertical structure of convection during dry and wet summer monsoon epochs over the western ghats. *Journal of Geophysical Research: Atmospheres, 124*(3), 1352–1369. https://doi.org/10.1029/2018JD028960

Vani, K. G., Mohan, G. M., Hazra, A., Pawar, S. D., Pokhrel, S., Chaudhari, H. S., Konwar, M., Saha, S. K., Mallick, C., Das, S. K., Deshpande, S., Ghude, S. D., Domkawale, M., Rao, S. A., Nanjundiah, R. S., & Rajeevan, M. (2022). Evaluation and usefulness of lightning forecasts made with lightning parameterization schemes coupled with the WRF model. *Weather and Forecasting, 37*(5), 709–726. https://doi.org/10.1175/WAF-D-21-0080.1

Vincent, P., Larochelle, H., Lajoie, I., Bengio, Y., & Manzagol, P.-A. (2010). Stacked denoising autoencoders: Learning useful representations in a deep network with a local denoising criterion. *Journal of Machine Learning Research, 11*, 3371–3408.

Wang, J., Sheng, Z., Zhou, B., & Zhou, S. (2014). Lightning potential forecast over Nanjing with denoised sounding-derived indices based on SSA and CS-BP neural network. *Atmospheric Research, 137*, 245–256. https://doi.org/10.1016/J.ATMOSRES.2013.10.014

Wilks, D. S. (2006). Statistical methods in the atmospheric sciences. *Angewandte Chemie International Edition, 6*(11), 951–952.

Williams, E., Chan, T., & Boccippio, D. (2004). Islands as miniature continents: Another look at the Land-ocean lightning contrast. *Journal of Geophysical Research D: Atmospheres, 109*(16). https://doi.org/10.1029/2003JD003833.

Williams, E., Guha, A., Boldi, R., Christian, H., & Buechler, D. (2019). Global lightning activity and the hiatus in global warming. *Journal of Atmospheric and Solar-Terrestrial Physics, 189*, 27–34. https://doi.org/10.1016/j.jastp.2019.03.011

Williams, E., & Montanyà, J. (2019). A closer look at lightning reveals needle-like structures. *Nature, 568*(7752), 319–320. https://doi.org/10.1038/d41586-019-01178-7

Williams, E., Mushtak, V., Rosenfeld, D., Goodman, S., & Boccippio, D. (2005). Thermodynamic conditions favorable to superlative thunderstorm updraft, mixed phase microphysics and lightning flash rate. *Atmospheric Research, 76*(1–4), 288–306. https://doi.org/10.1016/j.atmosres.2004.11.009

Williams, E. R. (1985). Large-scale charge separation in thunderclouds. *Journal of Geophysical Research, 90*(D4), 6013–6025. https://doi.org/10.1029/jd090id04p06013

Williams, E. R. (1988). The electrification of thunderstorms. *Quarterly Journal of the Royal Meteorological Society, 107*(452), 277–298. https://doi.org/10.1002/qj.49710745202

Williams, E. R. (1989). The tripole structure of thunderstorms. *Journal of Geophysical Research, 94*(D11). https://doi.org/10.1029/jd094id11p13151.

Williams, E. R. (2005). Lightning and climate: A review. *Atmospheric Research, 76*(1–4), 272–287. https://doi.org/10.1016/j.atmosres.2004.11.014

Williams, E. R. (2009). The global electrical circuit: A review. *Atmospheric Research, 91*(2–4), 140–152. https://doi.org/10.1016/j.atmosres.2008.05.018

Williams, E. R., Zhang, R., & Rydock, J. (1991). Mixed-phase microphysics and cloud electrification. *Journal of the Atmospheric Sciences, 48*(19), 2195–2203. https://doi.org/10.1175/1520-0469(1991)048<2195:MPMACE>2.0.CO;2

Wilson, J. W., Crook, N. A., Mueller, C. K., Sun, J., & Dixon, M. (1998). Nowcasting thunderstorms: A status report. *Bulletin of the American Meteorological Society, 79*(10), 2079–2099. https://doi.org/10.1175/1520-0477(1998)079%3c2079:NTASR%3e2.0.CO;2

Wu, F., Cui, X., Zhang, D.-L., Liu, D., & Zheng, D. (2016). SAFIR-3000 lightning statistics over the Beijing Metropolitan Region during 2005–07. *Journal of Applied Meteorology and Climatology, 55*(12), 2613–2633. https://doi.org/10.1175/JAMC-D-16-0030.1

Xayasouk, T., Lee, H., & Lee, G. (2020). Air Pollution Prediction Using Long Short-Term Memory (LSTM) and Deep Autoencoder (DAE) Models. *Sustainability, 12*(6), 2570. https://doi.org/10.3390/su12062570

Yair, Y., Lynn, B., Price, C., Kotroni, V., Lagouvardos, K., Morin, E., Mugnai, A., & del Carmen Llasat, M. (2010). Predicting the potential for lightning activity in Mediterranean storms based on the Weather Research and Forecasting (WRF) model dynamic and microphysical fields. *Journal of Geophysical Research, 115*(D4), D04205. https://doi.org/10.1029/2008JD010868

Zeng, R., Zhuang, C., Zhou, X., Chen, S., Wang, Z., Yu, Z., & He, J. (2016). Survey of recent progress on lightning and lightning protection research. *High Voltage, 1*(1), 2–10. https://doi.org/10.1049/hve.2016.0004

Ziegler, C. L., & Macgorman, D. R. (1994). Observed lightning morphology relative to modeled space charge and electric field distribution in a tornadic storm. *Journal of the Atmospheric Sciences, 51*(6), 833–851.

Zipser, E. J., Cecil, D. J., Liu, C., Nesbitt, S. W., & Yorty, D. P. (2006). Where are the most intense thunderstorms on earth? *Bulletin of the American Meteorological Society, 87*(8), 1057–1072. https://doi.org/10.1175/BAMS-87-8-1057

Severe Storms Forecasting and Forecast Demonstration Project Over the Indian Region

Soma Sen Roy, Pradeep Sharma, and M. Mohapatra

Abstract To address the requirement for improvement of forecasting of thunderstorms over the Indian subcontinent a special project was started since 2005 during the summer season. The present chapter provides details of developments since 2017, undertaken as a part of the Severe Thunderstorm Observation and Regional Modelling (STORM) program. The Forecast Demonstration Project of the STORM program as coordinated by India, is conducted during March to June every year. As a part of this project, there has been rapid expansion of observation tools as well as mesoscale model products that forecast weather associated with thunderstorms such as lightning, gusty winds and hail. Correspondingly there have been major improvements in forecasting quality in short range to nowcast scale with quantitative forecasts for thunderstorms being issued round the clock at three hour intervals for all districts of India and more than a thousand stations. Using the same metrics and a constant verification dataset, significant improvement is noted in all scores, both in short range as well as nowcasts scale for all subdivisions of India and all months of the FDP period (March to June). However, there are consistently higher forecast detection errors in short range scale over the arid western parts of India. On a monthly scale, all India short range thunderstorm forecast skill is generally lower in March as compared to April and May and June. However upon detailed analysis, it reflects a practical aspect of the forecast methodology adopted by an operational forecaster vis-à-vis the gradual increase in atmospheric moisture as the season progresses. With the increase in moisture in the lower troposphere as the season progresses, the probability of maturation of cumulus cells into cumulonimbus stage significantly increases and time taken to do so, decreases. The slight deterioration in scores in 2022 and 2023 compared to previous years may be ascribed to the unfavourable middle tropospheric conditions resulting in subdued thunderstorm activity over the Indian region compared to previous years.

S. Sen Roy (✉) · P. Sharma · M. Mohapatra
India Meteorological Department, Lodi Road, New Delhi, India
e-mail: senroys@gmail.com

S. Das and W-K. Tao (eds.), *Severe Storms*,
https://doi.org/10.1007/978-981-97-7075-5_19

1 Introduction

The Indian subcontinent is susceptible to various types of extreme weather hazards, including rainfall and associated flooding, thunderstorms accompanied by hail, lightning, wind squalls, downbursts and dust storms as well as heat and cold waves. According to the National Crime Records Bureau (NCRB) report, the high population density over India and lack of preparedness, causes more than 7000 human casualties every year in association with severe weather over the Indian region (https://ncrb.gov.in/en/adsi-reports-of-previous-years). In addition to human casualties, there are severe economic losses (De et al., 2005),huge agricultural losses (Rao et al., 2014; FAO, 2016; Chattopadhyay et al., 2017) and destruction of livestock and wildlife associated with occurrence of severe weather events (Narwade et al., 2014).

Forecasting such severe weather events over a tropical region like India is a challenge. The conditional instability of the lower troposphere (Williams & Renno, 1993) and the local vertical direction being almost perpendicular to the earth's axis of rotation so close to the equator (Sobel, 2012), contribute to the more spontaneous nature of weather events with consequent reduction of forecast skill in longer lead times. Most of these events are meso beta—meso gamma in scale (spatial scale of a few km to few tens of km) and have a lifetime of a few minutes to a few hours.

The increase in surface temperature as a result of climate change in the past few decades has an accelerating impact on global frequency of occurrence of severe weather. The change in temperature has a direct impact on the hydrological cycle as predicted by the Clausius-Clapeyron relation leading to a "dry-drier, wet-wetter" paradigm of weather (Held & Soden, 2006). According to trend analysis of occurrence of severe weather over India during recent years, extreme precipitation events are generally increasing in a contiguous region extending from the northwestern Himalayas in Kashmir through most of the Deccan Plateau in the south and decreasing values in the eastern part of the Gangetic Plain, northeastern states of India and parts of western Himalayas (Sen Roy & Balling, 2004). Studies also indicate a gradually increasing trend in frequency of thunderstorm days over east central and northeast Indian regions in the last 20 years where they are especially severe (Sen Roy & Sen Roy, 2021). The increasing trend in the past two decades is also reflected in the satellite based lightning observations over the Indian subcontinent (Qie et al., 2021). This increasing trend in severe weather has increased the vulnerability of the public and also increases the challenges for a weather forecaster (Scher & Messori, 2019).There is low-to-medium confidence of future intensification of severe convective storms over south Asia in a climate change scenario according to the latest IPCC report (Seneviratne et al., 2021).

Globally, in order to provide reliable predictions of weather and climate extremes on short and long time scales to reduce potential risks and damages, the WCRP Extremes Grand Challenge (Zhang et al., 2014; Alexander et al., 2016) is organized around four overarching research themes: Document (focusing on observational requirements), Understand (focusing on the relative roles of different spatial scales and their interactions), Simulate (focusing on model reliability and improvement), and Attribute (focusing on unraveling the contributors to extreme events). The Severe Weather Forecasting Program (SWFP https://community.wmo.int/activity-areas/severe-weather-forecasting-programme-swfp) of the World Meteorological Organization (WMO) was developed in 2006 to strengthen the capacity of WMO Members to deliver improved forecasts and warnings of severe weather in order to save lives and livelihoods and protect property and infrastructure and also develop capacity for impact-based forecasts and warning services for improved decision-making in collaboration with Public Weather Services Programme (PWS).

It is notable that the National Crime Records Bureau (NCRB) report of India indicates that on an average, nearly 40% of the total annual human casualties over India on account of "forces of nature" are associated with thunderstorms and associated lightning (https://ncrb.gov.in/en/adsi-reports-of-previous-years). This amounts to more than 2800 deaths every year, with an increasing trend in the past few years (Mahapatra et al., 2018).

Considering all the above, there is a need to augment the observation, modeling, research, forecasting and warning services for thunderstorms. With a goal to improve the forecasting of thunderstorms, especially during the March to May period when they are especially severe, the Government of India (Department of Science and Technology) started the nationally coordinated Severe Thunderstorm Observation and Regional Modeling (STORM) program in 2005. It was the first comprehensive observational and modeling effort to improve understanding and prediction of severe thunderstorms during the period of March to May every year (Mohanty et al., 2006, 2007). In view of the commonality of the weather systems affecting them, the STORM program was expanded later under the South Asian Association for Regional Cooperation (SAARC), to cover other South Asian countries—Bangladesh, Nepal and Bhutan in 2009, Pakistan and Afghanistan in 2012 and Sri Lanka and Maldives in 2013 (Das et al., 2014). The focus of the program was multi country additional observations for potential probable areas of thunderstorms during Intense Observation Periods (IOP) to better analyse the weather systems. A tri-weekly STORM Bulletin was generated, delineating thunderstorm prone areas over the Indian subcontinent for the subsequent two to three days. Simultaneously, the Commonwealth Games of 2010 first brought forth the need for a service for operational location specific accurate forecast of convective weather with short lead times and intensity all over the Indian region to address the rapidly evolving tropical weather. In view of the requirement, three hourly location specific nowcasting of thunderstorms, squall and hail storm was initiated under the aegis of STORM program, for 120 cities across India by Meteorological Centres of IMD in 2013 (Ray et al., 2015).

Since 2017, the STORM program has shifted focus to severe weather only over the Indian region during the period from March to June (Sen Roy et al., 2019, 2021). In this phase of the Forecast Demonstration Project (FDP) STORM program, the India Meteorological Department has taken the lead role to meet the growing requirements of Indian public for location specific, impact based warnings for severe weather related to thunderstorms. The purpose of this chapter is to discuss the major initiatives and findings from this phase of the FDP STORM program. The following section provides details of the new initiatives undertaken as a part of this phase of the STORM program and verification of the thunderstorm forecasts being generated as a part of this new endeavour. The last section of this chapter details the way forward towards improvement in analysis, forecast and dissemination of forecasts of thunderstorms and associated weather over the Indian region.

2 New Initiatives During the FDP STORM Program from 2017

IMD has put in place a new integrated system for thunderstorm forecasting and nowcasting to meet the growing demand for location specific information of severe weather and also handle an expected increase in extreme weather events in a climate change scenario. The program—FDP STORM addresses the targets of the WCRP Extremes Grand Challenge—Document, Understand and Simulate severe weather events and Attribute reasons to extreme events. Simultaneously, this program seeks to improve the forecasting and dissemination mechanism in order to better serve the public for severe weather forecasting.

Instead of thrice weekly bulletin earlier, a twice-daily "Severe Weather Advisory" Bulletin is being issued from the National Weather Forecasting Centre (NWFC), IMD (New Delhi) about potential areas for occurrence of severe weather 24 h in advance and updated in the afternoon. The different kinds of severe weather are delineated as polygons on the map of India after discussions with all the Meteorological Centres of IMD during a daily Video Conference held at 1030 IST. The bulletin optimizes upon the larger, subdivision scale weather forecasts issued in medium range (1–5 days) time scale. The detailed weather expected for a region is mentioned in the accompanying text in the Severe Weather Advisory Bulletin. The Severe Weather advisory is issued for the following phenomena: (a) Regions of thunderstorm occurrence and accompanying weather in the form of gusty wind (b) thunderstorm accompanied by damaging cloud to ground lightning (c) thunderstorm accompanied by hailstorm (d) thunderstorm with duststorm (e) intensity of rainfall spell.

While severe weather forecasts are generated throughout the year, special effort is made every year to introduce new products as well as analyse thunderstorms in greater detail during the period of March to June every year, when they are especially severe. A special severe weather "**FDP STORM Bulletin**" is generated daily. In addition to a summary of the actual weather that occurred during the past 24 h, there is a mesoscale weather summary in the bulletin for convection forecast during the upcoming 48 h over the Indian region. The summary analyzes global and mesoscale model forecasts of convection, severe weather indices and morning observations of temperature, moisture and wind to provides partial and temporal guidance for occurrence and intensity of thunderstorms. While this aspect of forecast methodology for severe weather events is similar to that of other global weather forecasting centres, the short temporal and spatial scale of development of weather over the Indian region has brought about further refinement to the methodology. District level nowcasts are issued by all State Meteorological Centres for all districts of India, as well as for about 1200 locations all over India, round the clock at three hourly intervals or less. This list of locations includes all the district headquarters all over India as well as towns and cities that are important for commercial or tourism purposes. The nowcasts are more detailed and quantitative than those given in the Severe Weather Advisory and provide the period of occurrence of the weather as well as its associated impact. These Nowcast warnings are updated on the IMD website at 3 hourly intervals at https://mausam.imd.gov.in/responsive/stationWiseNowcast.php. Additionally, district level nowcasts have been started for all the 735 districts of India from April 2018 (https://mausam.imd.gov.in/responsive/districtWiseNowcast.php). The impact expected due to the severe weather has also been added to the nowcast warnings in terms of colour codes following WMO Technical Note, 2015 (WMO, 2015) and NDMA guidelines (https://ndma.gov.in/images/pdf/Draft-Guidelines-thunderstorm.pdf). The categories for which nowcast warnings are issued are:

(i) ***Green colour*** (No severe weather)
(ii) ***Yellow colour*** (Light rain: < 5 mm/h/Light snow < 5 cm/h/Light Thunderstorms with maximum surface wind speed up to 40 kmph/Slight dust storm with wind speed up to 40 kmph and visibility is less than 1,000 m but more than 500 m due to dust/Low (<30%) probability of cloud to ground lightning occurrence)
(iii) ***Orange colour*** (Moderate rain: 5–15 mm/h/Moderate snow: 5–15 cm/h/ Moderate Thunderstorms with maximum surface wind speed between 41 and 61 kmph (In gusts)/Moderate dust storm with wind speed between 41 and 61 kmph and visibility between 200 and 500 m due to dust/Moderate (30–60%) probability of cloud to ground lightning occurrence)
(iv) ***Red colour*** (Heavy rain: > 15 mm/h/Heavy snow: > 15 cm/h/Severe Thunderstorms with maximum surface wind speed between 62 and 87 kmph (In gusts)/ Very Severe Thunderstorms with maximum surface wind speed > 87 kmph (In gusts)/Thunderstorms with Hail/Severe dust storm with surface wind speed (in gusts) exceeding 61 kmph and visibility is less than 200 m due to dust/High (>60%) probability of cloud to ground lightning occurrence).

Figure 1 is a sample map of the Day 1 Severe Weather Advisory issued by IMD while Fig. 2 displays the increase in the number of stations every year. Figure 3 displays a snapshot of station level and district level nowcast warnings issued at three hour intervals by various meteorological centres of IMD. When severe weather is expected, for maximum effectiveness of the warning, detailed SMS/WhatsApp messages and e-mails are issued to district collectors, State Disaster Management Authorities and local administration of the district concerned apart from print and electronic media. This expansion of nowcast services has become possible on account of many new products becoming available in recent years. This includes expansion of the Doppler radar network and AWS network throughout India and countrywide real time information of lightning flashes from the ground based lightning networks of the Indian Institute of Tropical Meteorology (IITM) and the Indian Air Force. This has permitted expansion of the nowcast service to stations not within the observation zone of the DWR network.

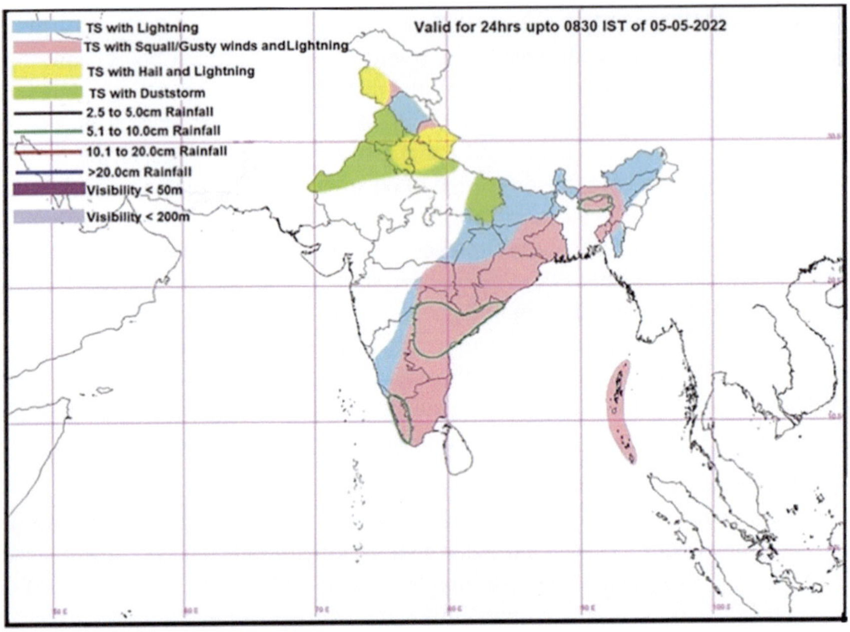

Fig. 1 24 h Severe weather advisory issued for 04 May 2022

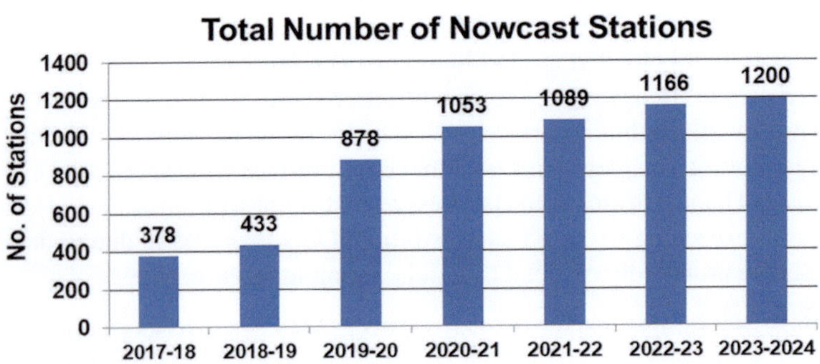

Fig. 2 Annual increase in the number of Nowcast stations

In April and May of 2018, there were wide spread devastating thunderstorms throughout India. The first of these storms happened on 11 April 2018, killing 19 people across North India. The next such incident happened during 02–03 May 2018. Thunderstorms with high velocity down drafts, swept across parts of North India and more than 134 people died and over 400 people were injured. In Uttar Pradesh, Agra district was the worst hit, accounting for at least 43 deaths and injuries to 51 others. In neighbouring Rajasthan, about 35 people died and over 209 were injured after winds downed more than 8,000 electricity poles and uprooted hundreds of trees. Eleven people were killed in Telangana, six were killed in Uttarakhand and

(a) (b)

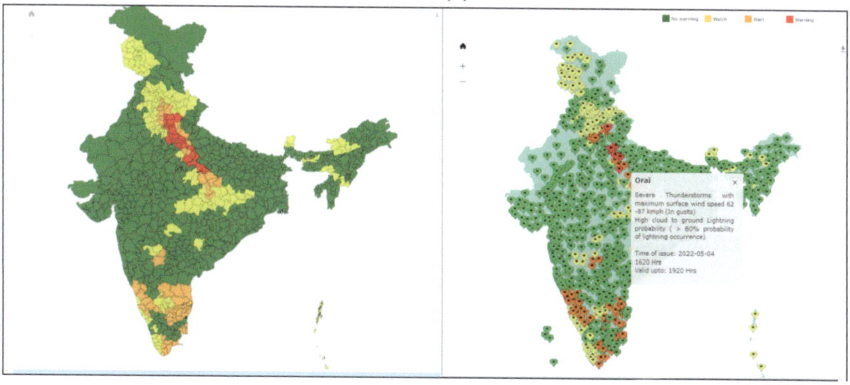

Fig. 3 **a** Snapshot of district level nowcasts at 1630 IST of 04 May 2022. **b** Snapshot of station level nowcasts at 1630 IST of 04 May 2022

two in Punjab. Nearly 100 people were also injured in Telangana, Uttarakhand and Punjab. Following these thunderstorms, there was damage to infrastructure. Power supply was cut off in many areas as squall uprooted trees and snapped electricity cables. At least 20,000 electric poles were uprooted and 2,500 electric transformers were damaged during the two days. More than 1,800 houses were damaged and 800 cattle were also killed (source: https://www.hindustantimes.com/india-news/thunderstorms-on-may-2-3-killed-134-injured-over-400-in-5-states-home-ministry/story-LjD8RCkaEEeo8cVlGWmINN.html). The third such event happened on 13 May 2018. On this day, duststorms and thunderstorms wreaked havoc in Uttar Pradesh, West Bengal, Andhra Pradesh and the national capital of Delhi.18 people died and 28 people were injured in Uttar Pradesh due to thunderstorm and hail, while 12 people including four children were killed in West Bengal, 9 in Andhra Pradesh, and 2 in Delhi. At several places in north India including Delhi, high-velocity winds uprooted trees and affected road, rail and air services (source: http://www.newindianexpress.com/nation/2018/may/13/thunderstorms-lightning-wreak-havoc-in-delhi-uttar-pradesh-west-bengal-and-andhra-pradesh-states-1814281.html).

In order to address the lacunae in the weather forecasting capabilities that caused these events, the newly formed Ministry of Earth Sciences focused on developing various mesoscale models to forecast lightning and other weather hazards associated with thunderstorms. This includes the National Centre for Medium Range Weather Forecasting regional unified model (NCUM-R) 4 km forecast for gusty winds, lightning, fog and air quality (Sandeep et al., 2021) and the Dynamic Lightning Parameterization (DLP scheme) based on cloud top height defined by the radar reflectivity factor in the Weather Research and Forecasting (WRF) model (Mohan et al., 2021).

Along with these developments, new Standard Operating Procedures (SOP) have been put in place throughout the country for prediction of thunderstorms in short range to nowcast scale (Sen Roy et al., 2021). These SOPs are dynamic and are periodically reviewed to update with respect to the new products that become available for severe weather detection and prediction.

All these developments have greatly increased the quality, outreach and effectiveness of severe weather warnings issued by IMD.

3 Verification of Thunderstorm forecasts Over India

A significant measure of improvement of skill is to objectively verify the forecasts and compare it with the verification scores of previous years. According to the WMO, verification needs to be done for two main purposes, administrative and scientific. At the administrative level, it is necessary to demonstrate the impact of the new developments described above, on the FDP STORM program in terms of improved operational forecasting services in a consistent manner during the past few years through some objective measure of forecast quality (WMO-UNESCO, 2014). At the scientific level, the main need is to establish the consistent, systematic spatial and temporal pattern of measurement of accuracy of thunderstorm forecasts for meteorological centres across India. In this section the thunderstorm forecast accuracy in terms of categorical scores is calculated in short range (24 h Severe Weather Guidance at subdivisional level) and nowcast scale (nowcasts for individual stations valid for 3 h and issued every three hours) every year, going back to 2016 when the new forecast system was put into place.

The thunderstorm data reported by observatories of IMD and the Indian Air Force are used for verification of the weather forecasts. WMO defines thunderstorms as one or more sudden electrical discharges, manifested by a flash of light (lightning) and a sharp or rumbling sound (thunder) (WMO, 2021). Of the network of 558 IMD observatories, only 141 are of Class I type, which are manned 24 h a day and send three hourly observations through the Global Telecommunication System (GTS), following WMO standards for recording present and past weather at and around a station in SYNOP code (WMO, 2011). The information is available on GTS throughout the world. These observations are supplemented by the significant weather report in the daily digital MMR (Monthly Meteorological Register) reports of these observatories. This information is stored in the National Data Centre of IMD at Pune. A combination of both datasets was available for verification during the period of seven FDP seasons (March to June) from 2016 to 2022.

In view of the mesoscale nature of thunderstorm activity over the Indian region, there is considerable argument to supplement the observatory reports with remotely sensed observations of thunderstorms. One of the proposed supplementary observations, with comparable length of uniform data (from 2016 to 2022 and available round the clock), is ground based lightning observations of the World Wide Lightning Network (WWLN). However, the low density of sensors of the WWLN network around the globe severely affects the lightning strike detection efficiency below lightning current amplitude of ~30 kA (Jacobson et al., 2006). Various studies have noted the 40 dBZ threshold in radar reflectivity to correspond to convective type of precipitation (Churchill & Houze, 1984; Steiner et al., 1995; Tokay & Short, 1996; Falk, 1997, Romatschke et al., 2010;). However, the current network of DWRs in India (at 250 km range) does not fully cover the Indian region, leaving some unsampled areas in the thunderstorm observations. Also on account of beam curvature away from the earth and beam broadening, thunderstorm cells which are far away from the radar site are not adequately sampled. Consequently, the present network is not dense enough for uniform validation of nowcasts and forecasts of thunderstorms. Also most of these DWRs have been installed in phases since 2014. The year wise installation of DWRs in India is shown in Fig. 4. Hence long term validation of the entire country using this data source is not possible. In order to maintain a uniform methodology of verification throughout the period of 2016–2022, and match the results with the verification methodology of previous years, only the data from the observatories of IMD and the Indian Air Force is considered for verification. This data is supplemented by the non realtime reports from class II and III observatories and data from the Indian Air Force. A total number of about 230 daily reporting stations are used for verification of the nowcasts during the March to June period.

The 24 h forecast is verified on a subdivision scale. Subdivision is a meteorologically homogenous part of India, which may comprise part of or a complete state or multiple small states. In order to verify the 24 h Severe Weather Advisory at subdivision level; if any observatory or news media reports the occurrence of thunderstorms in a location within the area of the subdivision delineated by the IOP, the forecast for the subdivision for the day is considered as correct. For verification purposes, a yes–no criterion (2×2 configuration table) is applied for occurrence-non-occurrence of thunderstorms. This data is used to compute various categorical skill scores in terms of (a) Probability of detection (POD) which measures the success of the forecast in correctly predicting the occurrence of thunderstorm, (b) False Alarm Ratio (FAR) which measures the number of false alarms per total number of thunderstorm predictions. (c) Critical Success Index (CSI) which is defined as the ratio of the number of hits (correct thunderstorm forecasts) to the number of events which occurred plus the number of false alarms. (d) Equitable Threat Score (ETS) which is a modification to the CSI, takes into account the number of correct forecasts of events (hits) that would be expected purely due to chance. The CSI is somewhat sensitive to the climatology of the event, tending to give poorer scores for rare events. It should be used in combination with other contingency table statistics (e.g., POD, FAR). The ETS is sensitive to hits and does not distinguish the source of forecast error and penalizes both misses and false alarms in the same way and is better for rare events (https://hwt.nssl.noaa.gov/Spring_2012/SkillScoresDescriptionSE2012.docx). For a good forecast, the FAR should approach zero and the rest of the variables should approach unity.

Similar verification metrics are calculated for the three hourly nowcasts issued by the State Meteorological Centres of IMD. However in this case, nowcast for thunderstorm is matched to the actual occurrence of thunderstorm over the Nowcast station. Since the number of nowcast stations is many more than reporting IMD observatories, the verification results for nowcasts for these stations are considered representative

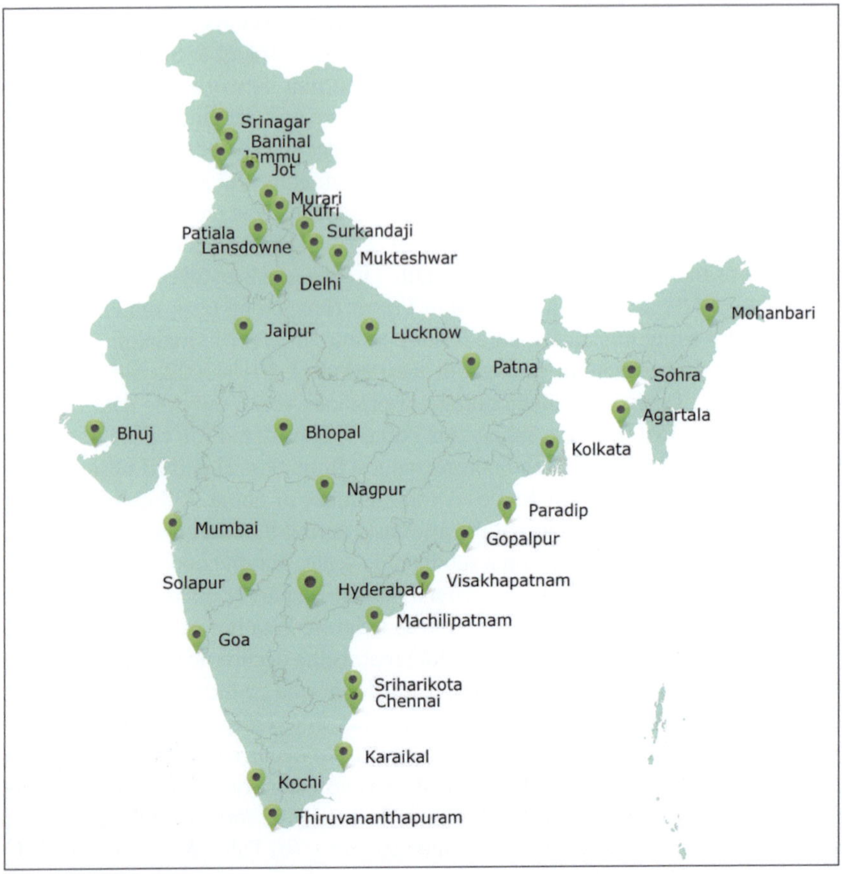

Fig. 4 The current Doppler radar network of India

of the skill of the larger nowcasting dataset. This is one of the limitations of this study. Media reports do not reliably give this information and hence are not used for verifying the three hourly nowcasts unless reported from multiple sources. While it may be argued that this is not a complete verification of all nowcasts issued by the respective Meteorological Centre, it may be considered an unbiased sample of the entire nowcast process. Also, it may be noted that this is a varying sample size, since every year, more and more stations are being brought under the Nowcast framework and the sample size is increasing annually. Attempt is being made to utilize the crowd sourced data in future for verification of nowcasts over all the nowcast stations.

3.1 *Verification of 24 h Advisory for Occurrence of Thunderstorms Over India*

Figure 5 displays the all India categorical verification scores during the March to June period-from 2016 to 2023 for 24 h thunderstorm advisory. As may be noted, there is systematic 200%improvement in the all India Probability of Detection (POD) scores, from 0.3 in 2016 to 0.86 in 2023. The False Alarm Ratio (FAR) score has remained nearly constant between 0.3 and 0.4. The Critical Success Index (CSI) score has also had a systematic improvement through the years with accuracy more than doubling in 2023 (0.61) over its value in 2016 (0.27). The Equitable Threat Score (ETS) score indicates that there was a 150% improvement in the forecast accuracy for rare events, increasing from 0.14 in 2016 to 0.34 in 2023. However, the improvement in error is not uniform throughout the country. Figure 6a–h display the spatial variability of the verification scores at subdivision scale in 2023 compared to the same scores in 2016. As may be noted, POD score for most of the subdivisions of west, central and peninsular India and the Himalayan subdivisions of Jammu and Kashmir and Arunachal Pradesh was below 0.4, with scores < 0.2 over the subdivisions of central and peninsular India in 2016 (Fig. 6a). POD score significantly improved in all the subdivisions by 2023, with scores > 0.8 in most subdivisions except parts of Gujarat, Maharashtra, Jammu & Kashmir, Bihar and Andaman and Nicobar Islands, where the score was 0.6–0.8 (Fig. 6b). This was a consistent pattern through the years, as is seen in Fig. 5. FAR score also significantly improved during this period (Fig. 6c for 2016 and Fig. 6d for 2023). Over most subdivisions, the score was between 0.2 and 0.6 in 2016 which improved to between 0.2 and 0.4 in 2023. However, scores were significantly poor for Lakshadweep Islands and Arunachal Pradesh in 2023. CSI score was < 0.4 over most of India in 2016 with slightly higher score over the subdivisions in northwest and adjoining west central India (Himachal Pradesh, Haryana, West Uttar Pradesh and West Madhya Pradesh) (Fig. 6e). The CSI scores significantly improved in all the subdivisions by 2023, with scores > 0.6 in most subdivisions of north, central, east, northeast and peninsular India. The CSI scores were poorer for Gujarat, Konkan & Goa, Arunachal Pradesh, Andaman & Nicobar and Lakshadweep (Fig. 6f). This was also a consistent pattern through the years, as is seen in Fig. 5. ETS Scores also improved from <0.2 in 2016 (Fig. 6g) to <0.4 in 2023 for most of the subdivisions (Fig. 6h). The poorest ETS scores were for Gujarat Region, South Interior Karnataka, Arunachal Pradesh and Lakshadweep subdivisions.

Figure 7a–d display the month wise evolution of the all India categorical scores from March to June during the period of 2016–2023. A consistent trend from 2016 to 2023, is the gradual improvement in the month wise forecasts (as noted from the scores) through the years. Another interesting result is the general pattern of improvement of forecasts from March to April and May and deterioration thereafter in June. The deterioration is less during recent years.

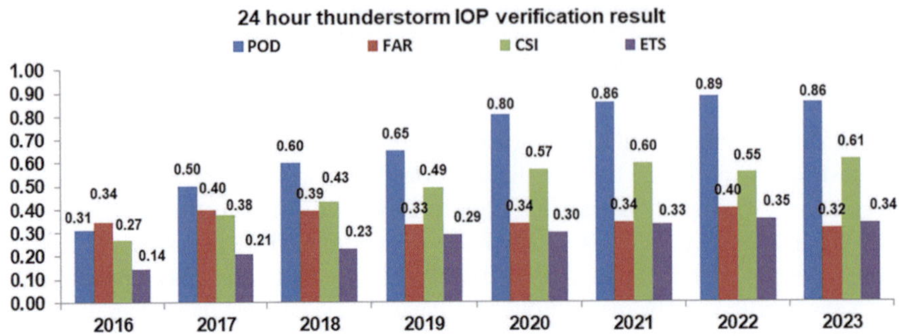

Fig. 5 24 h thunderstorm forecast verification result for the entire FDP season of 2016–2023

3.2 *Verification of 3 h Nowcast of Thunderstorms Over India*

Figure 8 displays the all India verification scores from 2016 to 2022 for 3 hourly thunderstorm nowcast. As may be noted, there is systematic improvement in the all India POD scores—from 0.74 in 2016 to 0.89 in 2020 with slight deterioration thereafter in 2021 and 2022 (0.83). The FAR score also significantly reduced in 2017 (0.37) as compared to 2016 (0.64) before increasing up to 2022 (0.57). The CSI score has also had a systematic improvement through the years up to 2020, with accuracy almost doubling (0.58) over its value in 2016 (0.32) before deteriorating slightly during 2021 and 2022 to (0.39).The ETS score showed similar trend with accuracy almost doubling in 2020 (0.56) over its value in 2016 (0.29) before deteriorating during 2021 and 2022 (0.37).Despite year-to-year variability, all scores indicated significant improvement up to 2022.

As mentioned in Sect. 2, the station level nowcasts were not started uniformly across all subdivisions at the same time. Hence the temporal evolution of spatial accuracy cannot be verified for all subdivisions for the same period. However, the statewise verification score during 2022 is displayed in Fig. 9a–d. POD score over most of the subdivisions was >0.6 in 2022, except Uttar Pradesh, Kerala, Lakshadweep and Telangana. FAR score is better for west and northwest India and east peninsular India, but is >0.8 over Kerala, Lakshadweep, Vidarbha, Uttar Pradesh and Northeast India. CSI and ETS scores show similar patterns with the worst score over the plains of north India, Maharashtra, Telengana and Kerala.

Figure 10a–d display the month wise evolution of the all India categorical scores for 3 hourly nowcasts from March to June during the period of 2016–2022. In 2016, POD score deteriorated from March (0.67) to April (0.36) and improved thereafter up to June (0.75), while FAR score was 0.6 in March, deteriorated slightly in April but improved slightly thereafter during the rest of the FDP season up to June (0.52). By 2022, the POD score was similar in March as compared to 2016 (0.65) and improved consistently up to June (0.89) while FAR score had an improving trend throughout the season. This trend of gradual improvement of FAR scores as the season progressed is seen as a persistent trend through the years except 2022 when scores deteriorated

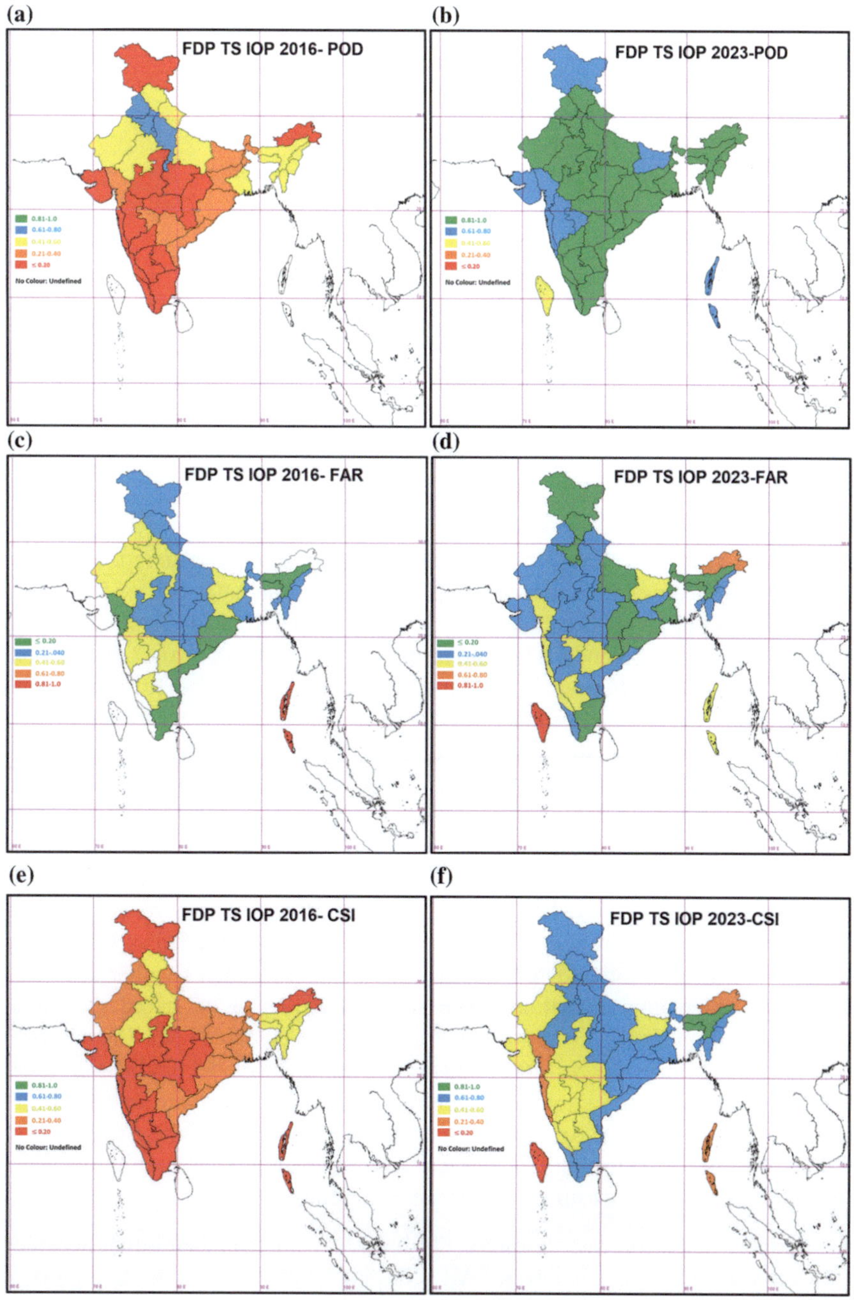

Fig. 6 **a** POD scores for 24 h TS IOP of Indian subdivisions in 2016. **b** POD scores for 24 h TS IOP of Indian subdivisions in 2023. **c** FAR scores for 24 h TS IOP of Indian subdivisions in 2016. **d** FAR scores for 24 h TS IOP of Indian subdivisions in 2023

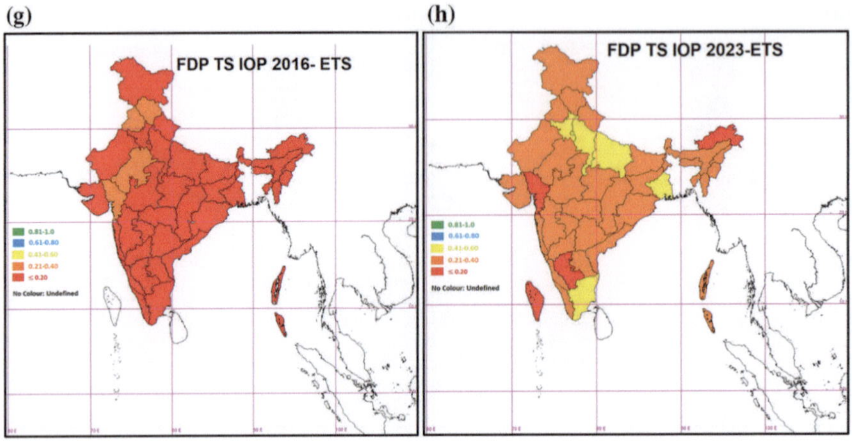

Fig. 6 (continued)

as compared to 2021 in all months. The change in these scores is also reflected in the change in CSI score, with a deteriorating trend during the latter half of the season replaced by an improving trend till the end of the season during all years. ETS scores also had an improving trend till the end of the season during all years. Overall, both CSI and ETS scores deteriorated in March but improved over the 2016 score with the progress of the season from 2016 to 2022.

4 Discussions and Conclusions

The present purpose of the FDP STORM program is to focus on improvement of the operational forecasting of thunderstorms over the Indian region. This aims towards accurate quantitative forecasts of thunderstorms and associated weather at district scale and with temporal resolution of 1–3 h over the Indian region. In order to implement these new initiatives, a multi-pronged approach was adopted involving improvement in observation resources, customization of mesoscale model products for thunderstorm prediction, and implementation of standard operating procedures for all aspects of severe weather. An important aspect of improvement of this service is to evaluate the forecasts and highlight the lacunae to target future scope of improvement. The major findings of this study may be summarized as follows:

(1) In short range forecast timescale (24 h forecast of thunderstorms), accuracy is lower (POD score is lower, FAR score is higher) over the western parts of India (Gujarat and west coast of India and interior Maharashtra), the western Himalayan region and the maritime islands (Andaman and Nicobar Islands). Previous studies indicate that during this period of the year, thunderstorm

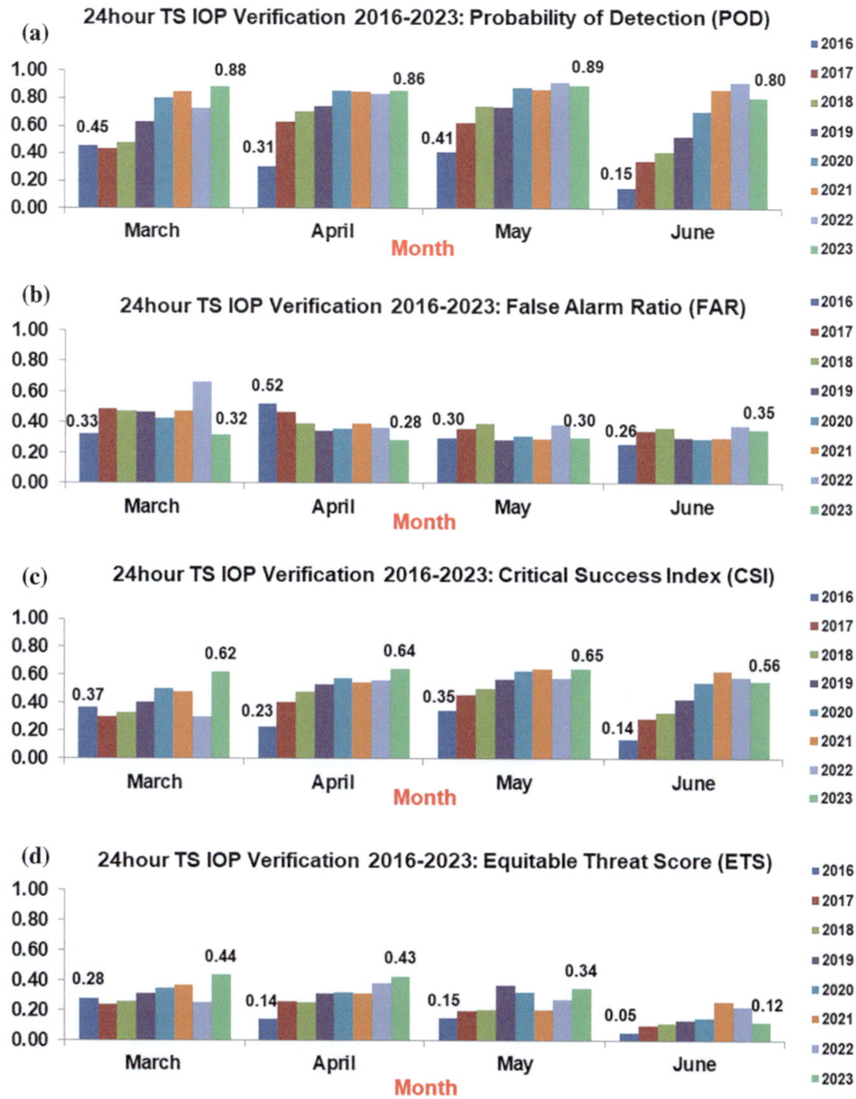

Fig. 7 **a** Month wise evolution of the all India POD scores from March to June during the period of 2016–2023. **b** Month wise evolution of the all India FAR scores from March to June during the period of 2016–2023. **c** Month wise evolution of the all India CSI scores from March to June during the period of 2016–2023. **d** Month wise evolution of the all India ETS scores from March to June during the period of 2016–2023

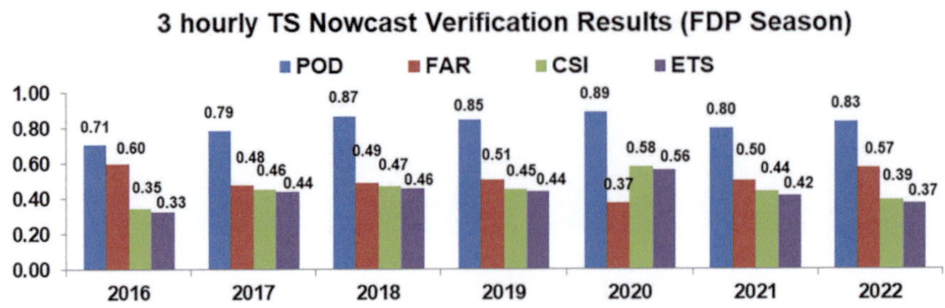

Fig. 8 Three hourly thunderstorm nowcast verification result for the entire FDP season of 2016–2022

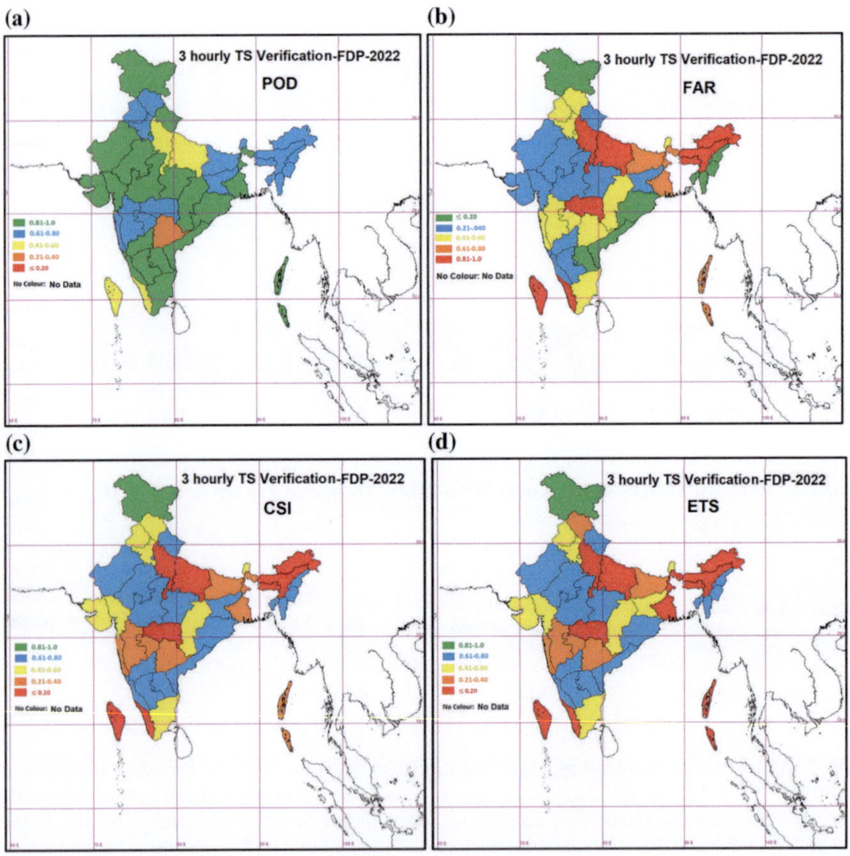

Fig. 9 **a** POD scores for 3 hourly nowcast over Indian states in 2022. **b** FAR scores for 3 hourly nowcast over Indian states in 2022. **c** CSI scores for 3 hourly nowcast over Indian states in 2022. **d** ETS scores for 3 hourly nowcast over Indian states in 2022

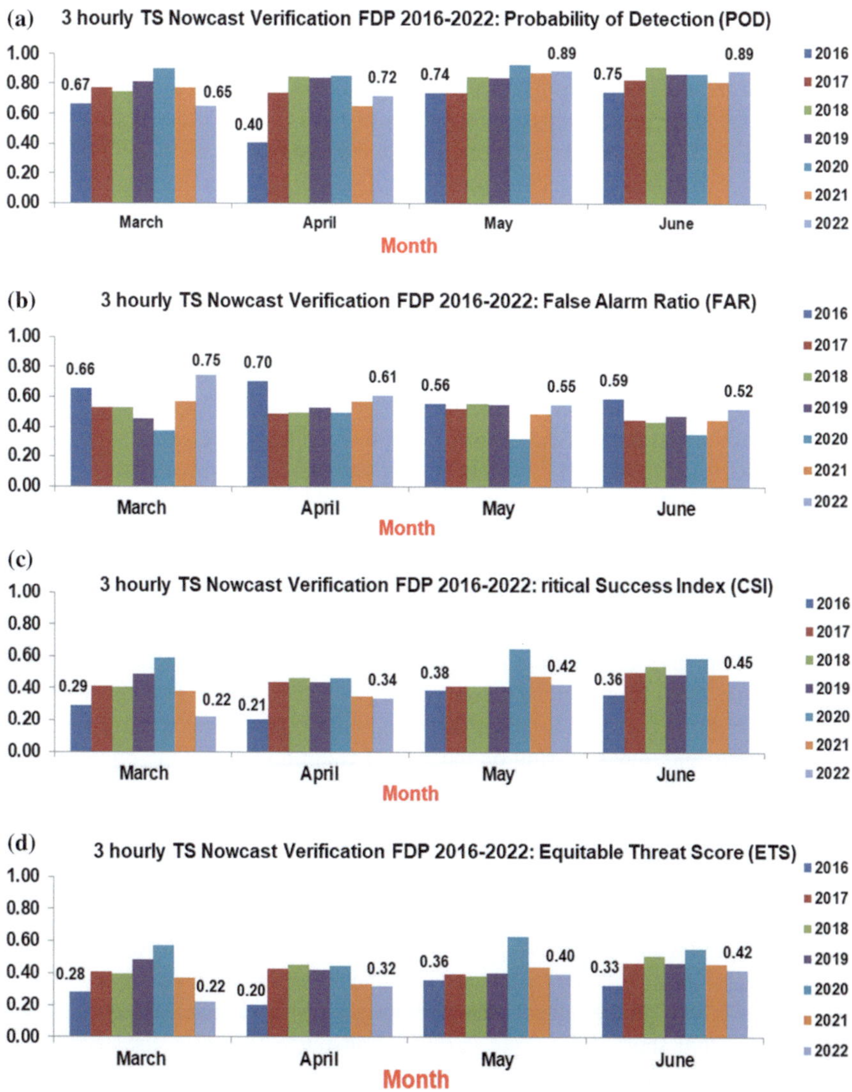

Fig. 10 **a** Month wise evolution of the all India 3 hourly nowcast POD scores from March to June during the period of 2016–2022. **b** Month wise evolution of the all India 3 hourly nowcast FAR scores from March to June during the period of 2016–2022. **c** Month wise evolution of the all India 3 hourly nowcast CSI scores from March to June during the period of 2016–2022. **d** Month wise evolution of the all India 3 hourly nowcast ETS scores from March to June during the period of 2016–2022

activity is subdued over the west Indian region on account of the strong anticyclone in the middle levels over the Arabian Sea (Srinivasan et al., 1973). For similar reasons, false alarms were higher in 2022 as compared to 2021 especially over east India, on account of the anomalously strong anticyclone during the first half of the FDP season which subdued the thunderstorm activity (FDPSTORM Report, 2022) and also during June in 2023. Under such conditions, predictions for thunderstorm occurrence made on the basis of lower atmospheric conditions, often fail due to unsupportive middle tropospheric conditions. The small spatial scale of the Andaman and Nicobar islands indicates that the predicted thunderstorm activity often occurs over the sea and goes unrecorded, thereby decreasing the accuracy of this region. Accuracy is higher in short-range time scale but poorer in nowcast time scale, over east and northeast India and western Himalayan region.

(2) On a monthly scale, all India short range thunderstorm forecast CSI score and ETS score are generally lower in March and June as compared to April and May. However upon detailed analysis, it is noted that POD as well as FAR score is lower in June as compared to March. This reflects a practical aspect of the forecast methodology adopted by an operational forecaster vis-à-vis the gradual increase in atmospheric moisture as the season progresses. With the increase in moisture in the lower troposphere as the season progresses, the probability of maturation of cumulus cells into cumulonimbus stage significantly increases and time taken to do so, decreases (Zhang & Klein, 2010). Hence forecasters may be erring on the side of imprudence while giving thunderstorm forecasts in March, and may be erring on the side of caution while giving thunderstorm forecasts in June.

(3) The monthly pattern of all India verification scores in nowcast timescales followed a similar pattern to short-range verification scores. The deterioration in scores during March and April of 2022 and June of 2023 compared to previous years may be ascribed to the subduedthunderstorm activity over the Indian region during the period on account of the active mid level anticyclone over parts of the Indian mainland, which increased forecast inaccuracy.

(4) However, some of the categorical scores for nowcast scale had a slightly different pattern from short-range scale. Unlike short-range forecasts, the verification scores of nowcasts over most of the subdivisions of north and east Indian plains and northeast Indian plains were poor. On the other hand, scores were poor for Maharashtra in short-range as well as nowcast scale. Scores in nowcast scale were better for west India, where short-range forecast skill was poor.

(5) Also, unlike the 24 h forecasts, the three hourly nowcasts generally improved from March to June, which is a consistent feature through the years.

It may be kept in mind that unlike the short-range time scale where forecasts are issued daily on the basis of an all Indian consensus, nowcasts are issued and verified at three hour intervals by the respective state meteorological offices. The persistent regional trends across multiple meteorological centres, points to lacunae in observations used for verification. It also points to the requirement for better understanding of thunderstorm characteristics over these regions in order to improve the forecast quality. With increasing public awareness of the effect of climate change on extreme weather events, there is urgent requirement for improvement of these services.

As mentioned in the introduction section, not only do thunderstorms have a devastating effect for humans, livestock, wildlife and agriculture over India, they also are a huge drain to the state exchequer. Due to their short lifetimes, mesoscale area of occurrence and fast temporal evolution over India, monitoring and forecasting these phenomena in a way that has an impact on the weather forecasting aspect is an evolving field. While monitoring these weather systems has a long history, this has gained impetus in recent years with the technological advancement in monitoring and forecasting of these systems as well as improvement of communication methods.

The nowcast warnings are now available for use by the general public through web based weather apps such as MAUSAM, UMANG and MEGHDOOT. The other major initiative in recent years is the Common Alert Protocol (CAP) as recommended by WMO and as implemented under the leadership of the National Disaster Management Authority (Mohapatra et al., 2022). The areas selected for warning under the Common Alert Protocol are automatically warned through their mobile phones.IMD has launched a public outreach initiative through a website for crowdsourcing of weather information where citizens will be able to record realized weather and can upload their local weather conditions (Chug et al., 2021).

Since thunderstorm is a highly localized event, the three hourly forecasts at district level are not location specific enough to take immediate action. In this context, the Indian Institute of Tropical Meteorology (IITM) has launched 'Damini', a free mobile-based application that can warn people about lightning in their vicinity at least 30–45 min before it strikes.

The newest initiative towards detailed analysis of thunderstorms is the **Thunderstorm Testbed Project** over east India, planned for the five year period of 2021–2026. The project is being organized by the Ministry of Earth Sciences comprising of scientists of IMD, IITM, NCMRWF for planning in-depth field experiments and numerical experiments with available and augmented instrumentation and computing facilities. This will be a first experiment of its kind in India. The equipment on the ground will be set up in the contiguous area of north Odisha, east Jharkhand and south West Bengal. Simultaneously a multi-agency thunderstorm project is proposed through joint collaboration of the Indian Space Research Organization, Defence Research and Development Organization and the Ministry of Earth Sciences in Balasore in East India. The observation equipment will comprise ground based observations, remote sensing measurements as well as drone measurements of the thunderstorms during their life cycle including formation, growth, movement and dissipation. All these developments have greatly increased the quality, outreach and effectiveness of impact based severe weather warnings issued by IMD.

References

Alexander, L., Zhang, X., Hegerl, G., Seneviratne, S., Behrangi, A., Fischer, E., Martius, O., Otto, F., Sillmann, J., & Vautard, R. (2016, June). *Implementation plan for WCRP grand challenge on understanding and predicting weather and climate extremes–the "Extremes Grand Challenge"*. Version.

Chattopadhyay, N., Devi, S. S., John, G., & Choudhari, V. R. (2017). Occurrence of hail storms and strategies to minimize its effect on crops. *Mausam, 68*(1), 75–92.

Chug, S., Nath, S., Srivastava, K., & Singh, S. L. (2021). Recent advances in communication infrastructure of IMD for early warning system. *Mausam, 72*(4), 935–946.

Churchill, D. D., & HouzeJr, R. A. (1984). Development and structure of winter monsoon cloud clusters on 10 December 1978. *Journal of the Atmospheric Sciences, 41*(6), 933–960.

Das, S., Mohanty, U. C., Tyagi, A., Sikka, D. R., Joseph, P. V., Rathore, L. S., Habib, A., Baidya, S. K., Sonam, K., & Sarkar, A. (2014). The SAARC STORM: A coordinated field experiment on severe thunderstorm observations and regional modeling over the South Asian Region. *Bulletin of the American Meteorological Society, 95*(4), 603–617.

De, U. S., Dube, R. K., & Rao, G. P. (2005). Extreme weather events over India in the last 100 years. *The Journal of Indian Geophysical Union, 9*(3), 173–187.

Falk, K. W. (1997). Techniques for issuing severe thunderstorm and tornado warnings with the WSR-88D Doppler radar (pp. 4–97). NOAA Technical memorandum SR-185. Retrieved May 13, 2020, from https://repository.library.noaa.gov/view/noaa/6360/noaa_6360_DS1.pdf.

F.A.O. (2016). Losses from climate-related disasters in agricultural sectors. FAO. https://www.fao.org/3/i6486e/i6486e.pdf.

FDP Report—Pre-Monsoon Thunderstorms during 2022: A Report. (2023). No. MoES/IMD/FDP/Storm Report/01(2023)/13.

Held, I. M., & Soden, B. J. (2006). Robust responses of the hydrological cycle to global warming. *Journal of Climate, 19*(21), 5686–5699.

Jacobson, A. R., Holzworth, R., Harlin, J., Dowden, R., & Lay, E. (2006). Performance assessment of the world wide lightning location network (WWLLN), using the Los Alamos sferic array (LASA) as ground truth. *Journal of Atmospheric and Oceanic Technology, 23*(8), 1082–1092.

Mahapatra, B., Walia, M., & Saggurti, N. (2018). Extreme weather events induced deaths in India 2001–2014: Trends and differentials by region, sex and age group. *Weather and Climate Extremes, 21*, 110–116.

Mohapatra, M., Kumar, N., Sharma, M., Devi, S. S., Senroy, S., Devi, K. S., & Singh, K. K. (2022). Short to medium range weather forewarning system in India. In *Social and Economic Impact of Earth Sciences* (pp. 1–29). Springer Nature Singapore.

Mohan, G. M., Vani, K. G., Hazra, A., Mallick, C., Chaudhari, H. S., Pokhrel, S., Pawar, S. D., Konwar, M., Saha, S. K., Das, S. K., & Deshpande, S. (2021). Evaluating different lightning parameterization schemes to simulate lightning flash counts over Maharashtra, India. *Atmospheric Research, 255*, 105532.

Mohanty, U. C., et al. (2006). Weather summary during pilot experiment of severe thunderstorms—Observations and regional Modeling (STORM) programme—2006 (13 April to 31 May, 2006). Report 177 of Department of Science & Technology, Government of India, Technology Bhawan, New Delhi.

Mohanty, U. C., et al. (2007). *Weather summary during pilot experiment of Severe Thunderstorms Observations and Regional Modeling (STORM) programme—2007*. Report 179 of Department of Science & Technology, Government of India, Technology Bhawan, New Delhi.

Narwade, S., Gaikwad, M. C., Fartade, K., Pawar, S., Sawdekar, M., & Ingale, P. (2014). Mass mortality of wildlife due to hailstorms in Maharashtra, India. *Bird Populations, 13*, 28–35.

Qie, K., Qie, X., & Tian, W. (2021). Increasing trend of lightning activity in the South Asia region. *Science Bulletin, 66*(1), 78–84.

Rao, V. U. M., Rao, B. B., Sikka, A. K., Rao, A. S., Singh, R., & Maheswari, M. (2014). Hailstorm threat to Indian agriculture: A historical perspective and future strategies.

Ray, K., Bandopadhyay, B. K., & Bhan, S. C. (2015). Operational nowcasting of thunderstorms in India and its verification. *Mausam, 66*(3), 595–602.
Romatschke, U., Medina, S., & Houze, R. A., Jr. (2010). Regional, seasonal, and diurnal variations of extreme convection in the South Asian region. *Journal of Climate, 23*, 419–439.
Sandeep, A., Jayakumar, A., Sateesh, M., Mohandas, S., Prasad, V. S., & Rajagopal, E. N. (2021). Assessment of the efficacy of lightning forecast over India: A diagnostic study. *Pure and Applied Geophysics, 178*, 205–222.
Scher, S., & Messori, G. (2019). How global warming changes the difficulty of synoptic weather forecasting. *Geophysical Research Letters, 46*(5), 2931–2939.
Seneviratne, S.I., Zhang, X., Adnan, M., Badi, W., Dereczynski, C., Di Luca, A., Vicente-Serrano, S. M., Wehner, M., & Zhou, B. (2021). Weather and climate extreme events in a changing climate (Chap. 11).
Sen Roy, S., & Balling, R. C., Jr. (2004). Trends in extreme daily precipitation indices in India. *International Journal of Climatology: A Journal of the Royal Meteorological Society, 24*(4), 457–466.
Sen Roy, S., & Roy, S. S. (2021). Spatial patterns of long-term trends in thunderstorms in India. *Natural Hazards, 107*(2), 1527–1540.
Sen Roy, S., Mohapatra, M., Tyagi, A., & Roy Bhowmik, S. K. (2019). A review of nowcasting of convective weather over the Indian region. *Mausam, 70*(3), 465–484. Retrieved May 13, 2020, from https://metnet.imd.gov.in/mausamdocs/17035_F.pdf. WMO.
Sen Roy, S., Sharma, P., Sen, B., Sathi Devi, K., Sunitha Devi, S., Gopal, N. K., Kumar, N., Mishra, K., Katyar, S., Singh, S. P., & Balakrishnan, S. (2021). A new paradigm for short-range forecasting of severe weather over the Indian region. *Meteorology and Atmospheric Physics, 133*, 989–1008.
Sobel, A. H. (2012). Tropical weather. *National Education Knowledge, 3*(12), 2.
Srinivasan, V., Ramamurthy, K., Nene, Y. R. (1973, December). India Meteorological department forecasting manual. Part-III: Discussion of Typical Synoptic Weather Situations; Deputy Director General of Meteorology (Forecasting) Poona, FMU Report No. III-2.2.
Steiner, M., HouzeJr, R. A., & Yuter, S. E. (1995). Climatological characterization of three-dimensional storm structure from operational radar and rain gauge data. *Journal of Applied Meteorology, 34*(9), 1978–2007.
Tokay, A., & Short, D. A. (1996). Evidence from tropical raindrop spectra of the origin of rain from stratiform versus convective clouds. *Journal of Applied Meteorology, 35*(3), 355–371.
Williams, E., & Renno, N. (1993). An analysis of the conditional instability of the tropical atmosphere. *Monthly Weather Review, 121*(1), 21–36.
WMO, UNESCO. (2014). Forecast verification for the African severe weather forecasting demonstration projects. *World Meteorological Organization Geneva Switzerland, 38*.
World Meteorological Organization. (2011). Manual on codes—International codes, volume I.1, annex II to the WMO technical regulations: part A—Alphanumeric codes, World Meteorological Organization (WMO), 2019 edition edn. Retrieved January 26, 2023, from https://library.wmo.int/index.php?lvl=notice_display&id=13617#.Y9JrgOxBw6E.
World Meteorological Organization. (2021). *Technical regulations, volume I: General meteorological standards and recommended practices. secretariat of the World Meteorological Organization.* Retrieved January 26, 2023, from https://library.wmo.int/index.php?lvl=notice_display&id=14073.
Zhang, Y., & Klein, S. A. (2010). Mechanisms affecting the transition from shallow to deep convection over land: Inferences from observations of the diurnal cycle collected at the ARM Southern Great Plains site. *Journal of Atmospheric Science, 67*(9), 2943–2959.
Zhang, X., Hegerl, G., Seneviratne, S., Stewart, R., Zwiers, F., & Alexander, L. (2014). *WCRP grand challenge: Understanding and predicting weather and climate extremes* (Technical report, World Climate Research Program). http://www.wcrp-climate.org/images/documents/grand_challenges/GC_Extremes_v2.pdf. White Paper.

How Long Could Winter Storms Last? Assessing Cold-Season Event-Based Extreme Precipitation Events Over the Conterminous U.S.

Toshi Matsui and David M. Mocko

Abstract This study examines cold-season event-based extreme precipitation statistics over the conterminous U.S. using radar-gauge merged 1/8th-degree hourly precipitation data from the North American Land Data Assimilation System (NLDAS) Phase 2 datasets. Analysis is conducted during Northern Hemisphere cold seasons (October to March) from 1997 to 2015. At each 1/8th-degree grid, event-based statistics were derived for storm duration, total accumulated precipitation, and duration of dry period. Histograms of event-based statistics are used to estimate the thresholds for *rare* (above 99th percentile) and *very rare* (above 99.9th percentile) events. The *very rare* extreme events exceed 10 cm of accumulated precipitation over 68 h in Northwest region. Climatology maps for the rare long-lasting winter storm events depicts unique distributions of these events. Further, time series of these statistics maps revealed that Northern Great Plains, Midwest, and Northeast regions of the United States experienced above-normal extreme precipitation accumulation events after 2006. However, these events are neither accompanied with prolonged duration of precipitation events nor dry periods, as hypothesized in previous.

T. Matsui (✉)
Mesoscale Atmospheric Processes Laboratory, Code 612, NASA/Goddard Space Flight Center, Greenbelt, MD 20771, USA
e-mail: Toshihisa_Matsui-1@nasa.gov

Earth System Science Interdisciplinary Center, University of Maryland, College Park, MD 20742, USA

D. M. Mocko
Hydrological Sciences Laboratory, NASA Goddard Space Flight Center, Greenbelt, MD, USA

Science Applications International Corporation, McLean, VA, USA

S. Das and W-K. Tao (eds.), *Severe Storms*,
https://doi.org/10.1007/978-981-97-7075-5_20

1 Introduction

During the last decade, the Arctic has warmed approximately twice as rapidly as has the entire Northern Hemisphere due to increased greenhouse gases and feedback processes (Stroeve et al., 2012). As a result, the temperature gradient between the equator and Arctic regions has been reduced. Decreasing Arctic sea ice has had a central role in this Arctic Amplification (e.g., Screen and Simmonds, 2010). Francis and Vavrus (2012) discussed that Arctic Amplification induced weakened zonal winds and increased the amplitude of Rossby waves, which favor persistent midlatitude weather patterns during the cold seasons, potentially inducing prolonged dry-period or precipitation extremes.

Liu et al. (2012) also linked Arctic Amplification to midlatitude blocking patterns across the U.S., which increased the likelihood of winter extremes. For long term, Changnon et al. (2006) also found an upward trend in snowfall extremes over the Midwest, East, and Northeast in the U.S. An overall upward trend of winter extreme in Changnon et al. (2006) indicates a climate shift in large-scale circulations that favor prolonged dry period as well as intensified winter storms discussed in Liu et al. (2012) and Francis and Vavrus (2012).

The Third National Climate Assessment (Melillo et al., 2014; available online at: http://nca2014.globalchange.gov) found similar increasing trends in total annual and heavy event precipitation in the Northeast and Midwest. Annual total precipitation changes for 1991–2012 compared to the 1901–1960 average showed the highest increase in precipitation over the Northeast, Midwest, and parts of the Northern Great Plains (Peterson et al., 2013). The Northeast and Midwest also show the highest increases in the heaviest 1% of daily precipitation from 1958 to 2012 (Karl et al., 2009). The NOAA U.S. Climate Extremes Index (2016) also shows an increase of extreme one-day precipitation over Conterminous U.S (CONUS) in recent decades (EPA, 2016).

Due to dormant status of vegetation and limited solar insolation, land surface is easily saturated during winter so that winter extreme precipitation could result in prolonged flooding event (citation). Therefore, this study examines event-based statistics of cold-season precipitation accumulation, precipitation duration, and dry period uniquely derived from an hourly radar-gauge assimilated precipitation dataset over a recent 18 years of record across the CONUS. Although the time record is relatively short as compared with the aforementioned previous studies, this study intends to provide unique insight of precipitation extremes by examining gridded hourly rainfall. This manuscript is organized as follow. Section 2 describe data and analysis method. Section 3 discusses histogram of extreme precipitation and non-precipitation events for each region of CONUS, their spatial distributions, and interannual variability. Section 4 summarizes the results.

2 Method

The motivation of this study is to create unique event-based precipitation statistics from the North American Land Data Assimilation System (NLDAS) (Mitchell et al., 2004; Cosgrove et al., 2003). NLDAS rainfall data is an hourly 1/8th-degree resolution rainfall rate dataset covering the conterminous U.S. and parts of Canada and Mexico from 25 to 53 North. Over CONUS, the data is essentially a merged gauge-radar product produced by temporally disaggregating a CPC PRISM-adjusted daily rain gauge analysis using hourly weights from NCEP Stage II Doppler radar precipitation estimates. NLDAS Phase 2 data (Xia et al., 2012) is used in this study, which uses CMORPH (NOAA CPC MORPHing Technique; Joyce et al., 2004) precipitation estimates (available since December 2002) from satellite microwave observations for the temporal disaggregation if the Stage II radar data is not available. These gaps in the radar data, from lack of radar coverage to equipment maintenance, average around 13% for the total area of the conterminous U.S. If both the Stage II radar data and the CMORPH data are not available, the 2.0 $\times$ 2.5-degree hourly CPC Hourly Precipitation Dataset (HPD; Higgins et al., 1996) is used for the temporal disaggregation. North American Regional Reanalysis (NARR) model-simulated precipitation is used if all three other products are unavailable. No matter the method of the disaggregation, the total daily rainfall amounts remain the CPC PRISM-adjusted gauge-based analysis values.

In a previous study, NLDAS-2 precipitation was used to construct hourly diurnal cycles of warm-season precipitation (Matsui et al. 2010). This study integrates hourly NLDAS precipitation from 1997 to 2015 during cold seasons (Oct 1996–Mar 1997 to Oct 2014–Mar 2015) periods, since the impacts of Arctic Amplification are more apparent in this season (Liu et al., 2012, Francis and Vavrus, 2012). Although the NLDAS-2 hourly datasets extend back to Jan 1979 and continue to be produced operationally at NOAA, only the period after 1996 are considered for this study. The reason is that Stage II observations were not widely available before mid-1996, and significant breakpoints of NLDAS-2 sub-daily precipitation frequency and intensity were found in mid-1996 (Ferguson and Mocko, 2017). Thus, only the period from October 1996 is used in this study, when the Stage II radar estimates are primarily used for temporal disaggregation.

Gridded hourly precipitation is used to determine event-based precipitation duration, here after denoted as continuous wet period (WetP), and simultaneously accumulated precipitation (AccumP) is measured for each WetP event. Meanwhile, continuous non-precipitating dry period (DryP) is defined between each WetP events. These statistics are derived from October 1_{st} to March 31st for each winter. On Mar 31st, the WetP or DryP become discontinued. This event separation is strict in that even a one-hour discontinuity of precipitation is separated into different precipitation events.

Previously, observational records of hourly precipitation record are relatively sparse and limited, and so are analyses of sub-daily or event-based precipitation extremes (Westra et al., 2014). This study highlights event-based statistics by using hourly precipitation rate, which is important to consider dependence on duration on extreme precipitation events (Westra et al., 2014). Unlike the study in Changnon et al. (2006), this study considers not only snowfall but also total measured precipitation (including both rain and snow) events during cold seasons.

3 Results

3.1 Histogram of Extreme Events

18-year event-based AccumP, WetP, and DryP are used to construct histograms for the entire CONUS, as well as over National Climate Assessment (NCA) regions, including Northwest (NW), Southwest (SW), Great Plains (divided for this study into Northern Great Plains (NGP) and Southern Great Plains (SGP)), Midwest (MW), Northeast (NE), and Southeast (SE). These histograms are used to define "rare" (above 99th percentile) and "very rare" (above 99.9th percentile) events by constructing accumulated probability density functions from the histogram. These rare and very rare events represent 99th and 99.9th percentile of event occurring for different NCA regions.

Figure 1 shows histograms of AccumP for CONUS and the different NCA regions with corresponding rare and very rare thresholds values. All histograms are exponentially distributed with different slopes, and extended beyond 45 cm per single continuous event. Among the different NCA regions, NGP shows the most narrowly distributed histogram up to 18 cm, while NW shows the mostly broadly distributed histogram beyond 45 cm. This simply expresses that the NW region most often experiences very rarely occurring extreme AccumP events, and vice versa in the NGP region. Thresholds of rare and very rare events strongly vary over the different NCA regions. For example, the wettest regions, NW, shows 3.1 cm and 10.5 cm of AccumP, while the driest region, NGP, shows 1.6 and 3.6 cm of AccumP for rare and very rare cases. Over the entire CONUS, they are 3.5 and 8.6 cm, respectively. The CONUS values are also used as threshold of event mapping later.

Similarly, Fig. 2 shows histograms of WetP for CONUS and other NCA regions. The histograms are extended up to 288 h (12 days) period. We found small but periodic spikes at every 24 h increment in CONUS and the individual NCA regions. This would be an artifact of the NLDAS precipitation product. This is most likely related to events when the Stage II radar-estimated daily total precipitation is zero, but the CPC gauge-based analysis has non-zero precipitation for that day. In this case, the temporal disaggregation spreads the precipitation evenly over the entire day (24 h period). This results in the artifact that precipitation may be slightly more likely to start or stop exactly on 24 h periods in the NLDAS-2 dataset.

Nevertheless, this histogram shows unique variability of WetP among different NCA regions. The most broadly distributed histogram is NW, the wettest region, similar to the AccumP distribution. Two top very rare WetP regions are NW (68 h) and SW (58 h), where atmospheric river often generate long-lasting precipitation events

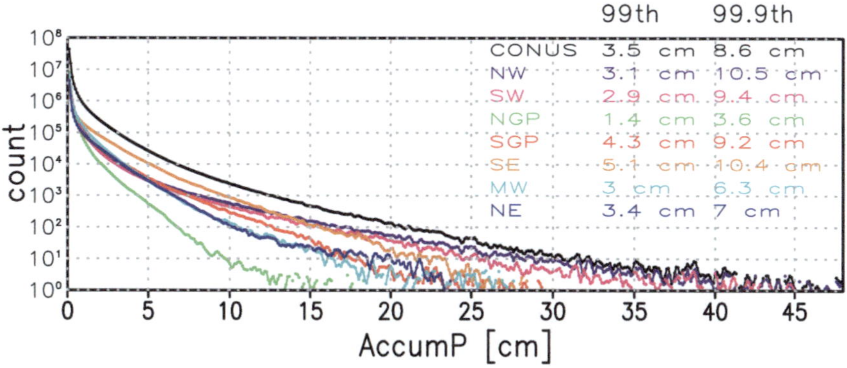

Fig. 1 18-year histogram of accumulated precipitation (AccumP) for over the entire CONUS, as well as National Climate Assessment (NCA) regions, including Northwest (NW), Southwest (SW), Northern Great Plains (NGP), Southern Great Plains (SGP), Midwest (MW), Northeast (NE), and Southeast (SE). (Spatial depiction of the NCA regions are shown in Fig. 4). Tables describe 99th (rare, left column) and 99.9th percentile (very rare, right column) events for each NCA region

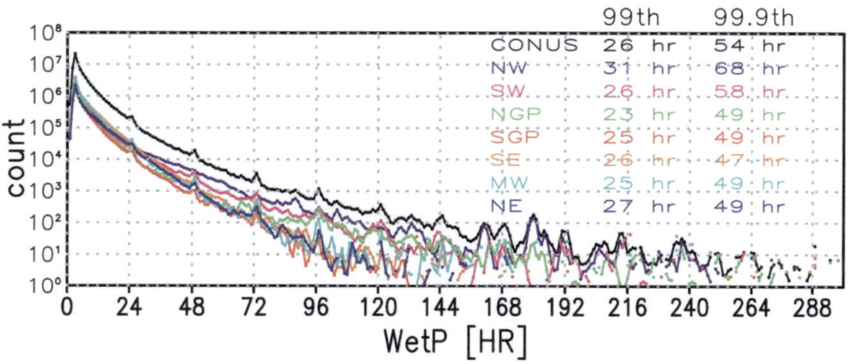

Fig. 2 Similar to Fig. 1, but using continuous wet period (WetP)

(Cobb et al., 2022). However, NGP's WetP is also extended broadly beyond 288 h, although the AccumP histogram is much narrowly distributed (Fig. 1). This is simply because cold-season rare precipitation events in NGP could be most likely characterized by long-lasting snowfall accumulation. In comparison with AccumP, thresholds of rare and very rare events between NGP, SGP, SE, MW, and NE resemble very closely each other, which are about 1-day and 2-day periods, respectively. Over the entire CONUS, thresholds of rare and very rare events are 26 and 54 h, respectively, which will be used to create spatial map of these events.

Figure 3 shows histograms of DryP events for CONUS and the different NCA regions. The histograms are extended beyond 2880 h (~four months). The most broadly distributed histogram appears in the SW, where more frequent dry spells create desert and semi-desert environment. SW is also characterized with frequent

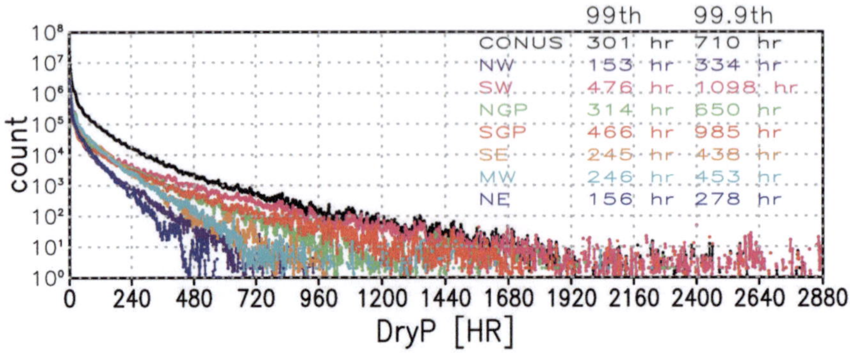

Fig. 3 Similar to Fig. 1, but using continuous dry period (DryP)

persistent rainfall events simultaneously (Fig. 2), because this region encompasses both coastal and semi-arid zones (discussed in the next section). On the other hand, NW (153 h in rare event) and NE (156 h in rare event) have shortest DryP among the NCA regions. Overall, in comparison with WetP, DryP histograms exhibit more variability than WetP does. Thresholds of rare and very rare events over CONUS are 301 and 710 h, respectively. Again CONUS thresholds will be used to create spatial map of these events.

3.2 Climatology Map of Rare Precipitation Events

Using the previous threshold values of the CONUS rare and very rare events, unique climatology maps are constructed for examining spatial variability of such extreme precipitation event frequencies (Figs. 4, 5 and 6). In these maps, event frequencies are normalized per year; i.e., describing how many events could happen per each cold season in the climatological sense.

Figure 4 describes distributions of rare (above 99th percentile) and very rare (above 99.9th percentile) AccumP events. The coastal zones of the West Coast in the NW and SW regions receive the highest frequencies of both rare and very rare AccumP events, peaking in the Cascade Range. These are mostly associated with strong winter-time precipitation events due atmospheric rivers, often called the "pineapple express", due to moisture transport from the central Pacific near the Hawaiian Islands (e.g., Higgins et al., 2000). There are relatively high frequencies over the eastern portion of SGP, southern portions of MW, eastern portions of NE, and over much of the SE, peaking around Louisiana. The highest frequency of very rare AccumP events appears in Olympic Mountain in NW, where more than 8.5 cm continuous precipitation events likely happen ~10 times in each cold season. As seen in Fig. 2, these very rare events are historically recorded up to 45 cm-accumulated rain. The distribution maps also depict that NGP receives the least frequent rare and very rare AccumP events (Fig. 1).

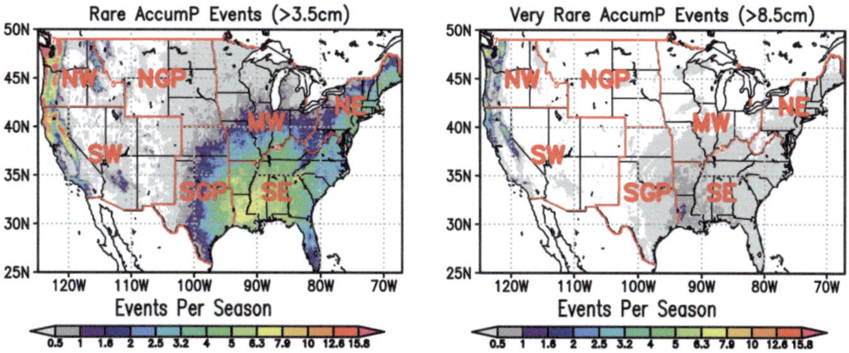

Fig. 4 Spatial distributions of rare (CONUS 99th percentile) and very rare (CONUS 99.9th percentile) events of accumulated precipitation (AccumP). Note that the NCA Great Plains region is divided into Northern (NGP) and Southern (SGP) in this study

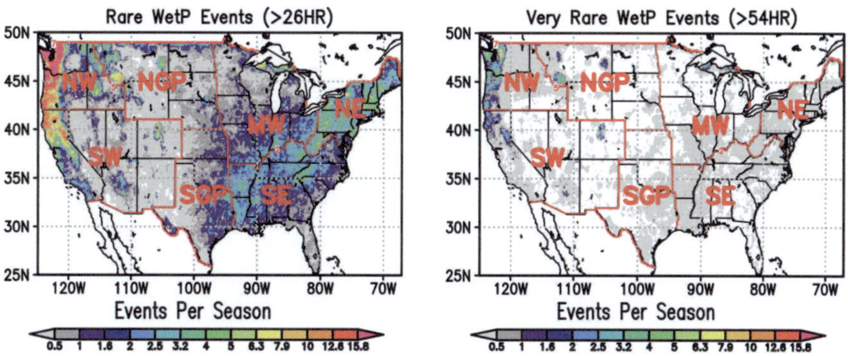

Fig. 5 Similar to Fig. 4, but using continuous precipitation period (WetP)

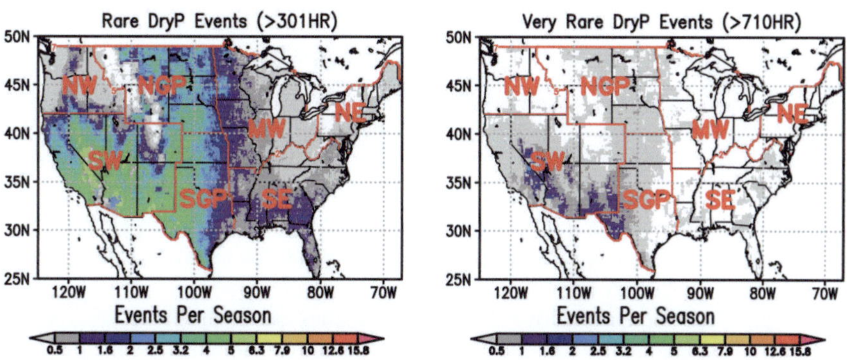

Fig. 6 Similar to Fig. 4, but using continuous dry period (DryP)

Figure 5 describes the distribution of rare and very rare WetP events. Similar to the AccumP map, the West Coast receives the highest frequencies of both rare and very rare AccumP events. In this map, several isolated high-frequency spots are found in NW, SW, and NGP. These isolated spots are clearly draped over the high mountain ridge, e.g., the Rocky Mountains in Colorado. However, it should be noted that NLDAS precipitation often lacks coverage of the radar and gauge networks for these high-mountain elevations. Uncertainties in gridded precipitation datasets, especially over mountainous regions, including NLDAS, were explored through the use of ensemble estimates over CONUS by Newman et al. (2015). However, the result shown in this study closely agrees with the snow-storm climatology map constructed with weather stations from 1901 to 2001 by Changnon et al. (2006).

There are also isolated high-WetP spots along the Lake Superior, which is associated with lake-effect snowstorms (e.g., Eichenlaub, 1970). Spatial distributions of WetP are quite different from those of AccumP over SE and NE. High frequencies of rare AccumP events are more concentrated toward coastlines, while more frequent rare WetP events appear along the Appalachian Mountains. These events are mostly characterized by Nor'easter events (Zielinski, 2002). The highest frequency of very rare WetP events appears around Glacier Peak in the northern Cascade Range, where 54 h continuous precipitation events likely happen ~10 times in each cold season.

Figure 6 describes the spatial distribution of rare and very rare DryP events. These are the counterpart of the WetP maps so that high frequencies of rare DryP events (>301 h or 12.5 day) appear in the SW, NGP, and SGP regions, excepting mountain ranges. These represent the broadly distributed histogram in Fig. 3. On the other hand, NE and NW have the lowest frequencies of such dry spells during the cold season. The highest frequency of very rare DryP events (>710 h or >29.6 day) appears around Death Valley in southern California. Again, as seen in Fig. 3, these very rare DryP events are recorded beyond 2880 h (~4 months) continuous period.

3.3 Interannual Variability of Rare Precipitation Events

In this section, event-based AccumP, WetP, and DryP histograms are constructed similar to Fig. 1, but separately for each Oct-Mar season, and the histogram of each year is subtracted from the 18-year mean histogram so that time series of histogram anomalies are derived. As a result, Figs. 7, 8 and 9 show interannual variability of the rare events histogram (units: number of events per cold season) over the CONUS as well as the each NCA region.

In Fig. 7, the largest positive AccumP anomalies occur during 2005–2007 in CONUS. These anomalies are contributed from positive anomalies from almost all NCA regions, except the NGP region. It should be noted that due to overwhelming sample number of precipitation events over NW, the CONUS time series is strongly dragged by the results in NW; i.e., the anomalies time series appears to resemble between CONUS and NW. In the NGP, MW, and NW regions, there is an increasing tendency in rare and very rare AccumP events after around 2006. Note that these

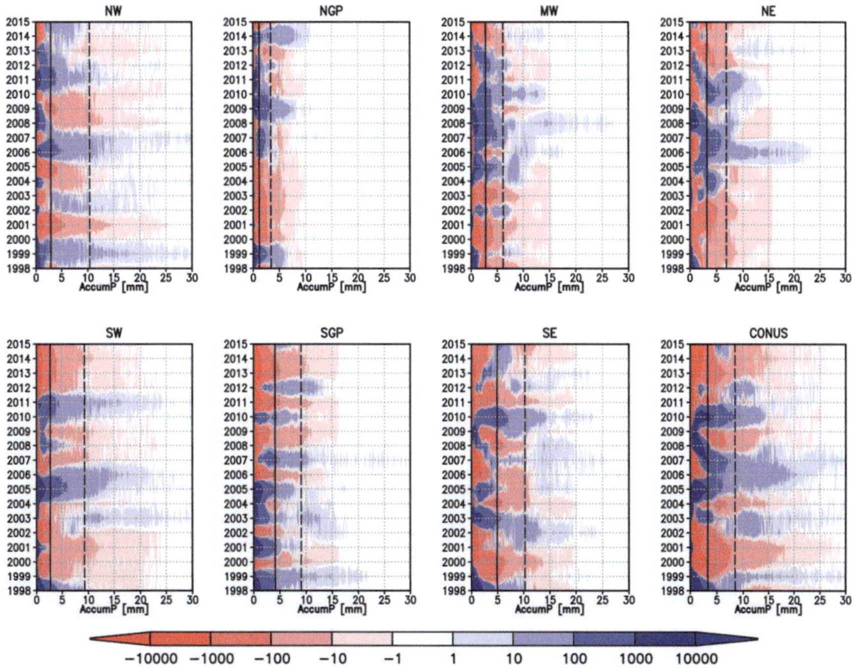

Fig. 7 Time series of AccumP histogram anomalies from 1998 to 2015 for CONUS and the NCA regions. Solid and dotted lines represent rare (99th percentile) and very rare (99.9th percentile) event thresholds for each NCA region. Unit is the number of events per cold season

correspond to an abnormal decline 1000–500 hPa thickness difference and 500 hPa zonal wind, as measured as important Arctic Amplification index (Francis and Vavrus 2012), which correspond to the positive anomalies of extreme AccumP events in these years.

Figure 8 depicts time series of WetP anomalies. Year 2000 dominates the positive WetP anomalies in rare and very rare events, which is largely contributed from NW and NGP. In similar manner, the anomalies time series between CONUS and NW appears to resemble closely each other due to dominant sampling number in NW. It is interesting that more frequent positive WetP anomalies after 2005 in NGP, MW, and NE, similar to those of AccumP. However, these positive anomalies are mostly contributed neither from rare nor very rare events. Thus, this contrast between AccumP and WetP anomaly time series suggests that rare and very rare AccumP events in these regions are not always contributed from longer-persisting precipitation events, but from relatively shorter-lasting (<24 h) intense precipitation events.

In addition to the quality of the NLDAS precipitation product over mountainous areas, one large complication is that the response of orographic precipitation extremes to a changing climate responds differently than precipitation extremes over flat terrain due to fundamental physical drivers associated with a combination of mountain wave dynamics and midlatitude jet stream (O'Gorman, 2015). Thus, this induces

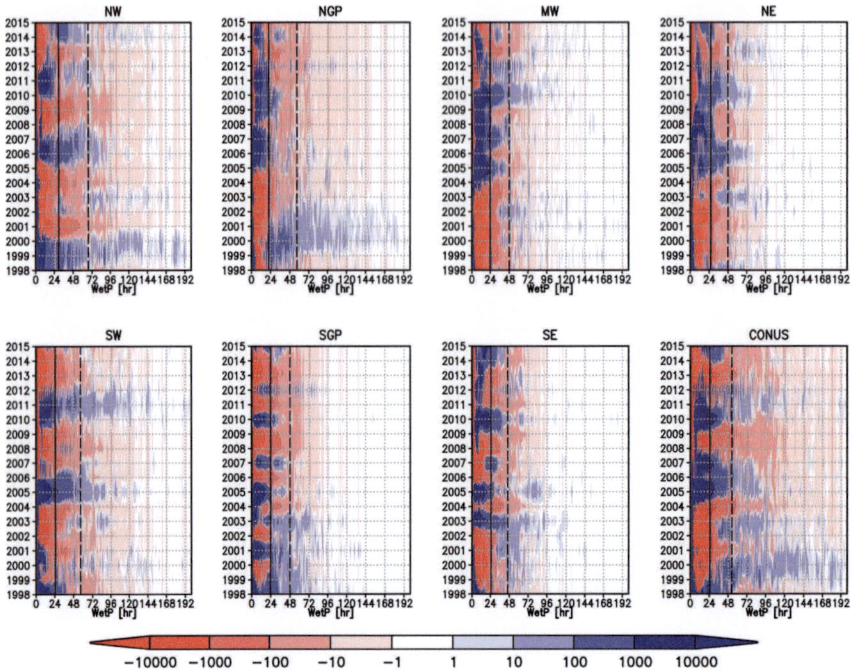

Fig. 8 Similar to Fig. 7, but using WetP

uncertainties in analyzing the trends or variability of our results in the western portion of CONUS regions.

Figure 9 describes time series of DryP anomalies. The largest positive DryP anomalies occur during 2006–2008. These positive anomalies are largely contributed from those of NGP and SGP. These anomalous years are somewhat coincident with the largest positive AccumP anomalies (Fig. 7, CONUS). In other words, in these years, CONUS experienced abnormally heavy precipitation as well as abnormally long dry periods during cold seasons.

Unlike WetP time series, there are no consistent tendencies of positive strong dry-period anomalies in MW and NE after 2006. This indicates that more frequent intense cold-season precipitation events are not evidently accompanied with extended extreme dry-period (non-precipitation) tendencies due to prolonged weather conditions, as indicated by Francis and Vavrus (2012). In fact, NGP exhibits different tendencies in that more (less) frequent shorter-term (longer-term) dry periods appeared after 2008, and vice versa before 2008; i.e., extreme cold-season dry-period events are happening less frequently after 2008.

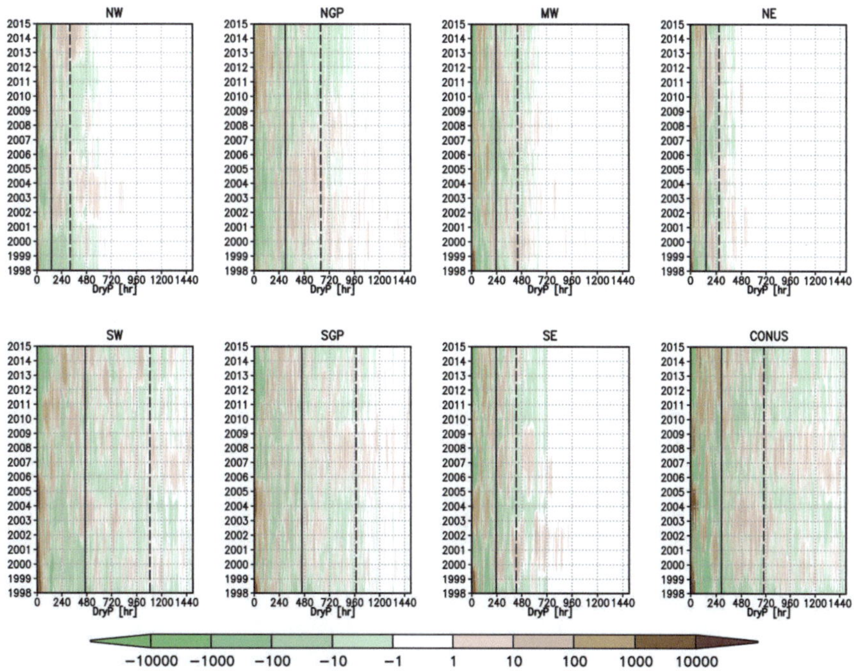

Fig. 9 Similar to Fig. 7, but using DryP

4 Summary and Discussion

This study examines gridded event-based precipitation statistics in order to characterize spatio-temporal variability of extreme rare precipitation and dry-period events over the CONUS and different NCA regions. In contrast to most of the previous studies that used daily precipitation (extensively reviewed by O'Gorman 2015) or snow-storm criteria (Changnon et al. 2006), the NLDAS-2 hourly precipitation product revealed detailed distributions of event-based frequencies of precipitation variability in time and space.

We have quantified extreme events by accounting for the 99th percentile and 99.9th percentile of event-based precipitation statistics for CONUS and NCA regions. The spatial distributions of these extreme (rare and very rare) events are summarized in the spatial distributions over CONUS, and qualitatively agree with previous studies (Changnon et al., 2006, Higgins et al., 2000, Zielinski et al. 2002).

Further, we have identified the temporal variability of these extreme events, and found that NGP, MW, and NE regions experienced more extreme precipitation events after 2006, corresponding to the very strong Arctic Amplification index as measured by Francis and Vavrus (2012). However, these events are not accompanied with prolonged duration of precipitation events. In the same years, MW and NE does not have clear feature of more frequent long-lasting dry-period events, too. In fact, NGP experienced less frequent extreme dry-period events after 2009.

A simple theoretical framework suggests that snowfall extremes tend to occur around −2 °C to −4 °C (O'Gorman 2015). Therefore, if amplified slow-moving Rossby waves (midlatitude jet stream) create more situations of this snow-extreme optimal temperature, the theory suggests additional likelihood of increase in snow extremes. But, such attempt is beyond the scope of this manuscript.

El Niño Southern Oscillation (ENSO) and the North Atlantic Oscillation (NAO)/ Arctic Oscillation (AO) phenomena are well-known climate modes that affect cold-season precipitation patterns over CONUS. For example, El Niño (La Niña) induces a more frequent south-ward (north-ward) storm track in the southern (northern) CONUS (e.g., Redmond and Koch, 1991; Smith and O'Brien 2001). Positive (negative) NOA induces cold (warm) temperature and more snowy (warm rain) situations over eastern CONUS. However, the observed time record does not simply explain the cold-season precipitation extreme variability exploited in this study due to rather short-term period (18 years). Thus, both Arctic Amplification as well as dominant climate modes must be considered when projecting future cold-season precipitation extremes in the future.

Nevertheless, this study provides detailed insight of extreme precipitation statistics using accurate high-resolution precipitation datasets. In addition to foci of this paper, these findings and datasets will be also useful for examining the performance of convection-permitting climate simulations (Prein et al. 2016) and/or verification of satellite high-resolution precipitation datasets (Huffman et al., 2015).

Acknowledgements NLDAS project is a collaboration project among several groups and funded with support from the NOAA Climate Program Offices's Modeling, Analysis, Predictions, and Projections (MAPP) Program. The NLDAS-2 forcing data (https://doi.org/10.5067/THUF4J1RLSYG) used in this effort were acquired as part of the activities of NASA's Science Mission Directorate and are archived and distributed by the Goddard Earth Sciences (GES) Data and Information Services Center (DISC). We thank Brian Cosgrove and Charles Alonge for leading the development of the NLDAS Phase 2 rainfall product.

References

Changnon, S., Changnon, D., & Karl, T. (2006). Temporal and spatial characteristics of snowstorms in the contiguous United States. *Journal of Applied Meteorology and Climatology, 45*, 1141–1155. https://doi.org/10.1175/JAM2395.1

Cobb, A., Ralph, F. M., Tallapragada, V., Wilson, A. M., Davis, C. A., Monache, L. D., Doyle, J. D., Pappenberger, F., Reynolds, C. A., Subramanian, A., Black, P. G., Cannon, F., Castellano, C., Cordeira, J. M., Haase, J. S., Hecht, C., Kawzenuk, B., Lavers, D. A., Murphy, M. J., Jr., Parrish, J., Rickert, R., Rutz, J. J., Torn, R., Wu, X., & Zheng, M. (2022). *Atmospheric river reconnaissance 2021: A review, weather and forecasting (published online ahead of print 2022)*. Retrieved Feb 18, 2023, from https://journals.ametsoc.org/view/journals/wefo/aop/WAF-D-21-0164.1/WAF-D-21-0164.1.xml.

Cosgrove, B. A., et al. (2003). Real-time and retrospective forcing in the North American Land Data Assimilation System (NLDAS) project. *Journal of Geophysical Research, 108*(D22), 8842. https://doi.org/10.1029/2002JD003118

Eichenlaub, V. (1970). Lake effect snowfall to the Lee of the great lakes: Its role in Michigan. *Bulletin of the American Meteorological Society, 51*, 403–412. https://doi.org/10.1175/1520-0477(1970)051%3c0403:LESTTL%3e2.0.CO;2

Ferguson, C. R., & Mocko, D. M. (2017). Diagnosing an artificial trend in NLDAS-2 afternoon precipitation. *Journal of Hydrometeorology.* In press. https://doi.org/10.1175/jhm-d-16-0251.1.

Francis, J. A., & Vavrus, S. J. (2012). Evidence linking Arctic amplification to extreme weather in mid-latitudes. *Geophysical Research Letters, 39*, L06801. https://doi.org/10.1029/2012GL051000

Higgins, R. W., Janowiak, J., & Yao, J. P. (1996). A gridded hourly precipitation data base for the United States (1963–1993) (47pp.). NCEP/CPC Atlas No. 1, U.S. Department of Commerce.

Higgins, R., Schemm, J., Shi, W., & Leetmaa, A. (2000). Extreme precipitation events in the Western United States related to tropical forcing. *Journal of Climate, 13*, 793–820. https://doi.org/10.1175/1520-0442(2000)013%3c0793:EPEITW%3e2.0.CO;2

Huffman, G. J., Bolvin, D. T., Braithwaite, D., Hsu, K., Joyce, R., Kidd, C., Nelkin, E. J., & Xie, P. (2015). *NASA global precipitation measurement (GPM) integrated multi-satellite retrievals for GPM (IMERG).* Algorithm Theoretical Basis Document, version 4.5, 26 pp. Available online at http://pmm.nasa.gov/sites/default/files/document_files/IMERG_ATBD_V4.5.pdf

Joyce, R. J., Janowiak, J. E., Arkin, P. A., & Xie, P. P. (2004). CMORPH: A method that produces global precipitation estimates from passive microwave and infrared data at high spatial and temporal resolution. *J. Hydrometeor., 5*, 487–503.

Karl, T. R., Melillo, J. T., & Peterson, T. C. (Eds.). (2009). *Global Climate Change Impacts in the United States* (189pp.). Cambridge University Press.

Liu, J., Curry, J. A., Wang, H., Song, M., & Horton, R. M. (2012). Impact of declining Arctic sea ice on winter snowfall. *Proceedings of the National Academy of Sciences, 109*(11), 4074–4079.

Matsui, T., Mocko, D. M., Lee, M.-I., Tao, W.-K., Suarez, M. J., & Pielke, R. A., Sr. (2010). Ten-year climatology of summertime diurnal rainfall rate over the conterminous U.S. *Geophysical Research Letters, 37*, L13807. https://doi.org/10.1029/2010GL044139.

Melillo, J. M., Richmond, T. C., & Yohe, G. W. (Eds.). (2014). *Climate change impacts in the United States: The third national climate assessment* (841pp.). U.S. Global Change Research Program. https://doi.org/10.7930/J0Z31WJ2.

Mitchell, K. E., et al. (2004). The multi-institution North American Land Data Assimilation System (NLDAS): Utilizing multiple GCIP products and partners in a continental distributed hydrological modeling system. *Journal of Geophysical Research, 109*, D07S90. https://doi.org/10.1029/2003JD003823.

Newman, A. J., Clark, M. P., Craig, J., Nijssen, B., Wood, A., Gutmann, E., Mizukami, N., Brekke, L., & Arnold, J. R. Gridded ensemble precipitation and temperature estimates for the contiguous United States. *Journal of Hydrometeorology, 16*, 2481–2500 (2015). https://doi.org/10.1175/JHM-D-15-0026.1

NOAA (National Oceanic and Atmospheric Administration). (2016). *U.S. climate extremes index.* http://www.ncdc.noaa.gov/extremes/cei.

O'Gorman, P. A. (2015). Precipitation extremes under climate change. *Curr. Clim. Change. Rep., 1*, 49. https://doi.org/10.1007/s40641-015-0009-3

Peterson, T. C., et al. (2013). Monitoring and understanding changes in heat waves, cold waves, floods, and droughts in the United States: State of knowledge. *Bulletin of the American Meteorological Society, 94*, 821–834. https://doi.org/10.1175/BAMS-D-12-000066.1

Prein, A. F., Ramussen, R. M., Ikeda, K., Liu, C., Clark, M. P., & Holland, G. J. (2016). The future intensification of hourly precipitation extremes. *Nature Climate Change.* https://doi.org/10.1038/nclimate3168.

Redmond, K. T., & Koch, R. W. (1991). Surface climate and streamflow variability in the western United States and their relationship to large-scale circulation indices. *Water Resources Research, 27*, 2381–2399. https://doi.org/10.1029/91WR00690

Screen, J. A., & Simmonds, I. (2010). The central role of diminishing sea ice in recent Arctic temperature amplification. *Nature, 464*, 1334–1337. https://doi.org/10.1038/nature09051

Smith, S. R., & O'Brien, J. J. (2001). Regional snowfall distributions associated with ENSO: Implications for seasonal forecasting. *Bulletin of the American Meteorological Society, 82*, 1179–1192 (2001). https://doi.org/10.1175/1520-0477(2001)082<1179>2.3.CO;2

Stroeve, J. C., Serreze, M. C., Holland, M. M., Kay, J. E., Maslanik, J., & Barrett, A. P. (2012). The Arctic's rapidly shrinking sea ice cover: A research synthesis. *Climate Change, 110*, 1005–1027. https://doi.org/10.1007/s10584-011-0101-1

U.S. Environmental Protection Agency. (2016). *Climate change indicators in the United States* (Fourth ed.). EPA 430-R-16-004. http://www.epa.gov/climate-indicators.

Westra, S., Fowler, H. J., Evans, J. P., Alexander, L. V., Berg, P., Johnson, F., Kendon, E. J., Lenderink, G., & Roberts, N. M. (2014). Future changes to the intensity and frequency of short-duration extreme rainfall. *Reviews of Geophysics, 52*, 522–555. https://doi.org/10.1002/2014RG000464

Xia, Y., et al. (2012). Continental-scale water and energy flux analysis and validation for the North American Land Data Assimilation System project phase 2 (NLDAS-2): 1. Intercomparison and application of model products. *Journal of Geophysical Research, 117*, D03109. https://doi.org/10.1029/2011JD016048.

Zielinski, G. (2002). A classification scheme for winter storms in the Eastern and Central United States with an emphasis on nor'easters. *Bulletin of the American Meteorological Society, 83*, 37–51. https://doi.org/10.1175/1520-0477(2002)083%3c0037:ACSFWS%3e2.3.CO;2

Part 3: Severe Storms in the Changing Climate, Societal Impacts & Disaster Risk Reduction

Climate Change and Monsoon Rainstorms

R. Krishnan, Kalik Vishisth, D. C. Ayantika, K. M. Sumit, and Ramesh Vellore

Abstract This study presents a brief analysis of observed changes in monsoon rainstorms for the period 1959–2021 over Western India (WI), particularly suitable for the Indian states of Gujarat and Maharashtra, using the gridded rainfall dataset from India Meteorological Department (IMD) and ECMWF Reanalysis v5 (ERA5).This analysis also uses the Integrated Multi-satellitE Retrievals for GPM (IMERG) precipitation dataset for the period 2000–2021. The results of our study indicate that the duration of monsoon rainstorms over WI have risen since 1990s, together with a distinct enhancement in rainstorm days occurring mostly during the years of positive Indian Ocean Dipole (pIOD) events in the tropical Indian Ocean, e.g., a remarkably long rainstorm event persisted for 17 days (i.e., 26 July–11 August) during the extreme pIOD of 2019. The sustenance of monsoon rainstorm days over WI during pIOD events is supported by the background large-scale circulation anomalies associated with intensified cross-equatorial transport of moisture from the south-eastern tropical Indian Ocean and strengthening of cyclonic vortices in the lower and mid-tropospheric levels over the Indian subcontinent. While state-of-the-art climate model projections suggest higher likelihood of extreme pIODs in the future, they also raise several implicative questions on the role of extreme pIODs on organized monsoon convection, scale interactive processes, among others, on the future monsoon rainstorms in the backdrop of increasing water vapour in a globally warming climate. Addressing these challenging scientific questions warrants comprehensive modelling and diagnostic investigations in the future.

Keywords Indian summer monsoon · Rainstorms over Western India (WI) · Climate change · Rainstorm days · Positive Indian Ocean Dipole (pIOD)

R. Krishnan (✉) · K. Vishisth · D. C. Ayantika · K. M. Sumit · R. Vellore
Indian Institute of Tropical Meteorology, Pune 411008, India
e-mail: krish@tropmet.res.in

S. Das and W-K. Tao (eds.), *Severe Storms*,
https://doi.org/10.1007/978-981-97-7075-5_21

1 Introduction

There is robust scientific evidence that human-caused climate change has driven detectable changes in the global water cycle since the mid-20th century (IPCC, 2021). In particular, increase of atmospheric moisture due to global warming has led to an overall increase in daily mean precipitation intensities in a majority of land regions during the recent few decades (Douville et al., 2021). While the magnitude of intensification of heavy precipitation events due to global warming is similar to the increase in atmospheric water holding capacity of about 7% per 1 °C of warming, changes in atmospheric circulation patterns and moisture transport can significantly alter the dynamics of extreme precipitation events at regional scales (Douville et al., 2021).

The summer monsoon rainy season over the Indian region, which is typically from June through September, contributes to nearly 80% of the total annual rainfall over a large part of the country. It is recognized that increases in the Indian summer monsoon precipitation due to global warming were counteracted by precipitation decreases due to cooling caused by anthropogenic aerosol emissions over the 20th century (Douville et al., 2021 and references therein). Additionally, several studies have reported a significant rising trend in the frequency and intensity of extreme heavy rain events over the Indian region, particularly over central India, since the mid-20th century (e.g., Goswami et al., 2006; Rajeevan et al., 2008, Dash et al., 2009; Kulkarni et al., 2020; Krishnan et al., 2016; Roxy et al., 2017 and others). Furthermore, long-term trends in the extreme rain events over the Indian region exhibit significant spatial variations (Ghosh et al., 2012; Kishtawal et al., 2010; Krishnamurthy et al., 2009; Pattanaik & Rajeevan, 2010; Paul et al., 2018). In a recent study, Nikumbh et al. (2019) have reported an increasing trend in the average size of spatially collocated rainfall extremes over India, since 1980, with large-sized extreme rainfall events (area $\geq$ 70,000 km^2) often associated with monsoon LPS and precursory planetary-scale surface pressure patterns.

Monsoon cyclonic disturbances in the form of low pressure systems (LPS) and depressions, during the southwest monsoon months of June–September, often cause sustained heavy rainfall along and near their tracks as they move west and north-westward from the Bay of Bengal across the Indian subcontinent (Rao, 1976). These heavy rainfall causing synoptic disturbances during the southwest monsoon season are referred to as monsoon rainstorms. Heavy rainfall associated with monsoon rainstorms can cause flooding of rivers, especially in the catchments that lie in the rainstorm area. Dhar et al. (1984) documented one of the most severe rainstorms of India, during 1–5 July 1941, associated with extraordinarily heavy rains in the coastal areas of South Gujarat, north Konkan and adjoining places. This rainstorm event caused severe flooding of rivers, heavy destruction and loss of life, livestock, agricultural crops and damage to property; while the landslides in the Western Ghats led to massive disruption of road and rail traffic between Pune and Mumbai for nearly 10 days (Dhar et al., 1984). As seen from the analysis of Dhar et al. (1984), the centre of this rainstorm was at Dharampur in the Surat district of the Indian state of Gujarat which recorded the highest one-day rainfall of 987 mm on 2 July 1941 (Fig. 1a), while the heavy rains resulted from a monsoon depression that moved west-northwestward from the Bay of Bengal during this period (Fig. 1b). Severe monsoon rainstorms are also known to occur in many parts of northwest India, including north Gujarat, Rajasthan, Punjab, Himachal Pradesh and Jammu and Kashmir, due to a combination of meteorological conditions viz., (a) well-marked low pressure system (b) cyclonic circulation extending up to mid-tropospheric levels (c) off-shore trough along the west coast extending from north Maharashtra to Kerala (d) upper air troughs in the subtropical westerlies laying over north Pakistan and neighbourhood (Kulkarni et al., 1996).

Dhar and Nandargi (1993, 1995) examined severe rainstorms over the Indian region using daily rainfall based on station observations for the period 1880–1990, following the broad criteria that ***(a)*** They should have a closed isohyetal pattern of point rainfall maximum $\geq$ 200 mm on day 1 and the rainstorm centre should have a cumulative rainfall maximum $\geq$ 250 mm and $\geq$ 300 mm on days 2 and 3, respectively ***(b)*** The extent of rain area around the heavy rain centre should be of the order of 50,000 km^2 or more for each of the 1-, 2-, and 3-day durations of the rainstorms. Note that a rain centre could be any station point or any point in the isohyetal analysis with highest rainfall within the rainstorm area for that duration. Their analysis showed that the heavy-rain centres of severe rainstorms have preferred areas of zones, where

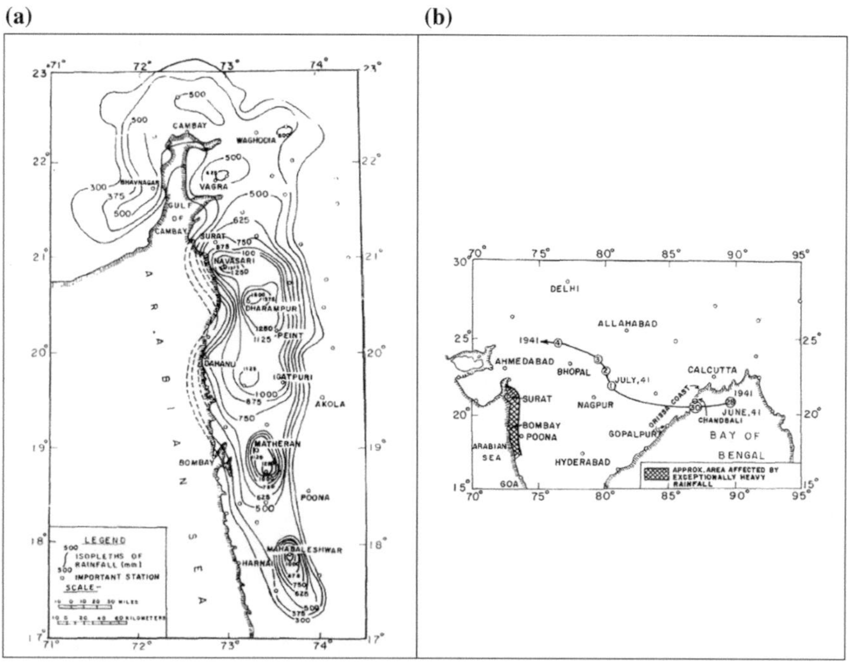

Fig. 1 **a** Isohyetal pattern of the 1–5 July 1941 rainstorm over the south Gujarat–north Konkan region of India. **b** Track of the July 1941 monsoon depression. Adapted from Dhar et al. (1984)

they frequently tend to occur. They also reported that 231 severe rainstorms, which affected India during this 110 year period (1880–1990), occurred in four rainstorm zones over India (Dhar & Nandargi, 1995).

A study on the temporal variation of rainstorms over India was carried out by Sagar et al. (2017) using the IMD high-resolution (0.25° [latitude] × 0.25° [longitude]) daily gridded rainfall dataset for the period 1951–2015. They identified rainstorms using the criteria that the rainfall at the centre ≥ 125 mm day^{-1}, the rainstorm area is at least 50,000 km^2 where rainfall ≥ 25 mm day^{-1} and the rainstorms sustain for at least two consecutive days. Sagar et al. (2017) reported statistically significant increasing trends in the frequency and duration of rainstorms over northern parts of India during this period. They also noted positive correlations between the duration of rainstorms and atmospheric water vapour over the Indian region and north Pacific; plus linkages between the rainstorm days and large-scale moisture transport variations from the tropical Indian Ocean towards the Indian landmass associated with cold sea surface temperature (SST) anomalies in the eastern equatorial Indian Ocean.

With the observed rising trend in the frequency of localized heavy rainfall occurrences over India (Kulkarni et al., 2020), several parts of the country have been experiencing enhanced risk of floods since the mid-20th century (Mujumdar et al., 2020). A list of major flood events that occurred over India during 2005–2019, including the devastating floods of Uttarakhand (2013), Kerala (2018, 2019) and several others, is detailed by Mujumdar et al. (2020). During 2022 (i.e. last year), the north-eastern Indian state of Assam witnessed heavy rains in May and June leading to massive flooding in several areas of the state, which are discussed in the National Remote Sensing Centre (NRSC) report by Shiva Prasad Sharma and Mohan (2022). However, the dominant catastrophe of 2022 in South Asia turned out to be infamous Pakistan floods (Mallapati, 2022). This flood event in Pakistan appears to have resulted from a combination of incessant heavy monsoon rains, excessive runoff from glacier melt in the northern mountainous areas caused by intense pre-monsoon heatwaves, an unusual depression that moved into the region plus a strong La Niña event that intensified the South Asian monsoon circulation (Mallapati, 2022).

This article focuses on a specific class of monsoon rainstorm events over Western India (68°E–76°E, 15°N–26°N), especially in the regions of Gujarat and Western Maharashtra, where heavy rainfall is often associated with the development of mid-tropospheric cyclones (MTC) (see Miller and Keshavamurty, 1968, hereafter MK68) together with significant contributions from westward moving monsoon LPS originating over the Bay of Bengal (Choudhury et al., 2018 and references therein). Several studies have examined the dynamical aspects of MTCs (e.g., MK68, Krishnamurti & Hawkins, 1970, Mak, 1975, 1983, Carr, 1977, Jayanthi et al., 2006, Routray et al., 2010, Choudhury et al., 2018, Kushwaha et al., 2021). It is also recognized that MTC events are accompanied by a steady build-up of water vapour in the mid-tropospheric levels that is favourable for heavy rains during the later stages of the MTC (Choudhury et al., 2018). This is an important point for understanding the observed changes in rainstorm events over Western India (WI) in a changing climate, especially given that water vapour increases at a rate of about 7% per 1 °C of warming.

2 Materials and Methods

The datasets used in this study include the India Meteorological Department (IMD) high-resolution (0.25° longitude × 0.25° latitude) gridded rainfall dataset (Pai et al., 2014) for the period 1959–2021 and the Integrated Multi-satellitE Retrievals for GPM (IMERG) precipitation dataset for the period 2000–2021 (Huffman et al., 2020). Also employed in this analysis are daily winds and specific humidity, constructed from 6-hourly fields at 23 vertical levels (1000–200 hPa), for the period 1959–2021 using the ERA5 dataset (Hersbach et al., 2018). In addition, sea surface temperature (SST) from the HadISST dataset (Rayner et al., 2003) and 2-m surface air temperature from ERA5 are used in the present analysis. We have identified rainstorm events over the domain (68°E–76°E, 15°N–26°N) covering the regions of Gujarat and Western Maharashtra in WI for the period (1959–2021) using the IMD gridded rainfall dataset. The rainstorm events over WI were identified following the criteria similar to Sagar et al. (2017) that the rainfall at the centre $\geq$ 125 mm day^{-1}, the rainstorm area is at least 50,000 km^2 where rainfall $\geq$ 25 mm day^{-1} and the rainstorms sustain for at least two consecutive days. The rain days for a particular rainstorm event correspond to the duration (or the number of days) for which the event lasts. By applying the above criteria, we identified 166 rainstorm events over WI during the 63-year period (1959–2021). It is noted that the total number of rain storm days for the 166 events is 532. During this 63-year period, there were 7 cases of long (duration $\geq$ 7 days) monsoon rainstorm events over WI, which are given in Table 1. Also shown in Table 1 are the duration of these 7 monsoon rainstorm events and the accumulated rainfall spatially averaged over WI (Indian land points only).Information about the dates of all the 166 rainstorm events (i.e., 532 days) is provided in the following github site: https://github.com/kalik-hub/Severe-Rainstorm/blob/main/all_rainstorm_events.txt.

Table 1 List of long monsoon rainstorm events (duration $\geq$ 7 days) over WI during 1959–2021. Also shown are the duration of the events and accumulated rainfall (mm) spatially averaged over WI (Indian land points only). Note that the IMD gridded rainfall dataset is constructed using station data only over the Indian land region (Pai et al., 2014)

No.	Long monsoon rainstorm events (duration $\geq$ 7 days)	Duration (days)	Accumulated rainfall over WI (mm)
1	20 July 1959–29 July 1959	10	198
2	10 August 1963–16 August 1963	07	124
3	13 August 1992–19 August 1992	07	109
4	04 July 2000–10 July 2000	07	113
5	25 June 2005–04 July 2005	10	235
6	05 August 2006–12 August 2006	08	233
7	26 July 2019–11 August 2019	17	447

3 Results

The time-series of the number of rainstorm events over WI, during the June–September (JJAS) monsoon rainy season, for the period (1959–2021) is shown in Fig. 2a. The time-series in Fig. 2b is the number of rain days per JJAS season over this 63-year period. It is noted that on an average ~ 2.6 rainstorm events per JJAS season occurred over WI during 1959–2021, while more than 4 events per JJAS season (i.e., mean + 1 std.dev) occurred during the following 18 years (i.e., 1959, 1964, 1970, 1975, 1976, 1977, 1981, 1983, 1984, 1990, 1991, 1994, 2006, 2007, 2010, 2015, 2016 and 2019). On an average, there are about 8.4 rain days per JJAS season associated with rainstorms over WI (Fig. 2b). It is interesting to note that the number of rain days associated with rainstorms over WI appear to have increased since the early 1990s, with a maximum of 26 rain days during the 2019 JJAS season. It is seen that the number of rain days per JJAS season during the 8 years (1959, 1976, 1994, 1997, 2005, 2006, 2007, 2019) indicated by circles in Fig. 2b, exceed the (mean + 1 std.dev) threshold of 13.5 days per year. The total number of rain days for the above 8 cases is 145.

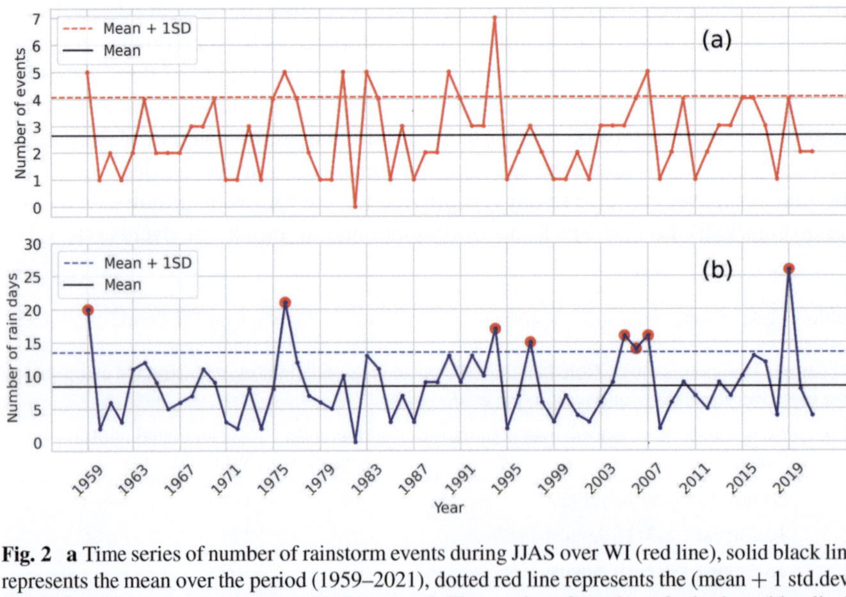

Fig. 2 **a** Time series of number of rainstorm events during JJAS over WI (red line), solid black line represents the mean over the period (1959–2021), dotted red line represents the (mean + 1 std.dev) threshold of 4 rainstorm events per JJAS season. **b** Time series of number of rain days (blue line). Black line is the mean and the dotted blue line is the (mean + 1 std.dev) threshold of 13.5 days per JJAS season. The red circles correspond to the 8 years when the duration of rainstorms per JJAS season exceeded the 13.5 days threshold

3.1 Composites Maps of Rainfall and Circulation Anomalies

The map of rainfall anomaly composited based on all the 166 rainstorm events (i.e., 532 days) during 1959–2021, using the IMD gridded rainfall dataset, is shown in Fig. 3a. An anomalous increase in rainfall can be seen over WI. The rainfall anomaly is maximum (>30 mm day^{-1}) over the Indian state of Gujarat and extends southward along the Western Ghats and west coast of India into the states of Maharashtra, Goa, Karnataka and Kerala. We also note an increase in the rainfall anomaly over northwest India and along a south-eastward branch extending from the Gujarat region into central and eastern India. However, the magnitude of rainfall is considerably reduced over eastern and central parts of India, as compared to WI. It is interesting to note negative anomalies extending over north-central and north-eastern India and parts of the Indo-Gangetic plains (IGP) and the south-eastern parts of Peninsular India, indicating a decrease of monsoon rainfall over these areas. The map in Fig. 3b shows the rainfall anomaly composite based on the 145 days corresponding to the 8 years when the duration of rain days per JJAS season exceeded the 13.5 days threshold. It can be seen that the overall spatial pattern of rainfall anomaly in Fig. 3b is similar to Fig. 3a, but with slightly enhanced magnitude of the positive and negative anomalies.

We also compared the rainfall anomaly patterns between the IMD and IMERG datasets for the common period 2000–2021 (Fig. 4). For this comparison, we considered 72 rain days corresponding to the 4 years (2005, 2006, 2007 and 2019) when the number of rain days per JJAS season exceeded the 13.5 days threshold. It is seen from Fig. 4 that the overall pattern of rainfall anomalies over the Indian land region is very similar in the two datasets. Additionally, the IMERG anomaly shows a band

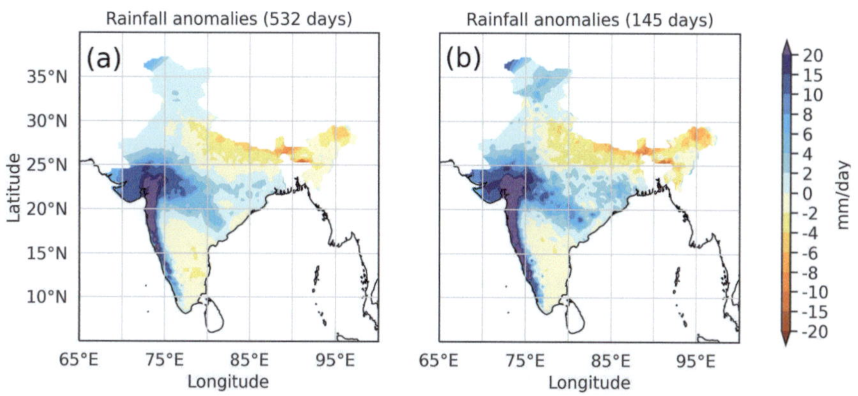

Fig. 3 **a** Composite map of daily IMD rainfall anomalies (mm d^{-1}) for all the rainstorm days (532 days) for the period 1959–2021. **b** Composite map of daily IMD rainfall anomalies (mm d^{-1}) for the 145 rainstorm days (145 days) during the years when the number of raindays per JJAS season exceeded the 13.5 days threshold. The daily rainfall anomalies are constructed, at every grid-point, by subtracting the daily climatological values from the daily rainfall gridded data. The daily climatological values are derived using the daily rainfall for the period 1959–2021

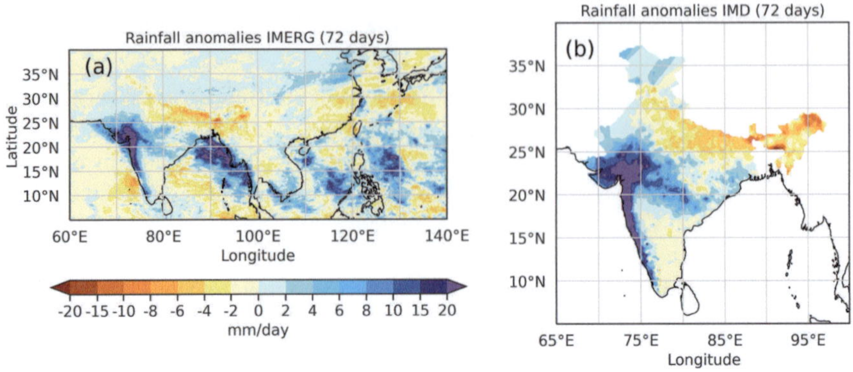

Fig. 4 **a** Composite of daily IMERG rainfall anomalies (mm d^{-1}) based on 72 raindays corresponding to the years (2005, 2006, 2007 and 2019). Note that the IMERG data is available from the year 2000 onward. **b** Same as in **a** except for the IMD dataset. The daily rainfall anomalies for the IMERG dataset are constructed, at every grid-point, by subtracting the daily climatological values from the daily rainfall gridded data. The daily climatological values are derived using the daily rainfall for the period 2000–2021

of increased rainfall extending south-eastward from the northern Bay of Bengal into parts of Myanmar, Thailand, Vietnam and South China Sea. Also seen in Fig. 4a are negative anomalies over the south-eastern Arabian Sea and the east coast of India, with the latter extending south-eastward towards to the equatorial eastern Indian Ocean. The main point from Fig. 4b is that the rainfall increase associated with WI rainstorms is characterized by a large-scale (>5000 km) pattern of enhanced monsoon rainfall that extends south-eastward from Gujarat across central and eastern India, Bay of Bengal into Myanmar, Thailand, Vietnam and South China Sea and is flanked by reduced rainfall to the north and south.

The composite map of precipitable water (PWAT) and mid-tropospheric wind anomalies (Fig. 5a), based on the 145 rain days corresponding to the 8 years when the total number of rain days exceeded the 13.5 days threshold, shows a prominent increase in atmospheric water vapour over a large area covering the South and Southeast Asian monsoon regions, with maximum increase over northwest India. Also seen in Fig. 5a is a dipole pattern of positive (negative) PWAT anomalies over the western (eastern) regions of the near-equatorial Indian Ocean. The mid-tropospheric wind anomalies show an intensified monsoon cross-equatorial flow, a large-scale cyclonic circulation anomaly extending longitudinally between 60°E and 100°E across the Indian subcontinent with the core cyclonic anomaly located over northwest India. The 850 hPa wind anomaly composite shows an intensified monsoon-cross equatorial flow over the Indian Ocean with a prominent cyclonic anomaly over northwest India (Fig. 5b). It may be noted that the monsoon westerly wind anomalies to the north of 10°N exhibit a large zonal extent stretching eastward from the Somali region, across the Arabian Sea, Indian landmass, Bay of Bengal, Southeast Asia, up to the tropical west Pacific. It turns out that a majority of these 8 years, except 1959, 1976 and 2005, coincided with the evolution of a positive Indian Ocean Dipole (pIOD) event in the Indian Ocean (Saji et al., 1999; Yamagata et al., 2004; Meyers et al., 2007; Behera et al., 2008, Ratna et al., 2021). Here it is also important to mention that the 2019 pIOD event has been among the strongest pIODs on record (Ratna et al., 2021), even as the WI region experienced the maximum rainstorm days during 1959–2021 (Fig. 2b). The composite map of sea surface temperature (SST) anomaly, based on the 8 years when the total number of rain days exceeded the 13.5 days threshold, shows a pattern of cooler-than-normal (warmer-than-normal) anomalies in the equatorial eastern (central-western) Indian Ocean that is typical of pIOD events (e.g., Saji et al., 1999, Yamagata et al., 2004). The intensified monsoon cross-equatorial flow in Fig. 5b is consistent with the observed strengthening of the large-scale summer monsoon circulation during pIOD events, as reported by several previous studies (e.g., Behera et al., 1999, Ashok et al., 2001, Krishnan & Swapna, 2009, Krishnan et al., 2011, Prajeesh et al., 2022).

3.2 Temporal Variations of Water Vapour and Circulation

Here we briefly discuss the temporal variations of the JJAS mean water vapour and circulation parameters, spatially averaged over a broader region (60°E–80°E, 15°N–30°N) that covers the central-eastern Arabian Sea and areas in western, central and northwest India, during the post-satellite period 1979–2021, using the ERA5 dataset (Hersbach et al., 2020).The time-series of mid-level and low-level relative vorticity averaged over this broad region are shown in Fig. 6a, b, respectively. It is interesting to note that the time-series of mid-level relative vorticity shows a prominent increasing trend ~ $0.05 \times 10^{-6}\ s^{-1}\ yr^{-1}$ during this 43-year period, although the trend in the low-level relative-vorticity time-series is insignificant. It is also noteworthy that the mid-level relative vorticity (Fig. 6a) and PWAT (Fig. 6c) significantly enhanced during the 6 years 1994, 1997, 2005, 2006, 2007 and 2019 when the total number of rain days per JJAS season exceeded the 13.5 days threshold. It is also important to note that a majority of these 6 years, except 2005, coincided with pIOD events. It is also known that the large-scale circulation anomalies over the tropical Indian Ocean during pIOD events are conducive for the generation and sustenance of long-lived monsoon low pressure systems (LPS) (Krishnan et al., 2011). We will take up this point for discussion later.

It is interesting to notice from Fig. 6c that the atmospheric moisture content over this region of western, central and northwest India increased at a rate of ~0.1 mm yr^{-1} (statistically significant $p < 0.05$) during 1979–2021, while the surface air temperature over this region increased at a rate of ~0.016 °C yr^{-1} (Fig. 6d) which is qualitatively indicative of the response of water vapour to climate warming. Observations also indicate that the Indian Ocean SST, averaged over a large domain (30°S–30°N, 40°–120°E), has experienced a significant warming trend (~0.06 °C $decade^{-1}$) from 1950 to 2020 and a maximum inter-decadal fluctuation around 1940 (Yang et al., 2022). The scatter-plot of the relationship between the PWAT and surface-air temperature variations over this broad region is shown in Fig. 7. While it is seen that the

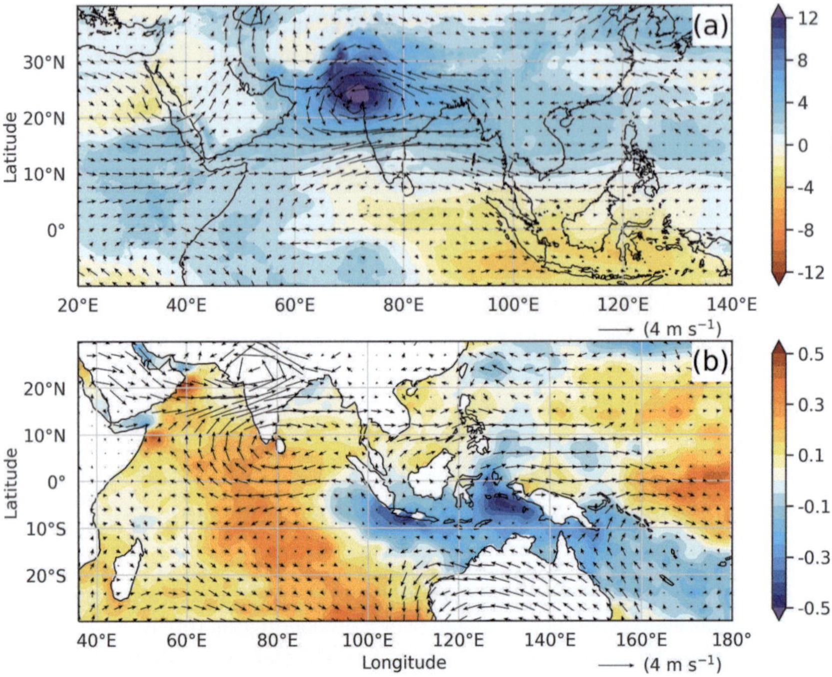

Fig. 5 **a** Composite map of daily anomalies of precipitable water [i.e., vertically integrated specific humidity from surface to 300 hPa] (PWAT, mm) and circulation (m s^{-1}) at 500 hPa based on the 145 raindays for the 8 years when the number of rain days per JJAS season exceeded the 13.5 days threshold. **b** Composite map of the JJAS sea surface temperature (SST °C) anomalies, based on the 8 years with number of raindays per JJAS season exceeded the 13.5 days threshold. Also shown in **b** are the composite of daily anomalies of 850 hPa winds (m s^{-1}) for the above mentioned 145 rain days. The daily anomalies of PWAT, 500 and 850 hPa winds are constructed by subtracting the corresponding daily climatological values computed for the period 1979–2021. The seasonal anomalies of SST in **b** are constructed by subtracting the seasonal climatological mean for the period 1950–2022

PWAT shows a modest increase with rising temperatures, approximately 3.6% per°C temperature change, there is also considerable spread in the PWAT values around the regression line pointing to the large water vapour variability. Observed changes in the atmospheric moisture content over the Indian monsoon region and adjoining areas during 1979–2021 are known to be dominated by substantial interannual variations and not just limited to the water vapour increase in a globally warming climate following the Clausius-Clapeyron thermodynamical constraint (see Douville et al., 2022).

The correlation coefficients of the detrended time-series of rainstorm days over WI with the detrended time-series of PWAT, specific humidity at 500 hPa and relative vorticity at the lower and mid-tropospheric levels, respectively for the period (1959–2021) are found to be statistically significant ($p < 0.05$) (see Table 2). Additionally, it is also important to recognize that the variability of the monsoon hydrological

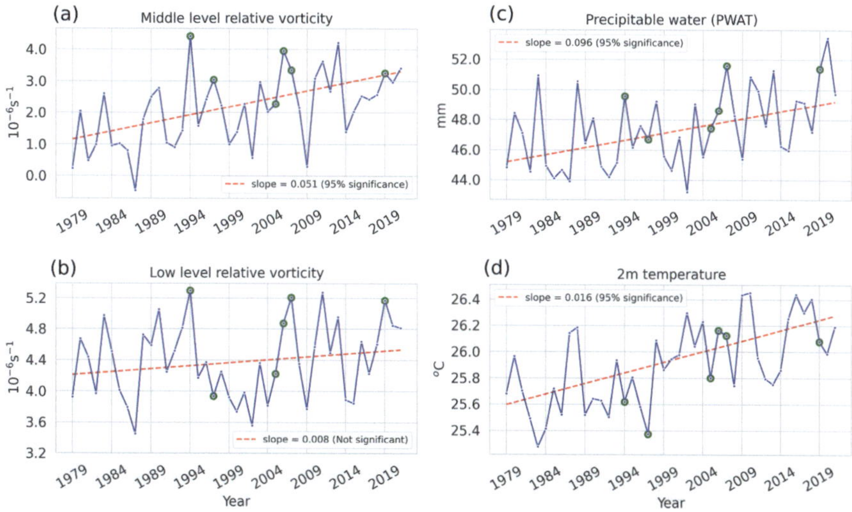

Fig. 6 Time series averaged over 60°E–80°E, 15°N–30°N. **a** Middle level (averaged between 550 and 650 hPa) relative vorticity (s^{-1}), **b** Low level (averaged between 850–1000 hPa) relative vorticity (s^{-1}), **c** PWAT **d** 2 m surface air temperature. The red lines correspond to linear trends of the time-series, computed using non-parametric Mann–Kendall test. The slopes of the linear trends and the corresponding 95%significance levels are mentioned. The green circles correspond to the 6 years when the total number of rain days per JJAS season exceeded the 13.5 days threshold

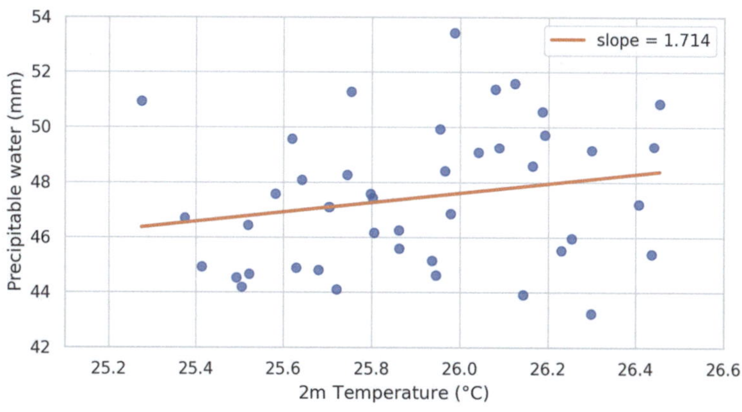

Fig. 7 Scatter plot of 2 m surface air temperature (°C) and precipitable water (mm) over the region 60°E–80°E, 15°N–30°N. The slope of the regression line, computed using the student's t-test, is statistically insignificant ($p = 0.16$)

Table 2 Pearson correlation coefficient (R value) and the corresponding significance level (p value) between the detrended time series of the duration of rainstorms (days) over WI and the detrended time-series of low-level (1000–850 hPa) vorticity(s^{-1}), mid-level (650–550 hPa) vorticity(s^{-1}), low-level specific humidity (g kg^{-1}), mid-level specific humidity (g kg^{-1}) and PWAT (mm)

	R value (Pearson correlation coefficient)	p value
Low-level vorticity	0.39	0.002
Mid-level vorticity	0.29	0.02
Low-level specific humidity	0.27	0.03
Mid-level specific humidity	0.31	0.01
Precipitable water	0.32	0.01

cycle, including the monsoon circulation, precipitation and evaporation over the land and oceanic regions are influenced by various modes of climate variability such as the IOD and El Nino-Southern Oscillation (ENSO), as well as other drivers such as natural and anthropogenic aerosols, land-use land-cover changes and so on (see Krishnan et al., 2021; Dhara & Krishnan, 2022).

3.3 *Discussion*

A particularly noteworthy point from the present study is the enhancement of rainstorm days over WI during pIOD events. Krishnan et al. (2011) noted that the strengthening of monsoon cross-equatorial moisture transport from the south-eastern tropical Indian Ocean to the Bay of Bengal induced by pIOD events, together with the enriched barotropic instability of the monsoonal winds, facilitates large-scale conditions that promote the activity of long-lived (>5 days) westward moving monsoon depressions and low-pressure systems (LPS) with their tracks extending from the Bay of Bengal into northwest India. They reported that the frequency of long-lived monsoon LPS tends to be relatively higher by about 12% during pIOD events as compared to non-pIOD years. In this connection, it is also interesting to note that several peaks in the time-series of mid-level relative vorticity over WI viz., 1994, 1997, 2006, 2007 and 2019 coincide with pIOD events (Fig. 6a).

To gain deeper insight about the dynamics of long-lasting rainstorms over WI, we examined composite maps of rainfall, PWAT and 500 hPa circulation anomalies for the long-lived rainstorm events (duration $\geq$ 7 days) during 2000, 2005, 2006 and 2019(see Table 1), which are shown in Fig. 8. Note that this composite also includes the 2019 record-breaking rainstorm event of 2019 which lasted for 17 days (Table 1).A large-scale well-organized rain belt, with northwest-southeast tilt structure, can be seen extending south-eastward from Gujarat into Myanmar coast and further downstream into Vietnam and tropical west Pacific (Fig. 8a). Such large-scale organized rain belts, with a horizontal scale ~ 7200 km, during active phases

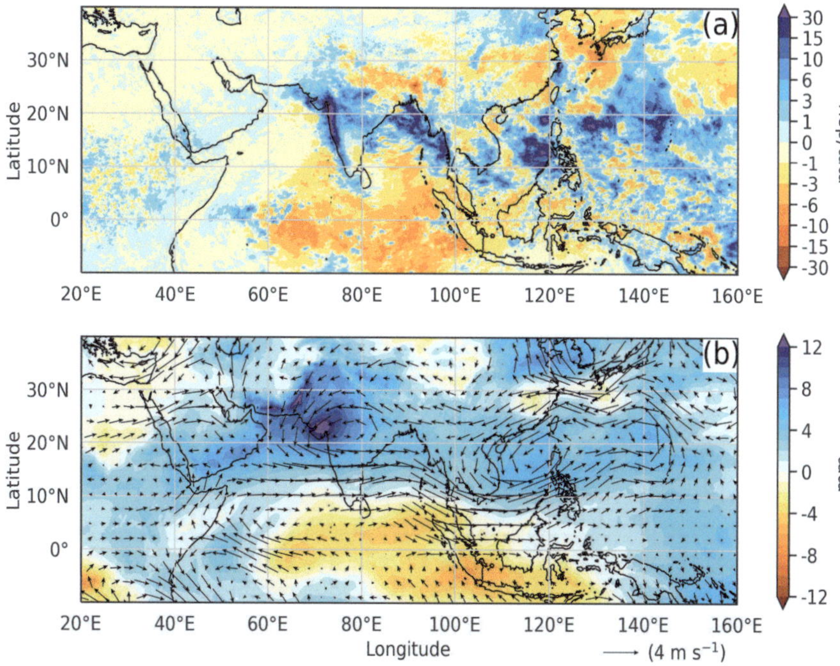

Fig. 8 **a** Composite map of rainfall anomalies (mm d^{-1}) using IMERG data based on the rainstorm events over WI that lasted for more than 7 days during 2000, 2005, 2006 and 2019 (Table 1). **b** Same as **a**, except for PWAT (mm) and 500 hPa circulation (m s^{-1}) using ERA5 reanalysis data. The daily anomalies of PWAT and 500 hPa winds are constructed by subtracting the corresponding daily climatological values computed for the period 1979–2021. The daily IMERG rainfall climatology is computed for the period 2000–2021

of the Indian summer monsoon are known to be composed of smaller meso-scale convective systems (MCS) (Choudhury & Krishnan, 2011).

We also note high rainfall over Gujarat, west coast of India and Central India, whereas rainfall is suppressed areas over the Himalayan foothills and north-eastern India. The composite map of 500 hPa winds and PWAT (Fig. 8b) shows a large-scale mid-tropospheric cyclonic circulation anomaly with enhanced water vapour that extends eastward from northwest India into Southeast Asia, Philippines, and tropical west Pacific. It has been observed that the development of mid-tropospheric cyclonic vortices (MTCs) and the augmentation of mid-tropospheric moisture over WI and northern Arabian Sea is often accompanied by heavy rainfall amounts over Gujarat, Western Maharashtra and adjoining areas (see MK68, Choudhury et al., 2018). Also the formation of MTCs is tied to the slow northward propagation of summer monsoon organized rain belts, which are enriched in stratiform-type precipitation with top-heavy latent heating that favours generation of mid-tropospheric cyclonic vortices (Choudhury et al., 2018). Basically the positive vertical gradient of latent heating from organized stratiform precipitating systems (Johnson and Houze, 1987), with maximum in the mid-troposphere levels, promotes generation of mid-tropospheric cyclonic vortices through enhanced mid-level convergence and vorticity stretching (Choudhury et al., 2018; Mapes, 1993).

Thus longer duration rainstorms over WI appear to be supported under the influence of mid-tropospheric cyclonic circulation and enrichment of precipitable water over the region. The life time of MTCs over WI can vary from a few days to about 2–3 weeks, which can be inferred from the comprehensive list of MTC events compiled by Choudhury et al. (2018). Based on this compiled list of MTC events, the average lifetime of MTCs over WI is estimated to be around 9 days. Here, it is also worth mentioning that Miller and Keshavamurthy (1968) presented a detailed observational analysis of MTCs that developed over the northeastern Arabian Sea during the period 25 June to 14 July1963. The composite analysis in their study is for the period 2–10 July, 1963, when the large-scale features of the MTC remained quasi-stationary. Here, it must be noted that Choudhury et al. (2018) did not consider short-lived MTCs (<5 days) in their analysis. Further studies are required to determine more precisely the life time of MTCs.

Extreme rainfall over WI and south western coast of India can also occur due to westward moving monsoon low pressure systems and depressions from the Bay of Bengal (Kottayil et al., 2022). Conversely, MTCs can also form over WI in association with westward moving monsoon depressions (see MK68, Carr, 1977; Choudhury et al., 2018). However, the individual roles of the MTCs and migrating monsoon depressions on the WI rainstorms are not yet clearly understood. Addressing this problem will require a separate in-depth analysis in the future.

Lastly, it is interesting to note that the large-scale cyclonic circulation pattern in Fig. 8b essentially comprises of two continental scale cyclonic anomalies—one over the Indian subcontinent and the other over Southeast Asia and tropical west Pacific. Our understanding suggests that the large-scale mid-tropospheric cyclonic anomalies can be interpreted as a quasi-stationary response to latent heating by the organized monsoon precipitation systems (see Choudhury & Krishnan, 2011). It is also realized that the evolution of monsoon rainstorms within this large-scale envelope of cyclonic circulation anomaly involves complex interactive feedbacks among atmospheric circulation, moisture and MCS heating fields, with multiple spatial and temporal scales. Further research is required to unravel these complex scale interactions of the monsoon rainstorms, which is beyond the scope of the present article.

4 Summary

It is seen from the present analysis that the duration of monsoon rainstorm events over western India (WI), as inferred from the number of rain days per JJAS season, has risen since 1990s; with a distinct increase in the rainstorm days observed during pIOD events. On the one hand, there is significant warming noted in surface air temperatures over a broader domain (60°E–80°E, 15°N–30°N), covering the central-eastern

Arabian Sea and areas in western, central and northwest India during 1979–2021, but the associated increase of PWAT over this region is found to be rather modest (~3.6% per°C temperature increase), due to large interannual variability. On the other hand, prominent increases are noted in PWAT and relative vorticity at the low and mid-tropospheric levels during the pIOD events of 1994, 1997, 2006, 2007 and 2019. According to the study by Cai et al. (2014), which assessed the influence of global warming on pIOD events using an ensemble of Coupled Model Intercomparison Project phase 5 (CMIP5) models, it has been reported that the frequency of extreme pIOD events is projected to increase by almost a factor of three, from one event every 17.3 years over the twentieth century to one event every 6.3 years over the twenty-first century. However, a more detailed analysis is needed to confirm the link between rainstorm days over WI and pIOD events in the future. Nevertheless, if indeed extreme pIODs such as the 2019 event become more frequent in the future, it raises several key research questions, such as:

(1) What is the influence of extreme pIODs on the frequency, intensity and duration of monsoon rainstorms in the future?
(2) What are the relative roles of extreme pIODs versus elevated levels of atmospheric water vapour in sustaining monsoon rainstorms?
(3) What is the role of global warming on the large-scale organization of mesoscale convective systems, monsoon LPS and MTCs in the future?
(4) How will climate change influence monsoon rainstorms in different regional monsoon systems of the world?

Unravelling these scientific issues warrants further investigations using high-resolution climate models, diagnostics based on satellite and reanalysis data, and the application of innovative analysis approaches, including AI/ML techniques.

Acknowledgements IITM is funded by the Ministry of Earth Sciences (MoES), Government of India. RK is thankful to Dr. M. Rajeevan, Former Secretary, MoES for the insightful scientific discussions during TROPMET 2022 at IISER, Bhopal.

References

Ashok, K., Guan, Z., & Yamagata, T. (2001). Impact of the Indian Ocean dipole on the relationship between the Indian monsoon rainfall and ENSO. *Geophysical Research Letters, 28*, 4499–4502. https://doi.org/10.1029/2001GL013294

Behera, S. K., Krishnan, R., & Yamagata, T. (1999). Unusual ocean-atmosphere conditions in the tropical Indian Ocean during 1994. *Geophysical Research Letters, 26*, 3001–3004. https://doi.org/10.1029/1999GL010434.

Behera, S. K., Luo, J.-J., & Yamagata, T. (2008). Unusual IOD event of 2007. *Geophysical Research Letters, 35*, L14S11. https://doi.org/10.1029/2008GL034122

Cai, W., Santoso, A., Wang, G., et al. (2014). Increased frequency of extreme Indian Ocean Dipole events due to greenhouse warming. *Nature, 510*, 254–258. https://doi.org/10.1038/nature13327

Carr, F. H. (1977). Mid-tropospheric cyclones of the summer monsoon. *Pure and Applied Geophysics, 115*, 1383–1412. https://doi.org/10.1007/BF00874415

Choudhury, A. D., & Krishnan, R. (2011). Dynamical response of the South Asian Monsoon trough to latent heating from stratiform and convective precipitation. *Journal of Atmospheric Science, 68*, 1347–1363.

Choudhury, A. D., Krishnan, R., Ramarao, M. V. S., Vellore, R., Singh, M., & Mapes, B. (2018). A phenomenological paradigm for midtropospheric cyclogenesis in the Indian Summer Monsoon. *Journal of Atmospheric Science, 75*, 2931–2954.

Dash, S., Kulkarni, M. A., Mohanty, U., & Prasad, K. (2009). Changes in the characteristics of rain events in India. *Journal of Geophysical Research (atmospheres), 114*, D10109.

Dhar, O. N., Kulkarni, A. K., & Mandal, B. N. (1984). The most severe rainstorm of India—A brief appraisal. *Hydrological Sciences Journal, 28*(2), 219–229. https://doi.org/10.1080/02626668409490935

Dhar, O. N., & Nandargi, S. (1993). The zones of severe rainstorm activity over India. *International Journal of Climatology, 13*, 301–311.

Dhar, O. N., & Nandargi, S. (1995). On some characteristics of severe rainstorms of India. *Theoretical and Applied Climatology, 50*(3–4), 205–212.

Dhara, C., & Krishnan, R. (2022). Advancing South Asian monsoon climate change projections: Challenges and opportunities. *The Journal of Indian Geophysical Union, 26*(4), 245–257.

Douville, H., Raghavan, K., Renwick, J, Allan, R. P, Arias, P. A., Barlow, M., Cerezo-Mota, R., Cherchi, A., Gan, T. Y., Gergis, J., Jiang, D., Khan, A., Pokam Mba, D., Rosenfeld, D., Tierney, T., & Zolina, O. (2021). Water cycle changes. In: V. Masson-Delmotte, P. Zhai, A. Pirani, S. L. Connors, C. Péan, S. Berger, N. Caud, Y. Chen, L. Goldfarb, M. I. Gomis, M. Huang, K. Leitzell, E. Lonnoy, J. B. R. Matthews, T. K. Maycock, T. Waterfield, O. Yelekçi, R. Yu, & B. Zhou (Eds.), *Climate Change 2021: The Physical Science Basis. Contribution of Working Group I to the Sixth Assessment Report of the Intergovernmental Panel on Climate Change.* Cambridge University Press.

Douville, H., Qasmi, S., Ribes, A., & Bock, O. (2022). Global warming at near-constant tropospheric relative humidity is supported by observations. *Communications Earth & Environment, 3*, 237. https://doi.org/10.1038/s43247-022-00561-z

Ghosh, S., Das, D., Kao, S.-C., & Ganguly, A. R. (2012). Lack of uniform trends but increasing spatial variability in observed Indian rainfall extremes. *Nature Climate Change, 2*, 86–91.

Goswami, B. N., Venugopal, V., Sengupta, D., Madhusoodanan, M. S., & Xavier, P. K. (2006). Increasing trend of extreme rain events over India in a warming environment. *Science, 314*, 1442–1445.

Hersbach, H., Bell, B., Berrisford, P., et al. (2020). The ERA5 global reanalysis. *Quarterly Journal of the Royal Meteorological Society, 146*, 1999–2049.https://doi.org/10.1002/qj.3803.

Huffman, G. J., Bolvin, D. T., Braithwaite, D., Hsu, K. -L., Joyce, R. J., Kidd, C., Nelkin, E. J., Sorooshian, S., Stocker, E. F., Tan, J., Wolff, D. B., & Xie, P. (2020). Integrated multisatellite retrievals for the global precipitation measurement (GPM) mission (IMERG). In: V. Levizzani, C. Kidd, D.B. Kirschbaum, C.D. Kummerow, K. Nakamura, F.J. & Turk (Eds.), *Satellite Precipitation Measurement* (Vol. 1, pp. 343–353). Springer International Publishing, Cham. https://doi.org/10.1007/978-3-030-24568-9_19

IPCC. (2021). Summary for policy makers in climate change 2021: The physical science basis. In V. Masson- Delmotte, P. Zhai, A. Pirani, S.L. Connors, C. Péan, S. Berger, N. Caud, Y. Chen, L. Goldfarb, M.I. Gomis, M. Huang, K. Leitzell, E. Lonnoy, J.B.R. Matthews, T.K. Maycock, T. Waterfield, O. Yelekçi, R. Yu, & B. Zhou (Eds.), *Contribution of Working Group I to the Sixth Assessment Report of the Intergovernmental Panel on Climate Change.* Cambridge University Press. https://doi.org/10.1017/9781009157896.001.

Kishtawal, C. M., Niyogi, D., Tewari, M., Pielke, R. A., & Shepherd, J. M. (2010). Urbanization signature in the observed heavy rainfall climatology over India. *International Journal of Climatology, 30*, 1908–1916. https://doi.org/10.1002/joc.2044

Kottayil, A., Xavier, A., Xavier, P., Koovekkallu, P., & Mohankumar, K. (2022). Evolution of large-scale factors influencing extreme rainfall over south western coast of India. *International Journal of Climatology, 42*(8), 4164–4178. https://doi.org/10.1002/joc.7455

Krishnamurthy, C. K. B., Lall, U., & Kwon, H.-H. (2009). Changing frequency and intensity of rainfall extremes over India from 1951 to 2003. *Journal of Climate, 22*, 4737–4746.
Krishnamurti, T. N., & Hawkins, R. (1970). Mid-tropospheric cyclones of the southwest monsoon. *Journal of Applied Meteorology, 9*, 442–458.
Krishnan, R., Ayantika, D. C., Kumar, V., & Pokhrel, S. (2011). The long-lived monsoon depressions of 2006 and their linkage with the Indian Ocean dipole. *International Journal of Climatology, 31*, 1334–1352.
Krishnan, R., Sabin, T. P., Vellore, R., et al. (2016). Deciphering the desiccation trend of the South Asian monsoon hydroclimate in a warming world. *Climate Dynamics, 47*, 1007–1027. https://doi.org/10.1007/s00382-015-2886-5
Krishnan, R., Sabin, T. P., Sagar, A., Swapna, P., & A. D. Choudhury. (2021). South Asian Monsoon climate change projections. In M. N. Rajeevan, J. Santos (Eds.), *CLIVAR Exchanges: India's Monsoon Mission*. WCRP.
Krishnan, R., & Swapna, P. (2009). Significant influence of the boreal summer monsoon flow on the Indian Ocean response during dipole events. *Journal of Climate, 22*, 5611–5634.
Kulkarni, A. K., Mandal, B. K., & Sangam, R. B. (1996). Analysis of severe rainstorms over Gujarat and Punjab during the 1993 monsoon. *International Journal of Climatology, 16*, 35–47.
Kulkarni, A., et al. (2020). Precipitation changes in India. In R. Krishnan, et al. (Eds.), *Assessment of Climate Change over the Indian Region—A Report of the Ministry of Earth Sciences, Government of India* (Chap. 3, pp. 47–72). Springer. https://link.springer.com/book/10.1007/978-981-15-4327-2.
Kushwaha, P., Sukhatme, J., & Nanjundiah, R. (2021). A global survey of midtropospheric cyclones. *Monthly Weather Review, 149*, 2737–2753.
Jayanthi, N., Rajeevan, M., Srivastava, A. K., Devi, S., Roy Bhowmik, S. K., & Hatwar, H. R. (2006). *Monsoon 2006: A report*. Synoptic Meteorology, No. 4/2006, Indian Meteorological Department (103pp.).
Johnson, R. H., & Houze, R. A. (1987). Precipitating cloud systems of the Asian monsoon. In C.-P. Chang & T. N. Krishnamurti (Eds.), *Monsoon Meteorology* (pp. 298–353). Oxford University Press.
Mallapati, S. (2022). Why are Pakistan's floods so extreme this year? *Nature*. https://doi.org/10.1038/d41586-022-02813-6
Mak, M.-K. (1975). The monsoonal mid-tropospheric cyclogenesis. *Journal of Atmospheric Science, 32*, 2246–2253.
Mak, M.-K. (1983). A moist baroclinic model for monsoonal mid-tropospheric cyclogenesis. *Journal of Atmospheric Science, 40*, 1154–1162.
Mapes, B. E. (1993). Gregarious tropical convection. *Journal of Atmospheric Science, 50*, 2026–2037.
Meyers, G., McIntosh, P., Pigot, L., & Pook, M. (2007). The years of El Niño, La Niña, and interactions with the Tropical Indian Ocean. *Journal of Climate, 20*, 2872–2880. https://doi.org/10.1175/JCLI4152.1
Miller, F. R., & Keshavamurthy, R. N. (1968). *Structure of an Arabian Sea Summer Monsoon System*. Meteorological Monographs (Vol. 1, 94pp.). East–West Center Press.
Mujumdar, M., Bhaskar, P., Ramarao, M. V. S., Uppara, U., Goswami, M., Borgaonkar, H., Chakraborty, S., Ram, S., et al. (2020). Droughts and floods. In Krishnan, R., et al. (Eds.), *Assessment of Climate Change over the Indian Region—A Report of the Ministry of Earth Sciences, Government of India* (pp. 117–141). Springer. https://link.springer.com/book/10.1007/978-981-15-4327-2.
Nikumbh, A. C., Chakraborty, A., & Bhat, G. S. (2019). Recent spatial aggregation tendency of rainfall extremes over India. *Science and Reports, 9*, 10321. https://doi.org/10.1038/s41598-019-46719-2
Pai, D. S., Sridhar, L., Badwaik, M. R., Rajeevan, M. (2014). Analysis of the daily rainfall events over India using a new long period (1901–2010) high resolution (0.25° × 0.25°) gridded rainfall dataset. *Climate Dynamics*. https://doi.org/10.1007/s00382-014-2307-1.

Pattanaik, D. R., & Rajeevan, M. (2010). Variability of extreme rainfall events over India during southwest monsoon season. *Meteorological Applications, 17*, 88–104. https://doi.org/10.1002/met.164

Paul, S., Ghosh, S., Mathew, M., Devanand, A., Karmarkar, S., & Niyogi, D. (2018). Increased spatial variability and intensity of extreme monsoon rainfall due to urbanization. *Scientific Reports, 8*, 3918. https://doi.org/10.1038/s41598-018-22322-9

Prajeesh, A. G., Swapna, P., Krishnan, R., Ayantika, D. C., Sandeep, N., Manmeet, S., Aditi, S., & Sandip, I. (2022). The Indian summer monsoon and Indian Ocean Dipole connection in the IITM Earth System Model (IITM-ESM). *Climate Dynamics, 58*, 5–6. https://doi.org/10.1007/s00382-021-05999-z

Rajeevan, M., Bhate, J., & Jaswal, A. K. (2008). Analysis of variability and trends of extreme rainfall events over India using 104 years of gridded daily rainfall data. *Geophysical Research Letters, 35*, L18707. https://doi.org/10.1029/2008GL035143

Rao, Y. P. (1976). Southwest Monsoons. *Meteorological Monograph, Synoptic Meteorology* (pp. 1–367). No. 1, India Meteorological Department.

Ratna, S. B., Cherchi, A., Osborn, T. J., Joshi, M., & Uppara, U. (2021). The extreme positive Indian Ocean Dipole of 2019 and associated Indian summer monsoon rainfall response. *Geophysical Research Letters, 48*, e2020GL091497. https://doi.org/10.1029/2020GL091497

Rayner, N. A., Parker, D. E., Horton, E. B., Folland, C. K., Alexander, L. V., Rowell, D. P., Kent, E. C., & Kaplan, A. (2003). Global analyses of sea surface temperature, sea ice, and night marine air temperature since the late nineteenth century. *Journal of Geophysical Research, 108*, 4407, D14. https://doi.org/10.1029/2002JD002670

Roxy, M. K., Ghosh, S., Pathak, A., Athulya, R., Mujumdar, M., Murtugudde, R., Terray, P., & Rajeevan, M. (2017). A threefold rise in widespread extreme rain events over central India. *Nature Communications, 8*, 708. https://doi.org/10.1038/s41467-017-00744-9

Routray, A., Mohanty, U. C., Niyogi, D., Rizvi, S., & Osuri, K. K. (2010). Simulation of heavy rainfall events over Indian monsoon region using WRF-3DVAR data assimilation system. *Meteorology and Atmospheric Physics, 106*, 107–125. https://doi.org/10.1007/s00703-009-0054-3

Sagar, S. K., Rajeevan, M., & Vijaya Bhaskara Rao, S. (2017). On increasing monsoon rainstorms over India. *Natural Hazards, 85*, 1743–1757. https://doi.org/10.1007/s11069-016-2662-9

Saji, N. H., Goswami, B. N., Vinayachandran, P. N., & Yamagata, T. (1999). A dipole mode in the tropical Indian Ocean. *Nature, 401*, 360–363. https://doi.org/10.1038/43854

Shiva Prasad Sharma, S. V., & Mohan, A. (2022). Satellite based analysis on Assam floods—Flood mapping and monitoring in Assam State. Disaster Management Support Group. National Remote Sensing Centre (NRSC), Indian Space Research Organization, Government of India.

Yamagata, T., Behera, S. K., Luo, J.-J., et al. (2004). *Coupled Ocean-atmosphere variability in the Tropical Indian Ocean. Earth's climate* (pp. 189–211). American Geophysical Union (AGU).

Yang, Y.-M., et al. (2022). Increased Indian Ocean-North Atlantic Ocean warming chain under greenhouse warming. *Nature Communications, 13*, 3978. https://doi.org/10.1038/s41467-022-31676-8

Severe Storms in Changing Climate Scenario with a Distinctive Emphasis on South Asia and India

Jagabandhu Panda, Debashis Paul, Ankan Sarkar, Asmita Mukherjee, Ipshita Bhasi, Geo Tom, Sushant Das, Nitin Lohan, Debjyoti Roy, and Sushil Kumar

Abstract The severe storms are the disturbed state of the atmosphere and could come in the forms of thunderstorms, squall lines, cloudbursts, tornadoes, hail, tropical cyclones, typhoons, hurricanes, windstorms, dust and sand storms, winter storms, blizzards, etc. They can be understood and analyzed through meteorological observations and derived information. Besides, they could be represented through a numerical weather model to study the associated physical processes, meteorological characteristics and dynamics. However, the advancement in understanding indicates a role of large scale synoptic conditions and ocean and land feedback in governing these regional and local storms. Also, they show climatological variability and a connection to the changing climate. While the whole globe experiences such storms depending upon their geographical relevance, severity depends upon the prevailing meteorological conditions, and strength of the land and ocean feedbacks. South Asia and India experiences most of these storms and the associated characteristics are found to be altering with a changing climate in this region. Based on the geographical relevance, some of these storms exhibited primarily in South Asia and India are described and discussed in this chapter.

1 Introduction

Storms are the significant disruptions created in the natural environment of any planetary object. The typical names include tornadoes, hail, thunderstorms, heavy precipitation events, tropical cyclones, windstorms, dust and sand storms, winter

J. Panda (✉) · D. Paul · A. Sarkar · A. Mukherjee · I. Bhasi · G. Tom · S. Das · D. Roy
Department of Earth and Atmospheric Sciences, National Institute of Technology Rourkela, Rourkela 769008, Odisha, India
e-mail: pandaj@nitrkl.ac.in

N. Lohan · S. Kumar
Department of Applied Mathematics, Gautam Buddha University, Greater Noida 201312, UP, India

S. Das and W-K. Tao (eds.), *Severe Storms*,
https://doi.org/10.1007/978-981-97-7075-5_22

storms, blizzards, etc. Intensity of such disruptions determines their severity. For instance, higher intensity tropical cyclones could be considered as severe cyclones, typhoons, hurricanes, etc. And higher intensity thunderstorms could form tornadoes, multi-cell thunderstorms, squall lines, etc. In Indian context, different intensity thunderstorms and lightning events, and tropical cyclones are well known, though sand and dust storms, and heavy rainfall events are prominent at several places based on the geographical relevance. In this chapter, different types of storms and the associated characteristic features are briefly described, while highlighting their types, formation mechanisms, anatomy and classifications, associated climatological significance, impact and mitigation. However, tropical cyclones and thunderstorms are primarily emphasized since they are the major storms, which impact the life of common people in India followed by heavy rainfall events and dust storms.

2 Thunderstorms and Lightning

Convective weather phenomena, such as tornadoes, hailstorms, thunderstorms, and lightning, are considered as the world's most dangerous hazards. Thunderstorm events are marked by thunder, lightning, hail, rain, as well as high surface winds, occurring in cumulonimbus clouds (Araghi et al., 2016). These extreme occurrences are responsible for a significant proportion of consequences, often including damaged houses, destroyed crops, uprooted trees, felled power lines, damaged aircrafts and automobiles, and human fatalities (Bhardwaj et al., 2017). Globally, the rate of thunderstorm recurrence is relatively higher, reporting around 1000 events every hour near the tropics (Williams, 2005; Mahanta and Yamane 2020). The African area experiences the most intense thunderstorms (Harel and Price 2020) of the world. The Indian subcontinent mostly experiences intense thunderstorm occurrences during the pre-monsoon or early monsoon months (i.e., March-June), when they are accompanied by violent squalls. In north-west India, it is mostly connected to western disturbances, while those in the northern and eastern India are associated with deep convective processes (Srivastava and Ray 1999). Some of the most severe thunderstorm events are observed in the eastern (West Bengal, Odisha and Jharkhand) and north-eastern (Arunachal Pradesh, Assam, Meghalaya, Manipur, Mizoram, Nagaland and Tripura) states of India during the pre-monsoon season (predominantly in the months of April and May) as favorable conditions (hot and humid climate) for the genesis exists over these regions (Sahu et al., 2022; Chandrasekar 2022). One of the types of thunderstorms coming from North-western direction being associated with hails, and gusts occur over the eastern parts of India and Bangladesh in the pre-monsoon months and are called Nor'westers or Kalbaisakhi. The Nor'westers cause severe causalities and damages to properties worth millions of dollars (De et al., 2005) and perhaps surpassing the losses associated with the TCs in a decadal scale. To perceive the socio-economic significance, the Severe Thunderstorm Observation and Regional Modeling (STORM) program, a comprehensive observational and modeling attempt to enhance understanding and prediction of severe thunderstorms at national level, was instituted through the Department of Science and Technology (DST), Government of India in 2005. The study proposed that the assimilation of radar data into the model and use of collective satellite and radar data along with modeling could help in the advancement of thunderstorm forecasting techniques (Das et al., 2014). Thunderstorm events with high financial impact include cases worldwide such as Europe (Doswell, 2003), Australia (Allen and Allen 2016), Canada (Sills and Joe 2019), USA (Eblen et al., 1990), etc. One of the notable event over northern parts of India that occurred on May 2, 2018 killed more than 120 people and injured about 300, was due to severe dust and thunderstorms (https://www.indiawaterportal.org/articles/storms-india-science-severity).

Thunderstorms are often considered as the most frequent source of lightning. A minor thunderstorm system typically generates a lightning strike every 20–30 s for nearly an hour spanning over an area of ~ 100–300 km. While, the larger storm systems may generate more than one lightning strike every second across regions a hundred times or more expansive. Lightning strikes are transient, strong electric flash with lengths extending up to several kilometers. Lightning events are observed when the atmosphere acquires electrical energy sufficiently big enough for the accompanying electric fields to produce electrical breakdown of the air (Uman, 2012), where the conditional instability may play a significant role (Williams and Renno 1993). There are approximately nine million lightning discharges occurring daily worldwide. According to the Accidental Deaths and Suicides report by the National Crime Record Bureau (http://ncrb.gov.in), every year more than 2,000 casualties are recoded due to lightning strikes in India. Thus, the lightning accompanied by severe thunderstorms could create havoc in any area by impacting the atmospheric composition as well as causing the accidental deaths, localized floods, etc.

The thunderstorm activities are highly modulated by the local atmospheric processes. The studies suggest that the changes in the land use characteristics feedback to the regional atmospheric processes can encourage the formation and persistence of thunderstorms (Rath et al., 2021, 2022). In major urban agglomerations, a significant enhancement in the intensity and frequency of thunderstorms are observed, however the mid-sized cities account for a less pronounced influence, signifying that climatological factors are partly regulating the urban effects on thunderstorms (Ashley et al., 2012). In large cities, the urban heat island (UHI) effect can trigger and accelerate the deep convection, which can lead to severe weather events. In the absence of dominant convective inhibition, the excess temperature of the UHI becomes the predominant factor that triggers the convection (Vraciu, 2023). Besides, UHI can impact the thermal and dynamical characteristics and thereby determines the location of thunderstormsand precipitation distribution (Lin et al., 2008). The thermal perturbations and higher surface roughness over the urban areas elevate the vertical uplift and moisture convergence, leading to increase in water vapor and the storms thus generate drags during the peak precipitation period that further leads to shifted precipitation center and stagnation in the major urban areas (Huang et al., 2022; Li et al., 2016b).

2.1 Anatomy, Classification and Triggering Mechanisms

Thunderstorms possess enormous vertical extent where the upper limit exists near the tropopause while their bases are at the top of the boundary layer (Stull, 2015). It is widely known that convective instabilities, adequate low-level moisture, severe wind shear, and a dynamic lifting mechanism that may discharge the instability significantly, all trigger thunderstorm events (Kessler, 1982). Also, the vertical shear of the ambient winds must match the amount of the convective instability in order for a huge convective cloud to grow properly (Asnani, 1992).The thunderstorms usually exhibit a cellular structure (Fig. 1). There are variations in the structure of these cells which may include only updrafts or downdrafts, or both. A thunderstorm typically comprises cells at various phases of the life cycle. While all storms must start as a single cell, the simple unicellular form is very uncommon after the storm reaches the wet, thundery stage (Byers, 1949). In general, thunderstorm cells pass through three distinct phases in a life cycle viz., the cumulus, mature and dissipating stage. The development of a thunderstorm (i.e., cumulus stage) occurs when warm, humid air rises in a conditionally unstable atmosphere.

As the ascending unstable air cools adiabatically, cumulus clouds develop. Initial clouds may last for just a few minutes before rapidly dissipating, but they serve a crucial role in the genesis of thunderstorms by carrying water vapor from the surface to the middle troposphere. The cumulus-stage clouds may reach greater heights at a pace of ~5–25 ms^{-1} (Knupp and Cotton 1985). Cumulus stage often lasts up to ten minutes and is not associated with precipitation, adverse weather, lightning, or thunder (Chandrasekar, 2022). The initiation of the mature stage is

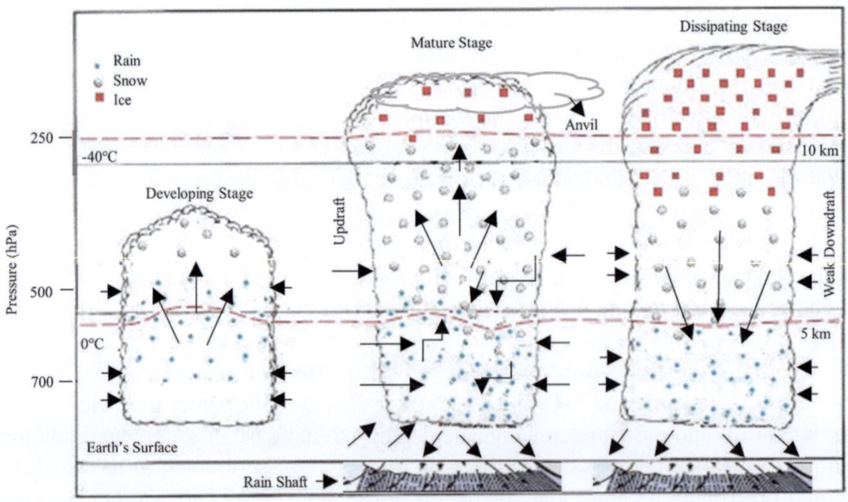

Fig. 1 Schematic diagram of a typical ordinary thunderstorm depicting its life cycle i.e., cumulus stage, mature stage, and dissipating stage

marked by severe rainfall events accompanied by hailstorm, thunder and lightning that instigates a downdraft. A mature thunderstorm's stem has a cloudy primary updraft tower surmounted by a vertical wind (VWS) bubble which when collides with the tropopause, forms the anvil (Stull, 2015). During the mature stage, updrafts are observed inside the cloud while downdrafts occur outside the cloud. This stage generally lasts for nearly ten to twenty minutes. The final or the dissipating stage is characterized by the occurrence of downdrafts across the base of the cloud. Towards the end of this stage, the amount of precipitation begins to recede and the sky starts to become clearer. Complex thunderstorms may have multiple updraft and downdraft zones to create supercell thunderstorms. And supercell thunderstorms are characterized by a quasi-steady spinning updraft and are the most powerful types, and they might be referred as a mesocyclone.

Thunderstorms are broadly classified into (1) basic storms, (2) mesoscale convective systems (MCS) and (3) supercells. Basic storms include weakly forced, multicell and orographic thunderstorms which may occur at the top of the boundary layer or at airmass boundaries. Multi-cell thunderstorms also fall under this category where different cells coexist at different stages of their cycle, which can be detected by weather radars based on their distinct precipitation cores and wind. Single or multi-cell thunderstorms stimulated by mountains generate orographic thunderstorms. MCS types are multi-cell storms, often associated with narrow line of thunderstorms and heavy precipitation. They require broad convective instability and wind shear over a deep layer of the atmosphere, triggered by ridges, synoptic systems or fronts. MCSs include squall lines, bow echoes, mesoscale convective complex and mesoscale convective vortex.

Certain atmospheric (pre-storm) precursors are required for the genesis of a severe thunderstorm such as (1) high humidity in the boundary layer, (2) non-local conditional instability, (3) significant wind shear, and (4) a lifting mechanism of boundary-layer air in the atmosphere (Stull, 2015). During the formation stage, thermals ascend heated humid air parcels in an unstable atmospheric boundary layer. The ascending parcel usually reaches the statically stable layer at the top of the mixed layer and stops ascending without initiating a thunderstorm. If the air parcel ascends a small distance above capping inversion, it would cool and drop back into the mixed layer (ML) due to its negative buoyancy. However, a trigger mechanism forces the boundary-layer air above the capping inversion, despite negative buoyancy. This results in cooling down of the air parcel till it becomes saturated at the lifting condensation level (LCL). The trigger mechanism continues to drive the air parcel beyond its LCL, causing water vapor to condense and form clouds, transforming latent heat into sensible heat. Above the free convection height (LFC), positively buoyant air parcels continue to ascend further and accelerate. This positive buoyancy powers thunderstorms and may create severe and intense updrafts to act as a trigger mechanism (Colquhoun, 1987). The air parcel ultimately reaches an elevation called equilibrium level (EL), where it is colder than the adjacent air. The thunderstorm cloud-base is located at the LCL and the anvil top is at the EL. For violent thunderstorms, EL may be located above the tropopause.

2.2 Thermodynamic Characteristics

The vertical air motion in moist convective condition rely on the convective available potential energy (CAPE). The primary hypothesis connecting a temperature-dependent CAPE to the wet bulb potential temperature, θ_w, was popular in studies concerning the tropical region (Williams and Renno 1993). Much of the CAPE in the tropics is located at higher elevations, where it does little to stimulate mixed phase microphysics (Williams et al., 2005).Despite CAPE being similar over warm tropical waters to that over land (Williams and Renno, 1993), the lightning activity is more by an order of magnitude (Williams and Stanfill 2002). On the other hand, when taking into account CAPE-θ_w correlations, the average maximum θ_w values over land are higher than the ocean by 1–2 °C, which is 1000–2000 J/kg (Williams and Renno, 1993). Thus, the air temperature above land/ocean has adapted to land/ocean surface conditions. At higher latitudes, the troposphere is often significantly out of equilibrium with the surface air (Doswell & Bosart, 2001), which results in very unstable weather conditions and powerful electrical discharges (Williams, 2001). The second issue with the CAPE-based concept corresponds to the parcel theory, where any assessment of CAPE is considered incorrect by virtue of the nature of turbulent convection. Even for the greatest extratropical supercells, the maximum air speeds predicted by CAPE and parcel theory are far higher than measurements (Williams and Stanfill 2002). Thus, any conjectures built upon the parcel theory, where vertical velocity plays a central role may be unreliable. The study of Williams and Stanfill (2002) looked at how to make updraft speeds different between land and sea while using the same CAPE, and Williams et al. (2005) re-examined the hypothesis that both the width of the boundary layer updrafts and the subsequent moist convection are proportional to the cloud base height (CBH). If typical 500 m CBHs are present in the ocean, the updrafts will be narrower, the convective ‘bubbles’ will be smaller, and the conversion efficiency of CAPE to updraft kinetic energy would be lower. Therefore, convection and updrafts in continental convection with 2–5 times greater CBHs will be more powerful (Williams and Stanfill 2002), and CBH has a significant impact on the convective organization (McCaul and Cohen 2002).Besides CAPE and CIN (Moncrieff and Miller 1976; Colby, 1984; Williams and Renno 1993), there are several other indices, which signify the thermodynamics of thunderstorms, viz., BOYD, BRN, LI, KI, SHOW, SWEAT, CTI, VTI, TTI, HI (Boyden, 1963; Weisman and Klemp 1982, 1984; Means, 1952; DeRubertis, 2006; George, 1960; Showalter, 1953; Miller, 1967, 1975; Litynska et al., 1976; Dhawan et al., 2008; Tyagi et al., 2011; Sahu et al., 2020), etc.

3 Tropical Cyclones

Tropical cyclone (TC) is a warm core synoptic weather system that develops over warm tropical waters with counter clockwise (clockwise) winds around a low-pressure center in the northern (southern) hemisphere. The World Meteorological Organization (WMO) characterizes a TC consisting of a vortex circulation of tropical origin with a synoptic scale diameter, approximately circular in shape, minimum surface pressure less than 900 hPa with maximum sustained winds of at least 64 knots or 118 km/h accompanied with well-organized deep convective systems and often supported by thunderstorms. The seven global ocean basins serve as breeding grounds for TCs. Among them, the North Indian Ocean (NIO), consisting of two major sub-basins, namely the Arabian Sea (AS) and the Bay of Bengal (BOB), accounts for 6–7% of global TCs with an yearly average of 3–5.

A cyclonic disturbance (CD) with relatively higher intensity is called as hurricane in the Atlantic and the north-eastern Pacific Oceans, TC in the Southern Pacific and the north and south Indian Oceans, typhoon in the western North Pacific, Willy Willies around Australia, Baguio around the Philippine Islands, and Taifu around Japan. Usually, the Saffir Simpson hurricane wind (SSHW) scale is adopted to classify the CDs into five categories based on the intensities (Taylor et al., 2010), viz., category one that includes CDs with a maximum sustained wind (MSW) in the range 119–153 km/h, two (MSW 154–177 km/h), three (MSW 178–208 km/h), four (MSW 209–251 km/h), and five (MSW 252 km/h or higher). However, over NIO, the India Meteorological Department (IMD) classification scheme is followed (http://www.rsmcnewdelhi.imd.gov.in). Theyare of seven categories, i.e., depression or D (MSW 17–27 knots or 8.74–13.89 ms^{-1}), deep depression or DD (MSW 28–33 knots or 14.4–16.97 ms^{-1}), cyclonic storm or CS (MSW 34–47 knots or 17.49–24.17 ms^{-1}), severe cyclonic storm or SCS (MSW 48–63 knots or 24.69–32.41 ms^{-1}), very severe cyclonic storm or VSCS (MSW 64–89 knots or 39.92–45.78 ms^{-1}), extremely severe cyclonic storm or ESCS (MSW 90–119 knots or 46.3–61.22 ms^{-1}), and super cyclonic storm or SuCS (MSW > 120 knots or > 61.73 ms^{-1}). Sometimes the depressions and deep depressions are considered as 'depressions' only. The ESCS and SuCS can be taken together to be termed as highly intensified cyclonic storm or HICS. Similarly, SCS and VSCS can be roughly considered as SCS only.

3.1 Formation and Structure

There are multiple factors involved in the formation of TCs, which involves both ocean and atmosphere to govern the genesis, development, propagation, and dissipation. SST is one of the most critical factors that governs the formation and strengthening of TCs. For instance, Roy Chowdhury et al. (2020), and Singh et al., (2019a, 2019b), indicated that SST plays a pivotal role in the genesis of NIO tropical cyclones and their climatological activity. Notably, the threshold temperature necessary for the formation and development of TCs is 26°C. An increased SST emanates the heat flux transmitting from the ocean to the atmosphere, involving increased convection and creating favorable conditions for TC genesis. The Low vertical wind shear (VWS) between the upper and lower troposphere must be minimal for cyclogenesis (Hawkins and Imbembo 1976). The strength of NIO TCs is determined by the high upper ocean heat content (UOHC) as well (Maneesha et al., 2015). The UOHC largely influences the life cycle and intensity of TCs across the NIO basin (Sharma and Ali, 2014). Because of the continual supply of latent heat fluxes from the ocean's surface to the atmosphere, TCs with a higher support from UOHC are more likely to sustain, accelerate and strengthen (Shay et al., 2000). The warm ocean waters ensure sufficient evaporation and moisture supply that provides necessary latent heat. An increased humidity level in the mid-troposphere will lead to the formation of cumulonimbus clouds. If dry air is entrained, it will inhibit cumulonimbus cloud formation. However, when the relative humidity of mid-tropospheric air <50–60%, there will be no occurrence of cumulonimbus convection, even in ocean areas. Also, the Coriolis parameter (f) must exceed a certain critical value, i.e., a value beyond 5° latitudes, for the development of cyclone vortex. It is observed that 65% of the world's TCs develop between 10° and 20° latitudes.

TC development involves four stages, i.e., formative stage, immature or developing stage, mature stage, and decaying stage. The formative stage does not consist of well-organized circulation; the center of the circulation is not well defined (eye not visible), but the presence of spiral bands of cumulus clouds can be seen. During this stage, strong surface wind is seen away from the center of the system (Fig. 2). The structural aspects of a TC can be visualized through remote sensing-based observational platforms in a proper manner (Panda et al., 2023), though can be re-constructed through numerical simulations as well.

In the developing stage, the convective cloud cluster begins to organize in a tight spiral band around minimum sea level pressure (MSLP), which decreases below 1000 hPa. This tends to the intensification of the system indicated by the increase in surface wind speed near the center that leads to clear visibility of the eye of the cyclone through satellite images. The convective clouds form into spiral bands around the eye as the TC develops. The mature stage is manifested when the intensification is reduced, and the TC attends a quasi-steady state, which is characterized by high variability in MSLP and MSW. The TC attains the maximum horizontal extent, which is infused with cumulous cloud-organized deep convective spiral bands. The diameter of the eye also tends to fluctuate during this stage due to varying wind shear along dynamic and thermodynamic parameters. The decaying stage indicates the termination of the organized deep convection in different spiral bands around the TC, leading to dissipation. The stage is characterized by the dissipation of the warm core, which is determined by the highest measured temperature in a TC that is observed between 700–500 hPa and 350–200 hPa.

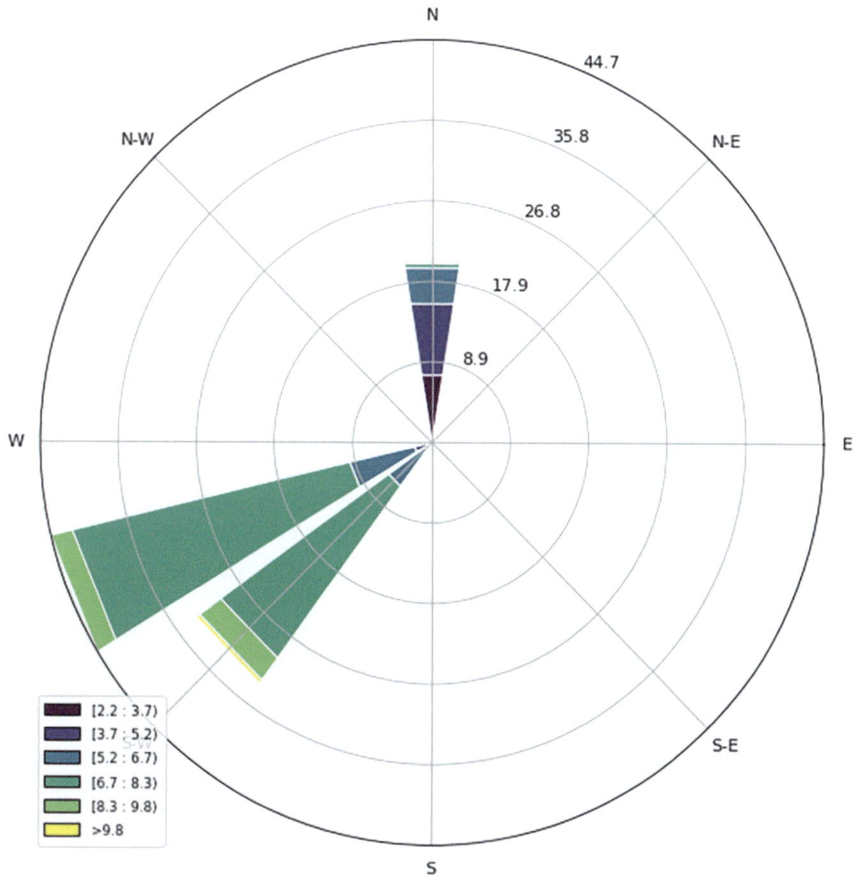

Fig. 2 Wind rose diagram displaying wind direction occupying a particular quadrant i.e., SW. Maximum Sustainable Wind is for Gati (21 November, 2020) cyclone originating over Arabian Sea and is observed through the SCATSAT

3.2 Tropical Cyclone Climatology

The average frequency, season, and strength of TCs varies from basin to basin. The recent rise in the frequency of severe TCs in major tropical ocean basins, particularly the North Atlantic and western Pacific basins, has stirred a significant discussion about the impact of climate change. Several large-scale environmental factors such as vorticity, relative humidity (Kuleshov et al., 2009), VWS (Camargo et al., 2007b), etc., contribute to the TC activity and hence climatology (Singh et al., 2016, 2019a, 2019b, 2020a, 2020b; Kumar et al., 2021, 2023; Mohanty et al., 2012; Mohapatra et al., 2015; Bielli et al., 2021; Paul et al., 2022; Panda et al., 2023). For instance, accumulated cyclone energy (ACE), a parameter that combines the number, lifetime, and intensity of TCs occurring in a basin over a given period of time, may serve as an indicator concerning TC climatology (Camargo and Sobel 2005; Kumar et al., 2023). Emanuel (2005) defined the power dissipation index (PDI) to be the sum of the maximum one-minute sustained wind speed cubed at six-hourly intervals for all periods when the TC is at least of tropical storm strength. Klotzbach and Gray (2006) reported that there has been a large increase in ACE over the North Atlantic and a large decline in the Northeast Pacific during the past decade. Also, SST plays a vital role in the prediction of the fluctuation of PDI and ACE of TCs. And a statistically significant relationship was observed between SSTs and ACE with Category 4–5 hurricanes for the North Atlantic and Northeast Pacific. The tropical Atlantic SST coefficient remained favorable for both PDI and ACE (Villarini and Vecchi 2012) and is higher than the tropical SST coefficient.

Over NIO, the climatology indicates a decreasing trend of TC frequency in the current warming period, contrasting the pre-warming time frame (Singh et al., 2019a, 2019b, 2020a). The TC activity over southern and northern BOB regions is found to be decreasing in the warming climate scenario. However, the southern sector of BOB hosts mostly severe systems, and the middle sector hosts most TCs. TC activity over the eastern sector of AS shows a considerable surge during the warming climate scenario. An increasing SST, sea surface wind, relative humidity at the middle troposphere, and potential evaporation factor (PEF) are favorable in the formation of intensified storms during the warming climate scenario. A large difference in the temperature anomaly between the atmosphere and ocean is perceived to play a key role in modulating the enhanced intensity of TCs during the current warming scenario. The SST range of 27.5–29.5 °C and the prevailing circulation pattern is helping to boost the middle and upper tropospheric humidity to result in enhanced SST, PEF, and relative humidity and support the intensification of TCs through a feedback mechanism. Similarly, the landfall pattern observed in the current warming scenario is supported by the SST distribution and circulation pattern over the NIO region (Singh et al., 2019b). A consequent distinct pattern in the rainfall contribution by the NIO cyclonic disturbances was realized over India (Singh et al., 2020b) as well.

3.3 Significance of ENSO, and Other Climate Modes

Climate modes such as El Niño Southern Oscillation (ENSO) can play a profound and prominent role in changing the ocean and atmospheric environments, which effectively modulates global TC activity (Chan, 1985; Camargo et al., 2010; Wang et al., 2010a; Aiyyer and Thorncroft 2011; Camargo and Hsiang 2015; Zheng et al., 2015). Owing to atmospheric teleconnections, the effects of ENSO can be felt beyond the Pacific, which modulates TC activities globally. Bell et al. (2014) emphasized the influence of ENSO on TC activity at a global scale (Fig. 3). They have shown that during the El-Niño years, a drop in TC activity is observed over the Atlantic, whereas TCs are found to form near the international date line over both the eastern and western North Pacific. Similarly, Wang et al. (2010a) have shown that ENSO, along with inter-decadal Pacific oscillation, has a very strong and positive influence on global TC activity.

The Western North Pacific generally experiences normal or above normal TC activity during the El-Niño years, whereas during the La-Niña developing years, it experiences normal or below normal activity (Chan, 1985). Central and Eastern Pacific is the second most active TC basin in the world, where on an average 19 named storms occur every year (Neumann, 1993). A higher TC activity was observed over the central pacific during both 2015 El-Niño season as well as 2016 La-Niña season (Boucharel et al., 2016; Murakami et al., 2017). This indicates significance of various environmental factors that are responsible for TC activity over the Central and Eastern Pacific that are mainly related to ENSO.

Gray (1984) illustrated that during the boreal summer of El-Niño years, North Atlantic experiences lesser TC activity compared to the La-Niña years. However, La Niña events were linked to a stronger Walker circulation and unusually strong

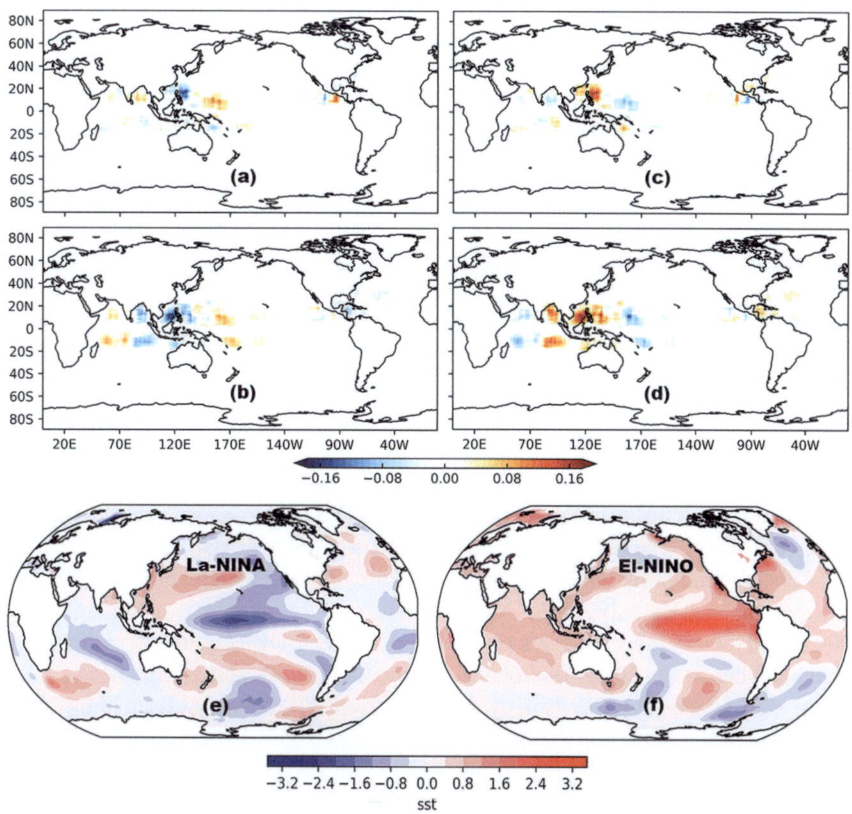

Fig. 3 Global TC density composite during El-Niño years **a** pre-monsoon **b** post-monsoon. Similarly, global TC density composite during La-Niña years **c** pre-monsoon **d** post-monsoon. The period taken is 1980–2020. Figure **e** represents 1998 winter (big La Niña season)

easterly winds in the upper levels of the Caribbean. During El-Nino years, higher VWS caused due to eastward shift of the Walker circulation and weaker upper tropospheric winds contributes to lesser TC activity; whereas, enhanced Walker circulation along with remarkably stronger upper tropospheric easterly winds over the Caribbean contributes to enhanced TC activity over North Atlantic during the La-Niña years (Gray et al., 1994; Goldenberg and Shapiro 1996; Tartaglione et al., 2003; Camargo et al., 2007a; Klotzbach, 2011; Patricola et al., 2014; Klotzbach et al., 2017). With the availability of improved and quality controlled datasets post-satellite era, various studies (Chand and Walsh 2009; Kuleshov et al., 2009; Vincent et al., 2011; Dowdy and Kuleshov 2012; Ramsay et al., 2012; Magee et al., 2017; Panda et al., 2023) noted a strong linkage between TC activity and ENSO.

The effect of ENSO on TC activity over NIO is mostly observed during the post-monsoon season. BOB experiences lesser intense TCs during the El-Niño years (Singh et al., 2000), however stronger and intense TCs are likely to form during the La-Niña years. It is mostly attributed to weaker VWS and the presence of more moisture over the central and northern BOB (Felton et al., 2013; Mahala et al., 2015). However, not much difference is observed in TC number between both scenarios (Li et al., 2016a). During the pre-monsoon, La-Niña tends to increase lower tropospheric cyclonic vorticity and mid-tropospheric humidity in the northern BOB, enhancing TC activity. However, higher VWS over the southern BOB, on the other hand, tends to suppress TC activity (Balaguru et al., 2016).

Similar to ENSO, Indian Ocean Dipole (IOD) plays a very significant role in modulating TC activity over the NIO basin. During a positive IOD event, the SST anomalies reach a maximum value during autumn, which significantly influences the TC activity over the NIO basin during post-monsoon (Saji et al., 1999). However, enhanced TC activity is observed over the BOB sub-basin during the negative phase of IOD (Kripalani and Kumar 2004). On some occasions, ENSO and IOD both act together, complementing each other. El-Niño (La-Niña) events, when paired with positive (negative) IOD, decrease (increase) the TC activity. Similarly, Pacific Decadal Oscillation (PDO) can change how ENSO impacts BOB TC activity during the post-monsoon (Girishkumar et al., 2015) too. Both El-Niño and La-Niña have different effects. This difference is bigger when PDO is in its warm phase compared to the cold phase. The MJO and associated equatorial waves also called as CCEWs often create conditions which is conducive for cyclogenesis (Bessafi and Wheeler 2006; Frank and Roundy 2006). Various studies in the past have documented the effect of MJO and other CCEWs on TC genesis over BOB (Liebmann et al., 1994; Krishnamohan et al., 2012; Tsuboi and Takemi 2014; Fosu and Wang 2015).

The significance of CCEWs in cyclogenesis over NIO basin has been highlighted by Schreck et al. (2012) and Frank and Roundy (2006) with distinct findings. Several studies also indicate that Madden Julian Oscillations (MJO) modulate the TC genesis most among all the CCEWs (Schreck et al., 2012; Li and Zhou 2018). An example of such system could be the TC Nargis (Fig. 4) triggered by a Rossby wave (Kikuchi et al., 2009). Similarly, the shift of MJO convection in the NIO toward the northeast supplements the quick growth of TCs (Kemball-Cook and Wang 2001). The significant factors that contribute to such intra-seasonal oscillations for enhancing

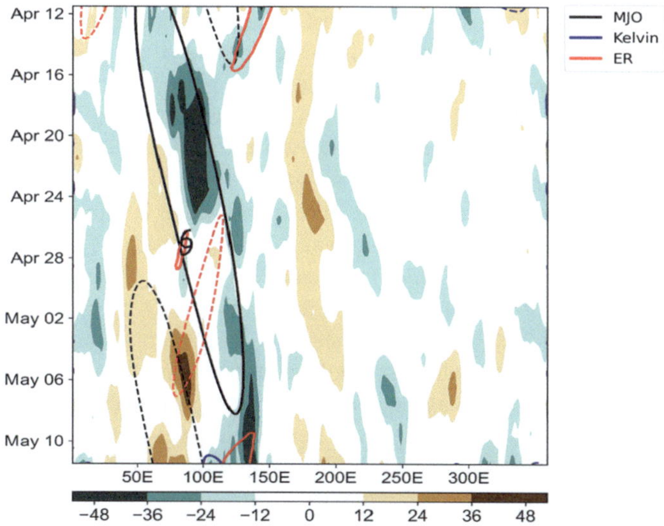

Fig. 4 Hovmoller diagram of average (15°S –15°N) anomalous OLR (shaded) overlaid by anomalies filtered for Kelvin (blue contours), Madden–Julian oscillation (black contours) and Equatorial Rossby (red contours) waves during TC Nargis (2008). TC symbols indicates genesis time and longitude of TC Nargis

TC activity include vertical velocity, relative vorticity, VWS (Wang and Moon 2017; Moon et al., 2018), etc. TC activity over NIO can also be enhanced by Inter-Decadal Pacific Oscillation (IPO) along with quasi-biennial oscillation, which creates conditions similar to a La-Niña event (Xiang and Wang 2013; Wahiduzzaman et al., 2019).

4 Tornadoes

Tornadoes, one of the most destructive storms worldwide, are defined as the fast-spinning narrow columns of air (with an average diameter in hundreds of meters) that are either suspended from or positioned under the base of a thunderstorm cloud. Generally, most of them rotate anticlockwise in the northern hemisphere, and a few rotate clockwise, which is quite rare (Davies-Jones, 1995). Moreover, these circular systems, which move rapidly often appear as a condensation funnel or a whirling cloud of debris and dust mounting from the ground (Kessler, 1970). This recognizable funnel shape signifies the low air pressure that predominates inside the vortex. They can be of different sizes and forms, like twisted rope, cylindrical, and sometimes huge dark funnels like an elephant's trunk hanging from a cumulonimbus cloud. Most tornadoes usually last just a few minutes and travel along paths typically ~ 7 km long. However, some thunderstorms can produce a series of tornadoes that last for two hours or more and travel at least hundreds of kilometers in the form of a tornado outbreak. These tornado families are considered as one of the deadliest storms in terms of damage and killing of people. In general, tornadoes go through several stages as they develop. The first stage is known as the dust-whirl stage, when a small funnel frequently extends downward from the base of a thunderstorm, and dust is seen whirling upward from the ground. The tornado intensifies in the organizing phase. When a tornado is mature, the damage is typically at its worst when the funnel reaches its widest point and is nearly vertical. The wind speed in this phase can reach up to ~140 ms^{-1} and cause significant damage (Boruff et al., 2003). The shrinking stage of a tornado is distinguished by an overall drop in the funnel's breadth, an increase in the funnel's tilt, and a shortening of the damage path at the surface. It is typically stretched out into the shape of a rope during the last stage, which is referred to as the decay stage. It typically undergoes significant deformation before eventually dissipating. Even though these stages are the general characteristics of most tornadoes, minor ones may only go through the organizing stage and sometimes even bypass the mature stage and enter the decline stage. The Fujita scale categorizes tornadoes based on rotating wind speed and storm damage (Fujita, 1971). Generally, the majority of tornadoes fall into the F0 and F1 categories (weak). Only a small percentage of them are classified as stronger than F3 (severe), and there may be only one F5 tornado recorded annually.

Tornadoes may occur everywhere in the globe, but no nation encounters more tornadoes than the United States, which averages more than thousands in a year (Goliger and Milford 1998; Doswell, 2003). In USA, the Central Plains are most prone to tornadoes, often known as tornado belt. Since tornadoes are more likely to form in the presence of strong VWS and vertical and horizontal moisture and temperature gradients, their frequencies are seen to be higher in the summer and spring than in the winter. Also, they tend to occur more frequently in the afternoon, when the surface air is the most unstable. However, the tornado-prone zones are expanding outside the well-known tornado belt areas (Dixon et al., 2011; Agee et al., 2016; Moore, 2018). Also, severe tornado days with tightly packed clusters of tornadoes could possibly be linked to climate change (Brooks et al., 2014; Elsner et al., 2015). In India, tornadoes are very uncommon in general. However, sometimes they can form in the eastern (specifically West Bengal and Odisha) and northeast parts of this country (De et al., 2005; Singh, 1981). Although tornadoes do not usually occur in northwest parts of India, there have been a few instances (Litta et al., 2010) too.

5 Sand and Dust Storms

Dust storms, which frequently happen in arid and semi-arid areas, are considered events in which blowing dust causes visibility drop (Shao et al., 2011; Lancaster, 2014). These storms occur when strong winds lift large amounts of sand and dust from the dry soils of the arid and semi-arid regions into the atmosphere through the thunderstorms or convective updrafts. They can considerably influence the key elements of the earth's system, such as the atmosphere, biosphere, hydrosphere, lithosphere, and cryosphere (Goudie and Middleton, 2006). The world's most significant dust storms primarily occur in the Northern Hemisphere, particularly in the well-known 'Dust Belt' region (Middleton, 2019; Muhs et al., 2014). It is a vast dry terrain stretching from the Sahara, the Middle East, and Central Asia to northwest India, along with Mongolia and China. In contrast, despite hotspots of dust storms in Central Australia, Southern Africa, and South America, drylands in the Southern Hemisphere constitute a far lower contributor to severe dust and sand storm events (Middleton and Kang 2017). Among all these, Sahara is estimated to provide more than 50% of the world's total dust emissions, followed by China and Central Asia (~20%) (Goudie, 2014). However, dust storms seem to be happening more frequently in some places worldwide, but not everywhere. Some predictions indicate that this will continue due to rising land-use demands and declining soil moisture levels brought on by climate change (Goudie, 2020). The United Nations General Assembly (UNGA) passed a resolution in 2015 called "Combatting sand and dust storms" (A/RES/70/195) in response to this recent upsurge in dust storms (Ganor et al., 2010; Hsu et al., 2012; Notaro et al., 2015). After that, in 2016, the UN Environment Assembly-2 (UNEA-2) also adopted a resolution calling on nations to take appropriate policy steps to address the problems posed by sand and dust storms.

Generally, dust storms take place all around the year, but their frequency and severity vary over a range of timescales. In the north-west India, Pakistan, and Afghanistan, convective dust storms form during the pre-monsoon season (Joseph et al., 1980). For instance, one such storm happened in winter of 2022 is illustrated in Fig. 5, that is derived from NOAA-NASA satellite. These storms are locally known as "Aandhi". Due to the exceptionally high temperatures and little moisture content in the lowest atmosphere over northwest India, severe thunderstorms form with high bases above the ground (~3–4 km). These storms with violent winds further produce dust storms by removing loose fine-grained dust particles from dry and unconsolidated surfaces. They primarily originate if the near-surface wind reaches beyond the emission threshold. Many parameters, such as precipitation, soil moisture, vegetation, and snow cover, significantly impact this threshold (Marticorena, 2014; Parolari et al., 2016). These factors are also responsible for significant regional, seasonal, and inter-annual variations in dust emissions. However, the recent human-induced land degradation or extended desertification could make these storms more vulnerable in the context of climate change. Dust storms are now getting much more attention since they affect not only the originating dryland areas but also the downwind places far from the source due to the long-range transport. And the particles' size, shape, density, and prevailing wind patterns mainly decide how the particles move via the various modes of transportation (Youlin et al., 2002). The dust episodes have extensive effects on soil properties, nutrient dynamics, and biogeochemical cycles in both marine and terrestrial ecosystems (Middleton, 2017). They are also associated with numerous social and economic hazards, such as traffic situations and flight delays, crop damage due to the removal of nutrient-rich elements from the topsoil,

Fig. 5 Satellite imagery illustrating a dust storm over Arabian Sea and parts of Indian region (https://earthobservatory.nasa.gov/images/149378/dust-storm-envelops-the-arabian-sea). The imagery is for late January 2022 and is derived from visible infrared imaging radiometer suite of NOAA-NASA satellite

severe health issues such as cardiovascular and respiratory diseases, and even water contamination.

6 Hail Storms

When thunderclouds with strong updraft velocities ($\geq$15 ms^{-1}) reach a sufficient height and contain significant amounts of supercooled water droplets and ice crystals to sustain the growth of hailstones, they become hailstorms (Fig. 6; Knight and Knight 2001). The hailstones form the storm's core when the embryos come into contact with supercooled liquid drops and produce ice crystals on their surfaces. The growth can range from "dry growth", where all accumulated water freezes onto the surface of the hailstone, to "wet growth", in which a portion of the water is allowed to stay liquid before freezing (Pruppacher and Klett 2010). Often, numerous layers of various growth types may be observed in the dissected hailstones. The number of hail embryos present can influence the size of the hailstones since more hail embryos mean greater competition for the limited amount of supercooled water (Knight and Knight 2001). A hailstone also needs enough time to become the size of a golf ball, with the length of time determined by the strength of the updraft and the embryo's path through the storm. Notably, the primary updraft's width and tilt are crucial for hailstone generation (Foote, 1984; Dennis and Kumjian 2017). The formation of rather large hailstones is connected with a widespread, moderate-strength storm updraft in which hail embryos can be suspended while they expand. On the other hand, significant accumulations of small hails may occur when there is an abundance of supercooled liquid and a high number of embryos without adequate updraft strength or growing time. The damage-causing hail is more likely to happen in supercell-type thunderstorms or coordinated multicellular convective motions, both of which need moderate to high wind shear to develop (Nelson, 1987). The last key factor impacting maximum hailstone size is how much hailstones melt when they hit the ground and fall below the melting level height (Fraile et al., 2003). Small hailstones melt faster than giant hailstones, which are large enough to fall more quickly due to their heavy weight. However, the hailstones may be trapped in a powerful updraft as they slowly drop, then be taken back down to start the process all over again.

India is one of the countries where hailstorms frequently occur (approximately 29 days per year), with moderate to severe hail (Nizamuddin, 1993). In contrast to other tropical stations in the northern hemisphere, hailstorms occur in India and Bangladesh during the winter and pre-monsoon seasons, with hardly any occurrences following the arrival of the south-west monsoon. In winter, low-level available potential energy and moisture combined with upper-level baroclinic instability brought on by western disturbances induce instability, which encourages the occurrence of winter hailstorms (Das, 2022). There have been reports of hail as big as tennis balls, mangoes, and lemons. A study by Eliot (1899) reported that appreciable numbers of hailstorms in India produced hailstones having a diameter of at least 3 cm. The observational and modeling studies indicate possible rise in hailstorm frequency in Australia and Europe and fall in East Asia and North America, while

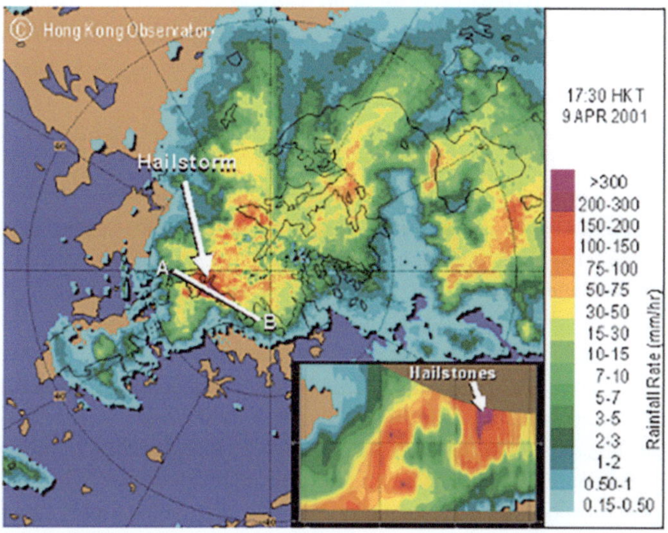

Fig. 6 Radar imagery illustrating an active hail storm over Hong Kong, China in the late afternoon hours. The imagery is for early April 2001 and is taken from Hong Kong Observatory (https://www.hko.gov.hk/en/wxinfo/radars/radar_gallery/storms.htm)

hail severity would increase across most of the places. The anthropogenic climate change may increase the likelihood of hailstorms and allow for the development of larger hailstones by enhancing low-level moisture and convective instability.

7 Squall Line

Multi-cell thunderstorms occasionally appear as a line and are found ahead of some cold fronts, described as a squall line. According to Bluestein and Jain (1985) and Klimowski et al. (2003), the squall lines could be identified based on certain criteria, viz., a length of 50 km or more with length to breadth ratio >5:1, persisting for 30 min or more, the first occurrence of the required length, and length to width ratio determines the start of the squall line. They are incorporated with stronger winds, heavy precipitation, lightning, hailstorms and tornadoes. Thus, they are regarded as a severe hazard to aviation because of the turbulence accompanied by them. Notably, there is no proven evidence correlating the cold front aloft with squall lines. Also, the presence of VWS with a component normal to the cold front defines the second one (Bollay and Beers 1945). The convection above and nearer the cold front is inhibited by the subsiding air over the frontal surface, although the thunderstorms developing in the forefront of the cold front are moved further away as they perforate the high winds above.

The development of a squall line is generally recognized by the identification of thunderstorms with significant wind shifts and increased pressure. The squall lines advance relatively uniformly with an average speed of ~64 km/h. Occasionally, the apparent speed may exceed 96 km/h (Brunk, 1953). Hocker and Basara (2008) analyzed the spatial climatology of squall lines to find a frequent increase during April, and a maximum number of occurrences in May over Oklahoma, USA, whereas significant fraction of the storms (i.e., ~63%), observed during the month from April to June. During the south-west monsoon season, when the dry easterly winds aloft are replaced by moist winds, a line of thunderstorms advances westward over central Africa, showing similar patterns as that of the American squall line (Hamilton et al., 1945). Based on a study in eastern China, the squall lines start to emerge in March and peak in July, and their diurnal fluctuation exhibits two minor peaks in the early morning and early afternoon and a significant peak in the early evening. The squall lines are moving eastward at a pace of 14.4 m per second with a dominating southwest-northeast orientation (Meng et al., 2013). An extremely severe squall line reported in the USA on July 14, 1948, was accountable for a loss of about 9,000,000 dollars due to a significant wind speed ~161 km/h or higher (Brunk, 1953). An investigation of the fatal tornadic and non-tornadic convective wind events along with the storm morphological characteristics across the United States concluded that the second-most fatal non-tornadic convective storm types were squall lines and clusters of cells, each accounting for 19% of fatalities (Schoen and Ashley 2011). Conversely, Brunk (1953) stated tornadoes generally happen before the formation of a squall line and, moreover, are commonly found when there is no recognizable squall line. This essentially encapsulates the current perception based on a survey of tornado research initiatives (Davies-Jones et al., 2001); supercells seem to pose the greatest tornado threat, causing a significant portion of the tornadoes, 1200 on average, to occur within the United States each year.

8 Heavy Rainfall Events

According to a report published by the Intergovernmental Panel on Climate Change (IPCC) titled "Changes in Climate Extremes and their Impact on the Natural Physical Environment," climate change causes alterations in the frequency, intensity, spatial extent, duration, and timing of weather and climate extremes, which can result in unprecedented extremes (Field et al., 2012). The frequency of heavy precipitation would rise in many parts of the world in the twenty-first century too. The most immediate consequence of heavy rain is the risk of flooding. This danger is heightened in metropolitan settings, where non-permeable pavement drives water into sewage systems. Some of the recent extreme flood situations to happen globally are Flooding in Jackson, Mississippi in the summer of 2022, Tennessee flood in August 2021, etc.

The highly populated South Asian subcontinent is one of the most susceptible places in the globe to the effects of natural catastrophes due to the monsoon variability (Singh et al., 2014; Dash et al., 2009, 2011). In recent decades, variability has increased with a progressive drop in monsoon circulation, but a tremendous increase in heavy rain events (Wang and Ding 2006; Turner and Annamalai 2012; Mishra et al., 2012). Roxy et al. (2017) found a threefold for increase in extreme rain events over central India during 1950–2015.

8.1 Cloudbursts

A cloudburst or rain gust is a quick, intense rainfall over a localized area in a relatively short time frame (e.g., Fig. 7). According to IMD, the cloudburst phenomenon is defined as high intensity precipitation, often 100 mm/h, across a limited area in a short time frame. Mountainous regions are particularly prone to cloudbursts. Due to the mountainous topography, the states of Himachal Pradesh, Jammu and Kashmir, and Uttarakhand in India, and Nepal are most impacted. The flash floods that occur along with cloudbursts are mostly to blame for the destruction of property, failure of communication systems, and human casualties. Recent years have seen a rise in reports of "cloudburst" related oddities in and near the Indian Himalayas' southern rim (Dimri et al., 2017). The majority of these occurrences involve unforeseen heavy precipitation. With its steep, unstable inclines, the Himalayan orography creates the ideal environment for a cloudburst occurrence to cause flash floods or landslides.

Cloudburst is an example of cumulonimbus convection that occurs when steep orography, marked moist thermodynamic instability, and deep, quick dynamic lifting

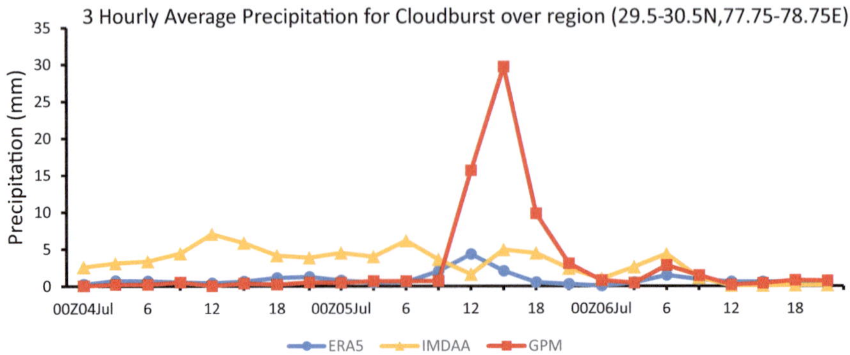

Fig. 7 Three hourly average precipitation for cloudburst over Haridwar, India. The comparison of rainfall from the ERA5 global analyses, IMDAA regional analyses and GPM observations is depicted to highlight higher precipitation in a relatively shorter duration

are present. According to Orlanski (1975), cloudburst occurrences take place at the mesoscale (2–20 km) and might be challenging to identify from thunderstorms. CAPE required for a cloudburst is released by the orographic lifting of moist unstable air, but research on the intricate interactions between cloud dynamics and orographic dynamics is still in its infancy. Cloudburst clouds can be 1–10 km in size. Their vertical range may also extend 10–15 km from low to high atmospheric levels.

However, in Indian scenario, cloudburst events are commonly mentioned in several studies, and researches have emphasized mostly on case studies, including reconstruction from the geomorphic signature of the hindcast (Hobley et al., 2012; Juyal, 2010; Gupta et al., 2013; Thayyen et al., 2013; Shrestha et al., 2015; Knös et al., 2022). The majority of cloudburst incidents take place during the monsoon season, with a higher frequency in July and August (Bhan et al., 2004). These episodes may have been triggered by westward-moving cyclonic circulations in the middle troposphere (500 hPa) over the Tibet-Ladakh region during an intense monsoon (Bhan et al., 2015). Sikka et al. (2015) and Ray et al. (2015) proposed that one of the primary causes of the catastrophic floods that occurred over Uttarakhand in 2013 and Jammu and Kashmir in 2014 was the interaction of low-level westward moving monsoonal systems and eastward moving mid-tropospheric westerly trough. The societal impact of cloudburst could be significantly minimized if high-resolution measurements (~10 km) of atmospheric parameters and vertical profiles are supplied through mesonet observations such as Radiosonde, Automatic Weather Station, and Doppler weather radar. Also, educating and preparing local officials to issue brief warnings would significantly aid in disaster diminution.

9 Future Climate Scenarios Concerning Severe Storms and Extreme Events

The climate scenarios incorporated with the Coupled Model Inter comparison Project Phase 5 (CMIP5) global models are the representative concentration pathways (RCPs) describing the greenhouse gas concentration trajectories, which were adopted for the IPCC assessment report 5 (AR5). The RCPs represent the possible extent of radiative forcing values and there are in total a set of four pathways, viz., RCP2.6, RCP4.5, RCP6.0, RCP8.5, which consider 2.6,4.5, 6.0, and 8.5 W/m^2 of radiative forcing respectively. Climate scenarios from the CMIP6 global models are the latest ones, which were incorporated in the IPCC assessment report 6 (AR6). The shared Socio economic pathways (SSPs) in CMIP6 framework cover a broader range of greenhouse gas and pollutant considerations compared to the previous RCP scenarios. The CO_2 emission scenarios with and without climate change mitigation measures as well as new low CO_2 emission pathways were incorporated.

An assessment of the projected changes in the severe days over South Asia in the mid and far future under RCP 2.6 and 8.5 scenarios during the pre (March–May or MAM) and post (September–November or SON) monsoon period has been performed by Glazer et al. (2021) using the RegCM model mainly for the CORDEX domain (Fig. 8). The detection of severe day is based on both CAPE and the vertical shear between the wind at the surface and the wind in the mid-troposphere ~6 km. Glazer et al. (2021) followed Seeley and Romps (2015) and Trapp et al. (2009) to attribute the weather conditions to be severe in terms of CAPE and vertical shear. Large increases in severe days are noticed mainly over Bangladesh and parts over eastern coast of India in both near and far future under both the RCP scenarios mostly during MAM. However, in the RCP8.5, there is decrease in the severe days over the Indo-Gangetic Plains during MAM. During SON, increase in severe days (~1–2 days) are noticed over the parts of central and northern India in the near and far future periods in RCP8.5 scenarios. Both CAPE and vertical shear responses in both the seasons to climate warming can potentially affect the spatio-temporal patterns of severe days over South Asia.

An investigation on the changes in the annual cycle of TCs over different ocean basins has been performed by Torres-Alavez et al. (2021) using the RegCM model emphasizing the CORDEX domain and the mid and far future under RCP2.6 and RCP8.5 scenarios. The tracking of TCs is based on feature-tracking algorithm TRACK following Hodges (1995, 1999). For NIO, there is a projected increase in occurrence of TCs in the far future period especially in RCP8.5 scenarios with significant changes occurring in the months of November and December. Over other basins including North Atlantic, there is consistent decrease in the occurrence of TCs, whereas there is increase in the occurrence over Eastern Pacific and Northwest Pacific in the mid to far future under RCP8.5 scenario.

Several approaches have been adopted to analyze the projected variabilities in storms and extreme weather events for the historic and near future time frames. For instance, a multi-model assessment of the future changes in the climate and weather

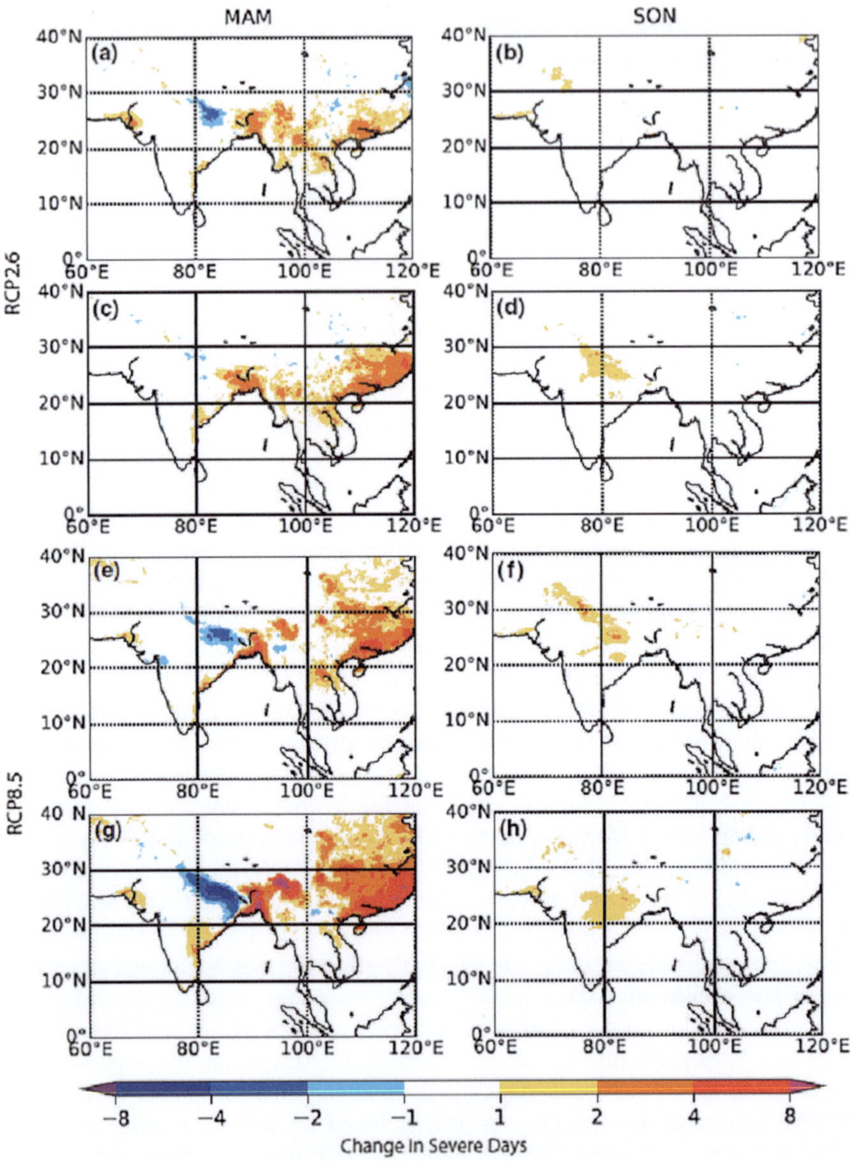

Fig. 8 Spatial distribution of changes in the severe days relative to historical period (1995–2014) at the mid (2041–2060) and far (2080–2099) future time periods over the South Asian region in the CORDEX-CORE simulations using RegCM model. The left panel is for March to May and the right panel is for September to November months. The figure is adopted from Glazer et al. (2021)

events is considered in this chapter for the low (SSP2.6), moderate (SSP4.5), and high (SSP8.5) emission scenarios to depict the projections at the end of the century (2066–2100) in comparison to a historical time frame (1980–2014). The analysis displays a rise in frequency of precipitation and extreme wind events in various regions, especially in the higher emission scenario (e.g. Fig. 9).

The models with greater skill in simulating climate events projected a small rise in the number of droughts and heat waves in Eurasia, whereas a greater number of heavy precipitation and strong winds everywhere under all emission scenarios (Ridder et al., 2022). Almazroui et al. (2021) analyzed the projected variabilities by employing four indices viz., annual maximum of maximum temperature, extreme heat wave days' frequency, annual maximum consecutive five-day precipitation and consecutive dry days under four SSP scenarios. The results indicated a rise in frequency and intensity of temperature and precipitation extremes over the land; specifically, extra tropical regions account for a notable increase in the intensity of hottest days as compared

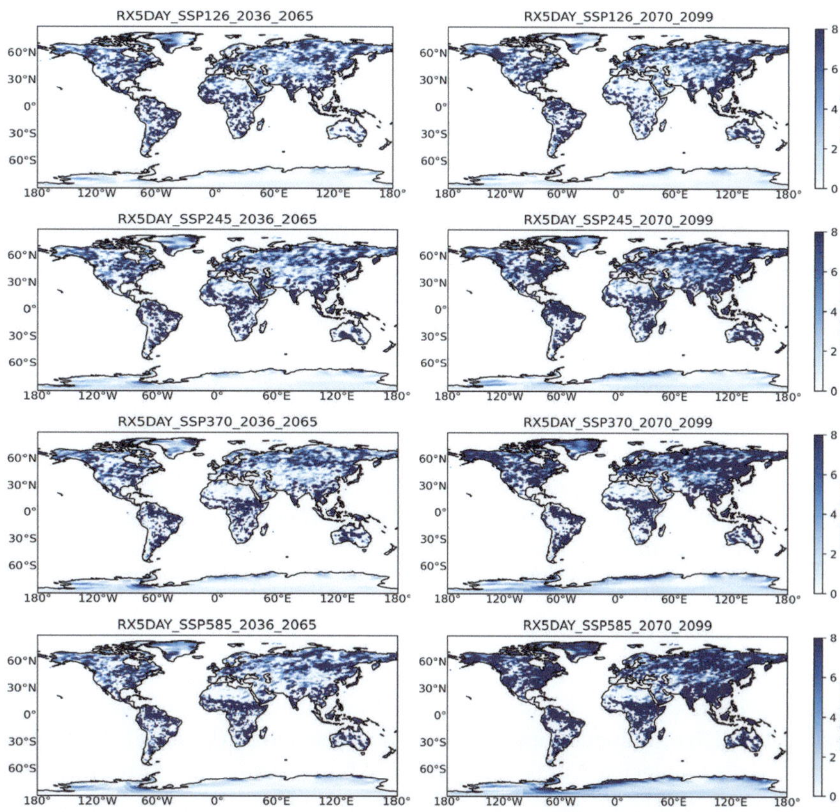

Fig. 9 Spatial distribution of future changes in Annual maximum consecutive 5-day precipitation (in mm), for scenarios (SSP2.6, SSP4.5, SSP7.0 and SSP8.5) and for the two future time slices (2036–2065 and 2070–2099) in comparison to the reference period (1985–2014) using GFDL –ESM4 model. The figure is reproduced by following Vogel et al. (2020)

to tropics. Conversely, the frequency of hottest days is predicted to increase over the tropical region. The regions accompanied under the Asian monsoon zones become more vulnerable towards the extreme flash flood events and Brazil, the Mediterranean, South Africa and Austria became more hazardous in terms of drought frequency. A multi-model ensemble approach on the projected occurrence of hot, dry and extreme event's clusters at different level of global warming also compliments the previous findings i.e., the Northeast, Mediterranean region and south eastern regions were under substantial risk of occurrence of extreme events (Vogel et al., 2020). Besides, the spatial variability in the annual maximum of consecutive 5-day precipitation (Fig. 6) and maximum temperature (Fig. 10) for the near and far future conditions in reference to 1985–2014 period using GFDL-ESM4 model also shows consistency with the previous findings. Notably, majority of the studies suggested an increment in the intensity as well as frequency of the extreme weather phenomena in the near future, which may further increase with the higher emission scenario (SSP8.5).

The response of the convective events in a global warming scenario were assessed using variables such as CAPE, CIN, VWS, storm relative helicity (SRH) and covariate indices (Lepore et al., 2021). The sensitivity of the convective variables as function of global average temperature were considered and the results reveal an increase in the favorable convective environments which leads to a frequency increase on the order of 5–20% per °C rise in global temperature. The impact of predominant dust storm activities on the air quality of china were investigated and projected for a time frame of 2020–2100 (Li et al., 2022). The result reveals reduction in the dust storm frequency in the future for SSP8.5 scenario, while a fluctuating decadal variability was observed in the SSP2.6 scenario.

The future projected variabilities in the intensity and frequency of TCs in the warming climate scenario is discussed in many literature. For instance, Lockwood et al. (2022) investigated the interrelation of sea level rise (SLR) to TC activities using 26 CMIP6 models by incorporating two major TC modulating factors, viz., potential intensity (PI) and VWS. Their study inferred that SLR exhibits a stronger positive correlation with PI change, whereas a negative correlation to VWS changes realized over majority of western North Atlantic and North West Pacific, under the higher emission scenario. An assessment of the characteristics of landfalling TCs in a pseudo global warming framework was implemented over South East Asia with perturbed conditions under SSP7.0 (Tran et al., 2022) that reveals the TCs to be 8% more intense, 2.8% more faster and would become relatively smaller at landfall by the end of 21 century. A location shift in the landfalling TCs would occur towards northward by extending the track more towards the inland. An estimation of the global TC activity in response to the increasing greenhouse emissions was studied with a downscaled high resolution coupled ocean atmosphere model (Emanuel, 2021). The results revealed an increase in both severity and frequency of TCs correlating with the increase in greenhouse gases. The trend is significantly weighted to the Northern Hemisphere chiefly on the higher latitudes.

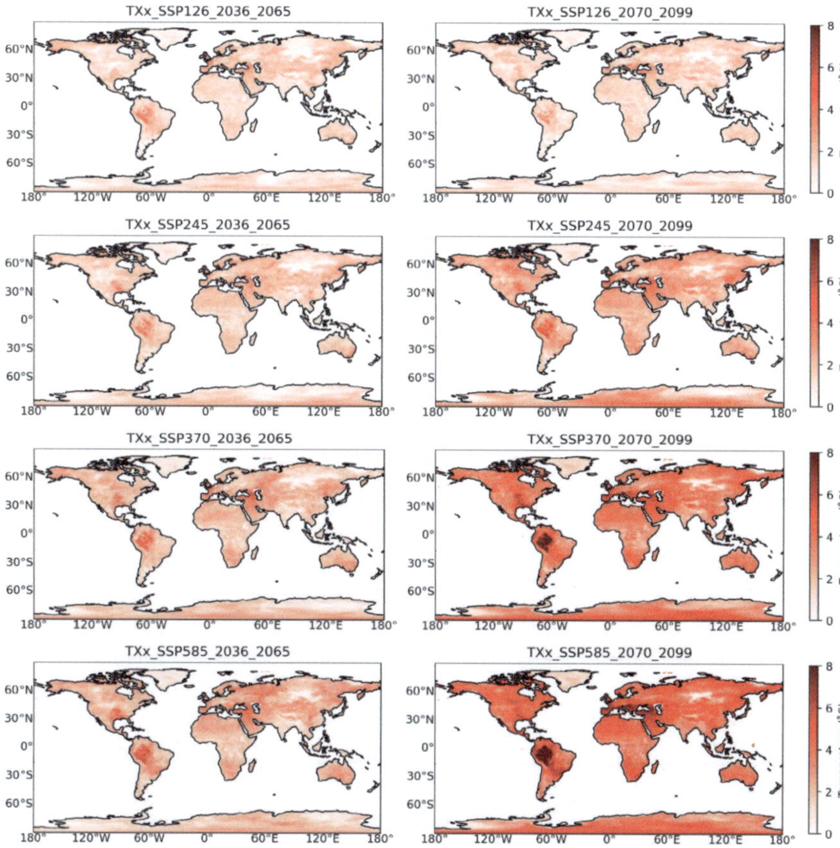

Fig. 10 Spatial distribution of future changes in Annual maximum of maximum temperature (in °C), for scenarios various SSP scenarios (SSP2.6, SSP4.5, SSP7.0 and SSP8.5) and for the two future time slices (2036–2065 and 2070–2099) as a comparison to the reference period (1985–2014) using GFDL-ESM4 model. The figure is reproduced by following Vogel et al. (2020)

10 Impacts of Severe Convective Storms and Mitigation

10.1 Impact of Thunderstorms and Analogous Extreme Events

The thunderstorms, which are considered hazardous natural phenomena generate torrential rainfall, strong gusty winds and hail storms. Violent rainfall, tornadoes, cloudbursts and flooding events are usually the repercussions of supercell thunderstorms (Smith et al., 2001). They cause loss of lives and property damage and could be regarded as a global threat to the societal and economic sectors. The response time obtained to reduce the risks incorporated with the thunderstorms is minimal and raises the destructions to an erratic scale. Shi and Cui (2012) evaluated the impacts and hazards of meteorological disaster events in Shanghai, China's biggest populated urban city. The rainstorm-induced consequences interrupted the regular routines of the urban society, and the accumulation of rain water led to flash flood events intruding on the traffic and inundating properties and residential sectors, resulting in economic losses. The number of rainstorm days exhibits a slight rise with a linear rate of 0.2 days per 10 years. According to the historical data, amongst all the causalities linked to the meteorological disasters in Shanghai, 85 out of the 312 deaths were caused by thunderstorms and induced a direct economic loss of 13 million CNY. Tornados account for the most significant number of causalities; about 1,134 and 10,546 buildings collapsed, produced an economic loss of 650 million CNY. The rainstorms and accompanied flood events generated the least number of causalities and mortalities, even though leads to an economic loss of about 273 million CNY. Also, the lightning accompanied with thunderstorms were accountable for a significant amount of fatalities per year. Globally, the lightning hazards induce about 24,000 deaths and 240,000 injuries per year (Holle and Lopez 2003) and in Germany about 1000 people are injured by lightning per year (Hinkelbein et al., 2013). In United States, statistics indicates 3,645 lightning deaths and an average of 75.9 per year fatalities occurred during 1959–2006 (Ashley and Gilson 2009).

The air transport sector facilitates vital connectivity at regional, national and international levels and time-efficient, convenient methods. However, adverse weather events' impacts make it a precarious disruptive sector. The thunderstorms and associated risks threaten life and properties, reported much more property damage in the United States compared to hurricanes. In 2011 about 47 billion dollars of loss had been reported, and often more mortalities than hurricanes (Re, 2011). The majority of the thunderstorm-related losses in the United States had a normalized loss of more than 250 million dollars per occurrence, according to a time series study of the severe thunderstorm-forcing activities for the years 1970–2009 (Sander et al., 2013). In the continental United States, the meteorological aviation routine weather reports combined with Federal Aviation Administration's Operations Network data, the meteorological impacts on the commercial aviation sector were analyzed by Goodman and Griswold (2019). As per the analysis, thunderstorm-induced lightning, icing, turbulence, wind shear, microbursts and severe rainfall are regarded as hazardous threats to the aviation sector. In addition, delays and cancellations inconvenience the passengers, affect the efficiency of airports and subsequently lead to higher economic losses. At the Hartsfield–Jackson Atlanta International Airport, the freezing precipitation, snow and thunderstorms accelerate the annual impacts on working efficiency and similar conditions are observed in the cases of Miami International Airport and John F. Kennedy International Airport in New York (Goodman and Griswold 2019) too.

During the outbreak of thunderstorms, airborne allergens, including pollen, near-ground-level particles, and fungal spores concentration rises, subsequently increase the risk of exposure in vulnerable individuals (D'Amato et al., 2019; Darvall et al., 2018; Marks et al., 2001). Epidemic thunderstorm asthma specifies the appearance of acute Asthma after a thunderstorm. In 2016 November 21, a significant thunderstorm asthma epidemic was reported in Melbourne, Australia, regarded as the enormous health impact incorporated with a storm ever recorded in history. During the storm event, hospital admissions rose by about 1000%, exceeding the health infrastructural capacity to worsen the situation (Price et al., 2021). A comprehensive investigation conducted on the occurrence of thunderstorms and associated asthma illness in Atlanta, Georgia, for the years 1993–2003 reveals 24,350 out of 2,15,832 cases occurred on the following days of thunderstorms (Grundstein et al., 2008). A study in Canada identifies that about 5–17% of the total outdoor visits were for Asthma and shortness of breath during a thunderstorm day as compared to a non-stormy day (Wardman et al., 2002). In southern and Eastern parts of England, a significant interrelationship observed between epidemics of Asthma and distribution of thunderstorms (Higham et al., 1997). The heavy winds associated with the storms set off an enormously higher amount of bioaerosols into the air, inciting a physiological response in sensitive individuals (Wardman et al., 2002). The projections of a rise in thunderstorms and heavy rainfall events leading to a rise in aeroallergens concentrations could increase the threat to the sensitive population.

An assessment of the damages and risks incorporated with crop production in Pakistan by a parametric analysis disclosed that the regions which underwent severe storms, particularly the hail storm, significantly destructed the standing crops, whereas thunderstorms and wind storms led to remarkable damages and subsequently weakened yields (Elahi et al., 2022). A contemplative analysis of the destruction of food crops in Vietnam indicates that most losses were incorporated with droughts and rainstorms (Simelton et al., 2015). The majority of the studies estimated the impacts of climate change-induced severe weather events as a threat to food security and forecasted a considerable reduction in the global crop yield (Kang et al., 2009). The impacts induced by a weather event don't possess identical magnitude, which may vary with time and space and according to the societal, economic and physical capacities of the population. A study on the socio-economic status and severe regional storm-induced impacts in Bangladesh indicated that the livelihood and environment, crop production and yields and other resources were extremely affected by the unexpected flash floods and severe thunderstorms (Karmakar et al., 2015). The plant resources, household properties, agricultural crops and fisheries were adversely affected, and even environmental degradation was identified. Correspondingly, the decrease in agricultural production broke food security and caused stress among the livelihood of the inhabitants belonging to the coastal zone.

The hail storms, tornados, lightning and flash floods accompanied by the thunderstorms produce substantial damages, causalities and loss of properties in the Indian subcontinent during the pre-monsoon season. The frequency of thunderstorm events is almost 30–40 in some parts of the northeastern parts of India during pre-monsoon months (Tyagi, 2007). Although a minimum number of thunderstorm days occurred during March, where the mean number of thunderstorm days does not exceed more than 6–8, and on the other hand, almost 14–16 days of thunderstorm events were experienced during the month of May. As per the IMD historical records, approximately 29 hail days per year with moderate to severe intensity were reported (Nizamuddin, 1993). In the state of Bihar, enormous damage to properties and loss of lives occurred due to a thunderstorm event in April 21, 2015. Similarly, the thunderstorm in Haryana that occurred in April 2015 was responsible for a monetary loss of about 1092 Crore INR. In March 2015, the hailstorm in Nagpur and in Uttar Pradesh accounted for an economic loss more than 1000 crore INR. The hailstorm occurred in Warangal, Andhra Pradesh, on April 2002, accounted for about 1550 crore INR (Das, 2015). Also, the lightening induced causalities are predominant among the major natural calamities and accountable for a higher percentage of mortality during the past decades. As per the records, an average of 1,775 deaths were reported per annum and recognized as the major damaging natural phenomena in terms of mortality (Illiyas et al., 2014). Since India is developing, the financial losses and expected economic burden from future extreme weather events could create an economic imbalance in the country. Chakravarty et al. (2021) investigated the life cycle of the dust storm occurred on May 30, 2014, and decline in air quality over the Indian Capital, New Delhi and adjoining parts of Indo-Gangetic planes. Ambient temperature considerably dropped to 32 °C from 42 °C during the thunderstorm events due to downdrafts, and the density of PM_{10} and $PM_{2.5}$ rose to a magnitude of 950 and 550 $\mu g/m^3$ respectively in the said event. The increased concentration of aerosol particles in the atmosphere induced a major threat to individuals who are sensitive to allergens and respiratory illnesses.

10.2 Mitigation Measures Concerning Thunderstorms and Analogous Weather

The fatalities, causalities and losses in environmental, societal and economic sectors due to regional weather events are a global concern in the warming climate scenario. In general, mitigation is regarded as the actions to be undertaken to reduce the vulnerability. The governments, policymakers, and national and international stakeholders formulate risk reduction approaches and mitigation strategies to minimize the damage associated with severe weather hazards. For the implementation of adaptable strategic mitigation frameworks, the International organizations of Red Cross and Red Crescent Societies grouped the disasters as a natural and non-natural triggers. Internationally, this distinction is helpful in developing risk reduction policies and mitigation plans in a discrete manner by considering the sources of extreme environmental events. Initially, the unfocussed risk reduction schemes for minimizing vulnerability by avoiding hazards and damages were executed for quick assistance in disasters and calamity scenarios. The refocused schemes with an emphasis on the necessity of mitigation were implemented during the International Decade for Natural Disaster Risk Reduction (IDNDR), regarded as a new initiative, strengthening the policy and efforts to reduce disaster risk at the global level.

The second largest country of the world, Canada is susceptible to a wide range of weather and weather-induced hazards including tornadoes, hurricanes, hail storms, blizzards, storm surges, ice storms and floods, making the population, natural and socio-economic environments prone to impairment. The hailstorms and approximately eighty tornadoes are annually recorded in southern Ontario, southeastern Quebec, and in the Prairie Provinces of Manitoba, Saskatchewan and Alberta (McBean and Henstra 2003). The hurricanes and storm surges on the Atlantic coast (Bruce, 2002) and the frequently occurring winter storms have put the livelihood and socio-economic sectors of the country under threat. The Government established a national disaster mitigation strategy (NDMS) combined with the Department of Public Safety and Emergency Preparedness by consultation with provinces, territories, and other stakeholders in order to magnify Canada's capacity to mitigate disasters and create disaster-resilient communities. Canada's Emergency framework is based on a teamwork approach in which provincial, territorial and local level governments have responsibilities for managing the emergencies according to the scale of impact. When a disaster exceeds the level of capacity of the local-level authorities, a first-level response is a transfer to the territorial and provisional governments for support. If the severity exceeds the capacities further, the Government of Canada supports the management actions. The Department of public safety and emergency preparedness is an authority within the government of Canada primarily focusing on mitigation-related initiatives and identifying the best possible ways to collaborate with non-governmental stakeholders to reduce Government efforts (Hwacha, 2005).

The dominant organization in the United States for disaster mitigation, the Federal Emergency Management Agency (FEMA), invested more than 1 million dollars in single research to evaluate and calculate the savings brought about by its hazard mitigation initiatives. Conventionally, the risk reduction could be comprised of avoidance, mitigation and preparedness. The relocations of settlements from a highly hazard-prone region refer to avoidance, the reinforcement of the infrastructures to reduce damage or overcome the disaster constitute mitigation and evacuation of the population and strategies to reduce the post-disaster impacts refer to preparedness (Ganderton, 2005). The benefit–cost analysis types of studies were generally undertaken to aware the public, authorities and stakeholders regarding the significance of mitigation efforts in the reduction of loss and damages. It seeks to provide insight into the losses and damages that society would suffer in the absence of mitigation. The potential benefits obtained from an efficient mitigation scheme include a reduction in loss of lives and causalities, damages to properties, a reduction in business interventions, a decrease in expenses for disaster relief by government and non-governmental organizations, and a lessening of the fear and caution about the hazards. Implementing risk reduction and mitigation strategies without scientific support, research and consultation from the experts will not provide long-term benefits; instead, a viable mitigation framework consistent with future hazardous scenarios is needed to be implemented.

In the changing climate scenario, minimizing the degree of vulnerability refers to adaptation that is achieved by enhancing capacity building and making the population resilient to cope with future severe weather scenarios. Ji and Lee (2021) investigated the outcomes of the U.S. Hazard Mitigation Grant Program (HMGP), which was executed to enhance disaster resilience at the community level. It especially intended to reduce the risk, damages and losses from the upcoming hazardous disasters by implementing several mitigation projects regionally in areas, exposed and vulnerable to naturally occurring hazards. As per the natural hazard loss data, concern to the US provinces that obtained HMGP funds during the disaster between 2010 and 2015 probably undergo reduced property damages in future hazardous scenarios compared to their counterparts. Several developing countries implemented land use policies as a part of their sustainable urban planning. Improving the resilience of vulnerable communities is a top priority in effective mitigation efforts. In Rabaul, the Papua New Guinea government provided land to the entire volcano victims at the safest locations (IDNDR, 1996) and relocation of the population from landslide and flood-prone areas in Lima, Peru (Leandro and Miranda 2000). Taiwan and Korea implemented combined private–public sector land acquisition and planning policies aimed to enlarge the contribution of serviced land (Pugh, 1995). A collaborative attempt at land acquisition in Sao Paulo, Brazil, and the urban reform in Bangladesh seek to control the illegal settlements, reform slum settlements and improve access to land. The sustainable mitigation policies are enhanced by adopting suitable shelter design, appropriate construction methods and usage of building materials. Different technical solutions are provided by the United Nations Disaster Relief Co-ordinator based on the class of hazard, physical situations of the settlements and accessible resources. The technical solutions include promoting the usage of water-resistant materials and waterproof plastering techniques, elevating basements from the flooding levels in flood-prone regions, avoiding low-pitched and flat, lightweight roofs, avoiding roof overhangs, a stronger connection from the basement to roof, evading empty-spaces under the basements, etc.

To mitigate the risks incorporated with the aviation sector, a satellite-based real-time thunderstorm now-casting viz., Cb-TRAM scheme for the tracking, detection and forecasting is developed. Complimented with the Numerical Weather Forecasting schemes, the probable locations for the development of thunderstorms in current and future times could be identified. The use of Cb-TRAM information in the cockpit enables the pilots capable of executing strategic flight planning one hour in advance rather than only minutes in advance, based on the out-of-window view and the constrained range of the onboard radar (Forster et al., 2016). The Epidemic thunderstorm asthma (ETSA) conditions were validated as a health hazard followed by the occurrence of the thunderstorms. Mitigation of the severity of asthmatic illness measure constitutes recognizing the risk variables to enable prevention interventions to be directed to individuals who are at maximum risk, preventative measures taken by individuals, as well as major systems-based strategies including forecast alert systems and environmental health literacy (Chatelier et al., 2021).

Similarly, the sand and dust storms generate hazards in several ways. The entrainment of the dust particles leads to soil loss, nutrient and fertilizer loss, the transport leads to air pollution causing respiratory problems, eye infection and disease transmission, and its deposition create crop growth problems, drinking-water contamination, reduction of solar power potential, etc. (Middleton and Kang 2017). The mitigation efforts comprise agronomic practices for reducing soil erosion utilizing live plants or crop remains to shield soil from the wind. A study in Sahelian Africa indicated that retaining 2% of agricultural land crop residue cover decreases the possibility of wind erosion by at least a factor of three (Toure et al., 2011). Wind-breakers and shelterbelts could be constructed to minimize the erosive force of the wind, i.e., fences or walls positioned at a right angle to erosive winds, or windbreaks made from trees or shrubs can function as shelter belts (Chendev et al., 2015; Wang et al., 2010a).

10.3 Mitigation Strategies of Government of India for Extreme Weather Impacts

The ministry of home affairs of the Government of India established the national disaster management program in 1993 and formulated the national disaster management authority in 1995 that initially worked as an organization focused on rescue and relief operations till the 2000s. Based-on the awareness regarding the consequences of the devastating impacts imparted by the super cyclone in 1999, the government proposed a national framework by combining the economic and environmental factors for the systematic and enhanced response and mitigation of natural disasters. A severe cyclonic storm that formed in the Bay of Bengal and hit the coast of Andra Pradesh near Kakinada in 1996 caused enormous loss and damage to life and property. Similarly, a TC in 1977, regarded as the worst cyclonic event accompanied by storm surges resulted in loss of lives of about 750 people, and affected more than 700,000 people with a total loss of about 5375 cores (Shanmugasundaram et al., 2000). Therefore, the Indian code of practice recommended constructing buildings and structures in coastal regions with a wind load more than 50 m/s (Standard, 1987). In regards to hazardous conditions similar to 1996 and 1999 TCs, and based on the post disastrous surveys conducted by the Structural Engineering Research Centre (SERC), Madras, several mitigation strategies are suggested including increasing the lateral strength of buildings and structures to resist cyclonic forces and vertical strength to withstand dead and live loads, construction of hipped type roof structure in cyclone hazard-prone locations to resist the uplifting forces during stronger winds, concrete restraining strips over the tiles, etc.

The state of Odisha, India, established Odisha State Disaster Management Authority (OSDMA) after the super cyclonic storm in 1999. in order to reduce the threats and to implement mitigation and risk reduction strategies. The authority is aimed to promote capacity building and coordinate disaster management activities by bringing together international to local level stakeholders. In collaboration with the Indian Red Cross society, they built 203 multi-purpose cyclonic shelters, 10 km from the Odisha coastline (OSDMA, 2016). Besides, the buildings that cleared the safety norms in the public sector, such as schools, community centres and temples also been facilitated as cyclone shelters (OSDMA, 2013). In the scenario of increasing natural disasters, Government of India implemented a Disaster Management Act in 2005 to define the duties and responsibilities of stakeholders in improving disaster risk reduction strategies and prompting capacity building to mitigate the impacts of forthcoming disasters. For instance, IMD early warning system provided a timely warning, i.e., five days before the landfall of the TC Phailin enabling the mass evacuation of the population who were under risk (about 1.2 million people in total) that reduced the severity of the impacts to a significant level (UNDP-GEAS, 2013). In order to improve and enhance the resilience of the vulnerability community, a five-year resilience program were implemented from 2011 to 2015. This comprehensive effort enabled the population and community to be capable of facing the future anticipated impacts and cope with the changing risks. A cluster approach was adopted by connecting the population possessing similar geographical, physical and socio-economic characteristics to disseminate the benefits to a larger population. As the population is heterogeneous in nature, a strategic mitigation approach is built up from the community level, where the population exhibits homogeneity up to a certain level.

In order to lead and adopt a comprehensive and integrated approach to disaster management, the National Disaster Management Authority (NDMA), headed by the Prime Minister, State Disaster Management Authorities (SDMAs), governed by the Chief Ministers, and District Disaster Management Authorities (DDMAs), supervised by the District Collector, District Magistrate, or Deputy Commissioner, were established. The NDMA is the responsible authoritative agency for implementing policies, plans and guidelines for timely and essential responses. The SDMA establish standards, oversee the state Plan's execution, and suggest funding allocations for preparedness and mitigation strategies. They are also responsible for identifying the vulnerable regions in different parts of the state to diverse disasters and designing methods for their prevention and mitigation. The district-level disaster management authority has the responsibility to create disaster management plans at the district level (Das, 2012). The multi-level disaster management approach is effective in developing and coordinating risk reduction, adaptation plans, and mitigation strategies efficiently, besides implementing preparedness practices and building up a climate-resilient community.

The agricultural sector is highly vulnerable to severe weather events. The country experienced anomalous and extensive events of thunderstorms during the years 2015 and 2020 (Jaybhaye, 2022). Hail, frozen and solid form of precipitation causes substantial losses to properties and growing crops. The management of mitigation and adaptation strategies requires interdisciplinary knowledge, technological inventions and suitable interventions. The Agromet Advisory Bulletin (AAB) provide site-specific information and suggestions within pre- and post-hail storm events to increase the advantages and reduce the losses in agricultural production. AAB is regarded as an effective aid in weather-smart agricultural technologies to reduce agriculture risks and create resilient agricultural production. The communication regarding the forecasted thunderstormsand hailstorms is disseminated through multimedia platforms (Jaybhaye et al., 2018). Elahi et al. (2022) estimated the adaptation and mitigation strategies adopted to alleviate the risks associated with extreme weather events. The analysis revealed that the farmers who adopted contemporary management procedures, could able to increase the wheat production remarkably. It indicates a community-level multidimensional approach can effectively minimize the losses and safeguard the population and properties from the devastating impacts of the future disasters in a changing climate scenario.

11 Concluding Remarks

The severe thunderstorms, cyclones, and heavy rain events like extreme weather scenarios impact the life in Earth significantly. The scientific advancement helps in understanding the different aspects of these extreme weather events. Consequently, the understanding leads to better forecasting to save the lives of people through disaster management and helps in the policy making. It is important to further the knowledge in view of their climatology or long-term variability since the prevailing warming climate has capability to alter their characteristic features. In the current warming environmental scenario, furthering the understanding is quite important for accurate forecasting while considering the impact of changing climate, as it coincides with the aspiration of socio-economic advancement of many countries including India. This chapter is expected to fulfill the objective of providing sufficient knowledge starting from the basics to the impact of changing climate on the considered severe storms so that the information might help in the future policy making and disaster management.

References

Agee, E., Larson, J., Childs, S., & Marmo, A. (2016). Spatial redistribution of US tornado activity between 1954 and 2013. *Journal of Applied Meteorology and Climatology, 55*(8), 1681–1697.

Aiyyer, A., & Thorncroft, C. (2011). Interannual-to-multidecadal variability of vertical shear and tropical cyclone activity. *Journal of Climate, 24*(12), 2949–2962. https://doi.org/10.1175/2010JCLI3698.1

Allen, J. T., & Allen, E. R. (2016). A review of severe thunderstorms in Australia. *Atmospheric Research, 178*, 347–366.

Almazroui, M., Saeed, F., Saeed, S., Ismail, M., Ehsan, M. A., Islam, M. N., ... & Nadeem, I. (2021). Projected changes in climate extremes using CMIP6 simulations over SREX regions. *Earth Systems and Environment, 5*(3), 481–497.

Araghi, A., Adamowski, J., & Jaghargh, M. R. (2016). Detection of trends in days with thunderstorms in Iran over the past five decades. *Atmospheric Research, 172*, 174–185.

Ashley, W. S., & Gilson, C. W. (2009). A reassessment of US lightning mortality. *Bulletin of the American Meteorological Society, 90*(10), 1501–1518.

Ashley, W. S., Bentley, M. L., & Stallins, J. A. (2012). Urban-induced thunderstorm modification in the Southeast United States. *Climatic Change, 113*, 481–498.

Asnani, G. C. (1992). *Tropical Meteorology* (Vol. 2, pp. 829–833, 852).

Balaguru, K., Leung, L. R., Lu, J., & Foltz, G. R. (2016). A meridional dipole in pre monsoon Bay of Bengal tropical cyclone activity induced by ENSO. *Journal of Geophysical Research: Atmospheres, 121*, 6954–6968. https://doi.org/10.1002/2016JD024936

Bell, R., Hodges, K., Vidale, P. L., Strachan, J., & Roberts, M. (2014). Simulation of the global ENSO–tropical cyclone teleconnection by a high-resolution coupled general circulation model. *Journal of Climate, 27*(17), 6404–6422. https://doi.org/10.1175/JCLI-D-13-00559.1

Bessafi, M., & Wheeler, M. C. (2006). Modulation of south Indian Ocean tropical cyclones by the Madden-Julian oscillation and convectively coupled equatorial waves. *Monthly Weather Review, 134*, 638–656. https://doi.org/10.1175/MWR3087.1

Bhan, S. C., Devrani, A. K., & Sinha, V. (2015). An analysis of monthly rainfall and the meteorological conditions associated with cloudburst over the dry region of Leh (Ladakh), India. *Mausam, 66*(1), 107–122.

Bhan, S. C., Paul, S., & Kharbanda, K. L. (2004). Cloud bursts in Himachal Pradesh. *Mausam, 55*(4), 712–713.

Bhardwaj, P., Singh, O., & Kumar, D. (2017). Spatial and temporal variations in thunderstorm casualties over India. *Singapore Journal of Tropical Geography, 38*(3), 293–312.

Bielli, S., Barthe, C., Bousquet, O., Tulet, P., & Pianezze, J. (2021). The effect of atmosphere-ocean coupling on the structure and intensity of tropical cyclone Bejisa in the Southwest Indian Ocean. *Atmosphere, 12*(6), 688.

Bluestein, H. B., & Jain, M. H. (1985). Formation of mesoscale lines of precipitation: Severe squall lines in Oklahoma during the spring. *Journal of the Atmospheric Sciences, 42*(16), 1711–1732.

Bollay, E., & Beers, N. R. (Eds.). (1945). *Handbook of Meteorology*. McGraw-Hill.

Boruff, B. J., Easoz, J. A., Jones, S. D., Landry, H. R., Mitchem, J. D., & Cutter, S. L. (2003). Tornado hazards in the United States. *Climate Research, 24*(2), 103–117.

Boucharel, J., Jin, F.-F., England, M. H., & Lin, I. I. (2016). Modes of hurricane activity variability in the Eastern Pacific: Implications for the 2016 season. *Geophysical Research Letters, 43*,. https://doi.org/10.1002/2016GL070847

Boyden, C. J. (1963). A simple instability index for use as a synoptic parameter. *Meteorological Magazine, 92*, 198–210.

Brooks, H. E., Carbin, G. W., & Marsh, P. T. (2014). Increased variability of tornado occurrence in the United States. *Science, 346*(6207), 349–352.

Bruce, J. (2002). Consequence Analysis of Storm Surge in the Charlottetown, Prince Edward Island Area. Office of Critical Infrastructure Protection and Emergency Preparedness=Bureau de la protection des infrastructures essentielles et de la protection civile.

Brunk, I. W. (1953). Squall lines. *Bulletin of the American Meteorological Society, 34*(1), 1–9.

Byers, H. R. (1949). Structure and dynamics of the thunderstorm. *Science, 110*(2856), 291–294.

Camargo, S. J., & Hsiang, S.M. (2015). Tropical cyclones. In M. Chaves, M. Ghil, & J. Urrutia-Fucugauchi (Eds.), *Extreme Events: Observations, Modeling, and Economics* (pp. 303–342). American Geophysical Union (AGU). https://doi.org/10.1002/9781119157052.ch18.

Camargo, S. J., & Sobel, A. H. (2005). Western North Pacific tropical cyclone intensity and ENSO. *Journal of Climate, 18*(15), 2996–3006.

Camargo, S. J., Emanuel, K. A., & Sobel, A. H. (2007a). Use of a genesis potential index to diagnose ENSO effects on tropical cyclone genesis. *Journal of Climate, 20*(19), 4819–4834.

Camargo, S. J., Emanuel, K. A., & Sobel, A. H. (2007b). Use of a genesis potential index to diagnose ENSO effects on tropical cyclone genesis. *Journal of Climate, 20*(19), 4819–4834. https://doi.org/10.1175/JCLI4282.1

Camargo, S. J., Sobel, A. H., Barnston, A. G., & Klotzbach, P. J. (2010). The influence of natural climate variability on tropical cyclones, and seasonal forecasts of tropical cyclone activity. In J. C. I. Chan & J. D. Kepert, (Eds.), *Global Perspectives on Tropical Cyclones* (pp. 325–360). https://doi.org/10.1142/9789814293488_0011.

Chakravarty, K., Vincent, V., Vellore, R., Srivastava, A. K., Rastogi, A., & Soni, V. K. (2021). Revisiting Andhi in northern India: A case study of severe dust storm over the urban megacity of New Delhi. *Urban Climate, 37*, 100825.

Chan, J. C. (1985). Tropical cyclone activity in the northwest Pacific in relation to the El Niño/ Southern Oscillation phenomenon. *Monthly Weather Review, 113*(4), 599–606.

Chand, S. S., & Walsh, K. J. E. (2009). Tropical cyclone activity in the Fiji region: Spatial patterns and relationship to largescale circulation. *Journal of Climate, 22*, 3877–3893.

Chandrasekar, A. (2022). *Basics of Atmospheric Science*. EBook ISBN: 9789391818258, ISBN Print: 9789391818241. PHI Learning Pvt. Ltd.

Chatelier, J., Chan, S., Tan, J. A., Stewart, A. G., & Douglass, J. A. (2021). Managing exacerbations in thunderstorm asthma: Current insights. *Journal of Inflammation Research, 14*, 4537.

Chendev, Y. G., Sauer, T. J., Hernandez Ramirez, G., & Burras, C. L. (2015). History of east European Chernozem soil degradation; protection and restoration by tree windbreaks in the Russian steppe. *Sustainability, 7*(1), 705–724.

Colby, F. P., Jr. (1984). Convective inhibition as a predictor of convection during AVE-SESAME II. *Monthly Weather Review, 112*(11), 2239–2252.

Colquhoun, J. R. (1987). A decision tree method of forecasting thunderstorms, severe thunderstorms and tornadoes. *Weather and Forecasting, 2*(4), 337–345.

D'Amato, G., Annesi-Maesano, I., Cecchi, L., & D'Amato, M. (2019). Latest news on the relationship between thunderstorms and respiratory allergy, severe Asthma, and deaths from Asthma. *Allergy, 74*(1), 9–11.

Darvall, J. N., Durie, M., Pilcher, D., Wigmore, G., French, C., Karalapillai, D., ... & Bellomo, R. (2018). Intensive care implications of epidemic thunderstorm asthma. *Critical Care and Resuscitation*, 20(4), 294–303.

Das, P. (2012). Disaster management in India: Policy review and institutional structure. *Asiapac Journal of Social Sciences, 4*, 37–52.

Das, S. (2022). Challenges in predicting extreme weather events over the South Asian Region. In *Extreme Natural Events* (pp. 51–106). Springer.

Das, S., Mohanty, U. C., Tyagi, A., Sikka, D. R., Joseph, P. V., Rathore, L. S., & Sarkar, A. (2014). The SAARC STORM: A coordinated field experiment on severe thunderstorm observations and regional modeling over the South Asian Region. *Bulletin of the American Meteorological Society, 95*(4), 603–617.

Das, Y. (2015). Some aspects of thunderstorm over India during pre-monsoon season: A preliminary report-I. *Journal of Geosciences and Geomatics, 3*(3), 68–78.

Dash, S., Kulkarni, M. A., Mohanty, U., & Prasad, K. Changes in the characteristics of rain events in India. *Journal of Geophysical Research: Atmospheres. 114*, D10109 (2009).

Dash, S., Nair, A. A., Kulkarni, M. A., & Mohanty, U. (2011). Characteristic changes in the long and short spells of different rain intensities in India. *Theoretical and Applied Climatology, 105*, 563–570.

Davies-Jones, R. (1995). Tornadoes. *Scientific American, 273*(2), 48–57.

Davies-Jones, R., Trapp, R. J., & Bluestein, H. B. (2001). Tornadoes and tornadic storms. In *Severe Convective Storms* (pp. 167–221). American Meteorological Society.

De, U. S., Dube, R. K., & Rao, G. P. (2005). Extreme weather events over India in the last 100 years. *Journal of Indian Geophysical Union, 9*(3), 173–187.

Dennis, E. J., & Kumjian, M. R. (2017). The impact of vertical wind shear on hail growth in simulated supercells. *Journal of the Atmospheric Sciences, 74*(3), 641–663.

DeRubertis, D. (2006). Recent trends in four common stability indices derived from US radiosonde observations. *Journal of Climate, 19*(3), 309–323.

Dhawan, V. B., Tyagi, A., & Bansal, M. C. (2008). Forecasting of thunderstorms in pre-monsoon season over northwest India. *Mausam, 59*(4), 433–444.

Dimri, A. P., Chevuturi, A., Niyogi, D., Thayyen, R. J., Ray, K., Tripathi, S. N., … Mohanty, U. C. (2017). Cloudbursts in Indian Himalayas: A review. *Earth-Science Reviews, 168*, 1–23.

Dixon, P. G., Mercer, A. E., Choi, J., & Allen, J. S. (2011). Tornado risk analysis: Is Dixie Alley an extension of Tornado Alley? *Bulletin of the American Meteorological Society, 92*(4), 433–441.

Doswell, C. A., III. (2003). Societal impacts of severe thunderstorms and tornadoes: Lessons learned and implications for Europe. *Atmospheric Research, 67*, 135–152.

Doswell, C. A., & Bosart, L. F. (2001). Extratropical synoptic-scale processes and severe convection. In *Severe Convective Storms* (pp. 27–69). American Meteorological Society.

Dowdy, A., & Kuleshov, Y. (2012). An analysis of tropical cyclone occurrence in the Southern Hemisphere derived from a new satellite-era data set. *International Journal of Remote Sensing, 33*, 7382–7397.

Eblen, L. H., Ladd, J. W., & Hicks, T. M. (1990). Severe thunderstorm forecasting: an operational review. NOAA Technical Memorandum NWS SR-130. Available at https://repository.library.noaa.gov/view/noaa/6332

Elahi, E., Khalid, Z., Tauni, M. Z., Zhang, H., & Lirong, X. (2022). Extreme weather events risk to crop-production and the adaptation of innovative management strategies to mitigate the risk: A retrospective survey of rural Punjab, Pakistan. *Technovation, 117*, 102255.

Eliot, J. (1899). Hailstorm in India during the period 1883–1897 with a discussion on their distribution. *Indian Meteorological Memoirs, 6*, 237–315.

Elsner, J. B., Elsner, S. C., & Jagger, T. H. (2015). The increasing efficiency of tornado days in the United States. *Climate Dynamics, 45*(3), 651–659.

Emanuel, K. (2005). Increasing destructiveness of tropical cyclones over the past 30 years. *Nature, 436*(7051), 686–688. https://doi.org/10.1038/nature03906

Emanuel, K. (2021). Response of global tropical cyclone activity to increasing CO 2: Results from downscaling CMIP6 models. *Journal of Climate, 34*(1), 57–70.

Felton, C. S., Subrahmanyam, B., & Murty, V. S. N. (2013). ENSO-modulated cyclogenesis over the Bay of Bengal. *Journal of Climate, 26*(24), 9806–9818. https://doi.org/10.1175/JCLI-D-13-00134.1

Field, C. B., Barros, V., Stocker, T. F., Qin, D., Dokken, D. J,. Ebi, K. L., Mastrandrea, M. D., Mach, K. J., Plattner, G.-K., Allen, S. K., Tignor, M., Midgley, P. M., & IPCC. (2012). *Managing the Risks of Extreme Events and Disasters to Advance Climate Change Adaptation, A Special Report of Working Groups I and II of the Intergovernmental Panel on Climate Change* (p. 582). Cambridge University Press.

Foote, G. B. (1984). A study of hail growth utilizing observed storm conditions. *Journal of Applied Meteorology and Climatology, 23*(1), 84–101.

Forster, C., Ritter, A., Gemsa, S., Tafferner, A., & Stich, D. (2016). Satellite-based real-time thunderstorm nowcasting for strategic flight planning en route. *Journal of Air Transportation, 24*(4), 113–124.

Fosu, B. O., & Wang, S. Y. S. (2015). Bay of Bengal: Coupling of pre-monsoon tropical cyclones with the monsoon onset in Myanmar. *Climate Dynamics, 45*(3), 697–709.

Fraile, R., Castro, A., López, L., Sánchez, J. L., & Palencia, C. (2003). The influence of melting on hailstone size distribution. *Atmospheric Research, 67*, 203–213.

Frank, W. M., & Roundy, P. E. (2006). The role of tropical waves in tropical cyclogenesis. *Monthly Weather Review, 134*, 2397–2417.

Fujita, T. T. (1971). Proposed characterization of tornadoes and hurricanes by area and intensity (No. NASA-CR-125545).

Ganderton, P. T. (2005). 'Benefit–cost analysis' of disaster mitigation: Application as a policy and decision-making tool. *Mitigation and Adaptation Strategies for Global Change, 10*(3), 445–465.

Ganor, E., Osetinsky, I., Stupp, A., & Alpert, P. (2010). Increasing trend of African dust, over 49 years, in the eastern Mediterranean. *Journal of the Geological Society Australia, 115*(D7), D07201.

George, J. J. (1960). *Weather Forecasting for Aeronautics*. Academic. https://doi.org/10.1016/C2013-0-12567-6, ISBN: 978–1–4832–3320–8.

Girishkumar, M. S., Suprit, K., Vishnu, S., Prakash, V. T., & Ravichandran, M. (2015). The role of ENSO and MJO on rapid intensification of tropical cyclones in the Bay of Bengal during October–December. *Theoretical and Applied Climatology, 120*, 797–810.

Glazer, R. H., Torres-Alavez, J. A., Coppola, E., Giorgi, F., Das, S., Ashfaq, M., & Sines, T. (2021). Projected changes to severe thunderstorm environments as a result of twenty-first century warming from RegCM CORDEX-CORE simulations. *Climate Dynamics, 57*, 1595–1613.

Goldenberg, S. B., & Shapiro, L. J. (1996). Physical mechanisms for the association of El Niño and West African rainfall with Atlantic major hurricane activity. *Journal of Climate, 9*, 1169–1187. https://doi.org/10.1175/1520-0442(1996)009%3c1169:PMFTAO%3e2.0.CO;2

Goliger, A. M., & Milford, R. V. (1998). A review of worldwide occurrence of tornadoes. *Journal of Wind Engineering and Industrial Aerodynamics, 74*, 111–121.

Goodman, C. J., & Griswold, J. D. S. (2019). Meteorological impacts on commercial aviation delays and cancellations in the continental United States. *Journal of Applied Meteorology and Climatology, 58*(3), 479–494.

Goudie, A. S. (2014). Desert dust and human health disorders. *Environment International, 63*, 101–113.

Goudie, A. S. (2020). Dust storms and human health. In *Extreme Weather Events and Human Health* (pp. 13–24). Springer.

Goudie, A. S., & Middleton, N. J. (2006). *Desert Dust in the Global System.* 288pp. ISBN: 978-3-540-32354-9, 978-3-642-06890-4, 978-3-540-32355-6. Springer Science & Business Media. https://doi.org/10.1007/3-540-32355-4

Gray, W. M. (1984). Atlantic seasonal hurricane frequency. Part I: El Niño and 30 mb quasi biennial oscillation influences. *Monthly Weather Review, 112*, 1649–1668. https://doi.org/10.1175/1520-0493(1984)112%3C1649:ASHFPI%3E2.0.CO;2

Gray, W. M., Landsea, C. W., Mielke, P. W., & Berry, K. J. (1994). Predicting Atlantic basin seasonal tropical cyclone activity by 1 June. *Weather and Forecasting, 9*, 103–115. https://doi.org/10.1175/1520-0434(1994)009%3c0103:PABSTC%3e2.0.CO;2

Grundstein, A., Sarnat, S. E., Klein, M., Shepherd, M., Naeher, L., Mote, T., & Tolbert, P. (2008). Thunderstorm associated asthma in Atlanta, Georgia. *Thorax, 63*(7), 659–660.

Gupta, V., Dobhal, D. P., & Vaideswaran, S. C. (2013. August 2012 cloudburst and subsequent flash flood in the Asi Ganga, a tributary of the Bhagirathi river, Garhwal Himalaya, India. *Current Science*, 249–253.

Hamilton, R. A., Archbold, J. W., & Douglas, C. K. M. (1945). Meteorology of Nigeria and adjacent territory. *Quarterly Journal of the Royal Meteorological Society, 71*(309–310), 231–264.

Harel, M., & Price, C. (2020). Thunderstorm trends over Africa. *Journal of Climate, 33*(7), 2741–2755.

Hawkins, H. F., & Imbembo, S. M. (1976). The structure of a small, intense hurricane—Inez 1966. *Monthly Weather Review, 104*(4), 418–442.

Higham, J., Venables, K., Kupek, E., Bajekal, M., & Kopek, E. (1997). Asthma and thunderstorms: Description of an epidemic in general practice in Britain using data from a doctors' deputising service in the UK. *Journal of Epidemiology and Community Health, 51*(3), 233.

Hinkelbein, J., Spelten, O., & Wetsch, W. A. (2013). Lightning strikes and lightning injuries in prehospital emergency medicine. Relevance, Results, and Practical Implications. *Der Unfallchirurg, 116*(1), 74–79.

Hobley, D. E., Sinclair, H. D., & Mudd, S. M. (2012). Reconstruction of a major storm event from its geomorphic signature: The Ladakh floods, 6 August 2010. *Geology, 40*(6), 483–486.

Hocker, J. E., & Basara, J. B. (2008). A 10-year spatial climatology of squall line storms across Oklahoma. *International Journal of Climatology: A Journal of the Royal Meteorological Society, 28*(6), 765–775.

Hodges, K. I. (1995). Feature tracking on the unit sphere. *Monthly Weather Review, 123*(12), 3458–3465.

Hodges, K. I. (1999). Adaptive constraints for feature tracking. *Monthly Weather Review, 127*(6), 1362–1373.

Holle, R. L., & Lopez, R. E. (2003, September). A comparison of current lightning death rates in the US with other locations and times. In *International Conference on Lightning and Static Electricity* (pp. 16–18).

Hsu, N. C., Gautam, R., Sayer, A. M., Bettenhausen, C., Li, C., Jeong, M. J., … Holben, B. N. (2012). Global and regional trends of aerosol optical depth over land and ocean using Sea WiFS measurements from 1997 to 2010. *Atmospheric Chemistry and Physics, 12*(17), 8037–8053.

Huang, X., Wang, D., Li, L. Z., Li, Q., & Zeng, Z. (2022). Modeling urban impact on Zhengzhou storm on July 20, 2021. *Journal of Geophysical Research: Atmospheres, 127*(22), e2022JD037387.

Hwacha, V. (2005). Canada's experience in developing a national disaster mitigation strategy: A deliberative dialogue approach. *Mitigation and Adaptation Strategies for Global Change, 10*(3), 507–523.

Illiyas, F. T., Mohan, K., Mani, S. K., & Pradeepkumar, A. P. (2014). Lightning risk in India: Challenges in disaster compensation. *Economic and Political Weekly*, 23–27.

IDNDR (1996). *Cities at risk; making cities safer… before disaster strikes.* International Decade for Natural Disaster Reduction. Available at https://www.undrr.org/publication/cities-risk-making-cities-safer-disaster-strikes

Jaybhaye, P. (2022). Mitigation and adaptation strategies of plants against hailstorm under changing climate. *Plant Defense Mechanisms, 123.*

Jaybhaye, P. R., Deore, N. H., & Shinde, P. B. (2018). Agromet advisory bulletin-a weather smart agriculture technology option for adaptation and mitigation of changing climate. *International Journal of Current Microbiology and Applied Sciences, 7*(2), 2644–2653.

Ji, H., & Lee, D. (2021). Disaster risk reduction, community resilience, and policy effectiveness: The case of the Hazard Mitigation Grant Program in the United States. *Disasters, 45*(2), 378–402.

Joseph, P. V., Raipal, D. K., & Deka, S. N. (1980). ANDHI, the convective dust storm of northwest India. *Mausam, 31*, 431–442.

Juyal, N. (2010). Cloud burst-triggered debris flows around Leh. *Current Science, 99*(9), 1166–1167.

Kang, Y., Khan, S., & Ma, X. (2009). Climate change impacts on crop yield, crop water productivity and food security–A review. *Progress in Natural Science, 19*(12), 1665–1674.

Karmakar, S., Quadir, D. A., & Mannan, M. A. (2015). Trends in maximum temperature and thunderstorms, their correlation and impacts on the livelihood of Bangladesh. *The Atmosphere, 5*(1), 113–129.

Kemball-Cook, S., & Wang, B. (2001). Equatorial waves and air-sea interaction in the boreal summer intra seasonal oscillation. *Journal of Climate, 14*, 2923–2942.

Kessler, E. (1970). Tornadoes. *Bulletin of the American Meteorological Society, 51*(10), 926–937.

Kessler, E. (1982). *Thunderstorm Morphology and Dynamics* (Vol. 2). US Department of Commerce, National Oceanic and Atmospheric Administration, Environmental Research Laboratories.

Kikuchi, K., Wang, B., & Fudeyasu, H. (2009). Genesis of tropical cyclone Nargis revealed by multiple satellite observations. *Geophysical Research Letters, 36*, L06811. https://doi.org/10.1029/2009GL037296

Klimowski, B. A., Bunkers, M. J., Hjelmfelt, M. R., & Covert, J. N. (2003). Severe convective windstorms over the northern high plains of the United States. *Weather and Forecasting, 18*(3), 502–519.

Klotzbach, P. J. (2011). The influence of El Niño-southern oscillation and the Atlantic multi-decadal oscillation on Caribbean tropical cyclone activity. *Journal of Climate, 24*, 721–731. https://doi.org/10.1175/2010JCLI3705.1

Klotzbach, P. J., & Gray, W. M. (2006). Causes of the unusually destructive 2004 Atlantic basin hurricane season. *Bulletin of the American Meteorological Society, 87*(10), 1325–1334.

Klotzbach, P. J., Saunders, M. A., Bell, G. D., & Blake, E. S. (2017). North Atlantic seasonal hurricane prediction: Underlying science and an evaluation of statistical models. *Climate Extremes: Patterns and Mechanisms, Geophysical Monograph Series, 226*, 315–328. https://doi.org/10.1002/9781119068020.ch19

Knight, C. A., & Knight, N. C. (2001). Hailstorms. In *Severe Convective Storms* (pp. 223–254). American Meteorological Society.

Knös, D., Karagiorgos, K., Haas, J., Blumenthal, B., Nyberg, L., & Halldin, S. (2022). Cloudburst-disaster modelling—A new open-source catastrophe model. *International Journal of Disaster Risk Reduction, 67*, 102679.

Knupp, K. R., & Cotton, W. R. (1985). Convective cloud downdraft structure: An interpretive survey. *Reviews of Geophysics, 23*(2), 183–215.

Kripalani, R. H., & Kumar, P. (2004). Northeast monsoon rainfall variability over south peninsular India vis-à-vis the Indian Ocean dipole mode. *International Journal of Climatology: A Journal of the Royal Meteorological Society, 24*(10), 1267–1282.

Krishnamohan, K. S., Mohanakumar, K., & Joseph, P. V. (2012). The influence of Madden Julian oscillation in the genesis of north Indian Ocean tropical cyclones. *Theoretical and Applied Climatology, 109*(1), 271–282.

Kuleshov, Y., Chane Ming, F., Qi, L., Chouaibou, I., Hoareau, C., & Roux, F. (2009). Tropical cyclone genesis in the Southern Hemisphere and its relationship with the ENSO. *Annales Geophysicae, 27*, 2523–2538.

Kumar, S., Panda, J., Paul, D., & Guha, B. K. (2023). Impact of environmental variables on the North Indian Ocean tropical cyclones radial parameters. *Climate Dynamics, 60*(3–4), 813–830.

Kumar, S., Panda, J., Singh, K., Guha, B. K., & Kant, S. (2021). Structural characteristics of North Indian Ocean tropical cyclones during 1999–2017: A scatterometer observation-based analysis. *Theoretical and Applied Climatology, 143*, 227–240.

Lancaster N. (2014). Aeolian processes. *Reference Module in Earth Systems and Environmental Sciences*. https://doi.org/10.1016/B978-0-12-409548-9.09126-0, 978-0-12-409548-9.

Leandro, C., & Miranda, L. (2000). Lima preparing communities for disaster. *Habitat Debate, 6*(2), 25–26.

Lepore, C., Abernathey, R., Henderson, N., Allen, J. T., & Tippett, M. K. (2021). Future global convective environments in CMIP6 models. *Earth's Future, 9*(12), e2021EF002277.

Li, H., Sriver, R. L., & Goes, M. (2016a). Modeled sensitivity of the Northwestern Pacific upper-ocean response to tropical cyclones in a fully coupled climate model with varying ocean grid resolution. *Journal of Geophysical Research: Oceans, 121*(1), 586–601.

Li, J., Hao, X., Liao, H., Yue, X., Li, H., Long, X., & Li, N. (2022). Predominant type of dust storms that influences air quality over northern China and future projections. *Earth's Future, 10*(6), e2022EF002649.

Li, R. C., & Zhou, W. (2018). Revisiting the intraseasonal, interannual and interdecadal variability of tropical cyclones in the western North Pacific. *Atmospheric and Oceanic Science Letters, 11*(2), 198–208.

Li, X. X., Koh, T. Y., Panda, J., & Norford, L. K. (2016b). Impact of urbanization patterns on the local climate of a tropical city, Singapore: An ensemble study. *Journal of Geophysical Research: Atmospheres, 121*(9), 4386–4403.

Liebmann, H., Hendon, H., & Glick, J. D. (1994). The relationship between tropical cyclones of the western Pacific and Indian Oceans and the Madden-Julian Oscillation. *Journal of the Meteorological Society of Japan. Series II, 72*(3), 401–412. https://doi.org/10.2151/jmsj1965.72.3_401.

Lin, C. Y., Chen, W. C., Liu, S. C., Liou, Y. A., Liu, G. R., & Lin, T. H. (2008). Numerical study of the impact of urbanization on the precipitation over Taiwan. *Atmospheric Environment, 42*(13), 2934–2947.

Litta, A. J., Mohanty, U. C., & Bhan, S. C. (2010). Numerical simulation of a tornado over Ludhiana (India) using WRF-NMM model. *Meteorological Applications: A Journal of Forecasting, Practical Applications, Training Techniques and Modelling, 17*(1), 64–75.

Litynska, Z., Parfiniewicz, J., & Piwkowski, H. (1976). The prediction of air mass-thunderstorms and hails (pp. 128–130). WMO Interpretation of Broad-Scale NWP Production for Local Forecasting Purposes (SEE N 77–18671 09–47).

Lockwood, J. W., Oppenheimer, M., Lin, N., Kopp, R. E., Vecchi, G. A., & Gori, A. (2022). Correlation between sea-level rise and aspects of future tropical cyclone activity in CMIP6 models. *Earth's Future, 10*(4), e2021EF002462.

Magee, A. D., Verdon-Kidd, D. C., Diamond, H. J., & Kiem, A. S. (2017). Influence of ENSO, ENSO Modoki, and the IPO on tropical cyclogenesis: A spatial analysis of the southwest Pacific region. *International Journal of Climatology, 37*, 1118–1137.

Mahala, B. K., Nayak, B. K., & Mohanty, P. K. (2015). Impact of ENSO and IDO on tropical cyclone activity in the Bay of Bengal. *Natural Hazards, 75*, 1105–1125. https://doi.org/10.1007/s11069-014-1360-8

Mahanta, R., & Yamane, Y. (2020). Climatology of local severe convective storms in Assam. *India. International Journal of Climatology, 40*(2), 957–978.

Maneesha, K., Sadhuram, Y., & Prasad, K. V. S. R. (2015). Role of upper ocean parameters in the genesis, intensification and tracks of cyclones over the Bay of Bengal. *Journal of Operational Oceanography, 8*(2), 133–146.

Marks, G. B., Colquhoun, J. R., Girgis, S. T., Koski, M. H., Treloar, A. B. A., Hansen, P., ... & Car, N. G. (2001). Thunderstorm outflows preceding epidemics of Asthma during spring and summer. *Thorax, 56*(6), 468-471.

Marticorena, B. (2014). Dust production mechanisms. In *Mineral Dust* (pp. 93–120). Springer.

McBean, G. A., & Henstra, D. (2003). *Climate change, natural hazards and cities.* Institute for Catastrophic Loss Reduction. ISBN 0-9732213-9-9. Available at https://www.iclr.org/wp-content/uploads/PDFS/climate_change_natural_hazards_and_cities.pdf

McCaul, E. W., Jr., & Cohen, C. (2002). The impact on simulated storm structure and intensity of variations in the mixed layer and moist layer depths. *Monthly Weather Review, 130*(7), 1722–1748.

Means, L. L. (1952). Stability index computation graph for surface data. Unpublished manuscript available from F. Sanders, 9.

Meng, Z., Yan, D., & Zhang, Y. (2013). General features of squall lines in East China. *Monthly Weather Review, 141*(5), 1629–1647.

Middleton, N. (2019). Variability and trends in dust storm frequency on decadal timescales: Climatic drivers and human impacts. *Geosciences, 9*(6), 261.

Middleton, N. J. (2017). Desert dust hazards: A global review. *Aeolian Research, 24*, 53–63.

Middleton, N., & Kang, U. (2017). Sand and dust storms: Impact mitigation. *Sustainability, 9*(6), 1053.

Miller, R. C. (1967). *Notes on Analysis and Severe-Storm Forecasting Procedures of the Military Weather Warning Center* (Vol. 200). Air Weather Service (MAC), United States Air Force.

Miller, R. C. (1975). *Notes on Analysis and Severe-Storm Forecasting Procedures of the Air Force Global Weather Central* (Vol. 200). AWS.

Mishra, V., Smoliak, B. V., Lettenmaier, D. P., & Wallace, J. M. (2012). A prominent pattern of year-to-year variability in Indian summer monsoon rainfall. *Proceedings of the National Academy of Sciences of the United States of America, 109*, 7213–7217.

Mohanty, U. C., Osuri, K. K., Pattanayak, S., & Sinha, P. (2012). An observational perspective on tropical cyclone activity over Indian seas in a warming environment. *Natural Hazards, 63*(3), 1319–1335. https://doi.org/10.1007/s11069-011-9810-z

Mohapatra, M., Geetha, B., Balachandran, S., & Rathore, L. S. (2015). On the tropical cyclone activity and associated environmental features over North Indian Ocean in the context of climate change. *Journal of Climate Change, 1*(1–2), 1–26.

Moncrieff, M. W., & Miller, M. J. (1976). The dynamics and simulation of tropical cumulonimbus and squall lines. *Quarterly Journal of the Royal Meteorological Society, 102*(432), 373–394.

Moon, J. Y., Wang, B., Lee, S. S., & Ha, K. J. (2018). An intra seasonal genesis potential index for tropical cyclones during Northern Hemisphere summer. *Journal of Climate, 31*(22), 9055–9071. https://doi.org/10.1175/JCLI-D-18-0515.1

Moore, T. W. (2018). Annual and seasonal tornado trends in the contiguous United States and its regions. *International Journal of Climatology, 38*(3), 1582–1594.

Muhs, D. R., Prospero, J. M., Baddock, M. C., & Gill, T. E. (2014). Identifying sources of Aeolian mineral dust: Present and past. In *Mineral Dust* (pp. 51–74). Springer.

Murakami, H., Vecchi, G. A., Delworth, T. L., Wittenberg, A. T., Underwood, S., Gudgel, R., et al. (2017). Dominant role of subtropical Pacific warming in extreme eastern Pacific hurricane seasons: 2015 and the future. *Journal of Climate, 30*, 243–264. https://doi.org/10.1175/JCLI-D-16-0424.1

Nelson, S. P. (1987). The hybrid multicellular–supercellular storm—An efficient hail producer. Part II. General characteristics and implications for hail growth. *Journal of the Atmospheric Sciences, 44*(15), 2060–2073.
Neumann, C. J. (1993). Global guide to tropical cyclone forecasting. WMO/TC-No. 560, Report No. TCP-31, World Meteorological Organization, Geneva, Switzerland.
Nizamuddin, S. (1993). Hail occurrences in India. *Weather, 48*(3), 90–92.
Notaro, M., Yu, Y., & Kalashnikova, O. V. (2015). Regime shift in Arabian dust activity, triggered by persistent Fertile Crescent drought. *Journal of the Geological Society Australia, 120*(19), 10229–10249.
Orlanski, I. (1975). A rational subdivision of scales of atmospheric processes. *Bulletin American Meteorology Social, 56*, 527–530.
OSDMA. (2013). State management plan Odisha. Odisha State Disaster Management Authority, Odisha. Available at https://health.odisha.gov.in/sites/default/files/2020-02/Disaster_Management.pdf
OSDMA. (2016). Multipurpose cyclone shelters. Retrieved May 16, 2016, from http://www.osdma.org/ViewDetails.aspx?vchglinkid=GL007&vchplinkid=PL040&vchslinkid=SL013.
Panda, J., Paul, D., & Kumar, S. (2023). Characteristics of tropical cyclones through remote sensing-based observational platforms (Chap. 18). In Singh & Tiwari (Eds.), *Atmospheric Remote Sensing* (pp. 325–354). Elsevier. https://doi.org/10.1016/B978-0-323-99262-6.00001-8.
Parolari, A. J., Li, D., Bou-Zeid, E., Katul, G. G., & Assouline, S. (2016). Climate, not conflict, explains extreme Middle East dust storm. *Environmental Research Letters, 11*(11), 114013.
Patricola, C. M., Saravanan, R., & Chang, P. (2014). The impact of the El Niño-Southern Oscillation and Atlantic Meridional Mode on seasonal Atlantic tropical cyclone activity. *Journal of Climate, 27*, 5311–5328. https://doi.org/10.1175/JCLI-D-13-00687.1
Paul, D., Panda, J., & Routray, A. (2022). Ocean and atmospheric characteristics associated with the cyclogenesis and rapid intensification of NIO super cyclonic storms during 1981–2020. *Natural Hazards*, 1–29.
Price, D., Hughes, K. M., Thien, F., & Suphioglu, C. (2021). Epidemic thunderstorm asthma: lessons learned from the storm down-under. *The Journal of Allergy and Clinical Immunology: In Practice, 9*(4), 1510–1515.
Pruppacher, H. R., & Klett, J. D. (2010). Microstructure of atmospheric clouds and precipitation. *Microphysics of Clouds and Precipitation*, 10–73.
Pugh, C. (1995). The role of the World Bank in housing. Housing the urban poor: policy and practice in developing countries, 34–92.
Ramsay, H. A., Camargo, S. J., & Kim, D. (2012). Cluster analysis of tropical cyclone tracks in the Southern Hemisphere. *Climate Dynamics, 39*, 897–917.
Rath, S. S., Panda, J., & Mohanty, S. (2021). Urban transition and its impact on mesoscale weather: A review. *Methods and Applications of Geospatial Technology in Sustainable Urbanism*, 187–216.
Rath, S. S., Panda, J., & Sarkar, A. (2022). Distinct urban land cover response to meteorology in WRF simulated pre-monsoon thunderstorms over the tropical city of Kolkata. *Meteorology and Atmospheric Physics, 134*(4), 76.
Ray, K., Bhan, S.C. and Bandopadhyay, B.K., 2015. The catastrophe over Jammu and Kashmir in September 2014: a meteorological observational analysis. Current Science, pp.580–591.
Re, M. (2011). *Topics Geo, natural catastrophes 2010, analyses, assessments. Positions.* Available at https://www.preventionweb.net/publication/topics-geo-natural-catastrophes-2010-analyses-assessments-positions
Ridder, N. N., Ukkola, A. M., Pitman, A. J., & Perkins-Kirkpatrick, S. E. (2022). Increased occurrence of high impact compound events under climate change. *NPJ Climate and Atmospheric Science, 5*(1), 3.
Roxy, M. K., Ghosh, S., & Pathak, A. (2017). A threefold rise in widespread extreme rain events over central India. *Nature Communications, 8*, 708. https://doi.org/10.1038/s41467-017-00744-9

Roy Chowdhury, R., Prasanna Kumar, S., Narvekar, J., & Chakraborty, A. (2020). Back-to-back occurrence of tropical cyclones in the Arabian Sea during October–November 2015: Causes and responses. *Journal of Geophysical Research: Oceans, 125*(6), e2019JC015836.

Sahu, R. K., Dadich, J., Tyagi, B., Vissa, N. K., & Singh, J. (2020). Evaluating the impact of climate change in threshold values of thermodynamic indices during pre-monsoon thunderstorm season over Eastern India. *Natural Hazards, 102*(3), 1541–1569.

Sahu, R., Tyagi, B., Vissa, N., & Mohapatra, M. (2022). Pre-monsoon thunderstorm season climatology of convective available potential energy (CAPE) and convective inhibition (CIN) over eastern India. *Mausam, 73*(3), 565–586.

Saji, N. H., Goswami, B. N., Vinayachandran, P. N., & Yamagata, T. (1999). A dipole mode in the tropical Indian Ocean. *Nature, 401*(6751), 360–363. https://doi.org/10.1038/43854

Sander, J., Eichner, J. F., Faust, E., & Steuer, M. (2013). Rising variability in thunderstorm-related US losses as a reflection of changes in large-scale thunderstorm forcing. *Weather, Climate, and Society, 5*(4), 317–331.

Schoen, J. M., & Ashley, W. S. (2011). A climatology of fatal convective wind events by storm type. *Weather and Forecasting, 26*(1), 109–121.

Schreck, C. J., Molinari, J., & Aiyyer, A. (2012). A global view of equatorial waves and tropical cyclogenesis. *Monthly Weather Review, 140*, 774–788.

Seeley, J. T., & Romps, D. M. (2015). The effect of global warming on severe thunderstorms in the United States. *Journal of Climate, 28*(6), 2443–2458.

Shanmugasundaram, J., Arunachalam, S., Gomathinayagam, S., Lakshmanan, N., & Harikrishna, P. (2000). Cyclone damage to buildings and structures—a case study. *Journal of Wind Engineering and Industrial Aerodynamics, 84*(3), 369–380.

Shao, Y., Wyrwoll, K. H., Chappell, A., Huang, J., Lin, Z., McTainsh, G. H., ... & Yoon, S. (2011). Dust cycle: An emerging core theme in Earth system science. *Aeolian Research, 2*(4), 181–204.

Sharma, N., & Ali, M. M. (2014). Importance of ocean heat content for cyclone studies. *Oceanography: Open Access*, 1–4.

Shay, L. K., Goni, G. J., & Black, P. G. (2000). Effects of a warm oceanic feature on Hurricane Opal. *Monthly Weather Review, 128*(5), 1366–1383.

Shi, J., & Cui, L. (2012). Characteristics of high impact weather and meteorological disaster in Shanghai, China. *Natural Hazards, 60*(3), 951–969.

Showalter, A. K. (1953). A convective index as an indicator of cumulonimbus development. *Journal of Applied Meteorology, 5*, 839–846.

Shrestha, P., Dimri, A. P., Schomburg, A., & Simmer, C. (2015). Improved understanding of an extreme rainfall event at the Himalayan foothills–a case study using COSMO. *Tellus a: Dynamic Meteorology and Oceanography, 67*(1), 26031.

Sikka, D. R., Ray, K., Chakravarthy, K., Bhan, S. C., & Tyagi, A. (2015). Heavy rainfall in the Kedarnath valley of Uttarakhand during the advancing monsoon phase in June 2013. *Current Science*, 353–361.

Sills, D. M., & Joe, P. I. (2019). From pioneers to practitioners: A short history of severe thunderstorm research and forecasting in Canada. *Atmosphere-Ocean, 57*(4), 249–261.

Simelton, E., Dam, B. V., & Catacutan, D. (2015). Trees and agroforestry for coping with extreme weather events: Experiences from northern and central Viet Nam. *Agroforestry Systems, 89*(6), 1065–1082.

Singh, D., Tsiang, M., Rajaratnam, B., & Diffenbaugh, N. S. (2014). Observed changes in extreme wet and dry spells during the South Asian summer monsoon season. *Natural Climate Change, 4*, 456–461.

Singh, K., Panda, J., & Kant, S. (2020a). A study on variability in rainfall over India contributed by cyclonic disturbances in warming climate scenario. *International Journal of Climatology, 40*(6), 3208–3221.

Singh, K., Panda, J., & Mohapatra, M. (2020b). Robustness of best track data and associated cyclone activity over the North Indian Ocean region during and prior to satellite era. *Journal of Earth System Science, 129*, 1–20.

Singh, K., Panda, J., & Rath, S. S. (2019a). Variability in landfalling trends of cyclonic disturbances over North Indian Ocean region during current and pre-warming climate. *Theoretical and Applied Climatology, 137*(1), 417–439.

Singh, K., Panda, J., Osuri, K. K., & Vissa, N. K. (2016). Progress in tropical cyclone predictability and present status in the North Indian Ocean region. *Recent Developments in Tropical Cyclone Dynamics, Prediction, and Detection, 1*, 193.

Singh, K., Panda, J., Sahoo, M., & Mohapatra, M. (2019b). Variability in tropical cyclone climatology over North Indian Ocean during the period 1891 to 2015. *Asia-Pacific Journal of Atmospheric Sciences, 55*(2), 269–287.

Singh, O. P., Khan, T. M. A., & Rahman, M. S. (2000). Changes in the frequency of tropical cyclones over the North Indian Ocean. *Meteorology and Atmospheric Physics, 75*, 11–20.

Singh, R. (1981). On the occurrence of tornadoes and their distribution in India. *Mausam, 32*(3), 307–314.

Smith, J. A., Baeck, M. L., Zhang, Y., & Doswell, C. A., III. (2001). Extreme rainfall and flooding from supercell thunderstorms. *Journal of Hydrometeorology, 2*(5), 469–489.

Srivastava, A. K., & Ray, K. S. (1999). Role of CAPE and CINE in modulating the convective activities during April over India. *Mausam, 50*(3), 257–262.

Standard, I. (1987). Code of practice for design loads (other than earthquake) for buildings and structures. *Part II, imposed loads (second revision), IS-875-1987*.

Stull, R. (2015). *Practical meteorology. An Algebra-based survey of atmospheric science*. Available at https://openlibrary-repo.ecampusontario.ca/jspui/handle/123456789/405

Tartaglione, C. A., Smith, S. R., & O'Brien, J. J. (2003). ENSO impact on hurricane landfall probabilities for the Caribbean. *Journal of Climate, 16*, 2925–2931. https://doi.org/10.1175/1520-0442(2003)016%3c2925:EIOHLP%3e2.0.CO;2

Taylor, H. T., Ward, B., Willis, M., & Zaleski, W. (2010). *The saffir-simpson hurricane wind scale*. Available at https://mail.hwn.org/media/pdf/sshws.pdf

Thayyen, R. J., Dimri, A. P., Kumar, P., & Agnihotri, G. (2013). Study of cloudburst and flash floods around Leh, India, during August 4–6, 2010. *Natural Hazards, 65*(3), 2175–2204.

Torres-Alavez, J. A., Glazer, R., Giorgi, F., Coppola, E., Gao, X., Hodges, K. I., ... & Sines, T. (2021). Future projections in tropical cyclone activity over multiple CORDEX domains from RegCM4 CORDEX-CORE simulations. *Climate Dynamics, 57*, 1507–1531.

Toure, A. A., Rajot, J. L., Garba, Z., Marticorena, B., Petit, C., & Sebag, D. (2011). Impact of very low crop residues cover on wind erosion in the Sahel. *CATENA, 85*(3), 205–214.

Tran, T. L., Ritchie, E. A., Perkins-Kirkpatrick, S. E., Bui, H., & Luong, T. M. (2022). Future changes in tropical cyclone exposure and impacts in Southeast Asia From CMIP6 pseudo-global warming simulations. *Earth's Future, 10*(12), e2022EF003118.

Trapp, R. J., Diffenbaugh, N. S., & Gluhovsky, A. (2009). Transient response of severe thunderstorm forcing to elevated greenhouse gas concentrations. *Geophysical Research Letters*, *36*(1).

Tsuboi, A., & Takemi, T. (2014). The interannual relationship between MJO activity and tropical cyclone genesis in the Indian Ocean. *Geoscience Letters, 1*(1), 1–6.

Turner, A. G., & Annamalai, H. (2012). Climate change and the South Asian summer monsoon. *Nature Climate Change, 2*, 587–595.

Tyagi, A. (2007). Thunderstorm climatology over Indian region. *Mausam, 58*(2), 189–212.

Tyagi, B., Naresh Krishna, V., & Satyanarayana, A. N. V. (2011). Study of thermodynamic indices in forecasting pre-monsoon thunderstorms over Kolkata during STORM pilot phase 2006–2008. *Natural Hazards, 56*(3), 681–698.

Uman, M. A. (2012). *Lightning*. Courier Corporation.

UNDP-GEAS. (2013). *Cyclone Phailin in India: early warning and timely actions saved lives*. Retrieved May 16, 2016, from http://www.unep.org/pdf/UNEP_GEAS_NOV_2013.pdf.

Villarini, G., & Vecchi, G. A. (2012). North Atlantic power dissipation index (PDI) and accumulated cyclone energy (ACE): Statistical modeling and sensitivity to sea surface temperature changes. *Journal of Climate, 25*(2), 625–637.

Vincent, E. M., Lengaigne, M., Menkes, C. E., Jourdain, N. C., Marchesiello, P., & Madec, G. (2011). Interannual variability of the South Pacific convergence zone and implications for tropical cyclone genesis. *Climate Dynamics, 36*, 1881–1896.

Vogel, M. M., Hauser, M., & Seneviratne, S. I. (2020). Projected changes in hot, dry and wet extreme events' clusters in CMIP6 multi-model ensemble. *Environmental Research Letters, 15*(9), 094021.

Vraciu, C. V. (2023). In what conditions an urban heat island can initiate deep convection? Theoretical estimations. *Theoretical and Applied Climatology*, 1–13.

Wahiduzzaman, M., Oliver, E. C. J., Klotzbach, P. J., Wotherspoon, S. J., & Holbrook, N. J. (2019). A statistical seasonal forecast model of North Indian Ocean tropical cyclones using the quasi-biennial oscillation. *International Journal of Climatology, 39*(2), 934–952. https://doi.org/10.1002/joc.5853

Wang, B., & Ding, Q. (2006). Changes in global monsoon precipitation over the past 56 years. *Geophysical Research Letters, 33*, L06711.

Wang, B., & Moon, J. Y. (2017). An anomalous genesis potential index for MJO modulation of tropical cyclones. *Journal of Climate, 30*(11), 4021–4035.

Wang, B., Yang, Y., Ding, Q., Murakami, H., & Huang, F. (2010a). Climate control of the global tropical storm days (1965–2008). *Geophysical Research Letters, 37*, L07704. https://doi.org/10.1029/2010GL042487

Wang, X. M., Zhang, C. X., Hasi, E., & Dong, Z. B. (2010b). Has the Three Norths Forest Shelterbelt Program solved the desertification and dust storm problems in arid and semiarid China? *Journal of Arid Environments, 74*(1), 13–22.

Wardman, A. E., Stefani, D., & MacDonald, J. C. (2002). Thunderstorm-associated asthma or shortness of breath epidemic: A Canadian case report. *Canadian Respiratory Journal, 9*(4), 267–270.

Weisman, M. L., & Klemp, J. B. (1982). The dependence of numerically simulated convective storms on vertical wind shear and buoyancy. *Monthly Weather Review, 110*(6), 504–520.

Weisman, M. L., & Klemp, J. B. (1984). The structure and classification of numerically simulated convective storms in directionally varying wind shears. *Monthly Weather Review, 112*(12), 2479–2498.

Williams, E. R. (2001). The electrification of severe storms. In *Severe Convective Storms* (pp. 527–561). American Meteorological Society.

Williams, E. R. (2005). Lightning and climate: A review. *Atmospheric Research, 76*(1–4), 272–287.

Williams, E., & Renno, N. (1993). An analysis of the conditional instability of the tropical atmosphere. *Monthly Weather Review, 121*(1), 21–36.

Williams, E., & Stanfill, S. (2002). The physical origin of the land–ocean contrast in lightning activity. *Comptes Rendus Physique, 3*(10), 1277–1292.

Williams, E., Mushtak, V., Rosenfeld, D., Goodman, S., & Boccippio, D. (2005). Thermodynamic conditions favorable to superlative thunderstorm updraft, mixed phase microphysics and lightning flash rate. *Atmospheric Research, 76*(1–4), 288–306.

Xiang, B., & Wang, B. (2013). Mechanisms for the advanced Asian Summer Monsoon onset since the mid-to-late 1990s. *Journal of Climate, 26*, 1993–2009. https://doi.org/10.1175/JCLI-D-12-00445.1

Youlin, Y., Squires, V., & Qi, L. (2002). Global alarm: dust and sandstorms from the world's drylands. United Nations Convention to Combat Desertification (UNCCD).

Zheng, Z.-W., Lin, I. I., Wang, B., Huang, H.-C., & Chen, C. H. (2015). A long neglected damper in the El Niño–typhoon relationship: A 'Gaia-like' process. *Scientific Reports, 5*, 11103. https://doi.org/10.1038/srep11103

Historical Trend and Future Projection of Extreme Precipitation in Nepal from High-Resolution CMIP6 Data

Shankar Sharma, Piyush Dahal, Suresh Marahatta, Archana Shrestha, Vishnu Prasad Panday, and Binod Pokharel

Abstract Understanding the spatial and temporal characteristics of precipitation and its extremes for the past and future periods can help identify the driving potential of extreme events, especially in the context of climate change. This study presents the spatial and temporal variation of precipitation on a seasonal and annual scale and their change in future periods (2015–2100) in response to historical periods (1980–2014), using bias-corrected and downscaled nine Coupled Model Intercomparison Project -sixth phase (CMIP6) ensemble members under mid- (SSP245) and high- (SSP585) emission scenarios over Nepal. The precipitation trend and related precipitation extremes (ten indices), distribution, and future changes were also assessed. The bias-corrected CMIP6 models slightly underestimated the observed precipitation amount but generally captured the observed precipitation pattern over the country. The spatial distribution of mean precipitation is similar between the historical and the projected periods, with higher precipitation likely to occur in the future, particularly under SSP585 scenarios. In the historical period, precipitation was found to be increasing in the monsoon season only, which is likely to continue in the future period at a rate of 2.3 mm/year and 5.8 mm/year under SSP245 and SSP585, respectively. The decreasing trend of winter precipitation in the historical period will continue in future periods, whereas precipitation in pre- and post-monsoon will likely increase in future periods. Further, all extreme precipitation indices except for consecutive dry

S. Sharma · P. Dahal · S. Marahatta · B. Pokharel (✉)
RECHAM Consult (Pvt.) Ltd., Kathmandu, Nepal
e-mail: binod.pokharel@cdhm.tu.edu.np

S. Marahatta · B. Pokharel
Central Department of Hydrology and Meteorology, Tribhuvan University, Kirtipur, Nepal

A. Shrestha
Department of Hydrology and Meteorology, Government of Nepal, Kathmandu, Nepal

V. P. Panday
Department of Civil Engineering, Institute of Engineering, Pulchowk Campus, Pulchowk, Lalitpur, Nepal

Center for Water Resources Studies (CWRS), Institute of Engineering, Tribhuvan University, Kirtipur, Nepal

S. Das and W-K. Tao (eds.), *Severe Storms*,
https://doi.org/10.1007/978-981-97-7075-5_23

days were found to be increasing, with higher frequency and intensity projected in the future. The increase in high-intensity and wetness-related precipitation extremes are related to the increasing precipitation in the monsoon season, while the increase in dry days-related extremes may be related to reduced winter precipitation. The findings of the present study can help understand an early snapshot of future precipitation and extreme event scenarios, which can serve as a reference for developing strategies to mitigate climate-related risks in the region.

Keywords CMIP6 · Precipitation · Extreme · Projection · Nepal

1 Introduction

Extreme precipitation, defined as intense and heavy rainfall or snowfall events, can have significant impacts on society and the environment. Extreme precipitation events can have significant impacts on many different sectors, including agriculture (Li et al., 2019), ecosystem (Knapp et al., 2008), natural disasters (Dezfuli, 2020), energy (Wasti et al., 2022), transportation (Jia et al., 2022), property and infrastructure (Frame et al., 2020), and health (Curriero et al., 2001). Though extreme precipitation events are affecting a number of different sectors, they have attracted the most interest due to the particular threat to human life. For example, extreme precipitation-induced floods caused half a million fatalities in the world from 1980 to 2009 (Doocy et al., 2013). In 2015, heavy rainfalls in Eastern China during the period of June 21–30 caused widespread flooding that collapsed 46,600 houses and affected 5,380,000 people (Wang & Gu, 2016). Extreme precipitation-triggered debris flows from 1950 to 2011 causing 77,779 fatalities from 213 events (Dowling & Santi, 2014). An extreme precipitation event between January 23 and 25, 2020, in southeast Brazil caused cascading effects of flooding and landslides, causing the loss of 56 lives and the displacement of numerous individuals (Dalagnol et al., 2022).

One important ingredient for the occurrence of extreme precipitation is the availability of atmospheric moisture (Min et al., 2011; Tabari, 2020). Global warming is expected to increase the amount of atmospheric moisture, resulting in increased extreme precipitation events (Held & Soden, 2006; Soden & Held, 2006; Trenberth et al., 2005). Changes in precipitation are mainly governed by the change in atmospheric moisture as described in the Clausius-Clapeyron (C–C) relation, which states that a warmer atmosphere can hold more water vapor, increasing at a rate of about seven percent per degree Celsius (Dong et al., 2022; Pall et al., 2007; Sugiyama et al., 2010; Trenberth et al., 2003). However, studies suggest that this change is site-specific and mainly depends on intensity (Tabari, 2020; Zhan et al., 2017).

The significant volume of studies suggests that precipitation is changing in response to anthropogenic global warming (Fischer & Knutti, 2016; Haerter et al., 2010; Lenderink & van Meijgaard, 2008; Myhre et al., 2019; Papalexiou & Montanari, 2019; Sun et al., 2021; Utsumi et al., 2011), and it is projected that extreme precipitation events will increase in proportion to their intensity in the future too (Ge et al., 2019; Kharin et al., 2013; Wehner, 2020). However, there exists non-uniformity in the behavior of extreme events across the globe, where some regions exhibit more sensitivity than others (Papalexiou & Montanari, 2019; Shi et al., 2021; Westra et al., 2013). For example, observed increases in frequency and intensity of extreme precipitation, in 1984–2013 relative to 1951–1980, are generally larger over Europe and Japan than over the USA and Australia (Myhre et al., 2019). Further, (Min et al., 2011) found that human-induced climate change intensified extreme precipitation events in two-thirds of observed data in Northern Hemisphere land areas.

The impacts of changes in extreme precipitation in the Himalayas can be wide-ranging and potentially severe (Zhang et al., 2017). There is evidence that the Himalayas are experiencing changes in extreme precipitation as a result of a changing climate (Goswami et al., 2006; Guhathakurta et al., 2017; Nandargi & Dhar, 2011; Zhan et al., 2017). In the Himalayas, a region with complex topography and a varied climate, understanding changes in extreme precipitation is of particular importance. The topography of the Himalayas also plays a role in extreme precipitation events (Dimri et al., 2017; Medina et al., 2010). Besides this, several other factors have been identified as potential drivers of the observed changes in extreme precipitation in the Himalayas. One factor is the influence of the Asian monsoon, which is known to affect precipitation patterns in the region (Vellore et al., 2016). Studies have shown that the monsoon is intensifying due to climate change, leading to increased extreme precipitation events (Turner & Annamalai, 2012). Another driver of extreme precipitation in the Himalayas is the El Niño-Southern Oscillation (ENSO) (Kumar et al., 2006). ENSO can affect the monsoon circulation and the frequency and intensity of extreme precipitation events in the Himalayas (Pokharel et al., 2019).

To evaluate the characteristics of extreme precipitation events, scientists use a set of standardized indices. These indices enable the consistent evaluation of changes in the frequency, intensity, and duration of extreme precipitation events over time, thereby providing valuable insights into the potential impacts of these events on various sectors. The present study entails an analysis of the extreme precipitation indices as defined by the Expert Team on Climate Change Detection and Indices (ETCCDI) (Zhang et al., 2011). The ETCCDI extreme precipitation indices provide a standardized and consistent approach for the evaluation of extreme precipitation events and have been widely used to evaluate extreme precipitation around the world (Diaconescu et al., 2016; Westra et al., 2013). These indices include several variables such as the number of heavy precipitation events, maximum daily precipitation amount, consecutive wet days, simple daily intensity index, etc.

The study of future changes in extreme events is crucial for developing effective adaptation and mitigation strategies. Prediction of such extreme weather events is still filled with uncertainties; a proper assessment of likely future trends would help in setting up infrastructure for disaster preparedness. Future changes in regional precipitation are mainly predicted using climate model data. In the present study, we examine projected changes in extreme precipitation over the Himalayas in the NASA Earth Exchange Global Daily Downscaled Projections archive (NEX-GDDP-CMIP-6) (Thrasher et al., 2022). This archive contains a set of bias-corrected high-resolution climate projections that have been generated using the latest generation of climate models, known as Coupled Model Intercomparison Project Phase 6 (CMIP6), for historical and future changes (1980–2100) in temperature, precipitation, and other climate variables on a daily basis. Based on these data, we summarized the current state of knowledge on extreme precipitation in the changing climate in Nepal.

2 Material and Methods

2.1 Study Area

Nepal is a Himalayan country with a diverse and complex climate system, covering an area of 147,516 km^2. Nepalis on the border between China and India and extends 885 km from east to west and 140–250 km from north to south. Within a short latitudinal distance, Nepal is characterized by unique topographic ranges from 60 m above sea level (low-elevation tarai plains) to 8848.8 masl (the highest peak, Mt. Everest). This rapid elevation change creates a variety of microclimates, ranging from tropical to sub-tropical in the lowlands to polar and tundra in the high mountains (Karki et al., 2016). The precipitation in the country is heavily influenced by the South Asian Monsoon and Westerly system, resulting in different precipitation patterns that determine the four climatic seasons: pre-monsoon (March–May), monsoon (June–September), post-monsoon (October–November), and winter (December-February) (Hamal et al., 2020a; Nayava, 1975). The monsoon season is extensively wet, with about 80% of the total annual precipitation, while winter precipitation in the form of snow occurs in high-elevation areas, particularly in the western hills (Hamal et al., 2020a; Sharma et al., 2020a). This heavy rainfall, combined with the region's complex topography, often results in a high risk of landslides and flash floods (Talchabhadel et al., 2018; Talchabhadel et al., 2022). The country's traditional farming system heavily relies on the monsoon season, with July–August receiving the highest precipitation, amounting to half of the total annual precipitation.

2.2 Datasets

The reference precipitation dataset used in this study is the Asian Precipitation—Highly-Resolved Observational Data Integration Towards Evaluation (APHRODITE), which covers the period 1950–2014 with a 0.25° × 0.25° spatial resolution. APHRODITE is generated from archives of surface meteorological stations from the south Asian region, including Nepal, which has been subject to extensive quality control measures (Yatagai et al., 2012). Some previous studies, for example, Sunilkumar et al. (2019), Sharma et al. (2020a) and Hamal et al. (2020a) suggested an adequate representation of the regional precipitation over Nepal and recommended it as the best alternative to gauge observation in Nepal. For the current study, APHRODITE precipitation daily data from 1980 to 2014 is used to validate the CMIP6 product over Nepal.

The Coupled Model Intercomparison Project (CMIP) is a collaborative international effort to evaluate and improve global climate models (GCMs) used for climate projections (Taylor et al., 2012). The sixth phase—CMIP6, includes updated versions of the GCMs used in the previous CMIP5 project and new models and model configurations (Eyring et al., 2016). These updated models incorporate the latest scientific understanding of the Earth's climate system and improvements in computational resources and modeling techniques to better our understanding of the underlying processes driving climate change, including the role of clouds, aerosols, and biogeochemical cycles (Try et al., 2022). The NEX-GDDP-CMIP6 dataset is the bias-corrected downscaled version of GCM, which was developed using 0.25-degree historical daily average data for various meteorological parameters, including maximum and minimum temperature, precipitation, near-surface humidity, radiation, and wind speed (Thrasher et al., 2022). This CMIP6 ensemble member consists of historical and projection simulations covering the time period from 1960 to 2100. The historical simulation covers the time period from 1960 to 2014, while the projection simulation covers the period from 2015 to 2100. The GCM dataset was bias-corrected using the Global Meteorological Forcing Dataset (GMFD) from the Terrestrial Hydrology Research Group at Princeton University as an observation source for Land Surface Modeling. The observational dataset was then merged with reanalysis data of varying spatial resolutions (0.25, 0.5, and 1.0°) and temporal resolutions (3-hourly, daily, and monthly) (Sheffield et al., 2006). The algoisrithm-generated CDFs for the GMFD data and the historical GCM simulations were compared to develop the quantile map. Further, based on this map, GCM values were translated to corresponding GMFD values, and the algorithm was used to adjust future climate projections by checking the probability quantile and accepting the corresponding observed climatology as an adjusted prediction, by removing the original GCM biases. Finally, temperature and precipitation climate trends are added back to the adjusted GCM climate fields. To assess precipitation and its extreme variability in Nepal, we obtained the bias-corrected versions of 22 CMIP6 model experiments from https://www.nccs.nasa.gov/services/data-collections/land-based-products/nex-gddp-cmip6. We used historical simulations from 1980 to 2014 and projected periods from 2015 to 2100 under two Shared Socioeconomic Pathways (SSPs): SSP245 and SSP585. The SSP245 (SSP585) is an update of the CMIP5 scenario RCP4.5 (RCP8.5), with an additional radiative forcing of 4.5 (8.5) W/m^2 by the year 2100, with now combined with socioeconomic reasons (Thrasher et al., 2022).As, these scenarios form an essential part of climate change research and assessment and represents the medium (SSP245) and high (SSP585) pathway of future greenhouse gas emissions. The SSP245 extrapolates the past and current global development into the future, whereas SSP858 is based on an intensified exploitation of fossil fuel resources with a high percentage of coal and an energy-intensive lifestyle globally. This scenario assumes that climate protection measures are being taken. Table 1 shows the list of bias-corrected CMIP6 ensemble members used in our study.

Table 1 Bias-corrected CMIP6 model details used in the current study

Source ID	Institution ID	Resolution	Study period historical (projected)
ACCESS-ESM1-5	Australian Commonwealth Sc. and Indust. Res. Org. (Australia)	0.25° × 0.25°	1980–2014 (2015–2100)
BCC-CSM2-MR	Beijing Climate Center (China)	0.25° × 0.25°	1980–2014 (2015–2100)
CNRM-ESM2-1	Centre National de Recherche Meteorologiques (France)	0.25° × 0.25°	1980–2014 (2015–2100)
GFDL-ESM4	NOAA—Geoph. Fluid Dyn. Lab. (USA)	0.25° × 0.25°	1980–2014 (2015–2100)
INM-CM4-8	Institute for Numerical Mathematics (Russia)	0.25° × 0.25°	1980–2014 (2015–2100)
INM-CM5-0	Institute for Numerical Mathematics (Russia)	0.25° × 0.25°	1980–2014 (2015–2100)
IPSL-CM6A-LR	Institute Pierre Simon Laplace (France)	0.25° × 0.25°	1980–2014 (2015–2100)
MIROC-ES2L	JAMSTEC (Japan)	0.25° × 0.25°	1980–2014 (2015–2100)
MPI-ESM1-2-HR	MPI-M DWD DKRZ	0.25° × 0.25°	1980–2014 (2015–2100)

2.3 *Methodology*

The comparison of bias-corrected CMIP6 ensemble and APHRODITE precipitation was performed over the historical period (1980–2014), which corresponds to the mutual time coverage. For the comparison, following the previous study by Chen et al. (2021), Hamal et al. (2020b), Sharma et al. (2020b), we have used metrics such as the mean bias (MB), the root mean square error (RMSE), and the Pearson correlation coefficient (R). The MB and RMSE measure the average deviation (difference) and magnitude of error between bias-corrected CMIP6 models and APHRODITE, while R measures the strength and direction of the linear relationship. The MB is close to zero, and the smaller RMSE values show the better performance of the models. The R-value is generally bounded between –1 and 1, with an optimal value of 1 as a perfect correlation. All the comparison was performed on grid-to-grid as both CMIP6 and APHRODITE have the same spatial resolution. The values in Table 1 show the statistical metrics calculated between CMIP6 bias-corrected models and APHRODITE data, and out of 22 ensemble models, nine best-performing models are selected for further analysis (low MB, RMSE, and high correlation). Using the best-performing CMIP6 models over Nepal, Fig. 1 compares the CMIP6 ensemble mean and APHRODITE precipitation at spatial and temporal scales from 1980 to 2014. Mean annual precipitation in Fig. 1a shows the high precipitation (above 2500 mm) location are concentrated on mid-elevation areas of the central region and high-elevation areas of the eastern region. Based on the monthly precipitation series from observations and model outputs, we calculated the bias in mean precipitation totals over the study area (Fig. 1b). The bias in the CMIP6 ensemble mean in Fig. 1b shows positive (negative) biases in those areas where precipitation is lower (higher), while other areas show biases between +25 and −25%. The low precipitation areas i.e., high-elevation areas of the central region shows the higher positive bias (Fig. 1a, b). APHRODITE, being a gauge-based precipitation product, its accuracy relies on a limited number of stations in Nepal due to the availability of high-quality observation data. It generally underestimates the actual precipitation of Nepal (Sharma et al., 2020a) which might be related to the higher positive bias in CMIP6 data. Although there is a slight change in the high precipitation area, the overall precipitation pattern remains the same. The annual cycle and monthly distribution of precipitation in Fig. 1c, d show an identical performance with APHRODITE, indicating that the historical bias-corrected CMIP6 dataset can represent the observed precipitation distribution over Nepal.

After the validation, we computed the mean precipitation distribution by averaging the selected model (as an ensemble mean) for the historical and projected periods under SSP245 and SSP585. The temporal distribution is also computed by averaging the precipitation in Nepal. Further, we have also calculated the change in precipitation and extremes by dividing the study period into the short-term (2025–2054) and mid-term (2055–2084) in reference to the historical period. Further, the Mann–Kendall (MK) statistical test (Kendall, 1938) was used to assess the monotonic upward or downward trend, and Sen's slope estimator (Sen, 1968) was used to calculate the magnitude of trend for seasonal and annual precipitation and extreme time series.

We have computed ten extreme precipitation indices as suggested by the CCl/WCRP/JCOMM Expert Team on Climate Change Detection and Indices (ETCCDI)

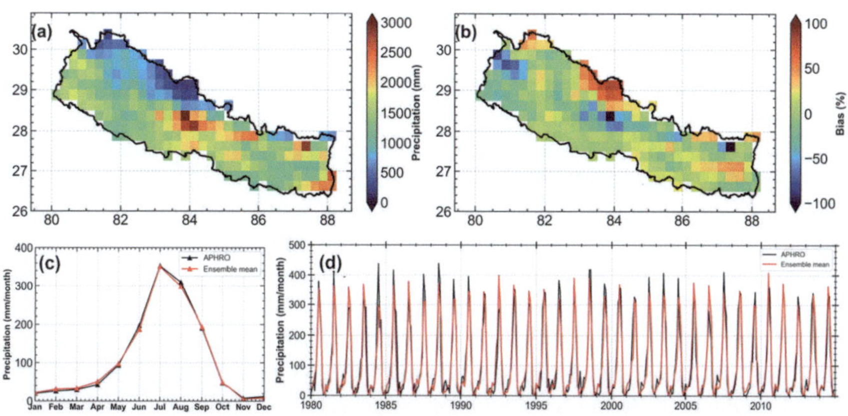

Fig. 1 Study Area Nepal, **a** APHRODITE mean annual precipitation during 1980–2014. **b** Percentage bias in mean ensemble member of nine bias-corrected CMIP6 models, with APHRODITE precipitation data, **c** monthly precipitation cycle in the ensemble mean and APHRODITE, **d** monthly precipitation distribution in CMIP6 ensemble members and APHRODITE data during 1980–2014

(Zhang et al., 2011). Precipitation intensity-related extreme events were calculated at each grid using daily datasets by selecting ten indicators for each grid: consecutive dry days (CDD), consecutive wet days (CWD), heavy precipitation days (R10mm), extreme precipitation days (R20mm), one-day maximum precipitation total (Rx1day), annual total precipitation (PRCPTOT), simple precipitation intensity index (SDII), total wet days (RR1mm), and Total precipitation from events above the 95th (R95pTOT) and 99th (R99pTOT) percentile of wet day amounts. These indices include frequency and duration-based extreme precipitation indices. Details of these extreme indices are presented in Table 2. R10mm and R20mm days represent the total days of heavy ($\geq$10 mm) and extreme precipitation ($\geq$20 mm) events in the year, respectively. R10mm, R25mm, SDII, Rx1day, R95pTOT, and R99pTOT are recognized as pivotal indices denoting intense precipitation events; these particular events are frequently the precursors to phenomena such as flash floods and landslides.Consecutive dry days (CDD) and consecutive wet days (CWD) denote the longest consecutive dry and wet days with $<$1 and $\geq$1 mm rainfall, respectively. These duration indices indicate wet and dry days, highlighting their importance in the field of agriculture (Casanueva Vicente et al., 2014).These indices were adopted from the Expert Team on Climate Change Detection, Monitoring and Indices (ETCCDI), jointly developed by the World Meteorological Organization (WMO) and Commission for Climatology, and the Research Programme on Climate Variability and Predictability.

3 Results

3.1 Spatio-Temporal Variation of Precipitation and Its Trend

Figure 2 shows the spatial distribution of mean precipitation during the historical (1980–2014) and projected period (2015–2100) under SSP245 and SSP585 scenarios. As it captured the characteristics of observed precipitation (Fig. 1), projected precipitation under both scenarios also shows a similar distribution but with higher precipitation amounts. The highest precipitation areas (above 2000 mm/year) in bias-corrected CMIP6 are located in the central and eastern parts of the country, while low precipitation (about below 1000 mm/year) is observed over the high-elevation areas of the central and western regions. The complex mountain geometry and wind patterns result in diverse precipitation distributions (Karki et al., 2017), which may not be accurately captured by the CMIP6 models and result in changes in the location of high and low precipitation areas. Additionally, the estimation of orographic precipitation in model datasets is affected by the model resolution and other physical processes (Fallah et al., 2019; Xu et al., 2019), which may create uncertainty in the model projections. In the high-elevaiton areas, coarse-resolution simulations smooth the land surface and fail to represent topography-induced effects and orographic precipitation (Chen et al., 2021; Wang et al., 2020). Furthermore, both

Table 2 Definition of the precipitation indices used in this study

Short name	Long name	Definition	Plane language description	Units
R10mm	Number of heavy rain days	Number of days when Prec ≥ 10 mm	Days when rainfall is at least 10 mm	days
R20mm	Number of heavy rain days	Number of days when Prec ≥ 20 mm	Days when rainfall is at least 20 mm	days
CDD	Consecutive dry days	Maximum annual number of consecutive wet days (Prec < = 1 mm)	Longest dry spell	days
CWD	Consecutive wet days	Maximum annual number of consecutive wet days (Prec ≥ 1 mm)	Longest wet spell	days
RR1	Total rain days	Annual total precipitation days (when Prec ≥ 1 mm)	Total wet days	days
SDII	Daily Prec intensity	Annual total Prec divided by the number of wet days (when total Prec ≥ 1 mm)	Average daily wet-day rainfall intensity	mm/day
PRCPTOT	Annual total wet-day Prec	Sum of daily Prec ≥ 1 mm	Total wet-day rainfall	mm
Rx1day	Max 1-day Prec	Maximum 1-day Prec total	Maximum amount of rain that falls in one-day	mm
R95pTOT	Total precipitation from events above the 95th percentile of wet day amounts	The sum of precipitation in days where daily precipitation exceeds the 95th percentile threshold value	Annual total precipitation due to events exceeding the 95th percentile value	mm
R99pTOT	Total precipitation from events above the 99th percentile of wet day amounts	The sum of precipitation in days where daily precipitation exceeds the 99th percentile threshold value	Annual total precipitation due to events exceeding the 99th percentile value	mm

scenarios (SSP245 and SSP585) show a very identical pattern of mean precipitation in the projected period, but it is likely to observe higher precipitation amounts in the future. Under higher emission scenarios, the precipitation amount will be higher than in moderate emission scenarios and historical data, which may result in the frequent occurrence of moderate to extreme precipitation in the future. It is worth noting that there is no shift in high or low precipitation areas in the future (as of the reference period), even though there is high precipitation.

The temporal distribution of seasonal and annual precipitation and its changes in reference to historical data over Nepal are analyzed in Fig. 3. The precipitation distribution in the country is very heterogeneous, with the highest precipitation during the monsoon season and the lowest in the winter season. The precipitation in the country is dominated by the monsoon season, as it contributes 80% to the annual

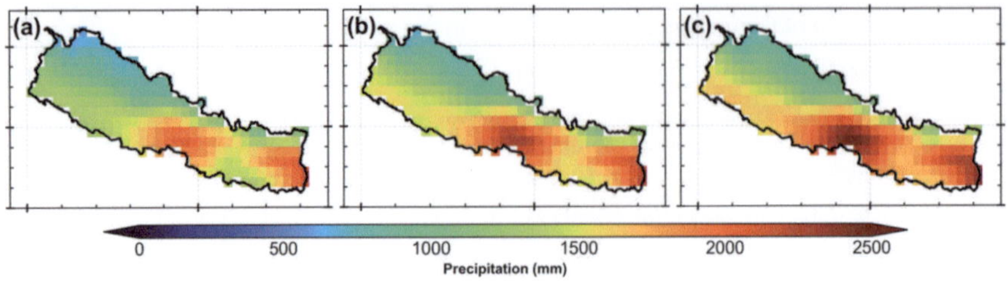

Fig. 2 Spatial distribution of historical and projected nine ensembles of mean precipitation during **a** the historical period (1980–2020), **b** SSP245 (2015–2100), **c** SSP585 (2015–2100)

precipitation under the influence of the South Asian Summer Monsoon (SASM) (Hamal et al., 2021; Sharma et al., 2020a), while the winter season is affected by western disturbances (Hamal et al., 2020a). As the winter season in Nepal contributes only ~5% to the annual precipitation, it is a very important season for agriculture and glacier accumulation. The temporal distribution of winter precipitation and its future change are presented in Fig. 3a, b. In the historical period, winter precipitation ranges from 30 mm/year to 90 mm/year, however, the precipitation was in decreasing trend (−0.1 mm/year). The pattern is also likely to continue in the future period, as both scenarios showa downward trend (not significant) in the future period (Fig. 3a, Table 3). The negative change in Fig. 3b indicates that the winter season in the future will be drier than in the historical period. The decrease in winter precipitation in the future may increase the drought risk and shrink the glaciers in the Himalayan region, which may create food insecurity and water shortages in the country. Further, precipitation distribution in the monsoon season shows an increasing trend and it is likely to rise in the future with higher rates (Fig. 3c). The monsoon precipitation was found to be increasing at a rate of 0.24 mm/year in the historical period, which is expected to rise by 2.32 and 5.82 mm/year under SSP245 and SSP585 (Fig. 3c). The decreasing winter precipitation and increasing monsoon precipitation in the future will increase the number of extreme precipitation events over the country, such as frequent droughts in the winter and severe floods and landslides in the monsoon season. The positive change in monsoonal precipitation as compared to the historical period, with higher changes in the 80s under both SSPs (Fig. 3d). The winter and monsoon precipitation before the 2060s are very similar in both scenarios. As the monsoon season contributes the majority of the annual precipitation, the annual precipitation in Fig. 3d reveals significant variation between historical and projected periods, with similar variation under SSP245 and SSP585 until 2060 (Fig. 3e). However, there will be a considerable increase in precipitation amount under SSP585 compared to SSP245 after 2060. The mean annual precipitation under SSP245 remains below 1600 mm/year, while it exceeds 1600 mm/year under SSP585 after 2060. The change in precipitation amount in Fig. 3f reveals that both scenarios show a change of less than 7% in the 2050s, which increases to over 12% in the 2080s compared to the historical period (1980–2010) under both SSPs. The difference in precipitation changes

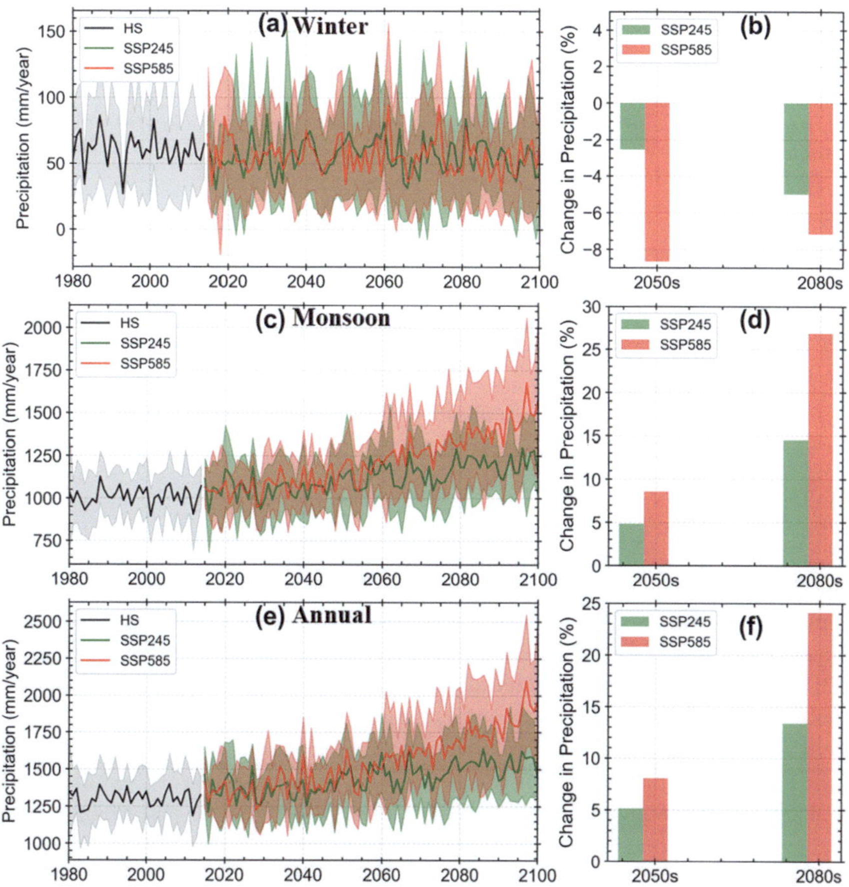

Fig. 3 Ensemble means precipitation distribution for historical and projected periods and changes in precipitation amount under SSP245 and SSP585 as compared to historical data for **a**, **b** winter season, **c**, **d** summer monsoon season, and **e**, **f** annual. The shaded area and dashed line show the inter-modal (9 bias-corrected models) standard deviation. The bargraph shows the change in the precipitation during early future (2025–2054) and mid-future (2055–2084)

between SSP245 and SSP585 is only 3% in the 2050s, but it increases to more than 10% in the 2080s.

3.2 *Extreme Precipitation Indices*

Nepal is experiencing frequent extreme precipitation events, impacting the country's overall economy (Karki et al., 2017; Talchabhadel et al., 2018). Here, we evaluated the temporal distribution of nine extreme precipitation indices and their changes over

Table 3 Seasonal trend in CMIP6 ensemble means for historical (1980–2014) and projection period (2015–2100)

Season	Historical (mm/year)	SSP245 (mm/year)	SSP585 (mm/year)
Pre-monsoon (March–May)	−0.04	0.18*	0.67**
Monsoon (June–September)	0.24	2.32**	5.82**
Post-monsoon (October–November)	−0.18	0.17**	0.52**
Winter (December–February)	−0.1	−0.03	−0.05
Annual	−0.07	2.72**	7.06**

*Note**, ** denote that the values are statistically significant at the 95 and 99% confidence levels, respectively

Nepal from 1980 to 2100 using an ensemble mean of bias-corrected CMIP6 datasets (as indicated by the black line). In the historical period, average heavy precipitation days in the ensemble mean (inter-model, shaded area) range between 5 and 6 (4 and 8) events per year (Fig. 4a), which is even lower for extreme precipitation, which ranges from 2 to 3 (2 to 4) events in a year (Fig. 4c). The duration-based indices, i.e., CDD and CWD, also showed considerable interannual variations with longer dry spells than wet spells, indicating that the country has suffered from frequent drought spells (Fig. 4e, g). Some previous studies based on observed data also mentioned some severe drought events during this period (Hamal et al., 2021; Sharma et al., 2021). The total number of rainy days (RR1) ranges between 15 and 17 days, with inter-model variation between 12 and 17 days from 1980 to 2014 (Fig. 4i). A similar interannual variation was also observed for SDII, PRCPTOT, and Rx1day. All the extreme precipitation shows a nominal downward trend (not significant), except for rainy days, which show an upward trend over the study area. Some previous extreme precipitation-based studies also noted that there is no distinct trend over the study areas (Sharma et al., 2020c).

The extreme precipitation for the future period under SSP245 and SSP585 scenarios (as indicated by green and red lines) is presented in Fig. 4. The inter-annual precipitation variation under both scenarios is identical before the 2060s, and the distribution of all extreme indices is also very similar. The average heavy precipitation (extreme) precipitation days in the future are increasing at a rate of 0.13 (0.009) and 0.03 (0.023) per year under SSP245 and SSP585 scenarios, respectively (Fig. 4a, c). The upward trend in heavy and extreme precipitation might be related to the increase in monsoonal precipitation over the country. Similarly, for heavy and extreme precipitation events, the upward trend is expected to be observed for both dry (Fig. 4e) and wet spells (Fig. 4g). The number of dry spells is higher than that of wet spells, however, the inter-annual variation is quite similar in dry spells under both scenarios. The decreasing trend of winter and the increasing trend of monsoon precipitation might be related to the upward trends for dry and wet spells,

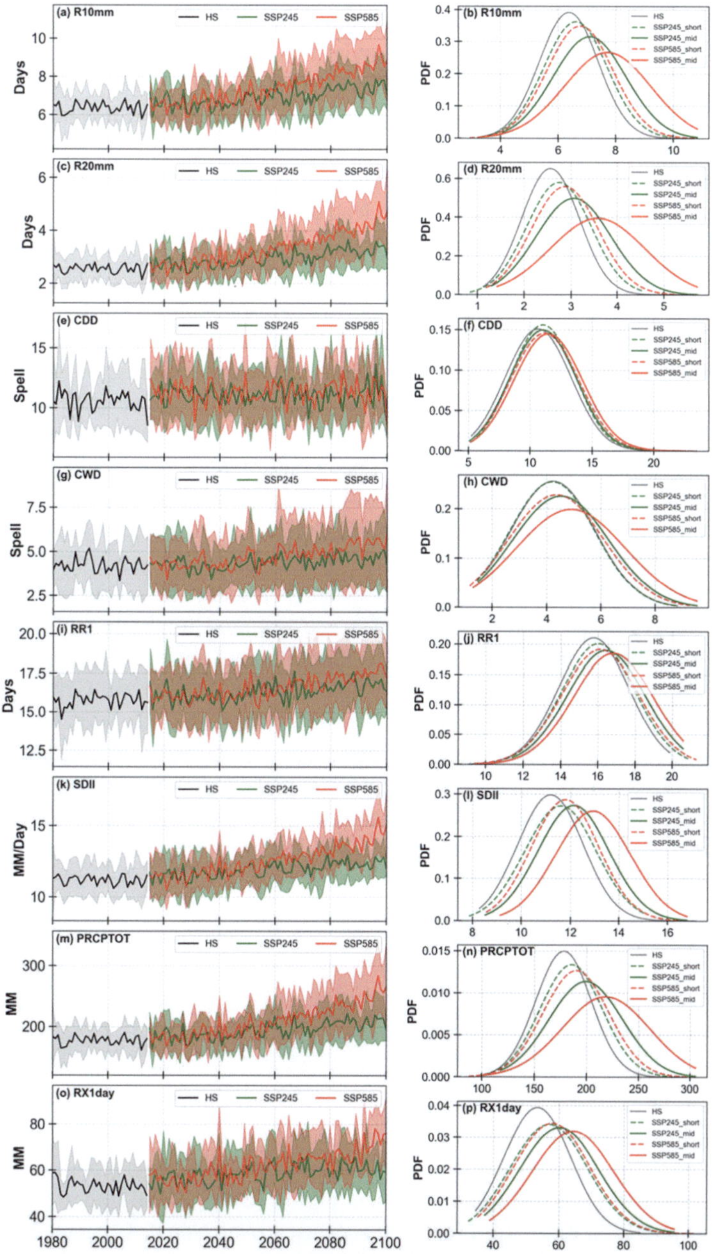

Fig. 4 Temporal distribution of historical and projected extreme precipitation indices over Nepal (left panel), probability density function (PDF) for extreme precipitation indices for the historical and projected period (right panel).Indices are in panel (**a**, **b**) R10mm, (**c**, **d**) R20mm, (**e**, **f**) CDD, (**g**, **h**) CWD, (**i**, **j**) RR1, (**k**, **l**) SDII, (**m**, **n**) PRCPTOT, (**o**, **p**) RX1day. Green and red dash and continuous line present for the early (2025–2054) and mid-future (2055–2084) under SSP245 and SSP585, respectively. The shaded area shows the inter-model standard deviation

respectively, over Nepal. The number of rainy days (Fig. 4i), and total precipitation (Fig. 4m) are also expected to increase under both scenarios. Moreover, the number of continuous wet days will incline faster than dry days, and the precipitation of a single day or total rainy days will rise. Both the amount of extremely heavy precipitation and the total precipitation are going to increase, which is more likely to result in extreme precipitation events in the future. It is interesting to note that, dry spell increment during the winter season can lead to drought events and may exacerbate air quality and fire events in the future (Pokharel et al., 2023).The increase in total precipitation coupled with a decrease in the number of rainy days may contribute to an elevated risk of floods and landslides across the country (Karki et al., 2020; Pokharel et al., 2019; Talchabhadel et al., 2018, 2022).

Based on the probability distribution in the right panel of Fig. 4, it is projected that the duration of heavy and extreme precipitation will increase in the future as the mode of duration for the future will shift to the higher duration compared to the historical period. A higher duration of heavy and extreme precipitation events are observed in the 80s under both scenarios as compared to the 50s. The probability density function in Fig. 4f indicates that the duration of dry spells is almost similar for the 50s and 80s to the historical period, but the duration of wet spells is higher in the 80s (Fig. 4f, h). The number of rainy events is found to be higher under SSP585 for the 80s (Fig. 4j). A significant difference is found for SDII with marginally higher intensity for the later period 80s than 50s and historical period (Fig. 4l). Additionally, we have also calculated the total precipitation on wet (R95pTOT) and very wet (R99pTOT) days for historical and projected periods (Fig. 5). The time series (panels 5a and 5c) indicate that historical data remains relatively consistent for both indices, with higher precipitation totals in the wet precipitation events. Both SSP245 and SSP585 scenarios suggest an upward trend in the future, with the latter predicting more pronounced events after 2080. The probability distribution functions (panels 5b and 5d) further highlight the notable shift in the duration of wet and very wet days in mid future than in the near future.

4 Conclusion

Nepal, the Himalayan country, is one of the regions with frequent flood disasters and is quite sensitive to global warming. This study investigated the precipitation pattern and their extremes using observation and nine bias-corrected CMIP6 models for the historical (1980–2014) and project period (2015–2100) under intermediate (SSP245) and high emission (SSP585) scenarios over Nepal. Changes [the early-future (2025–2054), mid-future (2055–2084)] in the precipitation and ten different precipitation extreme indices are also accessed for the project period in reference to the historical period over the country. The MK statistical test and Sen's Slope estimator were performed on the annual, seasonal, and extreme precipitation time series to present the trend and magnitudes of precipitation and extremes.

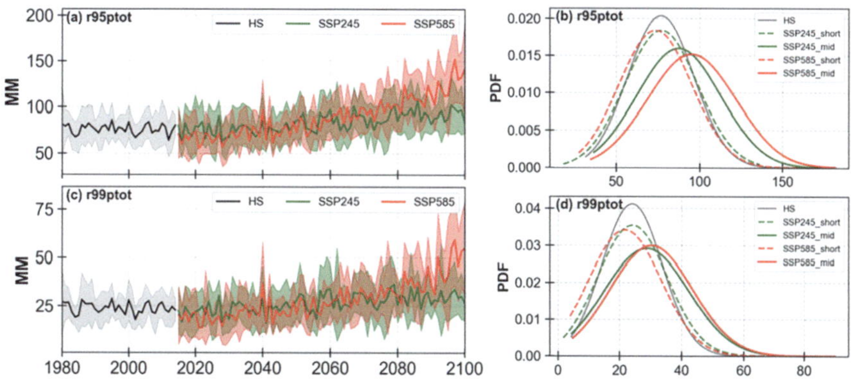

Fig. 5 Temporal distribution of historical and projected extreme precipitation indices over Nepal (left panel), probability density function (PDF) for (**a, b**) R95pTOT and (**c, d**) R99pTOT. Green and red dash and continuous line present for the early (2025–2054) and mid-future (2055–2084) under SSP245 and SSP585, respectively. The shaded area shows the inter-model standard deviation

Bias-corrected CMIP6 ensemble mean shows a slight shift in high and low precipitation areas but generally captured the APHRODITE precipitation pattern (low MB, RMSE, and high correlation) between 1980 and 2014 over Nepal. The winter precipitation shows a non-significant downward trend for the historical period, which is likely to continue for future periods under both SSP245 and SSP585 scenarios. Precipitation in the monsoon season was found to be significantly increasing at a rate of 0.24 mm/year in the historical period, which is expected to accelerate with higher rates under both SSPs. The precipitation in the pre-and post-monsoon was in a decreasing trend for the historical period but significantly increasing for the projected period under both scenarios. As the monsoon season is likely to rise, annual precipitation also shows a similar trend of 2.72 mm/year and 6.96 mm/year for the project period under SSP245 and SSP585 scenarios. Further, an increase in the precipitation amount was found to be more evident after the 2060s.

Several different precipitation indices (R10mm, R20mm, CWD, CDD, RR1, SDII, PRCPTOT, RX1day, R95PTOT, R99PTOT) used in this study showed an upward trend, among which only CDD has a non-significant upward trend under all scenarios. The intensity and frequency of extreme precipitation events in Nepal have increased over time and are expected to increase at higher rates in the future. The increment of high-intensity and wetness-related precipitation extremes for future periods might be related to the increasing trend of the monsoon period, whereas the increment of dry days-related extremes is related to decreased winter precipitation. Moreover, in the context of ongoing climate change, this study can assist in developing strategies to combat climate-related threats and high-intensity related precipitation extremes and their resulting losses, as part of preparedness and adaptation efforts in response to climate change.

Acknowledgements This research is supported by the Research Grants to Faculty Appointed through Research Excellence based Open Competition Under the Higher Education Reforms Project (HERP DLI-7B).This research is funded by the National Priority Research Grants (team led by Dr. Binod Pokharel) and Collaborative Research Grants (team led by Prof. Dr. Vishnu Prasad Pandey) from the Research Directorate of Research Coordination and Development Council, Tribhuvan University.We would like to express our sincere gratitude and appreciation to the scientists at the APHRODITE team and NASA Center for Climate Simulation for providing the precipitation and bias-corrected downscaled CMIP6 data, respectively.

Funding This research is funded by the Research Grants to Faculty Appointed through Research Excellence based Open Competition Under the Higher Education Reforms Project (HERP DLI-7B), the National Priority Research Grants (team led by Dr. Binod Pokharel) and Collaborative Research Grants (team led by Prof. Dr. Vishnu Prasad Pandey) from the Research Directorate of Research Coordination and Development Council, Tribhuvan University.

Data Availability Statement The bias corrected CMIP6 data for Nepal can be freely obtained from https://www.nccs.nasa.gov/services/data-collections/land-based-products/nex-gddp-cmip6. APHRODITE gridded precipitation data can be freely accessed from http://aphrodite.st.hirosaki-u.ac.jp/products.html.

Conflict of Interest The authors declare that they have no conflict of interest.

References

Casanueva Vicente, A., Rodríguez Puebla, C., Frías Domínguez, M. D., & González Reviriego, N. (2014). Variability of extreme precipitation over Europe and its relationships with teleconnection patterns.

Chen, Y., et al. (2021). Spatial performance of multiple reanalysis precipitation datasets on the southern slope of central Himalaya. *Atmospheric Research, 250*, 105365.

Curriero, F. C., Patz, J. A., Rose, J. B., & Lele, S. (2001). The association between extreme precipitation and waterborne disease outbreaks in the United States, 1948–1994. *American Journal of Public Health, 91*, 1194–1199.

Dalagnol, R., et al. (2022). Extreme rainfall and its impacts in the Brazilian Minas Gerais state in January 2020: Can we blame climate change? *Climate Resilience and Sustainability, 1*.

Dezfuli, A. (2020). Rare atmospheric river caused record floods across the middle east. *Bulletin of the American Meteorological Society, 101*, E394–E400.

Diaconescu, E. P., Gachon, P., Laprise, R., & Scinocca, J. F. (2016). Evaluation of precipitation indices over North America from various configurations of regional climate models. *Atmosphere-Ocean, 54*, 418–439.

Dimri, A. P., et al. (2017). Cloudbursts in Indian Himalayas: A review. *Earth-Science Reviews, 168*, 1–23.

Dong, T., Zhu, X., Deng, R., Ma, Y., & Dong, W. (2022). Detection and attribution of extreme precipitation events over the Asian monsoon region. *Weather and Climate Extremes, 38*, 100497.

Doocy, S., Daniels, A., Murray, S., & Kirsch, T. D. (2013). The human impact of floods: A historical review of events 1980–2009 and systematic literature review. *PLoS Currents*.

Dowling, C. A., & Santi, P. M. (2014). Debris flows and their toll on human life: A global analysis of debris-flow fatalities from 1950 to 2011. *Natural Hazards, 71*, 203–227.

Eyring, V., Bony, S., Meehl, G. A., Senior, C. A., Stevens, B., Stouffer, R. J., & Taylor, K. E. (2016). Overview of the Coupled Model Intercomparison Project Phase 6 (CMIP6) experimental design and organization. *Geoscientific Model Development, 9*, 1937–1958.

Fallah, A., Rakhshandehroo, G. R., Berg, P., & Orth, S. O. R. (2019). Evaluation of precipitation datasets against local observations in Southwestern Iran. *International Journal of Climatology*.

Fischer, E. M., & Knutti, R. (2016). Observed heavy precipitation increase confirms theory and early models. *Nature Climate Change, 6*, 986–991.

Frame, D. J., Wehner, M. F., Noy, I., & Rosier, S. M. (2020). The economic costs of Hurricane Harvey attributable to climate change. *Climatic Change, 160*, 271–281.

Ge, F., et al. (2019). Risks of precipitation extremes over Southeast Asia: Does 1.5 °C or 2 °C global warming make a difference? *Environmental Research Letters, 14,* 044015.

Goswami, B. N., Venugopal, V., Sengupta, D., Madhusoodanan, M. S., & Xavier, P. K. (2006). Increasing trend of extreme rain events over India in a warming environment. *Science, 314*, 1442–1445.

Guhathakurta, P., Pai, D. S., & Rajeevan, M. N. (2017). Variability and trends of extreme rainfall and rainstorms. In M. N. Rajeevan, & S. Nayak (Eds.), *Observed Climate Variability and Change over the Indian Region* (pp. 37–49). Springer.

Haerter, J. O., Berg, P., & Hagemann, S. (2010). Heavy rain intensity distributions on varying time scales and at different temperatures. *Journal of Geophysical Research, 115*, D17102.

Hamal, K., Sharma, S., Baniya, B., Khadka, N., & Zhou, X. (2020a). Inter-annual variability of winter precipitation over Nepal coupled with ocean-atmospheric patterns during 1987–2015. *Frontiers in Earth Science, 8*, 161.

Hamal, K., et al. (2020b). Evaluation of MERRA-2 precipitation products using gauge observation in Nepal. *Hydrology, 7*, 40.

Hamal, K., Sharma, S., Pokharel, B., Shrestha, D., Talchabhadel, R., Shrestha, A., & Khadka, N. (2021). Changing pattern of drought in Nepal and associated atmospheric circulation. *Atmospheric Research*, 105798.

Held, I. M., & Soden, B. J. (2006). Robust responses of the hydrological cycle to global warming. *Journal of Climate, 19*, 5686–5699.

Jia, L., Yang, S., Wang, W., & Zhang, X. (2022). Impact analysis of highways in China under future extreme precipitation. *Natural Hazards, 110*, 1097–1113.

Karki, R., Talchabhadel, R., Aalto, J., & Baidya, S. K. (2016). New climatic classification of Nepal. *Theoretical and Applied Climatology, 125*, 799–808.

Karki, R., Schickhoff, U., Scholten, T., & Böhner, J. (2017). Rising precipitation extremes across Nepal. *Climate, 5*, 4.

Karki, R., ul Hasson, S., Gerlitz, L., Talchabhadel, R., Schickhoff, U., Scholten, T., & Böhner, J. (2020). Rising mean and extreme near-surface air temperature across Nepal. *International Journal of Climatology, 40*, 2445–2463.

Kendall, M. G. (1938). A new measure of rank correlation. *Biometrika, 30*, 81–93.

Kharin, V. V., Zwiers, F. W., Zhang, X., & Wehner, M. (2013). Changes in temperature and precipitation extremes in the CMIP5 ensemble. *Climatic Change, 119*, 345–357.

Knapp, A. K., et al. (2008). Consequences of more extreme precipitation regimes for terrestrial ecosystems. *BioScience, 58*, 811–821.

Kumar, K. K., Rajagopalan, B., Hoerling, M., Bates, G., & Cane, M. (2006). Unraveling the mystery of Indian monsoon failure during El Niño. *Science, 314*, 115–119.

Lenderink, G., & van Meijgaard, E. (2008). Increase in hourly precipitation extremes beyond expectations from temperature changes. *Nature Geoscience, 1*, 511–514.

Li, Y., Guan, K., Schnitkey, G. D., DeLucia, E., & Peng, B. (2019). Excessive rainfall leads to maize yield loss of a comparable magnitude to extreme drought in the United States. *Global Change Biology, 25*, 2325–2337.

Medina, S., Houze, R. A., Kumar, A., & Niyogi, D. (2010). Summer monsoon convection in the Himalayan region: Terrain and land cover effects. *Quarterly Journal of the Royal Meteorological Society*, n/a-n/a.

Min, S.-K., Zhang, X., Zwiers, F. W., & Hegerl, G. C. (2011). Human contribution to more-intense precipitation extremes. *Nature, 470*, 378–381.

Myhre, G., et al. (2019). Frequency of extreme precipitation increases extensively with event rareness under global warming. *Scientific Reports, 9*, 16063.

Nandargi, S., & Dhar, O. N. (2011). Extreme rainfall events over the Himalayas between 1871 and 2007. *Hydrological Sciences Journal, 56*, 930–945.

Nayava, J. L. (1975). Climates of Nepal. *The Himalayan Review, 1*, 9–12.

Pall, P., Allen, M. R., & Stone, D. A. (2007). Testing the Clausius-Clapeyron constraint on changes in extreme precipitation under CO2 warming. *Climate Dynamics, 28*, 351–363.

Papalexiou, S. M., & Montanari, A. (2019). Global and regional increase of precipitation extremes under global warming. *Water Resources Research, 55*, 4901–4914.

Pokharel, B., Wang, S. Y. S., Meyer, J., Marahatta, S., Nepal, B., Chikamoto, Y., & Gillies, R. (2019). The east–west division of changing precipitation in Nepal. *International Journal of Climatology, 40*, 3348–3359.

Pokharel, B., et al. (2023). Amplified drought trends in Nepal increase the potential for Himalayan wildfires. *Climatic Change, 176*, 17.

Sen, P. K. (1968). Estimates of the regression coefficient based on Kendall's Tau. *Journal of the American Statistical Association, 63*, 1379–1389.

Sharma, S., Hamal, K., Khadka, N., & Joshi, B. B. (2020a). Dominant pattern of year-to-year variability of summer precipitation in Nepal during 1987–2015. *Theoretical and Applied Climatology, 142*, 1071–1084.

Sharma, S., Khadka, N., Hamal, K., Shrestha, D., Talchabhadel, R., & Chen, Y. (2020b). How accurately can satellite products (TMPA and IMERG) detect precipitation patterns, extremities, and drought across the Nepalese Himalaya? *Earth and Space Science, 7,* e2020EA001315.

Sharma, S., Khadka, N., Hamal, K., Baniya, B., Luintel, N., & Joshi, B. B. (2020c). Spatial and temporal analysis of precipitation and its extremities in seven provinces of Nepal (2001–2016). *Applied Ecology and Environmental Sciences, 8*, 64–73.

Sharma, S., et al. (2021). Projected drought conditions over southern slope of the Central Himalaya Using CMIP6 models. *Earth Systems and Environment*, 1–11.

Sheffield, J., Goteti, G., & Wood, E. F. (2006). Development of a 50-year high-resolution global dataset of meteorological forcings for land surface modeling. *Journal of Climate, 19*, 3088–3111.

Shi, X., Chen, J., Gu, L., Xu, C.-Y., Chen, H., & Zhang, L. (2021). Impacts and socioeconomic exposures of global extreme precipitation events in 1.5 and 2.0 °C warmer climates. *Science of the Total Environment, 766*, 142665.

Soden, B. J., & Held, I. M. (2006). An assessment of climate feedbacks in coupled ocean-atmosphere models. *Journal of Climate, 19*, 3354–3360.

Sugiyama, M., Shiogama, H., & Emori, S. (2010). Precipitation extreme changes exceeding moisture content increases in MIROC and IPCC climate models. *Proceedings of the National Academy of Sciences, 107*, 571–575.

Sun, Q., Zhang, X., Zwiers, F., Westra, S., & Alexander, L. V. (2021). A global, continental, and regional analysis of changes in extreme precipitation. *Journal of Climate, 34*, 243–258.

Sunilkumar, K., Yatagai, A., & Masuda, M. (2019). Preliminary evaluation of GPM-IMERG rainfall estimates over three distinct climate zones with APHRODITE. *Earth and Space Science, 6*, 1321–1335.

Tabari, H. (2020). Climate change impact on flood and extreme precipitation increases with water availability. *Scientific Reports, 10*, 13768.

Talchabhadel, R., Karki, R., Thapa, B. R., Maharjan, M., & Parajuli, B. (2018). Spatio-temporal variability of extreme precipitation in Nepal. *International Journal of Climatology, 38*, 4296–4313.

Talchabhadel, R., Sharma, S., Khadka, N., Hamal, K., Karki, S., & Thapa, B. R. (2022). An outlook on the applicability of satellite precipitation products for monitoring extreme precipitation events in Nepal Himalaya. *Weather*.

Taylor, K. E., Stouffer, R. J., & Meehl, G. A. (2012). An overview of CMIP5 and the experiment design. *Bulletin of the American Meteorological Society, 93*, 485–498.

Thrasher, B., Wang, W., Michaelis, A., Melton, F., Lee, T., & Nemani, R. (2022). NASA global daily downscaled projections, CMIP6. *Scientific Data, 9*, 262.
Trenberth, K. E., Fasullo, J., & Smith, L. (2005). Trends and variability in column-integrated atmospheric water vapor. *Climate Dynamics, 24*, 741–758.
Trenberth, K. E., Dai, A., Rasmussen, R. M., & Parsons, D. B. (2003). The changing character of precipitation. *Bulletin of the American Meteorological Society, 84*, 1205–1218.
Try, S., Tanaka, S., Tanaka, K., Sayama, T., Khujanazarov, T., & Oeurng, C. (2022). Comparison of CMIP5 and CMIP6 GCM performance for flood projections in the Mekong River Basin. *Journal of Hydrology: Regional Studies, 40*, 101035.
Turner, A. G., & Annamalai, H. (2012). Climate change and the South Asian summer monsoon. *Nature Climate Change, 2*, 587–595.
Utsumi, N., Seto, S., Kanae, S., Maeda, E. E., & Oki, T. (2011). Does higher surface temperature intensify extreme precipitation?: TEMPERATURE AND EXTREME RAINFALL. *Geophysical Research Letters, 38*, n/a-n/a.
Vellore, R. K., et al. (2016). Monsoon-extratropical circulation interactions in Himalayan extreme rainfall. *Climate Dynamics, 46*, 3517–3546.
Wang, L., & Gu, W. (2016). The Eastern China flood of June 2015 and its causes. *Science Bulletin, 61*, 178–184.
Wang, Y., et al. (2020). Synergy of orographic drag parameterization and high resolution greatly reduces biases of WRF-simulated precipitation in central Himalaya. *Climate Dynamics, 54*, 1729–1740.
Wasti, A., Ray, P., Wi, S., Folch, C., Ubierna, M., & Karki, P. (2022). Climate change and the hydropower sector: A global review. *WIREs Climate Change, 13.*
Wehner, M. F. (2020). Characterization of long period return values of extreme daily temperature and precipitation in the CMIP6 models: Part 2, projections of future change. *Weather and Climate Extremes, 30*, 100284.
Westra, S., Alexander, L. V., & Zwiers, F. W. (2013). Global increasing trends in annual maximum daily precipitation. *Journal of Climate, 26*, 3904–3918.
Xu, X., Frey, S. K., Boluwade, A., Erler, A. R., Khader, O., Lapen, D. R., & Sudicky, E. (2019). Evaluation of variability among different precipitation products in the Northern Great Plains. *Journal of Hydrology: Regional Studies, 24*, 100608.
Yatagai, A., Kamiguchi, K., Arakawa, O., Hamada, A., Yasutomi, N., & Kitoh, A. (2012). APHRODITE: Constructing a long-term daily gridded precipitation dataset for Asia based on a dense network of rain gauges. *Bulletin of the American Meteorological Society, 93*(9), 1401–1415.
Zhan, Y.-J., et al. (2017). Changes in extreme precipitation events over the Hindu Kush Himalayan region during 1961–2012. *Advances in Climate Change Research, 8*, 166–175.
Zhang, X., et al. (2011). Indices for monitoring changes in extremes based on daily temperature and precipitation data. *Wires Climate Change, 2*, 851–870.
Zhang, X., Zwiers, F. W., Li, G., Wan, H., & Cannon, A. J. (2017). Complexity in estimating past and future extreme short-duration rainfall. *Nature Geoscience, 10*, 255–259.

Evolution of Storms: From Solar to Severe

Aditi Upadhyay and A. P. Dimri

Abstract Climate change is increasing the likelihood of severe storms. Attribution of any such single event on climate change is difficult, but warming has increased evaporation, intensifying the droughts and frequency of heavy rainfall or snowfall events. The decrease in temperature gradients and hence corresponding potential over different regions has changed the wind circulation patterns, affecting the weather. This chapter addresses the climate/weather/storm conditions from their origin to their evolutionary viewpoint. The role of the very processes/forces that has led to the formation of the Earth's atmosphere and its current climate trend are assumed to be trivial now from the energy budget perspective, and their role in climate change is highly debated. However, from its evolution perspective, they are still significant and require an understanding of climate history and natural causes. The Sun and the heat flux emanating from the Earth's surface acts as an energy source. The influence of the Sun and its variabilities (fluctuations in electromagnetic radiations and particles during different phases of the solar cycle and solar storms/flares) on climate conditions has been discussed. This also encapsulates the effect of Galactic cosmic rays and cloud-radiation interaction. The particle precipitation over the polar region can heat that region and impact the North Atlantic Oscillations (NAO) and Arctic Oscillation (AO) indices, which are representative of changes in the regional weather conditions. The solar cycle signals are also manifested in tropospheric winds. Furthermore, the effect of the increase in atmospheric CO_2 content and ocean acidification on temperature increase and vis-e-versa, a feedback loop, has also been discussed from the perspective of climate changes and extreme weather conditions.

Keywords Solar storm · Cloud-radiation interaction · Particle precipitation · Geomagnetism · Extreme weather condition

---

A. Upadhyay (✉)
Applied Physics Laboratory, The Johns Hopkins University, Laurel, MD, USA
e-mail: aditiu147@gmail.com; Aditi.Upadhyay@jhuapl.edu

A. P. Dimri
Indian Institute of Geomagnetism, Navi Mumbai, India

School of Environmental Science, Jawaharlal Nehru University, 110067 New Delhi, India

S. Das and W-K. Tao (eds.), *Severe Storms*,
https://doi.org/10.1007/978-981-97-7075-5_24

Acronyms

AO	Arctic Oscillation
CIR	Corotating Interaction Region
CME	Coronal Mass Ejection
CZ	Convection Zone
GCR	Galactic Cosmic Rays
GLE	Ground Level Enhancement (related to solar energetic particles)
GPS	Global Positioning System
GSFC	Goddard Space Flight Centre
IPCC	Intergovernmental Panel on Climate Change
NAO	North Atlantic Oscillations
NASA	National Aeronautics and Space Administration
NCEP	National Centers for Environmental Prediction
NOAA	National Oceanic and Atmospheric Administration
RADAR	Radio detection and ranging
SDO	Solar Dynamic Observatory
SOHO	Solar and Heliospheric Observatory
SRES	Special Report on Emission Scenarios
SSI	Spectral Solar Irradiance
TSI	Total Solar Irradiance
UV	Ultra Violet

Activity Indices (indicate the response and conditions of the magnetosphere and ionosphere)

DST	Disturbed storm time index, low latitude
Kp, Ap, ap	Planetary index, mid-latitude
AE, AL, AU	Auroral electrojet index, auroral regions

1 Introduction

Human society and our biomes are most tragically affected by the extreme events, which by definition are the ones which lies between 5 and 10% of the highest or lowest of the historical measurements. During such events the atmospheric conditions like temperature, pressure, precipitation rank above a cut-off point close to the upper or lower limits of the distribution of historical observations. The events are also categorized by their deviation from the mean, the recurrence interval, or the likelihood of occurrence. One way that individuals perceive climate change is through extreme weather (Howe et al., 2014). Extreme occurrences contrast with long-term climate change trends, which appear abstract, far away, slow, and complex. Extreme events are sudden, take place right now, and are very apparent. The issues which is now cropping up and catching all the headlines is incrementaly more of the natural disaster/climate events are falling in this category or it can be rephrased as-the intensities, duration, frequencies and the destructive capabilities of the natural calamities have increased. Due to the warming of climate system, there have been changes in the severe weather events' observed frequency, severity, and duration. Climate models have also been used to mimic such variations in extreme weather, and some of the causes are known. For instance, it is anticipated that warming would increase the possibility of exceptionally hot days and nights. Moreover, it is anticipated that warming would increase atmospheric moisture, which might increase the frequency of heavy precipitation and snowfall events, and evaporation, which may aggravate droughts. There are strong scientific arguments for studying the causation of extreme weather events. The fundamental physics that govern the functioning of the climate system and the wide-ranging effects of the fast increase of greenhouse gases on the climate system are well studied. Nonetheless, there is still much to learn about how certain weather occurrences are impacted by the changing climate. It is increasingly challenging to assess how much climate change affects a certain weather or event. It entails taking into account a wide range of potential natural and man-made variables (such as large-scale circulation, internal mechanisms of climate variability, anthropogenic climate change, and aerosol impacts) that interact to create the unique circumstances of an event. Since these events are uncommon, there are only limited examples of them available in the past at any specific site. To understand the role of natural and anthropogenic variables on changing climate and the occurrence characteristics of the extreme weather events, a comparatively new field of science, event attribution is being applied. The capacity of scientists to reliably anticipate and estimate future weather and climatic conditions will be improved by improved attribution, and eventually by prediction of severe occurrences, which will show an even more nuanced and comprehensive knowledge of the climate system. The confidence in the capabilities for extreme events to natural and anthropogenic climate changes varies for every particular event. A list of that is provided in Table S1 of National Academies of Sciences et al. (2016). In this chapter we aim to understand the attribution of natural variability towards the extreme events and especially due to the Sun. The Sun is a variable star and the primary source of energy for the Earth and its climate system. To understand its role in climate change, we require the knowledge and answer regarding: (a) What is solar variability (short and long time scales)? (b) How do the solar and terrestrial connections work? (c) How does the Earth's climate system respond to them? We broadly aim to answer these questions in this chapter.

2 The Sun

The Sun is our closest star, and its proximity piques our attention. It is the primary energy source for all life forms on Earth. It contains 99.86% of the mass of the solar system and is located at its centre. Many cultures (ex. Hindu, Japanese, Greek, Syrian, Finnish, Roman, etc.) revere the sun and regard it as a deity; evidence of this can be found throughout history. However, astronomers claim that it is a typical star. It is a middle-aged G type main sequence star (G2V) based on its location on the Hertzsprung-Russel diagram of surface temperature against luminosity. It is commonly referred as a yellow dwarf. The temperature inside the Sun is of the order of 10^7 K, and density is approximately 10^5 kg/m^3, proving to be a favourable region for ionized gas molecules to undergo thermonuclear reactions and fuse to form heavier nuclei. The movement of heated, ionized gases in the outer layers of the Sun causes electrical currents to occur. It is generally accepted that electrical currents function as a magnetic dynamo inside the Sun and produce the magnetic field. Photographs of the Sun revels dark spots on visible region of the solar atmosphere (Photosphere). These are fibrillated structures having a sunken, bowl-like form, called Sunspots. They have higher magnetic field (10^3–10^4 Gauss) and lower temperatures (4000 K) as compared to the rest of the photospheric surface. They are a manifestation of the magnetic field and dynamic nature of the Sun. They occur in pairs of opposite magnetic polarity with periodicity of approximately 11 years and polarity reversal in 22 years. Therefore, Sun's magnetic dynamo has a 22-year cycle. The magnetic north pole is located in the northern hemisphere during the first half of the cycle, while the magnetic south pole is located in the southern hemisphere. During solar maxima, the peak of the sunspot cycle, the magnetic poles switch positions such that the magnetic north is now positioned in the southern hemisphere. The sunspots are the active regions of the Sun and are responsible for release of powerful energy bursts called flares, coronal mass ejection etc. which will be discussed later.

The ionized particles (electrons, protons, and ions) leave the Sun's surface and the plasma extends into the space as the solar corona is incredibly hot. Their thermal energy lies in the 1.5–10 keV range. The term 'solar wind' refers to this flow of particles immersed in the remnant magnetic field of the Sun. With respect to distance and direction from the Sun, the density, temperature, and speed change. When the particles get to the Earth's surface, they accelerate and usually reach a speed between 250 and 750 km/s. The solar wind can be divided in two states: fast and slow solar wind. The classification is mainly based on the bulk velocity. The slow solar wind travels with speeds less than 350 km/s, and if the speed is greater than 600 km/s, it will be termed as fast solar wind. In addition to the speed disparity, there is a difference in the plasma parameters and location of the Sun from where the particles are launched. The particles of the slow solar wind originate from the equatorial regions (having closed magnetic field lines), whereas in fast solar wind, they are leaving through the coronal hole (region of open field lines, unipolar open magnetic field line regions). When the two kinds of streams interact, a compressional region is formed with a shock front which is called 'co-rotating interaction region'. The plasma particles in these regions get accelerated and the shock-related turbulence and magnetic compression areas can deflect cosmic rays. When planets come in the path of CIR, rapid fluctuations in the solar wind plasma and the geomagnetic disturbances are observed.

3 Sun-Earth Interaction

The Earth is one of the eight planets revolving around the Sun. Some of the planets also have their own intrinsic magnetic field like Sun, although the source of it is different. The interplay of the planet's rotation and the movement of inner conducting material (molten rocks and metal or liquified metallic gases) results in the formation of planetary magnetic fields by the geodynamo process. Its requirements are that the rotation of the planet must be considerably fast and its core must have an energy source which could keep driving the conductive inner molten fluid in convective motion. Mercury rotates so slowly that its magnetic field is feeble. Venus doesn't appear to have a significant field since its molten interior doesn't seem to have much convection. Although it once had one, Mars has no measurable magnetic field now because of the planet's interior has solidified. The Jovian planets have extremely strong magnetic fields. Although, to a small amount, the currents flowing in the plasma surrounding the planets also contribute to the total magnetic field.

The solar wind plasma is highly conducting and the magnetic field is frozen in it. Due to the frozen-in flux condition, the particles of the solar wind cannot directly penetrate into the region of planetary magnetic field. The magnetized solar wind drapes around the planets and other terrestrial bodies in its path. It creates compression and *bowshock* in the sunward direction and elongated *magnetotail* downstream in the anti-sunward direction. A pictorial representation of these regions can be found in Fig. 1. The interaction between the planetary magnetic field and the solar wind leads to the formation of magnetosphere, which is a bullet shaped cavity in which the planets lie and the space in it is dominated by the planet's magnetic field (Baumjohann and Treumann, 2012).

3.1 The Earth's Magnetosphere

The Earth's magnetosphere is divided into different regions based on the particle temperature, density, particle species and their movement. These regions are magnetosheath, magnetopause, ring current, radiation belts, plasmasphere, plasma sheet, magnetotail lobe etc., and they are bounded by current systems (Baumjohann and Treumann, 2012). Some of the regions are shown in Fig. 1. The radiation belts are formed around planets having a stable dipolar field. The magnetic field captures the energetic particles mostly coming from the solar wind and holds them around. The Earth has two radiation belts and sometimes a third one is also seen. Inner belt lies

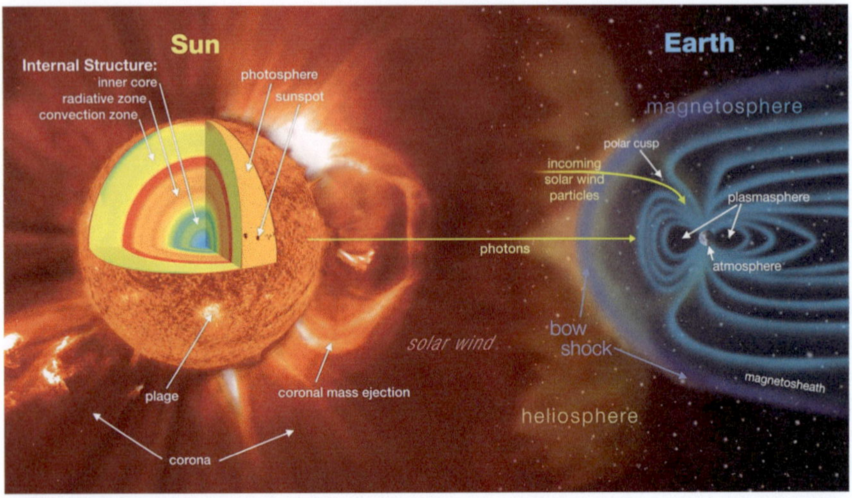

Fig. 1 Depiction of the Sun–Earth system and their interactions that influence the space weather. The significant regions that we have discussed in the text are shown in the figure for reference. Picture credit: GSFC/NASA (*Source* https://svs.gsfc.nasa.gov/30481)

between lies between 0.2 and 2 R_E. Here, R_E is the Earth radii ($\approx$ 6400 km). The distances at such scales are measured in terms of multiple of Earth radii. It has electrons of few KeV (1 eV = 11,600 K) energies and protons of 100 MeV energies. The outer belt lies in 3–8 R_E. It has electrons of energies in the range 0.1–10 MeV. Inside it lies the plasmasphere. The plasmasphere is torus in shape and lies deep within the magnetosphere and extends till 4 R_E. It is linked to mid latitude ionosphere and co-rotates with Earth. It comprises of cool and dense plasma and ions of ionospheric origin. Across these two systems flows the ring current, which is a large scale current system flowing westward in approximate circles around earth in a radial distance of several R_E. Beneath the plasmsphere lie the ionosphere and the atmosphere.

Although the particles of solar origin cannot directly penetrate into the Earth's magnetosphere, but the radiations can. On an average Sun gives out 10^{17} W of energy in form of radiations. The Sun behaves like a blackbody of effective temperature of 5800 K and emits radiations in the range of Ultraviolet (<400 nm) to Infrared regions (>1000 nm). The Earth-Sun distance or time of the year, solar activity, the angle at which the sun's rays enter the Earth's atmosphere, geographical location, the weather, and the level of air pollution all have a significant impact on the intensity of the solar spectrum that is received on the Earth's surface. The atmosphere is a blanket of different gases which are held around the Earth's surface by the virtue of its gravitational pull. When the energetic radiations reach the Earth, they can ionize the particles at the top of our atmosphere resulting in ions and free electrons. This weakly ionized region is called the ionosphere and it is mostly ionized due to the Sun's UV rays, X-rays, corpuscular radiations and also the cosmic rays from the outer universe. The ionosphere is spread between 60 and 1000 km altitudes. Based on the region's plasma density profiles, it is further divided into D, E and F layers. The ionized layers can conduct electricity and affects the propagation of radio waves. Any fluctuations in the particles and radiations coming from the Sun are first intercepted by the magnetosphere and the ionosphere systems (Kivelson and Russell, 1995).

3.2 Solar Activity and Solar Storms

The surface of the Sun is not as quiet as it appears. The outer layer contains huge convections cells (500 kms) called granulations. They frequently form and disappear within 5 min durations. Sometimes they have longer lifetimes up to one day, which we call supergranulations. The knowledge of the solar structure and its activities have increased by the advancement in the theories and magnetohydrodynamic modelling, high resolution ground observations, and from satellite missions like Skylab, SOHO, SDO, Parker Solar Probe, etc. Observations of the Sun in different frequency filters or coronagraph can reveal the pictures of the atmosphere at different altitudes. With Hα filters, we can observe the chromosphere with sunspots, flares and filaments etc. These have lifetimes between few minutes to few days. The Sun has no surface features that endure for any discernible amount of time. Helioseismology has also revealed that the solar structure experiences oscillations with amplitudes of merely a few cm/sec and different periodicities.

The sunspots, which are the active regions of the sun as previously discussed, give rise to huge erupting bubbles from the surface known as **coronal mass ejection (CME).** It carries with it tonnes of coronal matter and embedded magnetic field and travel with speed upto 3000 km/s. Twisting in the field lines in the strong and stressed magnetic flux regions leads to the process of magnetic reconnection. It leads to sudden explosive release of electromagnetic energy in form of **solar flare** and accompanied with enormous quantities of plasma which is CME. If it is moving with high speed, shock waves are formed at its front. The radiative instabilities in the outer region can also lead to formation of **prominences**. They are large sheets of dense plasma which looks like a thin and dark filament in the Hα photos. Their length is around 2×10^7 m and remains stable for months. These structures can also become unstable and erupt. More interesting phenomenon of coronal heating and solar wind acceleration are happening around the sun which reveals its continuous activity and instabilities within it. The long term variations in the solar activity and solar wind parameters follow the 11-year solar cycle (Kivelson and Russell, 1995, NOAA).

3.3 Solar Wind-Magnetosphere-Ionosphere System

The solar wind, magnetosphere and ionosphere acts a unified system that is driven by the transmission of energy and momentum from the solar wind to the magnetosphere and then to the ionosphere (Lyon, 2000). The transfer of energy between solar wind and magnetosphere takes place by the process of compression of the earth's magnetic field lines and magnetic reconnection. The explosive release of energy and the propagating disturbances in the solar wind directed towards the earth can bring geomagnetic disturbances. The high solar activity can disturb the magnetospheric configuration and current systems. Various satellite and ground based instruments like magnetometers, langmuir probe, GPS, RADARs, ionosonde, and riometers etc. are used to observed the fluctuations in the magnetic and electric field and changes in configurations of ionosphere and magnetosphere. When the earth observes a CME, more particles are injected in the radiation belt and ring current by the night side reconnection. The energy in the ring current increases which creates depression in the magnetic field as observed from the near-equatorial region ground based magnetometers. This is called **Geomagnetic Storm**. High density and southward orientation of solar wind can also cause **Magnetospheric Substorm**. This is localized high latitude phenomenon during which huge particle precipitation from the polar region takes places. Substorms lead to westward auroral electrojet currents and display of auroras.

An assessment of the solar and magnetospheric activity can be made by observing changes in terrestrial magnetic field and current systems. Scientists have created a number of indices to provide a measure of the activity in order to analyse the fluctuations in the geomagnetic field. Magnetic field fluctuations in the regular current systems are caused by frequent changes in solar radiation. There are also irregular current systems, which can be brought on by interactions between the magnetosphere and ionosphere, the magnetosphere itself, or the solar wind and the magnetosphere. These erratic current systems cause changes in the geomagnetic field, which are accessed by calculating magnetic activity indices. These measurements help in tracking geomagnetic activity, which develops when the magnetosphere and ionosphere react to solar forcing. A few of the indices regularly used to comprehend and recognise instances of space weather are: DST (disturbed storm time index, low latitude) (Dessler and Parker, 1959). Kp, Ap, ap (planetary index, mid-latitude) (Bartels, 1949) and AE, AL, AU (auroral electrojet, auroral regions) (Davis and Sugiura, 1966).

3.4 Role of the Atmosphere

The atmosphere has a crucial role in defining the local terrestrial solar irradiance as it affects how much radiation or sunlight from the top of the atmosphere reaches the ground. The atoms and molecules in the air selectively absorb or scatter light of particular wavelengths. A photon of light provides all of its energy to an atom or molecule through the process of absorption. To achieve this, it elevates an electron to a higher energy level, absorbing photon in the process. Therefore, it converts electromagnetic energy into heating atmosphere.

Ozone (O_3) is the main absorber of the energy at UV wavelengths below 400 nm. Water vapour (H_2O) and molecular oxygen (O_2) exhibits strong absorption bands for light in the mid-infrared wavelengths (till 3000 nm), and red and near infrared region (600–800 nm), respectively. Above around 2500 nm, carbon dioxide (CO_2) starts to have a role in the absorption of light. Incoming solar energy is dispersed into different directions by molecules in the atmosphere through the process of scattering. The molecules and/or particles divert the photons from their initial path. Mie scattering and Rayleigh scattering are the two forms of scattering that are most common in the atmosphere. Rayleigh scattering is mostly caused by extremely tiny dust particles and gas molecules like oxygen and nitrogen in the upper atmosphere. The reason why the sky appears blue during the day and red at sunrise and sunset is because Rayleigh scattering preferentially scatters blue light. Mie scattering is mostly caused by water vapour and big dust/pollutant particles in the lower atmosphere. Mie scattering gives the fog and clouds white color. The terrestrial spectrum can also be further altered by regional factors including the local climate and pollution levels.

4 Natural Factors in Climate Change

The summers have gotten hotter, and the rainfall pattern has been haphazard lately. We are observing a higher frequency of extreme weather events, and all have been attributed to the change in the climate. The climate, or the long-term averages of weather conditions, is undoubtedly changing. This has literally become a burning issue (since the heat waves started striking various places around the globe). However, we do not know what it all can be pointed to. Much debate and research have been going on this topic, and numerous factors are at play. If the temperature keeps rising as it is , it will affect the precipitation, and these changes will be reflected in the crop yields, water supplies, human health, and various ecosystems.

The natural climate formation factors which are:

(a) Heliophysical and astrophysical factors: impact of Sun (luminosity, activity, and gravitational influence) and other solar system bodies
(b) Geophysical factors: peculiarities of Earth as a planet—its mass, size, rotation, electric and magnetic fields, and inner processes like volcanic and tectonic activities
(c) Circulation factors: redistribution of energy and matter within the system of atmosphere, hydrosphere, lithosphere, and biosphere.

4.1 How Does Sun Affect the Climate?

The solar wind with the continuous solar electromagnetic radiations governs the formation and pattern of the magnetosphere, atmosphere, and climate. The Sun's activity varies from time scales of seconds to billions of years. The total solar irradiance (TSI) observed from Earth shows a variability of 0.15% from solar maxima to solar minima. Variations in hard electromagnetic emissions across the solar cycle bring changes in the ionospheric layers, thermospheric temperatures and atmospheric heights. The lack of the dominant 11 years solar cycles between around 1640 and 1700 (Eddy, 1976), together with the nearly complete absence of sunspots themselves, is almost as conspicuous as their existence. This Maunder minimum, when there were essentially no sun spots on the solar surface, is a great illustration of a grand minimum of solar activity and example of effect of solar activity on climate. It just so happened that during this minima, a period of extreme cold and suffering coincided what is known as the Little Ice Age. In Europe, during this time the Thames River frequently froze over and mountain glaciers developed in the valleys. Other than the 11-year solar cycle, scientists have also hypothesized the presence of other patterns like the Gleissberg and Suess cycle. However, these cycles are very long-term (70–100 and 210 years, respectively) and consist of several 11-year solar cycle. Any contribution coming from them will be several orders less than the 0.15% variation in the energy reaching the Earth. When the magnetic energy associated with the sunspots is released, intense burst of radiations and/or accelerated particles come out with it. If these bursts are directed towards the Earth, they are capable of bringing about temporary changes in the Earth's magnetospheric-ionospheric system. Changes in emitted EM radiations (gamma to radio range) during high solar activity brings about modification in the temperature, circulation and chemistry of the upper atmosphere and the stratosphere. While the increase in the particle densities and energies (during CMEs and shocks) create geomagnetic disturbances. They transmit energies to the magnetosphere and that, via auroral activity (eV or MeV), is imparted to the upper polar atmosphere (Lean, 1997). The energetic particles entering from the polar cusp penetrates into the atmosphere and causes additional ionisation. These activities affect the temperature, chemistry and circulation of the thermosphere. Furthermore, with the changes in the heliospheric structure and composition, the amount of Galactic Cosmic Rays (GCRs; GeV energies) reaching Earth also modulates. These particles also affect the chemistry and the ionisation of the upper atmosphere.

The consequence of the incoming energetic particles in the Polar Regions from various processes and their loss via ionisation is atmospheric heating, modification in atmospheric electricity and chemistry. The majority of ions produced in the middle atmosphere are $N2^{+}$ and $O2^{+}$, and the electron produced are either lost in recombination or forms a negative ion on interaction with a neutral molecule. The most critical chemical reactions taking place in that region involves the ozone layer. Ozone destroying radicals (NO_x and HO_x) are generated by the incident proton which causes a chain reaction and accelerate the reduction process by a factor of 2–4. During geomagnetic active conditions, the auroral oval gets extended in the southward direction (in the Northern hemisphere and opp. in the Southern hemisphere), leading to particle precipitation in the mid-latitude atmosphere. Due to the latitudinal exchange and transport processes, ozone concentrations in the equatorial regions also show reduction. The latitudinal asymmetry in the solar EM radiations and energetic particles reaching Earth cause a latitudinal temperature gradient in the thermosphere generating a regular meridional circulation. The particle precipitation in the high latitudes acts as a heating source. Therefore, in the Polar Regions, additional secondary circulation cells evolve, enabling the transfer of heat towards lower latitudes. During high speed solar winds and active geomagnetic conditions, the heating in the Polar Regions increases because of increased particle precipitation. This leads to expansion of the secondary cells and suppresses the thermally driven circulation. Depending upon the solar activity, closed zonal circular patterns, vortices in the stratosphere and thermosphere also exists. All these affect the seasons and weather patterns on Earth and play a crucial role in the seasonal appearance of the ozone hole.

Through the above mentioned studies and indirect evidences, relationship between solar activity and terrestrial climate is generally accepted and has been confirmed to have existed for past 7000 years. However, the role of active geomagnetic events to modify the weather conditions and bringing about any changes in the global climate has long been speculated, since 1801 (Herschel, 1801), but is still controversial. On the basis of statistical grounds and energy considerations, discussions in support of these relations have been criticised (Pittock, 1978). Considering the changes in the TSI over one solar cycle, the amount of variation in the energy received by the Earth is 0.17 W/m^2 or 0.07%. For climate change, the long term changes in the radiative forcing are required, which when calculated from TSI, have small values and are insignificant for the energetic consideration (0.017 W/m^2/decade only). However, the present rate of radiative forcing due to greenhouse gases is 0.03 W/m^2. The temperature rise as per these calculations should only be 0.07 K, indicating more complex processes are involved and intricate and extensive calculations are required to predict it correctly (Change, 2007, Hofmann et al., 2009, OHLICH, 2006).

Although the changes in TSI over the solar cycle is modest, from the spectral solar irradiance (SSI) changes at ultra-violet (UV) and extreme UV (EUV) wavelengths vary from 7 to 100%, which can result in changes in middle and upper atmosphere. For instance, westerly winds observed at the Earth's surface during winters are twice as strong during sunspot minima as compared to sunspot maxima. For active geomagnetic conditions also drastic fluctuations in UV and EUV are observed. The energetic particles emanating from the Sun during those times can also change the climate system. Climate re-analysis data has shown that the surface air temperatures exhibit a difference (4 K) between the periods of high geomagnetic activity to then when they were few (Seppälä et al., 2009). The mechanisms for such significant changes are still not well known.

Earth is also exposed to energetic particles within the Galaxy. These particles are accelerated to energies even greater than GeV by supernova remnants etc. and can penetrate the heliosphere and magnetosphere and reach Earth's upper atmosphere (Büsching et al., 2005). These GCRs can interact with atmospheric molecules and atoms and causes ionization in the upper and middle atmosphere; they are the main source of ionization below 40–35 kms. They also cause the generation of cosmogenic nuclides, e.g., 10Be and14C. Many publications have studied the ground-level enhancement (GLE, measured using neutron monitors) of solar energetic particles and GCRs and their effects on atmospheric ionization (Calisto et al., 2013, Mironova and Usoskin, 2014). A rapid reduction in the GCR is observed during active geomagnetic times. Hence, no straightforward relationship between GLE and the ionization effect in the polar region is observed. It was noticed that the fluctuation in air ionisation brought on by cosmic ray might potentially have an impact on the optical transparency of the atmosphere by altering the generation of aerosols or the transition of water between its various phases (Svensmark, 2000). Dorman and Dorman (2004) have looked at the potential statistical relationships between the solar activity cycle and the related long-term Cosmic ray intensity fluctuations with aspects of climate change. The cosmic rays have been also known to bring changes in the global temperatures by altering the world's cloud cover through ionisation (Carslaw et al., 2002). There have been many successful attempts in finding out the signatures of the solar cycle in the stratospheric and tropospheric temperatures and winds, and the tropical circulation. A response in tropospheric zonally averaged temperature and winds has been observed according to a study using the NCEP/National Centre for Atmospheric Research reanalysis data set (Haigh et al., 2005). The midlatitude jets become stronger and move poleward in the solar maxima years. Because of the weak signal, the limited amount of data, and notably the significant inaccuracies in estimations of vertical velocities, it is challenging to estimate the 11-year solar signal in tropical circulations. However, Labitzke and Van Loon (1995) hypothesized that the Hadley cell in which there is generalized upwelling at equatorial latitudes and descent in the subtropics was greater at solar maxima in their examination of station radiosonde data from the tropics and subtropics. Using the data of the NCEP vertical velocities, (Kodera, 2004) reported a decrease in near-equatorial convective activity during solar maxima and an enhancement of off equatorial convection in the Indian monsoon. In the northern hemisphere, the North Atlantic Oscillation (NAO) and Arctic Oscillation (AO) are the two main modes of climatic variability. The former describes opposing pressure anomalies between the North Atlantic -European regions and while the latter describes opposing anomalies between middle to high latitude Northern hemisphere. These modes of variabilities can influence the temporal and spatial weather patterns. The long-term influence of geomagnetic activity on NAO index has been studied and significant correlation active geomagnetic condition and NAO are found (Bucha, 2014).

4.2 Other Natural Factors

Other natural factors affecting the global climate include the plate tectonics. The gradual movement of continents over millions of years affects volcanic activity and surface erosion, both of which have an impact on the atmosphere's composition.

Another important aspect is the presence of atmospheric gases that exist naturally and are referred to as greenhouse gases because they warm the surface of the Earth. They absorb emitted heat energy and reflect it back towards the surface. The temperature of Earth is also impacted by explosive volcanic activity, such the eruption of the Mount Pinatubo during 1991 in the Philippines. These eruptions release aerosols into the stratosphere, where they can stay for up to a year and cool the Earth's surface by reflecting some of the incoming sunlight back into space. According to Milutin Milankovitch, cyclical changes in the geometry of Earth's orbit around the Sun affects the earth's climate on the time scales of millions of years. The position and movement of the planet have a significant impact on the overall climate. The seasons are brought about by the earth's present 23.5-degree tilt to the plane of its orbit as it revolves around the sun. The amount of solar radiation that Earth receives is strongly influenced by the cycles/periodicities in the eccentricity, precession, and obliquity of the Earth's orbit. According to this theory, the Milankovitch cycles, as they are now called, affect the climate globally and are responsible of the long term gradual warming and cooling of the planet.

5 Anthropogenic Factors in the Climate Change

An important area of scientific research now centres on the problem of anthropogenic climate change. Not a day goes by without the discussions of the potential origins of climate change and its effects in the media and social platforms. Today, thousands of scientists employed by government organisations, research institutions, academic institutions, and weather bureaus are involved in climate research. In 1988, the World Meteorological Organization and the United Nations Environment Programme set up the Intergovernmental Panel on Climate Change (IPCC) to coordinate the results of this international study conducted by the involvement of thousands of scientist globally. The purpose of this organisation is to evaluate data, conduct research, and compile the results into reports. The IPCC publishes climate change assessment reports in 1990, 1996, 2001 and 2007. For anybody researching the issue of global warming, these studies have grown to be an essential information source. According to the IPCC's most recent report (Pachauri et al., 2007), published in 2007, the phenomena of global warming is undeniable and that, between 1906 and 2005, the earth's average temperature increased by 1.3F (0.74C). The IPCC also discovered that the probability of this warming being caused completely by natural processes were fewer than 5%, making human action the most likely cause of it. The report states that carbon dioxide emissions rose by 70% between 1970 and 2004 and are now increasing at a progressively faster rate. Extreme weather events, such as storms, floods, and droughts, which are predicted to have less than a 10% probability of occurring in any one region, are becoming more frequent, according to scientific assessments on climate change. Even little modifications to typical weather patterns can result in severe climate through feedback loops and the unintended consequences of remote occurrences. However, the pace of growth in extreme weather conditions remains unpredictable due to the small amount of historical data that scientists have to draw on in order to make their estimates. It is also note-worthy that the projections of future climate implicates that no particular weather occurrence can be connected to the phenomena of global warming. In the IPCC 2007 report, the global average temperature rise is 'very likely' due to the anthropogenic factors. Here 'very likely' is parallel to a 90% confidence level. When both human and natural forcings are taken into account, climate model simulations are more consistent with known changes and trends in global near-surface temperatures throughout the 20th century (Hegerl et al., 2007). Numerous studies have identified the impact of various variables on changes in a range of climate indicators (Stott et al., 2010), with human impacts dominating in recent decades. Thus, these evaluations raise confidence in the predictions of future climate change resulting from rising greenhouse gas concentrations made by climate models. According to a variety of alternative scenarios for future anthropogenic emissions, the forecasts for climate change by 2100 likely ranges (>66%) from 1.1 to 2.9 K for the low-emission B1 SRES scenario to 2.4–6.4 K for the high-emission A1FI SRES scenario (Meehl, 2007). Plans to lessen this anticipated warming and its potential effects have so far focused on reducing CO_2 and other greenhouse gas emissions. The IPCC evaluated studies that looked at how natural forcing variables affected historical climate, but no systematic analysis of the potential effects of changes in natural external forcing elements in the future has been done. The expected anthropogenic warming may be increased or decreased by natural external climatic forcing variables. There has been speculation that a hypothetical decline in solar activity (Abreu et al., 2008) might aid in climate change mitigation efforts or "buy time" for adaptation to the anticipated changes in the climate (McKie, 2006). Therefore, it is quite interesting to investigate what probable changes in solar activity may occur in the future and what potential effects on the climate might result.

6 Summary

In this chapter we have discussed the solar variabilities at different time scales, the Sun-earth connection, and the how Earth system responds to the fluctuations in the Sun. This has enabled us to understand the weightage of natural and anthropogenic factors which plays a role in the sudden increase in the global warming and the extreme weather conditions we have been witnessing in the recent decades. In brief, the Sun controls the Earth's climate and weather. Incoming solar radiation and outgoing infrared radiations balance lead to provide dimensions to climate and weather. The Earth is lying in the protective shield of bullet shaped magnetospheric cavity called magnetosphere and the atmosphere. They prevent the harmful short wavelength radiations and energetic particles to reach the ground, which are detrimental to existence of life. However, the Sun is by far the largest supplier of energy to the Earth's surface. The protection offered by the Earth's magnetosphere- atmosphere and the warmth given by the Sun's radiation are two elements essential for the existence of life. Any change in the Sun's radiative output has an impact on the energy balance of the Earth's surface and atmosphere, which has an impact on our climate. The Ultraviolet radiations control the stratospheric chemistry mostly by the formation and destruction of ozone. Any modification in the spectrum of the radiations coming from the Sun, especially in the higher frequency part (UV), can impact the climate. The Sun may have an impact on Earth's climate in additional, more complicated ways, such as by modifying the amount of cosmic rays flux reaching earth, which has been suggested to enhance the covering by low-lying clouds. Radiative balance play a decisive role in determining cloud property leading to positive and negative feedback mechanism which are imperatively key. There are indices (quantitative representation of different influencing factors/activities) discussed which influence the global climate and weather and researchers have shown based on these indices about the fluctuation during solar minima and maximum. Indices representing the Sun's fluctuations viz., DST, Kp, AE, AL etc. are instrumental in determining/deciding on role of sun's emission on earth. Subtle changes in Earth's orbit around the Sun have been known to cause the past ice ages. Representation of these balances will determine the intensity of weather. For a better visualization of the process which affects the earth's climate, a schematic is shown in Fig. 2. In this figure, all the methods and channels from which the solar variability, both in irradiance and radiation, effects the Earth's climate are laid out. Some of these are long term variations and requires long period datasets, proxies, and modelling to understand. The variations in the Solar dynamo effects the convection zone and surface fields and eventually brings deviation in the solar irradiance (and indirectly the GCRs because of the associated changes in the magnetic field) up to the decadal and centennial times scales. Their corresponding variations in the Earth's temperature, clouds, ozone layer, winds, precipitation and its patterns, and NAO etc. have been observed. Sunspot number, faculae or CME, changes in the Solar magnetic flux or field lines are manifestation of solar activity which eventually disturbs and reconfigures the geomagnetic field and effects the incoming radiations from solar, galactic and inter-galactic sources. The cosmic ray have been known to form the cosmogenic isotopes which affect cloud formation. The Total Solar Irradiance (which is the integration of the solar irradiance over all incoming wavelengths) is responsible for the formations of atmospheric layers like thermosphere, mesosphere, stratosphere and troposphere (as shown in Fig. 2), and variations in it and its absorption by the ozone (involved in the radiative heating of the stratosphere) and oceans, can modulate the heat input and, therefore, bring long-term changes in Earth system. The low energy Solar Energetic Particles are also known to affects the thermosphere through inducing ionospheric currents and particle precipitation. In addition to this, it is also shown in the Fig. 2 that human activities, volcanic eruption, and changes in the Earth's orbit (Milankovitch cycle) affects the weather and climate. It is now understood that Earth's temperature fluctuations are cyclical in nature and have regularly happened without human intervention. The climate of the Earth has changed far more and more frequently in the past than it does now. Over the course of its billions of years of existence, the Earth has frequently experienced periods of tremendous solar flares and tectonic disasters.

Numerous asteroid attacks have hit the Earth, including one that is supposed to have killed off the dinosaurs. However, the warming that we have witnessed over

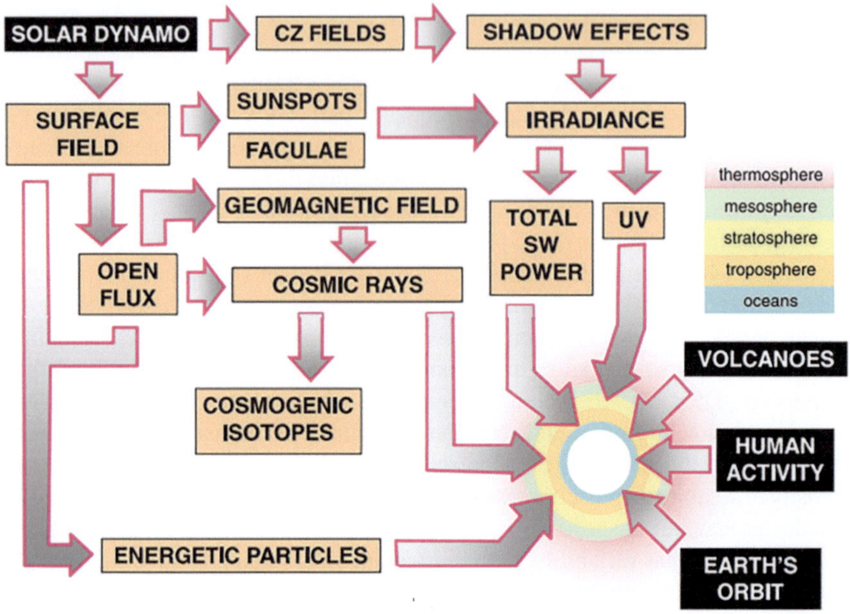

Fig. 2 The schematic shows different climate forcings of the Earth's atmosphere and the factors that affect the forcing associated with solar variability. The CZ refers to convection zone (Picture credit: Gray et al. 2010)

the recent years is too rapid to be linked to fluctuation in Earth's orbit and too large to be caused by solar activity. As per the United Nations Intergovernmental Panel on Climate Change (IPCC) 2007 report, the long and short-term variations in solar activity play only a very minute role in Earth's climate. Warming from increased levels of human-produced greenhouse gases is actually many times stronger than any effects due to recent variations in solar activity. However, it is not entirely anthropogenic, model prediction report that 5% contribution is from the natural factors. Natural causes can be used to explain the climate's gradual and inevitable fluctuation. The human component undoubtedly has an impact on these developments, but it is not a major, worldwide factor. According to the scientist, we still do not fully comprehend the process causing global warming. In addition to satisfying the public's curiosity, extensive, thorough scientific analysis of the relationship between extreme weather events and changes in the climate system can be a useful tool for emergency managers, regional planners, and policy makers at all levels of government by outlining the potential risks of such events in the future. Understanding and confidence in the various risk calculations that underpin many of the society's building codes, water, land, health, and food management, insurance, transport networks, and many other aspects of daily life can be gained from a thorough understanding of the attribution of extreme weather events in the context of a changing climate.

7 Discussion

Climate includes natural elements such as temperature, pressure, humidity, wind, and precipitation. These elements exhibit variations both over time and across different locations. The chemical composition of the atmosphere, Sun-earth distance, and fluctuation in the Earth's orbit also affect the climate. Moreover, each region on the planet possesses its unique climatic and weather conditions, influenced by factors like latitude, elevation, local topography, and proximity to large bodies of water, including oceans. In the past, meteorologists predominantly considered extreme weather as a natural phenomenon resulting from an odd combination of natural elements of the climate. However, the changing composition of the atmosphere due to human activities might be influencing global climate patterns and contributing to anthropogenic extreme weather events. During the last century, our planet has warmed 1.33F and the level of carbon dioxide in the atmosphere has risen from 280 ppm to about 380 ppm in the last two centuries. Taking the accelerating rise in CO_2 levels into account, in the next half century, it is predicted that the earth could warm another 2–3F (Streissguth and Mann, 2011). Higher sea surface temperatures as a result would lead to more powerful weather systems that support hurricanes and tropical storms. The hydrologic cycle would be amplified and strengthened by increased evaporation and water vapor in the atmosphere. Future social, economic, and political repercussions will be severe if extreme weather gets harsher and spreads more frequently. Heat waves, droughts, hurricanes, and floods, will hinder the ability of less developed countries to provide food and other necessities to their people. Due to competition over diminishing natural resources like fertile soil, energy, and water, these people will become more reliant on aid organizations and international donors for these products, which will lead to political and military confrontation on a global scale. As farmers and rural residents who are unable to maintain their families move to metropolitan regions to increase their prospects of survival, migration from impacted areas may also take place. This movement may overload already over crowded cities, making it difficult for them to provide citizens with food, water, shelter, security and health care. While there may be substantial costs associated with addressing and resolving global warming, there may also be huge consequences associated with inaction. The United States saw 58 weather-related disasters between 1980 and 2004 which has cost at least 1 billion US Dollars in damages, and these extreme weather events are occurring increasingly often. It will be necessary to develop new techniques for conserving water for residential, agricultural, and irrigation use. Coastal communities will need to build flood diversion channels and raise seawalls to defend themselves against flooding. Systems for medical care must be ready for the introduction of tropical illnesses into previously uninhabited regions. Humans need to figure out how to regulate or reduce greenhouse gas emissions in order to prevent a future tipping point. They can do this by a global accord, the development of new technology that reduces the need to burn fossil fuels, or a combination of the two. Inaction on climate change is the simplest solution, but it's also the most hazardous (Streissguth and Mann, 2011).

References

Abreu, J., Beer, J., Steinhilber, F., Tobias, S., & Weiss, N. (2008). For how long will the current grand maximum of solar activity persist? *Geophysical Research Letters, 35*(20).

Bartels, J. (1949). The standardized index, Ks, and the planetary index, Kp. *IATME Bulletin, 97*(12b), 0001.

Baumjohann, W., & Treumann, R. A. (2012). *Basic Space Plasma Physics*. Imperial College Press.

Bucha, V. (2014). Geomagnetic activity and the north Atlantic oscillation. *Studia Geophysica Et Geodaetica, 58*(3), 461–472.

Büsching, I., Kopp, A., Pohl, M., Schlickeiser, R., Perrot, C., & Grenier, I. (2005). Cosmic-ray propagation properties for an origin in supernova remnants. *The Astrophysical Journal, 619*(1), 314.

Calisto, M., Usoskin, I., & Rozanov, E. (2013). Influence of a carrington-like eventon the atmospheric chemistry, temperature and dynamics: Revised. *Environmental Research Letters, 8*(4), 045010.

Carslaw, K., Harrison, R., & Kirkby, J. (2002). Cosmic rays, clouds, and climate. *Science, 298*(5599), 1732–1737.

Change, C. (2007). IPCC fourth assessment report. *The Physical Science Basis, 2*, 580–595.

Davis, T. N., & Sugiura, M. (1966). Auroral electrojet activity index AE and its universal time variations. *Journal of Geophysical Research, 71*(3), 785–801.

Dessler, A. J., & Parker, E. N. (1959). Hydromagnetic theory of geomagnetic storms. *Journal of Geophysical Research, 64*(12), 2239–2252.

Dorman, L. I., & Dorman, L. I. (2004). Cosmic ray influence on planetary cloud covering and long term climate change. In *Cosmic Rays in the Earth's Atmosphere and Underground (Norwegian Register for Scientific Journals and Series)* (pp. 591–624).

Eddy, J. A. (1976). The maunder minimum: The reign of louis xiv appears to have been a time of real anomaly in the behavior of the sun. *Science, 192*(4245), 1189–1202.

Gray, L. J., Beer, J., Geller, M., Haigh, J. D., Lockwood, M., Matthes, K., Cubasch, U., Fleitmann, D., Harrison, G., & Hood, L., et al. (2010). Solar influences on climate. *Reviews of Geophysics, 48*(4), 18.

Haigh, J. D., Blackburn, M., & Day, R. (2005). The response of tropospheric circulation to perturbations in lower-stratospheric temperature. *Journal of Climate, 18*(17), 3672–3685.

Hegerl, G. C., Zwiers, F. W., Braconnot, P., Gillett, N. P., Luo, Y., Orsini, J. A. M., Nicholls, N., Penner, J. E., Stott, P. A., & Allen, M., et al. (2007). Understanding and attributing climate change. In *IPCC WGI Fourth Assessment Report*.

Herschel, W. (1801). Xiii. observations tending to investigate the nature of the sun, in order to find the causes or symptoms of its variable emission of light and heat; with remarks on the use that may possibly be drawn from solar observations. *PhilosophicalTransactions of the Royal Society of London, 91*, 265–318.

Hofmann, D. J., Butler, J. H., & Tans, P. P. (2009). A new look at atmospheric carbon dioxide. *Atmospheric Environment, 43*(12), 2084–2086.

Howe, P. D., Boudet, H., Leiserowitz, A., & Maibach, E. W. (2014). Mapping the shadow of experience of extreme weather events. *Climatic Change, 127*(2), 381–389. https://doi.org/10.1007/s10584-014-1253-6

Kivelson, M. G., & Russell, C. T. (1995). *Introduction to Space Physics*. Cambridge University Press.

Kodera, K. (2004). Solar influence on the Indian Ocean monsoon through dynamical processes. *Geophysical Research Letters, 31*(24).

Labitzke, K., & Van Loon, H. (1995). Connection between the troposphere and stratosphere on a decadal scale. *Tellus A, 47*(2), 275–286.

Lean, J. (1997). The sun's variable radiation and its relevance for earth. *Annual Reviewof Astronomy and Astrophysics, 35*(1), 33–67.

Lyon, J. G. (2000). The solar wind-magnetosphere-ionosphere system. *Science, 288*(5473), 1987–1991.
McKie, R. (2006). The cooling sun brings relief to sweltering earth.
Meehl, G. A. (2007). Global climate projections in climate change 2007. In *The Physical Science Basis. Contribution of Working Group I to the Fourth Assessment Report of the IPCC* (pp. 747–846).
Mironova, I., & Usoskin, I. (2014). Possible effect of strong solar energetic particle events on polar stratospheric aerosol: A summary of observational results. *Environmental Research Letters, 9*(1), 015002.
National Academies of Sciences Engineering Medicine, et al. (2016). *Attribution of Extreme Weather Events in the Context of Climate Change*. National Academies Press.
NOAA. Coronal mass ejection. https://www.swpc.noaa.gov/phenomena/coronal-mass-ejections.
Ohlich, C. (2006). Solar irradiance variability since 1978. *Space Science Reviews, 125*, 53–65.
Pachauri, R. K., Reisinger, A., et al. (2007). *IPCC fourth assessment report*. IPCC.
Pittock, A. B. (1978). A critical look at long-term sun-weather relationships. *Reviewsof Geophysics, 16*(3), 400–420.
Seppälä, A., Randall, C., Clilverd, M. A., Rozanov, E., & Rodger, C. (2009). Geomagnetic activity and polar surface air temperature variability. *Journal of Geophysical Research: Space Physics, 114*(A10).
Stott, P. A., Gillett, N. P., Hegerl, G. C., Karoly, D. J., Stone, D. A., Zhang, X., & Zwiers, F. (2010). Detection and attribution of climate change: a regional perspective. *Wiley Interdisciplinary Reviews: Climate Change, 1*(2):192–211.
Streissguth, T., & Mann, M. E. (2011). *Extreme weather*. Greenhaven Press.
Svensmark, H. (2000). Cosmic rays and earth's climate. In *Cosmic Rays and Earth Proceedings of an ISSI Workshop*, 21–26 March 1999, Bern, Switzerland (pp. 175–185). Springer.

Socio-Economic Implications of Severe Storms in the Changing Climatic Scenarios: A Case of South Asia

Jagdish Jadhav, Pramod Kamble, Devesh Sharma, Hemlata Manglani, Gyana Ranjan Panda, and Priya Rajput

Abstract The present study aims to highlight the socio-economic implications of severe storms in a changing climate scenario in South Asia. To explore these impacts, the study first outlined the vulnerability of South Asia to climate change and severe storms. The study further investigated the socio-economic consequences of severe storms and discussed loss in human and infrastructure, implications on water, agriculture, and livestock sector, and issues of migration, resettlement, and rehabilitation. The socio-economic consequences highlighted that countries like Cambodia and Lao People's Democratic Republic are facing an average annual loss of 5–7% of their Gross Domestic Product. Several billion people in countries like India and China live in danger every year due to severe storms like tropical cyclones. Some implications of severe storms are being seen in forest areas and mangrove growth. In countries such as Bhutan, India, China, and Vietnam, forest cover has increased over time, while in countries such as Myanmar, Cambodia, and the Democratic People's Republic of Korea, it has declined. Due to severe storms, the highest number of newly displaced people was seen in South Asia, in which India and Bangladesh had the highest migration. The study also attempts to bring relevance to Traditional Ecological Knowledge (TEK) to Asian indigenous communities to

J. Jadhav (✉)
Department of Social Work, Central University of Rajasthan, NH-08, Bandarsindri, Kishangarh, Ajmer, Rajasthan 305 817, India
e-mail: jagdishjadhav_sw@curaj.ac.in

P. Kamble
Department of Environment Science, Central University of Rajasthan, Ajmer, India

D. Sharma
Department of Atmospheric Sciences, Central University of Rajasthan, Ajmer, India

H. Manglani
Department of Economics, Central University of Rajasthan, Ajmer, India

G. R. Panda
Department of Public Policy, Law and Governance, Central University of Rajasthan, Ajmer, India

P. Rajput
Department of Social Work, Central University of Rajasthan, Ajmer, India

S. Das and W-K. Tao (eds.), *Severe Storms*,
https://doi.org/10.1007/978-981-97-7075-5_25

respond to climate change and severe storms. Although many solutions for disaster management are available in India's ancient historical Maurya period and *Kautilya's Arthashastra*, present-day India became fully involved in disaster mitigation through the Yokohoma Plan, the Hyogo Framework, and the ongoing Sendai Framework. The final part of the study discussed the Disaster Management Act, of 2005, whose last mile of disaster preparedness and management is a product of defined planning, coordination, and implementation, and actual rescue and rehabilitation operations during disaster situations. The study recommends that local communities should consider practicing their indigenous knowledge in disaster risk reduction activities, including prevention, mitigation, preparedness, response, and recovery with support from international, national, and state-level mitigation and preparation strategies.

Keywords Severe storms · South Asia · Socio-economic implications · Traditional Ecological Knowledge (TEK) · Yokohama strategy · Hyogo framework · Sendai framework · Disaster management act 2005

1 Introduction

Climate change is having a severe impact on people's lives all across the world. Climate change is the long-term, widespread shifts in Earth's average temperature and precipitation patterns (NAPCC, 2008; Parikh & Gokam, 1992; Nema et. al., 2012). Extreme weather events like hurricanes, heat waves, and rising sea levels are becoming more often as a result of global warming. The average change in weather conditions such as changes in temperature, humidity, and precipitation are the cause of extreme events (Parikh and Gokam, 1992). In certain instances, ecosystems have been destroyed past the point of no return, irreversibly altering the livelihoods of the populations they support. Loss and Damage are terms used in climate diplomacy to describe these kinds of undesirable outcomes (Woetzel et al., 2020). 'Loss and Damage' refers to the inescapable societal and economic impacts of climate change (Woetzel et al., 2020). Many lives are lost, embankments are destroyed, the land is rendered unproductive and habitats are permanently altered as a result of these environmental changes at the global level.

South Asia region is considered as highest populous region in the world, and also a place for living of two-thirds of the world's poor live. The high statistics of impoverishment along with laggards' human development indices of the region broadly shaped and exacerbated by various factors marred by very frequent disasters mostly driven by extreme climatic events and severe storms. Though the region has adequate natural bases and biodiversity, its usability requires economic capacities which are frequently limited by recurrent occurrences of such events. The South Asian constituent countries are agriculture-dependent and their development capacity in the economic sense is treated as an "underdeveloped" category at par with sub-Saharan Africa leaving many in acute poverty. The region is most exposed to natural disasters and climate-induced vulnerabilities and barring Afghanistan, Nepal, and

Bhutan, other South Asian countries are highly exposed to severe storms such as cyclones, thunderstorms, hailstorms, dust storms, squall lines, tornadoes, cloudbursts, and lightning. The incidences of climate-induced natural disasters are frequent which has damaged the progress of the region which can be observed from the World Bank (2022) that nearly 64% of the world demography residing in the South Asian region are affected by severe floods in one way and other. To read further from the World Bank estimates (2022), the region is bracketed as the most vulnerable and climate-sensitive region as extreme climatic vulnerabilities affect half of the region's demography very often. As per the International Monetary Fund (IMF), 2019 estimates, such high incidences affect the economy of constituent countries and the projected losses in terms of the GDP per capita are higher than the global average. While considering the vulnerability and disaster risk in South Asia, various factors are attributed to its growth. The soaring population density along with increased urbanization, unemployment, poverty, and the rising degree of pollution and degradation of natural resources besides weak governance structure both at the national and local level can be considered as the factor acerbating the causes of vulnerability and disasters in almost all countries in South Asia region. This results in the loss of lives and economy in almost all countries of South Asia. Loss of lives, measured Mortality Risk Index (MRI) (UN, 2022) is highest in Bangladesh, India, Pakistan, and Sri Lanka due to severe storms, and its associated disasters like cyclones, floods, and landslides. Such extent of vulnerabilities and disaster risks have a bearing on the economic losses of those countries.

The Aftermath of Socio-economic Loss from Severe Storms

With a mean temperature of 1.39 °C above the average from 1981 to 2010, the year 2023 was the hottest on record in Asia. According to the World Meteorological Organisation (WMO), the effects of extreme weather and climate change on Asia in 2023 resulted in thousands of fatalities, millions of population displacements, and a hundred billion dollars in asset damage. Verkhoyansk, Russian Federation, had a temperature of 38.0 °C, which is likely the hottest temperature ever recorded thus far north of the Arctic Circle. There were also several other noteworthy heat extremes (WMO, 2021). South Asia had highly vigorous summer monsoons as many people were killed and relocated as a result of floods and landslides caused by this and the more common tropical cyclones (WMO, 2021). A recent multi-agency review organized by the World Meteorological Organisation found that environmental deterioration, increased food and water insecurity, and health problems make sustainable development an unattainable aim.

The ocean's heat content and the temperature of its surface have profound implications on marine life, ocean circulation, and the atmosphere–ocean nexus. Moreover, in 2020, average sea surface temperatures in the Indian, Pacific, as well as Arctic oceans all reached new highs. The Arabian Sea and other areas near Asia are seeing a tripling of their average ocean and sea surface temperatures (WMO, 2021). The quantity of sea ice in the Polar Regions is an important climatic indicator for both the region and the rest of the world. After the summer melt, the minimum ice extent in the Arctic that year (2020) was the second lowest on record since satellite measurements began in 1979. In the summer, there was no ice at all along the Northern Sea Route and the Eurasian shelf sea (WMO, 2021). Since the early 1990s, the average sea level has increased by 3.3 mm annually. Sea levels are rising significantly faster in the North Indian Ocean and the Northwestern Pacific Ocean than the worldwide average (WMO, 2021). Glaciers covering around 100,000 km2 are concentrated in high-mountain Asia, mostly in the Tibetan Plateau and Himalayas. It is the source of ten significant Asian rivers and has the highest concentration of ice outside of the Polar Regions. The existence and incomes of around 750 million people in the region would be impacted by the glacier mass decreasing by 20–40% by 2050 due to the rapid glacial retreat. The global sea level, local dangers (including landslides and avalanches), and regional water cycles are all significantly impacted by this. The expected decrease in glacier run-off has significant ramifications for both water security and ecosystems (WMO, 2021) (Fig. 1).

Climate change is a particularly crucial issue in South Asia due to the upsurge of extreme weather events and climatic variability in recent decades. It is well observed that Asian countries are extremely susceptible to extreme weather events that lead to socio-economic losses and damages due to changing climatic scenarios. Such events as extratropical and tropical cyclones, snow avalanches, intense rainfall, thunderstorms, and prolonged dry spells are expected to cause adverse effects in this part of world Asia. Table 1 shows the distribution of people affected due to storm events in various regions of Asia. Eastern and South Asia are regions with a high number of affected people.

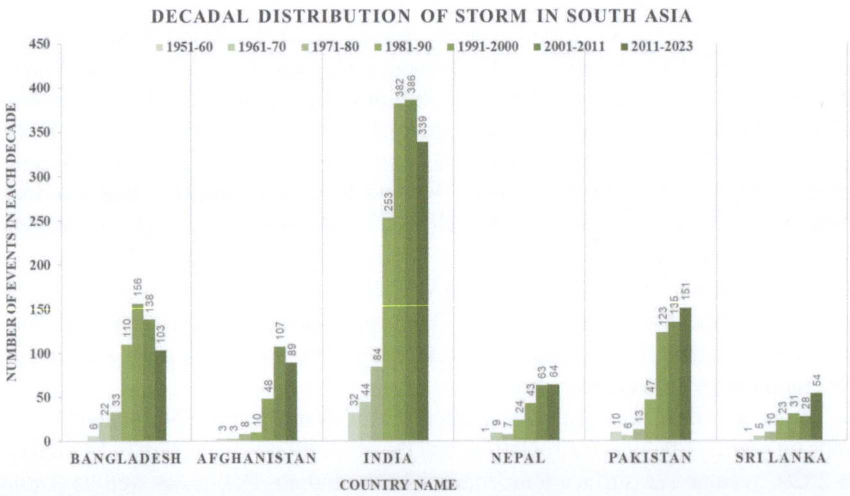

Fig. 1 Decadal distribution of storm events in South Asian countries. *Source* EM-DAT: the OFDA/CRED international disaster database

Table 1 Distribution of people affected in the Asian region due to storm events

Region	Injured	Affected	Deaths	Homeless
Central Asia	3454	10,610,458	4244	153,059
Eastern Asia	2,063,766	3,301,992,892	3,232,454	64,305,661
South-Eastern Asia	1,047,652	544,528,926	512,405	15,873,657
Southern Asia	2,212,008	3,167,211,779	2,789,155	64,593,275
Western Asia	781,931	55,250,279	121,770	2,484,623
Grand total	**6,108,811**	**7,079,594,334**	**6,660,028**	**147,410,275**

Source EM-DAT: the OFDA/CRED international disaster database

Table 2 Distribution of people affected in South Asian countries due to storm events

Country	Deaths	Injured	Homeless	Affected
Afghanistan	28,511	75,671	254,990	34,722,092
Bangladesh	650,696	**1,080,814**	14,462,180	457,266,960
Bhutan	314	30	1000	86,841
India	**1,766,388**	377,000	**34,897,377**	**2,447,215,693**
Maldives	376	2214	36,849	245,502
Nepal	27,651	31,448	258,439	17,493,217
Pakistan	122,058	249,403	9,663,295	122,227,623
Sri Lanka	41,074	204,673	4,528,065	27,779,408

Source EM-DAT: the OFDA/CRED international disaster database

The number of storm events that occurred in various decades, there is increased after the 1980s and continues to increase in recent years. Comparing the data after 1950, more events occurred in India followed by Bangladesh and Pakistan (Table 2).

1.1 Human and Infrastructure Loss

In Asia, many climate and weather-related evacuees are unable to return home, assimilate locally, or settle abroad, resulting in protracted displacement (WMO, 2021). More than 5,000 people were killed and 50 million were displaced throughout Asia due to flooding and storms in 2020 alone. This is proof that early warning systems in many Asian countries have been effective, as it is far lower than the annual average over the past twenty years (158 million people affected and roughly 15,500 fatalities) (WMO, 2021). According to ESCAP, tropical cyclones, floods, and droughts cause an estimated several hundred billion dollars in Average Annual Loss (AAL) per year. About US$ 238 billion in losses were predicted for China, US$ 87 billion for India, and US$ 83 billion for Japan (WMO, 2021). Bhutan is facing a potential loss of 18%, Nepal 13%, India 10%, and Pakistan 10% in 2100. Such extent of vulnerabilities and disaster risks have a bearing on the economic losses of those countries. The following section explains the AAL to disasters as % of GDP in constituent countries (ESCAP, 2020) (Table 3).

The above figures show that most countries are prone to economic losses and the gravity of losses is typically felt in smaller economies than the larger economies like India. Pakistan Bhutan, Nepal, Afghanistan, and Bangladesh are considered to be bearing more economic losses than other countries like Sri Lanka and Maldives. India however still faces 3.34% of the GDP as part of the AAL induced by extreme disasters. The following table explains the sectorial impacts of disasters (in USD million) in certain South Asian countries which are particularly felt in the Economic and Infrastructures sectors, productive sectors, and social sectors. This table indicates that disaster affects both social and economic sectors severely (Table 4).

The effects of global warming, which have been happening for several decades, create dire consequences for human society as well as wild and cultivated ecosystems. Climate change has the potential to severely disrupt agricultural systems, including

Table 3 Annual AAL of GDP in South Asian Countries

Country	Total AAL (Million of USD)	Total AAL (% of GDP)
India	87,617.60	3.34
Pakistan	16,029.40	6.77
Bhutan	163.95	6.92
Nepal	1663.35	7.65
Bangladesh	11,443.90	6.40
Afghanistan	1270.81	5.73
Srilanka	1704.01	2.06
Maldives	22.28	0.56

Source Asia Pacific disaster report (2019:1) ESCAP calculation based on probabilistic risk assessment

Table 4 Sectoral impacts of disasters in South Asia (in million USD)

Country	Broad prevalent disasters	Economic & infrastructure	Productive sectors	Social sectors
India	Floods, cyclones, & storms	6650.00	513.00	1418.95
Pakistan	Floods, cyclones & storms	1658.00	3631.00	1603.5
Nepal	Floods	710.2	890.65	2202.7
Bangladesh	Floods, cyclones & storms	230.7	513.0	1418.95
Srilanka	Landslides, floods	124.68	98.76	229.93

Source Asia Pacific disaster report (2019)

natural resources like soil and water, and their capacity to produce food, feed, fiber, and energy while also providing essential ecosystem services like water to people and a home for crop landraces, wild relatives, and pollinators (Nema et al., 2012). The costs associated with climate change can be both monetary and intangible. For example, when extreme heatwaves in India reduce wheat production, it can have a devastating effect on the income of many farmers and hence on the national economy as a whole. Non-economic loss and harm from climate change may include the loss of cultural traditions, indigenous knowledge, biodiversity, and ecological services (Köberle et al., 2021). More frequent and severe floods in South Asia are a potential outcome of climate change that might lead to the destruction of houses and other buildings (Woetzel et al., 2020). With global temperatures expected to rise, Asia particularly South Asia will be among the most hit areas.

1.2 Implications for the Water, Agriculture, and Livestock Sector

There has been a slowdown in efforts to improve food security and nutritional value. Estimates for 2020 put the number of undernourished individuals in Southeast Asia at 48.8 million, South Asia at 305.7 million and West Asia at 42.3 million. More than half of the world's total may be traced back to Asia (WMO, 2021). Water security and ecosystems in Asia face serious threats from melting glaciers and diminishing fresh-water resources, while declining coral reefs threaten food security (WMO, 2021). In 2019, over 74% of Asia's mangroves were found in Bangladesh (24%), Myanmar (19%), India (17%), and Thailand (14%). Bangladesh is a low-lying state that is vulnerable to tropical storms, and mangroves there have declined by 19% between 1992 and 2019 (WMO, 2021). However, mangroves are threatened by rising sea levels and ocean temperatures due to human activity, as well as by changes in precipitation and storm patterns.

Forests act as a "carbon sink," absorbing excess CO2 from the atmosphere. Bhutan, China, India, and Vietnam all had increases in forest cover between 1990 and 2018. However, there was a decrease in forest cover of 26% in Myanmar, 24% in Cambodia, and 12% in the Democratic People's Republic of Korea (WMO, 2021). Sand and dust storms are dangerous to people, animals, and crops. In the Arabian Peninsula, the broader Middle East, and Central Asia, the frequency and intensity of dust storms are on the rise due to land use land-cover changes, and climate-related variables. However, in 2020, the number of undernourished individuals in South Asia rose by 20%, whereas it increased by just 6% in Southeast Asia and West Asia. The issue was made worse by climate-related calamities (WMO, 2021).

1.3 Issue of Migration, Resettlement, and Rehabilitation

Displacement of people due to disasters and other risks is unabated and increasing with time. Ginnetti et al. (2015) write that the displacement figures in South Asia stand at 36% of the global displacement and this figure has remained unchanged since the 1970s. More people are displaced from their homes as a result of natural disasters than all the displacement brought on by war and violence combined. At least 23.7 million people were displaced in 104 countries and territories due to natural disasters in 2021, which is three times as many as the number of displaced individuals due to war and violence (see Fig. 2). East Asia and the Pacific, including South Asia, witnessed the highest number of newly displaced people due to natural disasters in 2020, with 4.9 million people forced to relocate in India, 780,000 in Vietnam, 749,000 in Indonesia, 1.27 million in Indonesia, and 99,000 in Bangladesh (IDMC, 2022). The entire region was hit by tropical cyclones, monsoon rains, and floods, but the largest displacement figures were recorded in Bangladesh.

From 9,500 in 2015 to an estimated 1.27 million in 2020, the annual number of Vietnamese people forced to relocate wildly fluctuates Mega-scale disasters are a major contributor to this variation since they are rare and unpredictable but have the potential to uproot large populations at once. For instance, in May 2020, Cyclone Amphan caused approximately five million people in Bangladesh, India, Myanmar, and Bhutan to flee their homes, making it the worst disaster-related displacement event on a worldwide scale that year. The variation has significant consequences for the planning and management of displacement, including the implementation of long-term solutions that prioritize the strength of vulnerable populations and the preparedness of shelters and schools in destination locations (UNESCO & UNU-IAS, 2023; IDMC, 2022). According to the UN Secretary-General's High-Level Panel

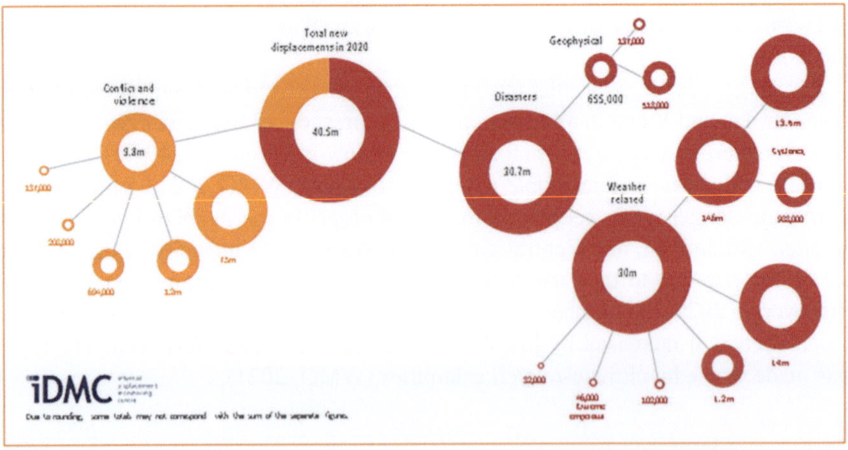

Fig. 2 Displacement in 2020: conflict & violence versus disasters & severe storms. *Source* IDMC (2022)

on Internal Displacement, "durable solutions" are achieved when people who have been uprooted within their own country no longer require their uprooting services and when they can exercise their rights without facing any prejudice as a result of their uprooting (UNESCO & UNU-IAS, 2023; IDMC, 2022)" (Table 5).

Climatic changes have suddenly appeared as a danger to biodiversity. Greenhouse gases including carbon dioxide, methane, CFCs, nitrogen oxides, and sulfur dioxide account for the vast bulk of the Earth's accumulated heat. From genes to species to communities to ecosystems, global warming may have far-reaching implications on all levels of biodiversity (Bellard et al., 2012; Parmesan, 2006). Positive and negative biological responses at the level of species and populations to climate change have been reported. Negative effects have been linked to declines in species populations on both local and global scales (Miller-Rushing et al., 2010; Montoya and Raffaelli, 2010; Dawson et al., 2011; Geyer et al., 2011), while positive effects have been associated with a noticeable increase in growth rate on both scales. Furthermore, many species are altering their seasonal migratory patterns due to climatic changes.

According to IUCN (2012) due to climatic changes, 33% of species are at risk of habitat changes and shifts, 29% due to extreme temperature, and 28% due to droughts (see Fig. 3). Similarly, complications arise due to flooding and storms (Staudinger et al., 2013). In May 2020, one of the strongest cyclones on record, Cyclone Amphan, struck the Sundarbans region between India and Bangladesh, forcing 2.4 million Indians and 2.5 million Bangladeshis to flee their homes (WMO, 2021). Panda (2010) estimated that climate change will lead to two different kinds of displacement in India; one within the country's territory due to an increase in drought, desertification, and water and food scarcity and the latter may attract migrants from the adjoining countries to cope with climate change (Gupta & Sharma, 2021; Chaturvedi and Doyle, 2010). For ages, India has acted as a refuge to many neighboring countries, even without being a signatory member of the international treaties on refugees

Table 5 Disaster affected vulnerable population

Year	Bangladesh	India	Indonesia	Tuvalu	Vietnam
2010	569,000	1,410,000	230	–	422,000
2011	4,000	1,430,000	2,000	–	230,000
2012	651,000	9,110,000	103,000	–	15,000
2013	1,160,000	2,090,000	316,000	–	1,040,000
2014	543,000	3,430,000	764,000	–	68,000
2015	531,000	3,530,000	169,000	5,430	9,550
2016	614,000	2,390,000	1,130,000	–	80,500
2017	939,000	1,350,000	199,000	–	633,000
2018	77,800	2,680,000	81,100	–	143,000
2019	4,090,000	5,020,000	168,000	–	88,700
2020	4,440,000	3,860,000	705,000	400	1,270,000

Source UNESCO & UNU-IAS (2023)

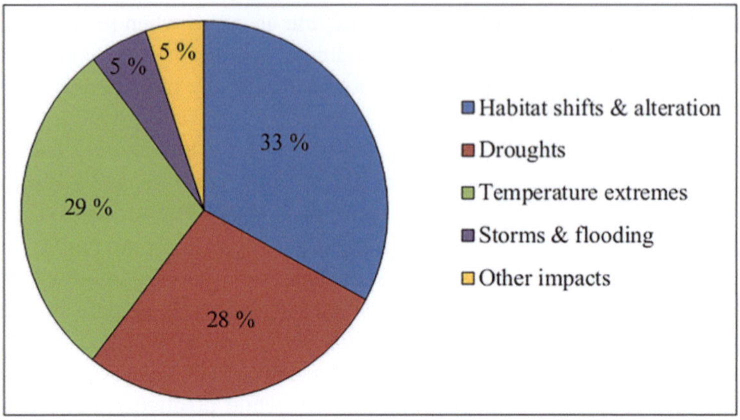

Fig. 3 Distribution of climate-related threats. *Source* IUCN, 2012. *Note* There can be overlays in species calculated amongst groups (IUCN, 2012) (Staudinger et al., 2013)

(Sawant and Sanjeev, 2022). However, the nation does acknowledge the right to refuge as provided for in Article 14 (1) of the UDHR.

2 Relevance of Traditional Ecological Knowledge (TEK)

For ages, the indigenous communities and local communities have been trying to escape the severity of such climatic changes and severe storms using TEK. This section attempts to bring the relevance of TEK as the first part underlined the concept of TEK and its controversies and the second part highlights TEK among South Asian indigenous communities to combat climate change and severe storms.

2.1 *Traditional Ecological Knowledge: Concept and Contradictions*

Traditional Ecological Knowledge (TEK) is known as the knowledge as well as experiences of different communities developed through their interaction with the natural environs. It is considered to be of great significance to attain the goal of Sustainable Development. TEK refers to the information and understanding gained via an in-depth study of a region that has been passed down orally (Huntington, 2000). Johannes, 1989 characterized TEK as the "environmental wisdom of traditional cultures". The knowledge among traditional people that aims to recognize the flow of ecosystems and live their lives while keeping a constructive relationship with it is called TEK (Freeman, 1992).

Hunn, 1993 argued the prominence of TEK as it is 'traditional', 'an ecological knowledge', and 'a knowledge'. According to him, just like traditions, TEK is developed over time in particular locations as adaptations in response to new demands and environmental changes, and it is the tried and true means for survival. TEK as an ecological knowledge includes numerous experiences of interaction with the local flora and fauna; their habitat and the ecology as a whole. As knowledge, TEK is not just an expression of its materialistic goods but a set of values and belief systems that prevails and is passed on orally to the next generations.

Contradictions exist between the foundations of TEK and other fields of knowledge, including Traditional Knowledge (TK), Indigenous Knowledge (IK), and its relevance to modern science. Although TEK is frequently associated with indigenous peoples, other people think it extends beyond them. Ballard (2001) has given the distinction between IK and TEK as an umbrella term that presents the understanding developed by different indigenous communities which are given different names and words according to their distinct culture and tradition. Whereas, TEK is considered to be part of a larger IK in which prominence is given to ecological knowledge (Dudgeon and Berkes, 2003). According to Kim et al. (2017), IK and TEK must be viewed as two distinct fields of knowledge while keeping in mind that TEK has its roots in IK.

The acceptance of TEK by scientists is still up for debate. Berkes (1993) differentiated between ecological knowledge and scientific knowledge as TEK is qualitative, intuitive, holistic, a mixture of mind and matter, moral, spiritual, observational, fact-finding through trial and error, and data generated through resource users over a long period. Haraway (2013) considered TEK as situated as it is the knowledge that lies within a specific geographical area or a community. Modern scientists consider TEK as a 'body of knowledge', whereas, for the indigenous people it is action-oriented i.e. to live according to that knowledge (McGregor, 2004). However, there is no denying the importance of TEK in comprehending, evaluating, and resolving the issues of the contemporary world (McGregor, 2005). TEK must be recognized by the scientific body of knowledge as it worked for people and communities for ages (Pierotti and Wildcat, 2000). McGregor, 2008 emphasized the need to link TEK and modern science to combating global environmental challenges.

2.2 Application of TEK in Combating and Adapting to Severe Storms

Many societies and communities around the world are utilizing TEK to fulfill their daily needs. It made them self-sufficient, overcame ecological severities, and pursued their customary modes of subsistence. It is necessary to pursue TEK in a localized manner since various cultures and societies employ various strategies to address similar issues. As a result, there is more interest in using TEK by indigenous people to fight climate change locally (Hosen et al., 2020; Kupika et al., 2019). Climate change detection, dealing with its repercussions, and adapting as a result demonstrate how TEK could promote global adaptation measures. This would support SDG 13's goal to combat climate change and SDG 15's goal to improve land life.

TEK is a significant way in which indigenous people exhibit their culture and is intrinsically tied to indigenous customs. The values they hold are what inform and guide their understanding of the effects of climate change on their communities and the world at large (Adger et al., 2013; Hosen et al., 2020). Because of this, indigenous people's TEK reactions to local-level climate changes provide major contributions to adaptation (Hosen et al., 2020), even though they are among the most negatively impacted by climate change. Rooted in an intimate familiarity with the land, indigenous communities are better able to weather environmental change and uncertainty because of their TEK (Hosen et al., 2020; Scaddan, 2008). Indigenous people make up less than five percent of the global population (Hosen et al., 2020; Ramos-Castillo et al., 2017), but they control about a quarter of the Earth's land area, which includes many of the planet's biodiversity hotspots (Hosen et al., 2020; Schuster et al., 2019). Therefore, they are in a prime position to protect the landscapes and ecosystems that are crucial for adapting to climate change (Hosen et al., 2020; Huang et al., 2018). Despite the significance of TEK to effective climate change-induced adaptation, it is often one of the most poorly understood parts of current efforts to deal with climate change (Hosen et al., 2020; Nalau et al., 2018) and receives huge criticism from natural sciences. Even with the speed with which globalization is progressing, indigenous peoples have proven time and again that they can sustainably co-evolve with the natural environment and offer ecological services. Institutional and societal prejudice has turned these groups into victims (Hosen et al., 2020; Salick and Byg, 2007).

Sacred groves are one of the best example of the conservation of natural resources exhibited by the indigenous people. Mass cultivation of medicinal plants could be the another best model for the same. The indigenous people coordinated their subsistence activities including farming, minor hunting, and fishing by monitoring and anticipating changes in the weather and other environmental factors. Traditional methods of weather prediction involved keeping an eye on things like sky color, temperature, moon phases, ripening of fruit trees, and animal movement (Hosen et al., 2020). According to the Lun Bawang, for instance, a crimson sunset sky indicates dry weather with no precipitation the next day. Since harvesting in the rain might cause crop damage and a subsequent reduction in productivity, however, this weather is excellent for gathering their lati'ba'. The Sa'ban and Penan, who grow upland rice, split the year into the "wet season" (October to February) and the "dry season" (March to September), with the former typified by considerable precipitation and the latter by drier conditions and higher temperatures. Upland rice farming relies significantly on the two seasons for optimal agronomic results. Land preparation is best done during the dry season, while strong rains during the wet season create ideal growing conditions for the crop. The Sa'ban claimed to have used moon phases as a planting calendar. They wait until the "bliin teng," also known as the "first quarter moon," appears in the sky on the night of the eighth month to plant their seeds. According to the scientific community, the first quarter moon is the wettest and most fruitful time of the month because of the influence of the tides on underground water supplies (Hosen et al., 2020; Harris & Summers, 2002).

3 Disaster Risk Reduction Policy: International and National Mechanisms

Since recorded human history, different severe storms like lightning strikes, tornadoes, straight-line winds, flash floods, floods, and hailstorms affect the natural and human-made environment. Considering the growing risk of different natural and human-made disasters, especially severe storms, nations are evolving with strong policy mechanisms backed with practical frameworks for preparedness, mitigation, and effective management. According to the United Nations International Strategy for Disaster Reduction (UNISDR), 2009, Disaster Risk Reduction is defined as:

> action taken to reduce the risk of disasters and the adverse impacts of natural hazards, through systematic efforts to analyze and manage the causes of disasters, including through avoidance of hazards, reduced social and economic vulnerability to hazards, and improved preparedness for adverse events.

This section aims to highlight the Disaster Risk Reduction Policies and Frameworks at the International and National levels with a special focus on severe storms.

3.1 Frameworks at the International Level

At the international level, some of the major steps taken for Disaster Risk Reduction can be seen as the Yokohama Strategy, 1994; The Hyogo Framework, 2005; and the Sendai Framework, 2015 (see Fig. 4). These international frameworks are comprised of certain direct and indirect points meant to protect human society, the natural environment, and the infrastructure in different ways against disasters.

The Yokohama Strategy for Action, 1994:

United Nations Member States met in May 1994, at the World Conference on Natural Disaster Risk Reduction in Yokohama, Japan. The conference was held

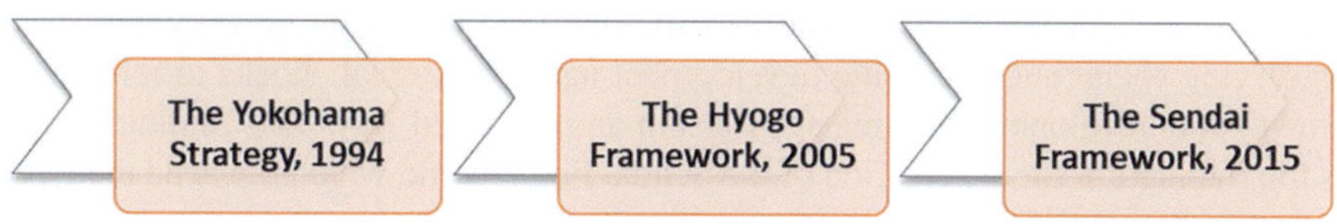

Fig. 4 International disaster risk reduction frameworks. *Source* Authors, 2023

to assess the progress made by the International Decade on Natural Disaster Reduction (IDNDR). The Yokohama Strategy emerged as the outcome of this conference. The strategy is also known as 'The Yokohama Strategy and Plan of Action for a Safer World' which provides guidelines for natural disaster prevention, preparedness, and mitigation. It includes 10 principles that cover: 'disaster assessment'; 'disaster prevention and preparedness'; 'development of strength and capabilities to prevent, reduce, and mitigate the disaster'; 'effective dissemination of early warning systems'; 'local level participation for prevention'; 'educating the communities'; 'sharing of technology'; 'merging environment sustainability with poverty alleviation'; and 'expressing strong policy determination for protecting the people and the national assets'. It also suggested a plan of action that suggested the member states shift from a 'relief-oriented approach' to 'mitigation and prevention.'

However, this strategy remained deficient in fulfilling its anticipated results as it was inefficient in dealing with the multiple aftereffects of disasters. Its analysis revealed that it lags in governance, risk identification, assessment and monitoring, knowledge management, reduction of risk factors, and preparedness for effective response and recovery (UNISDR, 2005). These gaps were further addressed in 2005 at the World Conference on Disaster Reduction in Kobe, Hyogo, Japan.

The Hyogo Framework for Action, 2005:

The Hyogo Framework for Action (HFA) was developed in 2005 to provide a more thorough approach for disaster management to emerge as a 'result-oriented plan of action' for 168 countries. These socioeconomic, and environmental strategies were planned for the next decade (2005–2015). HFA had outlined five key principles as considering disaster risk reduction as a national priority for enriched institutional implementation; identifying, assessing, and monitoring risks while refining early warning systems; nurturing a culture of safety and resilience through knowledge, innovation, and education; dropping the core risks; and improving disaster preparedness at all levels.

The Framework came out well in developing desired institutions, policies, and legislations on disaster risk reduction. However, the majority of nations have made insignificant progress in managing the fundamental disaster risk drivers. The integration of catastrophe risk factors into development decisions was generally not made solely by the institutional or policy frameworks. As a result, risk generation outpaced risk reduction in both higher- and lower-income countries, where vulnerability expanded more quickly than risk exposure.

After the HFA implementation, Member States acknowledged that their efforts had not resulted in reducing the material losses and financial repercussions. They came to the decision that national and international emphasis must switch from preserving social and economic development against external shocks to reforming growth and development to manage risks in an integrated way. This finding served as the foundation for the creation of the Sendai Framework, which replaced the HFA and was adopted in 2015.

The Sendai Framework for Action, 2015–2030:

During the Third National Conference on Disaster Risk Reduction in Sendai, Japan in 2015, the Sendai Framework for Action, 2015 was adopted. The framework set major priorities: build an understanding of disaster risk, strengthen disaster risk governance, investment in disaster risk reduction, and investment in recovery, rehabilitation, and recovery to 'build back better'. The Sendai Framework anticipates a significant decrease in the risk of disasters, as well as losses to lives, livelihoods, and health, as well as to the economic, physical, social, cultural, and environmental assets of people, businesses, communities, and nations (UNISDR, 2015a, 2015b).

The Sendai Framework outlines seven global targets a substantial reduction in global disaster mortality; a substantial reduction in the numbers of affected people; a reduction in economic losses with global GDP; a substantial reduction in disaster damage to critical infrastructure and a disruption of basic services, including health and education facilities; an increase in the number of countries with national and local disaster risk reduction strategies by 2020; enhanced international cooperation for developing countries; and increased access to multi-hazard early warning systems and disaster risk information and assessments (Wahlström, 2015). In the attempt to analyze the holistic approach and appropriateness of the Sendai Framework, its relationship with different dimensions like health, climate change, disability, gender, and livelihood are necessary to be examined. A variety of action plans, some of which are presented in Table 6, have also been developed for combating natural disasters which include various severe storms at the local and national levels.

Table 6 Plan of action at national and local levels

S. No.	Plan of action	Description
1	Framing of legislation & implementation plans	At various levels of government, every nation must create laws, policies, and programs to lessen the effects of natural disasters, especially severe storms. To address this, committees at the national and local levels must be included in the preparation of disaster risk assessment, mitigation, and emergency plans to foster cooperation and increase community resilience
2	Disaster resistant development	Development projects, such as the construction of infrastructure, should be undertaken with an evaluation of environmental and disaster risk
3	Public awareness, recognition, involvement & empowerment	Programs aimed at educating, informing, and fostering public opinion must be planned with increased media attention. It is important to give credit to the traditional knowledge, practices, and attitudes of the local communities to tackle disaster risks. Enhancing marginalized groups' empowerment and participation in decision-making is also vital

Source Authors, 2023

3.2 *Frameworks at the National Level*

India is one of the nations with the highest risk of severe storms and natural disasters including cyclones, floods, droughts, landslides, and earthquakes. India is located in the North Indian Ocean (NIO) Basin and has a long coastline of approximately 7,516 km of flat coastal terrain, a shallow continental shelf, and a high population density in its coastal areas (Barman et al., 2016). Severe tropical cyclones often strike 5,700 km of India's coastline, which includes 84 coastal districts that account for around 8% of the country's total area over 13 Coastal States and Union Territories (Minville, 2022). Because of the storm tidal effect in the region, Tropical Cyclones (TCs) have a comparatively greater destructive impact on India's east coast even though their frequency is the lowest in the world (7% of all TCs) in the North Indian Ocean (NIO), which includes the Bay of Bengal and the Arabian Sea. However, India has a long history of dealing with severe storms and natural disasters. Its disaster risk reduction strategies can be traced back to the Mauryan period. Kautlaya, in 'Arthashastra' (book on economics, politics, administration, etc.) has mentioned natural disasters and human-made disasters and the mechanisms to deal with them (Pohekar, 2023). In the current scenario, India is becoming more careful about disasters, subject to the international forums. With the Declaration of International Decade for Disasters, in 1989, the country realized the significance of Disaster Risk Reduction. India thereafter got fully involved in disaster mitigation via the Yokohoma plan, the Hyogo framework, and the ongoing Sendai framework. With a federal system of governance, India has defined responsibilities for both the state and the federal government. Before this, neither the central nor the state government had passed any laws addressing the management of disasters of all kinds in a comprehensive way (NDMD, 2002).

4 Disaster Management Act, 2005

In the year 2005, the Parliament of India enacted the Disaster Management Act, 2005 as a fundamental legislation. The Act lays out a thorough framework within which the state, district, and local level bodies are established and officials are designated to carry out their assigned tasks and responsibilities in the management of disasters. It not only specifies the national agencies and functionaries along with their powers and functions. The Act, as a whole, aims to establish a top-down model of disaster management in India by enforcing the central body in the formulation and implementation of the nation's disaster management policies and programs (Pandey, 2016).

The National Disaster Management Authority (NDMA) is set up at the highest level to coordinate "the enforcement and implementation of the policy and plan for disaster management." The National Executive Committee (NEC), which serves as the NDMA's executive branch, is responsible for, among other things, supporting "the

National Authority in the discharge of its functions and having the responsibility for implementing the National Authority's policies and plans and ensuring compliance with directives issued by the Central Government for disaster management in the country." The Act also states that every state must have its State Disaster Management Authority (SDMA) and State Executive Committee (SEC) to assist SDMA. This hierarchy further reaches to last mile. As per the Disaster Management Act, of 2005, the five levels of the Disaster Management Authority are as:

- NDMA: National Disaster Management Authority is headed by the Prime Minister.
- SDMA: State Disaster Management Authority is headed by the Chief Minister.
- DDMA: District Disaster Management Authority is headed by the Collector.

All these three authorities are responsible for providing precautionary warnings regarding disasters. They also act as planning, coordinating, and implementing bodies.

- TDMA: Tehsil Disaster Management Authority is headed by the Tehsildar.
- VLCC: Village Level Committee Chairperson, Sarpanch as the head.

These two authorities do the actual rescue and Rehabilitation work during the disaster situation. Therefore, the Disaster Management Act, of 2005 came out efficient for building institutions and mechanisms for disaster risk reduction in India with the inclusion of all stakeholders.

Especially in the context of severe storms different government agencies and organizations in India are contributing to its mitigation and preparedness. In the case of cyclones, the National Cyclone Risk Mitigation Project (NCRMP) was initiated in 2011 to focus on coastal states that are prone to cyclones to lessen the states' likelihood of cyclones and other hydro-meteorological hazards that affect coastal communities, as well as to improve state entities' ability to prepare for and respond to disasters (Lead, 2013). As part of the NDMA's multi-hazard mitigation program, it aims to develop a National Disaster Management communication network and to better comprehend various hazards. Early Warning Dissemination Systems, saline walls, underground electrical cabling, linking highways and bridges, and cyclone shelters are examples of some of the infrastructural benefits. A Hazard, Vulnerability, and Risk Assessment (HVRA) study was also carried out as part of the NCRMP Phase I, leading to the development of a Web-based Composite Risk Atlas (Web-CRA) using deterministic hazard and vulnerability analysis modeling ("ibid"). This Web-based tool, which can identify hotspots of high-vulnerability coastal areas for communities at risk, was commissioned in 2017. In its continuation, a Web-based Dynamic Composite Risk Atlas & Decision Support System (Web-DCRA & DSS) Tool for forecasting cyclone impacts over all four coastal states has been created under NCRMP Phase II housed at IMD, New Delhi. The Impact-based forecast represents a novel method for predicting "what the weather will do" in addition to "what the weather will be" (Mohapatra, et al., 2023). Disaster management authorities in India's cyclone-prone States and Union Territories (UTs) can use this information for proactive mitigation planning, including evacuation, contingency, and routing plans, as well as for allocating resources, including NDRF, SDRF, Fire and emergency services, and community volunteers/task forces, for dynamic response in real-time during a cyclonic event (Kumari, 2023).

To mitigate the impacts of a lightning strike IS 2309 Indian Standard: The Lightning-Code of Practise (Reaffirmed in 2005) defines the basic technical elements of lightning while emphasizing its principal implications (Trotsenko et al., 2021). It advises on proper material selection and acceptable engineering practices for specific scenarios such as establishments that contain explosives and temporary construction platforms. The Lightning Awareness Cell, established in June 2005 under the auspices of the Regional Energy Centre, Kerala promotes lightning protection awareness for local and technical people. Similarly, The Energy and Resources Institute (TERI) established the Lightning Awareness Cell in Guwahati in August 2005 and promoted lightning awareness among students, teachers, village panchayat members, and communities around the country. The Centre for Innovation in Science & Social Action (CISSA), Thiruvananthapuram, Kerala has collaborated with The Lightning Awareness Cell, Regional Energy Centre, Kerala, and South Asian Lightning Information and Resource Centre to conduct lightning protection workshops for various stakeholders.

Following the severe floods of 1954, the Government of India launched many initiatives and formed several committees to explore the country's flood concerns. The critical phases are as follows: (i) 'Policy Statement 1954' (ii) 'Flood High-Level Committee—1957'(iii) 'Policy declaration from 1958, Ministerial Committee on Flood Control—1964' (iv) 'Ministerial Committee on Floods and Relief—1972' (v) 'Rashtriya Barh Ayog (RBA) Working Group on Flood Control for Five-Year Plans—1980 National Water Policy (1987) National Commission for Integrated Water Resource Development Plan (1996)' (viii) 'National Water Policy (2002 revision)' 'National Disaster Management Authority (NDMA), 2005' and 'National Disaster Management Guidelines' (NDMG), 2008 (Sharma, 2003; Singh, 2004; Mohapatra & Singh, 2003; Goswami, 2003; Alam & Muzzammil, 2011). Additionally, the Government of India created the National Institute of Disaster Management (NIDM) as the principal institute for disaster management capacity building under the Disaster Management Act, of 2005 (Das, 2012). NIDM aims to strengthen disaster preventive and preparedness capability at all levels to establish a disaster-resilient India (Gupta, 2018). NIDM has been entrusted with major duties in human resource development, capacity building, training, research, documentation, and policy advocacy (Goel, 2007). NIDM offers technical assistance to state governments through the Disaster Management Centres (DMCs) of the Administrative Training Institutes of the States and Union Territories (ATI) (Parkash & Roy, 2014) and presently funds up to 30 of these facilities. Floods, earthquakes, cyclones, droughts, landslides, and industrial catastrophes are among the six being established as centers of competence in specialized risk management (Gupta, 2018; Verma et al., 2022).

Figure 5 represents the integrated wheel for disaster risk reduction. This integrated approach is highly focused on various dimensions like social, institutional, economic, environmental, and technological. There is a need to develop interdisciplinary approaches to have better solutions for disaster risk reduction. It is also

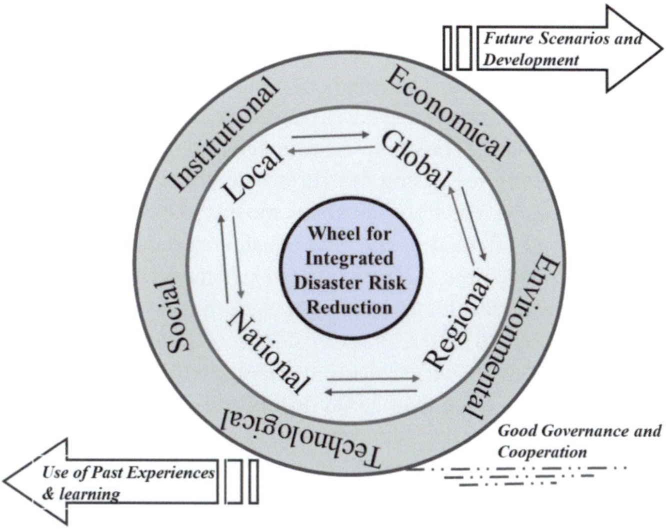

Fig. 5 Wheel for integrated disaster risk reduction. *Source* Authors, 2023

important to learn from past experiences and observations which can be used for future development. As a result, planning and approach development must take into account historical occurrences, learning, and potential future situations. Nowadays, there is a better understanding of future predictions with support of technological and computation support. Still, there is a scope to develop the local and national capacity of stakeholders to use the available information and disseminate it to the grassroots level with the support of grassroots institutions and NGOs. However, the smooth movement of the proposed wheel is highly dependent on the good governance and cooperation approach for any kind of disaster risk reduction including severe storms.

In the milieu of climate change and severe storms, "mitigation" and "adaptation" are two important areas to further move forward in the changing scenarios, severe storms, and for the sustainable development of society. The first is to look into further ways that technical advancement and intervention can be used to decrease greenhouse gas sources or improve sinks. Secondly, adaptation to act for existing or project climatic stimuli or their effects with the engagement of social scientists to ensure the last mile support.

5 Discussion

The effects of climate change and severe storms are evident everywhere, but they tend to be more severe in South Asia. As a consequence, it has resulted in "Loss and Damage", which are permanent effects on society and the economy. Due to its dense population, rapid urbanization, extreme poverty, and inadequate political systems, South Asia—which is home to two-thirds of the world's impoverished—faces many difficulties. Thousands of people lost their lives, millions were displaced, and significant asset damage was caused by the aftermath of severe storms from the past few decades. Threats to marine ecosystems and water security are significant due to melting ice, rising sea surface temperatures, and glacier retreat in high-mountain Asia. Severe storm incidents mostly strike South Asian countries, resulting in millions of affected people and considerable deaths. Since the 1980s, the frequency of these occurrences has increased substantially, increasing the socio-economic costs. The available losses and damages show how urgently South Asia has to address climate change to lessen its severe effects on people's lives, the economy, and the environment. Asia is experiencing increasing effects of climate change, as demonstrated in 2020 by more than 5,000 deaths and 50 million displaced people as a result of severe storms. The Average Annual Loss (AAL) of the economic cost is estimated to be in the hundreds of billions of dollars, with smaller countries in South Asia carrying a disproportionate share of the economic loss. Sectorial effects on infrastructure, agricultural, and social aspects highlight the complex aftereffects of climate-related disasters and severe storms beyond their economic implications.

Indigenous communities across the world find great value in Traditional Ecological Knowledge (TEK), which provides sustainable subsistence methods, ecological resilience, and self-sufficiency. The localized use of TEK serves as an example of adaptation to climate change, addressing common difficulties with a variety of approaches. Even though they make up fewer than 5% of the world's population, indigenous peoples care for 25% of the planet's land, which is essential for biodiversity. With its roots in traditional beliefs and a close relationship to the land, TEK enables indigenous groups to effectively manage environmental shifts. Sacred groves, a prime example of resource preservation, demonstrate the effectiveness of native methods. Institutional and social prejudice frequently marginalizes indigenous efforts, despite their shown capacity for sustainable co-evolution. This prevents TEK from receiving wider recognition and from being integrated into climate change measures. The present also work explores India's diverse strategy for reducing the impact of and being prepared for major storms, explaining programs like the National Cyclone Risk Mitigation Project and the National Disaster Management Communication Network. Along with highlighting initiatives in flood management and lightning protection, it also traces the development of committees and policies addressing these issues. The National Institute of Disaster Management's founding and its function in training and preparing for disasters are outlined. A comprehensive strategy that takes into account social, institutional, economic, environmental, and technical aspects is emphasized by the integrated wheel for disaster risk reduction. It emphasizes how crucial it is to use multidisciplinary methods, draw lessons from the past, and develop national and local skills for stakeholders. The need for mitigation and adaptation approaches in the face of severe storms and climate change is also emphasized.

6 Conclusion

The lack of resources in South Asia, as well as poor aid and emergency services, add to the negative health effects due to the occurrence of frequent extreme events in the region. Climate change has increased the frequency and severity of extreme events and there need to put extra effort into strengthening the community capacity, improvement in emergency services, and financial and medical aid to deal with this risk due to extreme events. Disaster mitigating plans should have measures to minimize loss of life and damage to property and other economic losses. There is also a great need for an international register of disaster case studies so that everyone can not only learn from their mistakes but also become aware of the positive aspects of dealing with particular incidents/disasters. An early warning system is not enough to deal with a tsunami. Disaster management plans are effective only when there is good coordination between grassroots people with support from higher organizations. It is important to educate the population about the dangers of extreme events like floods, cyclones, landslides, etc. Through the use of Indigenous knowledge, people can deal with a variety of hazards and disasters before, during and post disasters through their active participation and support during the events. With the use of better infrastructure and technologies, there is a high possibility to reduce the damage (social and financial) at various levels. There is also a need for a good network system for data observations to be further used in model simulation and forecasting. Data management and sharing with different organizations can play an important role in disaster management. It is recommended that disaster risk reduction stakeholders should consider integrating their knowledge of different disciplines and transforming local communities. Overall, there is a need for the conservation of traditional knowledge along with the application of recent technologies to manage disasters that are occurring frequently due to extreme events. Policies regarding disaster management need to be implemented through a well-managed plan to achieve sustainable development for local people.

References

Adger, W. N., Barnett, J., Brown, K., Marshall, N., & O'brien, K. (2013). Cultural dimensions of climate change impacts and adaptation. *Nature Climate Change, 3*(2), 112–117.

Alam, J., & Muzzammil, M. (2011). Flood disaster preparedness in Indian scenario. *International Journal of Recent Trends in Engineering & Technology, 5*(3), 34–38.

Ballard, M. (2001). *Flooding sustainable livelihoods of the Lake St. Martin First Nation: The need to enhance the role of gender and language in Anishinaabe knowledge systems.* (Unpublished Doctoral Dissertation). University of Manitoba.

Barman, N. K., Chatterjee, S., & Paul, A. K. (2016). *Coastal morphodynamics: Integrated spatial modeling on the deltaic Balasore coast*. Springer.

Bellard, C., Bertelsmeier, C., Leadley, P., Thuiller, W., & Courchamp, F. (2012). Impacts of climate change on the future of biodiversity. *Ecology Letters, 15*, 365–377. https://doi.org/10.1111/j.1461-0248.2011.01736.x

Berkes, F. (1993). Traditional ecological knowledge in perspective. *Traditional Ecological Knowledge: Concepts and Cases, 1*, 1–9.

Chaturvedi, S., & Doyle, T. (2010). Geopolitics of fear and the emergence of 'climate refugees': Imaginative geographies of climate change and displacements in Bangladesh. *Journal of the Indian Ocean Region, 6*(2), 206–222.

Das, P. (2012). Disaster management in India: Policy review and institutional structure. *Asiapac Journal of Socical Sciences, 4*, 37–52.

Dawson, T. P., Jackson, S. T., House, J. I., Prentice, I. C. (2011). Beyond predictions: Biodiversity conservation in a changing climate. *Science (New York, N.Y.), 332*. 53–58. https://doi.org/10.1126/science.1200303.

Dudgeon, R. C., & Berkes, F. (2003). Local understandings of the land: Traditional ecological knowledge and indigenous knowledge. In H. Selin (Ed.), *Nature Across* (pp. 75–96). Kluwer.

ESCAP UN (Economic and Social Commission for Asia and the Pacific). (2020). *The disaster risks cape across South and South-West Asia: Key takeaways for stakeholders*. ST/ESCAP/2879

Freeman, M. M. (1992). The nature and utility of traditional ecological knowledge. *Northern Perspectives, 20*(1), 9–12.

Geyer, J., Kiefer, I., Kreft, S., Chavez, V., Salafsky, N., Jeltsch, F., & Ibisch, P. L. (2011). Classification of climate-change-induced stresses on biological diversity. *Conservation Biology, 25*(4), 708–715.

Ginnetti, J., Lavell, C., & Franck, T. (2015). *Disaster-related displacement risk: Measuring the risk and addressing its drivers*. Published by the Internal Displacement Monitoring Centre, Norwegian Refugee Council Chemin de Balexert 7–9 CH-1219 Châtelaine (Geneva) Switzerland.

Gupta, A. K. (2018). Disaster governance and legal systems in India (pp. 39–60). Disaster risk governance in India and cross cutting issues.

Gupta, D., & Sharma, M. (2021). Chances of conflict on account of large-scale influx of climate refugees in India. In *Social-Ecological Systems (SES) from risks and insecurity to viability and resilience* (pp. 179–188).

Goel, S. L. (2007). *Disaster Administration and Management: Text and Case Studies*. Deep and Deep Publications.

Goswami, A. K. (2003). Water resource management in India—An overview. *Indian Journal of Public Administration, 49*(3), 245–253.

Haraway, D. (2013). Situated knowledges: The science question in feminism and the privilege of partial perspective. *Women, Science, and Technology, 3*, 455–472.

Harris, R. J., & Summers, W. (2002). *RJ Harris's Moon Gardening: The A-to-Z of the Cornish Head Gardener's Way of Producing Common Soft Fruits and Vegetables...: for Beginners at Gardening: for Beginners at Moon Gardening*. Really Useful Books.

Hosen, N., Nakamura, H., & Hamzah, A. (2020). Adaptation to climate change: Does traditional ecological knowledge hold the key? *Sustainability, 12*, 676. https://doi.org/10.3390/su12020676

Huang, C. W., McDonald, R. I., & Seto, K. C. (2018). The importance of land governance for biodiversity conservation in an era of global urban expansion. *Landscape and Urban Planning, 173*, 44–50.

Hunn, E. (1993). What is traditionell ecological knowledge. In N. Williams & G. Baines (Eds.), *Traditional Ecological Knowledge: Wisdom for Sustainable Development* (Vol. 13, p. 15). *Australian National University*.

Huntington, H. P. (2000). Using traditional ecological knowledge in science: Methods and applications. *Ecological Applications, 10*(5), 1270–1274.

IUCN (International Union for Conservation of Nature). (2012). IUCN Red list of threatened species. Version 2012.1. Retrieved September 17, 2012, from, www.iucnredlist.org.

Johannes, R. E. (1989). Introduction. In R. E. Johannes (Ed.), *Traditional Ecological Knowledge: A Collection of Essays* (pp. 5–9). IUCN (The World Conservation Union).

Kim, E.-J.A., Asghar, A., & Jordan, S. (2017). A critical review of Traditional Ecological Knowledge (TEK) in science education. *Canadian Journal of Science, Mathematics and Technology Education, 17*(4), 258–270. https://doi.org/10.1080/14926156.2017.1380866

Köberle, A. C., Vandyck, T., Guivarch, C., Macaluso, N., Bosetti, V., Gambhir, A., Tavoni, M., & Rogelj, J. (2021). The cost of mitigation revisited. *Nature Climate Change, 11*, 1035–1045.

Kumari, C. (2023). Inter-agency coordination in disaster management. *International Handbook of Disaster Research, 211.*

Kupika, O. L., Gandiwa, E., Nhamo, G., & Kativu, S. (2019). Local ecological knowledge on climate change and ecosystem-based adaptation strategies promote resilience in the Middle Zambezi Biosphere Reserve, Zimbabwe. *Scientifica.*

Lead, T. (2013). National Cyclone Risk Mitigation Project (NCRMP).

McGregor, D. (2004). Traditional ecological knowledge and sustainable development: Towards coexistence. *In the way of development: Indigenous peoples, life projects and globalization* (pp. 72–91).

McGregor, D. (2005). Traditional ecological knowledge: An Anishnabe woman's perspective. *Atlantis: Critical Studies in Gender, Culture & Social Justice, 29*(2), 103–109.

McGregor, D. M. D. (2008). Linking traditional ecological knowledge and western science: Aboriginal perspectives from the 2000 state of the lakes ecosystem Te of the lakes ecosystem conference. *The Canadian Journal of Native Studies, 28*(1), 139–158.

Miller-Rushing, A. J., Høye, T. T., Inouye, D. W., & Post, E. (2010). The effects of phenological mismatches on demography. *Philosophical Transactions of the Royal Society b: Biological Sciences, 365*(1555), 3177–3186.

Minville, G. (2022). *An Analysis of post-disaster recovery management in the 2016 and 2019 national disaster management plans of India* (Doctoral dissertation, Université d'Ottawa/University of Ottawa).

Mitra, A. P. (1996). Global change-greenhouse gas emission in India: methane budget estimates from rice fields based on data available up to 1995.

Mohapatra, M., Chauhan, A., Varshney, A., Gurjar, S., Bushair, M. T., Sharma, M., ... & K. Vani, G. (2023). Short to medium range impact based forecasting of heavy rainfall in India. *MAUSAM, 74*(2), 311–344.

Mohapatra, P. K., & Singh, R. D. (2003). Flood management in India. In *Flood problem and management in South Asia* (pp. 131–143).

Montoya, J. M., & Raffaelli, D. (2010). Climate change, biotic interactions and ecosystem services. *Philosophical Transactions of the Royal Society b: Biological Sciences, 365*(1549), 2013–2018.

Nalau, J., Becken, S., Schliephack, J., Parsons, M., Brown, C., & Mackey, B. (2018). The role of indigenous and traditional knowledge in ecosystem-based adaptation: A review of the literature and case studies from the Pacific Islands. *Weather, Climate, and Society, 10*(4), 851–865.

NAPCC (National Action Plan on Climate Change), Prime Minister's Council on Climate Change, Government of India. Retrieved June 30, 2008, from, http://pmindia.nic.in/Pg01-52.

NDMD (National Disaster Management Division). (2002). *Disaster management-the development perspective, an extract of the chapter in the tenth five year plan document (2002–2007).* Ministry of Home Affairs, Government of India.

Nema, P., Nema, S., & Roy, P. (2012). An overview of global climate changing in current scenario and mitigation action. *Renewable & Sustainable Energy Reviews, 16*, 2329–2336.

Panda, A. (2010). Climate refugees: Implications for India. *Economic and Political Weekly*, 76–79.

Pandey, R. K. (2016). Legal framework of disaster management in India. *ILI Law Review, Winter* (2016).

Parikh, J., & Gokarn, S. (1992). *Climate change and India's energy policy options* (No. 76).

Parkash, S., & Roy, T. S. (2014). Phalin cyclone 2013: Lessons Learnt. In *National Workshop Proceeding.*

Parmesan, C. (2006). Ecological and evolutionary responses to recent climate change. *Ecology and Evolution, 37*, 637–669.

Pierotti, R., & Wildcat, D. (2000). Traditional ecological knowledge: The third alternative (commentary). *Ecological Applications, 10*(5), 1333–1340.

Pohekar, P. D. (2023). Kautilya's antecedents, contemporary disaster resilience: Relevance of today's strengths and limitations. In *Fifth World Congress on Disaster Management* (Vol. IV, pp. 95–100). Routledge.

Ramos-Castillo, A., Castellanos, E. J., & Galloway McLean, K. (2017). Indigenous peoples, local communities and climate change mitigation. *Climatic Change, 140*, 1–4.

Salick, J., & Byg, A. (2007). *Indigenous peoples and climate change*. Tyndall Centre Publication.

Sawant, N. H., & Sanjeev, A. (2022). Climate refugees in India: A wake-up call for an inclusive policy. *India Quarterly, 78*(2), 371–380.

Scaddan, C. (2008). Climate change and indigenous peoples. *United Nations Permanent Forum on Indigenous Issues*. United Nations.

Schuster, R., Germain, R. R., Bennett, J. R., Reo, N. J., & Arcese, P. (2019). Vertebrate biodiversity on indigenous-managed lands in Australia, Brazil, and Canada equals that in protected areas. *Environmental Science & Policy, 101*, 1–6.

Sharma, M. K. (2003). Overview of flood Management in India. *Indian Journal of Public Administration, 49*(3), 520–538.

Singh, R. (2004). Practices and current status of flood forecasting and Management in India. *Emeritus Fellow, WRDTC*. Indian Institute of Technology, Roorkee, India.

Staudinger, M. D., Carter, S. L., Cross, M. S., Dubois, N. S., Duffy, J. E., Enquist, C., ... & Turner, W. (2013). Biodiversity in a changing climate: a synthesis of current and projected trends in the US. *Frontiers in Ecology and the Environment, 11*(9), 465-473.

Trotsenko, Y., Dixit, M., & Mykhailenko, V. (2021). Expression for calculation of lightning ground flash density for conditions of India. In *2021 IEEE International Conference on Modern Electrical and Energy Systems (MEES)* (pp. 1–5). IEEE.

UNISDR. (2005). Hyogo framework for action 2005–2015: Building the resilience of nations and communities to disasters. http://www.unisdr.org/files/1037_hyogoframeworkforactionenglish.pdf.

UNISDR (United Nations International Strategy for Disaster Reduction). (2015a). Sendai framework for disaster risk reduction 2015–2030. http://www.wcdrr.org/uploads/Sendai_Framework_for_Disaster_Risk_Reduction_2015-2030.pdf.

United Nations Disaster Risk Reduction. https://www.undrr.org/publication/hyogo-framework-action-2005-2015-building-resilience-nations-and-communities-disasters.

UNESCO and UNU-IAS. (2023). Asia-Pacific regional Synthesis, Climate Change, Displacement and the right to education. Published in 2023 by the United Nations Educational, Scientific and Cultural Organization, 7, place de Fontenoy, 75352 Paris 07 SP, France.

UNISDR. (2015b). *Sendai framework for disaster risk reduction 2015–2030*. UNISDR.

Verma, K., Kumar, S., Kumari, A., & Sharma, S. (2022). *Disaster Management in Dentistry*. Book Rivers.

Wahlström, M. (2015). New Sendai framework strengthens focus on reducing disaster risk. *International Journal of Disaster Risk Science, 6*, 200–201.

Woetzel, J., Pinner, D., & Samandari, H. (2020). Climate risk and response: Physical hazards and socioeconomic impacts.

Disaster Risk Reduction Policies in the Aftermath of Severe Storms for Asia and the Pacific Region

Sanjay Srivastava and Prangya Paramita Gupta

Abstract The recent years have witnessed the transformations in disaster risk reduction policies in the aftermath of severe storms worldwide. It introduced the paradigm of 'build back better' built on three key lessons:

- The **first** is preparing better to protect people for the severe cyclones. The evidence is unequivocal.
- The **second** is **building dedicated institutions such as national disaster management authorities, ministries, an exclusive department etc. for disaster related governance and strengthening of the national hydrometeorological originations in Asia and the Pacific.**
- The **third** is **putting communities at the center of the recovery and reconstruction process** as well as empowering at risk communities to manage disasters.

The paper summarises the key trends in policy responses and outlines pathways for adaptation and resilience to tropical cyclones which are on the rise both in terms of frequency and intensity.

1 Asia-Pacific World's Most Disaster Impacted Region

Asia and the Pacific is the most disaster-prone region in the world. Since 1970, disasters in Asia and the Pacific have killed over two million people—60% of the global death toll, that is 105 lives every day, or one life every 13 min (Asia-Pacific Disaster Report, 2023b). In the rest of the world the average number of fatalities per disaster was 189, but in Asia and the Pacific it was much higher at 338. As indicated

Storm here primarily refers to the cyclones.

S. Srivastava (✉) · P. P. Gupta
Disaster Risk Reduction, UN Economic and Social Commission for Asia and the Pacific, Bangkok 10200, Thailand
e-mail: srivastavas@un.org

S. Das and W-K. Tao (eds.), *Severe Storms*,
https://doi.org/10.1007/978-981-97-7075-5_26

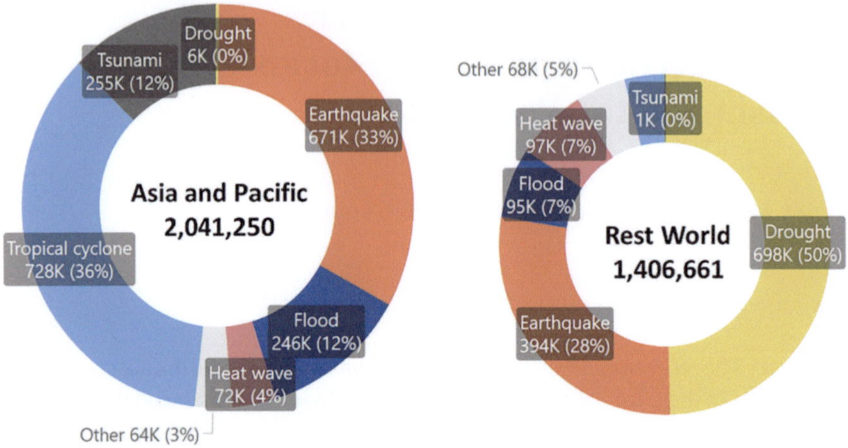

Fig. 1 Number of fatalities from disasters in the Asia–Pacific region and the rest of the world, 1970–2022. *Source* ESCAP calculations based on EM-DAT, accessed on 14 Feb 2023

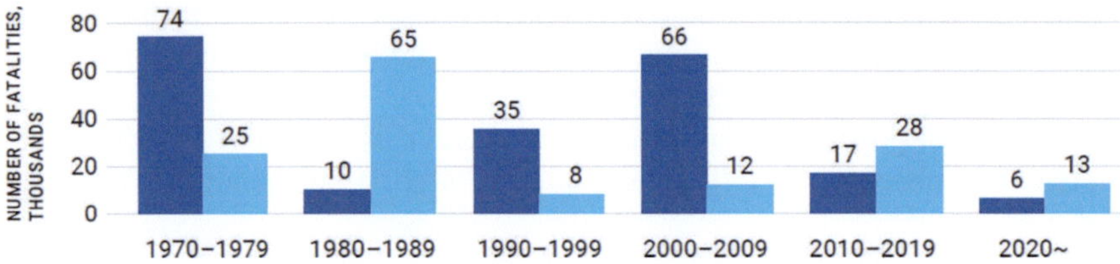

Fig. 2 Average number of fatalities by decade in the Asia–Pacific region and the rest of the world, 1970–2022. *Source* ESCAP calculations based on EM-DAT, accessed on 14 Feb 2023

in Fig. 1, the principal causes of natural disaster deaths in Asia and the Pacific were storms and earthquakes, followed by tsunami and floods. In the rest of the world the pattern was different: the death toll was lower, and the principal killer was drought especially in Africa, followed by earthquakes.[1]

There have been significant improvements, however, in the past decade from 1970 to 2022, the average loss of life per year in the Asia Pacific fell to 8 thousand, a significant improvement of almost 5-fold. When looking at the trend in the Asia–Pacific broken down by decades, it is possible to see a clearer downward trend in the number of fatalities, number of people affected and economic damages, especially since 2020 and onwards. Comparing the average of the year 2020 onwards to the immediate precedent decade, it is possible to see a drop of 64% for the number of fatalities, 58% for the number of people affected, and 43% for economic damage (Figs. 2, 3 and 4).

[1] Flood refers to Surface water flood; Drought referes to precipitation related droughts; Tropical cyclone refers to wind related Tropical cyclone; Heatwave refers to temperature related Heatwave; Earthquake referes to seismogenic earthquake; Tsunami refers to Earthquake triggered Tsunami.

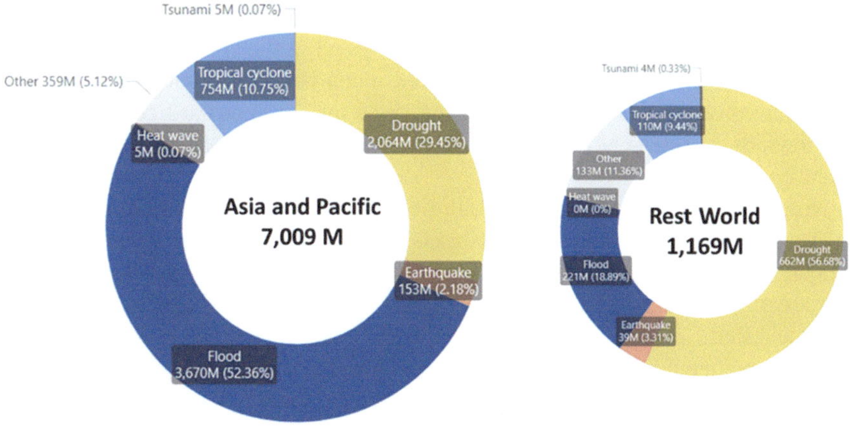

Fig. 3 Number of people affected from disasters in the Asia–Pacific region and the rest of the world, 1970–2022. *Source* ESCAP calculations based on EM-DAT, accessed on 14 Feb 2023

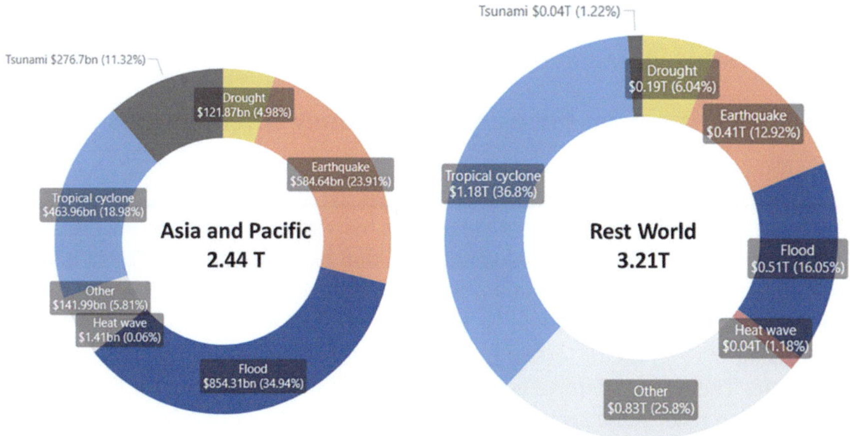

Fig. 4 Economic damage from disasters in the Asia–Pacific region and the rest of the world, 1970–2022. *Source* ESCAP calculations based on EM-DAT, accessed on 14 Feb 2023. [*Note* The economic damages of some disaster occurrences are not presented in the diagram due to data unavailability. ESCAP-Economic and Social Commission for Asia and the Pacific; EM-DAT—Emergency Events Database]

This drop is demonstrative of the efforts directed at enhancing the forecast, bringing awareness, and building resilience in the region. Accurate forecast and efficient early warning with considerable lead time in case of two recent Severe cyclones namely cyclone Mocha in the Bay of Bengal and cyclone Biparjoy in the Arabian sea saved thousands of lives despite their landfall in the densely populated coastal areas of India, Bangladesh and Myanmar. No fatality was reported in case of cyclone Biparjoy whereas only about two dozen deaths was reported in Myanmar for cyclone Mocha. In a similar scenario in 2008, cyclone Nargis caused more than 138,000 fatalities (ESCAP, 2023). In terms of trend in the Asia–Pacific compared to global average, Asia Pacific region fared much better at reducing the number of people affected and economic damage since the 2020 onward compared with the previous decade (EM-DAT—The International Disaster Database, 2021; SDG gateway—Data & Statistics, 2021).

Note: It is important to note that all the analysis mentioned in Sect. 1 is based on the EM-DAT emergency event datasets. The EM-DAT data is a unique global data on the disasters and their impacts on humans. It is compiled form heterogenous sources e.g., UN agencies, governments, and the International Federation of Red Cross and Red Crescent Societies as well as from non-governmental organizations, insurance companies, research institutes, and press agencies. Hence, the data often suffers from limitations such as regional reporting biases or political conditions despite thorough quality checks. Moreover, the quality of reporting is not constant over time. The "affected people," refers to people requiring immediate assistance, as well as displaced and evacuated people in many instances. Hence, the information should be interpreted with caution keeping in mind the heterogeneity and the biases in the data (Mahecha et al., 2020).

2 Sectoral Impacts: Focus Areas of Policy Interventions

The landscape of sectoral impacts varies across the subregions. In the Pacific, the social and productive sector incurs about 30% losses, while losses in infrastructure and cross cutting sectors stand at around 16%. Southeast Asia, almost 80% of disaster losses occur in the productive sectors, while in South and southwest Asia productive sector losses are relatively lower at around 60%, followed by social (25%) and infrastructure (13%) sector losses. Losses in the cross-cutting sectors are highest in the Pacific. Therefore, adaptive actions to build future resilience must be based on understanding risks at the sectoral level.

As per cent of country GDP for loss and damage and as a per cent of the total population for death and people affected, Vanuatu, Tonga, Fiji and Marshall Island are at higher risk than other countries in the Asia Pacific region. The risk is increasing over time. Vanuatu saw a two-fold increase in death and people affected from Tropical Cyclone Pam in 2015 to Tropical Cyclone Harold in 2020 and in Fiji, from Tropical cyclone Evan (2012) to Tropical cyclone Winston (2016). Loss and damage increased almost 10-fold. As climate change continues to increase the intensity and number of floods and cyclones, these negative trends are expected to continue. Only a complete transformation in how the region thinks about policies on adaptation and measures to mitigate the impacts can reverse these trends.

Policy actions are driven by impacts and risk visualization. Towards this, storms must be unpacked based on phases of action before (the preemptive actions on planned adaptation), during (anticipatory action a few days after a major hazard is forecast) and after their impact (contingency planning with fund accruing after

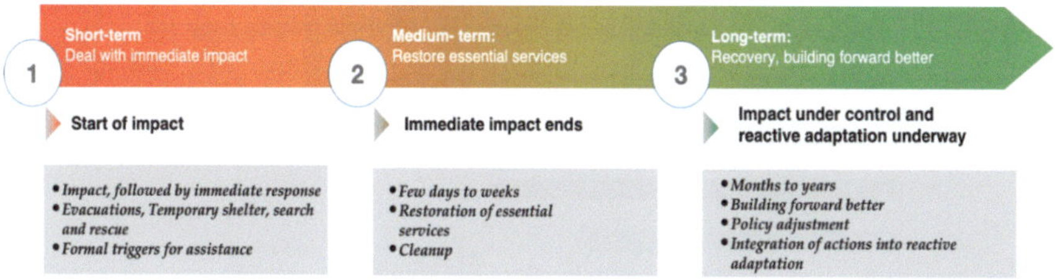

Fig. 5 Spectrum of actions in responding to climate impacts for rapid onset disasters (Adapted from Santiago Network 2023, Regional Scoping Workshop on Loss and Damage Under the Santiago Network: Latin America and Caribbean Region, Analysis of Workshop Discussions, 30 April 2023)

impact). Yet they must also be repacked into phases based on the nature and the actions required to anticipate, respond and recover. For rapid onset disasters such as storms, the phases which can be more directly quantifiable include (Fig. 5):

- Short term actions to response to immediate impacts
- Medium term response to recover essential services
- Long term reactive adaptation including reconstruction, climate proofing and building back better

3 Adaptation and Resilience Pathways for Addressing the Climate Crisis

Based on the three key lessons of the "building back better" the following section identifies the actions required under each key lessons.

3.1 Prepare Better to Protect People

UN Secretary-General António Guterres calls for 50% of the total share of climate finance to be spent on building resilience and adapting to the effects of a warming world (UN Climate Change News, 2021). ESCAP estimates the total climate adaptation costs for cascading hazards (natural and biological) under extreme climate change scenario (2040–2059) at US$ 270 billion for the Asia Pacific region (https://rrp.unescap.org/) of which US$ 68 billion would be for adapting to biological hazards (Asia-Pacific Disaster Report, 2021). Around 70% of these costs are in East and North-East Asia at $190 billion. These costs need to be considered alongside capacities to pay which vary from 1.4% of GDP for the Pacific small island developing States (SIDS), to less than 1% for South-East Asia and North and Central Asia. The adaptation costs vary substantially across the sub-region (Fig. 6).

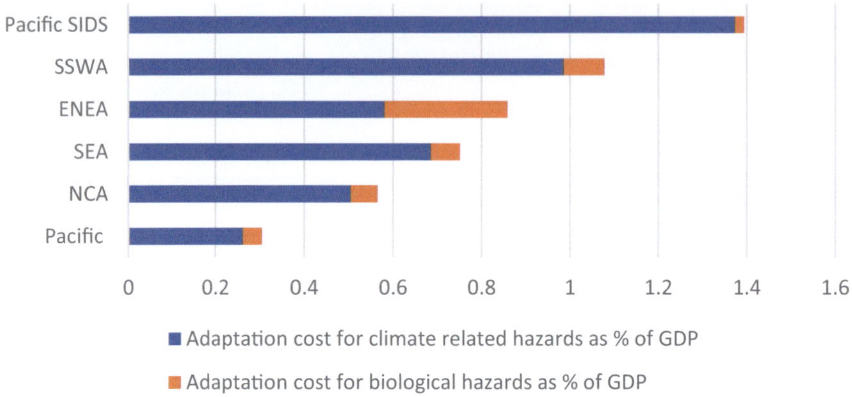

Fig. 6 Subregional adaptation costs for climate-related hazards and biological hazards, percentage of GDP

The risk profiles are quite diverse across the sub-regions. As a result, each subregion will have its own adaptation priorities, as illustrated in Fig. 7 based on categories established by the Global Commission on Adaptation: early-warning systems; climate-resilient infrastructure; improved dryland agriculture crop production; mangrove protection, and water security (Adapt now: A global call for leadership on climate resilience, 2019). In East and North-East Asia, the top adaptation priorities are making new infrastructure more resilient and strengthening early warning systems. In South and South-West Asia, the highest priorities are strengthening early warning systems and making new infrastructure resilient, followed by resilient water resource management improving drylands and protecting mangroves.

It's important to highlight that for individual Pacific SIDS, and for the LDCs, where economic assets are very exposed to natural hazards adaptation costs are disproportionately high. At the country level, the highest cost as per cent of GDP is recorded in the Pacific SIDS with 8.4% in Vanuatu and 6.8% in Tonga. In the Pacific SIDS, adaptation priorities derived from its unique riskscape include making water resource management resilient and improving dryland agriculture crop production followed by protecting mangroves, strengthening early warning systems, and making new infrastructure resilient. These matrices allow countries to prioritize adaptation action and investment in areas most applicable to their climate risk profiles. For example, adaptation priorities for Vanuatu includes improving dryland agriculture crop production and resilient water resource management followed by protecting mangroves, strengthening early warning systems and making new infrastructure resilient. This can be attributed to high exposure to drought and a major share of employed population that works in the agriculture sector. Further, investing in these priorities will ensure progress on achieving multiple sustainable development goals of the 2030 Agenda for Sustainable Development (Fig. 7).

Mangroves naturally protect the coastal regions from the impacts of storms and floodings by reducing the wave heights and preventing the generation of wind

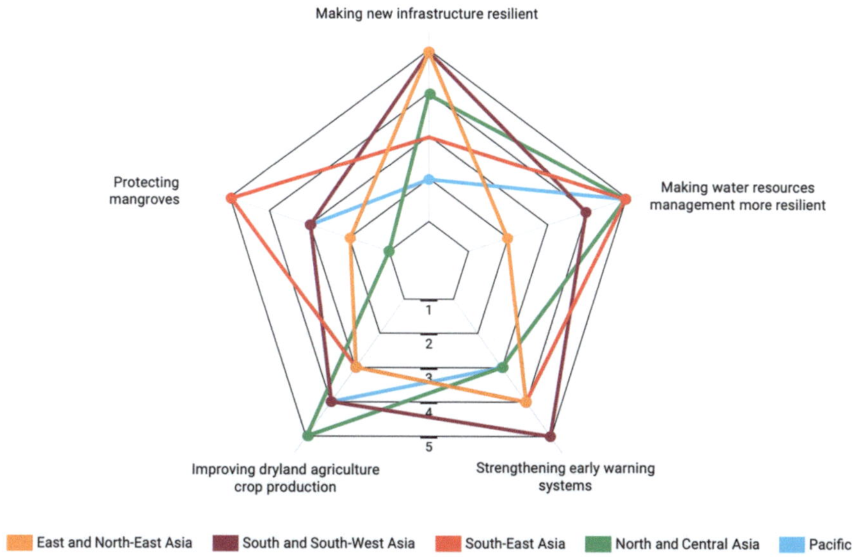

Fig. 7 Adaptation priorities for Asia and Pacific subregions (2040–2059)

waves. They are well recognized for adaptation, resilience and sustainable development strategies as well as advocated by many coastal nations in their recently submitted NDCs (UNFCCC Secretariat, 2021). It is estimated that without the existing mangroves, 15 million more people would be flooded annually across the world (Menéndez et al., 2020).

The extent of the protection depends on their density, height and width. Higher the width of the mangrove areas from the coast higher the protection. Studies indicate that up to 66% of wave height can be reduced by over 100 meters of mangroves and up to 100% by over 500 meters of mangroves. They reduce the flood depth and related loss of life caused by cyclones. Figure 8 describes how mangroves with different vegetation drag can attenuate wave height and sea level rise in the Pearl River Delta, China (Dominicis et al., 2023).

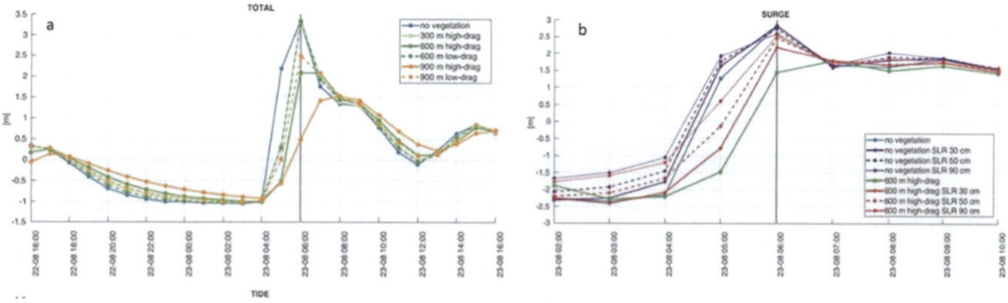

Fig. 8 Water levels during Typhoon Hato (**a**) and Sea Level Rise (**b**) with and without mangroves in Shenzhen Bay, China

Mangroves reduce the extent of flooding in low-lying coastal areas. An assessment of the coastal flood protection provided by the mangroves in Pagbilao, Philippines, reveals that without mangroves the extent of flooding is much wider and affects more people and property (Menéndez et al., 2019) (Fig. 9).

The economic benefit of mangroves under a cyclone and storm surge scenario is estimated by Menéndez et al. (2020). It shows that globally, population exposure, loss of properties and extent of flooding is significantly reduced in the "with mangrove" scenario compared to the "without mangrove" scenario (Fig. 10).

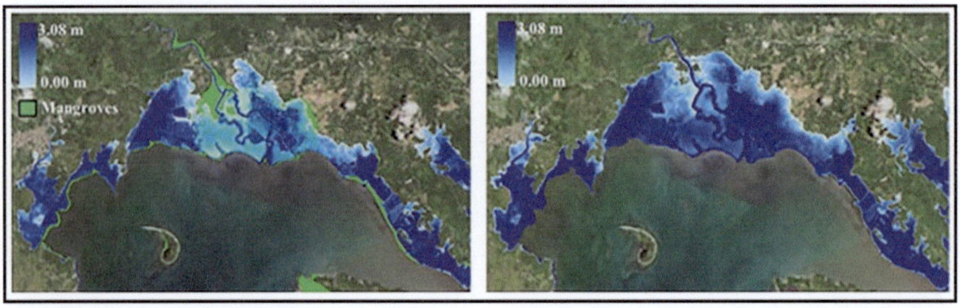

Fig. 9 Coastal flooding produced by 1-in-50 years tropical cyclone in Pagbilao—comparison between with and without mangroves

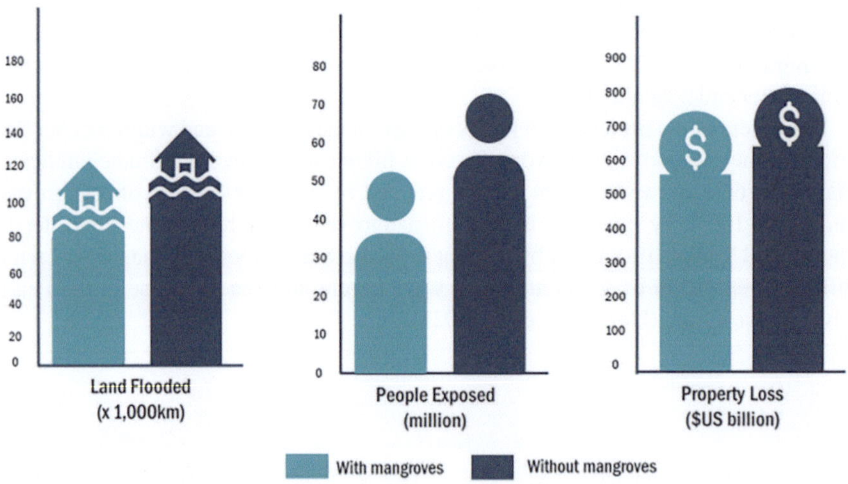

Fig. 10 Global annual benefit expected from flood protection (Menéndez et al., 2020)

3.2 *Building Dedicated Institutions*

Strengthening institutional capacities is essential particularly for regions prone to frequent cyclones and storms. Because of the complexity and nonlinearity in the atmospheric system, accurately predicting the track, positions and intensities of cyclones is a challenge. Numerical weather prediction (NWP) models and statistical models-based forecasts can provide advanced early warning for cyclones based on the data from weather stations, including coastal radars and ground-based stations. However, they require highly efficient infrastructure and skilled human resources to produce actionable information and effective communication channels among the stakeholders for preparedness and response. Developing transboundary collaboration with international organizations and agencies involved in cyclone and storm research and management can lead to the exchange of knowledge, data sharing, and access to global resources. And enhance the preparedness of the countries/regions with comparatively less equipped infrastructure for information generation.

The WMO/ESCAP Panel on Tropical Cyclones (PTC) is one such collaboration established half a century back in the Bay of Bengal and the Arabian Sea, has played critical role in many instances. The PTC promotes measures to improve tropical cyclone warning systems including dissemination of technical information on tropical cyclone research and forecasting operations to mitigate the socio-economic impacts of tropical cyclone-related disasters. The Regional Specialized Meteorological Centre (RSMC) at Indian Meteorological Department associated with the PTC had been able to forecast cyclone Mocha precisely four days in advance, which gave sufficient lead time for the PTC member country authorities to maneuver the coastal communities to safer areas. The RSMC monitors cyclones throughout the Northern Indian Ocean, from the shores of Oman in West Asia to Myanmar in Southeast Asia and works closely with 13 PTC members to facilitate the delivery of early warning products and services of tropical cyclones, often characterized by transboundary origins and impacts.

Public–private partnerships (PPPs) for cyclone early warning can be highly effective in improving the accuracy, reach, and timeliness of early warning systems. The Pacific Disaster Centre's (PDC) Disaster AWARE platform is one such example that serves the early warning and decision support needs of the public, as well as thousands of governmental and nongovernmental organizations around the world. It's also offered for business enterprise management needs.

3.3 *Putting Communities at the Centre of Recovery and Reconstruction*

Aftermath of Severe Storms follows a set of interventions that include humanitarian and post-cyclone recovery and reconstruction, which also aims at build back better. The Post-Disaster Rapid Assessment and similar disaster assessments are an essential

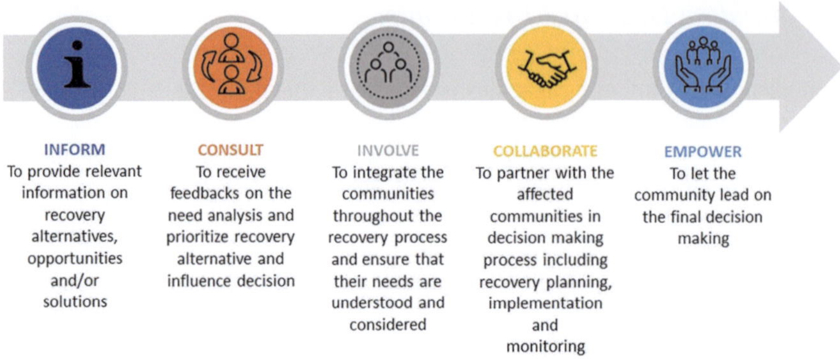

Fig. 11 The components of an effective community-based disaster recovery process. Adapted from UNDP, 2020

component of a successful disaster recovery framework (DRF) (Post-disaster rapid assessment: Tropical Cyclone Gita, 2018). The DRF helps to (i) develop a central vision for recovery; (ii) define guiding principles; (iii) identify primary recovery sectors; (iv) assess government capacity to manage recovery; (v) appoint an appropriate recovery leader; (vi) establish appropriate institutional arrangements; (vii) clarify institutional roles and responsibilities; and (viii) establish coordination mechanisms, conduct funding gap analysis, and mobilize funds. The DRF also outlines needs for the first three years of recovery as identified in this Post Disaster Rapid Assessment, including immediate needs, short-term needs, and medium-term needs.

In most of the post-disaster recovery process the top-down approach is undertaken which does not involve the participation of the affected communities. Although people-centered recovery can be more effective as it can identify the local priorities, enhance local capacities and skills for future, ensure ownership as well as maintain cultural integrity (UNDP, 2020). Figure 11 illustrates the components of an effective community-based disaster recovery process.

Economic and Social Commission for Asia and the Pacific (ESCAP) analyzed 48 Post-Disaster Needs Assessment (PDNAs) of disasters (1 drought event, 10 earthquakes, 16 floods, 16 tropical cyclones, 1 tsunami, and 1 volcanic eruption) occurring from 1999 to 2023 to examine the trends in impacts on mortality and morbidity and the social, infrastructure, productive and cross cutting sector losses (Asia-Pacific Disaster Report, 2023a). The overall results indicate stagnating action on measures to mitigate loss and damage to multiple sectors. The results show a decreasing trend of mortality and people affected from all disasters, trends in mortality for earthquakes is on the rise. A disaggregated analysis by three ESCAP subregions, note that the Pacific region shows increasing trends for loss and damage and deaths and people affected, while South and Southwest Asia shows a stark increase in loss and damage.

4 Building Back Better: Example of Best Practice of Post Cyclone Winston (Fiji)

Several innovative tools for prioritizing resilience interventions are already available. For example, the World Bank demonstrates how sustainability checklists can be applied to select and prioritize recovery and resilience interventions from existing national strategies, for building forward better from the pandemic. The checklist was applied to Fiji, which had already developed a nation-wide Capacity and Vulnerability Assessment (CVA) for natural hazards following the devastation of Cyclone Winston in 2016. This resulted in 125 possible interventions across 10 sectors: housing and land use, hazard management, transport, water, energy, health/education, environment, agriculture, fisheries, and social protection (Fig. 12) (How to build back better after the COVID-19 crises? A practical approach applied to Fiji, 2020).

Fiji is exposed to many natural hazards including cyclones, so any COVID-19 recovery plan must build resilience to future pandemics, natural hazards, as well as their potential interaction. The World Bank therefore looked at the CVA and found 63 interventions that could address both hazard and pandemic resilience. To develop an overall strategic plan that would most effectively strengthen resilience, the interventions were then assessed for their impacts on short term growth, long term growth, climate resilience, and decarbonisation. Comparing the results revealed priority interventions, that would have the strongest impacts at different time scales. This example demonstrates that short term stimulus packages, designed strategically, can also have long term benefits.

It is important to highlight that ecosystem of innovations has been built at sector-levels and then aggregated at national level with long term goals of resilience building and decarbonization. This presents a unique example of policy coherence through the window of opportunity available during post-crisis recovery.

5 Way Forward

Rapidly intensifying storms are becoming a new normal. While early warning systems have been helpful in saving lives through timely evacuation of at-risk communities before cyclones strike, the increasing economic impact of such events is a cause of grave concern. For instance, the economic impact of Amphan is estimated at $13 billion and $135 million in India and Bangladesh, respectively. The damages mostly relate to social infrastructure (housing, schools, and hospitals), physical infrastructure (energy, transport, water/sanitation, ICT) and agriculture (crops, livestock).

Why are the costs of climate change rising? Because of the exposure of critical infrastructure in high-risk areas. The Asia–Pacific Disaster Report 2019 (Asia-Pacific Disaster Report, 2019) has highlighted that high proportions of the existing critical infrastructures are located in multi-hazard risk hotspots, which when the natural

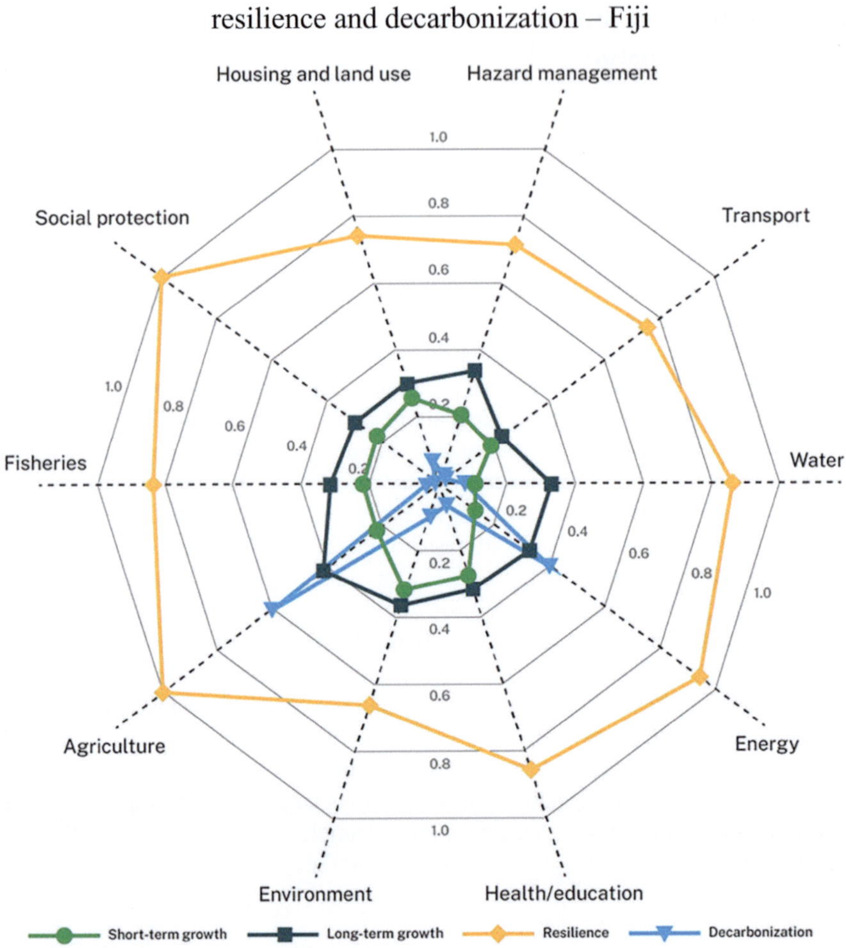

Fig. 12 How different interventions perform along different checklist categories with long term goal of achieving resilience and decarbonization—Fiji. *Source* World Bank 2020

disasters occur, lead to a significant disruption in economic activity. For instance, a third of energy power plants, ICT fibre-optic cables, and airports and over 42% of road infrastructure are located in multi-hazard risk hotspots. How can resilience be built in defending against climate change? This requires both monetary investments and policy actions. Building infrastructure that is climate-resilient (adaptation) and reducing the release of GHG emissions (mitigation) are two keys aspects, which can be strengthened through a coalition of stakeholders to move towards a low-carbon world.

References

Adapt now: A global call for leadership on climate resilience 2019. Global Center on Adaptation. https://gca.org/reports/adapt-now-a-global-call-for-leadership-on-climate-resilience/.

Asia-Pacific Disaster Report. (2019). *The Disaster Riskscape Across Asia-Pacific Pathways for resilience, inclusion and empowerment*. United Nations publication.

Asia-Pacific Disaster Report 2021. Resilience in a Riskier World—Managing systemic risks from biological and other natural hazards. United Nations publication, Bangkok. https://www.unescap.org/kp/2021/asia-pacific-disaster-report-2021.

Asia-Pacific Disaster Report 2023, ESCAP's flagship publication (under publication).

Asia-Pacific Disaster Report 2023. United Nations publication, Bangkok. https://www.unescap.org/kp/2023/seizing-moment-targeting-transformative-disaster-risk-resilience.

De Dominicis, M., Wolf, J., van Hespen, R., et al. (2023). Mangrove forests can be an effective coastal defence in the Pearl River Delta, China. *Communications Earth & Environment, 4*, 13. https://doi.org/10.1038/s43247-022-00672-7

EM-DAT—The International Disaster Database 2021. Available at https://www.emdat.be

ESCAP 2023. Benefits of 'Early Warnings for All' triples in vulnerable contexts. https://www.unescap.org/blog/benefits-early-warnings-all-triples-vulnerable-contexts.

How to build back better after the COVID-19 crises? A practical approach applied to Fiji 2020. https://www.worldbank.org/en/news/immersive-story/2020/09/17/how-to-build-back-better-after-the-covid-19-crisis-a-practical-approach-applied-to-fiji.

Mahecha, M. D., Guha-Sapir, D., Smits, J., Gans, F., & Kraemer, G. (2020). Data challenges limit our global understanding of humanitarian disasters triggered by climate extremes. In *Climate Extremes and Their Implications for Impact and Risk Assessment* (pp. 243–256). https://doi.org/10.1016/b978-0-12-814895-2.00013-6.

Menéndez, P., Losada, I. J., Torres-Ortega, S., Toimil, A., & Beck, M. W. (2019). Assessing the effects of using high-quality data and high-resolution models in valuing flood protection services of mangroves. *PLoS ONE, 14*(8), e0220941. https://doi.org/10.1371/journal.pone.0220941

Menéndez, P., Losada, I. J., Torres-Ortega, S., Narayan, S., & Beck, M. W. (2020). The global flood protection benefits of mangroves. *Science and Reports, 10*, 1–1. https://doi.org/10.1038/s41598-020-61136-6

Post-disaster rapid assessment: Tropical Cyclone Gita 2018. Government of Tonga. https://www.preventionweb.net/publication/post-disaster-rapid-assessment-tropical-cyclone-gita.

Risk and Resilience Portal, UNESCAP. https://rrp.unescap.org/.

SDG gateway—Data and Statistics 2021. https://www.unescap.org/stat/data.

Secretariat, UNFCCC. (2021). Nationally Determined Contributions Under the Paris Agreement: Synthesis Report by the Secretariat (UN Framework Convention on Climate Change, 2021). https://unfccc.int/NDCREG?gclid=Cj0KCQjw8e-gBhD0ARIsAJiDsaXAV_jNrWbHBB5v-Q_IwXdZt6lOfktCf6Fu0JT9LXu7zNLQZ6U-eJMaAsn1EALw_wcB.

UN Climate Change News 2021. António Guterres: 50% of All Climate Finance Needed for Adaptation. https://unfccc.int/news/antonio-guterres-50-of-all-climate-finance-needed-for-adaptation.

UNDP 2020. Guidelines for Community Participation in Disaster Recovery. https://reliefweb.int/report/world/guidelines-community-participation-disaster-recovery.

Epilogue

S. Das and W-K. Tao (eds.), *Severe Storms*, https://doi.org/10.1007/978-981-97-7075-5

Challenges in Prediction of Severe Storms

Someshwar Das and Wei-Kuo Tao

Abstract Severe Storms continue taking their toll on the human lives and economy even though there has been considerable advancement in the prediction models. This monograph is focused on local severe thunderstorms causing intense lightnings, strong winds and torrential rainfall, dust storms, hailstorms, tornadoes, and cloud-bursts. What causes the local severe storms so furious? How can we predict their intensity, location, and time of occurrence accurately at sufficient lead time? Will the severe storms be more frequent and intense in the changing climate? Such questions are addressed in this monograph in 3 parts containing 26 chapters by the leading scientists across the world.

Keywords Storms · Observations · Modelling · Assimilation · Prediction · Early warning · Disaster mitigation · Changing climate

1 Introduction

Severe thunderstorms occur everywhere globally. It is estimated that at any given time there are about 2000 thunderstorms taking place on the earth (http://www.nssl.noaa.gov/education/svrwx101/thunderstorms/). They severely disrupt the routine life, and have devastating impacts on agriculture and aviation, surfacetransportation, electricity, power, communication, essential services, and other socio-economic sectors. It is estimated that lightning due to severe storms kill more people than hurricanes, earthquakes and floods and is the biggest contributor to accidental deaths due to natural causes. It kills approximately 24,000 people around the world every year. In India, more than 2,500 people die due to thunderstorm and lightning every year

S. Das
Former Scientist-G/ Advisor, NCMRWF & IMD, Ministry of Earth Sciences, New Delhi, India

Former Professor, Central University of Rajasthan, Ajmer, India

W.-K. Tao
Emeritus Scientist, NASA/ Goddard Space Flight Center, Greenbelt, MD, USA

Chair Professor, National Central University, Taoyuan City, Taiwan

S. Das and W-K. Tao (eds.), *Severe Storms*, https://doi.org/10.1007/978-981-97-7075-5

(NDMA, 2018). A global survey on the mortality and economic losses due to storms (including tropical cyclones) indicates that 577,232 people died and economic losses of US Dollar 521 billion occurred during 1970–2019 (WMO, 2021). Storms had the highest human and economic toll around the globe. The IPCC (2021) projects an increase in the severity and frequency of severe storms in future due to the climate change.

As climate shocks become more common, better prediction and early warning is a critical need for decision-making, such as when to plan and when to flee. In March 2023, the UN has envisioned implementation of an effective system to protect every person on Earth by 2027, which is a huge task. According to UN, currently one-third of the people in the world, especially in Least Developed Countries and Small Island Developing States, are still not covered by early warning systems. Even the people covered sometimes do not get early warnings on time or in the proper manner to enable them to take mitigation measures such as moving to a safer place with their means of livelihood. "Systems that warn people of impending storms, are not a luxury but a cost-effective tool that saves lives, reduces economic losses, and provides a nearly tenfold return on investment", according to an UN website for the Early Warnings for all. These disasters displace three times more people than war. And the situation is getting worse, according to UN Secretary General at COP27. As of now, almost 50% of the countries do not have early warning systems. Evidence suggests that countries with good coverage of early warning systems have lower mortality rates compared to those lacking it (DTE, 2023).

Predicting the extreme weather events at precise location, intensity and lead time has been a challenge to the Meteorologists, especially in the developing countries due to lack of observations, resources, and the limit of the deterministic forecasts. However, observation system should be able to sense the changes in the structure of the atmosphere and the numerical weather prediction models should be able to provide accurate solutions to the future state of the atmosphere with usable skills. Accurate prediction of the severe storms requires observations of 3-dimensional structure of atmosphere at good temporal and spatial resolutions, assimilation of those observations in the numerical prediction model, applications of ensemble forecasting at cloud-resolving scale and high-performance computing resources (Das, 2022). The observations are required several times a day on routine basis from different sources such as satellites, radars, aircrafts, radiosonde balloons, automatic weather stations, surface meteorological observatories over land, and ocean from ships, and buoys.

After having obtained the forecasts from the models, another important step is to disseminate the warnings to the public at sufficient lead time and be prepared for rescue operations if there is a failure of the forecast in the aftermath of the event for disaster risk reduction and mitigation. Dissemination is done mainly via concerned government departments at the national, state and district levels. It is also done through print/digital/social media even to individual users through digital applications. Timely dissemination of early warnings to impacted communities is a major challenge, compounded by gaps in connectivity. Spreading awareness about steps to take during a disaster event also lacks focus (DTE, 2023).

2 What is the Severe Weather Forecast Demonstration Project (SWFDP)?

As the observation technology is improving rapidly, field campaigns are frequently organized to collect intensive observations on extreme weather events. Field observations of severe weather phenomena are collected to improve our understanding of their structures and life cycles and develop better models for forecasting the events.

The Severe Weather Forecasting Demonstration Project (SWFDP) of the World Meteorological Organization (WMO) is an initiative to strengthening capacity in National Meteorological and Hydrological Services (NMHSs) in developing and least developed countries including Small Island Developing States to deliver improved forecasts and warnings of severe weather to save lives, livelihoods, and property (https://public.wmo.int/en/programmes/severe-weather-forecasting-programme-swfp). SWFDP was initiated in 2006 to make the NWP products, including Ensemble Prediction System (EPS) products, of the most advanced Global Data-Processing and Forecasting System Centres available to all WMO Members. The Project made global-scale products available to Regional Specialized Meteorological Centres (RSMC) that integrate and synthesize them in order to provide daily guidance for short-range and medium-range forecasts of hazardous weather conditions and weather-related hazards to NMHSs in their geographical region. Thus, NMHSs are enabled to issue effective severe weather warnings to disaster management and civil protection authorities in their respective countries. Currently, SWFP covers over 80 developing countries.

Over the years several field experiments have been conducted both over the Indian region and outside. India Meteorological Department (IMD) had conducted three field experiments during 1929–1941 to study the outbreak of severe convective storms (IMD, 1944; Tyagi et al., 2012). Several field experiments have been conducted since then in USA and elsewhere to understand and predict the convective storms.

3 The SAARC STORM

The skill of forecasting severe storms depends on the type of storms, i.e., whether they are Squall line, light/moderate/or severe Nor'wester depending on their intensity (surface wind speed). The skill also depends on whether they are non-organized storm, a pulse storm, a line storm, or a supercell depending upon the maximum reflectivity, low level wind shear, vertically integrated liquid water (VIL) and the rotational track of the storm. Realizing the importance of these extreme weather events and their socioeconomic impact, the India Department of Science and Technology started the nationally coordinated Severe Thunderstorm Observation and Regional Modelling (STORM) program in 2005. It was a comprehensive observational and modelling effort to improve understanding and prediction of severe thunderstorms (STORM, 2005). The STORM program is a multiyear exercise and is quite complex in the formulation of its strategy for implementation. Considering that the thunderstorms occur at a spatial scale ranging from a few kilometers to a few hundred kilometers, it needs different surface-based observational platforms to be organized on a meso-net.

However, the weather knows no political boundaries. Since the neighbouring South Asian countries are also affected by the Nor'westers, the STORM program was expanded to cover the South Asian countries under the South Asian Association for Regional Cooperation (SAARC). The STORM program covered all the SAARC countries in three phases (Das et al., 2014). In the first phase, the focus was on Nor'westers that form over the eastern and northeastern parts of India, Bangladesh, Nepal, and Bhutan. In the second phase, the dry convective storms/dust storms and deep convection that occur in the western parts of India, Pakistan, and Afghanistan were investigated. Similarly, in the third phase, the maritime and continental thunderstorms over southern parts of India, Sri Lanka, and Maldives were investigated. Thus, overall, the SAARC-STORM program covered investigations about formation, modelling, and forecasting, including nowcasting of severe convective weather in the pre-monsoon season over South Asia.

4 Why Are Thunderstorms Furious?

The ferocity of thunderstorm depends on its damage potential in terms of the strength of wind, ability for a lightning strike, size of hail stones, and flash floods (Das, 2017). For a severe thunderstorm, the ingredients that must be present are moisture, instability, lift and strong speed and directional storm relative wind shear. Wind shear aids in the following: Tilting a storm (displacing updraft from downdraft), allows the updraft to sustain itself for a longer period of time, allows the development of a mesocyclone, and allows rotating air to be ingested into the updraft (tornado genesis). Severe storms also tend to have these characteristics over ordinary thunderstorms: higher CAPE, drier air in the middle levels of the atmosphere (convective instability), better moisture convergence, Baroclinic atmosphere, and more powerful lift.

4.1 Mechanism of Gale Wind (Sustained Surface Wind 63–87 km/h)

Wind is stronger when the pressure gradient is high. Strong winds experienced during a thunderstorm are due to the downdrafts beneath the cumulonimbus. Rain from the storm evaporates below the cloud, causing the air to cool beneath it. This cold air is heavy and crashes into the ground below. When it hits the ground, this cold air must turn sideways, and the result is strong winds. These winds are known as microbursts. Wind speeds in microbursts can exceed 160 km h^{-1} and cause significant damage even though they only last for 5–15 min. A typical thunderstorm is made up of a single cumulonimbus (CB) cloud. The CB cloud consists of strong vertical updrafts and downdrafts.

4.2 Mechanism of Lightning and Thunder

Lightning is caused by an electrical discharge of electrons moving very quickly from one place to another. These electrons move so quickly that they superheat the air around them causing it to glow. Thunderstorms are created by strong rising air currents called updrafts forming Cumulonimbus clouds. These updrafts create winds of about 80 km h^{-1} (or more) rising several miles into the air to form Cumulonimbus clouds. Warm updrafts and cooler weaker downdrafts create turbulence within thunderclouds resulting in the interaction of tiny water particles and microscopic ice crystals called hydrometeors. The continuous collisions of the water and ice create large and small particles. The different movement characteristics of these varyingly sized particles in turn create electrical charges within the cloud. Figure 2 (Chap. 7) illustrates the distribution of charges in a typical thunderstorm.

Observations suggest that smaller particles (less than 100 μm) acquire a positive charge, while larger particles gain a negative charge. Movement of air inside the cloud combined with the effects of gravity causes these differently sized water particles to separate. Updrafts carry the tiny particles to the top of the cloud and gravity pulls the larger water or ice particles towards the bottom of the cloud. This separation of smaller positively charged particles and the larger negatively charged particles create an electrical imbalance with an enormous electric potential of millions of volts across the storm cloud. The laws of physics requires this electrical difference to be neutralized. Lightning is formed as tremendous currents travel across the air to correct the huge imbalance of negative electrons created within the storm cloud. More detailed information on lightning can be found in http://www.aharfield.co.uk/lightning-protection-services/how-lightning-is-formed.

The bright glow from lightning is caused by the immense currents (about 200,000 amps) super heating the air it is traveling along to 3 ½ times hotter than the surface of the sun to the order of 20,000 °C. The sound of thunder is caused by the same superheated air expanding rapidly and creating a supersonic shock wave, which then decays to an acoustic wave as it propagates away from the lightning channel. As a lightning strike is not one continuous bolt of lightning, but a series of lightning bursts traveling along stepped leaders (typically 30 m in length) the sound created is not one single report or boom, but a series of them. On the return stroke, each stepped leader is discharged from the bottom-up super heating air and creating a shock wave at different altitudes. These differently timed shock waves at different altitudes give the rumble effect during a lightning strike. A series of secondary strokes can also travel along the same ionised channel as the initial return stroke making the thunderclap oscillate in volume as secondary shock waves are generated.

5 How Can We Model Thunderstorms?

A basic characteristic of the convective cloud models is that their governing equations are non-hydrostatic since the vertical and horizontal scales of convection are similar. Presently, mesoscale models having horizontal resolution $\leq$ 10 km are also used for simulation and prediction of regional weather systems. These models can be used for a wide variety of applications including simulation and prediction of severe storms and tropical cyclones. In order to adequately model the thunderstorms, the fundamental physical and mathematical laws governing the dynamics and thermodynamics of the clouds need to be implemented in a form which is physically meaningful, mathematically rigorous, and computationally efficient. Numerous physical factors linked to the dynamical and thermodynamical characteristics of the clouds need to be addressed. Many researchers have studied and developed numerical models of thunderstorms, squall lines and mesoscale convective systems (Zipser and LeMone, 1980; Weisman and Klemp, 1982; Cotton et al., 1986; Weisman et al., 1988; Tao et al., 1987, 1989; Rotunno et al., 1988; Tao and Simpson, 1993; Takemi and Satomura, 2000; Weisman and Rotunno, 2004; many others).

The modelling of storms depends on the requirement. The storm models may be one, two or three dimensional. Single Column Models (SCMs) or one-dimensional models are used for investigating individual physical processes in idealized condition. The SCMs can be considered as a single grid box of a climate model in which field observations may be used to test the parameterization of a physical process. Since the SCMs do not interact with the neighbouring grid columns, the effects of the surrounding grid boxes are specified in terms of the large-scale advective forcings to run the SCMs (Xu and Randall, Das et al., 1987; Liu et al., 2001).

Two-dimensional (2D) anelastic cloud model was developed to study cloud development under the influence of the surrounding environment (Ogura & Phillips, 1962). After Global Atmospheric Research Program's Atlantic Tropical Experiment (GATE), cloud ensemble modeling was developed to study the collective feedback of clouds on the large-scale tropical environment with the aim of improving cumulus parameterization in large-scale models (i.e., Soong and Tao 1980; Tao and Soong 1986; Tao et al., 1987; and many others), a quest that continues to this day.

6 Can We Forecast Severe Storms with Sufficient Accuracy?

The models discussed above are good for testing idealized simulations, process studies and development of new schemes. Forecasting of thunderstorm is done based on a combination of synoptic method, Doppler Weather Radar and Numerical Model products (Das et al., 2015, Das, 2017).

6.1 *Numerical Prediction of Thunderstorms:*

Numerical models can simulate all the phases of thunderstorm evolution. Thunderstorms are mesoscale phenomena. Mesoscale convective weather systems refer to those which are smaller than synoptic scale (about 1000 km) and larger than the cumulus scale (~1 km). Mesoscale models are designed for the simulation and prediction of mesoscale weather systems. Such models remain important for an operational numerical weather prediction center, because they can be run at very high resolution on a nested grid with a wide variety of options for the parameterization of physical processes. The global models do not have such privileges and, they are very expensive to run at high resolutions. Moreover, at finer resolution the mesoscale models are also capable of assimilating large number of high-resolution observations available from present day satellites and Doppler radars. The mesoscale models can be configured to run from global to cloud resolving scale for simulation of thunderstorms and cloud cluster properties. Since the early 1990s several important changes took place in mesoscale modelling. First was the introduction of nonhydrostatic dynamics into mesoscale models (e.g., Dudhia, 1993). The nonhydrostatic mesoscale models can be run at cloud resolving resolutions (~1 km or finer) without the restrictions of the hydrostatic assumption. This greatly increases the range of scientific problems to which the models can be applied. For example, at such resolution, mesoscale models can explicitly simulate convection and its interaction with the larger scale weather systems in a realistic way. However, the use of high-resolution mesoscale models for real-time NWP requires a tremendous number of computational resources. Presently, mesoscale models have been developed with a wide variety of flexibilities in terms of changing horizontal and vertical resolutions, nesting domains, and choosing options for different physical parameterization schemes, i.e. MM5, WRF, RAMS, ETA, ARPS, HIRLAM, etc. (Anthes, 1990; Case et al., 2002; Cotton et al., 1994; Dudhia, 1993; Mesinger, 1996; Toth, 2001). These models require initial and boundary conditions from a large-scale/global model and may be used for forecasting up to 72 h (skill can deteriorate beyond that). Such models can be run at cloud resolving scale for simulation of thunderstorms and cloud cluster properties.

6.2 *Ensemble Forecasting of Thunderstorm and Lightning Strike*

A single model may not be able to forecast accurately. Hence, for operational forecasting several models are run at one time—an ensemble. If each run produce similar outputs, the probability of the model forecast of an event is highest. If the runs look different, the probability of the model forecast is low. Another technique is to run the same model several times with varying initial weather conditions. This approach results in several predictions that produce a range of possible future weather outputs.

Probably the first real time regional ensemble prediction system was implemented at NCEP in 2001 based on a 10-member Short-Range Ensemble Forecasting (SREF). The NCEP-SREF emphasizes on both initial conditions (IC) and physics uncertainties. It uses multi-analyses, i.e., Ensemble Data Assimilation System (EDAS) and Global Data Assimilation System (GDAS), multi-LBCs (using NCEP global ensemble members), multi-model (Eta and RSM) and perturbing ICs (breeding approaches). Presently, the SPC (Storm Prediction Centre) SREF is constructed by post-processing 21 members of the NCEP SREF plus the 3-h time lagged, operational WRF-NAM (for a total of 22 members) each 6 h (03, 09, 15, and 21 UTC). Output is available at 3 h intervals through 87 h. The SPC ensemble post-processing focuses on diagnostics relevant to the prediction of SPC mission-critical high-impact, mesoscale weather including thunderstorms and severe thunderstorms, large scale critical fire weather conditions, and mesoscale areas of hazardous winter weather.

Bright et al. (2005) and Bright and Grams (2009) provide information on the SPC-SREF thunderstorm calibration technique with detailed verification results that show the forecasts to be both reliable and skilful. Bright et al. (2005) defined a Cloud Physics Thunder Parameter (CPTP) based on the (1) temperature at LCL (≥ -10 °C) to ensure presence of super cooled liquid water, (2) temperature at equilibrium level (≤ -20 °C) to ensure the cloud top is above the charge reversal zone, and (3) CAPE in the 0 to -20 °C layer $\geq$ 100–200 J kg^{-1} to ensure existence of sufficient vertical motion in mixed-phase region through the charge reversal temperature zone.

6.3 Rapid Update Cycle Data Assimilation

The first attempt to initialize a cloud-scale numerical model using radar observations for the simulation of a thunderstorm was made by Lin et al. (1993). In their study, wind from a dual-Doppler analysis, temperature and pressure from a thermodynamic retrieval, and rainwater from radar reflectivity observations were used to initialize a cloud-resolving numerical model. Sun and Crook (1997, 1998) developed a four-dimensional variational data assimilation (4DVAR) scheme for the initialization of a cloud-scale model with single-Doppler radar observations. Sun (2005) further explored the feasibility of numerical weather prediction of convective storms using detailed high-resolution observations from single-Doppler radar and the 4DVAR scheme developed by Sun and Crook (1997) based on Variational Doppler Radar Analysis System (VDRAS) for a real-data case of a supercell storm. The dry version of VDRAS has been implemented in a number of field projects, including the WMO Sydney 2000 Olympic Project (Crook and Sun 2002) and the Beijing 2008 Forecast Demonstration Project (B08FDP) in support of the Beijing Summer Olympics (Sun et al., 2010), to produce real-time low-level wind analysis for nowcasting of thunderstorms.

Although the fundamental data assimilation theory does not depend on the scale of interest, convective-scale data assimilation possesses several important differences from the large-scale. The techniques that have been applied to convective-scale assimilation are, (1) Single-Doppler retrieval and direct insertion of retrieved fields, (2) Successive correction, (3) Three-dimensional variational (3D-Var) technique, (4) 4D-Var, and (5) Ensemble Kalman Filter. The issues with the convective scale data assimilation are, (1) Processing Radar Data, (2) Combining Radar data with other types of observations, (3) Cycling frequency, (4) Nonlinearity in microphysics, (5) Single and Multiple Doppler Synthesis, (6) Estimation of observation and forecast errors, (7) Using Polarimetric Radar information, and (8) Low-level moisture and land-surface data assimilation. Details of these techniques may be referred from Sun et al. (2010), and Benjamin et al. (2016). Presently, NCEP (USA) uses an hourly updated assimilation and model forecast system called the Rapid Refresh (RAP)—Benjamin et al. (2016). It replaced the Rapid Update Cycle (RUC) in 2012. The RAP uses the Advanced Research version of the WRF-ARW and the Gridpoint Statistical Interpolation analysis system (GSI). RAP forecasts are generated every hour with forecast lengths going out 18 h. Multiple data sources go into the generation of RAP forecasts including commercial aircraft weather data, balloon data, radar data, surface observations, and satellite data.

7 What is the Skill of Forecasts?

Predicting the track of the movement of thunderstorm and its intensity by using a numerical model and quantify the skill of the model in predicting time and location of occurrence and the intensity poses a big challenge. As discussed earlier, a single deterministic model may not be able to provide skilful forecasts. Therefore, ensemble forecasting based on multi-models, multi-physics, multi-ICs and multi-LBCs are used for obtaining skilful forecasts.

In the past, the determination of several verification scores were the most common practice and the interpolated gauge measurements served as the verification data. These techniques are often marked as "traditional verification methods". They include the use of contingency table and the skill scores derived from the table or scalar error measures, like RMSE. The development of new verification methods started about ten years ago and was motivated by development of NWP models with high temporal and spatial resolutions, ensemble forecasting and data from radar and satellite measurements (e.g., Ebert and McBride, 2000; Casati et al., 2004, 2008; Ebert, 2008; Gilleland et al., 2009, Srivastava et al., 2011, Roy Bhowmik et al., 2011). As higher-resolution numerical models are now used to predict highly discontinuous fields, like convection, there is an increasing need of newer verification techniques (Casati et al., 2008; Gilleland et al., 2009). Newer verification techniques that attempt to better characterize model skill for discontinuous fields can be classified into one of four categories: neighbourhood (or fuzzy), scale-separation (or decomposition), feature based (or object based), and field deformation techniques (Gilleland et al., 2009). Roberts and Lean (2008) presented a spatial verification score where they directly compare fractional coverage of precipitation area over a certain threshold with areas surrounding observations and forecast. The traditional verification scores are often not appropriate measures of skill for high-resolution model forecasts of discontinuous fields (Gilleland et al., 2009), like strong convection.

8 Issues Discussed in This Monograph

For understanding a physically evolving system like a thunderstorm, we need to observe the system through various approaches, develop techniques to forecast its evolution in a short, medium, and long-time scales and develop policies for disaster mitigation in the aftermath of the event. These are discussed in three parts containing 26 chapters. The observational aspects of the severe storms through satellite, radar, aircraft, and ground-based network of stations are discussed in the first part consisting of 8 chapters. The numerical modelling and data assimilation techniques are discussed in the second part for the development of early warning system and finally the outlook of the severe storms in a changing climate, their socio-economic impacts and policies for disaster mitigation are discussed in the third part of the book.

8.1 Observational Aspects

8.1.1 Detection of Thunderstorms by Satellite

The Imager and Sounder on board the geostationary satellites like the INSAT-3D/R provides fine resolution vertical profiles of thunderstorm cells over India and neighbourhood, along with other polar satellite soundings. The INSAT-3DR imager provides imaging capability of the earth disc from geostationary altitude in one visible (0.52–0.72 μm) and five infrareds; 1.55–1.70 (SWIR), 3.80–4.00 (MIR), 6.50–7.00 (water vapour), 10.2–11.2 (TIR-1) and 11.5–12.5 (TIR-2) bands. The INSAT-3D/R carries multi-spectral imager channels (6 channel imager and 19 channel sounder payloads). It is staggered with a temporal frequency at every 15 min. Further, a web-based quick visualization and analysis tool for INSAT satellite data on a real-time basis is available. It is called Real-time Analysis of Products & Information Dissemination (RAPID). The resolution is nominally 1 km × 1 km for visible and SWIR bands, 4 km × 4 km for one MIR and both TIR bands and 8 km × 8 km for WV band.

The space-borne precipitation radar observations have enabled us to study characteristics of precipitation globally and statistically including those by the rainstorms (Chap. 1). The satellite-based thunderstorm monitoring has become popular for its enhanced spatial, spectral, and higher temporal resolution for near real-time thunderstorm detection and monitoring (Chap. 2). The satellite-based observations are also crucial for assimilation in the NWP models for accuracy of operational weather forecasting. Satellite observations are also the only practical alternative approach for locations with a lack of data.

Based on the accumulated space-borne precipitation radar observations, it is found that extremely heavy rainfalls do not often accompany extremely tall convection. There is an urgent need to quantify the change of precipitation characteristics associated with the global warming for adequate adaptation measure for the sustainable society. To this end, it is essential to understand mechanisms of heavy rainfalls in various environments and their modifications with response to the climate change.

8.1.2 What Can Be Expected from Radar?

Doppler Weather Radar (DWR) provides an effective means of observing severe storms and mesoscale convective systems (horizontal scales ~ 10–100 km), which produce damaging winds, extremely heavy rainfall, and hailfalls. Radial velocity measurements obtained by the DWR provide considerable insight on three-dimensional storm motion fields. Dual polarization observations allow significant improvements in the identification of hydrometeor types and estimations of precipitation rates (Chap. 3). They are also used for improved prediction of Hailstorms, Cloudburst and Heavy rainfall associated with the severe storms. The Microburst, Strong winds, Tornado and Hailstorms are monitored and predicted at least few hours in advance (nowcasting) by DWR (Chap. 4). The northeastern parts of India which receives the highest rainfall per day in the world (1003.6 mm) at place named Mawsynram and adjoining Bangladesh are prone to severe Nor'westers, tornadoes, torrential rainfall, and flash floods. Existing network of DWRs of Northeast India, Bangladesh, and Myanmar along with assimilation of these data in the NWP model can be a viable, effective solution to detect, monitor, and predict the severe storms in the northeast Indian subcontinent by international, domestic, and inter-organizational cooperation through open dialogues among stakeholders (Chap. 5).

The real time observations from both traditional national radar surveillance networks, and more specialized local radar networks that observe specific urban areas, now support a wide variety of data combination techniques. The resultant data sets will support continuing advancements in the assimilation of radar data into NWP models. The current research into machine learning techniques can be applied to weather radar data sets to further improve the anticipation and recognition of severe storms.

8.1.3 What is Suggested by the Field Experiment Involving Aircraft and Radars?

Based on the observations collected using aircrafts and Polarimetric Radars at ground during a field campaign named "The Cloud Aerosol Interactions and Precipitation Enhancement Experiment (CAIPEEX)" key information on the microphysical processes and organization of clouds and rainfall types were obtained (Chap. 6). It is shown that the key cloud dynamics and microphysics observations give pathways for improving the emerging global cloud-resolving models.

8.1.4 Electrical Characteristics of Thunderstorms

Electrical, microphysical, and dynamic characteristics occur almost simultaneously inside thunderstorms and interact with each other. Many studies in India have shown that some thunderstorms forming over India have inverted polarity charge structure. This inverted polarity charge structure is attributed to increased aerosol concentration or high liquid water content (Chap. 7). Laboratory and field experiments also have shown that electric forces in thunderstorms can significantly impact on rain formation processes. Studies show that electrical forces inside thunderstorms play a dominant role in modifying the microphysical characteristics and in initiation and strengthening. Results also show the changes in raindrop size distribution due to combined effect of increase in aerosol loading/pollution and electric field inside the clouds formed due to such aerosol loading. These observations demonstrate how electric field in thunderstorms could change the rainfall and its characteristics.

The above conclusions are drawn from measurements made at a few places only over India. Measurements of electric field at more place especially where the severe thunderstorms occur frequently could help in our understanding of effect electric fields on rainfall characteristics. This understanding can possibly improve the simulation of tropical rainfall in numerical models.

8.1.5 Raindrop Size Distributions in the Severe Storms

Understanding of the microphysical properties, particularly the Raindrop Size Distributions (RSD) in the severe storms and the cyclones is important for better precipitation estimations, modelling, and simulations of tropical storms. RSD is measured by ground based disdrometer. It can vary with geography and precipitation types. The RSD can vary between oceanic and continental storms. Studies have shown that the mass-weighted mean diameters of the RSD are higher due to more mid- and large-size drops. Study indicates that the RSDs are greater in the western Pacific (WP) as compared to the North Indian Ocean (NIO) basin (Chap. 8). A significant number of bigger drops in the WP Tropical Storms can be attributed to the association of intense precipitating clouds with surplus moisture in the WP.

8.2 Modelling, Assimilation, and Early Warning System

8.2.1 Numerical Model of Squall Line

Numerical modelling of the 3-dimensional structure of the thunderstorm is challenging, but very important for the development of early warning system. High-resolution (167 m) modelling of a linear-shaped Mesoscale Convective System (MCS) representing squall line is described in Chap. 9. Special attention is paid to the precise reproduction of fine-scale, complex topography based on the Digital

Elevation Model (DEM) dataset (50 m) in the high-resolution simulations. The model simulated the successive development of convective cells on the upstream side of a linear-shaped MCS and the stationarity of the linear-shaped MCS reasonably well. Key issues in this high-resolution modelling are to represent more precisely the complex and complicated nature of topography and to explicitly resolve convective motion. A notable feature for the generation of linear-shaped MCSs in their embedded environments is a very humid state not only in the lower troposphere but also in the middle-to-upper troposphere. Therefore, the processes leading to such humid conditions should be investigated in more detail. A multi-scale analysis of linear-shaped MCSs and their environmental properties is required to reveal mechanisms for the moistening of the free troposphere and the generation of linear-shaped MCSs.

8.2.2 Numerical Modelling of Hailstorms

An overview of hail and hailstorm characteristics, hail production mechanisms, and examples of various microphysical schemes currently employed by cloud-resolving and mesoscale models for use in hail research is described in Chap. 10. Validation is a critical component of hail modelling, so it is essential that microphysics schemes are systematically evaluated in their ability to produce realistic hail fields and characteristics as well as their time evolution relative to the storm dynamics. The most comprehensive approach is via systematic, statistical comparisons against radar observations, including polarimetric, not depleted by significant attenuation, which can sample entire convective systems, including convective cores. It is also important to note that other factors, namely grid resolution, can have a substantial impact on convective simulations in addition to the microphysics parameterization. Understanding how hail events may change in response to expected climate change is vital given the enormous economic risk to agriculture and property. Because hail sizes and number densities can vary through the lifecycles of propagating systems, statistical analyses of populations of hailstorms from continuous regional, seasonal simulations will likely be required to understand how characteristics such as system sizes and hail size distributions might change in a future climate, possibly presenting new threats to agriculture, transportation, and wildlife.

8.2.3 Observed and Simulated Characteristics of Severe Storms

Convective storms are often organized by multiple convective cells, and most of which develop to be a MCS that is characterized by an organized regions of deep convection across hundreds of kilometres horizontally over a time span exceeding 6 h. Severe convective storm generally occurs within MCS as an intense convective cell. The MCS are classified into 2 types: Linear MCS and Non-linear MCS. The Linear MCS are further classified into eight types according to the organizational mode, system-relative cell movement, and general features of the system (Bluestein and Jain, 1985). These eight types include trailing stratiform (TS) MCS, back-building/quasi-stationary (BB) MCS, trailing line/adjoining stratiform (TL/AS) MCS, leading stratiform (LS) MCS, parallel stratiform (PS) MCS, broken line (BL) MCS, embedded line (EL) MCS, and long line (LL) MCS. The Nonlinear (NL) MCS are unorganized strong convective echoes embedded within large areas of stratiform echoes. Chapter 11 presents a comprehensive analysis of convective storms in China.

Firstly, the Climatology of MCS is described, providing insights into their spatial and seasonal variations, and their organization patterns. The second part investigates the microphysical processes, structures, and characteristics that give rise to extreme precipitation events. The third part discusses the structure, vertical rotation characteristics, and evolutionary process of tornadoes. In the fourth part, the climatology, structure, and cloud microphysical characteristics of hailstorms are discussed.

8.2.4 Characteristics of Wind in the Boundary Layer During Severe Storms

Understanding the characteristics of winds associated with turbulence in the atmospheric boundary layer during severe storms is important for the safety of human life and properties. The boundary layer wind field during severe storm is described in Chap. 12 based on simulations using the Weather Research and Forecasting (WRF) model and a Large Eddy Simulation (LES) model.

8.2.5 Dynamics of Rainstorms Over Southern Slopes of the Himalayas

The Low-Pressure Systems (LPSs) such as monsoon lows, monsoon depressions, and mid-latitude troughs and their interaction create an environment that concentrates moist strong winds onto the Himalayan slopes and enhances forced uplift of the moist air, leading to heavy rainfall (Chap. 13). The locations of the LPSs and the mid-latitude troughs are strongly controlled by Intra-Seasonal Oscillations (ISOs) on larger spatiotemporal scales that create an environment suitable for the genesis and development of LPSs and troughs. To better understand and predict heavy precipitation events, we need to improve understanding of the controlling multiscale processes over the complex terrain of the Himalayas.

Precipitation tends to occur frequently in the afternoon and at night time at higher elevations. Understanding how interactions between the diurnal cycle and synoptic-scale disturbances might affect the occurrence of an extreme precipitation event warrants further investigation. To unravel the dynamics, we might need cloud-resolving model simulations with horizontal resolution better than 1 km capable of resolving the regional topography and its interaction with the large-scale atmospheric circulation, as well as advanced in situ observations such as those realized by ground precipitation radar. Additionally, further observational, and numerical studies are required to better understand how precipitation systems, which are strongly influenced by the land surface effects of both the Himalayan slopes and the Indian subcontinent, maintain the Himalayan glaciers at higher elevations.

8.2.6 Physical and Dynamic Characteristics of Severe Storms Over NE India and Bangladesh

The northeastern parts of India and adjoining Bangladesh experience many violent thunderstorms, which are also known as Nor'westers during the pre-monsoon period (March–May). These storms have great destructive potential and often gain the strength of a tornado accompanied by high magnitude squalls, gusty winds (60–150 km/h), torrential rains, lightning, and hails. They cause extensive damage to crops, infrastructures, and lives and livelihoods in the areas across which they traverse. Based on the simulations using the WRF model and ERA5 data, it is found that generally a low pressure is formed over the region producing strong south-westerly low-level winds. The wind scarry a plentiful of moisture from the Bay of Bengal towards the region, which vertically extended up to 200 hPa level. A line of discontinuity of warm and moist air from south and cold and dry air from north meet over the country. The strong westerly or north-westerly winds help to move the storm towards east, south-east direction. The area of occurrence of the severe storms are characterized by the high values of low-level relative vorticity ($>150 \times 10^{-5}\ s^{-1}$), upper level divergence ($>400\ s^{-1}$), vertical velocity (>12 m/s), CAPE (>3000 J/kg), low values of CINE (<50 J/kg), high values of cloud water mixing ratio (>30 gm/Kg), ice water mixing ratio (>80 gm/Kg), and reflectivity (>50 dBZ) vertically extending up to 150 hPa level with a core at 500 hPa level. The above physical and dynamical characteristics of the severe storms are commonly found over the region, which are the prerequisite for genesis of the severe storms (Chap. 14).

8.2.7 Assimilation of Data from Surface Mesonet, Radar, and Lidar for Prediction of Severe Storms

Data assimilation involves incorporating observational data into numerical models using variational or ensemble-based assimilation methods. These methods aim to find the best fit between the observations and the model's simulated state, effectively blending the observations with the model's background state. The assimilation process enables a more accurate representation of the initial atmospheric conditions and better captures the complex dynamics and microphysics associated with these convective systems, ultimately enhancing the skill and reliability of MCS forecasts.

Ground-based radar and lidar data offer valuable information about the structure and evolution of MCSs. Radar data, such as reflectivity and Doppler velocity, provide insights into precipitation patterns, storm intensity, and wind fields within the storms. Lidar data can provide vertical wind profiles. Chapter 15 illustrates that assimilating radar and lidar data into NWP models can enhance the representation of convective processes within MCSs, resulting in improved predictions of their track, intensity, and timing.

8.2.8 Rapid Update Cycle Data Assimilation for Storm Prediction

The Rapid Update Cycle (RUC) or High-Resolution Rapid Refresh (HRRR) is an hourly-updated, storm-resolving model Data Assimilation (DA) system used for accurately simulating the rapidly evolving convective storms by taking advantage of the latest weather observations. Chapter 16 demonstrates improvements in the forecasts by using a 1-hourly HRRR 4DVAR DA system with 1.5 km resolution convection permitting NCUM-R model (NCMRWF Unified—Regional model) by assimilating conventional, and satellite observations along with DWR radial velocity and reflectivity over Indian region.

8.2.9 Ensemble Forecasting of Severe Storms

The deterministic prediction has limits due to uncertainties in initial conditions, model formulations, and the chaotic nature of the atmosphere. These uncertainties, including inaccuracies in initial state estimation and model representation of atmospheric processes, make it challenging to quantify the individual contributions of these errors. To address this issue, an Ensemble Prediction System (EPS) is used to form a consistent probability distribution function (pdf) that describes uncertainty in forecasts. The combination of probability and determinism in forecasting is known as Ensemble Prediction.

Chapter 17 presents different ensemble forecast products based on the high resolution global (12 km) and regional (1.5 km) EPS of NCMRWF for prediction of severe weather events. The rainfall and lightning products obtained from EPS are used for early warning. Location specific EPSgrams and other forecasts products are made available globally under dedicated website of SWFDP project.

8.2.10 Lightning Forecasting and Utilization of AI/ML in Early Warnings

Lightning is considered as one of the biggest killers annually due to severe storms. Early warning of lightning is very difficult and challenging as the physical and dynamic processes leading to lightning are still poorly understood. Some attempts have been made recently to predict the lightning threats using the NWP models (Kumar et al., 2022), but accuracy of predicted lightning intensity (flash rate), time and location is still very poor.

Therefore, hybrid approach based on AI (Artificial Intelligence) -ML (Machine Learning) and NWP has been tested for lightning early warning system using the WRF model forecasts at 9 km resolution. The 24-h lead time forecast is used as input for a deep learning model with IITM Lightning Location Network (LLN) data as the target vector (Chap. 18). A Spearman rank correlation matrix is used to select input features for the AI. At the same time, Principal Component Analysis (PCA) was carried out to determine the hidden layer structure. The lightning-to-non-lightning event ratio is highly skewed, adding a challenge for any deep-learning model to learn the latent features of lightning. Hence, a novel two-autoencoder-based deep-learning model (C_{2AE}) is built, where positive autoencoder is trained on WRF output associated with lightning events and negative autoencoder on WRF output associated with non-lightning events. When tested on a dataset consisting of both lightning and non-lightning data, the C_{2AE} is able to predict lightning with statistically significant skills.

As the False Alarm rate in this approach is high, non-linear hidden layers may help the autoencoders learn hidden features. Further, the spatiotemporal information may have to be added to the dataset to identify the clustered lightning active regions. However, as the cost of misses is higher than that of false alarms for lightning, a carefully constructed deep learning algorithm with a simple architecture will have to be designed.

8.2.11 Severe Storms Forecasting and Forecast Demonstration Project (FDP)

The Forecast Demonstration Project of the STORM program is conducted during March to June every year in India. As a part of this project, there has been rapid expansion of observation tools as well as mesoscale model products that forecast weather associated with thunderstorms such as lightning, gusty winds, and hail. Correspondingly there have been major improvements in forecasting quality in short range to nowcast scale with quantitative forecasts for thunderstorms being issued round the clock at three-hour intervals for all districts of India and more than a thousand stations. Using the same metrics and a constant verification dataset, significant improvement is noted in all scores, both in short range and nowcasts scale for all subdivisions of India (Chap. 19). However, the skill scores vary with places and season of the year. There are consistently higher forecast detection errors in short range scale over the arid western parts of India.

The nowcast warnings are now available for use by the public through web-based weather apps such as MAUSAM, UMANG and MEGHDOOT. The other major initiative in recent years is the Common Alert Protocol (CAP) as recommended by WMO and as implemented by the National Disaster Management Authority (Mohapatra et al., 2022). The areas selected for warning under the Common Alert Protocol are automatically warned through their mobile phones. In addition, the Indian Institute of Tropical Meteorology (IITM) has launched 'Damini', a free mobile-based application that can warn people about lightning in their vicinity at least 30–45 min before it strikes.

IMD has launched a public outreach initiative through a website for crowdsourcing of weather information where citizens will be able to record realized weather and can upload their local weather conditions (Chug et al., 2021). The latest initiative towards detailed analysis of thunderstorms is the "Thunderstorm Testbed Project" over east India, planned for the five-year period of 2021–2026. The project is being planned for in-depth field experiments and numerical experiments with available and augmented instrumentation and computing facilities. All these developments have greatly increased the quality, outreach and effectiveness of impact based severe weather warnings issued by IMD.

8.2.12 How Long Could Winter Storms Last Over the Continental US?

The Northern Great Plains, Midwest, and Northeast regions of the United States experienced above-normal extreme precipitation accumulation events after 2006. However, these events are not accompanied with prolonged duration of precipitation events. The cold-season extreme precipitation events are analysed over the conterminous U.S. (CONUS) using radar-gauge merged 1/8th-degree hourly precipitation data from the North American Land Data Assimilation System (NLDAS) Phase 2 datasets. Analysis reveals that the *very rare* extreme events exceed 10 cm of accumulated precipitation over 68 h in Northwest region (Chap. 20).

8.3 Severe Storms in the Changing Climate, Societal Impacts and Disaster Risk Reduction

8.3.1 How Will Be the Monsoon Rainstorms in the Changing Climate

The IPCC (2021) report suggests that climate change caused by the human activities has driven detectable changes in the global water cycle since the mid-20th century. The global warming can cause increase of atmospheric moisture, which can lead to an overall increase in daily mean precipitation intensities in a majority of land regions during the recent few decades (Douville et al., 2021). While the increase in magnitude of intensification of heavy precipitation events due to global warming is about 7% per 1 °C of warming, changes in atmospheric circulation patterns and moisture transport can significantly alter the dynamics of extreme precipitation events at regional scales (Douville et al., 2021).

The results discussed in Chap. 21 indicate that the duration of monsoon rainstorms over Western India (WI) have risen since 1990s, together with a distinct enhancement in rainstorm days occurring mostly during the years of positive Indian Ocean Dipole (pIOD) events in the tropical Indian Ocean. The sustenance of monsoon rainstorm days over WI during pIOD events is supported by the background large-scale circulation anomalies associated with intensified cross-equatorial transport of moisture from the south-eastern tropical Indian Ocean and strengthening of cyclonic vortices in the lower and mid-tropospheric levels over the Indian subcontinent. While state-of-the-art climate model projections suggest higher likelihood of extreme pIODs in the future, they also raise several implicative questions on the role of extreme pIODs on organized monsoon convection, scale interactive processes, among others, on the future monsoon rainstorms in the backdrop of increasing water vapour in a globally warming climate.

Several key questions are raised such as; (1) What is the influence of extreme pIODs on the frequency, intensity and duration of monsoon rainstorms in the future? (2) What is the role of global warming on the large-scale organization of mesoscale convective systems, monsoon low pressure system (LPS) and mid tropospheric cyclone (MTC) in the future? (3) How will climate change influence monsoon rainstorms in different regional monsoon systems of the world? Unravelling these scientific issues warrants further investigations using high-resolution climate models, diagnostics based on satellite and reanalysis data, and the application of innovative analysis approaches, including AI/ML techniques.

8.3.2 What Will Be the Impact of Changing Climate on the Severe Storms Over South Asia?

The response of the convective events in a global warming scenario are assessed using variables such as CAPE, CINE, vertical wind shear (VWS), storm relative helicity (SRH) and covariate indices (Lepore et al., 2021). The sensitivity of the convective variables as function of global average temperature are considered and the results reveal an increase in the favorable convective environments which leads to a frequency increase on the order of 5–20% per °C rise in global temperature.

The climate scenarios incorporated with the Coupled Model Inter comparison Project Phase 6 (CMIP6) global models are the representative concentration pathways (RCPs) describing the greenhouse gas concentration trajectories, which are adopted for the IPCC assessment report 6 (AR6). The RCPs represent the possible extent of radiative forcing values and there are in total a set of four pathways, viz., RCP2.6, RCP4.5, RCP6.0, RCP8.5, which consider 2.6, 4.5, 6.0, and 8.5 W/m^2 of radiative forcing respectively. Furthermore, the shared Socio-economic pathways (SSPs) in CMIP6 framework cover a broader range of greenhouse gas and pollutant considerations compared to the previous RCP scenarios. The CO_2 emission scenarios with and without climate change mitigation measures as well as new low CO_2 emission pathways are incorporated. An assessment of the projected changes in the severe days over South Asia in the mid and far future under RCP 2.6 and 8.5 scenarios during the pre (March–May or MAM) and post (September–November or SON) monsoon period is performed by Glazer et al. (2021) using the RegCM model over the South Asian region in the CORDEX-CORE domain. Spatial distribution of changes in the severe days relative to historical period (1995–2014) at the mid (2041–2060) and far (2080–2099) future time periods are analyzed.

Results indicate large increases in severe days mainly over Bangladesh and parts over eastern coast of India in both near and far future under both the RCP scenarios mostly during MAM. However, in the RCP8.5, there is decrease in the severe days over the Indo-Gangetic Plains during MAM. During SON, increase in severe days (~1–2 days) are noticed over the parts of central and northern India in the near and far future periods in RCP8.5 scenarios.

Several approaches are adopted to analyze the projected variabilities in storms and extreme weather events for the historic and near future time frames. For instance, a multi-model assessment of the future changes in the climate and weather events is considered for the low (SSP2.6), moderate (SSP4.5), and high (SSP8.5) emission scenarios to depict the projections at the end of the century (2066–2100) in comparison to a historical time frame (1980–2014). The analysis displays a rise in frequency of precipitation and extreme wind events in various regions, especially in the higher emission scenario. The impact of predominant dust storm activities on the air quality of China were investigated and projected for a time frame of 2020—2100 (Li et al., 2022). The result reveals a reduction in the dust storm frequency in the future for SSP8.5 scenario, while a fluctuating decadal variability was observed in the SSP2.6 scenario. These issues are discussed in Chap. 22.

8.3.3 What is the Future Projection of Heavy Precipitation Over the Himalayas?

The Himalayas is a region with complex topography and a varied climate. There is evidence that the Himalayas are experiencing changes in extreme precipitation because of a changing climate. Studies have shown that the monsoon is intensifying due to climate change, leading to increased extreme precipitation events. ENSO can affect the monsoon circulation and the frequency and intensity of extreme precipitation events in the Himalayas. Extreme precipitation, defined as intense and heavy rainfall or snowfall events, can have significant impacts on society and the environment.

In Chap. 23, the spatial and temporal variation of precipitation on a seasonal and annual scale and their change in future climate (2015–2100) are assessed with reference to historical periods (1980–2014) using bias-corrected and downscaled CMIP6 ensemble members under mid- (SSP245) and high- (SSP585) emission scenarios over Nepal. Study indicates that the spatial distribution of mean precipitation is similar between the historical and the projected periods, with higher precipitation likely to occur in the future, particularly under SSP585 scenarios. The monsoon precipitation is likely to continue increasing in the future at a rate of 2.3 mm/year and 5.8 mm/year under SSP245 and SSP585, respectively. The precipitation during pre- and post-monsoon seasons will likely increase in future periods, whereas the winter precipitation is likely to decrease as in the historical dataset. Extreme precipitation indices except for consecutive dry days are found to be increasing, with higher frequency and intensity projected in the future.

8.3.4 What is the Role of Solar Storms on the Weather and Climate Extremes?

The influence of the Sun and its variability in electromagnetic radiation on Earth's weather and climate extremes during different phases of the solar cycle and solar storms is well known. The Galactic cosmic rays also influence cloud-radiation interaction. The particle precipitation over the polar region can heat that region and impact the North Atlantic Oscillations (NAO) and Arctic Oscillation (AO) indices, which causes changes in the regional weather conditions. The solar cycle signals are also manifested in tropospheric winds. Furthermore, there is a feedback loop between the effect of the increase in atmospheric CO_2 content and ocean acidification on temperature increase and vis-e-versa. Such natural variabilities can impact on the energy balance of the Earth's surface and atmosphere. Subtle changes in Earth's orbit around the Sun have been known to cause the past ice ages.

It is now understood that Earth's temperature fluctuations are cyclical in nature and have regularly occurred without human intervention. Numerous asteroids have hit the Earth, including one that is supposed to have killed off the dinosaurs since the existence of our planet. Higher sea surface temperatures due to global warming would lead to more powerful weather systems that support hurricanes and tropical storms. The hydrologic cycle would be amplified and strengthened by increased evaporation and water vapor in the atmosphere. Representation of these balances will determine the intensity of weather. These issues are discussed from the perspective of climate changes and extreme weather conditions in Chap. 24. The weightage of natural and anthropogenic factors which play a role in the sudden increase in global warming and the extreme weather conditions is also discussed in this chapter.

8.3.5 What Are the Socio-economic Impacts of Severe Storms and Strategies for Disaster Mitigations?

The fatalities, causalities, and losses in environmental, societal, and economic sectors due to severe weather events are a global concern in the warming climate scenario. In general, mitigation is regarded as the actions to be undertaken to reduce vulnerability. The governments, policymakers, and national and international stakeholders formulate risk reduction approaches and mitigation strategies to minimize the hazards associated with severe weather. For the implementation of adaptable strategic mitigation frameworks, the international organizations of Red Cross and Red Crescent Societies grouped the disasters as a natural and non-natural trigger (Chap. 22). Internationally, this distinction is helpful in developing risk reduction policies and mitigation plans in a discrete manner by considering the sources of extreme environmental events. Initially, the unfocussed risk reduction schemes for minimizing vulnerability by avoiding hazards and damages were executed for quick assistance in disasters and calamity scenarios. The refocused schemes with an emphasis on the necessity of mitigation were implemented during the International Decade for Natural Disaster

Risk Reduction (IDNDR), regarded as a new initiative, strengthening the policy and efforts to reduce disaster risk at the global level.

Many countries have established a national disaster mitigation strategy (NDMS) combined with the Department of Public Safety and Emergency Preparedness. For example, the USA has established the Federal Emergency Management Agency (FEMA). India has established the national disaster management authority (NDMA) in 1995. Many states Governments in India have established their own State Disaster Management Authority (SDMA). The risk reduction could be comprised of avoidance, mitigation, and preparedness. The relocations of settlements from a highly hazard-prone region refer to avoidance, the reinforcement of the infrastructures to reduce damage or overcome the disaster constitute mitigation and evacuation of the population and strategies to reduce the post-disaster impacts refer to preparedness. In the changing climate scenario, minimizing the degree of vulnerability refers to adaptation that is achieved by enhancing capacity building and making the population resilient to cope with future severe weather scenarios. Several developing countries implemented land use policies as a part of their sustainable urban planning. A collaborative attempt at land acquisition seeks to control the illegal settlements, reform slum settlements, and improve access to land. Sustainable mitigation policies are enhanced by adopting suitable shelter design, appropriate construction methods and usage of building materials. The technical solutions include promoting the usage of water-resistant materials and waterproof plastering techniques, elevating basements from the flooding levels in flood-prone regions, avoiding low-pitched and flat, lightweight roofs, avoiding roof overhangs, a stronger connection from the basement to roof, evading empty-spaces under the basements, etc.

To mitigate the risks incorporated with the aviation sector, a satellite-based real-time thunderstorm now-casting viz., Cb-TRAM (Tracking & Monitoring of Severe Convection) scheme for the tracking, detection and forecasting is developed. Complimented with the Numerical Weather Forecasting schemes, the probable locations for the development of thunderstorms in current and future times could be identified. The use of Cb-TRAM information in the cockpit enables the pilots to be capable of executing strategic flight planning one hour in advance rather than only minutes in advance, based on the out-of-window view and the constrained range of the onboard radar. The Epidemic thunderstorm asthma (ETSA) conditions is considered a health hazard followed by the occurrence of the thunderstorms. Mitigation of the severity of asthmatic illness measure constitutes recognizing the risk variables to enable prevention interventions to be directed to individuals who are at maximum risk, preventative measures taken by individuals, as well as major systems-based strategies including forecast alert systems and environmental health literacy.

Similarly, sand and dust storms generate hazards in several ways. The entrainment of the dust particles leads to soil loss, nutrient and fertilizer loss, the transport leads to air pollution causing respiratory problems, eye infection and disease transmission, and its deposition create crop growth problems, drinking-water contamination, reduction of solar power potential, etc. (Middleton and Kang 2017). The mitigation efforts comprise agronomic practices for reducing soil erosion utilizing live plants or crop remains to shield soil from the wind. Windbreakers and shelterbelts could be constructed to minimize the erosive force of the wind, i.e., fences or walls positioned at a right angle to erosive winds, or windbreaks made from trees or shrubs can function as shelter belts. The agricultural sector is highly vulnerable to severe weather events. Hail, frozen and solid form of precipitation causes substantial losses to properties and growing crops. The management of mitigation and adaptation strategies requires interdisciplinary knowledge, technological inventions, and suitable interventions. The Agromet Advisory Bulletin (AAB) setup in India provides site-specific information and suggestions within pre- and post-hailstorm events to increase the advantages and reduce the losses in agricultural production. Many such issues are discussed in Chaps. 22, 25 and 26.

9 Field Experiments on Thunderstorms

As the observation technology is improving rapidly, field campaigns are frequently organized to collect intensive observations of storms. Field observations of severe weather phenomena are collected to improve our understanding of their structures and life cycles and develop better models for forecasting the events. Over the years several field experiments have been conducted both over the Indian region and outside. IMD had conducted three field experiments during 1929–1941 to study the outbreak of severe convective storms (IMD, 1944; Tyagi et al., 2012). Several field experiments have been conducted since then in USA and elsewhere (Table 5) to understand and predict the convective storms.

Some of the most important experiments conducted on thunderstorms in USA are the VORTEX, BAMEX, and TELEX among others. The VORTEX project began in 1994, with the objective of explaining how tornadoes form (Wurman et al., 2012). The two-year field project resulted in ground-breaking data collection and led to several follow-up studies in the late 1990s. The VORTEX2 field project debuted in 2009 and continued through Spring 2010, with scientists hoping to understand how, when, and why certain supercells produce tornadoes. In 2016, NSSL researchers embarked on VORTEX Southeast, the latest field project to examine tornadic storms. In this research program, scientists sought to understand how environmental factors in the southeastern United States affect the formation, intensity, structure, and path of tornadoes in the region. That was also the first VORTEX experiment to emphasize societal impact, with social scientists studying how the community receives and responds to warnings.

The Bow Echo and Mesoscale Convective Vortex Experiment (BAMEX) was conducted over mid-America (Illinois, St. Louis, Missouri) between 20 May and 6 July 2003 (Davis et al., 2004). BAMEX concentrated on studying the life cycle of mesoscale convective systems (MCSs), particularly those producing severe surface winds (bow echoes) and those producing long-lived mesoscale convective vortices (MCVs) capable of initiating subsequent convection. BAMEX investigated competing hypothesis explaining the development of severe straight-line winds. The hypotheses are that they are caused by the production of negative buoyancy in the rear-inflow jet which is a function of the microphysical composition of the stratiform region, that they result from meso gamma scale vortices near the surface along the leading edge of the bow echo, and that they are produced by internal gravity waves produced from perturbations of the stable boundary layer by deep convection. BAMEX also investigated the process of MCV formation which had hitherto not been well observed and sought to understand how convection is initiated and organized in the vicinity of long-lived MCVs.

The Thunderstorm Electrification and Lightning Experiment (TELEX) was conducted in May–June 2003 and 2004 in central Oklahoma (MacGorman et al., 2008). The objective of the experiment was to study how lightning and other electrical storm properties depend on storm structure, updrafts, and precipitation formation. Measurements were taken by a lightning mapping array, polarimetric and mobile Doppler radars, and balloon-borne electric-field meters and radiosondes. The results showed that the interaction between cloud ice and riming graupel is an essential ingredient in electrifying storms, though it may not account for every region of storm charge.

10 Storm Chasing

Storm chasing is the pursuit of any severe weather condition, which can be for curiosity, adventure, scientific investigation, or for news or media coverage. While witnessing a tornado is generally the biggest objective for most chasers, many chase thunderstorms and delight in viewing cumulonimbus and related cloud structures, watching a barrage of hail and lightning. There are also a smaller number of storm chasers who intercept tropical cyclones and waterspouts. The storm chasing can have scientific objectives by collaborating with an university or government project. A few operate "chase tour" services, making storm chasing a niche for tourism. The first recognized storm chaser is David Hoadley (https://en.wikipedia.org/wiki/Storm_chasing) who is considered the pioneer storm chaser. He began chasing North Dakota storms in 1956, using data from area weather offices and airports. The most popular storm chasing-tour company was founded by Dr. Reed Timmer of the TV show "Storm Chasers". They specialized in tornado chasing led by an Oklahoma University Meteorologist. Local National Weather Service offices hold storm spotter training classes. Some offices collaborate to produce severe weather workshops oriented toward operational meteorologists. Storm chasers vary with regards to the amount of equipment used, some prefer a minimalist approach; where only basic photographic equipment is taken on a chase, while others use everything from satellite-based tracking systems and live data feeds to vehicle-mounted weather stations and hail guards.

11 Application of Artificial Intelligence (AI)/Machine Learning (ML) Techniques for Forecasting Severe Storms

Typically, 287 TB new data is archived every day at operational weather forecasting centres like the European Centre for Medium Range Weather Forecasts (ECMWF, October 2022; https://www.ecmwf.int/en/computing/our-facilities/data-handling-system). These data include global observations from different sources like surface and upper air observations, remote sensing platforms like the Radars, Satellites, etc., and model products. The Big Data generated for weather forecasting every day provides a good opportunity for the developers of the AI/ML models, which primarily use statistics, neural networks, data mining techniques to train a model.

However, the weather phenomena are physical systems, which are governed by the laws of Physics, hydrodynamics, and thermodynamics. They are scale dependent and have a life cycle. There are also many uncertainties in modelling the weather systems, particularly the convective systems like the thunderstorms, hailstorms, tornadoes, lightning, etc. due to their short life span, spatial scale, initial value problems, interaction with sub-grid scale processes, and chaos. Therefore, training a model based purely on the past data is a challenging task for the severe storm forecasting. Nevertheless, interests in using AI/ML in weather forecasting has been growing in recent years. Forecasters and emergency managers are making life-or-death decisions in short time periods and, if they are to use an AI/ML method to help streamline the process of data overload, they must fully trust the product. McGovern et al. (2023) provides a good review of the Machine Learning for Convective Weather.

11.1 Can AI/ML Methods Follow the Laws of Physics?

Ensuring that AI/ML models respect the laws of physics is a key unsolved challenge for AI/ML model development for all aspects of weather and climate forecasting. There are many small-scale processes that are critical to forecasting convection and must be simulated in a physics-based manner to ensure that the output is realistic. While many ML models can be trained based entirely on observational data, the models must also be constrained by the laws of physics, or they can learn patterns that cannot occur in the real world. With the goal of creating trustworthy ML, it is critical that the models respect the laws of physics both in their successful predictions and in their unsuccessful ones. Some methods already exist for incorporating physical laws into ML. These include custom loss functions, which punish the ML model for violating physical laws, and custom architectures, which use equations to enforce physical laws inside the model.

11.2 Uncertainty Quantification

AI/ML can be used to process hundreds of simulations of an event to identify common patterns, something that would be very challenging for a human to do. However, the promise of using AI/ML methods to identify fundamental new science discoveries in convective weather, such as identifying why one storm is tornadic and another similar storm never generates a tornado remains elusive. Sophisticated ML methods that are strongly tailored to the phenomena of interest may be required to advance our understanding of severe weather processes that have been studied for decades. This area is prime for additional research.

ML tornado applications have primarily focused upon detection and prediction, not characterization. Many relationships have been identified between tornado properties and storm mode, environment, and radar- and satellite-measured attributes. ML could exploit these and perhaps yet unknown relationships between Tornado attributes and regularly measured atmospheric variables to produce real-time estimates and short-term predictions of tornado intensity, damage, width, and duration.

Another challenge to training ML for severe weather is the cost (both economic and societal) for missing a forecast. While this cost exists to some degree in many traditional ML problems, it is literally often life-or-death in severe weather in a handwriting recognition system. The cost for missing an event even differs from tornado to tornado, with long-lived significant tornadoes passing through populated areas having much more of a penalty than brief squall line spin-ups in open fields. This poses a challenge because most current ML algorithms treat all reports as identical, not only in quality, but also potential impact during training and real-time prediction.

12 Unresolved Issues and Vision for the Next Decades

Clouds are the basic ingredients in the formation of all meteorological disasters. However, the clouds are barely predicted accurately in the models despite decades of analysing satellite and simulated data. Many questions remain. For example, how do cloud particles form, interact with aerosols and precipitate or re-evaporate? What determines the number of clouds in a given area? How do clouds interact with turbulence and visible and infrared radiation? Large thunderstorms pack a lot of energy in the form of updrafts that move, in extreme cases, 30–40 m s^{-1}. Models hardly account for these and other physical properties.

13 Summary and Conclusion

Thunderstorms form almost everywhere on the earth. The basic ingredients for their formation are moisture, instability, and lifting. Most of the damages from thunderstorms occur due to strong winds, lightning strike, hail stones, and flash floods. Numerical models of the storms range from Single Column model to one, two or three dimensional. The forecasting/nowcasting of thunderstorms are done based on the combination of synoptic methods, Doppler Radars, and numerical ensemble models with rapid update cycle data assimilation. The skill of forecasting the thunderstorms have spatial errors of about 40 km, temporal errors $\pm$ 30 min and intensity errors about -25% (underestimated) in 24 h (Sarkar et al., 2019).

Field experiments are conducted to improve our understanding of the structures, life cycles and develop better models for forecasting the thunderstorms. Many field campaigns are conducted in India since 1929, but most of them are conducted in USA in recent years to understand their mechanisms and improve forecasting skills. Storm Chasing has become an adventure, curiosity and hobby in addition to achieving scientific goals. While field campaigns on thunderstorms were conducted in India much before it started in USA, there is a need for more concerted efforts on collecting field observations of severe storms using latest equipment with focused scientific objectives, convective scale data assimilation and ensemble forecasting of the origin, track, lightning strike probability and intensity of the storm fury.

There is a need to establish special observation testbeds for severe storms, like the Nor'westers, Dust storms, Cloudbursts, etc., in the areas which are prone to frequent storms. Such testbeds should be equipped with modern equipment to collect intensive observations of the severe storms for 4–5 years. The observations could be used for understanding the physical and dynamical processes to study the life cycle of the storms and improve their representations in the numerical model. The observations could be assimilated into high resolution model with rapid update cycle for forecast demonstration projects and improving the accuracy of the early warning system. Ensemble model experiments could also be carried out to improve the accuracy and lead time of the forecasts.

While the frequent anomalous weather may be blamed to the effects of global warming and climate change, any such event is routed to the changes in the structure of the atmosphere. Our observation systems are supposed to sense any such change in the structure of the atmosphere and the weather and climate models are supposed to predict such changes with sufficient lead time, so that warnings may be issued to the public. Hence, it is a challenge to the Atmospheric Scientists.

Failure to observe the changes in the structure of the atmosphere implies that our observation network/technology is insufficient. The observation network is insufficient in many South Asian countries and in the remote areas and over the mountains. The discrete observation points have coarse resolution and hence, often they are unable to represent the sub-grid scale weather phenomena accurately. There is also a problem in accurately representing a continuous weather by the discrete observation points in space and time. These are somehow managed by the data assimilation. Many countries in South Asia do not even have a proper data assimilation system.

Weather is a result of the motion of air caused by a balance between various forces. This motion is described by the governing equations of the atmosphere based on the principles of hydrodynamics and thermodynamics. Some of the equations are nonlinear partial differential equations whose solutions are to be obtained numerically by making billions of calculations on the fastest supercomputers. So, regardless of the causes of extreme weather events, if we know the laws and the variables, why can't we make perfect forecasts even with a short lead time? The main reasons for this are (1) Inaccurate Initial Conditions, (2) Multiscale Interactions, (3) Chaos and Limit on Deterministic Predictability. Little change in the accuracy of the initial condition can cause large errors in the solutions of the model. There are uncertainties in the formulation of the governing equations. But even if the model was perfect, intrinsic nonlinearity of the atmosphere would restrict our ability to predict the weather to about two weeks, because there is a limit of the deterministic predictability due to chaos.

Despite the limitations in the observations and the modelling, the techniques of weather forecasting have continuously evolved in the past 3–4 decades. People have gradually started trusting the accuracy of weather forecasts as their skills have improved due to gradual improvements in the observation technology such as Satellites, Doppler Weather Radars, Automatic Weather Stations, etc., and improvements in our understanding of the atmosphere, data assimilation system and modelling techniques. However, more research is needed to improve the skills of forecasting the local severe storms to above 90% accuracy at 4–5 days lead time.

There is enough scope for further substantial improvement in the skill of prediction of localized severe weather systems. There is a need to develop suitable assimilation and modelling system for improved prediction of local severe storms. It may consist of utilization of observations of different types in data assimilation, model dynamics and physics, proper post processing, bias removal, ensemble and super-ensemble approaches. The products may be linked to a GIS system for better utilization in developing warning system for risk/vulnerability assessment, mitigation, and management strategies by various central and state agencies. The prediction of local severe storms can be achieved through an integrated approach combining field observations, research, and modelling.

References

Anthes, R. A. (1990). Recent applications of the Penn State/NCAR mesoscale model to synoptic, mesoscale, and climate studies. *Bulletin of the American Meteorological Society, 71*(11), 1610–1629.

Benjamin, S., Weygandt, S., Hu, M., Alexander, C., Smirnova, T., Olson, J., Brown, J., James, E., Dowell, D., Grell, G., Lin, H., Peckham, S., Smith, T., Moninger, W., & Manikin, G. (2016). A North American hourly assimilation and model forecast cycle: The rapid refresh. *Monthly Weather Review, 144*, 1669–1694. https://doi.org/10.1175/MWR-D-15-0242.1

Bluestein, H. B., & Jain, M. H. (1985). Formation of mesoscale lines of precipitation: Severe squall lines in Oklahoma during the spring. *Journal of Atmospheric Science, 42*, 1711–1732.

Bright, D. R. & Grams, J. S. (2009). Short range ensemble forecast (SREF) calibrated thunderstorm probability forecasts: 2007–2008 verification and recent enhancements. In *3rd Conference on Meteorological Applications of Lightning Data*. American Meteorological Society, 6.3.

Bright, D. R., Wandishin, M. S., Jewell, R. E., & Weiss, S. J. (2005). A physically based parameter for lightning prediction and its calibration in ensemble forecasts. In *Conference on Meteorological Applications of Lightning Data*. American Meteorological Society, CD-ROM, 4.3.

Casati, B., Ross, G., & Stephenson, D. B. (2004). A new intensity-scale approach for the verification of spatial precipitation forecasts. *Meteorological Appllcations, 11*, 141–154.

Casati, B., Wilson, L. J., Stephenson, D. B., Nurmi, P., Ghelli, A., Pocernich, M., Damtrath, U., Ebert, E. E., Brown, B. G., & Mason, S. (2008). Review, forecast verification: Current status and future directions. *Meteorological Applications, 15*, 3–18.

Case, J. L., Manobianco, J., Oram, T. D., Garner, T., Blottman, P. F., & Spratt, S. M. (2002). Local data integration over East-Central Florida using the ARPS data analysis system. *Weather and Forecasting, 17*(1), 3–26.

Chug, S., Nath, S., Srivastava, K., & Singh, S. L. (2021). Recent advances in communication infrastructure of IMD for early warning system. *Mausam, 72*(4), 935–946.

Cotton, W. R., Thompson, G., Mieike, P. W. (1994). Real-time mesoscale prediction on workstations. *Bulletin of the American Meteorological Society, 75*(3), 349–362.

Cotton, W. R., Tripoli, G., Rauber, R., & Mulvihill, E. (1986). Numerical simulation of the effects of varying ice crystal nucleation rates and aggregation processes on orographic snowfall. *Journal of Applied Meteorology, 25*, 1658–1680.

Crook, N. A., & Sun, J. (2002). Assimilating radar, surface, and profiler data for the Sydney 2000 forecast demonstration project. *Journal of the Seismological Society of Japan, 19*, 888–898.

Das, S. (2017). Severe thunderstorms observations and modelling—A review. *Vayumandal (bulletin of the Indian Meteorological Society), 43*(2), 1–29.

Das, S. (2022). Challenges in predicting extreme weather events over the South Asian Region. In Unnikrishnan et al. (Eds.), *Extreme Natural Events* (pp. 51–106). Springer, A. S. Center for Science & Technology of the Non-aligned and Other Developing Countries. https://doi.org/10.1007/978-981-19-2511-5_3.

Das, S., Johnson, D., & Tao, W.-K. (1999). Single-column and cloud ensemble model simulations of TOGA-COARE convective systems. *Journal of Meteorological Society of Japan, 77*(4), 803–826.

Das, S., Mohanty, U. C., & Sharma, O. P. (1987). Semiprognostic test of the Arakawa-Schubert cumulus parameterization during different phases of the summer monsoon. *Contributions to Atmospheric Physics, 60*(2), 255–275.

Das, S., Mohanty, U. C., Tyagi, A., Sikka, D. R., Joseph, P. V., Rathore, L. S., Habib, A., Baidya, S., Sonam, K., & Sarkar, A. (2014). The SAARC STORM—A coordinated field experiment on severe thunderstorm observations and regional modeling over the South Asian Region. *Bulletin of the American Meteorological Society, 95*, 603–617. https://doi.org/10.1175/BAMS-D-12-00237.1

Das, S., Sarkar, A., Das, M. K., Rahman, M. M., & Islam, M. N. (2015). Composite characteristics of nor'westers based on observations and simulations. *Atmospheric Research, 158–159*, 158–178. https://doi.org/10.1016/j.atmosres.2015.02.009

Das, S., Sud, Y. C., & Suarez, M. J. (1998). Inclusion of a prognostic cloud scheme with the relaxed Arakawa-Schubert cumulus parameterization; single column model studies. *Quarterly Journal of Royal Meterological Society, 124*, 2671–2692.

Davis, C., Atkins, N., Bartels, D., Bosart, L., Coniglio, M., Bryan, G., Cotton, W., Dowell, D., Jewett, B., Johns, R., Jorgensen, D., Knievel, J., Knupp, K., Lee, W.-C., McFarquhar, G., Moore, J., Pryzbylinski, R., Rauber, R., Smull, B., … Ziegler, C. (2004). The Bow-Echo and MCV Experiment (BAMEX): Observations and opportunities. *Bulletin of the American Meteorological Society, 85*, 1075–1093.

DTE. (2023). Down to earth. Retrieved November 1–15, 2023, from http://downtoearth.org.in.

Douville, H., Raghavan, K., Renwick, J, Allan, R. P., Arias, P. A., Barlow, M., Cerezo-Mota, R., Cherchi, A., Gan, T. Y., Gergis, J., Jiang, D., Khan, A., Pokam Mba, D., Rosenfeld, D., Tierney, T., & Zolina, O. (2021.) Water cycle changes. In V. Masson-Delmotte, P. Zhai, A. Pirani, S. L. Connors, C. Péan, S. Berger, N. Caud, Y. Chen, L. Goldfarb, M. I. Gomis, M. Huang, K. Leitzell, E. Lonnoy, J. B. R. Matthews, T. K. Maycock, T. Waterfield, O. Yelekçi, R. Yu, & B. Zhou (Eds.), *Climate Change 2021: The Physical Science Basis. Contribution of Working Group I to the Sixth Assessment Report of the Intergovernmental Panel on Climate Change.* Cambridge University Press.

Dudhia, J. (1993). A non-hydrostatic version of the Penn State/NCAR mesoscale model: Validation tests and simulation of an Atlantic cyclone and cold front. *Monthly Weather Review, 121*, 1493–1513.

Ebert, E. E. (2008). Fuzzy verification of high-resolution gridded forecasts: A review and proposed framework. *Meteorological Applications, 15*, 51–64.

Ebert, E. E., & McBride, J. L. (2000). Verification of precipitation in weather systems: Determination of systematic errors. *Journal of Hydrology, 239*, 179–202.

Gilleland, E., Ahijevych, D., Brown, B. G., Casati, B., & Ebert, E. E. (2009). Intercomparison of spatial forecast verification methods. *Weather Forecasting, 24*, 1416–1430.

Glazer, R. H., Torres-Alavez, J. A., Coppola, E., Giorgi, F., Das, S., Ashfaq, M., & Sines, T. (2021). Projected changes to severe thunderstorm environments as a result of twenty-first century warming from RegCM CORDEX-CORE simulations. *Climate Dynamics, 57*, 1595–1613.

IPCC. (2021). Summary for policymakers. In *Climate Change 2021: The Physical Science Basis. Contribution of Working Group I to the Sixth Assessment Report (AR-6) of the Intergovernmental Panel on Climate Change,* Cambridge University Press.

IMD. (1944). Nor'wester of Bengal (17pp.). India Meteorological Department Technical Note 10.

Kumar, A., Das, S., & Panda, S. K. (2022). Numerical simulation of a widespread lightning events over North India using an ensemble of WRF modeling system configurations. *Journal of Atmospheric and Solar-Terrestrial Physics*. https://doi.org/10.1016/j.jastp.2022.105984.

Lepore, C., Abernathey, R., Henderson, N., Allen, J. T., & Tippett, M. K. (2021). Future global convective environments in CMIP6 models. *Earth's Future, 9*(12), e2021EF002277.

Li, J., Hao, X., Liao, H., Yue, X., Li, H., Long, X., & Li, N. (2022). Predominant type of dust storms that influences air quality over northern China and future projections. *Earth's Future, 10*(6), e2022EF002649.

Lin, Y.-L., Farley, R. D., & Orville, H. D. (1993). Bulk parameterization of the snow field in a cloud model. *Journal of Climate Applied Meteorology, 22*, 1065–1092.

Liu, C., Moncrieff, M. W., Grabowski, W. W. (2001). Explicit and parameterized realizations of convective cloud systems in TOGA COARE. *Monthly Weather Review, 129*(7), 1689–1703.

MacGorman, D. R., Rust, W. D., Schuur, T. J., Biggerstaff, M. I., Straka, J. M., Ziegler, C. L., Mansell, E. R., Bruning, E. C., Kuhlman, K. M., Lund, N. R., Biermann, N. S., Payne, C., Carey, L. D., Krehbiel, P. R., Rison, W., Eack, K. B., & Beasley, W. H. (2008). TELEX: The thunderstorm electrification and lightning experiment. *Bulletin of the American Meteorological Society, 89*, 997–1018.

McGovern, A., Chase, R. J., Flora, M., Gagne, D. J., II., Lagerquist, R., Potvin, C. K., Snook, N., & Loken, E. (2023). A review of the machine learning for convective weather. *Artificial Intelligence for Earth Sciences*. https://doi.org/10.1175/AIES-D-22-0077.1

Mesinger, F. (1996). Improvements in quantitative precipitation forecasts with the eta regional model at the national centers for environmental prediction: The 48-km upgrade. *Bulletin of the American Meteorological Society, 77*(11), 2637–2650.

Middleton, N., & Kang, U. (2017). Sand and dust storms: Impact mitigation. *Sustainability, 9*(6), 1053.

Mohapatra, M., Kumar, N., Sharma, M., Devi, S. S., Senroy, S., Devi, K. S., & Singh, K. K. (2022). Short to medium range weather forewarning system in India. In *Social and Economic Impact of Earth Sciences* (pp. 1–29). Springer Nature Singapore.

NDMA. (2018). Guidelines for preparation of Action Plan—Prevention and management of thunderstorm & lightning/squall dust/hailstorm and strong winds. National Disaster management Authority (NDMA) (p. 39), Government of India.

Ogura, Y., & Phillips, N. A. (1962). Scale analysis of deep and shallow convection in the atmosphere. *Journal of Atmospheric Science, 19*, 173–179.

Roberts, N. M., & Lean, H. W. (2008). Scale-selective verification of rainfall accumulations from high resolution forecast of convective events. *Monthly Weather Review, 136*, 78–97.

Rotunno, R., Klemp, J. B., & Weisman, M. L. (1988). A theory for strong, long-lived squall lines. *Journal of Atmospheric Science, 45*, 463–485.

Roy Bhowmik, S. K., Sen Roy, S., Srivastava, K., Mukhopadhay, B., Thamp, S. B., Reddy, Y. K., Singh, H., Venkateswarlu, S., & Adhikary, S. (2011). Processing of Indian Doppler weather radar data for mesoscale applications. *Meteorology and Atmospheric Physics, 2011*, 133–147.

Sarkar, A., Das, S., & Dutta, D. (2019). Computation of skill of a mesoscale model in forecasting thunderstorm using Radar reflectivity. *Modeling Earth Systems and Environment, 5*, 443–454. https://doi.org/10.1007/s40808-018-0553-7

Srivastava, K., Gao, J., Brewster, K., Roy Bhowmik, S. K., Xue, M., & Gadi, R. (2011). Assimilation of Indian radar data with ADAS and 3DVAR techniques for simulation of a small-scale tropical cyclone using ARPS model. *Natural Hazards*. https://doi.org/10.1007/s11069-010-9640-4

Soong, S.-T., & Tao, W.-K. (1980). Response of deep tropical clouds to mesoscale processes. *Journal of Atmospheric Science, 37*, 2016–2036.

STORM. (2005). STORM science plan. India Department of Science and Technology Report (118pp.). www.imd.gov.in/SciencePlanofFDPs/STORM%20Science%20Plan.pdf.

Sun, J., Chen, M., & Wang, Y. (2010). A frequent-updating analysis system based on radar, surface, and mesoscale model data for the Beijing 2008 Forecast Demonstration Project. *Weather Forecasting, 25*, 1715–1735.

Sun, J. (2005). Convective-scale assimilation of radar data: Progress and challenges. *Quarterly Journal of Royal Meteorological Society, 131*, 3439–3463.

Sun, J., & Crook, N. A. (1997). Dynamical and microphysical retrieval from Doppler radar observations using a cloud model and its adjoint: I. Model development and simulated data experiments. *Journal of Atmospheric Science, 54*, 1642–1661.

Sun, J., & Crook, N. A. (1998). Dynamical and microphysical retrieval from Doppler radar observations using a cloud model and its adjoint: II. Retrieval experiments of an observed Florida convective storm. *Journal of Atmospheric Science, 55*, 835–852.

Takemi, T., & Satomura, T. (2000). Numerical experiments on the mechanisms for the development and maintenance of long-lived squall lines in dry environments. *Journal of Atmospheric Science, 57*, 1718–1740.

Tao, W.-K., & Simpson, J. (1993). The Goddard cumulus ensemble model. Part I: Model description. *Terrestrial Atmospheric and Oceanic Science, 4*, 35–72.

Tao, W.-K., Simpson, J., & McCumber, M. (1989). An ice-water saturation adjustment. *Monthly Weather Review, 117*, 231–235.

Tao, W.-K., Simpson, J., & Soong, S.-T. (1987). Statistical properties of a cloud ensemble: A numerical study. *Journal of Atmospheric Science, 44*, 3175–3187.

Tao, W.-K., & Soong, S.-T. (1986). A study of the response of deep tropical clouds to mesoscale processes: Three-dimensional numerical experiments. *Journal of Atmospheric Science, 43*, 2653–2676.

Toth, Z. (2001). Meeting summary: Ensemble Forecasting in WRF. *Bulletin of the American Meteorological Society, 82*(4), 695–698.

Tyagi, A., Sikka, D. R., Goyal, S., & Bhowmick, M. (2012). A satellite based study of pre-monsoon thunderstorms (Nor'westers) over eastern India and their organization into mesoscale convective complexes. *Mausam, 63*(1), 29–54.

Weisman, M. L., & Klemp, J. B. (1982). The dependence of numerically simulated convective storms on vertical wind shear and buoyancy. *Monthly Weather Review, 110*, 504–520.

Weisman, M. L., & Rotunno, R. (2004). "A theory for strong, long-lived squall lines" revisited. *Journal of Atmospheric Science, 61*, 361–382.

Weisman, M. L., Klemp, J. B., & Rotunno, R. (1988). Structure and evolution of numerically simulated squall lines. *Journal of Atmospheric Science, 45*, 1990–2013.

WMO. (2021). The *atlas of mortality and economic losses from weather, climate and water extremes (1970–2019). WMO-No. 1267,* World Meteorological Organization (WMO).

Wurman, J., Dowell, D., Richardson, Y., Markowski, P., Rasmussen, E., Burgess, D., Wicker, L., & Bluestein, H. B. (2012). The second verification of the origins of rotation in tornadoes experiment: VORTEX2. *Bulletin of the American Meteorological Society, 93*, 1147–1170.

Xu, K.-M., & Randall, D. A. (1996). Explicit simulation of cumulus ensembles with the GATE phase III data: Comparison with observations. *Journal of the Atmospheric Sciences, 53*(24), 3710–3736.

Zipser, E. J., & LeMone, M. A. (1980). Cumulonimbus vertical velocity events in GATE. Part II: Synthesis and model core structure. *Journal of Atmospheric Science, 37*, 24582–25469.

MIX
Papier aus verantwortungsvollen Quellen
Paper from responsible sources
FSC® C105338

If you have any concerns about our products, you can contact us on
ProductSafety@springernature.com

In case Publisher is established outside the EU, the EU authorized representative is:
Springer Nature Customer Service Center GmbH
Europaplatz 3, 69115 Heidelberg, Germany

Printed by Libri Plureos GmbH
in Hamburg, Germany